Lecture Notes in Engineering

Edited by C. A. Brebbia and S. A. Orszag

44

R. S. Edgar

Field Analysis and Potential Theory

Author
Robert Steel Edgar
1 Rosetta Terrace
Port Elliot
South Australia 5212

ISBN-13:978-3-540-51074-1 e-ISBN-13:978-3-642-83765-4
DOI: 10.1007/978-3-642-83765-4

Library of Congress Cataloging-in-Publication Data
Edgar, R. S.
Field analysis and potential theory / R. S. Edgar.
(Lecture notes in engineering ; 44)
Previously published by Defense Science and Technology
Organisation, Electronics Research Laboratory, Adelaide, c1985.
Includes bibliographical references and index.
ISBN-13:978-3-540-51074-1 (U.S.)
1. Electromagnetic fields – Mathematics. 2. Potential, Theory of.
I. Title. II. Series.
QC665.E4E34 1989
530.1'41–dc20 89-6365

PREFACE

"Electromagnetic theory is a peculiar subject. The peculiarity resides not so much in the stratification - superposed layers of electrostatics, magnetostatics, steady currents and time-varying fields - as in the failure that has attended all attempts to weld these layers into a logical whole . The lowest layer, electrostatics, defines certain concepts, such as $\overline{E}$, $\overline{D}$, ϕ, in a way that is generally satisfactory only for the static case. Yet the attempt is made to force these specialised definitions into the higher strata, with ad hoc modifications when necessary. The student, in looking through his text books on electromagnetics, can find general definitions only with difficulty, if at all; and even the most advanced treatises fail to present a rigorously logical development of the subject".

So wrote Moon and Spencer[1] some 30 years ago; and their criticism continues to be pertinent today.

More recently, a senior physicist of the National Bureau of Standards[2] has expressed his concern in similar terms:

"A logically consistent set of definitions of the electromagnetic field quantities is extremely difficult to find in the literature. Most text books either evade the problem or present definitions that are applicable only to special cases".

On a related issue R.W.P. King writes[3]

"--- where the purpose is to arrive at solutions of general or special problems and not follow an interesting historical development, it is best to learn how to proceed from the most general equations. Thus it is interesting, but immaterial for the man interested in learning to solve electrodynamical problems, to know by what winding and devious paths, by what bold leaps the Maxwell-Lorentz equations were finally formulated. If he nevertheless insists on seeing them 'derived' from other less general formulas such as Coulomb's law and Ampere's laws, he must content himself with the statement that this has never been done".

---

1. P. Moon and D.E. Spencer, "A Postulational Approach to Electromagnetism", Jour. Franklin Inst. 259, p. 293 (1955).

2. Chester H. Page, "Definitions of Electromagnetic Field Quantities", Am. J. Phys. 42, p. 490 (1974).

3. R.W.P. King, "Fundamentals of Electromagnetic Theory", p. 107, 2nd ed., Dover, New York (1963).

See also P. Duhem, "The Aim and Structure of Physical Theory", p. 200ff, tr. P. Wiener, Princeton University Press, New Jersey (1954).

Like many authors of advanced texts, King opts for a treatment which postulates Maxwell's equations, ab initio, and proceeds therefrom. However, this approach is not entirely satisfactory. The field vectors are not defined uniquely by Maxwell's equations and a rigorous development along conventional lines demands a distressing proliferation of postulates of one form or another (See Sec. 7.9).

One is therefore led to enquire whether an alternative treatment may not be available which admits of unequivocal definitions of the electromagnetic field quantities, which bypasses the unjustifiable extrapolations of the inductive method, and which demands fewer postulates than King's approach. Such a development does, in fact, exist and was presented as early as 1897 by Levi-Civita, who subsequently wrote[4]

> "We can find the essentials of Maxwell's theory even while starting from the classical laws. It is sufficient to complete them by the hypothesis that the actions at a distance are propagated with a finite velocity".

In other words, if ϕ and $\overline{A}$ represent appropriately-retarded forms of the scalar and vector potentials conventionally associated with time-invariant distributions of charge and current, and if $\overline{E}$ and $\overline{B}$ are defined by

$$\overline{E} = -\operatorname{grad}\phi - \frac{1}{c}\frac{\partial\overline{A}}{\partial t} \quad ; \quad \overline{B} = \operatorname{curl}\overline{A}$$

we can <u>deduce</u> the dynamical form of Maxwell's equations for a vacuum (in Gaussian units). The ancillary relationship

$$\operatorname{div}\overline{J} = -\frac{\partial\rho}{\partial t}$$

follows immediately from the postulate of the conservation of source strength or from the discrete physical model.

Now this approach is quite different from those discussed previously. It proposes to derive Maxwell's equations by mathematical manipulation of the space and time derivatives of potential functions defined in terms of scalar and vector source densities. Since such analysis, qua analysis, does not require that the source densities be based upon an electrical model, but only that they satisfy the equation of continuity, the proposed procedure transfers Maxwell's equations from the realm of physics to that of pure mathematics - to what may be called a branch of retarded potential theory.

In these circumstances the <u>physics</u> of electromagnetic theory is introduced through the constitutive relationships and the Lorentz force formula. The fact that these describe the interactions of charge complexes by expressions which involve the symbols of potential theory in no way requires that such symbols be necessarily given a physical interpretation.

---

4. Cited by A. O'Rahilly, "Electromagnetic Theory", p. 190, Dover, New York (1965).

This division of electromagnetics into pure and applied aspects brings with it a considerable simplification of fundamental principles.

The vector fields $\overline{E}$ and $\overline{B}$ are no longer defined by force effects on charge and current elements (with all the attendant difficulties posed by material media) but as analytical derivatives of the point functions ϕ and $\overline{A}$.

Div $\overline{B} = 0$, not because 'lines of $\overline{B}$ are always closed' (which is hardly ever true), but because $\overline{B} = \text{curl}\ \overline{A}$ and div curl $\equiv 0$.

Similarly, div $\overline{D} = 4\pi\rho$, neither by postulate nor by a misapplication of Coulomb's law, but because the equation div $\overline{E} = 4\pi\ (\rho - \text{div}\ \overline{P})$ follows from the expression for $\overline{E}$ in terms of ϕ and $\overline{A}$, irrespective of the relationship between $\overline{P}$ and $\overline{E}$.

Again, the analytical transformations which allow $\overline{B}$ and $\overline{H}$ to be expressed as the derivatives of either scalar or vector potential functions constitute the true basis of the 'pole' and 'whirl' models of magnetic material.

Finally, the displacement current term in Maxwell's generalisation of Ampère's work law in vacuo appears not as a postulate (to which arguments involving extrapolation from closed to open circuits ultimately reduce) but as an analytical consequence of the expansion of curl $\overline{B}$.

All of these matters are the province of potential theory; their only contact with the physical world lies in the identification of the mathematical source density functions such as ρ, $\overline{J}$, $\overline{P}$ and $\overline{M}$ with those deriving from the physical model when it is required, as an exercise in applied mathematics, to calculate the interaction between charge complexes by making use of the Lorentz formula or the relationships between $\overline{J}$ and $\overline{E}$, $\overline{P}$ and $\overline{E}$ (or $\overline{D}$) and $\overline{M}$ and $\overline{B}$ (or $\overline{H}$).

Should experimental evidence ever reveal a failure on the part of the Maxwell-Lorentz treatment to adequately describe some electrical phenomenon, Maxwell's equations would in no way be invalidated; their physical relevance, however, could (among other possibilities) be called into question. Thus, if experimental correlation were better served by the development of a ballistic theory (involving a modification of retardation kinematics) the latter theory would supersede the former in an electromagnetic context; as analytical developments they would be equally significant.

Our contention is identical with that expressed by Duhem[5] at the turn of the century:

> "A physical theory is not an explanation. It is a system of mathematical propositions deduced from a small number of principles which aim to represent as simply, as completely and as exactly as possible a set of experimental laws".

---

5. ibid. p. 19.

This document is concerned with the systematic development of retarded potential theory insofar as it is relevant to the study of classical electromagnetics[6]. To highlight its purely analytical nature overt mention of electricity and magnetism has been eschewed, although, for obvious reasons, standard symbolism has been retained.

Since a work of this type finds its most natural expression in terms of Gaussian units, no apology is offered for their adoption. A knowledge, on the part of the reader, of only the elements of vector analysis - amounting to little more than a familiarity with the addition, subtraction and multiplication of vectors - has been assumed.

R.S. Edgar

---

6. In dealing with doublet and whirl complexes (equivalent to polarised dielectrics and magnetised material) considerations are restricted to those complexes which are at rest in the system of coordinates in which the various field functions are evaluated.

NOTATION

The limited nature of available type face has led in part to the adoption of the following notation:

Vector quantities are represented by a bar over the associated symbol.

Complex quantities bear a tilde superscript while complex conjugates are starred in addition.

The dot in the scalar product is located at the foot of the component symbols rather than half-way up.

Unit vectors, other than $\bar{i}$, $\bar{j}$, $\bar{k}$, carry a circumflex superscript.

The symbol $\underline{R}$ is used on occasion to represent a region.

LIST OF TABLES

CONTENTS

CHAPTER 1

THE DIFFERENTIAL AND INTEGRAL CALCULUS OF VECTORS

CHAPTER 2

CURVILINEAR COORDINATE SYSTEMS

CHAPTER 3

GREEN'S THEOREM AND ALLIED TOPICS

CHAPTER 4

UNRETARDED POTENTIAL THEORY

CHAPTER 5

RETARDED POTENTIAL THEORY

CHAPTER 6

HELMHOLTZ'S FORMULA AND ALLIED TOPICS

CHAPTER 7

EXPONENTIAL POTENTIAL THEORY

APPENDICES

CHAPTER 1

THE DIFFERENTIAL AND INTEGRAL CALCULUS OF VECTORS

1.1 Scalar and Vector Fields

A collection of points, or a point set, is said to be connected if every pair of points in the set can be joined by a continuous curve composed wholly of points of the set. Such a point set is said to comprise a region. Thus a region may be associated with a straight line or a plane surface, or with a curved line or a curved surface, or with that portion of space included within a closed surface, but not with a set of discrete points or non-intersecting enclosures.

A point is an interior point of a given region of space if it is the centre of a sphere of non-zero radius which contains only points of that region; if the sphere contains no points of the region then its centre is exterior to that region. A point is a boundary point of a given region of space if every sphere of which it is the centre contains points both of that region and not of that region. The modification of these definitions to cover regions which comprise surfaces or lines is straightforward.

A region which is described in such a way that it contains only interior points is said to be open; if it contains all of its boundary points it is said to be closed. Thus the equation $x^2 + y^2 + z^2 < a^2$ defines an open region of space which is the interior of a sphere of radius a centred upon the origin of coordinates, while $x^2 + y^2 + z^2 \leq a^2$ defines the corresponding closed region.

If V(P) is a scalar point function defined throughout the region $\underline{R}$, ie if for every point P of $\underline{R}$ there is at least one value of V, then the points comprising $\underline{R}$ together with their associated scalar magnitudes are said to constitute a scalar field within $\underline{R}$. When there is only one value of V corresponding to each point of $\underline{R}$ the field is said to be single-valued; otherwise it is multiple-valued.

The function V(P) is said to be continuous at the point P_o in $\underline{R}$ if $V(P_o)$ has a definite finite value and if, for every positive number ε, no matter how small, it is possible to find a positive number δ such that for all $PP_o < \delta$ (where P is a point of $\underline{R}$ and PP_o is the distance between P and P_o), $|V(P) - V(P_o)| < \varepsilon$. When V(P) is continuous at every point of $\underline{R}$ it is said to be continuous in $\underline{R}$.

It is sometimes possible to divide a region $\underline{R}$ into a finite set of sub-regions such that a scalar point function which is undefined upon the boundaries of the sub-regions and is discontinuous through them is nevertheless continuous at the interior points of each sub-region and exhibits limiting values as any point of a boundary is approached from one side or another. The point function is then said to be piecewise continuous in $\underline{R}$.

A surface is said to be equi-scalar or level with respect to V when V has the same value at all points of the surface. By definition, two level surfaces corresponding to different scalar values cannot intersect at any point of a single-valued field.

If $\overline{F}(P)$ is a vector point function defined throughout the region $\underline{R}$, then the points of $\underline{R}$ and the associated vector quantities constitute a vector field in $\underline{R}$. This may be single or multiple-valued. The definition of continuity of $\overline{F}(P)$ is entirely analogous to that given above for the scalar field.

Unless stated to the contrary, the scalar and vector fields of the following pages will be assumed to be single-valued and continuous.

1.2 Directional Derivative of a Scalar Field

Gradient

Let $V = V(x,y,z)$ be a scalar point function having continuous partial derivatives $\frac{\partial V}{\partial x}$, $\frac{\partial V}{\partial y}$ and $\frac{\partial V}{\partial z}$ throughout a region of space $\underline{R}$,[1] and let P_o and P be two interior points of $\underline{R}$ whose coordinates are (x_o, y_o, z_o) and $(x_o + \Delta x, y_o + \Delta y, z_o + \Delta z)$.

Then

$$\begin{aligned} V(P)-V(P_o) &= V(x_o+\Delta x, y_o+\Delta y, z_o+\Delta z) - V(x_o, y_o+\Delta y, z_o+\Delta z) \\ &\quad + V(x_o, y_o+\Delta y, z_o+\Delta z) - V(x_o, y_o, z_o+\Delta z) \\ &\quad + V(x_o, y_o, z_o+\Delta z) - V(x_o, y_o, z_o) \end{aligned}$$

whence, by the mean-value theorem,

$$V(P) - V(P_o) = \Delta V = \left(\frac{\partial V}{\partial x}\right)_{\substack{x' \\ y_o+\Delta y \\ z_o+\Delta z}} \Delta x + \left(\frac{\partial V}{\partial y}\right)_{\substack{x_o \\ y' \\ z_o+\Delta z}} \Delta y + \left(\frac{\partial V}{\partial z}\right)_{\substack{x_o \\ y_o \\ z'}} \Delta z$$

where $x_o < x' < x_o + \Delta x$; $y_o < y' < y_o + \Delta y$; $z_o < z' < z_o + \Delta z$

Let $\overrightarrow{P_oP} = \Delta\overline{s} = \hat{\overline{s}}\Delta s = \overline{i}\Delta x + \overline{j}\Delta y + \overline{k}\Delta z$

where $\hat{\overline{s}}$ is a unit vector directed along $\overrightarrow{P_oP}$.

---

1. Unless stated to the contrary, continuity of the first partial derivatives will be taken to imply continuity of the function itself. Similarly, continuity of nth order partial derivatives will imply continuity of lower order derivatives and of the function.

The average rate of change of V along $\vec{P_oP}$ is given by

$$\frac{\Delta V}{\Delta s} = \left(\frac{\partial V}{\partial x}\right)_{\substack{x' \\ y_o+\Delta y \\ z_o+\Delta z}} \cos(sx) + \left(\frac{\partial V}{\partial y}\right)_{\substack{x_o \\ y' \\ z_o+\Delta z}} \cos(sy) + \left(\frac{\partial V}{\partial z}\right)_{\substack{x_o \\ y_o \\ z'}} \cos(sz)$$

where cos(sx), cos(sy) and cos(sz) are the direction cosines of $\hat{s}$, ie the cosines of the angles made by $\hat{s}$ with the positive x, y and z axes.

The rate of change of V at P_o in the direction defined by $\hat{s}$ is found by taking the limit of the above expression as $\Delta s \to 0$ while the direction cosines remain constant.

Since the first derivatives of V are continuous about P_o we obtain

$$\lim_{\Delta s \to 0} \frac{\Delta V}{\Delta s} = \left(\frac{dV}{ds}\right)_{P_o} = \left(\frac{\partial V}{\partial x}\right)_{P_o} \cos(sx) + \left(\frac{\partial V}{\partial y}\right)_{P_o} \cos(sy) + \left(\frac{\partial V}{\partial z}\right)_{P_o} \cos(sz)$$

or, in general,

$$\frac{dV}{ds} = \frac{\partial V}{\partial x}\cos(sx) + \frac{\partial V}{\partial y}\cos(sy) + \frac{\partial V}{\partial z}\cos(sz) \qquad (1.2\text{-}1)$$

Upon re-writing in the form of a scalar product this becomes

$$\frac{dV}{ds} = (\bar{i}\cos(sx) + \bar{j}\cos(sy) + \bar{k}\cos(sz)).\left(\bar{i}\frac{\partial V}{\partial x} + \bar{j}\frac{\partial V}{\partial y} + \bar{k}\frac{\partial V}{\partial z}\right)$$

$$= \hat{s} \,.\, \left(\bar{i}\frac{\partial V}{\partial x} + \bar{j}\frac{\partial V}{\partial y} + \bar{k}\frac{\partial V}{\partial z}\right)$$

$\frac{dV}{ds}$ is known as the directional derivative of V corresponding to the direction $\hat{s}$, and the vector point function $\bar{i}\frac{\partial V}{\partial x} + \bar{j}\frac{\partial V}{\partial y} + \bar{k}\frac{\partial V}{\partial z}$ is said to be the gradient (grad) of the scalar V, so that

$$\text{grad } V = \bar{i}\frac{\partial V}{\partial x} + \bar{j}\frac{\partial V}{\partial y} + \bar{k}\frac{\partial V}{\partial z} \qquad (1.2\text{-}2)$$

and

$$\frac{dV}{ds} = \hat{s}.\text{grad } V \qquad (1.2\text{-}3)$$

It is now seen that the magnitude of the directional derivative at a point is equal to the scalar component of grad V in the direction of $\hat{s}$ at that point, consequently grad V defines both in magnitude and direction the maximum positive rate of change of V. Further, grad V must be normal to the surface of constant V through any point, since only in this case is $\frac{dV}{ds}$ zero in all directions within the surface as required. This leads to an alternative definition:

$$\text{grad } V = \hat{n}\,\frac{dV}{dn} \tag{1.2-4}$$

where $\hat{n}$ is the unit vector normal to the surface of constant V through the point in question and directed towards larger positive values of V, and $\frac{dV}{dn}$ is the rate of change of V in this direction.

Since this definition in no way involves any coordinate system it follows that grad V, as expressed in (1.2-2), must be independent of the system of rectangular axes chosen. An analytical proof of this is given in Sec. 1.19.

A considerable simplification of expression may be effected by treating $\left(\bar{i}\,\frac{\partial}{\partial x} + \bar{j}\,\frac{\partial}{\partial y} + \bar{k}\,\frac{\partial}{\partial z}\right)$ as a differential vector operator, written as ∇ and called 'del' or 'nabla', which operates upon a scalar point function in accordance with the following convention:

$$\nabla V = \left(\bar{i}\,\frac{\partial}{\partial x} + \bar{j}\,\frac{\partial}{\partial y} + \bar{k}\,\frac{\partial}{\partial z}\right) V = \bar{i}\,\frac{\partial V}{\partial x} + \bar{j}\,\frac{\partial V}{\partial y} + \bar{k}\,\frac{\partial V}{\partial z} \tag{1.2-5}$$

Then

$$\text{grad } V = \nabla V \quad ; \quad \frac{dV}{ds} = \hat{s}.(\nabla V)$$

We may carry the notation further by treating ∇ as a formal vector in the scalar product $\hat{s}.\nabla$, in which case[2]

$$(\hat{s}.\nabla) = (\bar{i}\cos(sx) + \bar{j}\cos(sy) + \bar{k}\cos(sz)).\left(\bar{i}\,\frac{\partial}{\partial x} + \bar{j}\,\frac{\partial}{\partial y} + \bar{k}\,\frac{\partial}{\partial z}\right)$$

$$= \cos(sx)\,\frac{\partial}{\partial x} + \cos(sy)\,\frac{\partial}{\partial y} + \cos(sz)\,\frac{\partial}{\partial z} \tag{1.2-6}$$

---

2. More generally, if $\bar{F}$ is a vector point function, then

$$(\bar{F}.\nabla) = F_x\,\frac{\partial}{\partial x} + F_y\,\frac{\partial}{\partial y} + F_z\,\frac{\partial}{\partial z}$$

Then

$$(\hat{s}.\nabla)V = \cos(sx)\frac{\partial V}{\partial x} + \cos(sy)\frac{\partial V}{\partial y} + \cos(sz)\frac{\partial V}{\partial z}$$

whence

$$(\hat{s}.\nabla)V = \frac{dV}{ds} = \hat{s}.(\nabla V) \tag{1.2-7}$$

The differential form of (1.2-7) is

$$dV = ds(\hat{s}.\nabla)V = (d\bar{s}.\nabla)V \tag{1.2-8}$$

The finite difference form of this relationship is a series expansion of which $(\Delta\bar{s}.\nabla)V$ comprises the first term. This may be demonstrated by expanding $V(P)-V(P_o) = \Delta V$ in a Taylor series on the assumption that the 1st, 2nd .... partial and mixed derivatives of V are continuous. It may be shown that

$$\Delta V = \frac{\partial V}{\partial x}\Delta x + \frac{\partial V}{\partial y}\Delta y + \frac{\partial V}{\partial z}\Delta z + \frac{\partial^2 V}{\partial x^2}\frac{(\Delta x)^2}{2!} + \frac{\partial^2 V}{\partial y^2}\frac{(\Delta y)^2}{2!} + \frac{\partial^2 V}{\partial z^2}\frac{(\Delta z)^2}{2!}$$

$$+ \frac{\partial^2 V}{\partial x\partial y}\Delta x\Delta y + \frac{\partial^2 V}{\partial x\partial z}\Delta x\Delta z + \frac{\partial^2 V}{\partial y\partial z}\Delta y\Delta z + \text{-----}$$

where the derivatives are evaluated at P_o.

This may be put into the form

$$\Delta V = (\Delta\bar{s}.\nabla)V + \frac{1}{2!}(\Delta\bar{s}.\nabla)^2 V + \text{---} \tag{1.2-9}$$

where

$$(\Delta\bar{s}.\nabla)^2 = \left(\Delta x\frac{\partial}{\partial x} + \Delta y\frac{\partial}{\partial y} + \Delta z\frac{\partial}{\partial z}\right)^2$$

$$\equiv \sum (\Delta x)^2\frac{\partial^2}{\partial x^2} + 2\sum \Delta x\Delta y\frac{\partial^2}{\partial x\partial y}$$

1.3 Directional Derivative of a Vector Field

The directional derivative of the vector point function $\overline{F}$ in the direction defined by the unit vector $\hat{\overline{s}}$ is given at any point by the associated value of $\frac{d\overline{F}}{ds}$.

$$\frac{d\overline{F}}{ds} = \lim_{\Delta s\to 0}\frac{\Delta\overline{F}}{\Delta s} = \lim_{\Delta s\to 0}\left(\frac{\Delta\,\overline{i}F_x + \Delta\,\overline{j}F_y + \Delta\,\overline{k}F_z}{\Delta s}\right)$$

$$= \lim_{\Delta s\to 0}\left(\frac{\overline{i}\,\Delta F_x + \overline{j}\,\Delta F_y + \overline{k}\,\Delta F_z}{\Delta s}\right)$$

where F_x, F_y, F_z are rectangular components of $\overline{F}$, ie

$$\frac{d\overline{F}}{ds} = \overline{i}\,\frac{dF_x}{ds} + \overline{j}\,\frac{dF_y}{ds} + \overline{k}\,\frac{dF_z}{ds}$$

Then, provided that the first (partial) derivatives of $\overline{F}$ are continuous[3] at the point in question,

$$\frac{d\overline{F}}{ds} = \overline{i}(\hat{\overline{s}}.\nabla)\,F_x + \overline{j}(\hat{\overline{s}}.\nabla)\,F_y + \overline{k}(\hat{\overline{s}}.\nabla)\,F_z$$

$$= (\hat{\overline{s}}.\nabla)\,\overline{i}\,F_x + (\hat{\overline{s}}.\nabla)\,\overline{j}\,F_y + (\hat{\overline{s}}.\nabla)\,\overline{k}\,F_z$$

ie

$$\frac{d\overline{F}}{ds} = (\hat{\overline{s}}.\nabla)\overline{F} \qquad (1.3\text{-}1)$$

It should be noted that $(\hat{\overline{s}}.\nabla)\overline{F}$ cannot be replaced by $\hat{\overline{s}}.(\nabla\overline{F})$ since no meaning has been assigned to the direct operation of ∇ upon a vector. The differential form of (1.3-1) is

$$d\overline{F} = (d\overline{s}.\nabla)\overline{F} \qquad (1.3\text{-}2)$$

---

3. ie $\frac{\partial\overline{F}}{\partial x}$, $\frac{\partial\overline{F}}{\partial y}$, $\frac{\partial\overline{F}}{\partial z}$ are continuous (in which case the scalar components of $\overline{F}$ have continuous first derivatives).

More generally, (assuming continuity of the relevant partial derivatives) it follows from (1.2-9) that

$$\Delta\overline{F} = (\Delta\overline{s}.\nabla)\overline{F} + \frac{1}{2!}(\Delta\overline{s}.\nabla)^2\overline{F} + --- \qquad (1.3\text{-}3)$$

1.4 Differentiability of a Scalar Point Function

Gradient and Directional Derivative of Combined Scalar Fields

The scalar point function V is said to be differentiable at P_o if, in the notation of Sec. 1.2,

$$\left\{\Delta V - \left(\frac{\partial V}{\partial x}\right)_{P_o}\Delta x - \left(\frac{\partial V}{\partial y}\right)_{P_o}\Delta y - \left(\frac{\partial V}{\partial z}\right)_{P_o}\Delta z\right\}\Big/\Delta r \to 0 \qquad \text{as } \Delta r \to 0$$

where $\Delta r^2 = \Delta x^2 + \Delta y^2 + \Delta z^2$.

Thus, V is differentiable at P_o when the first partial derivatives of V are continuous about P_o, since in this case we may write

$$\Delta V = \left\{\left(\frac{\partial V}{\partial x}\right)_{P_o} + \varepsilon_1\right\}\Delta x + \left\{\left(\frac{\partial V}{\partial y}\right)_{P_o} + \varepsilon_2\right\}\Delta y + \left\{\left(\frac{\partial V}{\partial z}\right)_{P_o} + \varepsilon_3\right\}\Delta z$$

where $\varepsilon_1 \to 0$, $\varepsilon_2 \to 0$, $\varepsilon_3 \to 0$ as $\Delta x, \Delta y, \Delta z \to 0$.

This, however, is not a necessary condition[4].

At points where V is differentiable, the directional derivative is given for all $\hat{s}$ by

$$\frac{dV}{ds} = \hat{s}.\text{grad } V$$

where grad V is identified with $\overline{i}\,\frac{\partial V}{\partial x} + \overline{j}\,\frac{\partial V}{\partial y} + \overline{k}\,\frac{\partial V}{\partial z}$.

When V is not differentiable, grad V is no longer considered to exist (or is not considered to be a vector) although it may still be possible to evaluate $\overline{i}\,\frac{\partial V}{\partial x} + \overline{j}\,\frac{\partial V}{\partial y} + \overline{k}\,\frac{\partial V}{\partial z}$.

---

4. See H. and B.S. Jeffreys, "Methods of Mathematical Physics", Sec. 5.041,

For the case of two point functions, V_1 and V_2, we have

$$\bar{i}\frac{\partial}{\partial x}(V_1+V_2) + \bar{j}\frac{\partial}{\partial y}(V_1+V_2) + \bar{k}\frac{\partial}{\partial z}(V_1+V_2)$$

$$= \bar{i}\frac{\partial V_1}{\partial x} + \bar{j}\frac{\partial V_1}{\partial y} + \bar{k}\frac{\partial V_1}{\partial z} + \bar{i}\frac{\partial V_2}{\partial x} + \bar{j}\frac{\partial V_2}{\partial y} + \bar{k}\frac{\partial V_2}{\partial z}$$

wherever the partial derivatives are defined.

When V_1 and V_2 are also differentiable this may be replaced by

$$\text{grad } (V_1+V_2) = \text{grad } V_1 + \text{grad } V_2 \tag{1.4-1}$$

In addition,

$$\text{grad } V_1V_2 = V_1 \text{ grad } V_2 + V_2 \text{ grad } V_1 \tag{1.4-2}$$

$$\text{grad } (V_1/V_2) = (1/V_2) \text{ grad } V_1 - (V_1/V_2^2) \text{ grad } V_2 \quad (V_2 \neq 0) \tag{1.4-3}$$

ie the gradient operator obeys the usual rules of differentation when operating upon the sum, product and quotient of scalar point functions. Corresponding formulae hold for the directional derivatives, eg

$$\frac{d}{ds}(V_1V_2) = V_1 \hat{s} \cdot \text{grad } V_2 + V_2 \hat{s} \cdot \text{grad } V_1$$

EXERCISES

1-1. Derive an expression for the unit normal to the surface $3x^2yz+2y^2-z^2 = 13$ at the point (1,2,1)

Ans: $\dfrac{\bar{i}12+\bar{j}11+\bar{k}4}{\sqrt{281}}$

1-2. Show that the vector $\bar{i}x+\bar{j}y+\bar{k}z$ drawn from the origin of coordinates is normal to the surface $x^2+y^2+z^2$ = constant at the point of intersection.

1-3. Find the directional derivative of the scalar point function $4x^2y+3xz^3+5xyz$ in the direction $\bar{i}4+\bar{j}5-\bar{k}6$ at the point (1,2,3)

Ans: 6.50

1-4. Find the angle between the normals to the surfaces $x^2+3y^2+4z = 48$ and $2x+3xy+z^2 = 36$ at the point (1,3,5)

Ans: 65° 38'

1-5. Derive an expression for the unit tangent to the curve of intersection of the surfaces in the previous problem at the point (1,3,5)

Ans: $\bar{i}7+\bar{j}-\bar{k}8$ / [illegible]

1-6. Derive an expression for the directional derivative of the vector point function $\bar{i}4x^2y+\bar{j}3xz^3+\bar{k}5xyz$ in the direction of $\bar{i}4+\bar{j}5-\bar{k}6$ at the point (1,2,3)

Ans: $\dfrac{\bar{i}84-\bar{j}162+\bar{k}135}{\sqrt{77}}$

1-7. Prove that $(\bar{F}.\nabla)\bar{r} = \bar{F}$ where $\bar{r} = \bar{i}x+\bar{j}y+\bar{k}z$

1-8. If $V = V(u,v,w$ ----)

and $u = u(x,y,z)$
$v = v(x,y,z)$
$w = w(x,y,z)$

have continuous partial derivatives, prove that

$$\text{grad } V = \frac{\partial V}{\partial u}\text{grad } u + \frac{\partial V}{\partial v}\text{grad } v + \frac{\partial V}{\partial w}\text{grad } w + \text{----}$$

(see Sec. 2.4)

1-9. Confirm equation (1.2-9)

1-10. Let V be a scalar point function having continuous partial derivatives $\frac{\partial V}{\partial x}, \frac{\partial V}{\partial y}, \frac{\partial V}{\partial z}$ throughout a region of space. Prove that the partial derivatives $\frac{\partial V}{\partial x'}, \frac{\partial V}{\partial y'}, \frac{\partial V}{\partial z'}$ are likewise continuous throughout the region, where x', y', z' refer to another rectangular coordinate system of random orientation.

Hence show that the continuity of the partial derivatives of V in x, y, z is sufficient to ensure that

$$\bar{i}\frac{\partial V}{\partial x} + \bar{j}\frac{\partial V}{\partial y} + \bar{k}\frac{\partial V}{\partial z} = \bar{i}'\frac{\partial V}{\partial x'} + \bar{j}'\frac{\partial V}{\partial y'} + \bar{k}'\frac{\partial V}{\partial z'}$$

1-11. Make use of the relationships which exist between $\bar{i}$, $\bar{j}$, $\bar{k}$ and $\bar{i}'$, $\bar{j}'$, $\bar{k}'$ in two systems of rectangular coordinates (Sec. 1.19), to prove that the equality of the previous exercise subsists at a point in the presence of discontinuities of the partial derivatives, provided that V is differentiable at that point.

1-12. If $V = f(x,y,z)$

and

$$f(x,y,z) = \frac{x\,y\,z}{(x^2+y^2+z^2)^{3/2}} \qquad \left((x,y,z) \neq (0,0,0)\right)$$

$$f(0,0,0) = 0$$

show that $\bar{i}\frac{\partial V}{\partial x} + \bar{j}\frac{\partial V}{\partial y} + \bar{k}\frac{\partial V}{\partial z}$ is defined at the origin but grad V is not.

1.5 Scalar and Vector Line Integrals

Let PQ be a continuous curve (which will be designed Γ) and let $\bar{r}_o$, $\bar{r}_1$, --- $\bar{r}_i$ --- $\bar{r}_n$ be vectors[5] drawn from an arbitrary external origin to closely-spaced points on it (Fig. 1.1). The vectors $\bar{r}_1 - \bar{r}_o$, $\bar{r}_2 - \bar{r}_1$ --- comprise a set of directed chords which approximate the profile of the curve. The typical chord is shown in the figure; it is $\bar{r}_i - \bar{r}_{i-1} = \Delta\bar{r}_i$ and its positive sense is along the curve from P to Q.

If $\hat{s}_i$ is a unit vector directed along $\Delta\bar{r}_i$, then $\Delta\bar{r}_i = \hat{s}_i \, \Delta s_i$ where $\Delta s_i = |\Delta\bar{r}_i|$. Let the unit tangent to the curve at the end point of $\bar{r}_i$ be designated $\hat{t}_i$. Then $\hat{s}_i$ approaches $\hat{t}_i$ as $\Delta s_i \to 0$.

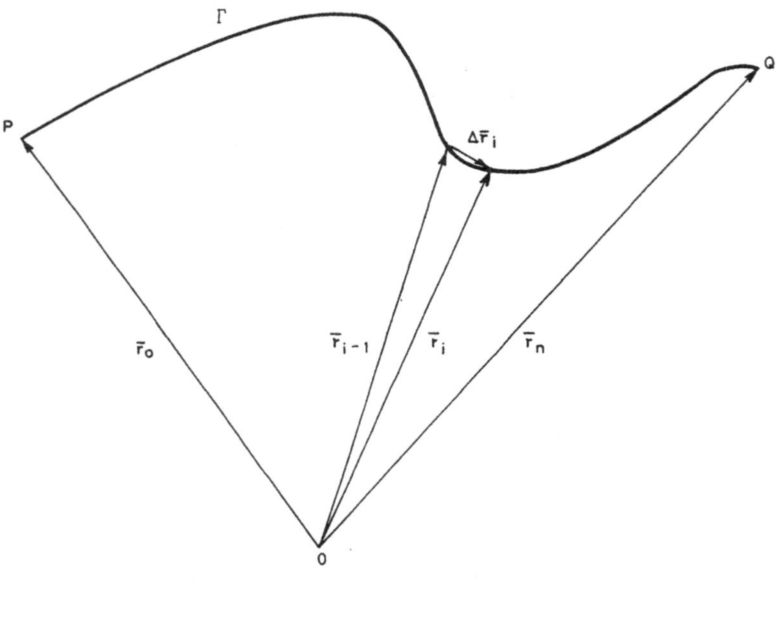

Fig. 1.1

5. These are known as position vectors since they serve to define the position of points on PQ relative to the external origin.

Suppose that V and $\overline{F}$ are bounded scalar and vector point functions which are piecewise continuous over PQ (where 'piecewise continuous' includes 'continuous')[6]. Then provided that PQ is regular[7], the following limits exist and serve to define four distinct types of line (or contour) integral:

(a)

$$\int_\Gamma V|d\overline{r}| = \lim_{\substack{n\to\infty \\ \Delta\overline{r}_i\to 0}} \sum_{i=1}^{n} V_i|\Delta\overline{r}_i| = \lim_{\substack{n\to\infty \\ \Delta s_i\to 0}} \sum_{i=1}^{n} V_i\,\Delta s_i = \int_\Gamma V\,ds \qquad (1.5\text{-}1)$$

where V_i is the value of V at any point of the element of arc intercepted by $\Delta\overline{r}_i$.

When V = 1 the integral defines the length of the curve between P and Q.

(b)

$$\int_\Gamma V\,d\overline{r} = \lim_{\substack{n\to\infty \\ \Delta\overline{r}_i\to 0}} \sum_{i=1}^{n} V_i\,\Delta\overline{r}_i = \lim_{\substack{n\to\infty \\ \Delta s_i\to 0}} \sum_{i=1}^{n} V_i\,\hat{s}_i\,\Delta s_i = \int_\Gamma V\,\hat{t}\,ds \qquad (1.5\text{-}2)$$

---

6. It has been assumed that any discontinuities of V or $\overline{F}$ which may be present can be accommodated within intervals of arbitrarily small total length. This applies equally to the line integrals considered below and, in equivalent form, to the surface and volume integrals of Secs. 1.6 and 1.7.

7. A regular curve consists of a finite number of non-intersecting regular arcs joined end to end. A regular arc is a set of points which, for some orientation of Cartesian axes, can be expressed as $y = f(x)$, $z = g(x)$ within an interval $a \le x \le b$, where $f(x)$ and $g(x)$ have continuous first derivatives. It follows that a regular curve has a continuously turning tangent at interior points of its component arcs; it is said to be piecewise smooth or piecewise differentiable. The curve may, of course, comprise only a single arc, in which case the unit vector tangent is continuous throughout.

It will be supposed, henceforth, that the curves under consideration are regular. For a detailed treatment the reader should consult O.D. Kellogg, "Foundations of Potential Theory," pp. 97-100, Frederick Ungar Publishing Co., n.d., New York.

With $\bar{\Delta r} = \bar{i}\Delta r_x + \bar{j}\,\Delta r_y + \bar{k}\Delta r_z$ this becomes

$$\int_\Gamma V\, d\bar{r} = \lim_{\substack{n\to\infty \\ \Delta\bar{r}_i\to\bar{0}}} \sum_{i=1}^{n} V_i\,(\bar{i}\,\Delta r_x + \bar{j}\,\Delta r_y + \bar{k}\,\Delta r_z)_i$$

Since the limiting value is independent of the mode of subdivision of the curve PQ during the limiting process, the above expression may be replaced by the sum of three individual limits, as shown below, where the associated n's and i's are unrelated[8].

$$\int_\Gamma V\, d\bar{r} = \lim_{\substack{n\to\infty \\ (\Delta r_x)_i\to 0}} \sum_{i=1}^{n} \bar{i}(V\Delta r_x)_i + \lim_{\substack{n\to\infty \\ (\Delta r_y)_i\to 0}} \sum_{i=1}^{n} \bar{j}(V\Delta r_y)_i + \lim_{\substack{n\to\infty \\ (\Delta r_z)_i\to 0}} \sum_{i=1}^{n} \bar{k}(V\Delta r_z)_i$$

$$= \bar{i}\int_\Gamma V\, dr_x + \bar{j}\int_\Gamma V\, dr_y + \bar{k}\int_\Gamma V\, dr_z \qquad (1.5\text{-}3)$$

(c)

$$\int_\Gamma \bar{F}.d\bar{r} = \lim_{\substack{n\to\infty \\ \Delta\bar{r}_i\to\bar{0}}} \sum_{i=1}^{n} \bar{F}_i.\Delta\bar{r}_i = \lim_{\substack{n\to\infty \\ \Delta s_i\to 0}} \sum_{i=1}^{n} \bar{F}_i.\hat{\bar{s}}_i\Delta s_i = \int_\Gamma F_t ds \qquad (1.5\text{-}4)$$

where $\bar{F}_i$ is the value of $\bar{F}$ at any point of the element of arc intercepted by $\Delta\bar{r}_i$.

---

8. This applies equally to the line integrals considered below, and to the surface and volume integrals of Secs. 1.6 and 1.7.

Further,

$$\int_\Gamma \bar{F}.d\bar{r} = \lim_{\substack{n\to\infty \\ \Delta\bar{r}_i\to\bar{0}}} \sum_{i=1}^{n} (F_x\Delta r_x+F_y\Delta r_y+F_z\Delta r_z)_i$$

$$= \lim_{\substack{n\to\infty \\ (\Delta r_x)_i\to 0}} \sum_{i=1}^{n} (F_x\Delta r_x)_i + \lim_{\substack{n\to\infty \\ (\Delta r_y)_i\to 0}} \sum_{i=1}^{n} (F_y\Delta r_y)_i + \lim_{\substack{n\to\infty \\ (\Delta r_z)_i\to 0}} \sum_{i=1}^{n} (F_z\Delta r_z)_i$$

$$= \int_\Gamma F_x dr_x + \int_\Gamma F_y dr_y + \int_\Gamma F_z dr_z \qquad (1.5\text{-}5)$$

This integral is known as the tangential line integral of $\bar{F}$ since the component of $\bar{F}$ tangential to the curve replaces V in (a).

(d)

$$\int_\Gamma \bar{F}\times d\bar{r} = \lim_{\substack{n\to\infty \\ \Delta\bar{r}_i\to\bar{0}}} \sum_{i=1}^{n} \bar{F}_i \times \Delta\bar{r}_i = \lim_{\substack{n\to\infty \\ \Delta s_i\to 0}} \sum_{i=1}^{n} \bar{F}_i \times \hat{s}_i\, \Delta s_i = \int_\Gamma (\bar{F}\times\hat{t})ds \qquad (1.5\text{-}6)$$

$$= \lim_{\substack{n\to\infty \\ \Delta\bar{r}_i\to\bar{0}}} \sum_{i=1}^{n} \{\bar{i}(F_y\Delta r_z-F_z\Delta r_y)_i + \bar{j}(F_z\Delta r_x-F_x\Delta r_z)_i + \bar{k}(F_x\Delta r_y-F_y\Delta r_x)_i\}$$

$$= \bar{i}\int_\Gamma (F_y dr_z-F_z dr_y) + \bar{j}\int_\Gamma (F_z dr_x-F_x dr_z) + \bar{k}\int_\Gamma (F_x dr_y-F_y dr_x) \qquad (1.5\text{-}7)$$

This is sometimes called the skew line integral of $\bar{F}$ between P and Q.

A line integral is said to be closed if the end points coincide, ie if the integration is carried out around a closed curve. In this case the integral sign is replaced by the symbol $\oint$ as in $\oint_\Gamma \overline{F} \times d\overline{r}$. The positive sense of integration around the curve must, of course, be stipulated when vector quantities are involved. The tangential line integral of $\overline{F}$ around the closed curve Γ, viz. $\oint_\Gamma \overline{F} \, . \, d\overline{r}$, is sometimes referred to as the circulation of $\overline{F}$ about Γ.

Although the general treatment of the integral calculus of scalar and vector fields does not require that definite integrals be evaluated, specific problems may, on the other hand, demand this. Under these circumstances it is necessary to reduce such integrals as appear in (a), (b), (c), (d) above to the form $\int_{x_1}^{x_2} f(x)\,dx$, or equivalent, so that integration may proceed in the usual way via the fundamental relationship:

$$\int_{x_1}^{x_2} f(x)\,dx = \phi(x_2) - \phi(x_1) \text{ where } \phi'(x) = f(x) \tag{1.5-8}$$

In this respect no difficulty is encountered with the scalar integrals $\int V\,ds$ and $\int F_t\,ds$ so long as V and F_t can be expressed as functions of s, since this establishes the required form of integral. In general, however, the procedure is more complicated.

Consider, for example, the evaluation of the tangential line integral of $\overline{F}$ along a curve lying wholly in the xy plane where $\overline{F} = \overline{F}(x,y)$. In this case (1.5-5) becomes

$$\int_\Gamma \overline{F}.d\overline{r} = \int_\Gamma F_x dr_x + \int_\Gamma F_y dr_y = \int_\Gamma f_1(x,y)dr_x + \int_\Gamma f_2(x,y)dr_y \tag{1.5-5(a)}$$

Since the equation of the curve relates x and y at each point, it is possible to express (1.5-5(a)) as

$$\int_\Gamma \overline{F}.d\overline{r} = \int_\Gamma g_1(x)dr_x + \int_\Gamma g_2(y)dr_y$$

But $|\Delta r_x| = |\Delta x|$ and $|\Delta r_y| = |\Delta y|$, hence

$$\int_\Gamma \bar{F}.d\bar{r} = \int_{x_P}^{x_Q} g_1(x)dx + \int_{y_P}^{y_Q} g_2(y)dy \qquad (1.5\text{-}5(b))$$

where the order of the limits of integration now carries that degree of responsibility for the sign of the result which was previously borne by the intrinsic sign of Δr_x and Δr_y.

(1.5-5(b)) is the form of integral required for evaluation since each integrand is a function of the integration variable alone.

For the case of a curve in space the Cartesian coordinates of each point are sometimes expressed as individual functions of some independent variable, say t. (Parametric representation). The appropriate form of (1.5-5) may then be shown to be

$$\int_\Gamma \bar{F}.d\bar{r} = \int_{t_P}^{t_Q} F_x \frac{dx}{dt}\,dt + \int_{t_P}^{t_Q} F_y \frac{dy}{dt}\,dt + \int_{t_P}^{t_Q} F_z \frac{dz}{dt}\,dt \qquad (1.5\text{-}5(c))$$

so long as the derivatives are finite throughout the interval.

Since the rectangular components of $\bar{F}$ are assumed to be functions of (x,y,z), they may be expressed over Γ as functions of t. The derivatives of x, y and z with respect to t are likewise functions of t so that (1.5-5(c)) may be written as

$$\int_\Gamma \bar{F}.d\bar{r} = \int_{t_P}^{t_Q} h_1(t)dt + \int_{t_P}^{t_Q} h_2(t)dt + \int_{t_P}^{t_Q} h_3(t)dt$$

which is the form required for evaluation.

1.6 Scalar and Vector Surface Integrals

Let S be a two-sided regular surface, either open or closed[9], and let its faces be divided into a system of elements designated ΔS_1, ΔS_2 --- ΔS_i --- ΔS_n. The positive sense of the vector normal at any point of the surface is taken to correspond to motion through the surface from the negative to the positive side. For a closed surface the outer side is conventionally chosen as positive; for an unclosed surface the choice is arbitrary.

We associate with the typical surface element ΔS_i the vector $\Delta\bar{S}_i = \bar{i}\,(\Delta S_x)_i + \bar{j}(\Delta S_y)_i + \bar{k}(\Delta S_z)_i$ where $(\Delta S_x)_i$, $(\Delta S_y)_i$ and $(\Delta S_z)_i$ are the projected areas of ΔS_i on the $\bar{i}$, $\bar{j}$ and $\bar{k}$ coordinate planes resp. (ie on the yz, xz and xy planes). The sign of the projected area is to be taken as positive if the vector normal at each point of ΔS_i makes an angle of less than 90° with the corresponding positive coordinate axis, and negative if the angle is greater than 90°. When there is a change of sign across an element the surface should be re-divided to remove the anomaly.

We now assign to ΔS_i the additional significance of $|\Delta\bar{S}_i|$, and define $\hat{\bar{n}}_i$ by means of $\Delta\bar{S}_i = \hat{\bar{n}}_i\,\Delta S_i$[10].

Then if V and $\bar{F}$ are bounded scalar and vector point functions which are piecewise continuous upon S, the following definitions apply.

(a)

$$\int_S V|d\bar{S}| = \lim_{\substack{n\to\infty \\ \Delta\bar{S}_i\to\bar{0}}} \sum_{i=1}^{n} V_i|\Delta\bar{S}_i| = \lim_{\substack{n\to\infty \\ \Delta S_i\to 0}} \sum_{i=1}^{n} V_i\,\Delta S_i = \int_S V\,dS \qquad (1.6\text{-}1)$$

---

9. A surface S is regular if it can be divided into a finite number of non-intersecting regular parts or 'faces', each of which can be represented, for some orientation of Cartesian axes, by $z = f(x,y)$, where x,y are the points of a region of the xy coordinate plane bounded by a closed regular curve and $f(x,y)$ has continuous first derivatives throughout the region. As a consequence, the unit vector normal to the surface is continuous at interior points of all faces but is not necessarily continuous upon their boundaries. (The latter may be shown to comprise regular curves.) S is then said to be piecewise smooth or piecewise differentiable. A surface is closed if each component arc of the regular curve which bounds any face ('edge' of a face) is common to two faces. Otherwise the surface is open. (See Kellogg, pp 105-12).

All surfaces to be treated subsequently will be assumed to be regular.

10. It will be evident from Ex. 1-24, p. 24 that as ΔS_i shrinks about an interior point of the element, so $\hat{\bar{n}}_i$ approaches the unit normal to the surface at that point.

where V_i is the value of V at any point of ΔS_i.

When V = 1 the integral defines the scalar area of the surface S.

(b)

$$\int_S V\,d\overline{S} = \lim_{\substack{n\to\infty \\ \Delta\overline{S}_i\to\overline{0}}} \sum_{i=1}^{n} V_i \Delta\overline{S}_i$$

$$= \lim_{\substack{n\to\infty \\ \Delta S_i\to 0}} \sum_{i=1}^{n} V_i \hat{\overline{n}}_i \Delta S_i = \int_S V\,\hat{\overline{n}}\,dS \tag{1.6-2}$$

$$= \lim_{\substack{n\to\infty \\ \Delta\overline{S}_i\to\overline{0}}} \sum_{i=1}^{n} V_i (\overline{i}\Delta S_x + \overline{j}\Delta S_y + \overline{k}\Delta S_z)_i$$

$$= \lim_{\substack{n\to\infty \\ (\Delta S_x)_i\to 0}} \sum_{i=1}^{n} \overline{i}(V\Delta S_x)_i + \lim_{\substack{n\to\infty \\ (\Delta S_y)_i\to 0}} \sum_{i=1}^{n} \overline{j}(V\Delta S_y)_i + \lim_{\substack{n\to\infty \\ (\Delta S_z)_i\to 0}} \sum_{i=1}^{n} \overline{k}(V\Delta S_z)_i \tag{1.6-3}$$

$$= \overline{i}\int_S V\,dS_x + \overline{j}\int_S V\,dS_y + \overline{k}\int_S V\,dS_z$$

When V = 1 the integral defines the vector area of the surface S.

(c)

$$\int_S \overline{F}.d\overline{S} = \lim_{\substack{n\to\infty \\ \Delta\overline{S}_i\to\overline{0}}} \sum_{i=1}^{n} \overline{F}_i.\Delta\overline{S}_i = \lim_{\substack{n\to\infty \\ \Delta S_i\to 0}} \sum_{i=1}^{n} \overline{F}_i.\hat{\overline{n}}_i\Delta S_i = \sum_S F_n dS \tag{1.6-4}$$

where $\overline{F}_i$ is the value of $\overline{F}$ at any point of ΔS_i.

Further,

$$\int_S \bar{F}.d\bar{S} = \lim_{\substack{n\to\infty \\ \Delta\bar{S}_i\to\bar{0}}} \sum_{i=1}^{n} (F_x\Delta S_x + F_y\Delta S_y + F_z\Delta S_z)_i$$

$$= \lim_{\substack{n\to\infty \\ (\Delta S_x)_i\to 0}} \sum_{i=1}^{n} (F_x\Delta S_x)_i + \lim_{\substack{n\to\infty \\ (\Delta S_y)_i\to 0}} \sum_{i=1}^{n} (F_y\Delta S_y)_i + \lim_{\substack{n\to\infty \\ (\Delta S_z)_i\to 0}} \sum_{i=1}^{n} (F_z\Delta S_z)_i$$

$$= \int_S F_x dS_x + \int_S F_y dS_y + \int_S F_z dS_z \qquad (1.6\text{-}5)$$

This integral is known as the normal surface integral of $\bar{F}$ over S, since the normal component of $\bar{F}$, viz. F_n, replaces V in the scalar surface integral (a).

(d)

$$\int_S \bar{F} \times d\bar{S} = \lim_{\substack{n\to\infty \\ \Delta\bar{S}_i\to\bar{0}}} \sum_{i=1}^{n} \bar{F}_i \times \Delta\bar{S}_i$$

$$= \lim_{\substack{n\to\infty \\ \Delta S_i\to 0}} \sum_{i=1}^{n} \bar{F}_i \times \hat{n}_i \, \Delta S_i = \int_S (\bar{F}\times\hat{n}) \, dS \qquad (1.6\text{-}6)$$

$$= \lim_{\substack{n\to\infty \\ \Delta\bar{S}_i\to\bar{0}}} \sum_{i=1}^{n} \bar{F}_i \times (\bar{i}\Delta S_x + \bar{j}\Delta S_y + \bar{k}\Delta S_z)_i$$

$$= \bar{i} \int_S (F_y dS_z - F_z dS_y) + \bar{j} \int_S (F_z dS_x - F_x dS_z) + \bar{k} \int_S (F_x dS_y - F_y dS_x) \qquad (1.6\text{-}7)$$

(e)

$$\int_S \bar{F}\, dS = \lim_{\substack{n\to\infty \\ \Delta S_i\to 0}} \sum_{i=1}^{n} \bar{F}_i \Delta S_i$$

$$= \lim_{\substack{n\to\infty \\ \Delta S_i\to 0}} \sum_{i=1}^{n} (\bar{i}F_x+\bar{j}F_y+\bar{k}F_z)_i \,\Delta S_i \tag{1.6-8}$$

$$= \lim_{\substack{n\to\infty \\ \Delta S_i\to 0}} \sum_{i=1}^{n} \bar{i}(F_x\Delta S)_i + \lim_{\substack{n\to\infty \\ \Delta S_i\to 0}} \sum_{i=1}^{n} \bar{j}(F_y\Delta S)_i + \lim_{\substack{n\to\infty \\ \Delta S_i\to 0}} \sum_{i=1}^{n} \bar{k}(F_z\Delta S)_i$$

$$= \bar{i}\int_S F_x dS + \bar{j}\int_S F_y dS + \bar{k}\int_S F_z dS \tag{1.6-9}$$

When integration is carried out over a closed surface, the integral symbol is written $\oint_S$ as in $\oint_S V\, d\bar{S}$.

Like their line integral counterparts, surface integrals such as (1.6-3), (1.6-5) and (1.6-7) are shorthand expressions for the limits of sums. For the purposes of evaluation these expressions are replaced by iterated integrals. Thus, (1.6-5) becomes

$$\iint F_x dydz + \iint F_y dzdx + \iint F_z dxdy \tag{1.6-5(a)}$$

where the appropriate limits of integration are to be supplied, and where F_x, F_y and F_z are expressed in terms of the integration variables alone, through the agency of the equation of the surface.

The iterated integrals are evaluated by the standard procedure of successive integration.

Before passing to the next subject it should be noted that the scalar area S of a surface has been defined by

$$S = \lim_{\Delta\bar{S}_i\to\bar{0}} \sum_{i=1}^{n} |\Delta\bar{S}_i| = \int_S |d\bar{S}|$$

while the vector area of the surface has been defined by

$$\bar{S} = \lim_{\Delta\bar{S}_i \to \bar{0}} \sum_{i=1}^{n} \Delta\bar{S}_i = \int_S d\bar{S}$$

In general, $\left|\int d\bar{S}\right| \neq \int \left|d\bar{S}\right|$; only in the case of a plane surface can we write $|\bar{S}| = S$.

1.7 Scalar and Vector Volume Integrals

Let the space within a closed regular surface (regular region) be divided into volume elements designated $\Delta\tau_1$, $\Delta\tau_2$ ---- $\Delta\tau_i$ --- $\Delta\tau_n$, and let these symbols also represent the volume of the elements. Then if V and $\bar{F}$ are bounded scalar and vector point functions which are piecewise continuous throughout the region, the associated scalar and vector volume integrals are given by

(a)

$$\int_\tau V d\tau = \lim_{\substack{n\to\infty \\ \Delta\tau_i \to 0}} \sum_{i=1}^{n} V_i \Delta\tau_i \tag{1.7-1}$$

where V_i is the value of V at any point of $\Delta\tau_i$.

(b)

$$\int_\tau \bar{F} d\tau = \lim_{\substack{n\to\infty \\ \Delta\tau_i \to 0}} \sum_{i=1}^{n} \bar{F}_i \Delta\tau_i \tag{1.7-2}$$

$$= \lim_{\substack{n\to\infty \\ \Delta\tau_i \to 0}} \sum_{i=1}^{n} \bar{i}(F_x\Delta\tau)_i + \lim_{\substack{n\to\infty \\ \Delta\tau_i \to 0}} \sum_{i=1}^{n} \bar{j}(F_y\Delta\tau)_i + \lim_{\substack{n\to\infty \\ \Delta\tau_i \to 0}} \sum_{i=1}^{n} \bar{k}(F_z\Delta\tau)_i$$

$$= \bar{i}\int_\tau F_x d\tau + \bar{j}\int_\tau F_y d\tau + \bar{k}\int_\tau F_z d\tau \tag{1.7-3}$$

where $\bar{F}_i$ is the value of $\bar{F}$ at any point of $\Delta\tau_i$.

Similar remarks apply to the integral form of (1.7-3) as apply to the surface integrals of the previous section.

It should be noted in connection with both surface and volume integrals that the shape of the elementary units (supposed regular) is immaterial so long as all dimensions decrease uniformly during the limiting process.

EXERCISES

1-13. If $\overline{F}$ is a function of the scalar variable w prove that

$$\overline{F} \cdot \frac{d\overline{F}}{dw} = F\frac{dF}{dw}$$

where $F = |\overline{F}|$.

Use this result to show that if the unit vector $\hat{a}$ is a function of time (t) then $\frac{d\hat{a}}{dt}$ is perpendicular to $\hat{a}$.

1-14. Let $\overline{F}$ be a function of time and let $V = V_o\Big/\left(1 - \frac{F^2}{\gamma^2}\right)^{\frac{1}{2}}$ where V_o and γ are constants.

If $\frac{d}{dt}(V\overline{F}) = \overline{G}$ prove that $\frac{dV}{dt} = \frac{\overline{G}.\overline{F}}{\gamma^2}$.

1-15. If $\overline{r}$ is the position vector of a particle relative to some fixed point, its velocity is given by $\overline{v} = \frac{d\overline{r}}{dt}$ and its acceleration by $\frac{d\overline{v}}{dt} = \frac{d^2\overline{r}}{dt^2}$.

Prove that

(a) $\overline{v} = v\hat{t}$

where $v = \frac{ds}{dt}$ and $\hat{t}$ is the unit tangent to the path. (s is distance along the path measured from some point in it.)

(b) $\frac{d\overline{v}}{dt} = \hat{t}\frac{dv}{dt} + \hat{n}\frac{v^2}{\rho}$

where ρ is the radius of curvature of the path and $\hat{n}$ is the unit normal to the path directed towards the centre of curvature.

1-16. In the above problem prove that the rate at which the position vector sweeps out (vector) area is given by

$$\frac{d\overline{S}}{dt} = \frac{1}{2}\left(\overline{r} \times \frac{d\overline{r}}{dt}\right)$$

Derive an expression for $\frac{d^2\overline{S}}{dt^2}$ and hence show that when the acceleration of a particle is directed towards a fixed point the position vector from that point moves in a plane and sweeps out area at a constant rate.

Ans: $\frac{d^2\overline{S}}{dt^2} = \frac{1}{2}\left(\overline{r} \times \frac{d^2\overline{r}}{dt^2}\right)$

1-17. If a particle moves under the influence of gravity and forces of constraint which are everywhere normal to the path, show that

$$\tfrac{1}{2}\, mv^2 + m\, g\, h = \text{constant}$$

where m = mass of particle

g = acceleration due to gravity

h = height of particle above some datum level

[Hint: Write down the equation of motion, multiply each side scalarly by $2\bar{v}$ and integrate the resulting equation with respect to time.]

1-18. Let r be the distance between two moving points. Show that $\frac{dr}{dt} = v_r$ where v_r is the radial component of the relative velocity $\bar{v}$ of the points. Show also that

$$\frac{d^2r}{dt^2} = a_r + \frac{1}{r}(v^2 - v_r^2)$$

where a_r is the radial component of the relative acceleration $\bar{a}$ of the points.

1-19. Use (1.5-5) to evaluate $\int_P^Q \bar{F}.d\bar{r}$

where $\bar{F} = \bar{i}\, f_1(x) + \bar{j}\, f_2(y) + \bar{k}\, f_3(z)$

and f_1, f_2 and f_3 are well-behaved[11] functions.

Hence show that $\oint \bar{F}.d\bar{r}$ is zero for all closed paths.

$$\text{Ans:} \quad \int_P^Q \bar{F}.d\bar{r} = [\phi_1(x)]_{x_P}^{x_Q} + [\phi_2(y)]_{y_P}^{y_Q} + [\phi_3(z)]_{z_P}^{z_Q}$$

where $\phi_1'(x) = f_1(x)$; $\phi_2'(y) = f_2(y)$; $\phi_3'(z) = f_3(z)$

11. ie the functions and their derivatives have the requisite degree of continuity for the problem in hand.

1-20. In the definitions of the scalar and vector line integrals given in Sec. 1.5, V_i and $\bar{F}_i$ were taken to be the values of V and $\bar{F}$ at any point of the arc of the integration curve intercepted by the rectilinear element $\Delta\bar{r}_i$. Show that if V and $\bar{F}$ are defined beyond the curve and are continuous in its neighbourhood, then the same values obtain for the line integrals if V_i and $\bar{F}_i$ are taken to be the values of V and $\bar{F}$ at any point of $\Delta\bar{r}_i$.

1-21. Given that

(a) $\bar{r}$ is the position vector relative to some fixed point

(b) $r = |\bar{r}|$

(c) f(r) is a single-valued, well-behaved function of r

(d) $\bar{a}$ is a constant vector point function

show that

$$\frac{f'(r)}{r}\,(\bar{a}.\bar{r})\bar{r}\,(\bar{r}.d\bar{r}) + f(r)\,(\bar{a}.\bar{r})\,d\bar{r} + f(r)\bar{r}\,(\bar{a}.d\bar{r}) = d\bar{F}$$

where $\bar{F}$ is some single-valued vector point function.

(In this case the left-hand expression is said to be an exact differential.)

Hence prove that for any closed curve

$$\oint\left(\frac{f'(r)}{r}\,(\bar{a}.\bar{r})\bar{r}(\bar{r}.d\bar{r})+f(r)(\bar{a}.\bar{r})d\bar{r}+f(r)\bar{r}(\bar{a}.d\bar{r})\right) = \bar{0}$$

Ans: $\bar{F} = f(r)\,(\bar{a}.\bar{r})\bar{r}$

1-22. Derive equation (1.5-5c) from first principles.

1-23. Show that the length of a curve in space between points whose abscissae are x_1 and x_2 may be expressed as

$$\int_{x_1}^{x_2}\left\{1 + \left(\frac{dy}{dx}\right)^2 + \left(\frac{dz}{dx}\right)^2\right\}^{\frac{1}{2}} dx$$

provided that at no point of the interval does the tangent to the curve lie in the yz plane.

If x, y and z are expressed in terms of a parameter t show that the equivalent integral becomes

$$\int_{t_1}^{t_2} \left\{\left(\frac{dx}{dt}\right)^2 + \left(\frac{dy}{dt}\right)^2 + \left(\frac{dz}{dt}\right)^2\right\}^{\frac{1}{2}} dt$$

so long as the derivatives are finite throughout the interval.

1-24. If a plane surface is randomly orientated with respect to rectangular axes, show that

(1) the associated vector area, $\bar{S} = \bar{i}S_x + \bar{j}S_y + \bar{k}S_z$, is directed along the normal to the surface and away from its positive side.

(2) the modulus of the vector area, $|\bar{S}|$, is equal to the scalar area of the surface (in the usual arithmetical sense).

In the light of these results and on the assumption that a curved surface may be treated, in the limit of sub-division, as a system of plane elements (see Ex.1-27. below), demonstrate that although vector area, and, ultimately, scalar area have been defined primarily in terms of Cartesian components, the surface integrals of Sec. 1.6 are nevertheless independent of any system of coordinates.

1-25. Show that for a non-plane surface

$$\left|\int_S d\bar{S}\right| \neq \int_S |d\bar{S}|$$

1-26. Prove that $\oint_S d\bar{S} \equiv 0$

1-27. Let the vertices of a small plane triangle of vector area $\Delta\bar{S}'$ lie in a spherical surface. Three planes, each containing one side of the triangle, cut the spherical surface in a curvilinear triangle whose vector area is $\Delta\bar{S}$. Use the result of Ex.1-26. to show that $|\Delta\bar{S}'-\Delta\bar{S}|/|\Delta\bar{S}'|\to 0$ as the dimensions of the triangle approach zero.

Suppose now that an open curved surface S is approximated by a polyhedral surface composed of contiguous triangles whose vertices lie in S, the outer edge of the polyhedral surface approximating the contour of S.

Then if $\overline{F}$ is a vector point function, continuous on and about S, show that

$$\int_S \overline{F}.d\overline{S} = \lim_{\substack{n\to\infty \\ \Delta\overline{S}'_i\to\overline{0}}} \sum_{i=1}^{n} \overline{F}_i.\Delta\overline{S}'_i$$

where $\overline{F}_i$ is the value of $\overline{F}$ at any point of the triangle $\Delta S'_i$.

1-28. In a homogeneous liquid at rest the pressure p is related to the height h above some datum level by the equation

$$p + \rho gh = \text{const}$$

where ρ = density of liquid

g = acceleration due to gravity

The pressure at any point acts equally in all directions.

Prove that the upthrust on a body of any shape, when completely immersed in the liquid, is equal to the weight of the liquid displaced.

Extend this to the case where the body projects through the horizontal plane of demarcation between two homogeneous liquids of different densities.

1-29. Let the scalar point functions U and V be finite, single valued and continuous in the region R, and let W be defined in R by

$$W_P = \int_O^P U\, dV$$

where O is a fixed point of R and the integration is carried out along a smooth curve joining O and P.

Show from first principles that, so long as U is a function of V, W is single-valued in R and grad W = U grad V.

1.8 Stokes's Theorem

Curl of a Vector Field

An open two-sided surface defines a boundary curve around which a closed line integral may be taken. The direction of integration is conventionally chosen as anti-clockwise in relation to an observer who sees the positive side of the surface; in other words, the positive sense of the normal at the surface and positive motion around the closed curve bear a right-handed screw relationship.

The closed tangential line integral of a vector point function $\overline{F}$ taken around the boundary Γ of a smooth open surface S may be shown to be equal to the normal surface integral over S of a related vector point function, so long as $\overline{F}$ has continuous first derivatives in a region of space containing S. This may be demonstrated as follows.

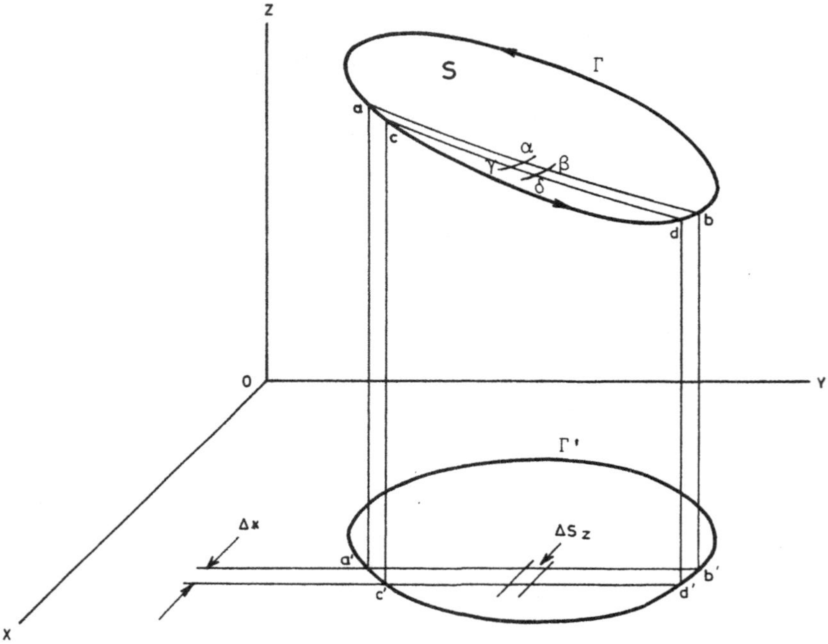

Fig. 1.2

In Fig. 1.2 the positive side of the open surface S is chosen arbitrarily as the side remote from 0 so that positive motion around the boundary Γ corresponds with the direction of the arrows. From (1.5-5) the tangential line integral of $\overline{F}$ around Γ is given by

$$\lim_{\substack{n\to\infty \\ (\Delta r_x)_i\to 0}} \sum_{i=1}^{n} (F_x \Delta r_x)_i + \lim_{\substack{n\to\infty \\ (\Delta r_y)_i\to 0}} \sum_{i=1}^{n} (F_y \Delta r_y)_i + \lim_{\substack{n\to\infty \\ (\Delta r_z)_i\to 0}} \sum_{i=1}^{n} (F_z \Delta r_z)_i \qquad (1.8\text{-}1)$$

where the summation is carried out around the entire curve.

We now proceed to transform the first term of (1.8-1) into a surface integral over S.

Let S be cut in the curves ab and cd by two planes drawn normal to the x axis, a distance Δx apart. The planes cut the projection of Γ on the xy coordinate plane, viz Γ', in a' and b', and c' and d'. The contribution to the first term of (1.8-1) from the elements of Γ intercepted by the planes may be written as $(F_x)_a\,(\overrightarrow{ac})_x + (F_x)_b\,(\overrightarrow{db})_x$. The orientations of $\overrightarrow{ac}$ and $\overrightarrow{db}$ relative to the positive direction of the x axis are such as to make this equal to $\{(F_x)_a - (F_x)_b\}\,\Delta x$, where Δx is intrinsically positive.

Now let a system of planes normal to the y axis be drawn through a and b and intermediate points, and let $\alpha\beta\delta\gamma$ be the typical surface element so defined. This will be designated ΔS. The projection of ΔS on the xy (or $\bar{k}$) coordinate plane, viz ΔS_z, is positive for the particular orientation of S adopted in the figure, because the positive normal at ΔS makes an angle of less than 90° with $\bar{k}$. ΔS_z is numerically equal to the product of Δx and the projection of $\alpha\beta$ on the xy plane. If Δy is the increment of the y coordinate on passing from β to α then it is clear that $\Delta y\Delta x = -\Delta S_z$. By projecting onto the xz plane we may show in a similar manner that $\Delta z\Delta x = +\Delta S_y$, where Δz is the increment of z on passing from β to α. (In this case the projected element of area will, in general, be non-rectangular.)

These two relationships continue to hold no matter how S may be distorted while the boundary remains fixed and the order of lettering from b to a remains $b\beta\alpha a$. Thus if S were so folded as to make Δy positive, then ΔS_z would be found to be negative because of the associated change of direction of the normal. The reader should convince himself of the validity of the contention by means of appropriate diagrams.

From considerations detailed in Sec. 1.2

$$(F_x)_\alpha - (F_x)_\beta = \frac{\partial F_x}{\partial y}\,\Delta y + \frac{\partial F_x}{\partial z}\,\Delta z$$

where Δy and Δz have the significance ascribed above, and the derivatives are referred to points in the neighbourhood of $\alpha\beta$ as dictated by the mean value theorem.

Then

$$(F_x)_a - (F_x)_b = \sum \left(\frac{\partial F_x}{\partial y}\,\Delta y + \frac{\partial F_x}{\partial z}\,\Delta z\right)$$

and

where the summations are carried out over all elements in the strip abdc.

Suppose now that Γ is divided entirely into pairs of elements such as ac and db by appropriately positioned planes drawn normal to the x axis, and that the strips so formed upon S are themselves divided by transverse planes as in the case of abdc.

Then

$$\sum_{\Gamma} F_x \Delta r_x = \sum_{S} \left(\frac{\partial F_x}{\partial z} \Delta S_y - \frac{\partial F_x}{\partial y} \Delta S_z \right)$$

and, as the number of intersecting planes is increased without limit while the spacing in each direction approaches zero,

$$\lim_{\Delta x \to 0} \sum_{\Gamma} F_x \Delta r_x = \lim_{\Delta S_y, \Delta S_z \to 0} \sum_{S} \left(\frac{\partial F_x}{\partial z} \Delta S_y - \frac{\partial F_x}{\partial y} \Delta S_z \right)$$

Since, in the limit, the partial derivatives are evaluated upon S itself, this is equivalent to[12]

$$\lim_{\substack{n \to \infty \\ (\Delta r_x)_i \to 0}} \sum_{i=1}^{n} (F_x \Delta r_x)_i = \int_S \left(\frac{\partial F_x}{\partial z} dS_y - \frac{\partial F_x}{\partial y} dS_z \right) \qquad (1.8\text{-}2)$$

equation (1.8-2) remains valid when the projection of Γ upon one or both coordinate planes is re-entrant. If, for example, Γ' is re-entrant, then some pair of adjacent planes drawn normal to the x axis will cut it in four, six --- places corresponding to two, three --- pairs of intercepted elements. However, since the line integral associated with each pair is equated to the surface integral taken over that portion of the strip between them (with sign dictated by currency) it is readily seen that the total line integral continues to be expressed in terms of the projected areas of S.

The transformation of the remaining terms of (1.8-1) may be effected in a similar manner. It is found that

$$\lim_{\substack{n \to \infty \\ (\Delta r_y)_i \to 0}} \sum_{i=1}^{n} (F_y \Delta r_y)_i = \int_S \left(\frac{\partial F_y}{\partial x} dS_z - \frac{\partial F_y}{\partial z} dS_x \right) \qquad (1.8\text{-}3)$$

---

12. See footnote, p. 50

and

$$\lim_{\substack{n\to\infty \\ (\Delta r_z)_i\to 0}} \sum_{i=1}^{n} (F_z \Delta r_z)_i = \int_S \left(\frac{\partial F_z}{\partial y}\, dS_x - \frac{\partial F_z}{\partial x}\, dS_y\right) \tag{1.8-4}$$

hence

$$\oint_\Gamma \bar{F}.d\bar{r} = \int_S \left\{\left(\frac{\partial F_z}{\partial y} - \frac{\partial F_y}{\partial z}\right) dS_x + \left(\frac{\partial F_x}{\partial z} - \frac{\partial F_z}{\partial x}\right) dS_y + \left(\frac{\partial F_y}{\partial x} - \frac{\partial F_x}{\partial y}\right) dS_z\right\}$$

This may be written as

$$\oint_\Gamma \bar{F}.d\bar{r} = \int_S (\text{curl } \bar{F}).d\bar{S} \tag{1.8-5}$$

where

$$\text{curl } \bar{F} \equiv \bar{i}\left(\frac{\partial F_z}{\partial y} - \frac{\partial F_y}{\partial z}\right) + \bar{j}\left(\frac{\partial F_x}{\partial z} - \frac{\partial F_z}{\partial x}\right) + \bar{k}\left(\frac{\partial F_y}{\partial x} - \frac{\partial F_x}{\partial y}\right)$$

(1.8-5) is known as Stokes's theorem.[13]

By treating ∇ as a formal vector in the operation $\nabla\times$ we obtain

$$\nabla \times \bar{F} = \left(\bar{i}\frac{\partial}{\partial x} + \bar{j}\frac{\partial}{\partial y} + \bar{k}\frac{\partial}{\partial z}\right) \times (\bar{i}F_x + \bar{j}F_y + \bar{k}F_z)$$

$$= \bar{i}\left(\frac{\partial F_z}{\partial y} - \frac{\partial F_y}{\partial z}\right) + \bar{j}\left(\frac{\partial F_x}{\partial z} - \frac{\partial F_z}{\partial x}\right) + \bar{k}\left(\frac{\partial F_y}{\partial x} - \frac{\partial F_x}{\partial y}\right)$$

or

$$\nabla \times \bar{F} = \text{curl } \bar{F} \tag{1.8-6}$$

This leads to the determinantal form which serves as a mnemonic:

---

13. The relationship holds for any regular, open surface since the integrals continue to exist when S is piecewise smooth.

$$\operatorname{curl} \bar{F} = \begin{vmatrix} \bar{i} & \bar{j} & \bar{k} \\ \frac{\partial}{\partial x} & \frac{\partial}{\partial y} & \frac{\partial}{\partial z} \\ F_x & F_y & F_z \end{vmatrix}$$

Alternative forms are[14]

$$\operatorname{curl} \bar{F} = \bar{i} \times \frac{\partial \bar{F}}{\partial x} + \bar{j} \times \frac{\partial \bar{F}}{\partial y} + \bar{k} \times \frac{\partial \bar{F}}{\partial z}$$

$$= \nabla F_x \times \bar{i} + \nabla F_y \times \bar{j} + \nabla F_z \times \bar{k}$$

When the surface of integration in (1.8-5) is a plane, we may write

$$\oint_\Gamma \bar{F}.d\bar{r} = \int_S (\operatorname{curl} \bar{F}).\hat{n}dS = \int_S (\operatorname{curl} \bar{F})_n dS$$

where $\hat{n}$ is constant.

Since the first derivatives of $\bar{F}$ are supposed to be continuous, $(\operatorname{curl} \bar{F})_n$ will be continuous over S. It then follows from the mean-value theorem (law of the mean) for integrals that

$$\oint_\Gamma \bar{F}.d\bar{r} = \Big((\operatorname{curl} \bar{F})_n\Big)_{P'} S$$

where P' is some point of the surface and S is its scalar area.

---

14. We have also

$$\operatorname{curl} \bar{F} = \nabla x \times \frac{\partial \bar{F}}{\partial x} + \nabla y \times \frac{\partial \bar{F}}{\partial y} + \nabla z \times \frac{\partial \bar{F}}{\partial z}$$

This form of expression carries over into general curvilinear coordinates and orthogonal surface curvilinear coordinates. (See Ex.2-15., p. 143 and p. 161). Similar remarks apply to the expression for gradient (p. 3) and divergence (p. 51).

If, now, the bounding contour shrinks about some point P_o within it (or upon it) so that all points of this curve approach P_o uniformly[15], then

$$\lim_{PP_o \to 0} \frac{1}{S} \oint_\Gamma \bar{F}.d\bar{r} = \left((\text{curl } \bar{F})_n\right)_{P_o} = ((\text{curl } \bar{F}).\hat{\bar{n}})_{P_o} \qquad (1.8\text{-}7)$$

where PP_o is the greatest distance between P_o and any point of the contour.

It follows that the circulation of $\bar{F}$ per unit plane area at a point is equal to the scalar component of curl $\bar{F}$ at that point in the direction of the positive normal to the plane; it consequently obeys the cosine law. It further follows that if the plane of integration is so orientated as to maximise the line integral per unit area positively, then this is the magnitude of curl $\bar{F}$, and the direction of curl $\bar{F}$ is the direction of the corresponding normal. It is, therefore, possible to express curl $\bar{F}$ (defined primarily as in (1.8-5)) in a form which is independent of any coordinate system, hence the rectangular Cartesian form must be invariant with respect to choice of rectangular axes. (See Sec. 1.19).

1.9 Alternative Approach to Stokes's Theorem

We will first derive an expression for the circulation of $\bar{F}$ around a small, not necessarily plane, closed curve.

The curve Γ is shown in Fig. 1.3, the positive sense of integration around it being indicated by the arrows.

P_o is a point in its vicinity and $\bar{r}$ is the position vector of any point of the curve relative to P_o. The closed curve Γ' and Γ'' are the projections of Γ on the xy and xz planes respectively, and P_o' and P_o'' are the corresponding projections of P_o.

The required expression for the line integral is given by (1.8-1). The first term of this may be transformed as follows.

As in the analysis of Sec. 1.8, let the simple surface S which spans Γ be cut in ab and cd by two planes set normal to the x axis at a distance Δx apart, and let the projections of ab and cd on the xy and xz coordinate planes be a'b', c'd' and a"b", c"d" respectively. The contribution of the intercepts ac and db to the first term of (1.8-1) is again taken as $\{(F_x)_a - (F_x)_b\}\ \Delta x$.

15. The shape which the contour assumes during the limiting process is immaterial so long as it comprises a simple closed curve, ie one which does not cut itself.

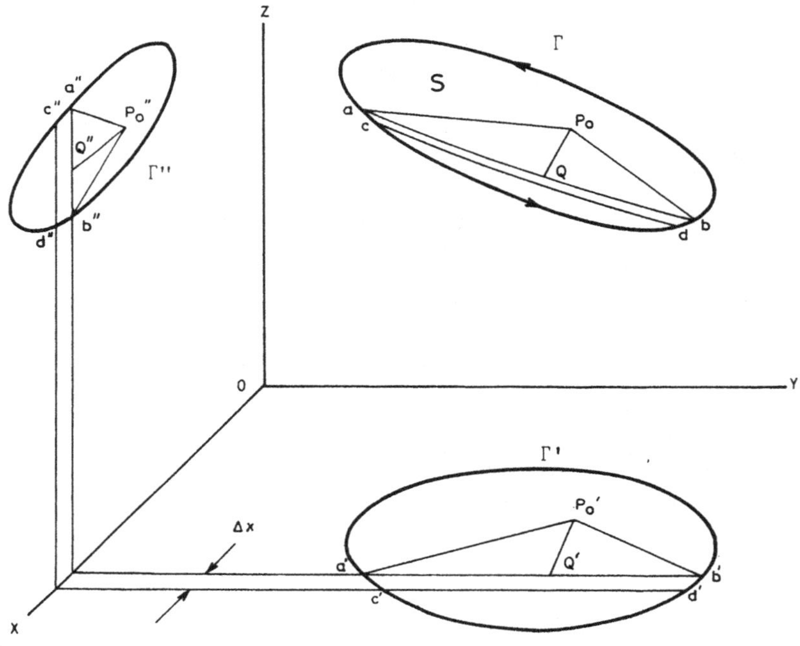

Fig. 1.3

On substituting F_x for V and $\bar{r}$ for $\Delta\bar{s}$ in (1.2-9), we obtain the value of F_x at a point whose position vector is $\bar{r}$ relative to P_o, in terms of F_x and its various derivatives (supposed continuous) at P_o.

$$F_x = (F_x)_{P_o} + ((\bar{r}.\nabla)F_x)_{P_o} + \text{---}$$

$$= (F_x)_{P_o} + r_x\left(\frac{\partial F_x}{\partial x}\right)_{P_o} + r_y\left(\frac{\partial F_x}{\partial y}\right)_{P_o} + r_z\left(\frac{\partial F_x}{\partial z}\right)_{P_o} + \text{---}$$

hence

$$(F_x)_a-(F_x)_b = \{(r_x)_a-(r_x)_b\}\left(\frac{\partial F_x}{\partial x}\right)_{P_o} + \{(r_y)_a-(r_y)_b\}\left(\frac{\partial F_x}{\partial y}\right)_{P_o}$$

$$+ \{(r_z)_a-(r_z)_b\}\left(\frac{\partial F_x}{\partial z}\right)_{P_o} + ---$$

where the rectangular components of $\bar{r}$ may be positive or negative depending upon the position of a and b relative to P_o.

It follows that

$$(F_x)_a-(F_x)_b = (P_o'Q'-P_o'Q')\left(\frac{\partial F_x}{\partial x}\right)_{P_o} + (-a'Q'-Q'b')\left(\frac{\partial F_x}{\partial y}\right)_{P_o}$$

$$+ (a''Q''+Q''b'')\left(\frac{\partial F_x}{\partial z}\right)_{P_o} + ---$$

where $P_o'Q'$, a'Q' etc represent positive length measures, so that

$$\{(F_x)_a-(F_x)_b\}\,\Delta x = \left(\frac{\partial F_x}{\partial z}\right)_{P_o} a''b''\Delta x - \left(\frac{\partial F_x}{\partial y}\right)_{P_o} a'b'\Delta x + ---$$

If now we suppose that Γ is divided entirely into pairs of elements such as ac and db by appropriately positioned planes and that the number of these planes increases without limit while their spacing everywhere approaches zero, then

$$\lim_{\substack{n\to\infty \\ (\Delta r_x)_i\to 0}} \sum_{i=1}^{n} (F_x\Delta r_x)_i = \left(\frac{\partial F_x}{\partial z}\right)_{P_o} \lim_{\Delta x\to 0} \sum_{S} a''b''\Delta x - \left(\frac{\partial F_x}{\partial y}\right)_{P_o} \lim_{\Delta x\to 0} \sum_{S} a'b'\Delta x + ---$$

where $\sum_S$ implies that the summation is carried out for all strips into which S is divided.

But $\lim\limits_{\Delta x\to 0}\sum\limits_{S} a''b''\Delta x$, which is essentially positive, is equal to the magnitude of the projected area of S on the xz coordinate plane[16]. Further, the projected area is itself positive as defined by the direction of integration around Γ. Hence this limit may be replaced by S_y. Similarly, the remaining limit may be replaced by S_z. These substitutions continue to be valid no matter how the surface may be folded so long as the periphery remains fixed. Since, in addition, the above equation is unaffected by possible re-entrance of Γ' or Γ", as discussed previously, we have in all cases

$$\lim_{\substack{n\to\infty \\ (\Delta r_x)_i\to 0}} \sum_{i=1}^{n} (F_x \Delta r_x)_i = \left(\frac{\partial F_x}{\partial z}\right)_{P_o} S_y - \left(\frac{\partial F_x}{\partial y}\right)_{P_o} S_z + ---$$

By similar constructions the second and third terms of (1.8-1) may be shown to be

$$\lim_{\substack{n\to\infty \\ (\Delta r_y)_i\to 0}} \sum_{i=1}^{n} (F_y \Delta r_y)_i = \left(\frac{\partial F_y}{\partial x}\right)_{P_o} S_z - \left(\frac{\partial F_y}{\partial z}\right)_{P_o} S_x + ---$$

and

$$\lim_{\substack{n\to\infty \\ (\Delta r_z)_i\to 0}} \sum_{i=1}^{n} (F_z \Delta r_z)_i = \left(\frac{\partial F_z}{\partial y}\right)_{P_o} S_x - \left(\frac{\partial F_z}{\partial x}\right)_{P_o} S_y + ---$$

whence

$$\oint_\Gamma \bar{F}.d\bar{r} = \left(\frac{\partial F_z}{\partial y} - \frac{\partial F_y}{\partial z}\right)_{P_o} S_x + \left(\frac{\partial F_x}{\partial z} - \frac{\partial F_z}{\partial x}\right)_{P_o} S_y + \left(\frac{\partial F_y}{\partial x} - \frac{\partial F_x}{\partial y}\right)_{P_o} S_z + ---$$

The remaining terms of the expansion comprise second and higher order derivatives multiplied by factors of the order of r^3 and higher. On writing these as γ we obtain

---

16. See footnote, p. 50

$$\oint_{\Gamma} \bar{F}.d\bar{r} = \left\{\bar{i}\left(\frac{\partial F_z}{\partial y} - \frac{\partial F_y}{\partial z}\right) + \bar{j}\left(\frac{\partial F_x}{\partial z} - \frac{\partial F_z}{\partial x}\right) + \bar{k}\left(\frac{\partial F_y}{\partial x} - \frac{\partial F_x}{\partial y}\right)\right\}_{P_o} .\bar{S} + \gamma \qquad (1.9\text{-}1)$$

where $\bar{S} = \bar{i}S_x + \bar{j}S_y + \bar{k}S_z$

Since $|\bar{S}|$ involves only the second power of r it is seen that the relative importance of the terms comprising γ decreases as the dimensions of the contour are reduced.

In the case of a plane contour of fixed orientation which shrinks about P_o in such a way that $PP_o \to 0$, where PP_o is the greatest distance between P_o and any point of the curve, (1.9-1) becomes

$$\lim_{PP_o \to 0} \frac{1}{S} \oint_{\Gamma} \bar{F}.d\bar{r} = \left\{\bar{i}\left(\frac{\partial F_z}{\partial y} - \frac{\partial F_y}{\partial z}\right) + \bar{j}\left(\frac{\partial F_x}{\partial z} - \frac{\partial F_z}{\partial x}\right) + \bar{k}\left(\frac{\partial F_y}{\partial x} - \frac{\partial F_x}{\partial y}\right)\right\}_{P_o} .\hat{n} \qquad (1.9\text{-}2)$$

where S is the scalar surface area within the contour and $\hat{n}$ is the unit positive normal to the surface.

This equation points to the existence of a vector which we call curl $\bar{F}$ and which is defined operationally in terms of the maximum circulation of $\bar{F}$ per unit plane area at a point, as discussed in the previous section. Its common Cartesian form in (1.9-2) and (1.9-1) allows us to re-write (1.9-1) as

$$\oint_{\Gamma} \bar{F}.d\bar{r} = (\text{curl } \bar{F})_{P_o} .\bar{S} + \gamma \qquad (1.9\text{-}3)$$

Alternatively, we may arrive at (1.9-3) directly from (1.9-1) by defining curl $\bar{F}$ primarily in terms of the Cartesian expression.

It remains to be shown that for contours of any size and shape (1.9-3) may be written without a remainder if the right-hand side is expressed as a surface integral.

To this end let the surface spanning the contour be divided into n elements, where the typical element is ΔS_i and its bounding curve is Γ_i (Fig. 1.4).

It will be seen that

$$\oint_{\Gamma} \bar{F}.d\bar{r} = \sum_{i=1}^{n} \oint_{\Gamma_i} \bar{F}.d\bar{r}$$

because of the cancellation of the pair of integrals associated with each

how the surface may be folded so long as all integrations are carried out right-handedly in relation to the local positive normal. The result is likewise independent of the degree of sub-division of S.

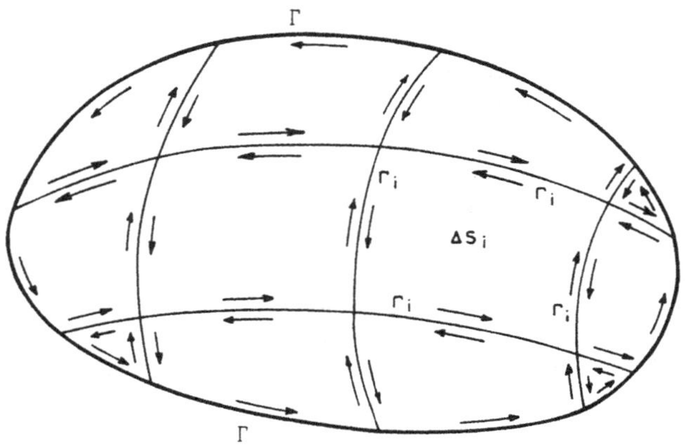

Fig. 1.4

On applying (1.9-3) to the typical surface element we get

$$\oint_{\Gamma_i} \bar{F}.d\bar{r} = (\text{curl } \bar{F})_{P_i}.\Delta\bar{S}_i + \gamma_i$$

where P_i is chosen to be some point of ΔS_i.

Then

$$\oint_{\Gamma} \bar{F}.d\bar{r} = \sum_{i=1}^{n} \{(\text{curl } \bar{F})_{P_i}.\Delta\bar{S}_i+\gamma_i\}$$

But

$$\frac{\gamma_i}{(\text{curl } \bar{F})_{P_i}.\Delta\bar{S}_i} \to 0 \quad \text{as} \quad \Delta\bar{S}_i\to\bar{0}$$

where it is supposed that all dimensions of ΔS_i decrease uniformly, so that

$$\frac{\sum_{i=1}^{n} \gamma_i}{\sum_{i=1}^{n} (\text{curl } \overline{F})_{P_i} . \Delta\overline{S}_i} \to 0 \quad \text{as} \quad n\to\infty \quad \text{and} \quad \Delta\overline{S}_i \to \overline{0}$$

hence

$$\oint_\Gamma \overline{F}.d\overline{r} = \lim_{\substack{n\to\infty \\ \Delta\overline{S}_i\to\overline{0}}} \sum_{i=1}^{n} (\text{curl } \overline{F})_{P_i} . \Delta\overline{S}_i$$

or

$$\oint_\Gamma \overline{F}.d\overline{r} = \int_S (\text{curl } \overline{F}).d\overline{S}$$

1.10 Application of Stokes's Theorem

We have been concerned as yet only with those surfaces having a single bounding curve. When, in addition, a surface is bounded internally, the appropriate form of Stokes's theorem can be developed anew along the lines indicated in Sec. 1.8 or 1.9. However, it is possible to transform a multiply bounded surface into one with a single bounding contour by means of a geometrical construction, and this permits of the direct application of the theorem developed above.

Fig. 1.5 represents a two-sided surface S with an outer contour Γ and two non-intersecting inner contours Γ_1 and Γ_2. Arbitrary non-intersecting curves ac and de are drawn upon the surface to join Γ to Γ_1 and Γ_1 to Γ_2. The single curve abacdefedgca (following the direction of the arrows) comprises a continuous contour of S. If the positive side of the surface faces the reader then right-handed integration will follow the sequence given above. Hence

$$\oint_{a.....a} \overline{F}.d\overline{r} = \int_S (\text{curl } \overline{F}).d\overline{S}$$

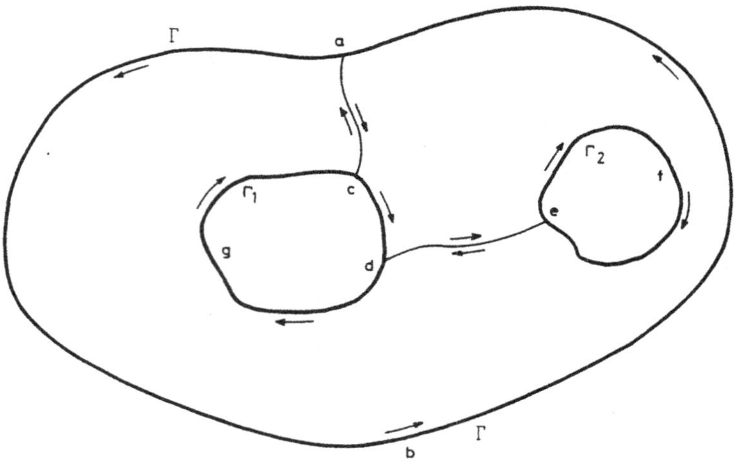

Fig. 1.5

It will be seen that those portions of the line integral relating to ac and de cancel in pairs because $\overline{F}$ has a unique value at each point of the curves and integration is carried out in both senses along them. There remain the integrals around Γ_1 and Γ_2. The above equation is therefore equivalent to

$$\oint_{\Gamma} \overline{F}.d\overline{r} + \oint_{\Gamma_1} \overline{F}.d\overline{r} + \oint_{\Gamma_2} \overline{F}.d\overline{r} = \int_S (\text{curl } \overline{F}).d\overline{S}$$

or, in general,

$$\oint_{\Gamma} \overline{F}.d\overline{r} = \int_S (\text{curl } \overline{F}).d\overline{S} - \sum_{h=1}^{n} \oint_h \overline{F}.d\overline{r} \qquad (1.10\text{-}1)$$

It is evident that the positive sense of integration around the outer contour is oppositely directed to that around the inner contours. This should be borne in mind when evaluating the above integrals.

The reader may prefer to arrive at the above result by dividing S into discrete parts by the addition of one or more curves such as ac in Fig. 1.5.

It is readily shown that the normal integral of curl $\overline{F}$ over any closed surface is zero if curl $\overline{F}$ is continuous upon the surface.

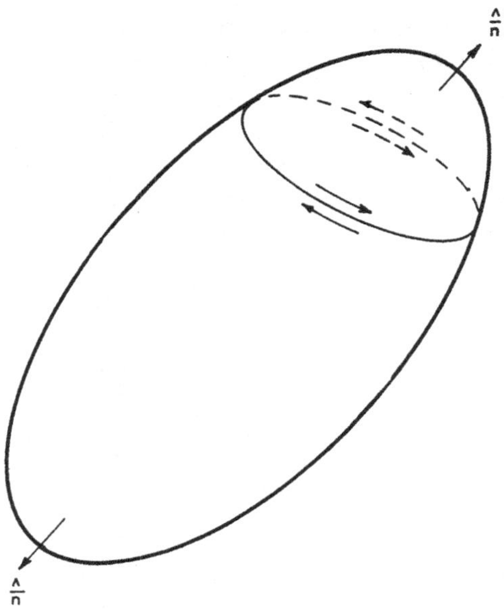

Fig. 1.6

Consider the closed surface of Fig. 1.6 upon which an arbitrary closed curve is drawn. The normal surface integral of curl $\overline{F}$ over the upper portion of the surface is equal to the line integral of $\overline{F}$ around the curve in the direction of the upper set of arrows (for a positive outward normal). The normal surface integral of $\overline{F}$ over the lower portion is equal to the line integral of $\overline{F}$ around the same curve but in the opposite sense (lower arrows). Since $\overline{F}$ has a unique value at each point of the curve the line integrals (and consequently the surface integrals) are equal and opposite, hence

$$\oint_S (\text{curl } \overline{F}).d\overline{S} \equiv 0$$

Alternatively, if a small portion of the closed surface were removed, the normal integral of $\overline{F}$ over the remainder would be equal to the line integral of $\overline{F}$ round the edge of the hole. The latter clearly approaches zero as the size of the hole is continuously reduced.

Stokes's theorem has been derived on the assumption that $\overline{F}$ and its first derivatives are continuous within the region of integration. If points of discontinuity are present it becomes necessary to exclude these from the region by means of additional bounding curves. Stokes's theorem may then be applied in the form (1.10-1).

Consider, for example, the two-dimensional[17] field defined by $\overline{F} = (-\overline{i}y+\overline{j}x)/(x^2+y^2)$. Within the xy plane which contains the origin of coordinates, $\overline{F}$ is seen to be a vector of magnitude $\frac{1}{r}$, where r is distance measured from the origin; it is perpendicular to the position vector drawn from the origin and is directed right-handedly in relation to the positive z axis. We find by differentiation that curl $\overline{F} = 0$ at all points of the plane except the origin, where it is undefined.

If a closed curve Γ is drawn in the plane in such a way that the origin is not included within it, a direct application of Stokes's theorem shows that the tangential line integral of $\overline{F}$ around Γ is zero. When Γ embraces the origin it is necessary to exclude this point from the surface of integration. Suppose that we do this by drawing about 0 a small circle of radius δ. Then from (1.10-1) we find that the line integral around Γ is equal and opposite to that around the circle. Since Γ was chosen arbitrarily it follows that all such curves have the same value of line integral, and since δ was chosen arbitrarily in relation to Γ it follows that the value of the line integral around the circle must be independent of δ. The latter is found, by direct computation, to have the value -2π, so that the line integral of $\overline{F}$ around Γ is equal to $+2\pi$.

1.11 The Irrotational Vector Field

A vector point function $\overline{F}$ is said to be irrotational or lamellar within a region R if the tangential line integral of $\overline{F}$ between every pair of points is independent of the path taken between them[18]. This implies that $\oint_\Gamma F.dr = 0$ for all closed curves within R.

It has been shown that the vanishing of curl $\overline{F}$ throughout a region is no guarantee that $\oint_\Gamma \overline{F}.d\overline{r} = 0$ unless Γ bounds an unbroken surface or one for which the sum of the line integrals over internal bounding cuves is zero[19].

A discussion of this restriction is simplified by the adoption of certain terms which will now be defined.

A region R is said to be simply connected or acyclic if every closed curve which can be drawn within it can be continuously contracted to a point without crossing any boundary of R. Such curves are said to be reducible. Alternative paths joining two points are said to be

---

17. A two-dimensional xy field exists throughout a region of space, and is defined by $F_x = f_1(x,y)$, $F_y = f_2(x,y)$, $F_z = 0$.

18. An irrotational field of force is said to be conservative.

19. Note, however, that most writers equate 'irrotational' with 'zero-curl'.

reconcilable if they can be continously deformed until one coincides with the other without passing beyond $\underline{R}$. If a region is not simply connected it is said to be multiply connected or cyclic.

Thus the region bounded by a simple closed surface is simply connected, and so is that bounded by two such surfaces when one encloses the other without touching it. But the closed region bounded by two concentric cylinders fitted with end plates is multiply connected as is that bounded by a torus and an enclosing sphere, since closed curves embracing the inner cylinder or the torus cannot be contracted to a point unless some part of them pass beyond the region.

It is possible to convert a multiply connected region into a simply connected region by the erection of one or more barriers (or 'cuts'). Thus in the case of the concentric cylinders, the barrier abcd shown in Fig. 1.7a is sufficient to prevent the setting up of an irreducible curve and to ensure that all paths between two points are reconcilable. In the case of the torus and sphere (shown sectionally in Fig. 1.7b) a diaphragm ef mounted in the central plane serves the same purpose.

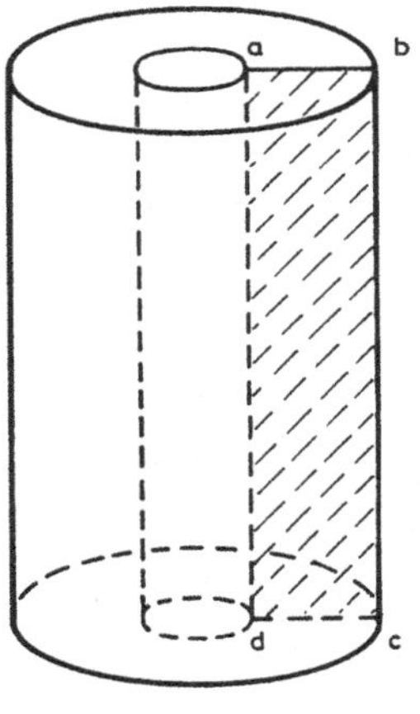

Fig. 1.7a

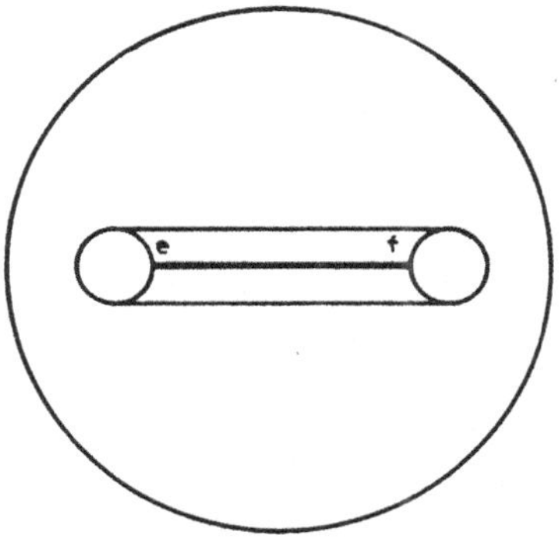

Fig.1.7b

The relevance of the above concepts to the present discussion lies in the fact that if a region has been so defined as to exclude discontinuities, the additional specification that it be simply connected is sufficient to ensure that the tangential line integral around any closed curve within the region may be evaluated by an integration of the curl over a surface lying entirely within the region. In particular, if $\overline{F}$ and its first derivatives are continuous and curl $\overline{F} = \overline{0}$ at all points of a simply connected region then $\oint_\Gamma \overline{F}.d\overline{r} = 0$ for all closed curves within the region and $\overline{F}$ is irrotational.

Theorem 1.11-1

If a scalar point function V is single-valued and has continuous first and second derivatives within a region R, and if $\overline{F}$ = grad V, then $\overline{F}$ is irrotational and curl $\overline{F} = \overline{0}$ within R.

Proof: Let P and Q be any two points of R and let Γ be a simple curve lying wholly within R and passing from P to Q.

Then

$$\int_\Gamma \overline{F}.d\overline{r} = \int_\Gamma \text{grad } V.d\overline{r} = \lim_{\substack{n\to\infty \\ \Delta\overline{r}_i\to\overline{0}}} \sum_{i=1}^{n} (\text{grad } V)_i.\Delta\overline{r}_i$$

Now $(\text{grad } V)_i.\Delta\overline{r}_i = \left(\frac{\partial V}{\partial x}\right)_i (\Delta r_x)_i + \left(\frac{\partial V}{\partial y}\right)_i (\Delta r_y)_i + \left(\frac{\partial V}{\partial z}\right)_i (\Delta r_z)_i$

where the derivatives are evaluated at some point of the arc of PQ intercepted by $\Delta\overline{r}_i$ (or on $\Delta\overline{r}_i$ itself - See Ex.1-20., p. 23).

But from the considerations of Sec. 1.2

$$\Delta V_i = \left(\frac{\partial V}{\partial x}\right)_\alpha (\Delta r_x)_i + \left(\frac{\partial V}{\partial y}\right)_\beta (\Delta r_y)_i + \left(\frac{\partial V}{\partial z}\right)_\gamma (\Delta r_z)_i$$

where α, β, γ are points in the neighbourhood of $\Delta\overline{r}_i$ which approach $\Delta\overline{r}_i$ as $\Delta\overline{r}_i\to\overline{0}$.

Then in virtue of the continuity of the partial derivatives we see that

$$\{\Delta V_i-(\text{grad } V)_i.\Delta\overline{r}_i\}/\Delta V_i\to 0 \quad \text{as} \quad \Delta\overline{r}_i\to\overline{0}$$

hence

$$\lim_{\substack{n\to\infty \\ \Delta\bar{r}_i\to\bar{0}}} \sum_{i=1}^{n} (\text{grad } V)_i.\Delta\bar{r}_i = \sum_{i=1}^{n} \Delta V_i$$

or $$\int_\Gamma \text{grad } V.d\bar{r} = V(Q) - V(P) \qquad (1.11\text{-}1)$$

It follows that the value of the tangential line integral of grad V between P and Q is independent of the path, hence $\bar{F}$ is irrotational within R.

From considerations leading to (1.8-7) if further follows that if, in addition, the first derivatives of $\bar{F}$ are continuous, then curl $\bar{F} = \bar{0}$. This can be proved alternatively by expanding the expression curl grad V in Cartesian coordinates.

$$\begin{aligned}\text{curl grad } V &= \nabla\times\nabla V \\ &= \left(\bar{i}\frac{\partial}{\partial x} + \bar{j}\frac{\partial}{\partial y} + \bar{k}\frac{\partial}{\partial z}\right) \times \left(\bar{i}\frac{\partial V}{\partial x} + \bar{j}\frac{\partial V}{\partial y} + \bar{k}\frac{\partial V}{\partial z}\right) \\ &= \bar{i}\left(\frac{\partial^2 V}{\partial y\partial z} - \frac{\partial^2 V}{\partial z\partial y}\right) + \bar{j}\left(\frac{\partial^2 V}{\partial z\partial x} - \frac{\partial^2 V}{\partial x\partial z}\right) + \bar{k}\left(\frac{\partial^2 V}{\partial x\partial y} - \frac{\partial^2 V}{\partial y\partial x}\right)\end{aligned}$$

whence

$$\text{curl grad } V \equiv \bar{0} \qquad (1.11\text{-}2)$$

since $\dfrac{\partial^2 V}{\partial y\partial z} = \dfrac{\partial^2 V}{\partial z\partial y}$ etc.

Theorem 1.11-2

If $\bar{F}$ is a continuous vector point function which is irrotational within the region R then $\bar{F}$ = grad V within R.

Proof: Taking any point O of the region as datum we define the scalar function V such that at any point P of the region

$$V(P) = \int_O^P \bar{F}.d\bar{r} = \int_O^P F_t ds$$

where the path of integration lies wholly within the region.

V will be single-valued because $\bar{F}$ is irrotational.

Then if Q is a point such that $\overrightarrow{PQ} = \Delta\bar{r} = \hat{s}\,\Delta s$ it follows from the mean-value theorem that

$$V(Q) - V(P) = \Delta V = (F_t)_{P'}\Delta s = (\bar{F}.\hat{s})_{P'}\Delta s$$

where P' is some point of PQ.

But from (1.11-1)

$$\Delta V = ((\text{grad } V)_t)_{P''}\Delta s = (\text{grad } V.\hat{s})_{P''}\Delta s$$

where P" is some point of PQ,

hence

$$(\bar{F}.\hat{s})_{P'} = (\text{grad } V.\hat{s})_{P''}$$

Since this holds independently of the magnitude of Δs we may let $\Delta s \to 0$, whence we find that

$$(\bar{F}.\hat{s})_P = (\text{grad } V.\hat{s})_P$$

or

$$(\bar{F} - \text{grad } V)_P.\hat{s} = 0$$

But this remains true for all orientations of PQ (ie of $\hat{s}$) so that

$$(\bar{F})_P = (\text{grad } V)_P$$

which is the required result.

EXERCISES

1-30. Derive (1.10-1) for a surface with a single internal boundary by means of an analysis similar to that of (a) Sec. 1.8 (b) Sec. 1.9.

1-31. By proceeding from first principles as in Sec. 1.8 derive the form of Stokes's theorem appropriate to a vector point function which is defined within a plane region and has no component normal to the plane (planar field). Hence show that if $V = V(x,y)$ and $U = U(x,y)$ are scalar functions of x and y which, together with their first derivatives, are continuous within the region of integration, then

$$\oint_{\Gamma} (V\, dr_y - U\, dr_x) = \int_{S_z} \left(\frac{\partial V}{\partial x} + \frac{\partial U}{\partial y}\right) dS_z$$

[This relationship is usually written in the form

$$\oint_{\Gamma} (V\, dy - U\, dx) = \iint_{S} \left(\frac{\partial V}{\partial x} + \frac{\partial U}{\partial y}\right) dx\, dy$$

and is known as Green's theorem in the plane. (See Ex.1-43, p. 63)]

1-32. Show that $\overline{F}$ is irrotational if

$$\overline{F} = \overline{i}\,(6xyz+5z^2) + \overline{j}\,(3x^2z+8y) + \overline{k}\,(3x^2y+10\,xz)$$

and find V such that $\overline{F} = \text{grad } V$

Ans: $V = 3x^2yz + 4y^2 + 5xz^2 + \text{const}$

1-33. For the two-dimensional or planar field defined by

$$\overline{F} = \overline{i}\,(4xy+6(x+y)-4) + \overline{j}\,(2x^2+6(x+y))$$

evaluate the line integral $\oint \overline{F}.d\overline{r}$ from (0,0) to (3,0) along the x axis, from (3,0) to (3,27) along $x = 3$, and from (3,27) to (0,0) along $y = x^3$, by (a) direct integration (b) application of Stokes's theorem.

Ans: 3174 - 3174 = 0

1-34. For the two-dimensional or planar field defined by

$$\overline{F} = \overline{i}x^2y + \overline{j}\,(2x+3y^2)$$

evaluate the line integral $\oint \overline{F}.d\overline{r}$ from (1,0) to (3,0) along the x axis, from (3,0) to (3,6) along x = 3, and from (3,6) to (1,0) along a straight line, by (a) direction integration (b) application of Stokes's theorem.

Ans: -22

1-35. The equations

$$x = a \sin \omega t \qquad y = a \cos \omega t \qquad z = bt$$

where a, b and ω are positive constants and t is an independent variable, represent a left-handed circular helix centred upon the z axis.

Determine the tangential line integral along the helix of the vector point function $\overline{F}$, as defined in Ex.1-32. above, between the points P and Q corresponding to $t = t_o$ and $t = t_o + \frac{2\pi}{\omega}$.

The path comprises one complete turn of the helix. Confirm that $\overline{F}$ is irrotational by computing the tangential line integral of $\overline{F}$ along the straight line PQ.

$$\text{Ans:} \quad \int_P^Q \overline{F}.d\overline{r} = \frac{6\pi a^3 b}{\omega}\sin^2 \omega t_o \cos \omega t_o + \frac{20\pi ab^2}{\omega}\left(t_o + \frac{\pi}{\omega}\right)\sin \omega t_o$$

1-36. Show that equation (1.8-5) continues to hold in the presence of an interior closed curve of discontinuity of $\overline{F}$ when the tangential component of $\overline{F}$ is continuous through this curve, provided that curl $\overline{F}$ is continuous at interior points of the two subregions and upon Γ, and the surface integral is understood to represent the limiting value of the sum of the integrals taken over the subregions as the line of discontinuity is approached from both sides.

1-37. Given that

(a) $\overline{F}$ and $\overline{G}$ are well-behaved vector point functions defined throughout the region $\underline{R}$

(b) for every open surface S of boundary Γ that can be contracted to a point without passing beyond $\underline{R}$

$$\oint_\Gamma \overline{F}.d\overline{r} = \int_S \overline{G}.d\overline{S}$$

1-38. Let $\underline{R}$ represent the unbounded region of the xy plane outside a circle centred upon the origin of coordinates. If

$$\bar{F} = \frac{-\bar{i}y + \bar{j}x}{x^2 + y^2} \quad \text{in} \quad \underline{R}$$

show that

$$\bar{F} = \text{grad } \tan^{-1} \frac{y}{x}$$

Prove by direct integration that

$$\oint_\Gamma \bar{F}.d\bar{r} = 2\pi \quad \text{or} \quad 0$$

according as Γ embraces the origin or not.

$\bar{F}$ is evidently non-irrotational although expressed everywhere in $\underline{R}$ as grad V. Discuss this

Ans: (1) Curl $\bar{F}$ is zero beyond the origin 0 but undefined at 0, and so admits of the possibility - as confirmed above - that $\oint_\Gamma \bar{F}.d\bar{r}$ be non-zero when Γ embraces 0.

(2) It follows that $\bar{F}$ cannot be expressed everwhere in $\underline{R}$ as the gradient of a single-valued function; if it could, $\bar{F}$ would necessarily be irrotational. It can be expressed as grad $\tan^{-1} \frac{y}{x}$ everwhere in $\underline{R}$ only because, corresponding to right-handed circulation about 0, $\tan^{-1} \frac{y}{x}$ can assume continuously increasing values with an increment of 2π per revolution.

(3) The erection of a radial barrier assures that Γ cannot embrace 0 and makes $\underline{R}$ simply connected. Correspondingly, $\tan^{-1} \frac{y}{x}$ can be expressed in single-valued form with a discontinuity of 2π across the barrier.

1.12 Flux Through a Closed Surface

The Divergence Theorem

Let S be an open, two-sided surface (Fig. 1.8a) and let $\overline{F}$ be a bounded vector point function, piecewise continuous upon S. Then the normal surface integral of $\overline{F}$ over S is known as the flux of $\overline{F}$ through S in the direction of the positive normal[20] at the surface.

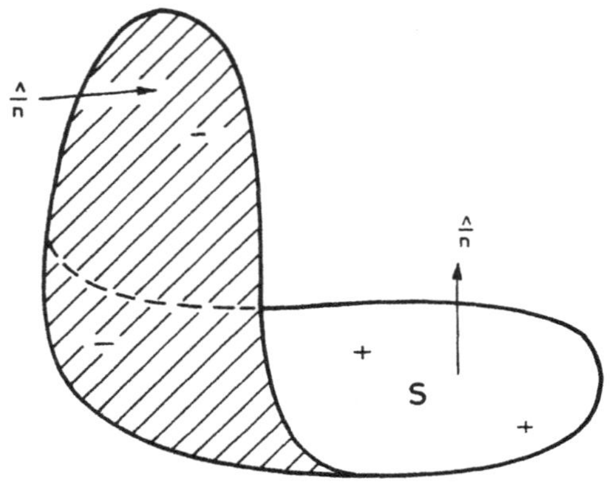

Fig. 1.8a

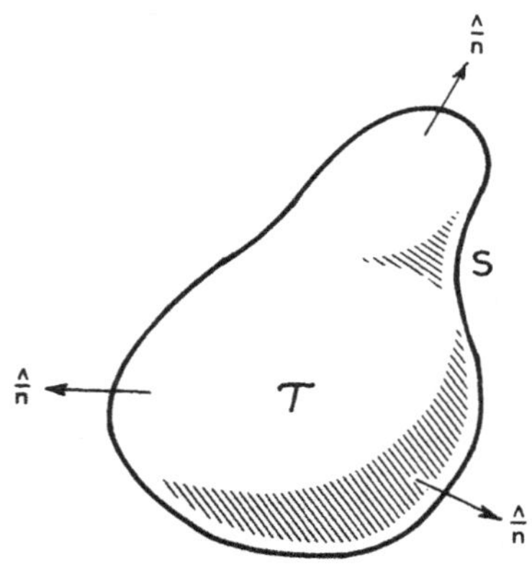

Fig. 1.8b

---

20. If a positive currency has been previously established for the boundary of S, this serves to define the positive side of S (or

When a region of space, τ, is bounded by a single closed surface S (Fig. 1.8b), and the outer side of S is chosen as positive in accordance with the convention mentioned previously, the flux so computed is an outwardly directed flux, and is commonly referred to as the flux of $\overline{F}$ out of the region τ.

It will now be shown that if $\overline{F}$ has continuous first derivatives at interior points of τ and upon its bounding surface (ie if $\overline{F}$ is continuously differentiable in the closed region τ), then the flux of $\overline{F}$ out of the region is equal to the volume integral, over τ, of a related scalar point function.

From (1.6-5) the normal surface integral of $\overline{F}$ over S is given by

$$\lim_{\substack{n\to\infty \\ (\Delta S_x)_i\to 0}} \sum_{i=1}^{n} (F_x \Delta S_x)_i + \lim_{\substack{n\to\infty \\ (\Delta S_y)_i\to 0}} \sum_{i=1}^{n} (F_y \Delta S_y)_i + \lim_{\substack{n\to\infty \\ (\Delta S_z)_i\to 0}} \sum_{i=1}^{n} (F_z \Delta S_z)_i \tag{1.12-1}$$

The first term of this expression may be transformed into a volume integral in the following way.

Let S be cut in the surface elements abcd and a'b'c'd' by a rectangular prism drawn parallel to the x axis (Fig. 1.9),[21] and let the sectional dimensions of the prism be Δy and Δz, where Δy and Δz are intrinsically positive. Since the positive normals are directed away from τ, it follows that $(\Delta S_x)_{abcd} = -\Delta y \Delta z$ and $(\Delta S_x)_{a'b'c'd'} = +\Delta y \Delta z$. The contribution of these elements to the summation may therefore be taken as $\{(F_x)_{a'} - (F_x)_a\} \Delta y \Delta z$.

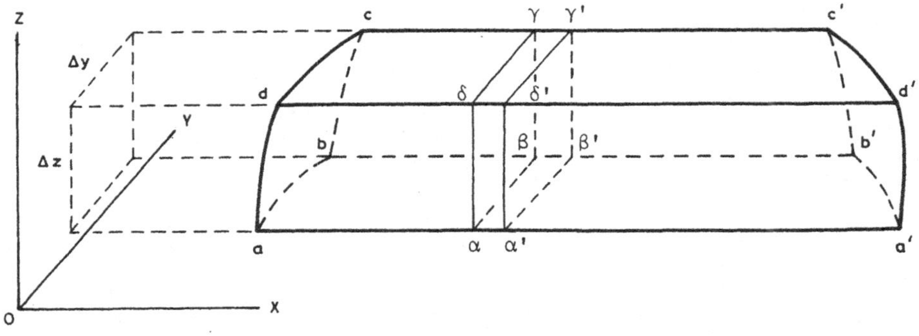

Fig. 1.9

---

21. It will be supposed, in the first instance, that S is non re-entrant, so that any prism cuts S in only two elements.

Suppose now that the prism is divided by a system of planes drawn normal to the x axis and passing through a, a' and intermediate points. Let the typical volume element so defined be $\alpha\beta\gamma\delta\alpha'\beta'\gamma'\delta'$. Then

$$(F_x)_{\alpha'} - (F_x)_\alpha = \frac{\partial F_x}{\partial x} \Delta x$$

where $\Delta x = \alpha\alpha'$ and the derivative is to be taken at some point of $\alpha\alpha'$.

Hence

$$(F_x)_{a'} - (F_x)_a = \sum \frac{\partial F_x}{\partial x} \Delta x$$

and

$$(F_x \Delta S_x)_{\substack{abcd \\ a'b'c'd'}} = \{(F_x)_{a'} - (F_x)_a\} \Delta y \Delta z = \sum \frac{\partial F_x}{\partial x} \Delta \tau$$

where $\Delta\tau = \Delta x \Delta y \Delta z$, and the summation includes all rectangular volume elements between the transverse planes drawn through a and a'.

If, now, S is divided entirely into pairs of surface elements by appropriately positioned prisms lying parallel to the x axis, and if those portions of the prisms intercepted by S are independently divided by planes drawn normal to the x axis, then

$$\sum_S F_x \Delta S_x = \sum \frac{\partial F_x}{\partial x} \Delta \tau$$

where the volume summation covers all rectangular elements defined by the intersection of the planes and prisms, and thereby approximates the volume enclosed by S.

If this mode of subdivision is maintained while the sectional dimensions of the prisms and the transverse plane spacings are continuously reduced, the above equation is replaced by[22]

---

22. This does not follow directly from the definition of the scalar volume integral given in Sec. 1.7, where it was supposed that the sum of all volume elements involved in the limiting process remains constant and equal to the enclosed volume. In the present instance the enclosed volume is taken to be the limit of the sum of the rectangular volume elements constructed during the limiting process, and the scalar integral under consideration is the limit of the sum of the quantities $\frac{\partial F_x}{\partial x} \Delta\tau$. However, the latter may be shown to be identical with the scalar volume integral as defined previously. Similar remarks apply to the derivation of the surface integrals in Secs. 1.8 and 1.9.

$$\lim_{\substack{n\to\infty \\ (\Delta S_x)_i\to 0}} \sum_{i=1}^{n} (F_x \Delta S_x)_i = \int_\tau \frac{\partial F_x}{\partial x}\, d\tau$$

This relationship remains valid when the surface is so folded as to be cut in more than two elements by one or more prisms. In this case the contributions from adjacent surface elements intercepted by a given prism are equated to the volume integral of $\frac{\partial F_x}{\partial x}$ between them, so that the overall form of the equation is unaffected.

Similar considerations show that the remaining terms of (1.12-1) may be equated to $\int_\tau \frac{\partial F_y}{\partial y}\, d\tau$ and $\int_\tau \frac{\partial F_z}{\partial z}\, d\tau$, so that

$$\oint_S \overline{F}.d\overline{S} = \int_\tau \left(\frac{\partial F_x}{\partial x} + \frac{\partial F_y}{\partial y} + \frac{\partial F_z}{\partial z}\right) d\tau \qquad (1.12\text{-}2)$$

The volume integrand of this equation is known as the divergence of $\overline{F}$. This is usually abbreviated to div $\overline{F}$, whence

$$\oint_S \overline{F}.d\overline{S} = \int_\tau \operatorname{div} \overline{F}\, d\tau \qquad (1.12\text{-}3)$$

where $\operatorname{div} \overline{F} \equiv \frac{\partial F_x}{\partial x} + \frac{\partial F_y}{\partial y} + \frac{\partial F_z}{\partial z}$

By treating ∇ as a formal vector in the operation $\nabla.$ we may write

$$\nabla.\overline{F} = \left(\overline{i}\,\frac{\partial}{\partial x} + \overline{j}\,\frac{\partial}{\partial y} + \overline{k}\,\frac{\partial}{\partial z}\right) . (\overline{i}F_x + \overline{j}F_y + \overline{k}F_z)$$

$$= \frac{\partial F_x}{\partial x} + \frac{\partial F_y}{\partial y} + \frac{\partial F_z}{\partial z}$$

hence $\nabla.\overline{F} = \operatorname{div} \overline{F}$ (1.12-4)

Alternative forms are

$$\operatorname{div} \overline{F} = \overline{i} . \frac{\partial \overline{F}}{\partial x} + \overline{j} . \frac{\partial \overline{F}}{\partial y} + \overline{k} . \frac{\partial \overline{F}}{\partial z}$$

$$= \overline{i}.\nabla F_x + \overline{j}.\nabla F_y + \overline{k}.\nabla F_z$$

(1.12-3) is known as the divergence theorem[23]. The names of Green, Gauss, and Ostrogradsky are associated with it.

Since the first derivatives of $\overline{F}$ are supposed to be continuous, div $\overline{F}$ will be continuous, and by the mean-value theorem for integrals

$$\oint_S \overline{F}.d\overline{S} = (\text{div }\overline{F})_{P'}\,\tau$$

where τ is the volume of the enclosure and P' is some point of it.

If S now shrinks about a point P_o within it or upon it, then

$$\lim_{PP_o \to 0} \frac{1}{\tau} \oint_S \overline{F}.d\overline{S} = (\text{div }\overline{F})_{P_o} \qquad (1.12\text{-}5)$$

where PP_o is the greatest distance between P_o and any point of S.

(No restriction is placed upon the shape of S during the limiting process).

Since the left-hand side of (1.12-5) does not involve any specific system of coordinates, the analytical form of div $\overline{F}$ given in (1.12-3) must be invariant with respect to choice of rectangular axes. An independent proof of this will be found in Sec. 1.19.

1.13 Alternative Approach to the Divergence Theorem

We will first derive an expression for the normal surface integral of $\overline{F}$ over a small closed surface S which embraces the point P_o.

Let $\overline{r}$ be the position vector relative to P_o of any point of the surface. Then from (1.2-9) the value of F_x at this point is given by

$$F_x = (F_x)_{P_o} + ((\overline{r}.\nabla)F_x)_{P_o} + ----$$

so that the first term of (1.12-1) becomes

23. The relationship holds for any closed regular surface S and the (regular) region τ which it bounds.

$$\lim_{\substack{n\to\infty \\ (\Delta S_x)_i\to 0}} \sum_{i=1}^{n} \{(F_x)_{P_o} + ((\bar{r}_i.\nabla)F_x)_{P_o} + ---\} (\Delta S_x)_i$$

$$= \lim_{\substack{n\to\infty \\ (\Delta S_x)_i\to 0}} \sum_{i=1}^{n} \left\{(F_x)_{P_o}(\Delta S_x)_i + \left(\frac{\partial F_x}{\partial x}\right)_{P_o}(r_x\Delta S_x)_i + \left(\frac{\partial F_x}{\partial y}\right)_{P_o}(r_y\Delta S_x)_i + \left(\frac{\partial F_x}{\partial z}\right)_{P_o}(r_z\Delta S_x)_i + ---\right\} \tag{1.13-1}$$

Let S be subdivided by prisms drawn parallel to the x axis as in the previous analysis (Fig. 1.9), and let the values of F_x to be associated with corresponding pairs of surface elements relate to points having the same y and z coordinates (such as a and a').

Then only the second term of (1.13-1) is finite because the derivatives, being referred to P_o, are constant, and

$$\sum_{i=1}^{n} (\Delta S_x)_i \quad , \quad \sum_{i=1}^{n} (r_y\Delta S_x)_i \quad \text{and} \quad \sum_{i=1}^{n} (r_z\Delta S_x)_i \quad \text{cancel in pairs.}$$

The second term may be written as

$$\left(\frac{\partial F_x}{\partial x}\right)_{P_o} \lim_{\substack{n\to\infty \\ (\Delta S_x)_i\to 0}} \sum_{i=1}^{n} (r_x\Delta S_x)_i$$

That portion of $\sum_{i=1}^{n} (r_x\Delta S_x)_i$ which derives from the paired surface elements shown in Fig. 1.9 is equal to $\{(r_x)_{a'} - (r_x)_a\}\,\Delta y\Delta z$, and this is the volume of the prism intercepted by the transverse planes through a and a'. It therefore follows from previous considerations that

$$\lim_{\substack{n\to\infty \\ (\Delta S_x)_i\to 0}} \sum_{i=1}^{n} (r_x\Delta S_x)_i = \tau$$

where τ is the volume of the enclosure, so that (1.13-1) reduces to

$$\left(\frac{\partial F_x}{\partial x}\right)_{P_o} \tau + ---$$

Similar constructions lead to corresponding expressions for the remaining terms of (1.12-1), whence

$$\oint_S \overline{F}.d\overline{S} = \left(\frac{\partial F_x}{\partial x} + \frac{\partial F_y}{\partial y} + \frac{\partial F_z}{\partial z}\right)_{P_o} \tau + \gamma \tag{1.13-2}$$

where γ comprises second and higher-order derivatives multiplied by factors of the order r^4 and higher. (1.13-2) is the expression which we set out to obtain.

If, now, S shrinks about P_o in such a way that all dimensions of τ approach zero, then

$$\lim_{\tau \to 0} \frac{1}{\tau} \oint \overline{F}.d\overline{S} = \left(\frac{\partial F_x}{\partial x} + \frac{\partial F_y}{\partial y} + \frac{\partial F_z}{\partial z}\right)_{P_o} \tag{1.13-3}$$

We see, therefore, that if $\overline{F}$ is continuous with continuous first derivatives throughout a region, the 'flux of $\overline{F}$ per unit volume at a point' defines a scalar function throughout the region which is independent of the limiting shapes of volume elements.

This is sometimes taken to be the primary definition of div $\overline{F}$.

In this case it remains to be shown that for singly bounded enclosures of any shape and size

$$\oint_S \overline{F}.d\overline{S} = \int_\tau \text{div}\,\overline{F}\,d\tau$$

Suppose that τ is divided into n volume elements by, say, three sets of orthogonal planes of equal spacings. Then the sum of the surface integrals of $\overline{F}$ associated with all volume elements is equal to the surface integral of $\overline{F}$ over S. This follows from the fact that those portions of the individual surface integrals associated with the common faces of adjacent elements are counted twice, but with reversed positive normals, and therefore cancel in the sum.

Hence

$$\oint_S \overline{F}.d\overline{S} = \sum_{i=1}^{n} \oint_{\Delta S_i} \overline{F}.d\overline{S}$$

where ΔS_i is the closed surface associated with the volume element $\Delta\tau_i$.

This equality continues to hold whatever the degree of subdivision.

From (1.13-3) and (1.13-2)

$$\oint_{\Delta S_i} \bar{F}.d\bar{S} = (\text{div } \bar{F})_{P_i} \Delta\tau_i + \gamma_i$$

where P_i is some point of $\Delta\tau_i$.

Then

$$\oint_S \bar{F}.d\bar{S} = \sum_{i=1}^{n} \{(\text{div } \bar{F})_{P_i} \Delta\tau_i + \gamma_i\}$$

But

$$\frac{\gamma_i}{(\text{div } \bar{F})_{P_i} \Delta\tau_i} \to 0 \quad \text{as} \quad \Delta\tau_i \to 0$$

where it is supposed that all dimensions of $\Delta\tau_i$ decrease uniformly, so that

$$\frac{\sum_{i=1}^{n} \gamma_i}{\sum_{i=1}^{n} (\text{div } \bar{F})_{P_i} \Delta\tau_i} \to 0 \quad \text{as} \quad n\to\infty \quad \text{and} \quad \Delta\tau_i \to 0$$

hence

$$\oint_S \bar{F}.d\bar{S} = \lim_{\substack{n\to\infty \\ \Delta\tau_i\to 0}} \sum_{i=1}^{n} (\text{div } \bar{F})_{P_i} \Delta\tau_i$$

or

$$\oint_S \bar{F}.d\bar{S} = \int_\tau \text{div } \bar{F} \, d\tau$$

1.14 Application of the Divergence Theorem

The divergence theorem has been formulated above for a closed region of volume integration bounded by a single surface. When more than one bounding surface is involved, as in Fig. 1.10, the appropriate form of the theorem may be developed from first principles along the lines indicated in Sec. 1.12 or 1.13[24].

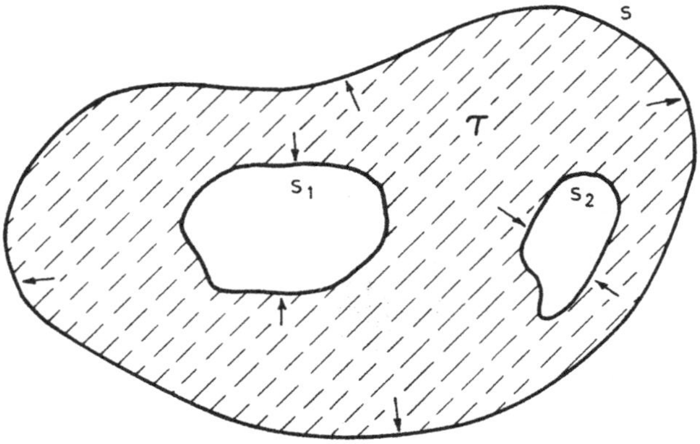

Fig. 1.10

Alternatively, a geometrical construction similar to that employed in the corresponding extension of Stokes's theorem may be utilised to derive the multiply bounded form from that already given.

The general form of the divergence theorem is

$$\int_{\tau} \operatorname{div} \overline{F}\, d\tau = \oint_{S} \overline{F}.d\overline{S} + \oint_{S_1} \overline{F}.d\overline{S} + \oint_{S_2} \overline{F}.d\overline{S} + --- \qquad (1.14\text{-}1)$$

where the surface integration is carried out over all closed surfaces bounding the region of volume integration τ, and the positive sense of the normal is everywhere directed away from τ[25].

---

24. The region need not be simply connected.

25. This definition of the positive sense of the normal is an extension of that adopted previously, where the region under consideration was bounded by a single surface only.

When bounding surfaces are also surfaces of discontinuity of $\overline{F}$, (1.14-1) is understood to apply to the region bounded by closed surfaces which lie just inside the surfaces of discontinuity.

The divergence theorem has been derived on the assumption that $\overline{F}$ and its first derivatives are continuous throughout τ. When τ contains points of discontinuity it is necessary that these be excluded from the integration space by means of closed surfaces, and the the theorem be applied in its multiply bounded form.

Consider, for example, the computation of the flux of $\overline{F}$ through a closed surface S which embraces the origin of coordinates, where

$$\overline{F} = \frac{\overline{r}}{r^3} = \frac{\overline{i}x + \overline{j}y + \overline{k}z}{(x^2+y^2+z^2)^{3/2}}$$

Div $\overline{F}$ is found to be zero for all non-zero values of r and is undefined at the origin.

Let the origin be excluded from the integration space by means of a small sphere of radius δ centred upon 0. Then (1.14-1) yields

$$\int_{\tau-\tau_\delta} \operatorname{div}\overline{F}\, d\tau = \oint_S \overline{F}.d\overline{S} + \oint_{S_\delta} \overline{F}.d\overline{S} = 0$$

The flux of $\overline{F}$ through S is seen to be equal and opposite to that through S_δ. But the flux through S_δ (which must be independent of δ since δ has been chosen arbitrarily) is readily shown, by direct integration, to have the value -4π. Hence the flux of $\overline{F}$ through S is equal to 4π.

1.15 The Solenoidal Vector Field

The field defined by a vector point function $\overline{F}$ is solenoidal (or circuital) within a region of space if $\oint \overline{F}.d\overline{S} = 0$ for every closed surface which can be drawn with the region.

The fact that div $\overline{F}$ may be zero within a region does not ensure that $\overline{F}$ is solenoidal within that region, since a surface of integration S may embrace a closed surface which bounds the region internally[26]. In this case it is the sum of the associated surface integrals which is zero rather than that taken over S alone. The reservation involved is clearly

26. Note, however, that most writers equate 'solenoidal' with 'zero-divergence'.

analogous to that encountered in the case of the irrotational vector field, where it was necessary to stipulate that the region be simply connected before a zero curl field could be identified as irrotational. In the present instance we may assert that so long as every closed surface can be contracted to a point without passing beyond the region[27] then the field is solenoidal if div $\overline{F} = 0$ at every point of it.

The flux through any open surface in a solenoidal vector field is dependent only upon its boundary.

In Fig. 1.11 the simple curve Γ is the common boundary of the open surfaces S_1, S_2 and S_3.

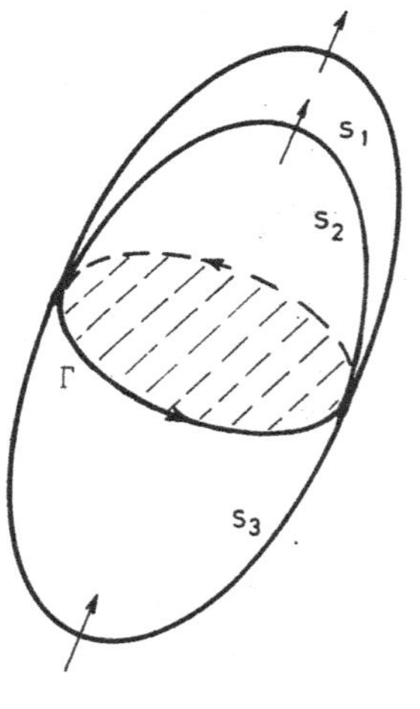

Fig. 1.11

Since the flux of $\overline{F}$ out of the region bounded by S_1 and S_2 is zero, it follows that the flux through S_1 is equal to that through S_2 if the positive sense through each surface is taken to correspond to the currency around Γ as defined by the arrows.

27. The region is then said to be aperiphractic.

If S_1 and S_2 should cut each other beyond Γ, then, since the fluxes through each are equal to that through S_3 in the direction shown, they continue to be equal to one another.

Theorem 1.15-1

If $\overline{F}$ and its first and second[28] derivatives are continuous throughout the region R, and if $\overline{V} = \text{curl}\ \overline{F}$, then $\overline{V}$ is solenoidal and $\text{div}\ \overline{V} = 0$ within R.

Proof: It has been shown in Sec. 1.10 that $\oint (\text{curl}\ \overline{F}).d\overline{S} = 0$ for every closed surface over which curl $\overline{F}$ is continuous. It follows that $\oint \overline{V}.d\overline{S} = 0$ for every closed surface within R, whence $\overline{V}$ is solenoidal within the region. But if $\overline{V}$ is solenoidal within R and has continuous first derivatives, then, from (1.12-5), $\text{div}\ \overline{V} = 0$ everywhere within R.

We may confirm that $\text{div}\ \overline{V} = 0$ by expanding div curl $\overline{F}$ in rectangular Cartesians. From (1.8-6) and (1.12-4)

$$\text{div curl}\ \overline{F} = \frac{\partial}{\partial x}\left(\frac{\partial F_z}{\partial y} - \frac{\partial F_y}{\partial z}\right) + \frac{\partial}{\partial y}\left(\frac{\partial F_x}{\partial z} - \frac{\partial F_z}{\partial x}\right) + \frac{\partial}{\partial z}\left(\frac{\partial F_y}{\partial x} - \frac{\partial F_x}{\partial y}\right)$$

ie

$$\text{div curl}\ \overline{F} \equiv 0 \qquad (1.15\text{-}1)$$

since

$$\frac{\partial^2 F_z}{\partial x \partial y} = \frac{\partial^2 F_z}{\partial y \partial x} \quad \text{etc.}$$

Theorem 1.15-2

If the vector point function $\overline{V}$ has continuous first derivatives and is solenoidal throughout the region of space R, then $\overline{V}$ may be expressed as curl $\overline{F}$ within R.

This will now be demonstrated for a region devoid of bounding surfaces.

---

28. In particular, the mixed derivatives should be continuous.

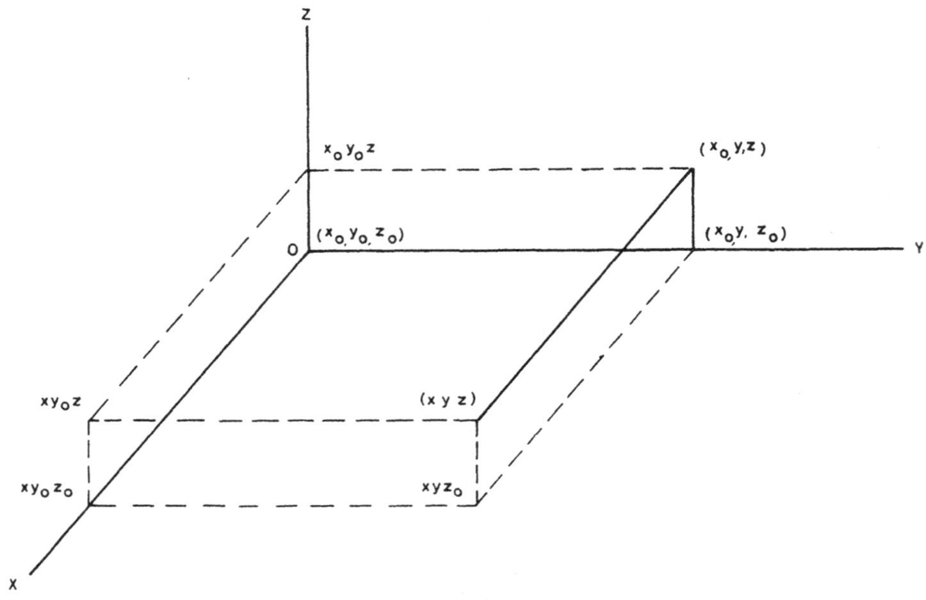

Fig. 1.12

Proof: Let rectangular axes be drawn through the fixed point (x_o, y_o, z_o) as in Fig. 1.12. It is evident that

$$V_x(x,y,z) = V_x(x_o,y,z) + \int_{x_o,y,z}^{x,y,z} \frac{\partial V_x}{\partial x}\, dx$$

But

$$V_x(x_o,y,z) = \frac{\partial}{\partial z} \int_{x_o,y,z_o}^{x_o,y,z} V_x\, dz$$

and

$$\int_{x_o,y,z}^{x,y,z} \frac{\partial V_x}{\partial x}\, dx = - \int_{x_o,y,z}^{x,y,z} \frac{\partial V_y}{\partial y}\, dx - \int_{x_o,y,z}^{x,y,z} \frac{\partial V_z}{\partial z}\, dx$$

∂V ∂V ∂V

Some consideration will show that

$$\int_{x_o,y,z}^{x,y,z} \frac{\partial V_y}{\partial y}\, dx = \frac{\partial}{\partial y} \int_{x_o,y,z}^{x,y,z} V_y\, dx \quad \text{and} \quad \int_{x_o,y,z}^{x,y,z} \frac{\partial V_z}{\partial z}\, dx = \frac{\partial}{\partial z} \int_{x_o,y,z}^{x,y,z} V_z\, dx$$

hence

$$V_x(x,y,z) = \frac{\partial}{\partial z} \int_{x_o,y,z_o}^{x_o,y,z} V_x\, dz - \frac{\partial}{\partial y} \int_{x_o,y,z}^{x,y,z} V_y\, dx - \frac{\partial}{\partial z} \int_{x_o,y,z}^{x,y,z} V_z\, dx$$

$$= \frac{\partial}{\partial y} \int_{x_o,y,z}^{x,y,z} (-V_y)\, dx - \frac{\partial}{\partial z} \left\{ \int_{x_o,y,z}^{x,y,z} V_z dx - \int_{x_o,y,z_o}^{x_o,y,z} V_x\, dz \right\}$$

Further,

$$V_y(x,y,z) = \frac{\partial}{\partial x} \int_{x_o,y,z}^{x,y,z} V_y\, dx$$

and

$$V_z(x,y,z) = \frac{\partial}{\partial x} \int_{x_o,y,z}^{x,y,z} V_z\, dx$$

It is now readily seen that by putting

$$F_x(x,y,z) = 0 \quad ; \quad F_y(x,y,z) = \int_{x_o,y,z}^{x,y,z} V_z\, dx - \int_{x_o,y,z_o}^{x_o,y,z} V_x\, dz \quad ;$$

$$F_z(x,y,z) = - \int_{x_o,y,z}^{x,y,z} V_y\, dx$$

We have

$$V_x(x,y,z) = \left(\frac{\partial F_z}{\partial y} - \frac{\partial F_y}{\partial z}\right)_{x,y,z} \quad ; \quad V_y(x,y,z) = \left(\frac{\partial F_x}{\partial z} - \frac{\partial F_z}{\partial x}\right)_{x,y,z} \quad ;$$

$$V_z(x,y,z) = \left(\frac{\partial F_y}{\partial x} - \frac{\partial F_x}{\partial y}\right)_{x,y,z}$$

ie $\overline{V} = \text{curl}\ \overline{F}$, where $\overline{F}$ has the Cartesian components specified above.

The above analysis is also applicable to a region bounded externally by a non-re-entrant surface, but complications arise when re-entrant or multiple bounding surfaces are present because of the discontinuities that may be introduced into the integration paths which define the components of $\overline{F}$. Nevertheless it can be shown, in the latter circumstance, that a well-behaved point function $\overline{F}$ exists, such that $\overline{V} = \text{curl}\ \overline{F}$ throughout the region over which $\overline{V}$ is defined. (See p. 344.) It is necessary that $\overline{V}$ be solenoidal and not merely divergence-free within any such region, since $\oint \text{curl}\ \overline{F}.d\overline{S} \equiv 0$ when $\overline{F}$ is well-behaved.

There are cases of interest in which $\overline{V}$ is undefined at some point - say 0, while div $\overline{V} = 0$ everywhere beyond 0. The above treatment, or an equivalent one (See Ex.1-49., p. 65), then permits of the derivation of a point function $\overline{F}$ such that $\overline{V} = \text{curl}\ \overline{F}$ at all points except those lying upon some coordinate line or surface containing 0, over which $\overline{F}$ is undefined. It may be possible to find an alternative expression for $\overline{F}$[29] which is well-behaved everywhere beyond 0, so long as $\overline{V}$ is solenoidal in all regions which exclude 0 (as in Ex.1-48., p. 64). On the other hand, if $\overline{V}$ is non-solenoidal in some region which excludes 0 (as in Ex.1-45., p. 64 and Ex. 2-20, p. 145), then $\overline{F}$ can never be well-behaved everywhere beyond 0.

EXERCISES

1-39. Derive (1.14-1) for a region bounded externally by S and internally by S_1, by means of an analysis similar to that of (a) Sec. 1.12 and (b) Sec. 1.13.

---

29. The addition of any gradient function to $\overline{F}$ leaves curl $\overline{F}$ unaffected. (Theorem 1.11-1)

1-40. Given that

(a) V and $\overline{F}$ are well-behaved scalar and vector point functions defined throughout the region of space $\underline{R}$

(b) for every closed surface S which may be contracted to a point without passing beyond $\underline{R}$

$$\oint_S \overline{F}.d\overline{S} = \int_\tau V\, d\tau$$

where τ is the space enclosed by S,

show that $V = \text{div}\,\overline{F}$ throughout $\underline{R}$.

1-41. Compute the value of the normal surface integral of $\overline{i}\,2x^2 + \overline{j}\,3xy + \overline{k}z$ over the closed surface formed by the planes $x = 0$, $x = 1$, $y = 0$, $y = 2$, $z = 0$, $z = 3$ by (a) direct integration (b) application of the divergence theorem.

Ans: 27

1-42. Show that the volume enclosed by a surface is given by $\frac{1}{3}\oint \overline{r}.d\overline{S}$ where $\overline{r}$ is the position vector of the typical surface element relative to an arbitrary origin.

1-43. Let Γ be a regular closed curve lying in the xy coordinate plane, and let the surface which it encloses be designated S_z. Suppose that this forms the base of a right cylinder of height h. Then if $V = V(x,y)$ and $U = U(x,y)$ are continuous scalar point functions with continuous first derivatives within and about the cylinder, and if $\overline{F}$ is a two-dimensional field defined by $\overline{F} = \overline{i}\,V + \overline{j}\,U$, show, by means of the divergence theorem, that

$$\oint_\Gamma \overline{F}.\hat{\overline{n}}'\,|d\overline{r}| = \int_{S_z} \text{div}\,\overline{F}\,dS_z$$

where $\hat{\overline{n}}'$ is the unit normal to $d\overline{r}$ directed away from S_z.

Hence show that

$$\oint_\Gamma (V\,dr_y - U\,dr_x) = \int_{S_z} \left(\frac{\partial V}{\partial x} + \frac{\partial U}{\partial y}\right) dS_z$$

[This result was obtained as a particular form of Stokes's theorem in Ex.1-31, p. 45. It is now seen to be a planar form of the divergence theorem (which is also known as Green's theorem) whence the name 'Green's theorem in the plane'.]

1-44. By an appropriate choice of V and U in Ex.1-43. above, express the area of the ellipse $\frac{x^2}{a^2} + \frac{y^2}{b^2} = 1$ as a line integral and compute its value.

Ans: πab

1-45. Prove that, for non-zero values of r, div $\overline{V} = 0$ where

$$\overline{V} = \frac{\overline{r}}{r^3} = \frac{\overline{i}x + \overline{j}y + \overline{k}z}{(x^2+y^2+z^2)^{3/2}}$$

and derive an expression for $\overline{F}$ such that $\overline{V}$ = curl $\overline{F}$. Confirm this by expanding curl $\overline{F}$.

Ans: $\overline{F} = \overline{i}0 + \overline{j}\frac{xz}{(y^2+z^2)r} - \overline{k}\frac{xy}{(y^2+z^2)r}$ beyond the x axis.

1-46. The solid angle subtended at an external point 0 by an open surface S is given by $\int_S \frac{\overline{r}}{r^3} \cdot d\overline{S}$, where $\overline{r}$ is the position vector of the typical surface element relative to 0. Show that all open surfaces, having the same simple closed bounding curve and a consistent sense of the normal based upon associated currency, subtend the same solid angle at 0, provided that no pair of surfaces enclose 0. Show further that if any two such surfaces enclose 0 and the outward normal is taken as positive, then the sum of the solid angles which they subtend is equal to 4π. [See also Sec. 3.4.]

1-47. Two concentric circles are centred upon the x axis and lie in a plane normal to it. The annulus defined by the circles subtends a certain solid angle at the origin. Utilise the result Ex.1-45. to express this solid angle in terms of line integrals of $\overline{F}$ around the circles. Evaluate these integrals by (a) Cartesian integration, (b) direct integration (noting first that $\overline{F}$ may be expressed simply in plane polar coordinates) and so show that the solid angle is given by $2\pi(\cos\theta_1 - \cos\theta_2)$ where θ_1 and θ_2 are the angles made with the x axis by the position vectors joining the origin to the inner and outer circles respectively.

Show also that the solid angle subtended at 0 by a circular disc centred upon and normal to the x axis cannot be evaluated directly, but only as a limiting case of the above expression, and explain the reason for this.

1-48. Let the vector point function $\overline{V}$ be defined by

$$\overline{V} = \frac{\overline{a} \times \overline{r}}{r^3}$$

where $\overline{a}$ is a constant vector point function and $\overline{r}$ is the position vector relative to the origin of coordinates.

Prove that div $\overline{V}$ is zero and $\overline{V}$ is solenoidal in all regions which exclude the origin, and show that $\overline{V}$ may be expressed outside the plane y = 0 as the curl of $\overline{F}$, where

$$\overline{F} = \overline{i}\,0 + \overline{j}\left\{\frac{a_z z}{y(y^2+z^2)^{\frac{1}{2}}} + \frac{a_x yx}{(y^2+z^2)r} - \frac{a_y}{(y^2)^{\frac{1}{2}}} + \frac{a_y}{r}\right\}$$

$$+ \overline{k}\left\{\frac{a_x zx}{(y^2+z^2)r} - \frac{a_z}{(y^2+z^2)^{\frac{1}{2}}} + \frac{a_z}{r}\right\}$$

By reducing curl $\overline{F}$ to the form

$$\overline{i}\left\{\frac{\partial}{\partial y}\left(\frac{a_z}{r}\right) - \frac{\partial}{\partial z}\left(\frac{a_y}{r}\right)\right\} + \overline{j}\left\{\frac{\partial}{\partial z}\left(\frac{a_x}{r}\right) - \frac{\partial}{\partial x}\left(\frac{a_z}{r}\right)\right\} + \overline{k}\left\{\frac{\partial}{\partial x}\left(\frac{a_y}{r}\right) - \frac{\partial}{\partial y}\left(\frac{a_x}{r}\right)\right\}$$

demonstrate that

$$\overline{V} = \text{curl}\left\{\overline{i}\,\frac{a_x}{r} + \overline{j}\,\frac{a_y}{r} + \overline{k}\,\frac{a_z}{r}\right\} = \text{curl}\,\frac{\overline{a}}{r}$$

Confirm that this holds at all points outside the origin.

1-49. $\overline{V}$ is an unbounded, solenoidal, vector point function which, together with its first derivatives, is everywhere continuous. Obtain a rectangular Cartesian expression for $\overline{F}$ such that $\overline{V}$ = curl $\overline{F}$ where F_y is everywhere zero, and confirm this by expanding curl $\overline{F}$.

Ans: $$F_x = \int_{x,y_o,z_o}^{x,y_o,z} V_y\,dz - \int_{x,y_o,z}^{x,y,z} V_z\,dy \quad ; \quad F_z = \int_{x,y_o,z}^{x,y,z} V_x\,dy \quad ;$$

or $$F_x = -\int_{x,y_o,z}^{x,y,z} V_z\,dy \quad ; \quad F_z = \int_{x,y_o,z}^{x,y,z} V_x\,dy - \int_{x_o,y_o,z}^{x,y_o z} V_y\,dx$$

1-50. Show that equation (1.12-3) continues to hold in the presence of an interior closed surface of discontinuity of $\overline{F}$ when the normal component of $\overline{F}$ is continuous through this surface, provided that div $\overline{F}$ is continuous at interior points of the two subregions and upon S, and the volume integral is understood to represent the limiting value of the sum of the integrals taken over the subregions as the surface of discontinuity is approached from both sides.

1.16 Expansion Formulae for Gradient, Curl and Divergence

(1) $\text{curl}(\overline{F}\pm\overline{G}\pm\overline{H}---) = \text{curl}\,\overline{F} \pm \text{curl}\,\overline{G} \pm \text{curl}\,\overline{H}---$ (1.16-1)

(2) $\text{div}(\overline{F}\pm\overline{G}\pm\overline{H}---) = \text{div}\,\overline{F} \pm \text{div}\,\overline{G} \pm \text{div}\,H---$ (1.16-2)

(3) $\text{curl}\,V\overline{F} = V\,\text{curl}\,\overline{F} + \text{grad}\,V \times \overline{F}$ (1.16-3)

(4) $\text{div}\,V\overline{F} = V\,\text{div}\,\overline{F} + \text{grad}\,V.\overline{F}$ (1.16-4)

In the above expressions V and $\overline{F}$, $\overline{G}$--- are differentiable scalar and vector point functions. Brackets have been omitted from the ultimate terms of equations (1.16-3) and (1.16-4) since only one reading is possible.

The relationships are readily proved by writing out the operands in rectangular components and differentiating in accordance with (1.8-6) and (1.12-4).

(1.16-1) and (1.16-2) also follow from (1.8-7) and (1.12-5) respectively.

(5) $\text{grad}(\overline{F}.\overline{G}) = (\overline{F}.\nabla)\overline{G} + (\overline{G}.\nabla)\overline{F} + \overline{F} \times \text{curl}\,\overline{G} + \overline{G} \times \text{curl}\,\overline{F}$ (1.16-5)

The operator $(\overline{F}.\nabla)$ has been defined in the footnote to p. 4. It should be noted that $(\overline{F}.\nabla)\overline{G}$ is sometimes written as $\overline{F}.\nabla\overline{G}$ because no meaning has been assigned to $\nabla\overline{G}$ and there can consequently be no ambiguity in the grouping[30]. (1.16-5) may be proved as follows

$$(\text{grad}(\overline{F}.\overline{G}))_x = \frac{\partial}{\partial x}(F_xG_x+F_yG_y+F_zG_z) = F_x\frac{\partial G_x}{\partial x} + F_y\frac{\partial G_y}{\partial x} + F_z\frac{\partial G_z}{\partial x} + ---$$

$$= F_x\frac{\partial G_x}{\partial x} + \left(F_y\frac{\partial G_x}{\partial y} + F_z\frac{\partial G_x}{\partial z} - F_y\frac{\partial G_x}{\partial y} - F_z\frac{\partial G_x}{\partial z}\right) + F_y\frac{\partial G_y}{\partial x} + F_z\frac{\partial G_z}{\partial x} + ----$$

$$= (\overline{F}.\nabla)\overline{G}_x + F_y(\text{curl}\,\overline{G})_z - F_z(\text{curl}\,\overline{G})_y + -----$$

$$= (\overline{F}.\nabla)G_x + (\overline{F} \times \text{curl}\,\overline{G})_x + (\overline{G}.\nabla)F_x + (\overline{G} \times \text{curl}\,\overline{F})_x$$

whence (1.16-5) follows.

(6) $\text{curl}(\overline{F}\times\overline{G}) = (\overline{G}.\nabla)\overline{F} - (\overline{F}.\nabla)\overline{G} + \overline{F}\,\text{div}\,\overline{G} - \overline{G}\,\text{div}\,\overline{F}$ (1.16-6)

---

30. $\nabla\overline{G}$ is assigned significance in dyadic notation, but we are concerned with this only in passing. (p. 592)

(7) $$\operatorname{div}(\overline{F}\times\overline{G}) = \overline{G}.\operatorname{curl}\overline{F} - \overline{F}.\operatorname{curl}\overline{G} \qquad (1.16\text{-}7)$$

(1.16-6) and (1.16-7) are proved in a manner similar to (1.16-5).

(8) $$\overline{F}\times(\overline{G}.\nabla)\overline{H} = ((\overline{G}\times\overline{F}).\nabla)\overline{H} + \overline{G}\times(\overline{F}.\nabla)\overline{H} + (\overline{F}\times\overline{G})\operatorname{div}\overline{H} + (\operatorname{curl}\overline{H})\times(\overline{F}\times\overline{G}) \qquad (1.16\text{-}8)$$

(9) $$\overline{F}.((\overline{G}.\nabla)\overline{H}) = \overline{G}.((\overline{F}.\nabla)\overline{H}) + (\overline{G}\times\overline{F}).\operatorname{curl}\overline{H} \qquad (1.16\text{-}9)$$

(1.16-8) and (1.16-9) hold at all points where $\overline{F}$ and $\overline{G}$ are defined and $\overline{H}$ is differentiable.

The relationships may be confirmed by routine expansion of the x components of the individual terms.

Formulae (1.16-1) to (1.16-7) are used frequently and, together with (1.4-1) to (1.4-3), should be committed to memory.

1.17 Deductions from Stokes's Theorem and the Divergence Theorem

A number of important integral transformations follow from the above theorems. These are stated and proved below.

Where volume integrals transform wholly or in part to surface integrals, the surface integration is carried out over all surfaces bounding the region of volume integration. Similarly, in those cases where surface integrals transform wholly or in part to line integrals, the line integration is carried out over all curves bounding the surface of integration. The sense of the positive normal for volume/surface transformations and the relation between the positive normal and boundary currency for surface/line transformations follow the conventions discussed previously.

The point functions involved in volume/surface transformations are supposed to be continuously differentiable throughout the closed region of space concerned, ie, they are supposed to have continuous first derivatives both at interior points of the region and upon its bounding surface (or surfaces)[31]. However, in the event that the point functions and their derivatives are continuous at interior points of a region and possess limits as the bounding surface S is approached along the interior normals, but are undefined upon S itself, the transformations relate the limiting values of surface and volume integrals for a closed region lying just inside S as S is approached at all points.

In general, the point functions involved in surface/line transformations are supposed to be continuously differentiable throughout a region of space which includes the surface of integration, although in certain cases the functions need be continuous only upon the surface itself (see

---

31. The derivatives at points of the surface will be 'single-ended' if the point function is not defined outside the enclosure.

Ex.1-31 p.45)[32]. Remarks similar to the above apply when a bounding curve is a line of discontinuity.

The subscripts previously appended to the integral signs will be omitted from now on, unless specially required, since the nature of the integration is sufficiently indicated by the associated integration variable and the presence of a closed or open integral sign.

(1)

$$\int d\bar{S} \times \text{grad } V = \oint V\, d\bar{r} \qquad (1.17\text{-}1)$$

Stokes's theorem with $\bar{F} = \bar{i}\, V$ yields

$$\int (\text{curl } \bar{i}\, V).d\bar{S} = \oint \bar{i}\, V.d\bar{r} = \oint V\, dr_x$$

But

$$\int (\text{curl } \bar{i}\, V)\, .\, d\bar{S} = \int \left(\bar{i}0+\bar{j}\, \frac{\partial V}{\partial z} - \bar{k}\, \frac{\partial V}{\partial y}\right) .\, d\bar{S}$$

$$= \int \left(\frac{\partial V}{\partial z}\, dS_y - \frac{\partial V}{\partial y}\, dS_z\right) = \int (d\bar{S} \times \text{grad } V)_x$$

hence

$$\int \bar{i}\, (d\bar{S} \times \text{grad } V)_x = \oint V\, \bar{i}\, dr_x$$

On putting $\bar{F} = \bar{j}\, V$ and $\bar{F} = \bar{k}\, V$ respectively, we obtain the additional equations

$$\int \bar{j}\, (d\bar{S} \times \text{grad } V)_y = \oint V\, \bar{j}\, dr_y$$

---

32. Transformations involving scalar and vector point functions which are defined only upon curved surfaces are treated in Ch.2.

$$\int \bar{k}\ (d\bar{S} \times \text{grad } V)_z = \oint V\ \bar{k}\ dr_z$$

whence, by addition, we obtain (1.17-1).

An alternative proof may be developed on the basis of the expansion (1.16-3).

Let $\bar{a}$ define a constant vector point function on and about the surface of integration. Then at each point of the surface

$$\text{curl } V\ \bar{a} = \text{grad } V \times \bar{a} \quad \text{since curl } \bar{a} = \bar{0}$$

By Stokes's theorem

$$\int (\text{curl } V\ \bar{a}).d\bar{S} = \oint V\ \bar{a}.d\bar{r}$$

Hence

$$\int d\bar{S}.(\text{grad } V \times \bar{a}) = \oint V\ \bar{a}.d\bar{r}$$

By interchanging the dot and cross of the triple scalar product we get

$$\int (d\bar{S} \times \text{grad } V).\bar{a} = \oint V\ \bar{a}.d\bar{r}$$

Some consideration will show that since $\bar{a}$ is constant this equation is equivalent to

$$\bar{a} . \int d\bar{S} \times \text{grad } V = \bar{a} . \oint V\ d\bar{r}$$

However, since $\bar{a}$ may have any direction the relationship can hold only if

$$\int d\bar{S} \times \text{grad } V = \oint V\ d\bar{r}$$

(1a)

$$\oint d\bar{S} \times \text{grad } V \equiv \bar{0} \qquad (1.17\text{-}2)$$

This follows from (1.17-1) by arguments similar to those employed in association with Fig. 1.6 to show that $\oint \text{curl } \bar{F}.d\bar{S} \equiv 0$.

(2)

$$\int \text{curl } \overline{F}\, d\tau = \oint d\overline{S} \times \overline{F} \tag{1.17-3}$$

This is known as Ostrogradsky's theorem.

We will prove it by showing that the x components of the two sides are equal for arbitrary axes.

Since

$$\int (\text{curl } \overline{F})_x \, d\tau = \int \left(\frac{\partial F_z}{\partial y} - \frac{\partial F_y}{\partial z}\right) d\tau$$

we may replace it by $\int \text{div } \overline{G}\, d\tau$ where

$$\overline{G} = \overline{i}0 + \overline{j}\, F_z - \overline{k}\, F_y$$

whence, by the divergence theorem,

$$\int (\text{curl } \overline{F})_x d\tau = \oint (\overline{i}0+\overline{j}F_z-\overline{k}F_y).d\overline{S} = \oint (F_z dS_y - F_y dS_z)$$

$$\text{ie} \int (\text{curl } \overline{F})_x d\tau = \oint (d\overline{S}\times\overline{F})_x$$

(1.17-3) leads directly to a definition of curl in terms of a surface integral per unit volume. (cf. (1.12-5) for divergence.)

$$\text{curl } \overline{F} = \lim_{\tau\to 0} \frac{1}{\tau} \oint d\overline{S} \times \overline{F} \tag{1.17-4}$$

(3)

$$\int \text{grad } V\, d\tau = \oint V\, d\overline{S} \tag{1.17-5}$$

On putting $\overline{F} = \overline{i}\, V$ in the divergence theorem we obtain

$$\int \text{div } \overline{F}\, d\tau = \int \frac{\partial V}{\partial x} d\tau = \oint V\, dS_x$$

$$\text{whence} \int \overline{i} \frac{\partial V}{\partial x} d\tau = \oint V\, \overline{i}\, dS_x$$

On putting $\overline{F} = \overline{j}\,V$ and $\overline{F} = \overline{k}\,V$ in succession we obtain similar equations which, when added to the above, yield (1.17-5).

It follows that

$$\text{grad } V = \lim_{\tau\to 0} \frac{1}{\tau} \oint V\, d\overline{S} \qquad (1.17\text{-}6)$$

(4)

$$\int (\overline{F} \times \text{grad } V)d\tau = \int V \text{ curl } \overline{F}\, d\tau + \oint V\, \overline{F} \times d\overline{S} \qquad (1.17\text{-}7)$$

This follows directly from a volume integration of (1.16-3) and a subsequent application of Ostrogradsky's theorem.

(5)

$$\int (\overline{F} \times \text{grad } V).d\overline{S} = \int V(\text{curl } \overline{F}).d\overline{S} - \oint V\, \overline{F}.d\overline{r} \qquad (1.17\text{-}8)$$

This follows directly from a normal surface integration of (1.16-3) and a subsequent application of Stokes's theorem.

(6)

$$\int \overline{F}.\text{grad } V\, d\tau = \int V(-\text{div } \overline{F})\, d\tau + \oint V\, \overline{F}.d\overline{S} \qquad (1.17\text{-}9)$$

This follows directly from a volume integration of (1.16-4) and a subsequent application of the divergence theorem.

(7)

$$\oint V \text{ grad } U.d\overline{S} = \int V \text{ div grad } U\, d\tau + \int \text{grad } V.\text{grad } U\, d\tau \qquad (1.17\text{-}10)$$

where the second derivatives of U are continuous.

(1.17-10) follows immediately from (1.17-9) when $\overline{F}$ is identified with grad U.

(8)

$$\oint (V \text{ grad } U - U \text{ grad } V).d\overline{S} = \int (V \text{ div grad } U - U \text{ div grad } V)\, d\tau \qquad (1.17\text{-}11)$$

where the second derivatives of V and U are continuous.

These two transformations are of considerable importance and are known, in the order in which they appear, as the first (asymmetrical) and second (symmetrical) form of Green's theorem[33].

(9)

$$\oint d\bar{r} \times \bar{F} = \int (d\bar{S}.\nabla)\bar{F} - \int \operatorname{div} \bar{F}\, d\bar{S} + \int d\bar{S} \times \operatorname{curl} \bar{F} \qquad (1.17\text{-}12)$$

We will confirm this for the x component.

$$\oint (d\bar{r}\times\bar{F})_x = \oint (dr_y F_z - dr_z F_y) = \int \{(d\bar{S} \times \operatorname{grad} F_z)_y - (d\bar{S} \times \operatorname{grad} F_y)_z\}$$

(from (1-17-1))

$$= \int \left(dS_z \frac{\partial F_z}{\partial x} - dS_x \frac{\partial F_z}{\partial z} - dS_x \frac{\partial F_y}{\partial y} + dS_y \frac{\partial F_y}{\partial x}\right)$$

$$= \int \left\{-\operatorname{div} \bar{F}\, dS_x + (d\bar{S}.\nabla)F_x + dS_y \left(\frac{\partial F_y}{\partial x} - \frac{\partial F_x}{\partial y}\right) - dS_z \left(\frac{\partial F_x}{\partial z} - \frac{\partial F_z}{\partial x}\right)\right\}$$

$$\text{ie} \quad \oint (d\bar{r}\times\bar{F})_x = \int (d\bar{S}.\nabla)F_x - \int \operatorname{div} \bar{F}\, dS_x + \int (d\bar{S} \times \operatorname{curl} \bar{F})_x$$

(9a)

$$\oint (d\bar{S}.\nabla)\bar{F} - \oint \operatorname{div} \bar{F}\, d\bar{S} + \oint d\bar{S} \times \operatorname{curl} \bar{F} \equiv \bar{0} \qquad (1.17\text{-}13)$$

This follows from (1.17-12) by arguments similar to those employed in association with Fig. 1.6 to show that $\oint \operatorname{curl} \bar{F}.d\bar{S} \equiv 0$.

(10)

$$\oint \bar{r} \times (d\bar{r}\times\bar{F}) = \int \bar{r}\times(d\bar{S}.\nabla)\bar{F} - \int \bar{r}\times \operatorname{div} \bar{F}\, d\bar{S} + \int \bar{r} \times (d\bar{S} \times \operatorname{curl} \bar{F}) + \int d\bar{S} \times \bar{F} \qquad (1.17\text{-}14)$$

where the origin of $\bar{r}$ is arbitrary.

This may be demonstrated as follows.

From the usual expansion for a triple vector product we have

---

33. Not to be confused either with Green's theorem of p. 45, or with the

$$\oint \bar{r} \times (d\bar{r}\times\bar{F}) = \oint \bar{r}.\bar{F}\, d\bar{r} - \oint \bar{F}\, \bar{r}.d\bar{r}$$

But from (1.17-1)

$$\oint \bar{r}.\bar{F}\, d\bar{r} = \int d\bar{S} \times \text{grad}\ \bar{r}.\bar{F}$$

and from (1.16-5)

$$\text{grad}\ \bar{r}.\bar{F} = (\bar{r}.\nabla)\bar{F} + (\bar{F}.\nabla)\bar{r} + \bar{r} \times \text{curl}\ \bar{F} + \bar{F} \times \text{curl}\ \bar{r}$$

$$= (\bar{r}.\nabla)\bar{F} + \bar{F} + \bar{r} \times \text{curl}\ \bar{F} \qquad \text{since curl}\ \bar{r} = \bar{0}$$

hence

$$\oint \bar{r}.\bar{F}\, d\bar{r} = \int d\bar{S} \times (\bar{r}.\nabla)\bar{F} + \int d\bar{S} \times \bar{F} + \int d\bar{S} \times (\bar{r} \times \text{curl}\ \bar{F})$$

Further,

$$\int (\text{curl}\ F_x\bar{r}).d\bar{S} = \oint F_x\bar{r}.d\bar{r} = \int (F_x\text{curl}\ \bar{r} + (\text{grad}F_x)\times\bar{r}).d\bar{S}$$

from (1.16-3)

$$= \int \nabla F_x \times \bar{r}\ .\ d\bar{S} = \int \bar{r} \times d\bar{S}\ .\ \nabla F_x \quad \text{by interchange of dot and cross}$$

whence

$$\oint \bar{F}\,\bar{r}\ .\ d\bar{r} = \int ((\bar{r}\times d\bar{S}).\nabla)\ \bar{F}$$

It follows that

$$\oint \bar{r} \times (d\bar{r}\times\bar{F}) = \int d\bar{S} \times (\bar{r}.\nabla)\bar{F} + \int d\bar{S} \times \bar{F} + \int d\bar{S} \times (\bar{r} \times \text{curl}\ \bar{F}) - \int ((\bar{r}\times d\bar{S}).\nabla)\bar{F}$$

But from (1.16-8) with $d\overline{S}$ substituted for $\overline{F}$, $\overline{r}$ substituted for $\overline{G}$ and $\overline{F}$ substituted for $\overline{H}$ we have

$$d\overline{S} \times (\overline{r}.\nabla)\overline{F} - ((\overline{r}\times d\overline{S}).\nabla)\overline{F} = \overline{r} \times (d\overline{S}.\nabla)\overline{F} + (d\overline{S}\times\overline{r}) \text{ div } \overline{F} + (\text{curl } \overline{F}) \times (d\overline{S}\times\overline{r})$$

hence

$$\oint \overline{r}\times(d\overline{r}\times\overline{F}) = \int \overline{r}\times(d\overline{S}.\nabla)\overline{F} - \int \overline{r}\times\text{div } \overline{F}\ d\overline{S} + \int(\text{curl } \overline{F})\times(d\overline{S}\times\overline{r}) + \int d\overline{S}\times(\overline{r}\times\text{curl } \overline{F}) + \int d\overline{S}\times\overline{F}$$

But expansion of the triple vector products shows that

$$(\text{curl } \overline{F}) \times (d\overline{S}\times\overline{r}) + d\overline{S} \times (\overline{r} \times \text{curl } \overline{F}) = \overline{r} \times (d\overline{S} \times \text{curl } \overline{F})$$

hence

$$\oint \overline{r} \times (d\overline{r}\times\overline{F}) = \int \overline{r} \times (d\overline{S}.\nabla)\overline{F} - \int \overline{r} \times \text{div } \overline{F}\ d\overline{S} + \int \overline{r} \times (d\overline{S} \times \text{curl } \overline{F}) + \int d\overline{S} \times \overline{F}$$

(11)

$$\int (\overline{F}.\nabla)\overline{G}\ d\tau = \int (-\text{div } \overline{F})\overline{G}\ d\tau + \oint \overline{G}\ \overline{F}.d\overline{S} \qquad (1.17\text{-}15)$$

This transformation may be derived from (1.17-9) by identifying V with G_x, G_y and G_z in turn, multiplying the respective equations by the unit vectors $\overline{i}$, $\overline{j}$ and $\overline{k}$, and adding.

(12)

$$\int (\overline{r}\times(\overline{F}.\nabla)\overline{G})\ d\tau = \int \overline{r} \times (-\text{div } \overline{F})\overline{G}\ d\tau + \oint (\overline{r}\times\overline{G})\overline{F}.d\overline{S} - \int (\overline{F}\times\overline{G})\ d\tau \qquad (1.17\text{-}16)$$

where the origin of $\overline{r}$ is arbitrary.

Since
$$(\overline{r}\times(\overline{F}.\nabla)\overline{G})_x = r_y(\overline{F}.\nabla)G_z - r_z(\overline{F}.\nabla)G_y$$

and

$$\text{div } G_z\overline{F} = G_z \text{ div } \overline{F} + (\overline{F}.\nabla)G_z$$

it is easily seen that

$$\int (\bar{r}\times(\bar{F}.\nabla)\bar{G})_x \, d\tau = \int (-\mathrm{div}\ \bar{F})(r_y G_z - r_z G_y) \, d\tau + \int (r_y \ \mathrm{div}\ G_z\bar{F} - r_z \ \mathrm{div}\ G_y\bar{F}) \, d\tau$$

But $\mathrm{div}\ r_y G_z \bar{F} = r_y \ \mathrm{div}\ G_z\bar{F} + G_z\bar{F}.\nabla\ r_y = r_y \ \mathrm{div}\ G_z\bar{F} + G_z F_y$

Similarly $\mathrm{div}\ r_z G_y \bar{F} = r_z \ \mathrm{div}\ G_y\bar{F} + G_y F_z$

hence

$$\int (r_y \ \mathrm{div}\ G_z\bar{F} - r_z \ \mathrm{div}\ G_y\bar{F}) \, d\tau = \oint (r_y G_z - r_z G_y)\bar{F}.d\bar{S} - \int (G_z F_y - G_y F_z) \, d\tau$$

whence

$$\int (\bar{r}\times(\bar{F}.\nabla)\bar{G})_x \, d\tau = \int (\bar{r}\times\bar{G})_x(-\mathrm{div}\ \bar{F}) \, d\tau + \oint (\bar{r}\times\bar{G})_x \bar{F}.d\bar{S} - \int (\bar{F}\times\bar{G})_x \, d\tau$$

from which (1.17-16) follows.

(13)

$$\int (\bar{F}.\nabla)\bar{G} \, d\tau = \int \bar{F} \ \mathrm{div}\ \bar{G} \, d\tau - \int (\bar{F} \times \mathrm{curl}\ \bar{G}) d\tau - \int (\bar{G} \times \mathrm{curl}\ \bar{F}) d\tau + \oint \bar{G} \times (d\bar{S}\times\bar{F}) \qquad (1.17\text{-}17)$$

From (1.16-5) it follows that

$$\int (\bar{F}.\nabla)\bar{G} \, d\tau = \int \mathrm{grad}\ \bar{F}.\bar{G} \, d\tau - \int (\bar{G}.\nabla)\bar{F} \, d\tau - \int (\bar{F} \times \mathrm{curl}\ \bar{G}) d\tau - \int (\bar{G} \times \mathrm{curl}\ \bar{F}) d\tau$$

But $\mathrm{div}\ F_x\bar{G} = F_x \ \mathrm{div}\ \bar{G} + \bar{G}.\nabla F_x$

whence

$$\oint \bar{F}\ \bar{G}.d\bar{S} = \int \bar{F} \ \mathrm{div}\ \bar{G} \, d\tau + \int (\bar{G}.\nabla)\bar{F} \, d\tau$$

Also, from (1.17-5)

$$\int \mathrm{grad}\ \bar{F}.\bar{G} \, d\tau = \oint \bar{F}.\bar{G} \, d\bar{S}$$

hence

$$\int (\overline{F}.\nabla)\overline{G}\, d\tau = \oint \overline{F}.\overline{G}\, d\overline{S} + \int \overline{F}\, \mathrm{div}\, \overline{G}\, d\tau - \oint \overline{F}\, \overline{G}.d\overline{S} - \int (\overline{F}\times \mathrm{curl}\, \overline{G})d\tau - \int (\overline{G}\times \mathrm{curl}\, \overline{F})d\tau$$

or

$$\int (\overline{F}.\nabla)\overline{G}\, d\tau = \int \overline{F}\, \mathrm{div}\, \overline{G}\, d\tau - \int (\overline{F}\times \mathrm{curl}\, \overline{G})d\tau - \int (\overline{G}\times \mathrm{curl}\, \overline{F})d\tau + \oint \overline{G}\times (d\overline{S}\times\overline{F})$$

(14)

$$\int (\overline{r}\times(\overline{F}.\nabla)\overline{G})\, d\tau = \int (\overline{r}\times\overline{F}\, \mathrm{div}\, \overline{G})\, d\tau - \int \overline{r}\times(\overline{F}\times \mathrm{curl}\, \overline{G})\, d\tau$$

$$- \int \overline{r}\times(\overline{G}\times \mathrm{curl}\, \overline{F})\, d\tau + \oint \overline{r}\times(\overline{G}\times(d\overline{S}\times\overline{F})) - \int (\overline{F}\times\overline{G})\, d\tau \qquad (1.17\text{-}18)$$

where the origin of $\overline{r}$ is arbitrary.

From (1.16-5) it follows that

$$\int \overline{r}\times(\overline{F}.\nabla)\overline{G}\, d\tau = \int \overline{r}\times \mathrm{grad}\, \overline{F}.\overline{G}\, d\tau - \int \overline{r}\times(\overline{G}.\nabla)\overline{F}\, d\tau - \int \overline{r}\times(\overline{F}\times \mathrm{curl}\, \overline{G})\, d\tau$$

$$- \int \overline{r}\times(\overline{G}\times \mathrm{curl}\, \overline{F})\, d\tau$$

Now $\qquad \mathrm{curl}\, \overline{r}\, \overline{F}.\overline{G} = \overline{F}.\overline{G}\, \mathrm{curl}\, \overline{r} + (\mathrm{grad}\, \overline{F}.\overline{G})\times\overline{r}$

hence

$$\int \mathrm{curl}\, \overline{r}\, \overline{F}.\overline{G}\, d\tau = -\int \overline{r}\times \mathrm{grad}\, \overline{F}.\overline{G}\, d\tau = \oint (d\overline{S}\times\overline{r})\overline{F}.\overline{G}$$

from (1.17-3)

Further, from (1.17-16), with $\overline{F}$ and $\overline{G}$ interchanged,

$$\int (\overline{r}\times(\overline{G}.\nabla)\overline{F})\, d\tau = \int \overline{r}\times(-\mathrm{div}\, \overline{G})\overline{F}\, d\tau + \oint \overline{r}\times\overline{F}\, \overline{G}.d\overline{S} - \int (\overline{G}\times\overline{F})\, d\tau$$

hence

$$\int (\overline{r}\times(\overline{F}.\nabla)\overline{G})\, d\tau = -\oint (d\overline{S}\times\overline{r})\overline{F}.\overline{G} + \int (\overline{r}\times\overline{F}\, \mathrm{div}\, \overline{G})\, d\tau - \oint (\overline{r}\times\overline{F})\overline{G}.d\overline{S}$$

$$+ \int (\overline{G}\times\overline{F})\, d\tau - \int \overline{r}\times(\overline{F}\times \mathrm{curl}\, \overline{G})\, d\tau - \int \overline{r}\times(\overline{G}\times \mathrm{curl}\, \overline{F})\, d\tau$$

But

$$(d\bar{S}\times\bar{r})\bar{F}.\bar{G} + (\bar{r}\times\bar{F})\bar{G}.d\bar{S} = (d\bar{S}\times\bar{F})\bar{r}.\bar{G} - \bar{G}(\bar{r}.d\bar{S}\times\bar{F})$$

as may be confirmed by expansion of the x components, hence

$$-\oint (d\bar{S}\times\bar{r})\bar{F}.\bar{G} - \oint (\bar{r}\times\bar{F})\bar{G}.d\bar{S} = -\oint (d\bar{S}\times\bar{F})\bar{r}.\bar{G} + \oint \bar{G}(\bar{r}.d\bar{S}\times\bar{F}) = \oint \bar{r}\times(\bar{G}\times(d\bar{S}\times\bar{F}))$$

whence

$$\int (\bar{r}\times(\bar{F}.\nabla)\bar{G})\,d\tau = \int (\bar{r}\times\bar{F}\ \mathrm{div}\ \bar{G})\,d\tau - \int \bar{r}\times(\bar{F}\times\mathrm{curl}\ \bar{G})\,d\tau - \int \bar{r}\times(\bar{G}\times\mathrm{curl}\ \bar{F})\,d\tau$$
$$+ \oint \bar{r}\times(\bar{G}\times(d\bar{S}\times\bar{F})) - \int (\bar{F}\times\bar{G})\,d\tau$$

EXERCISES

1-51. Confirm (1.16-8) and (1.16-9).

1-52. Prove that $\oint (\bar{F}\times\mathrm{grad}\ V).d\bar{S} = \int (\mathrm{grad}\ V.\mathrm{curl}\ \bar{F})\,d\tau$

[Hint: Expand div($\bar{F}$ × grad V) and apply the divergence theorem].

1-53. Prove that $\oint V\ \mathrm{curl}\ \bar{F}.d\bar{S} = \int (\mathrm{grad}\ V.\mathrm{curl}\ \bar{F})\,d\tau$

[Hint: Expand div(V curl $\bar{F}$) and apply the divergence theorem].

1-54. Demonstrate, independently of the results of the previous two exercises, that

$$\oint (\bar{F}\times\mathrm{grad}\ V).d\bar{S} = \oint V\ \mathrm{curl}\ \bar{F}.d\bar{S}$$

1-55. Show that

$$\int(\mathrm{grad}\ V\times\mathrm{grad}\ U).d\bar{S} = \oint V\ \mathrm{grad}\ U.d\bar{r} = -\oint U\ \mathrm{grad}\ V.d\bar{r}$$

1-56. Prove (1.17-3) by expanding div($\bar{F}\times\bar{a}$), where $\bar{a}$ is a constant vector point function, and applying the divergence theorem.

1-57. Prove (1.17-5) by expanding div V $\bar{a}$, where $\bar{a}$ is a constant vector point function, and applying the divergence theorem.

1-58. Show that curl $(\bar{\omega}\times\bar{r}) = 2\bar{\omega}$ where $\bar{r}$ is the position vector relative to an arbitrary origin, and $\bar{\omega}$ is a constant vector point function.

1-59. Make use of (1.16-3) to show that curl $\bar{r}\, f(r) = \bar{0}$ for all $f(r)$.

1-60. Prove that div $\bar{r}\, f(r) = 3\, f(r) + r\, f'(r)$.

1-61. Show that the vector area of an open surface with a boundary Γ is dependent only upon the shape and orientation of Γ.

The area may be shown to be equal to $\frac{1}{2}\oint_{\Gamma} \bar{r}\times d\bar{r}$ where the origin of $\bar{r}$ is arbitrary.

Demonstrate this by

(1) obtaining an expression for the area traced out by the position vector as its end point moves around Γ, and using the identity $\oint d\bar{S} \equiv \bar{0}$,

(2) expanding $\oint \bar{r}\times d\bar{r}$ in rectangular components and performing the associated integrations,

(3) substituting $\bar{r}$ for $\bar{F}$ in (1.17-12).

1-62. Show that $\int (d\bar{S}\times\nabla).\bar{F} = \int d\bar{S}.(\nabla\times\bar{F}) = \oint \bar{F}.d\bar{r}$

$$\text{where } (d\bar{S}\times\nabla) \equiv \begin{vmatrix} \bar{i} & \bar{j} & \bar{k} \\ dS_x & dS_y & dS_z \\ \frac{\partial}{\partial x} & \frac{\partial}{\partial y} & \frac{\partial}{\partial z} \end{vmatrix}$$

1-63. Prove that $\int (d\bar{S}\times\nabla)\times\bar{F} = \oint d\bar{r}\times\bar{F}$

1-64. Prove that $\int (d\bar{S}\times\nabla)V = \int d\bar{S}\times\nabla V = \oint V\, d\bar{r}$

1-65. Show that $\int ((\bar{a}.\nabla)\bar{F}).d\bar{S} = \bar{a}\,.\int \text{div}\,\bar{F}\, d\bar{S} + \bar{a}\,.\oint d\bar{r}\times\bar{F}$
where $\bar{a}$ is a constant vector point function.

1-66. Show that $\int ((\bar{a}\times\nabla)\times\bar{F}).d\bar{S} = \bar{a}.\int (\text{curl}\,\bar{F})\times d\bar{S} + \bar{a}.\oint d\bar{r}\times\bar{F}$

1-67. Let Γ denote a small, not necessarily plane , closed curve in the immediate vicinity of a point 0, and let $\overline{F}$ be a vector point function which, together with its first derivatives, is continuous everywhere in the neighbourhood of 0.

Make use of the approximation

$$\overline{F} \approx (\overline{F})_o + ((\overline{r}.\nabla)\overline{F})_o$$

where 0 is the origin of $\overline{r}$, to show that

$$\oint_\Gamma d\overline{r} \times \overline{F} \approx ((\overline{S}.\nabla)\overline{F})_o - \overline{S}(\text{div } \overline{F})_o + \overline{S} \times (\text{curl } \overline{F})_o$$

where $\overline{S}$ is the vector area of a surface whose contour is Γ.

Confirm this by working from (1.17-12).

[Hint: Expand $(d\overline{r}\times\overline{F})_x$ and carry out the associated plane surface integrations, noting carefully the currencies of the projections of Γ. Then bring the resulting expression into the required form.]

1-68. Derive Kelvin's generalisation of Green's theorem, viz.

$$\int W \text{ grad } V.\text{grad } U \, d\tau = \oint VW \text{ grad } U.d\overline{S} - \int V \text{ div}(W \text{ grad } U) \, d\tau$$

$$= \oint UW \text{ grad } V.d\overline{S} - \int U \text{ div}(W \text{ grad } V) \, d\tau$$

where W, V and U are well-behaved scalar point functions.

1-69. Prove that

$$\text{div}(\overline{F}.\nabla)\overline{G} = \overline{F}.\text{grad div } \overline{G} + \sum \text{grad } F_x.\text{grad } G_x + \sum (\text{grad } F_x \times \text{curl } \overline{G})_x$$

1.18 The Laplacian Operator ∇^2

The operator ∇^2 is defined by

$$\nabla^2 = \frac{\partial^2}{\partial x^2} + \frac{\partial^2}{\partial y^2} + \frac{\partial^2}{\partial z^2} \tag{1.18-1}$$

It may operate upon both scalar and vector point functions. When operating upon a well-behaved scalar point function it is equivalent to the operator div grad:

$$\text{div grad } V = \sum \frac{\partial}{\partial x}(\text{grad } V)_x = \frac{\partial^2 V}{\partial x^2} + \frac{\partial^2 V}{\partial y^2} + \frac{\partial^2 V}{\partial z^2}$$

ie $\text{div grad } V = \nabla.\nabla V = \nabla^2 V$ (1.18-2)

For operation upon a vector point function we have

$$\nabla^2\overline{F} = \frac{\partial^2\overline{F}}{\partial x^2} + \frac{\partial^2\overline{F}}{\partial y^2} + \frac{\partial^2\overline{F}}{\partial z^2}$$

But
$$\frac{\partial^2\overline{F}}{\partial x^2} = \frac{\partial^2}{\partial x^2}(\overline{i}F_x + \overline{j}F_y + \overline{k}F_z)$$

$$= \overline{i}\,\frac{\partial^2 F_x}{\partial x^2} + \overline{j}\,\frac{\partial^2 F_y}{\partial x^2} + \overline{k}\,\frac{\partial^2 F_z}{\partial x^2}$$

Similarly for $\frac{\partial^2\overline{F}}{\partial y^2}$ and $\frac{\partial^2\overline{F}}{\partial z^2}$

whence

$$\nabla^2\overline{F} = \overline{i}\,\nabla^2 F_x + \overline{j}\,\nabla^2 F_y + \overline{k}\,\nabla^2 F_z \qquad (1.18\text{-}3)$$

It follows that

$$(\nabla^2\overline{F})_x = \nabla^2 F_x = \text{div grad } F_x \quad \text{etc.} \qquad (1.18\text{-}4)$$

When $\overline{F}$, together with its first and second (mixed) derivatives are continuous, the operation is equivalent to grad div-curl curl, as may be shown by expanding in the basic Cartesian form.

$$(\text{grad div } \overline{F} - \text{curl curl } \overline{F})_x$$

$$= \frac{\partial}{\partial x}\left(\frac{\partial F_x}{\partial x} + \frac{\partial F_y}{\partial y} + \frac{\partial F_z}{\partial z}\right) - \left\{\frac{\partial}{\partial y}(\text{curl } \overline{F})_z - \frac{\partial}{\partial z}(\text{curl } \overline{F})_y\right\}$$

$$= \frac{\partial^2 F_x}{\partial x^2} + \frac{\partial^2 F_y}{\partial x \partial y} + \frac{\partial^2 F_z}{\partial x \partial z} - \left\{\frac{\partial}{\partial y}\left(\frac{\partial F_y}{\partial x} - \frac{\partial F_x}{\partial y}\right) - \frac{\partial}{\partial z}\left(\frac{\partial F_x}{\partial z} - \frac{\partial F_z}{\partial x}\right)\right\}$$

$$= \frac{\partial^2 F_x}{\partial x^2} + \frac{\partial^2 F_x}{\partial y^2} + \frac{\partial^2 F_x}{\partial z^2} \quad \text{since} \quad \frac{\partial^2 F_y}{\partial x \partial y} = \frac{\partial^2 F_y}{\partial y \partial x} \quad \text{etc.}$$

whence

$$\nabla^2\bar{F} = \text{grad div } \bar{F} - \text{curl curl } \bar{F} \tag{1.18-5}$$

It has been shown previously that the results of the operations grad, curl and div (defined by ∇, $\nabla\times$, and $\nabla.$) are independent of the set of right-handed rectangular axes chosen and this must be true of the double operations div grad, grad div and curl curl. It follows that ∇^2 is likewise invariant with respect to choice of axes when operating upon a scalar or vector point function.

1.19 Invariance of Grad, Div, Curl and ∇^2 With Respect to Choice of Rectangular Axes

Let $\bar{i}$, $\bar{j}$, $\bar{k}$ and $\bar{i}'$, $\bar{j}'$, $\bar{k}'$ be two right-handed rectangular systems of axes. The cosines of the angles between individual axes are set out in the following table.

	$\bar{i}$	$\bar{j}$	$\bar{k}$
$\bar{i}'$	l_1	m_1	n_1
$\bar{j}'$	l_2	m_2	n_2
$\bar{k}'$	l_3	m_3	n_3

(1.19-1)

On resolving $\bar{i}$, $\bar{j}$ and $\bar{k}$ in turn along the $\bar{i}'$, $\bar{j}'$, and $\bar{k}'$ axes we obtain

$$\begin{aligned} \bar{i} &= \bar{i}'l_1 + \bar{j}'l_2 + \bar{k}'l_3 \\ \bar{j} &= \bar{i}'m_1 + \bar{j}'m_2 + \bar{k}'m_3 \\ \bar{k} &= \bar{i}'n_1 + \bar{j}'n_2 + \bar{k}'n_3 \end{aligned} \tag{1.19-2}$$

To demonstrate the invariance of grad V it is necessary to show that

$$\bar{i}'\frac{\partial V}{\partial x'} + \bar{j}'\frac{\partial V}{\partial y'} + \bar{k}'\frac{\partial V}{\partial z'} = \bar{i}\frac{\partial V}{\partial x} + \bar{j}\frac{\partial V}{\partial y} + \bar{k}\frac{\partial V}{\partial z}$$

At points where V is well-behaved,

$$\frac{\partial V}{\partial x'} = \frac{\partial V}{\partial x}\frac{\partial x}{\partial x'} + \frac{\partial V}{\partial y}\frac{\partial y}{\partial x'} + \frac{\partial V}{\partial z}\frac{\partial z}{\partial x'}$$

ie $$\frac{\partial V}{\partial x'} = l_1\frac{\partial V}{\partial x} + m_1\frac{\partial V}{\partial y} + n_1\frac{\partial V}{\partial z}$$

Similarly $\frac{\partial V}{\partial y'} = l_2 \frac{\partial V}{\partial x} + m_2 \frac{\partial V}{\partial y} + n_2 \frac{\partial V}{\partial z}$

and $\frac{\partial V}{\partial z'} = l_3 \frac{\partial V}{\partial x} + m_3 \frac{\partial V}{\partial y} + n_3 \frac{\partial V}{\partial z}$

Hence $\bar{i}' \frac{\partial V}{\partial x'} + \bar{j}' \frac{\partial V}{\partial y'} + \bar{k}' \frac{\partial V}{\partial z'}$ may be brought into the form

$$(\bar{i}'l_1+\bar{j}'l_2+\bar{k}'l_3) \frac{\partial V}{\partial x} + (\bar{i}'m_1+\bar{j}'m_2+\bar{k}'m_3) \frac{\partial V}{\partial y} + (\bar{i}'n_1+\bar{j}'n_2+\bar{k}'n_3) \frac{\partial V}{\partial z}$$

and from (1.19-2) this is seen to be equal to

$$\bar{i} \frac{\partial V}{\partial x} + \bar{j} \frac{\partial V}{\partial y} + \bar{k} \frac{\partial V}{\partial z}$$

The proof of the invariance of div $\bar{F}$ requires that additional relationships be established between the direction cosines. Since $\bar{i}.\bar{i} = \bar{j}.\bar{j} = \bar{k}.\bar{k} = 1$ and $\bar{i}.\bar{j} = \bar{j}.\bar{k} = \bar{k}.\bar{i} = 0$ we have from (1.19-2)

$$\begin{aligned} l_1^2 + l_2^2 + l_3^2 &= 1 \\ m_1^2 + m_2^2 + m_3^2 &= 1 \\ n_1^2 + n_2^2 + n_3^2 &= 1 \end{aligned} \qquad (1.19\text{-}3)$$

and

$$\begin{aligned} l_1m_1 + l_2m_2 + l_3m_3 &= 0 \\ m_1n_1 + m_2n_2 + m_3n_3 &= 0 \\ n_1l_1 + n_2l_2 + n_3l_3 &= 0 \end{aligned} \qquad (1.19\text{-}4)$$

Now $F_{x'} = (\bar{i}F_x+\bar{j}F_y+\bar{k}F_z) \,.\, \bar{i}'$

$= l_1F_x + m_1F_y + n_1F_z$

hence $\frac{\partial F_{x'}}{\partial x'} = l_1 \frac{\partial F_x}{\partial x'} + m_1 \frac{\partial F_y}{\partial x'} + n_1 \frac{\partial F_z}{\partial x'}$

$$= l_1 \left\{ \frac{\partial F_x}{\partial x} l_1 + \frac{\partial F_x}{\partial y} m_1 + \frac{\partial F_x}{\partial z} n_1 \right\}$$

$$+ m_1 \left\{ \frac{\partial F_y}{\partial x} l_1 + \frac{\partial F_y}{\partial y} m_1 + \frac{\partial F_y}{\partial z} n_1 \right\}$$

$$+ n_1 \left\{ \frac{\partial F_z}{\partial x} l_1 + \frac{\partial F_z}{\partial y} m_1 + \frac{\partial F_z}{\partial z} n_1 \right\}$$

It may be shown similarly that

$$\frac{\partial F_{y'}}{\partial y'} = l_2 \left\{\frac{\partial F_x}{\partial x} l_2 + \frac{\partial F_x}{\partial y} m_2 + \frac{\partial F_x}{\partial z} n_2\right\}$$

$$+ m_2 \left\{\frac{\partial F_y}{\partial x} l_2 + \frac{\partial F_y}{\partial y} m_2 + \frac{\partial F_y}{\partial z} n_2\right\}$$

$$+ n_2 \left\{\frac{\partial F_z}{\partial x} l_2 + \frac{\partial F_z}{\partial y} m_2 + \frac{\partial F_z}{\partial z} n_2\right\}$$

and

$$\frac{\partial F_{z'}}{\partial z'} = l_3 \left\{\frac{\partial F_x}{\partial x} l_3 + \frac{\partial F_x}{\partial y} m_3 + \frac{\partial F_x}{\partial z} n_3\right\}$$

$$+ m_3 \left\{\frac{\partial F_y}{\partial x} l_3 + \frac{\partial F_y}{\partial y} m_3 + \frac{\partial F_y}{\partial z} n_3\right\}$$

$$+ n_3 \left\{\frac{\partial F_z}{\partial x} l_3 + \frac{\partial F_z}{\partial y} m_3 + \frac{\partial F_z}{\partial z} n_3\right\}$$

Upon adding the above expansions, collecting terms and substituting from (1.19-3) and (1.19-4) we obtain

$$\frac{\partial F_{x'}}{\partial x'} + \frac{\partial F_{y'}}{\partial y'} + \frac{\partial F_{z'}}{\partial z'} = \frac{\partial F_x}{\partial x} + \frac{\partial F_y}{\partial y} + \frac{\partial F_z}{\partial z}$$

as required.

Curl $\overline{F}$ may be shown to be invariant with respect to choice of axes by a similar analysis. It is necessary, for this purpose, to invoke three further sets of relationships between the direction cosines, based upon the equations $\overline{i} \times \overline{j} = \overline{k}$, $\overline{j} \times \overline{k} = \overline{i}$, $\overline{k} \times \overline{i} = \overline{j}$ and $\overline{i}' \times \overline{j}' = \overline{k}'$, $\overline{j}' \times \overline{k}' = \overline{i}'$, $\overline{k}' \times \overline{i}' = \overline{j}'$. The manipulations are left as an exercise for the reader.

As mentioned in Sec. 1.18, the invariance of the Cartesian forms of grad, curl and div is sufficient to ensure the invariance of $\nabla^2 V$ and $\nabla^2 \overline{F}$ via equations (1.18-2) and (1.18-5). This may be demonstrated independently by a direct transformation of the second derivatives (see Ex.1-72. and 1-73. below).

EXERCISES

1-70. Show that ∇V and $\nabla.\overline{F}$ remain unaffected when a left-handed system of axes is substituted for a right-handed system.

1-71. Show analytically that $\nabla \times \overline{F}$ is invariant with respect to choice of axes when working within either a right-handed or a left-handed system, but that it changes sign when one system is substituted for the other. $(\nabla\times(\nabla\times\overline{F})$ is consequently invariant under all conditions.)

1-72. Develop transformation formulae for the second derivatives of V and thereby show that

$$\nabla^2 V = \frac{\partial^2 V}{\partial x^2} + \frac{\partial^2 V}{\partial y^2} + \frac{\partial^2 V}{\partial z^2}$$

is invariant with respect to choice of axes and to transfer from a right-handed to a left-handed system.

1-73. Use the result of Ex.1-72. to show that

$$\nabla^2 \overline{F} = \overline{i}\,\nabla^2 F_x + \overline{j}\,\nabla^2 F_y + \overline{k}\,\nabla^2 F_z$$

is invariant to the same extent as $\nabla^2 V$.

1.20 Moving Systems and Time-Dependent Fields

1.20a Time rate of change of scalar or vector value at a point which moves in a time-dependent field

Let $V = V(x,y,z,t)$ be a scalar function of space and time which, together with its first space and time derivatives, is continuous both in space and time in the neighbourhood of x_o, y_o, z_o, t_o.

Suppose that corresponding to the times t_o and $t_o + \Delta t$ a moving point occupies the positions x_o, y_o, z_o and $x_o + \Delta x$, $y_o + \Delta y$, $z_o + \Delta z$. Then the change in the value of V at the point during this time interval is given by

$$\Delta V = V(x_o+\Delta x, y_o+\Delta y, z_o+\Delta z, t_o+\Delta t) - V(x_o,y_o,z_o,t_o)$$

$$= V(x_o+\Delta x, y_o+\Delta y, z_o+\Delta z, t_o+\Delta t) - V(x_o,y_o,z_o,t_o+\Delta t)$$

$$+ V(x_o,y_o,z_o,t_o+\Delta t) - V(x_o,y_o,z_o,t_o)$$

An appeal to Sec. 1.2 and the mean-value theorem reveals that

$$\Delta V = \left(\frac{\partial V}{\partial x}\right)_{\substack{x' \\ y_o+\Delta y \\ z_o+\Delta z \\ t_o+\Delta t}} \Delta x + \left(\frac{\partial V}{\partial y}\right)_{\substack{x_o \\ y' \\ z_o+\Delta z \\ t_o+\Delta t}} \Delta y + \left(\frac{\partial V}{\partial z}\right)_{\substack{x_o \\ y_o \\ z' \\ t_o+\Delta t}} \Delta z + \left(\frac{\partial V}{\partial t}\right)_{\substack{x_o \\ y_o \\ z_o \\ t'}} \Delta t$$

where $t_o < t' < t_o + \Delta t$ and x', y', z' have the meanings previously assigned,

whence it follows that

$$\lim_{\Delta t\to 0} \frac{\Delta V}{\Delta t} = \left(\frac{dV}{dt}\right)_{\substack{x_o \\ y_o \\ z_o \\ t_o}} = \left(\frac{\partial V}{\partial t} + \frac{\partial V}{\partial x}\frac{dx}{dt} + \frac{\partial V}{\partial y}\frac{dy}{dt} + \frac{\partial V}{\partial z}\frac{dz}{dt}\right)_{\substack{x_o \\ y_o \\ z_o \\ t_o}}$$

or, in general,

$$\frac{dV}{dt} = \frac{\partial V}{\partial t} + v_x \frac{\partial V}{\partial x} + v_y \frac{\partial V}{\partial y} + v_z \frac{\partial V}{\partial z}$$

ie

$$\frac{dV}{dt} = \frac{\partial V}{\partial t} + (\bar{v}.\nabla)V \qquad (1.20\text{-}1)$$

It will be seen that the partial derivative refers to the rate of change of V at a fixed point, while the total derivative refers to the rate of change of V at a point which moves with the velocity $\bar{v}$ and coincides with the fixed point at the instant under consideration.

If $\bar{F}$ is a vector function of space and time having the same degree of continuity as V, then

$$\frac{d\bar{F}}{dt} = \frac{\partial \bar{F}}{\partial t} + (\bar{v}.\nabla)\bar{F} \qquad (1.20\text{-}2)$$

This follows directly from a substitution of the rectangular components of $\bar{F}$ in (1.20-1), multiplication by the unit vectors $\bar{i}$, $\bar{j}$ and $\bar{k}$, and subsequent addition.

When V and $\bar{F}$ are invariant with respect to time the partial derivatives are equated to zero, and the above formulae reduce to

$$\frac{dV}{dt} = (\bar{v}.\nabla)V \qquad (1.20\text{-}1(a))$$

and

$$\frac{d\bar{F}}{dt} = (\bar{v}.\nabla)\bar{F} \qquad (1.20\text{-}2(a))$$

1.20b <u>Time rate of change of tangential line integral along a curve which moves in a time-dependent field</u>

In the following analysis the motion of the curve of integration will not be restricted to one of pure translation. Change of shape and of length will be permitted so long as the velocity vector is (a) continuous in time at every point of the curve, and (b) continuous along the curve at every point of time, in the interval under consideration. The vector field in which the curve moves, and its first space and time derivatives, are supposed to be continuous both in space and time.

Two configurations of the integration path (which is assumed to be open) are shown in Fig. 1.13a. The curve PQ corresponds to the time t_o while P'Q' corresponds to $t_o + \Delta t$. The positive sense of integration is shown by the arrows.

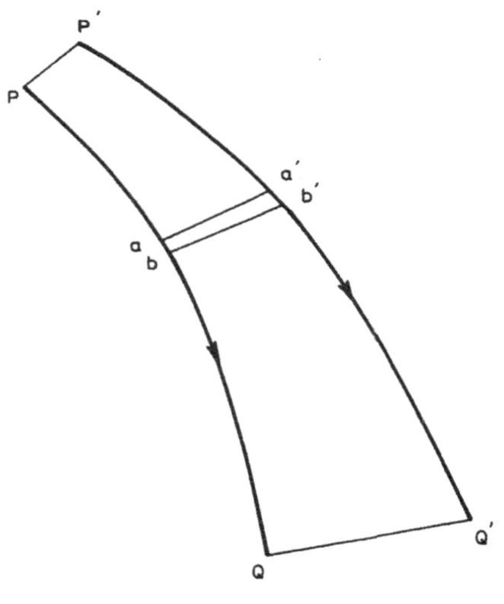

Fig. 1.13a

The change of line integral (say, ΔL) which takes place during the interval Δt is given by

$$\Delta L = \int_{P'}^{Q'} (\bar{F})_{t_o+\Delta t} \cdot d\bar{r} - \int_{P}^{Q} (\bar{F})_{t_o} \cdot d\bar{r}$$

This may be written as

$$\Delta L = \int_{P'}^{Q'} (\bar{F})_{t_o+\Delta t} \cdot d\bar{r} - \int_{P}^{Q} (\bar{F})_{t_o+\Delta t} \cdot d\bar{r} + \int_{P}^{Q} (\bar{F})_{t_o+\Delta t} \cdot d\bar{r} - \int_{P}^{Q} (\bar{F})_{t_o} \cdot d\bar{r}$$

(This expression is seen to involve integration along a path which does not coincide with the moving curve at the time specified for integration.)

It follows that

$$\frac{dL}{dt} = \lim_{\Delta t\to 0} \frac{1}{\Delta t}\left\{\int_{P'}^{Q'} (\bar{F})_{t_o+\Delta t} \cdot d\bar{r} - \int_{P}^{Q} (\bar{F})_{t_o+\Delta t} \cdot d\bar{r}\right\}$$

$$+ \lim_{\Delta t\to 0} \frac{1}{\Delta t}\left\{\int_{P}^{Q} (\bar{F})_{t_o+\Delta t} \cdot d\bar{r} - \int_{P}^{Q} (\bar{F})_{t_o} \cdot d\bar{r}\right\}$$

We now define the significance of the total and partial derivatives with respect to time in the present context by writing the above as

$$\left\{\frac{d}{dt}\int_{P}^{Q} \bar{F}.d\bar{r}\right\}_{t_o} = \lim_{\Delta t\to 0} \frac{1}{\Delta t}\left\{\int_{P'}^{Q'} (\bar{F})_{t_o+\Delta t} \cdot d\bar{r} - \int_{P}^{Q} (\bar{F})_{t_o+\Delta t} \cdot d\bar{r}\right\}$$

$$+\left\{\frac{\partial}{\partial t}\int_{P}^{Q} \bar{F}.d\bar{r}\right\}_{t_o} \qquad (1.20\text{-}3)$$

It is clear that

$$\frac{d}{dt}\int_{P}^{Q} \bar{F}.d\bar{r} = \int_{P}^{Q} \frac{d}{dt}(\bar{F}.d\bar{r}) \quad \text{and} \quad \frac{\partial}{\partial t}\int_{P}^{Q} \bar{F}.d\bar{r} = \int_{P}^{Q} \frac{\partial}{\partial t}(\bar{F}.d\bar{r})$$

Further, since partial differentiation involves a fixed path, it is permissible to write

$$\int_P^Q \frac{\partial}{\partial t}(\bar{F}.d\bar{r}) = \int_P^Q \frac{\partial \bar{F}}{\partial t} . d\bar{r}$$

the same line elements being used for successive integrations in time. This operation is not valid for the total derivative unless the motion of the curve is one of pure translation, in which case the individual vector elements are unaffected by the movement.

The limiting expression in (1.20-3) may be transformed into a line integral over PQ in the following way.

Suppose that a and b are closely-spaced points of PQ and that a' and b' are the corresponding points of P'Q'. (This means that those points of the moving curve coincident with a and b at time t_o are coincident with a' and b' at time $t_o + \Delta t$.) The rectilinear figure abb'a', shown enlarged in Fig. 1.13b, is composed of the triangles abb' and ab'a'. In general these triangles are not coplanar; their vector areas are given by $\frac{1}{2}\Delta\bar{r}_i \times \bar{v}_b\Delta t$ and $\frac{1}{2}\Delta\bar{r}_{i'} \times \bar{v}_a\Delta t$ when the positive sides of the surfaces face the reader. $\bar{v}_a$ and $\bar{v}_b$ are defined by $\bar{v}_a\Delta t = \overrightarrow{aa'}$ and $\bar{v}_b\Delta t = \overrightarrow{bb'}$, so that they represent mean velocities over the time interval Δt.

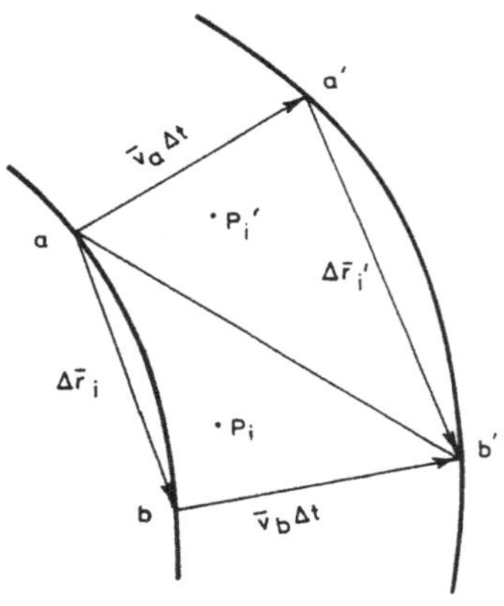

Fig. 1.13b

On applying Stokes's theorem to each of the triangles in turn we obtain

$$\int_a^b \bar{F}.d\bar{r} + \int_b^{b'} \bar{F}.d\bar{r} + \int_{b'}^{a} \bar{F}.d\bar{r} = \int_{abb'} (\text{curl } \bar{F}).d\bar{S}$$

and

$$\int_{b'}^{a'} \bar{F}.d\bar{r} + \int_{a'}^{a} \bar{F}.d\bar{r} + \int_a^{b'} \bar{F}.d\bar{r} = \int_{ab'a'} (\text{curl } \bar{F}).d\bar{S}$$

The value of $\bar{F}$ at each point will be taken as that obtaining at the time $t_o + \Delta t$.

By the mean-value theorem for integrals the sum of the above equations may be written as

$$\bar{F}_i.\Delta\bar{r}_i + \int_b^{b'} \bar{F}.d\bar{r} - \bar{F}_{i'}.\Delta\bar{r}_{i'} - \int_a^{a'} \bar{F}.d\bar{r} \tag{1.20-4}$$

$$= \tfrac{1}{2} (\Delta\bar{r}_i \times \bar{v}_b) \;.\; (\text{curl } \bar{F})_{P_i} \;\Delta t + \tfrac{1}{2} (\Delta\bar{r}_{i'} \times \bar{v}_a) \;.\; (\text{curl } \bar{F})_{P_{i'}} \;\Delta t$$

where $\bar{F}_i$ and $\bar{F}_{i'}$ are the values of $\bar{F}$ at certain points of ab and a'b', and P_i and $P_{i'}$ are points of the triangular surfaces abb' and ab'a'.

By writing $\Delta\bar{r}_{i'} = \Delta\bar{r}_i + \Delta\bar{v}_i \;\Delta t$ where $\Delta\bar{v}_i = \bar{v}_b - \bar{v}_a$ and employing (1.3-3) to express $(\text{curl } \bar{F})_{P_i}$ and $(\text{curl } \bar{F})_{P_{i'}}$ in terms of curl $\bar{F}$ and its derivatives at a neighbouring point, it is found that (1.20-4) may be replaced by

$$\bar{F}_i.\Delta\bar{r}_i + \int_b^{b'} \bar{F}.d\bar{r} - \bar{F}_{i'}.\Delta\bar{r}_{i'} - \int_a^{a'} \bar{F}.d\bar{r} \tag{1.20-5}$$

$$= \tfrac{1}{2} (\Delta\bar{r}_i \times \bar{v}_b) \;.\; (\text{curl } \bar{F})_b \;\Delta t + \tfrac{1}{2} (\Delta\bar{r}_i \times \bar{v}_a) \;.\; (\text{curl } \bar{F})_a \;\Delta t + \varepsilon_i$$

where

$$\varepsilon_i = \tfrac{1}{2}\,(\Delta\bar{r}_i\times\bar{v}_b)\;.\;((\overrightarrow{bP_i}.\nabla)\text{curl }\bar{F})_b\;\Delta t + \tfrac{1}{2}\,(\Delta\bar{r}_i\times\bar{v}_a)\;.\;((\overrightarrow{aP_{i'}}.\nabla)\text{curl }\bar{F})_a\;\Delta t$$

$$+ \tfrac{1}{2}\,(\Delta\bar{v}_i\times\bar{v}_a)\;.\;(\text{curl }\bar{F})_a\;(\Delta t)^2 + \text{----}$$

Since $\overrightarrow{bP_i}$ and $\overrightarrow{aP_{i'}}$ have magnitudes of the order $v_a\Delta t$ it is seen that second or higher powers of Δt appear implicitly or explicitly as multiplying factors in all component terms of ε_i.

Suppose now that PQ and P'Q' are divided into n matched pairs of elements such as ab and a'b' and that both sides of (1.20-5) are summed over all of these. This yields

$$\sum_{i=1}^{n} \bar{F}_i.\Delta\bar{r}_i + \bar{v}_Q\Delta t.\bar{F}_{QQ'} - \sum_{i=1}^{n} \bar{F}_{i'}.\Delta\bar{r}_{i'} - \bar{v}_P\Delta t.\bar{F}_{PP'}$$

$$= \tfrac{1}{2}\sum_{i=1}^{n} (\Delta\bar{r}_i\times\bar{v}_b).(\text{curl }\bar{F})_b\Delta t + \tfrac{1}{2}\sum_{i=1}^{n} (\Delta\bar{r}_i\times\bar{v}_a).(\text{curl }\bar{F})_a\Delta t + \sum_{i=1}^{n} \varepsilon_i$$

where $\bar{F}_{QQ'}$ and $\bar{F}_{PP'}$ are the values of $\bar{F}$ at certain points of the straight lines QQ' and PP', and where $\bar{v}_Q\Delta t = \overrightarrow{QQ'}$ and $\bar{v}_P\Delta t = \overrightarrow{PP'}$.

Then on interchanging the dots and crosses of the triple scalar products and taking limits as $n\to\infty$ and $\Delta\bar{r}_i\to\bar{0}$, we get

$$\int_P^Q \bar{F}.d\bar{r} + \bar{v}_Q\Delta t.\bar{F}_{QQ'} - \int_{P'}^{Q'} \bar{F}.d\bar{r} - \bar{v}_P\Delta t.\bar{F}_{PP'}$$

$$= \tfrac{1}{2}\int_P^Q (\bar{v}\times\text{ curl }\bar{F}).d\bar{r}\Delta t + \tfrac{1}{2}\int_P^Q (\bar{v}\times\text{ curl }\bar{F}).d\bar{r}\Delta t + \lim_{\substack{n\to\infty\\ \Delta\bar{r}_i\to\bar{0}}}\sum_{i=1}^{n}\varepsilon_i$$

or, with the subscript $t_o + \Delta t$ restored,

$$\int_{P'}^{Q'} (\bar{F})_{t_o+\Delta t}.d\bar{r} - \int_{P}^{Q} (\bar{F})_{t_o+\Delta t}.d\bar{r}$$

$$= \bar{v}_Q\Delta t.(\bar{F}_{QQ'})_{t_o+\Delta t} - \bar{v}_P\Delta t.(\bar{F}_{PP'})_{t_o+\Delta t} - \int_{P}^{Q} \{\bar{v}\times(\text{curl } \bar{F})_{t_o+\Delta t}\}.d\bar{r}\Delta t - \lim_{\substack{n\to\infty \\ \Delta\bar{r}_i\to\bar{0}}} \sum_{i=1}^{n} \varepsilon_i$$

where $\bar{v}_Q$, $\bar{v}_P$ and $\bar{v}$ continue to represent mean vector velocities over the interval Δt.

Hence

$$\lim_{\Delta t\to 0} \frac{1}{\Delta t} \left\{\int_{P'}^{Q'} (\bar{F})_{t_o+\Delta t}.d\bar{r} - \int_{P}^{Q} (\bar{F})_{t_o+\Delta t}.d\bar{r}\right\} = (\bar{v}_Q.\bar{F}_Q)_{t_o} - (\bar{v}_P.\bar{F}_P)_{t_o} \qquad (1.20\text{-}6)$$

$$- \int_{P}^{Q} (\bar{v} \times \text{curl } \bar{F})_{t_o}.d\bar{r}$$

since the components of $\frac{1}{\Delta t} \lim_{\substack{n\to\infty \\ \Delta\bar{r}_i\to\bar{0}}} \sum_{i=1}^{n} \varepsilon_i$ comprise the products of Δt, $(\Delta t)^2$ etc with expressions bounded in upper value.

It is clear that $\bar{v}_Q$, $\bar{v}_P$ and $\bar{v}$ as they appear in (1.20-6) are instantaneous velocities at the time t_o, and that $\bar{F}_Q$ and $\bar{F}_P$ are to be evaluated at Q and P.

The right-hand side of (1.20-6) may be replaced by

$$\int_{P}^{Q} \frac{d}{ds} (\bar{v}.\bar{F})_{t_o}\, ds - \int_{P}^{Q} (\bar{v} \times \text{curl } \bar{F})_{t_o} . d\bar{r}$$

On substituting this in (1.20-3) and dropping the t_o subscript we obtain the general relationship

$$\frac{d}{dt}\int_{P}^{Q} \bar{F}.d\bar{r} = \frac{\partial}{\partial t}\int_{P}^{Q} \bar{F}.d\bar{r} + \int_{P}^{Q} \frac{d}{ds} (\bar{v}.\bar{F})\, ds - \int_{P}^{Q} (\bar{v} \times \text{curl } \bar{F}).d\bar{r} \qquad (1.20\text{-}7)$$

When the motion is purely translational a scalar field $\bar{v}.\bar{F}$ may be generated from $\bar{F}$ and the unique vector $\bar{v}$. This field has continuous first derivatives in the neighbourhood of the integration path and allows us to write

$$(\bar{v}.\bar{F})_Q - (\bar{v}.\bar{F})_P = \int_P^Q (\text{grad } \bar{v}.\bar{F}).d\bar{r}$$

in which case (1.20-7) reduces to

$$\frac{d}{dt}\int_P^Q \bar{F}.d\bar{r} = \int_P^Q \left(\frac{\partial \bar{F}}{\partial t} + \text{grad } \bar{v}.\bar{F} - \bar{v} \times \text{curl } \bar{F}\right).d\bar{r} \qquad (1.20\text{-}8)$$

For the particular case of a closed curve in motion (1.20-7) is seen to become

$$\frac{d}{dt}\oint \bar{F}.d\bar{r} = \oint \left(\frac{\partial \bar{F}}{\partial t} - \bar{v} \times \text{curl } \bar{F}\right).d\bar{r} \qquad (1.20\text{-}9)$$

When the field is invariant with respect to time the general expression reduces to

$$\frac{d}{dt}\int_P^Q \bar{F}.d\bar{r} = \int_P^Q \frac{d}{ds}(\bar{v}.\bar{F})\, ds - \int_P^Q (\bar{v} \times \text{curl } \bar{F}).d\bar{r} \qquad (1.20\text{-}7(a))$$

(1.20-7) may be derived more easily, but in a less fundamental manner, by means of the following analysis.

Let $\Delta\bar{r}$ represent an element of the moving curve and let P be a particular point of it. Suppose that P is coincident with the fixed point P_o at the time t_o. Then, on the understanding that the expressions below are to be evaluated at $t = t_o$, we may write

$$\frac{d}{dt}((\bar{F})_P.\Delta\bar{r}) = \left(\frac{d}{dt}(\bar{F})_P\right).\Delta\bar{r} + (\bar{F})_{P_o} . \frac{d}{dt}(\Delta\bar{r})$$

$$= \left(\frac{\partial \bar{F}}{\partial t}\right)_{P_o} .\Delta\bar{r} + (((\bar{v})_{P_o}.\nabla)\bar{F})_{P_o}.\Delta\bar{r} + (\bar{F})_{P_o} . \frac{d}{dt}(\Delta\bar{r})$$

where $(\bar{v})_{P_o}$ is the velocity of P when coincident with P_o.

The vector $(\bar{v})_{P_o}$ and the vector point function $\bar{F}$ (at time t_o) define the field $(\bar{v})_{P_o}.\bar{F}$. On applying (1.16-5) to this we get

$$\text{grad } (\bar{v})_{P_o}.\bar{F} = ((\bar{v})_{P_o}.\nabla)\bar{F} + (\bar{v})_{P_o} \times \text{curl } \bar{F}$$

whence

$$\frac{d}{dt}((\bar{F})_P.\Delta\bar{r}) = \left(\frac{\partial\bar{F}}{\partial t}\right)_{P_o}.\Delta\bar{r} + (\text{grad}(\bar{v})_{P_o}.\bar{F})_{P_o}.\Delta\bar{r} - ((\bar{v})_{P_o} \times \text{curl } \bar{F})_{P_o}.\Delta\bar{r} + (\bar{F})_{P_o} . \frac{d}{dt}(\Delta\bar{r}) \tag{1.20-10}$$

It is readily shown from Fig. 1.13b that

$$\frac{d}{dt}(\Delta\bar{r}) = (\bar{v})_b - (\bar{v})_a = \Delta\bar{v}$$

where $\Delta\bar{r} = \vec{ab}$

Further, since $(\bar{v})_{P_o}.\bar{F}$ has continuous first space derivatives,

$$(\text{grad}(\bar{v})_{P_o}.\bar{F})_{P_o}.\Delta\bar{r} = \Delta((\bar{v})_{P_o}.\bar{F}) + \varepsilon_1 = (\bar{v})_{P_o}.\Delta\bar{F} + \varepsilon_1$$

where $\Delta(\ ) = (\ )_b - (\ )_a$ and where $\varepsilon_1/((\bar{v})_{P_o}.\Delta\bar{F}) \to 0$ as $\Delta\bar{r} \to \bar{0}$

Substitution in (1.20-10) then yields

$$\frac{d}{dt}((\bar{F})_P.\Delta\bar{r}) = \left(\frac{\partial\bar{F}}{\partial t}\right)_{P_o}.\Delta\bar{r} + (\bar{v})_{P_o}.\Delta\bar{F} + (\bar{F})_{P_o}.\Delta\bar{v} - ((\bar{v})_{P_o} \times \text{curl}\bar{F})_{P_o}.\Delta\bar{r} + \varepsilon_1$$

or

$$\frac{d}{dt}((\bar{F})_P.\Delta\bar{r}) = \left(\frac{\partial\bar{F}}{\partial t}\right)_{P_o}.\Delta\bar{r} + \Delta(\bar{v}.\bar{F}) - ((\bar{v})_{P_o} \times \text{curl}\bar{F})_{P_o}.\Delta\bar{r} + \varepsilon_1 + \varepsilon_2$$

On summing (1.20-11) over all elements of PQ the term $\Delta(\bar{v}.\bar{F})$ is replaced by $(\bar{v}.\bar{F})_Q - (\bar{v}.\bar{F})_P$; on taking limits as the number of elements is increased indefinitely and the magnitude of each approaches zero, we obtain

$$\frac{d}{dt}\int_P^Q \bar{F}.d\bar{r} = \int_P^Q \frac{\partial\bar{F}}{\partial t}\,.\,d\bar{r} + \int_P^Q \frac{d}{ds}(\bar{v}.\bar{F})\,ds - \int_P^Q (\bar{v}\times\text{curl}\bar{F}).d\bar{r}$$

1.20c <u>Time rate of change of flux through a surface which moves in a time-dependent field</u>

As in the previous analysis no restriction will be placed upon the nature of the motion, apart from the requirement that the velocity vector be continuous in time at every point of the surface, and continuous from point to point across the surface at every instant of time, within the interval under consideration. The vector field and its derivatives are supposed to exhibit the same degree of continuity as before.

Two configurations of the moving surface (which is assumed to be open) are shown as S and S' in Fig. 1.14a corresponding to the times t_o and $t_o + \Delta t$.

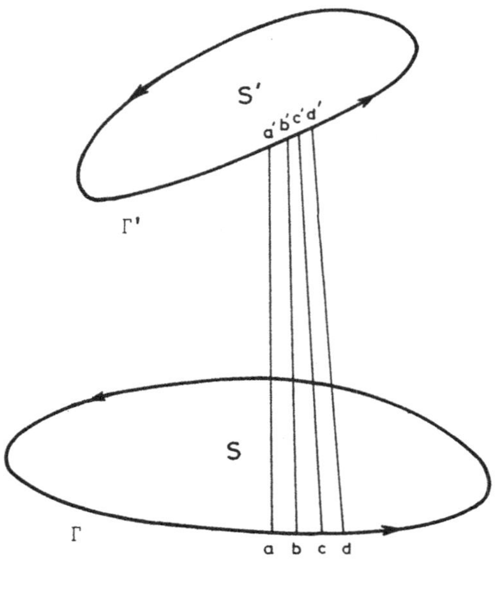

Fig. 1.14a

The change of flux during the interval Δt is given by

$$\Delta N = \int_{S'} (\overline{F})_{t_o+\Delta t}.d\overline{S} - \int_{S} (\overline{F})_{t_o}.d\overline{S}$$

where the currency around the contours Γ and Γ' define a consistent positive normal at S and S'.

This may be put into the form

$$\Delta N = \int_{S'} (\overline{F})_{t_o+\Delta t}.d\overline{S} - \int_{S} (\overline{F})_{t_o+\Delta t}.d\overline{S} + \int_{S} (\overline{F})_{t_o+\Delta t}.d\overline{S} - \int_{S} (\overline{F})_{t_o}.d\overline{S}$$

whence

$$\left(\frac{d}{dt}\int_{S} \overline{F}.d\overline{S}\right)_{t_o} = \underset{\Delta t\to 0}{\mathrm{Lim}}\ \frac{1}{\Delta t}\left\{\int_{S'} (\overline{F})_{t_o+\Delta t}.d\overline{S} - \int_{S} (\overline{F})_{t_o+\Delta t}.d\overline{S}\right\} + \left\{\frac{\partial}{\partial t}\int_{S} \overline{F}.d\overline{S}\right\}_{t_o} \qquad (1.20\text{-}11)$$

The limiting term must now be transformed into a surface integral over S.

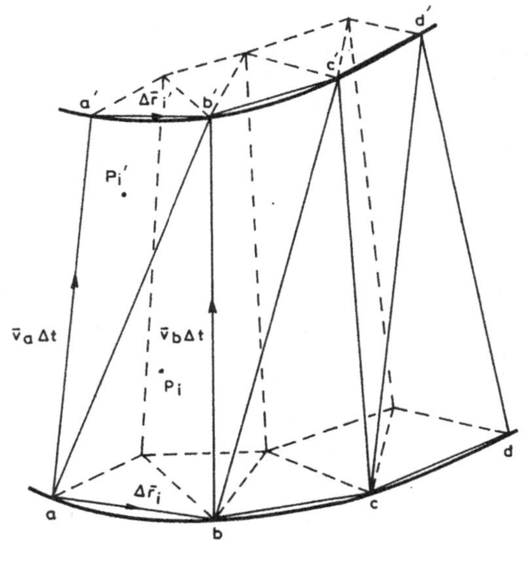

[Arrowheads have been placed in the centres of the associated vectors to ease congestion.]

Fig. 1.14b

The detailed treatment of the transformation is laborious and will not be set down here. It is suggested that the reader carry this out, as an exercise, along the following lines.

Let a,b,c,d -- be closely-spaced points of Γ and let a',b',c',d' -- be the corresponding points of Γ'. These points are joined as shown in Fig. 1.14b to form a series of contiguous triangular faces which approximate the curved surface traced out by the contour of the open surface during its movement. The surfaces S and S' are themselves approximated by polyhedral surfaces composed of two sets of triangular faces whose vertices are corresponding points lying in S and S'. Some of these are shown in the figure.

The divergence theorem is applied to each of the solid figures formed by joining the corresponding vertices of matching triangular end pieces and the result is summed over the whole system. Cancellation of the surface integral occurs over internal interfaces so that the final expression relates the external surface integral to the sum of the individual products of volume and divergence.

By invoking the mean-value theorem for integrals, together with (1.2-9) and (1.3-3), and by taking limits as the degree of subdivision is increased indefinitely, an equation is formed between

(a) the surface integrals of $\overline{F}$ over S and S' (see Ex.1-27. p. 24)

(b) the line integral $\oint_\Gamma d\overline{r} \times \overline{v}\,\Delta t.\overline{F}$ where $\overline{v}$ is the mean velocity of of the line element during the interval Δt

(c) the surface integral $\int_S \overline{v}\,\Delta t \operatorname{div} \overline{F}.d\overline{S}$ where $\overline{v}$ is the mean velocity of the surface element during the interval Δt

(d) a series of terms similar to (b) and (c) but involving space derivates of $\overline{F}$ and div $\overline{F}$ with multiplying factors $(\Delta t)^2$, $(\Delta t)^3$ ----.

Upon dividing the equation by Δt and taking limits as $\Delta t \to 0$ the factors $\overline{v}$, $\overline{F}$ and div $\overline{F}$, which were to be evaluated in the first instance at the time $t_o + \Delta t$, are finally referred to the time t_o, at which instant the moving surface and the surface of integration coincide. Coincidentally, the terms in (d) disappear and (1.20-11), with subscript t_o deleted, becomes

$$\frac{d}{dt}\int_S \overline{F}.d\overline{S} = \frac{\partial}{\partial t}\int_S \overline{F}.d\overline{S} + \int_S \overline{v} \operatorname{div} \overline{F}.d\overline{S} - \oint_\Gamma (\overline{v}\times\overline{F}).d\overline{r} \qquad (1.20\text{-}12)$$

When the motion of the surface is one of pure translation the vector field $\overline{v} \times \overline{F}$ is defined at all points where $\overline{F}$ is defined and has equal degrees of continuity. In this case (1.20-12) may be replaced by

$$\frac{d}{dt}\int_S \bar{F}.d\bar{S} = \int_S \left(\frac{\partial \bar{F}}{\partial t} + \bar{v}\,\text{div}\,\bar{F} - \text{curl}(\bar{v}\times\bar{F})\right).d\bar{S} \tag{1.20-13}$$

The particular case of a closed surface may be treated as a combination of two open surfaces. The line integral in (1.20-12) then cancels around the common bounding curve and we are left with

$$\frac{d}{dt}\oint_S \bar{F}.d\bar{S} = \frac{\partial}{\partial t}\oint_S \bar{F}.d\bar{S} + \oint_S \bar{v}\,\text{div}\,\bar{F}.d\bar{S} \tag{1.20-14}$$

In the case of a time-invariant field (1.20-12) simply reduces to

$$\frac{d}{dt}\int_S \bar{F}.d\bar{S} = \int_S \bar{v}\,\text{div}\,\bar{F}.d\bar{S} - \oint_\Gamma (\bar{v}\times\bar{F}).d\bar{r} \tag{1.20-12(a)}$$

An alternative derivation of (1.20-12), which is analogous to the second derivation of (1.20-7), will now be given in detail.

Let $\Delta\bar{S}$ represent an element of the moving surface and let P be a particular point of it. Suppose that P is coincident with the fixed point P_o at the time t_o. Then, on the understanding that the expressions below are to be evaluated at $t = t_o$, we may write

$$\frac{d}{dt}((\bar{F})_P.\Delta\bar{S}) = \left(\frac{d}{dt}(\bar{F})_P\right).\Delta\bar{S} + (\bar{F})_{P_o}.\frac{d}{dt}(\Delta\bar{S})$$

$$= \left(\frac{\partial\bar{F}}{\partial t}\right)_{P_o}.\Delta\bar{S} + ((\bar{v})_{P_o}.\nabla)\bar{F})_{P_o}.\Delta\bar{S} + (\bar{F})_{P_o}.\frac{d}{dt}(\Delta\bar{S})$$

where $(\bar{v})_{P_o}$ is the velocity of P when coincident with P_o.

The vector $(\bar{v})_{P_o}$ and the vector point function $\bar{F}$ (at time t_o) define the field $(\bar{v})_{P_o}\times\bar{F}$. On applying (1.16-6) to this we get

$$\text{curl}((\bar{v})_{P_o}\times\bar{F}) = -((\bar{v})_{P_o}.\nabla)\bar{F} + (\bar{v})_{P_o}\text{div}\,\bar{F}$$

whence

$$\frac{d}{dt}((\overline{F})_P.\Delta\overline{S}) = \left(\frac{\partial\overline{F}}{\partial t}\right)_{P_o}.\Delta\overline{S} + (\overline{v})_{P_o}(\text{div }\overline{F})_{P_o}.\Delta\overline{S} - (\text{curl}((\overline{v})_{P_o}\times\overline{F}))_{P_o}.\Delta\overline{S}$$

$$+ (\overline{F})_{P_o}.\frac{d}{dt}(\Delta\overline{S}) \tag{1.20-15}$$

Since the vector area of a closed surface is zero, it is easily seen that the rate of change of $\Delta\overline{S}$ with respect to time is equal to the rate at which the bounding curve of $\Delta\overline{S}$ sweeps out area. On taking account of the relationship between the positive normal at the surface and the currency of this curve it is found that

$$\frac{d}{dt}(\Delta\overline{S}) = \oint_{\Delta S} \overline{v}\times d\overline{r}$$

where the subscript indicates the element of area bounded by the curve around which the line integral is taken.

Further, since $(\overline{v})_{P_o}\times\overline{F}$ has continuous first space derivatives, it follows from (1.9-3) that

$$(\text{curl}((\overline{v})_{P_o}\times\overline{F}))_{P_o}.\Delta\overline{S} = \oint_{\Delta S}((\overline{v})_{P_o}\times\overline{F}).d\overline{r} - \gamma$$

where the ratio of γ to the line integral approaches zero as $\Delta\overline{S}\to\overline{0}$.

Substitution in (1.20-15) then yields

$$\frac{d}{dt}((\overline{F})_P.\Delta\overline{S}) = \left(\frac{\partial\overline{F}}{\partial t}\right)_{P_o}.\Delta\overline{S} + (\overline{v})_{P_o}(\text{div }\overline{F})_{P_o}.\Delta\overline{S}$$

$$+ \oint_{\Delta S}\{(\overline{F}\times(\overline{v})_{P_o}) + ((\overline{F})_{P_o}\times\overline{v})\}.d\overline{r} + \gamma \tag{1.20-16}$$

If, for the typical point of the curve bounding ΔS, we write

$$\overline{F} = (\overline{F})_{P_o} + \Delta\overline{F} \quad\text{and}\quad \overline{v} = (\overline{v})_{P_o} + \Delta\overline{v}$$

then the line integral in the above equation becomes

$$\oint_{\Delta S} ((\bar{F})_{P_o} \times (\bar{v})_{P_o}).d\bar{r} + \oint_{\Delta S} (\bar{F}\times\bar{v}).d\bar{r} - \oint_{\Delta S} (\Delta\bar{F}\times\Delta\bar{v}).d\bar{r} \qquad (1.20\text{-}17)$$

The first term of this expression is zero since $(\bar{F})_{P_o} \times (\bar{v})_{P_o}$ is constant, and the ratio of the third term to the second approaches zero as $\Delta\bar{S}\to\bar{0}$. On substituting this in (1.20-16) and summing over all elements of area, the term $\oint_{\Delta S} (\bar{F}\times\bar{v}).d\bar{r}$ is replaced by $\int_{\Gamma} (\bar{F}\times\bar{v}).d\bar{r}$ because of cancellation of the line integral over internal contours. If, now, limits are taken as the number of elements is increased indefinitely while the size of each approaches zero, we obtain

$$\frac{d}{dt}\int_S \bar{F}.d\bar{S} = \int_S \frac{\partial\bar{F}}{\partial t} . d\bar{S} + \int_S \bar{v} \text{ div } \bar{F}.d\bar{S} + \oint_{\Gamma} (\bar{F}\times\bar{v}).d\bar{r}$$

EXERCISES

1-74. Show that the rate of change of density at a fixed point in a field of fluid flow is given by

$$\frac{\partial\rho}{\partial t} = - \text{div}(\rho\bar{v})$$

Hence show that the rate of change of density at a point which moves with the fluid is given by

$$\frac{d\rho}{dt} = - \rho \text{ div } \bar{v}$$

1-75. Derive from first principles an expression for the rate of change of the line integral of a time-dependent field $\bar{F}$ along a straight line PQ, when PQ is stretched without change of orientation. Confirm this by making the appropriate substitutions in (1.20-7).

1-76. Show from first principles that the rate of change of the flux of the time-dependent field $\bar{F}$ through a plane surface S, which is stretched smoothly while remaining plane, is given by

$$\frac{d}{dt}\int_S \bar{F}.d\bar{S} = \frac{\partial}{\partial t}\int_S \bar{F}.d\bar{S} - \oint_{\Gamma} (\bar{v}\times\bar{F}).d\bar{r}$$

and confirm this by substitution in (1.20-12).

1-77. Write down (1.20-9) in terms of some vector point function $\overline{C}$ (rather than $\overline{F}$) and so derive (1.20-12) from (1.20-9) for the particular case: $\text{div}\ \overline{F} = 0$.

1-78. In the analysis which follows (1.20-10) confirm that

$$\varepsilon_1/((\overline{v})_{P_o}.\Delta\overline{F})\to 0 \qquad \text{as} \qquad \Delta\overline{r}\to\overline{0}$$

$$\varepsilon_2/\Delta(\overline{v}.\overline{F})\to 0 \qquad \text{as} \qquad \Delta\overline{r}\to\overline{0}$$

Show that the third term of (1.20-17) is one order smaller than the second, and that the ratio of the third to the second consequently tends to zero as $\Delta\overline{S}\to\overline{0}$.

1-79. Prove (1.20-7) for movement of a rigid contour in the following way. Since the movement may be reduced to one of translation through some point 0 and rotation with angular velocity $\overline{\omega}$ about an axis through 0 we may define a velocity field both on and off the contour by

$$\overline{v} = \overline{v}_o + (\overline{\omega} \times \overline{r})$$

where $\overline{v}_o$ is the velocity at 0 and $\overline{r}$ is the position vector of the point in question relative to 0.

By combining (1.20-2) and (1.16-5), with $\overline{G}$ replaced by $\overline{v}$, show that at each point of the contour

$$\frac{d\overline{F}}{dt} = \frac{\partial\overline{F}}{\partial t} + \nabla(\overline{v}.\overline{F}) - (\overline{F}.\nabla)\ \overline{v} - (\overline{v} \times \text{curl}\ \overline{F}) - (\overline{F} \times \text{curl}\ \overline{v})$$

$$= \frac{\partial\overline{F}}{\partial t} + \nabla(\overline{v}.\overline{F}) - (\overline{\omega} \times \overline{F}) - (\overline{v} \times \text{curl}\ \overline{F}) - (\overline{F} \times 2\overline{\omega})$$

Show further that

$$\frac{d}{dt}\ (\overline{F}.d\overline{r}) = \frac{d\overline{F}}{dt}\ .\ d\overline{r} + (\overline{F}.\overline{\omega} \times d\overline{r})$$

and combine these results to obtain

$$\frac{d}{dt}(\bar{F}.d\bar{r}) = \frac{\partial \bar{F}}{\partial t} . d\bar{r} + (\nabla(\bar{v}.\bar{F})).d\bar{r} - (\bar{v} \times \text{curl } \bar{F}).d\bar{r}$$

whence derive (1.20-7) by integration.

1-80. Make use of (1.17-15) to show that

$$\int_\tau \bar{v}\, d\tau = \int_\tau (-\text{div } \bar{v})\bar{r}\, d\tau + \oint_S \bar{r}\, \bar{v}.d\bar{S}$$

where $\bar{v}$ is a well behaved vector point function and $\bar{r}$ is the position vector measured from an internal or external origin.

Hence deduce that if a homogeneous incompressible fluid has no normal component of velocity upon a bounding surface the linear momentum of the enclosed fluid is zero.

1.21 Time Rates of Change of a Vector Quantity Referred to Coordinate Systems in Relative Motion

Let S and S' represent two rectangular systems of coordinates[34] whose unit vectors are $\bar{i}$, $\bar{j}$, $\bar{k}$ and $\bar{i}'$, $\bar{j}'$, $\bar{k}'$ respectively.

A vector quantity $\bar{F}$, observed in both systems, may be expressed as

$$\bar{F} = \bar{i}F_x + \bar{j}F_y + \bar{k}F_z \tag{1.21-1}$$

or

$$\bar{F} = \bar{i}'F_{x'} + \bar{j}'F_{y'} + \bar{k}'F_{z'}$$

If differentiation with respect to time is denoted by $\frac{d}{dt}$ in system S and by $\left(\frac{d}{dt}\right)'$ in system S', then, because $\bar{i}$, $\bar{j}$ and $\bar{k}$ are fixed in S and $\bar{i}'$, $\bar{j}'$ and $\bar{k}'$ are fixed in S',

$$\frac{d\bar{F}}{dt} = \bar{i}\frac{dF_x}{dt} + \bar{j}\frac{dF_y}{dt} + \bar{k}\frac{dF_z}{dt} \tag{1.21-3}$$

34. The use of S to denote both a surface and a coordinate system should not lead to any confusion.

and

$$\left(\frac{d\bar{F}}{dt}\right)' = \bar{i}'\left(\frac{\partial F_{x'}}{dt}\right)' + \bar{j}'\left(\frac{dF_{y'}}{dt}\right)' + \bar{k}'\left(\frac{dF_{z'}}{dt}\right)' \tag{1.21-4}$$

Since the time derivative of a scalar is the same in both systems

$$\frac{dF_x}{dt} = \left(\frac{dF_x}{dt}\right)' \quad \text{and} \quad \frac{dF_{x'}}{dt} = \left(\frac{dF_{x'}}{dt}\right)' \quad \text{etc.} \tag{1.21-5}$$

Suppose now that S and S' are in relative motion with the corresponding coordinates axes maintained parallel. Then $\bar{i}' = \bar{i}$ and $F_{x'} = F_x$ etc hence $\left(\frac{d\bar{F}}{dt}\right)' = \frac{d\bar{F}}{dt}$. This continues to hold for a random orientation of the two systems so long as the relative motion is one of pure translation. In this case the direction cosines of the axes, taken in pairs, remain constant in time, and (1.21-4) may be transformed into (1.21-3) by means of the formulae developed in Sec. 1.19.

The most general form of relative motion of S and S' involves rotation about some common instantaneous axis in addition to motion of translation. Suppose that, relative to S, S' has an angular velocity $\bar{\omega}$. Then it is easily shown that

$$\frac{d\bar{i}'}{dt} = \bar{\omega}\times\bar{i}'; \quad \frac{d\bar{j}'}{dt} = \bar{\omega}\times\bar{j}'; \quad \frac{d\bar{k}'}{dt} = \bar{\omega}\times\bar{k}'$$

hence, from (1.21-2),

$$\frac{d\bar{F}}{dt} = \sum \frac{d\bar{i}'}{dt}F_{x'} + \sum \bar{i}'\frac{dF_{x'}}{dt}$$

$$= \sum \bar{\omega}\times\bar{i}'F_{x'} + \sum \bar{i}'\left(\frac{dF_{x'}}{dt}\right)'$$

or $$\frac{d\bar{F}}{dt} = (\bar{\omega}\times\bar{F}) + \left(\frac{d\bar{F}}{dt}\right)' \tag{1.21-6}$$

Alternatively, from (1.21-1),

$$\left(\frac{d\bar{F}}{dt}\right)' = \sum\left(\frac{d\bar{i}}{dt}\right)'F_x + \sum \bar{i}\left(\frac{dF_x}{dt}\right)'$$

But $\left(\frac{d\bar{i}}{dt}\right)' = -\bar{\omega}\times\bar{i}$ because the angular velocity of S is $-\bar{\omega}$ relative to S', hence

$$\left(\frac{d\bar{F}}{dt}\right)' = \sum(-\bar{\omega}\times\bar{i}F_x) + \sum\bar{i}\frac{dF_x}{dt}$$

or $\left(\frac{d\bar{F}}{dt}\right)' = -(\bar{\omega}\times\bar{F}) + \frac{d\bar{F}}{dt}$

and this is identical with (1.21-6)

A relationship between the second time derivatives of $\bar{F}$ in the two systems may be found by differentiating (1.21-6).

$$\frac{d^2\bar{F}}{dt^2} = \left(\frac{d\bar{\omega}}{dt}\times\bar{F}\right) + \left(\bar{\omega}\times\frac{d\bar{F}}{dt}\right) + \frac{d}{dt}\left(\frac{d\bar{F}}{dt}\right)'$$

$$= \left(\frac{d\bar{\omega}}{dt}\times\bar{F}\right) + \left\{\bar{\omega}\times\left((\bar{\omega}\times\bar{F}) + \left(\frac{d\bar{F}}{dt}\right)'\right)\right\} + \left(\bar{\omega}\times\left(\frac{d\bar{F}}{dt}\right)'\right) + \left(\frac{d}{dt}\left(\frac{d\bar{F}}{dt}\right)'\right)'$$

The last two terms derive from the substitution of $\left(\frac{d\bar{F}}{dt}\right)'$ for $\bar{F}$ in equation (1.21-6) since this equation is unrestricted.

It follows that

$$\frac{d^2\bar{F}}{dt^2} = \left(\frac{d\bar{\omega}}{dt}\times\bar{F}\right) + (\bar{\omega}\times(\bar{\omega}\times\bar{F})) + \left(2\bar{\omega}\times\left(\frac{d\bar{F}}{dt}\right)'\right) + \left(\frac{d^2\bar{F}}{dt^2}\right)' \qquad (1.21\text{-}7)$$

where $\frac{d^2\bar{F}}{dt^2} = \bar{i}\frac{d^2F_x}{dt^2} + \bar{j}\frac{d^2F_y}{dt^2} + \bar{k}\frac{d^2F_z}{dt^2}$

and $\left(\frac{d^2\bar{F}}{dt^2}\right)' = \bar{i}'\left(\frac{d^2F_{x'}}{dt^2}\right) + \bar{j}'\left(\frac{d^2F_{y'}}{dt^2}\right) + \bar{k}'\left(\frac{d^2F_{z'}}{dt^2}\right)$

Correspondingly,

$$\left(\frac{d^2\bar{F}}{dt^2}\right)' = -\left(\frac{d\bar{\omega}}{dt}\times\bar{F}\right) + (\bar{\omega}\times(\bar{\omega}\times\bar{F})) - \left(2\bar{\omega}\times\frac{d\bar{F}}{dt}\right) + \frac{d^2\bar{F}}{dt^2} \qquad (1.21\text{-}8)$$

EXERCISES

1-81. By means of the transformation formulae of Sec. 1.19 show that $\left(\frac{d\bar{F}}{dt}\right)' = \frac{d\bar{F}}{dt}$ if the relative motion of the coordinate systems, S and S', is one of pure translation with random orientation of axes.

1-82. If the rectangular coordinate system S' has an angular velocity $\bar{\omega}$ relative to S, show that $\frac{d\bar{i}'}{dt} = \bar{\omega} \times \bar{i}'$.

1-83. Let O and O' be the origins of the rectangular coordinate systems S and S', and let P be a moving point whose position vectors relative to O and O' are $\bar{r}$ and $\bar{r}'$. Then if $\overrightarrow{OO'} = \bar{r}_o$ we may write

$$\bar{r} = \bar{r}_o + \bar{r}' = \bar{r}_o + \bar{i}'x' + \bar{j}'y' + \bar{k}'z'$$

where (x',y',z') are the coordinates of P in S'. By successive differentiation of this equation with respect to time in system S, show that,

$$\frac{d\bar{r}}{dt} = \frac{d\bar{r}_o}{dt} + (\bar{\omega} \times \bar{r}') + \left(\frac{d\bar{r}'}{dt}\right)'$$

and

$$\frac{d^2\bar{r}}{dt^2} = \frac{d^2\bar{r}_o}{dt^2} + \left(\frac{d\bar{\omega}}{dt} \times \bar{r}'\right) + (\bar{\omega} \times (\bar{\omega}\times\bar{r}')) + \left(2\bar{\omega} \times \left(\frac{d\bar{r}'}{dt}\right)'\right) + \left(\frac{d^2\bar{r}'}{dt^2}\right)'$$

These equations describe the relationships between the velocities and the accelerations of P in the two coordinate systems. The second equation is known as the Theorem of Coriolis.

1-84. Derive the results of Ex.1-83. by the substitution of an appropriate vector quantity in (1.21-6) and (1.21-7).

1.22 Complex Scalar and Vector Fields

Our considerations to date have been restricted to those scalar and vector fields which are characterised by point functions having real magnitudes. If f_1 and f_2 and $\bar{F}_1$ and $\bar{F}_2$ are such scalar and vector point functions and $j = \sqrt{-1}$, then $f_1 + jf_2$ and $\bar{F}_1 + j\bar{F}_2$ represent complex fields. When $\bar{F}_1$ and $\bar{F}_2$ are collinear $\bar{F}_1 + j\bar{F}_2$ has a definite direction and a complex magnitude; when non-collinear it may be treated as the sum of fields having real and imaginary magnitudes.

It is evident that

$$\frac{\partial^n}{\partial x^n}(f_1+jf_2) = \frac{\partial^n}{\partial x^n} f_1 + j \frac{\partial^n}{\partial x^n} f_2 \qquad \text{for } n = 1, 2, 3 \text{ ---}$$

and

$$\frac{\partial^n}{\partial x^n}(\overline{F}_1+j\overline{F}_2) = \frac{\partial^n}{\partial x^n} \overline{F}_1 + j \frac{\partial^n}{\partial x^n} \overline{F}_2 \qquad \text{for } n = 1, 2, 3 \text{ ---}$$

with similar equations for differentiation with respect to y, z and t. It follows that

$$\text{grad}(f_1+jf_2) = \text{grad } f_1 + j \text{ grad } f_2$$

$$\text{div}(\overline{F}_1+j\overline{F}_2) = \text{div } \overline{F}_1 + j \text{ div } \overline{F}_2$$

$$\text{curl}(\overline{F}_1+j\overline{F}_2) = \text{curl } \overline{F}_1 + j \text{ curl } \overline{F}_2$$

whence

$$\text{div grad}(f_1+jf_2) = \text{div grad } f_1 + j \text{ div grad } f_2$$

$$\text{grad div}(\overline{F}_1+j\overline{F}_2) = \text{grad div } \overline{F}_1 + j \text{ grad div } \overline{F}_2$$

$$\text{curl curl}(\overline{F}_1+j\overline{F}_2) = \text{curl curl } \overline{F}_1 + j \text{ curl curl } \overline{F}_2$$

We have also

$$\text{grad} \frac{1}{f_1 + jf_2} = - \frac{1}{(f_1+jf_2)^2} (\text{grad } f_1 + j \text{ grad } f_2)$$

If we write

$$f_1 + jf_2 = \tilde{f} \qquad g_1 + jg_2 = \tilde{g}$$

$$\overline{F}_1 + j\overline{F}_2 = \tilde{\overline{F}} \qquad \overline{G}_1 + j\overline{G}_2 = \tilde{\overline{G}}$$

then we may show that

$$
\begin{aligned}
\text{curl curl } \tilde{\bar{F}} &= \text{grad div } \tilde{\bar{F}} - \nabla^2 \tilde{\bar{F}} \\
\text{grad}(\tilde{f}\tilde{g}) &= \tilde{f} \text{ grad } \tilde{g} + \tilde{g} \text{ grad } \tilde{f} \\
\text{div}(\tilde{\bar{F}} \times \tilde{\bar{G}}) &= \tilde{\bar{G}}.\text{curl } \tilde{\bar{F}} - \tilde{\bar{F}}.\text{curl } \tilde{\bar{G}} \\
\text{grad}(\tilde{\bar{F}}.\tilde{\bar{G}}) &= (\tilde{\bar{F}}.\nabla)\tilde{\bar{G}} + (\tilde{\bar{G}}.\nabla)\tilde{\bar{F}} + \tilde{\bar{F}} \times \text{curl } \tilde{\bar{G}} + \tilde{\bar{G}} \times \text{curl } \tilde{\bar{F}} \\
\text{curl}(\tilde{\bar{F}} \times \tilde{\bar{G}}) &= (\tilde{\bar{G}}.\nabla)\tilde{\bar{F}} - (\tilde{\bar{F}}.\nabla)\tilde{\bar{G}} + \tilde{\bar{F}} \text{ div } \tilde{\bar{G}} - \tilde{\bar{G}} \text{ div } \tilde{\bar{F}}
\end{aligned}
$$

$$
\tilde{\bar{F}}.\{(\tilde{\bar{G}}.\nabla)\tilde{\bar{H}}\} = \tilde{\bar{G}}.\{(\tilde{\bar{F}}.\nabla)\tilde{\bar{H}}\} + (\tilde{\bar{G}} \times \tilde{\bar{F}}).\text{curl } \tilde{\bar{H}}
$$

Further,

$$
\begin{aligned}
\text{curl } \tilde{V} \tilde{\bar{F}} &= \tilde{V} \text{ curl } \tilde{\bar{F}} + (\text{grad } \tilde{V}) \times \tilde{\bar{F}} \\
\text{div } \tilde{V} \tilde{\bar{F}} &= \tilde{V} \text{ div } \tilde{\bar{F}} + (\text{grad } \tilde{V}).\tilde{\bar{F}} \\
\text{div}(\tilde{V} \text{ grad } \tilde{U}) &= \tilde{V} \nabla^2 \tilde{U} + \text{grad } \tilde{V} \,.\, \text{grad } \tilde{U}
\end{aligned}
$$

Each of the volume/surface/line integral relationships developed previously continues to hold for complex fields. Thus, in addition to the complex forms of Stokes's theorem, the divergence theorem and Ostrogradsky's theorem, we have, inter alia,

$$
\int d\bar{S} \times \text{grad } \tilde{V} = \oint \tilde{V} \, d\bar{r}
$$

$$
\frac{d\tilde{\bar{F}}}{dt} = \frac{\partial \tilde{\bar{F}}}{\partial t} + (\bar{v}.\nabla)\tilde{\bar{F}}
$$

$$
\frac{d}{dt} \int \tilde{\bar{F}}.d\bar{r} = \frac{\partial}{\partial t} \int \tilde{\bar{F}}.d\bar{r} + \int \frac{d}{ds} (\bar{v}.\tilde{\bar{F}}) \, ds - \int \bar{v} \times \text{curl } \tilde{\bar{F}}.d\bar{r}
$$

$$
\oint d\bar{r} \times \tilde{\bar{F}} = \int (d\bar{S}.\nabla)\tilde{\bar{F}} - \int \text{div } \tilde{\bar{F}} \, d\bar{S} + \int d\bar{S} \times \text{curl } \tilde{\bar{F}}
$$

If the notation Re {} is employed to denote the real part of a complex expression within the brackets, then it is seen that

$$\text{Re }\{\text{grad}(f_1+jf_2)\} = \text{grad Re }\{f_1+jf_2\}$$

$$\text{Re }\{\text{div}(\overline{F}_1+j\overline{F}_2)\} = \text{div Re }\{\overline{F}_1+j\overline{F}_2\}$$

$$\text{Re }\{\text{curl}(\overline{F}_1+j\overline{F}_2)\} = \text{curl Re }\{\overline{F}_1+j\overline{F}_2\}$$

and

$$\text{Re }\{\text{div grad}(f_1+jf_2)\} = \text{div grad Re }\{f_1+jf_2\}$$

$$\text{Re }\{\text{grad div}(\overline{F}_1+j\overline{F}_2)\} = \text{grad div Re }\{\overline{F}_1+j\overline{F}_2\}$$

$$\text{Re }\{\text{curl curl}(\overline{F}_1+j\overline{F}_2)\} = \text{curl curl Re }\{\overline{F}_1+j\overline{F}_2\}$$

Since the products of both complex scalars and complex vectors can be expressed as the sum of real and imaginary components, we have, in addition, such relationships as

$$\text{Re }\{\text{div}(\tilde{\overline{F}}\times\tilde{\overline{G}})\} = \text{div Re }\{\tilde{\overline{F}}\times\tilde{\overline{G}}\}$$

and

$$\text{Re }\{\text{div grad }\tilde{f}\tilde{g}\} = \text{div grad Re }\{\tilde{f}\tilde{g}\}$$

Similarly,

$$\text{Re }\left\{\frac{\partial}{\partial t}\,(\tilde{f}\tilde{g})\right\} = \frac{\partial}{\partial t}\,\text{Re }\{\tilde{f}\tilde{g}\}$$

$$\text{Re }\{\oint \tilde{\overline{F}}\times\tilde{\overline{G}}.d\overline{r}\} = \oint \text{Re }\{\tilde{\overline{F}}\times\tilde{\overline{G}}.d\overline{r}\}$$

$$\text{Re }\{\nabla^2(\tilde{\overline{F}}.\tilde{\overline{G}})\} = \nabla^2\ \text{Re }\{\tilde{\overline{F}}.\tilde{\overline{G}}\}$$

It should be noted that

$$\text{Re }\{\tilde{f}\}\ \text{Re }\{\tilde{g}\} \neq \text{Re }\{\tilde{f}\tilde{g}\}$$

$$\text{Re }\{\tilde{\overline{F}}\}\times\text{Re }\{\tilde{\overline{G}}\} \neq \text{Re }\{\tilde{\overline{F}}\times\tilde{\overline{G}}\}$$

$$\text{Re }\{\text{div }\tilde{\overline{F}}\}\ \text{Re }\{\text{div }\tilde{\overline{G}}\} \neq \text{Re }\{\text{div }\tilde{\overline{F}}\ \text{div }\tilde{\overline{G}}\}$$

Clearly, these inequalities do not conform to the basic relationships from which the previous equalities were derived, viz

$$\mathrm{Re}\,\{p(f_3+jf_4)\} = p\,\mathrm{Re}\,\{f_3+jf_4\}$$

and

$$\mathrm{Re}\,\{p(\overline{F}_3+j\overline{F}_4)\} = p\,\mathrm{Re}\,\{\overline{F}_3+j\overline{F}_4\}$$

where p is a linear operator.

It should also be borne in mind that the trigonometric functions are not linear operators. Thus it is easily shown by expansion in exponential form that

$$\begin{aligned}\sin\,(f+jg) &= \sin f \cos jg + \cos f \sin jg \\ &= \sin f \cosh g + j \cos f \sinh g\end{aligned}$$

CHAPTER 2

CURVILINEAR COORDINATE SYSTEMS

2.1 Curvilinear Coordinates

If three single-valued scalar point functions u, v and w are defined throughout a region of space $\underline{R}$, and have level surfaces (u = constant, v = constant, w = constant) which nowhere meet in a common curve or coincide, then we may associate with each point of $\underline{R}$ a triplet of values known as its curvilinear coordinates. These are the values assumed by u, v and w on the particular level surfaces which pass through the point[1].

The level surfaces need not be planar, and in this respect the curvilinear system differs from the familiar Cartesian system of coordinates where the level surfaces of x, y and z are composed of planes which lie parallel to the yz, xz and xy coordinate planes through the origin.

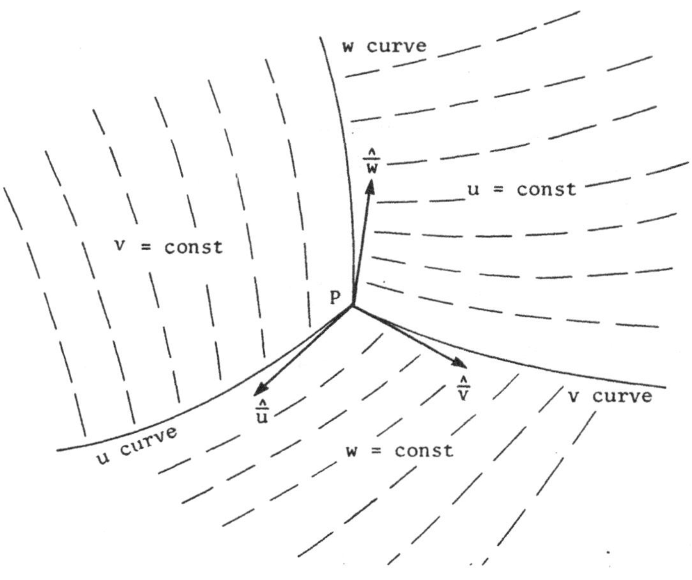

Fig. 2.1

---

1. For a detailed analysis of the relationship between Cartesian and curvilinear coordinates, see I.S. Sokolnikoff, "Advanced Calculus", Ch.12, McGraw-Hill, New York (1939).

In Fig. 2.1 the point P is shown as the intersection of level surfaces of u, v and w. These surfaces are known as the u, v and w coordinate surfaces through P. They meet pair-wise in coordinate curves. Thus the v and w surfaces meet in a curve along which v and w are constant and u varies. This is called a u curve. Similarly, the u and w coordinate surfaces meet in a v coordinate curve, and the v and u surfaces in a w curve.

The tangents to the coordinate curves through P define the coordinate axes at P. The unit vectors lying in the coordinate axes and directed towards increasing values of u, v and w are denoted by $\hat{u}$, $\hat{v}$ and $\hat{w}$. Unlike the unit vectors $\bar{i}$, $\bar{j}$ and $\bar{k}$, the vectors $\hat{u}$, $\hat{v}$ and $\hat{w}$ are, in general, functions of position because of variation of orientation.

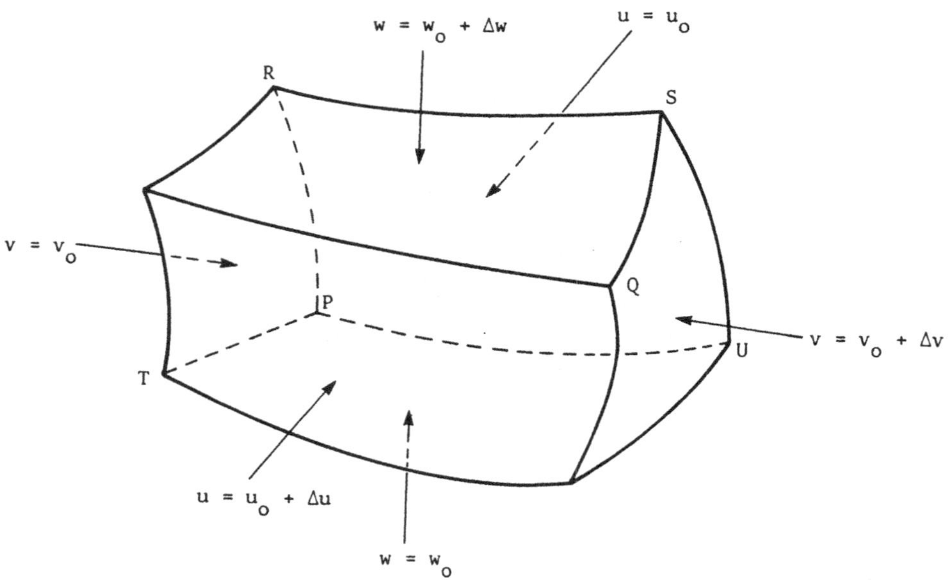

Fig. 2.2

In Fig. 2.2 the point P has been assigned the coordinates (u_o, v_o, w_o) and the point Q the coordinates $(u_o + \Delta u, v_o + \Delta v, w_o + \Delta w)$. The six associated level surfaces have been drawn and are seen to enclose a curvilinear volume element.

The curve PT is the intersection of the level surfaces $v = v_o$ and $w = w_o$. It is consequently a u coordinate curve, the increment of u between P and T being Δu.

If we denote $\overrightarrow{PT}$ by $(\Delta\bar{r})_u$ where $\bar{r}$ is the position vector relative to, say, the origin of rectangular coordinates, then we see that as $\Delta u \to 0$ the direction of $(\Delta\bar{r})_u$ approaches that of the coordinate axis at P.

Thus

$$\lim_{\Delta u \to 0} \frac{(\Delta\bar{r})_u}{\Delta u} = \hat{u}\, h_1$$

where h_1 is some positive scalar quantity.

On writing the above limit as $\dfrac{\partial\bar{r}}{\partial u}$ and extending the notation to motion along PU and PR, we get

$$\frac{\partial\bar{r}}{\partial u} = \hat{u}\, h_1 \quad ; \quad \frac{\partial\bar{r}}{\partial v} = \hat{v}\, h_2 \quad ; \quad \frac{\partial\bar{r}}{\partial w} = \hat{w}\, h_3 \qquad (2.1\text{-}1)$$

or

$$\frac{ds_u}{du} = h_1 \quad ; \quad \frac{ds_v}{dv} = h_2 \quad ; \quad \frac{ds_w}{dw} = h_3 \qquad (2.1\text{-}2)$$

where s_u, s_v and s_w represent distance measured along the u, v and w curves.

h_1, h_2 and h_3 are known as metrical coefficients. In the most general case each is a function of position.

It follows that

$$(\Delta\bar{r})_u = \overrightarrow{PT} \approx \hat{u}\, h_1\, \Delta u$$

$$(\Delta\bar{r})_v = \overrightarrow{PU} \approx \hat{v}\, h_2\, \Delta v$$

$$(\Delta\bar{r})_w = \overrightarrow{PR} \approx \hat{w}\, h_3\, \Delta w$$

The vector displacement between P and Q is given by

$$\Delta\bar{r} = \overrightarrow{PQ} = \overrightarrow{PR} + \overrightarrow{RS} + \overrightarrow{SQ} \approx \overrightarrow{PR} + \overrightarrow{PU} + \overrightarrow{PT}$$

The substitution of $\vec{PU}$ for $\vec{RS}$ and of $\vec{PT}$ for $\vec{SQ}$ may involve approximations both of magnitude and direction[2]. However, as the dimensions of the volume element approach zero the curvilinear coordinate surfaces approximate to a plane parallel system, so that we may write

$$\Delta\bar{r} = \hat{\bar{u}}\, h_1 \Delta u + \hat{\bar{v}}\, h_2 \Delta v + \hat{\bar{w}}\, h_3 \Delta w + \bar{\varepsilon}$$

where $|\bar{\varepsilon}|/|\Delta\bar{r}| \to 0$ as $\Delta u,\ \Delta v,\ \Delta w \to 0$

whence the differential form becomes

$$d\bar{r} = \hat{\bar{u}}\, h_1\, du + \hat{\bar{v}}\, h_2\, dv + \hat{\bar{w}}\, h_3\, dw \tag{2.1-3}$$

The scalar distance associated with this element of displacement is found from

$$\left.\begin{aligned} ds^2 = d\bar{r}.d\bar{r} &= h_1^2\, du^2 + h_2^2\, dv^2 + h_3^2\, dw^2 \\ &+ 2\, \hat{\bar{u}}.\hat{\bar{v}}\, h_1 h_2\, du\, dv + 2\, \hat{\bar{u}}.\hat{\bar{w}}\, h_1 h_3\, du\, dw \\ &+ 2\, \hat{\bar{v}}.\hat{\bar{w}}\, h_2 h_3\, dv\, dw \end{aligned}\right\} \tag{2.1-4}$$

It may be shown in a similar manner that the differential form of the scalar area of a surface element which lies within the u coordinate surface and is bounded by v and w coordinate curves is given by[3]

$$|dS_u| = |(\hat{\bar{v}}\times\hat{\bar{w}}) h_2 h_3\, dv\, dw| = \left| \left\{ \frac{\partial \bar{r}}{\partial v} \times \frac{\partial \bar{r}}{\partial w} \right\} dv\, dw \right| \tag{2.1-5}$$

Corresponding expressions hold for $|dS_v|$ and $|dS_w|$.

---

2. No such approximations are involved in the case of a rectangular or oblique Cartesian system.

3. The sign of dS_u will depend upon the sense of the positive normal at the element relative to $\hat{\bar{u}}$.

The magnitude of the volume element is likewise given by

$$d\tau = \left|(\hat{u}\times\hat{v}.\hat{w})\, h_1h_2h_3\, du\, dv\, dw\right|$$

or

$$d\tau = \left|\left(\frac{\partial\bar{r}}{\partial u}\times\frac{\partial\bar{r}}{\partial v}.\frac{\partial\bar{r}}{\partial w}\right)du\, dv\, dw\right| \qquad (2.1\text{-}6)$$

Since $\bar{r} = \bar{i}x + \bar{j}y + \bar{k}z$ it may also be expressed as a determinant:

$$d\tau = \begin{vmatrix} \frac{\partial x}{\partial u} & \frac{\partial y}{\partial u} & \frac{\partial z}{\partial u} \\ \frac{\partial x}{\partial v} & \frac{\partial y}{\partial v} & \frac{\partial z}{\partial v} \\ \frac{\partial x}{\partial w} & \frac{\partial y}{\partial w} & \frac{\partial z}{\partial w} \end{vmatrix} |du\, dv\, dw| \qquad (2.1\text{-}7)$$

(2.1-7) is often written as $J\left(\frac{x,y,z}{u,v,w}\right)|du\ dv\ dw|$ where $J\left(\frac{x,y,z}{u,v,w}\right)$ is known as the Jacobian of the functions $x = x(u,v,w)$; $y = y(u,v,w)$; $z = z(u,v,w)$.

When the coordinate axes are mutually perpendicular at all points the coordinates are said to be orthogonal. For a right-handed orthogonal system

$$\hat{u}.\hat{v} = \hat{u}.\hat{w} = \hat{v}.\hat{w} = 0$$

$$\hat{u}\times\hat{v} = \hat{w} \quad ; \quad \hat{v}\times\hat{w} = \hat{u} \quad ; \quad \hat{w}\times\hat{u} = \hat{v}$$

in which case

$$ds^2 = h_1^2\, du^2 + h_2^2\, dv^2 + h_3^2\, dw^2 \qquad (2.1\text{-}8)$$

$$|dS_u| = h_2h_3\, |dv\, dw| \qquad (2.1\text{-}9)$$

$$d\tau = h_1h_2h_3\, |du\, dv\, dw| \qquad (2.1\text{-}10)$$

The considerations of this section are best illustrated by reference to the two most commonly employed systems of curvilinear coordinates, viz cylindrical and spherical coordinates[4].

4. For more complex systems of curvilinear coordinates see J.A. Stratton, "Electromagnetic Theory", Sec. 1.18, McGraw-Hill, New York (1941).

2.2 Cylindrical Coordinates

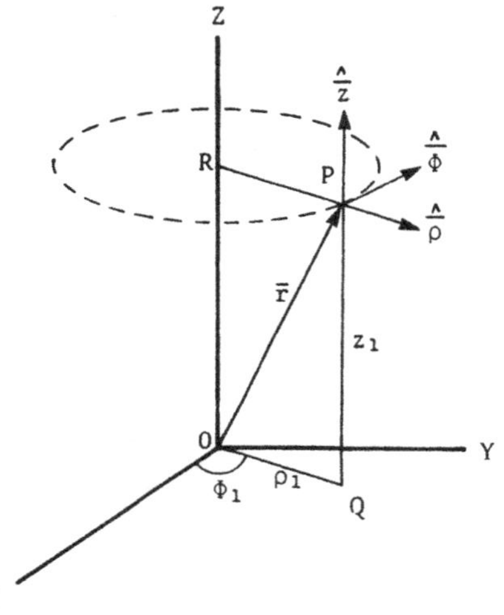

Fig. 2.3

Let Q be the projection of the point P on the xy coordinate plane through 0 (Fig. 2.3). If ρ_1 is the length of 0Q, and if Φ_1 is the angle made by $\overrightarrow{0Q}$ with the positive x axis when measuring right-handedly about the z axis, and if z_1 is the distance QP, then (ρ_1, Φ_1, z_1) are said to be the (circular) cylindrical coordinates of the point P.

By stipulating that $\rho_1 > 0$ and that $0 \leq \Phi_1 < 2\pi$ the coordinates of P are uniquely defined[5].

P is seen to be the point of intersection of the level surfaces $\rho = \rho_1$, $\Phi = \Phi_1$, $z = z_1$. The surface $\rho = \rho_1$ comprises a circular cylinder of radius ρ_1 centred on the z axis. The surface $\Phi = \Phi_1$ is a half-plane whose edge is the z axis. It makes an angle Φ_1 with 0X. The surface $z = z_1$ is a plane normal to the z axis and at a distance z_1 (measuring in the positive z direction) from the xy plane through 0.

The coordinate curve defined by the ρ and z surfaces is a circle through P centred on the z axis and normal to it. It is the curve traced out when Φ alone varies, and is known as the Φ curve through P. It will be seen that the ρ curve is RP produced, and the z curve QP produced in both directions.

---

5. Points on the z axis ($\rho = 0$) will be excluded from the following analyses. The z axis is said to comprise a singular line.

The coordinate axes through P are mutually perpendicular; cylindrical coordinates are consequently orthogonal. The unit vectors $\hat{\rho}$, $\hat{\Phi}$, $\hat{z}$ taken in that order form a right-handed set.

The relations between rectangular Cartesian and cylindrical coordinates may be written down by inspection.

$$x = \rho \cos \Phi \quad ; \quad y = \rho \sin \Phi \quad ; \quad z = z \tag{2.2-1}$$

$$\rho = (x^2+y^2)^{\frac{1}{2}} \quad ; \quad \Phi = \tan^{-1} \frac{y}{x} \quad ; \quad z = z \tag{2.2-2}$$

In this representation each coordinate of one set has been expressed entirely in terms of the coordinates of the other set (with the exception of the common z coordinate).

The position vector from 0 is given by

$$\bar{r} = \bar{i}x + \bar{j}y + \bar{k}z = \bar{i}\,\rho \cos \Phi + \bar{j}\,\rho \sin \Phi + \bar{k}z$$

hence[6]

$$\left.\begin{aligned} \frac{\partial \bar{r}}{\partial \rho} &= \bar{i} \cos \Phi + \bar{j} \sin \Phi = \hat{\rho}\, h_1 \\ \frac{\partial \bar{r}}{\partial \Phi} &= -\bar{i}\,\rho \sin \Phi + \bar{j}\,\rho \cos \Phi = \hat{\Phi}\, h_2 \\ \frac{\partial \bar{r}}{\partial z} &= \bar{k} = \hat{z}\, h_3 \end{aligned}\right\} \tag{2.2-3}$$

whence

$$h_1 = 1 \quad ; \quad h_2 = \rho \quad ; \quad h_3 = 1 \tag{2.2-4}$$

---

6. It should be particularly noted that when partial differentiation is carried out with respect to a variable of one set, the remaining variables <u>of the same set</u> are held constant. Thus whereas $\frac{\partial x}{\partial \rho}$ is the rate of change of x with respect to ρ while Φ and z remain constant, $\frac{\partial \rho}{\partial x}$ is the rate of change of ρ with respect to x while y and z remain constant. The readily-demonstrated equality of $\frac{\partial x}{\partial \rho}$ and $\frac{\partial \rho}{\partial x}$ does not, therefore, constitute an inconsistency.

and

$$\left.\begin{aligned} \hat{\underline{\rho}} &= \overline{i}\cos\Phi + \overline{j}\sin\Phi \\ \hat{\underline{\Phi}} &= -\overline{i}\sin\Phi + \overline{j}\cos\Phi \\ \hat{\underline{z}} &= \overline{k} \end{aligned}\right\} \qquad (2.2\text{-}5)$$

The mutually perpendicular nature of $\hat{\underline{\rho}}$, $\hat{\underline{\Phi}}$ and $\hat{\underline{z}}$ is confirmed by the fact that the scalar products, taken two at a time, are zero.

2.3 Spherical Coordinates

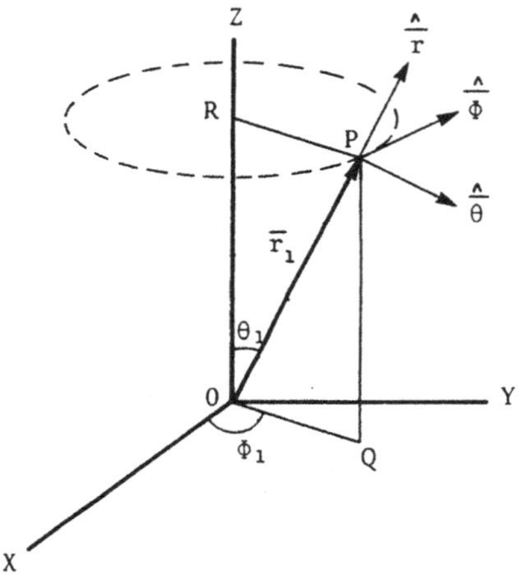

Fig. 2.4

Let Q be the projection of the point P on the xy plane through O (Fig. 2.4). If r_1 is the length of OP and if θ_1 is the angle made by $\overrightarrow{OP}$ with the positive z axis and if Φ_1 is the angle made by $\overrightarrow{OQ}$ with the positive x axis when measuring right-handedly about the z axis, then (r_1, θ_1, Φ_1) are said to be the spherical polar coordinates of P. By stipulating that $r_1 > 0$, $0 < \theta_1 < \pi$, $0 \leq \Phi_1 < 2\pi$ the coördinates of P are uniquely defined[7].

---

7. Points on the z axis ($r = 0$, or $\theta = 0, \pi$) will be excluded from the following analyses.

The coordinate surfaces through P are seen to be

(a) a sphere of radius r_1 centred upon 0

(b) a circular cone of half-angle θ_1 $\left(\text{or } \pi - \theta_1 \text{ if } \theta_1 > \frac{\pi}{2}\right)$ whose axis coincides with the z axis and whose vertex is at 0

(c) a half-plane whose edge is the z axis and which makes an angle Φ_1 with OX.

The coordinate curves are

(a) OP produced (r curve)

(b) a semi-circle of radius r_1 centred upon 0 and lying in the half-plane $\Phi = \Phi_1$ (θ curve)

(c) a circle of radius $r_1 \sin \theta_1$ normal to the z axis and passing through P (Φ curve).

The coordinates are clearly orthogonal; $\hat{r}, \hat{\theta}, \hat{\Phi}$ in that order are seen to form a right-handed set.

For points beyond the z axis

$$x = r \sin \theta \cos \Phi \quad ; \quad y = r \sin \theta \sin \Phi \quad ; \quad z = r \cos \theta \qquad (2.3\text{-}1)$$

$$r = (x^2+y^2+z^2)^{\frac{1}{2}} \quad ; \quad \theta = \cos^{-1} \frac{z}{(x^2+y^2+z^2)^{\frac{1}{2}}} \quad ; \quad \Phi = \tan^{-1} \frac{y}{x} \qquad (2.3\text{-}2)$$

Further,

$$\bar{r} = \bar{i}\, r \sin \theta \cos \Phi + \bar{j}\, r \sin \theta \sin \Phi + \bar{k}\, r \cos \theta$$

hence

$$\left.\begin{aligned}
\frac{\partial \bar{r}}{\partial r} &= \bar{i} \sin \theta \cos \Phi + \bar{j} \sin \theta \sin \Phi + \bar{k} \cos \theta = \hat{r}\, h_1 \\
\frac{\partial \bar{r}}{\partial \theta} &= \bar{i}\, r \cos \theta \cos \Phi + \bar{j}\, r \cos \theta \sin \Phi - \bar{k}\, r \sin \theta = \hat{\theta}\, h_2 \\
\frac{\partial \bar{r}}{\partial \Phi} &= -\bar{i}\, r \sin \theta \sin \Phi + \bar{j}\, r \sin \theta \cos \Phi = \hat{\Phi}\, h_3
\end{aligned}\right\} \qquad (2.3\text{-}3)$$

whence[8]

$$h_1 = 1 \quad ; \quad h_2 = r \quad ; \quad h_3 = r \sin\theta \tag{2.3-4}$$

and

$$\left.\begin{aligned} \hat{\bar{r}} &= \bar{i}\sin\theta\cos\Phi + \bar{j}\sin\theta\sin\Phi + \bar{k}\cos\theta \\ \hat{\bar{\theta}} &= \bar{i}\cos\theta\cos\Phi + \bar{j}\cos\theta\sin\Phi - \bar{k}\sin\theta \\ \hat{\bar{\Phi}} &= -\bar{i}\sin\Phi + \bar{j}\cos\Phi \end{aligned}\right\} \tag{2.3-5}$$

2.4 Line, Surface and Volume Integration in Cylindrical and Spherical Coordinates

2.4a Line integration in cylindrical and spherical coordinates

It follows from (2.2-3) and (2.2-4) that the differential form of the vector line element in cylindrical coordinates is given by

$$\left.\begin{aligned} d\bar{r} &= \hat{\bar{\rho}}\, h_1 d\rho + \hat{\bar{\Phi}}\, h_2 d\Phi + \hat{\bar{z}}\, h_3 dz \\ &\text{or} \\ d\bar{r} &= \hat{\bar{\rho}}\, d\rho + \hat{\bar{\Phi}}\, \rho\, d\Phi + \hat{\bar{z}}\, dz \end{aligned}\right\} \tag{2.4-1}$$

Hence

$$ds = (d\rho^2 + \rho^2 d\Phi^2 + dz^2)^{\frac{1}{2}} \tag{2.4-2}$$

(2.4-2) also follows from substitution of (2.2-4) in (2.1-8).

The scalar line integral of a point function V along a curve Γ is consequently given by

$$\int_\Gamma V\, ds = \int_\Gamma V\, (d\rho^2 + \rho^2 d\Phi^2 + dz^2)^{\frac{1}{2}} \tag{2.4-3}$$

For the purpose of computation this is replaced by

---

8. The metrical coefficients for cylindrical and spherical coordinates may be derived quite simply by the use of (2.1-2) and an appeal to the appropriate diagram. In more complicated systems the formal approach developed above may be required.

$$\int_\Gamma V\, ds = \int_\Gamma V \left\{\left(\frac{d\rho}{dt}\right)^2 + \rho^2 \left(\frac{d\Phi}{dt}\right)^2 + \left(\frac{dz}{dt}\right)^2\right\}^{\frac{1}{2}} |dt| \qquad (2.4\text{-}3(a))$$

when V and the coordinates of the curve can be expressed in terms of some parameter t. Alternatively,

$$\int_\Gamma V\, ds = \int_\Gamma V \left\{\left(\frac{d\rho}{d\Phi}\right)^2 + \rho^2 + \left(\frac{dz}{d\Phi}\right)^2\right\}^{\frac{1}{2}} |d\Phi| \qquad (2.4\text{-}3(b))$$

when, for example, V, ρ and z are known functions of Φ.

The corresponding expressions for the tangential line integral of $\overline{F}$ are

$$\int_\Gamma \overline{F}.d\overline{r} = \int_\Gamma (F_\rho d\rho + \rho F_\Phi d\Phi + F_z dz) \qquad (2.4\text{-}4)$$

$$\int_\Gamma \overline{F}.d\overline{r} = \int_\Gamma \left\{F_\rho \frac{d\rho}{dt} + \rho F_\Phi \frac{d\Phi}{dt} + F_z \frac{dz}{dt}\right\} dt \qquad (2.4\text{-}4(a))$$

$$\int_\Gamma \overline{F}.d\overline{r} = \int_\Gamma \left\{F_\rho \frac{d\rho}{d\Phi} + \rho F_\Phi + F_z \frac{dz}{d\Phi}\right\} d\Phi \qquad (2.4\text{-}4(b))$$

Equations (2.4-3) to (2.4-4b) reduce to their plane polar equivalents when the terms containing z are deleted.

In spherical coordinates

$$d\overline{r} = \hat{\overline{r}}\, dr + \hat{\overline{\theta}}\, r\, d\theta + \hat{\overline{\Phi}}\, r \sin\theta\, d\Phi \qquad (2.4\text{-}5)$$

$$ds = (dr^2 + r^2 d\theta^2 + r^2 \sin^2\theta\, d\Phi^2)^{\frac{1}{2}} \qquad (2.4\text{-}6)$$

The basic forms of $\int_\Gamma V\, ds$ and $\int_\Gamma \overline{F}.d\overline{r}$ are seen to be

$$\int_\Gamma V\, (dr^2 + r^2 d\theta^2 + r^2 \sin^2\theta\, d\Phi^2)^{\frac{1}{2}} \qquad (2.4\text{-}7)$$

$$\int_{\Gamma} (F_r dr + rF_\theta d\theta + r \sin\theta \, F_\phi d\phi) \tag{2.4-8}$$

whence the alternative forms follow directly.

2.4b Surface integration in cylindrical, spherical, and general surface curvilinear coordinates

The scalar area of a surface element which lies within a coordinate surface and is bounded by coordinate curves may be found by substitution of the appropriate coordinates and metrical coefficients in equations such as (2.1-9). For a cylindrical coordinate system this yields

$$|dS_\rho| = \rho|d\phi \, dz| \quad ; \quad |dS_\phi| = |d\rho \, dz| \quad ; \quad |dS_z| = \rho|d\rho \, d\phi| \tag{2.4-9}$$

There is, therefore, no difficulty in setting up any required form of integral when the surface of integration coincides with a coordinate surface. When this is not the case, but one coordinate upon the surface is a known function of the other two, we may proceed as follows.

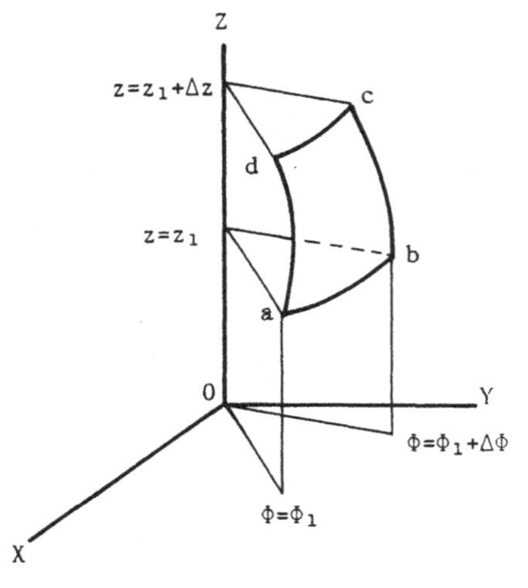

Fig. 2.5

Suppose that the equation of the surface of integration, S, takes the form $\rho = f(\phi,z)$. Fig. 2.5 shows the part of the surface intercepted by the coordinate planes $\phi = \phi_1$, $\phi = \phi_1 + \Delta\phi$, $z = z_1$, $z = z_1 + \Delta z$, where $\Delta\phi$ and Δz are positive increments of ϕ and z. The line ab is the intersection of S with the surface $z = z_1$, while ad is the intersection of S with $\phi = \phi_1$. If the end point of $\bar{r} = \bar{i}\rho \cos\phi + \bar{j}\rho \sin\phi + \bar{k}z$ is

$$\frac{d\bar{r}}{d\Phi} = \bar{i}\left\{\frac{\partial\rho}{\partial\Phi}\cos\Phi - \rho\sin\Phi\right\} + \bar{j}\left\{\frac{\partial\rho}{\partial\Phi}\sin\Phi + \rho\cos\Phi\right\}$$

(The total derivative sign is used here to distinguish the operation from $\left(\frac{\partial\bar{r}}{\partial\Phi}\right)$ in (2.2-3) where ρ, Φ and z are independent variables[9].)

It will be seen from the figure that since $\frac{d\bar{r}}{d\Phi}$ is the rate of change of $\bar{r}$ with Φ while z remains constant, the vector $\overrightarrow{ab}$ is given approximately by $\left(\frac{d\bar{r}}{d\Phi}\right)_a \Delta\Phi$. Similarly,

$$\frac{d\bar{r}}{dz} = \bar{i}\frac{\partial\rho}{\partial z}\cos\Phi + \bar{j}\frac{\partial\rho}{\partial z}\sin\Phi + \bar{k}$$

and $\overrightarrow{ad} \approx \left(\frac{d\bar{r}}{dz}\right)_a \Delta z$

It follows that in the limit as $\Delta\Phi$, $\Delta z \to 0$ the element of area intercepted by the planes becomes

$$(d\bar{S})_{\substack{\Phi_1,\Phi_1+\Delta\Phi \\ z_1,z_1+\Delta z}} = \left(\frac{d\bar{r}}{d\Phi}\times\frac{d\bar{r}}{dz}\right)_{\Phi_1,z_1} d\Phi\, dz$$

where the postive normal at the element makes an angle of less than 90° with $\hat{\rho}$.

On expanding the vector product and dropping the subscripts we obtain the general form

$$d\bar{S} = \left\{\bar{i}\left(\frac{\partial\rho}{\partial\Phi}\sin\Phi + \rho\cos\Phi\right) + \bar{j}\left(\frac{-\partial\rho}{\partial\Phi}\cos\Phi + \rho\sin\Phi\right) - \bar{k}\,\rho\,\frac{\partial\rho}{\partial z}\right\}|d\Phi\, dz|$$

or, from (2.2-5),

$$d\bar{S} = \left\{\hat{\rho}\,\rho - \hat{\Phi}\,\frac{\partial\rho}{\partial\Phi} - \hat{z}\,\rho\,\frac{\partial\rho}{\partial z}\right\}|d\Phi\, dz|$$

The scalar magnitude of the surface element is

---

9. Since $\bar{r}$ is also a function of z the notation is not a happy one. But we do not here consider the case in which Φ and z vary simultaneously.

$$\left\{\rho^2 + \left(\frac{\partial\rho}{\partial\Phi}\right)^2 + \rho^2\left(\frac{\partial\rho}{\partial z}\right)^2\right\}^{\frac{1}{2}} |d\Phi\ dz|$$

and the scalar surface integrals are given by

$$\int_S V\,dS = \iint_S V\left\{\rho^2 + \left(\frac{\partial\rho}{\partial\Phi}\right)^2 + \rho^2\left(\frac{\partial\rho}{\partial z}\right)^2\right\}^{\frac{1}{2}} |d\Phi\ dz| \qquad (2.4\text{-}10)$$

$$\int_S \overline{F}.d\overline{S} = \iint_S \left\{\rho F_\rho - F_\Phi \frac{\partial\rho}{\partial\Phi} - \rho F_z \frac{\partial\rho}{\partial z}\right\} |d\Phi\ dz| \qquad (2.4\text{-}11)$$

(2.4-11) represents the flux of $\overline{F}$ through S directed away from the z axis.

Corresponding expressions obtain when the surface is defined by $\Phi = g(\rho,z)$ or by $z = h(\rho,\Phi)$.

Surface integration in spherical coordinates may be carried out in a similar manner.

The elements of area which lie within the coordinate surfaces and are bounded by coordinate curves are seen to be

$$|dS_r| = r^2 \sin\theta\ |d\theta\ d\Phi| \quad ; \quad |dS_\theta| = r\sin\theta\ |dr\ d\Phi| \quad ; \quad |dS_\Phi| = r\ |dr\ d\theta| \qquad (2.4\text{-}12)$$

When the surface of integration is not a coordinate surface but r is a known function of (θ,Φ), the associated vector surface element is given by

$$d\overline{S} = \left\{\frac{d\overline{r}}{d\theta} \times \frac{d\overline{r}}{d\Phi}\right\} |d\theta\ d\Phi|$$

where the positive normal at the surface makes an angle of less than 90° with $\hat{\overline{r}}$.

The surface integrals are then found to be

$$\int_S V\,dS = \iint_S V\left\{\left(\frac{\partial r}{\partial\Phi}\right)^2 + \left(\frac{\partial r}{\partial\theta}\right)^2 \sin^2\theta + r^2\sin^2\theta\right\}^{\frac{1}{2}} r\ |d\theta\ d\Phi| \qquad (2.4\text{-}13)$$

$$\int \overline{F}.d\overline{S} = \iint \left\{rF_r \sin\theta - F_\theta \sin\theta\,\frac{\partial r}{\partial\theta} - F_\Phi \frac{\partial r}{\partial\Phi}\right\} r\ |d\theta\ d\Phi| \qquad (2.4\text{-}14)$$

(2.4-14) represents the flux of $\overline{F}$ through S directed away from the origin of cooordinates.

Similar expressions hold when θ is a known function of (r,Φ), or Φ is a known function of (r,θ).

Equations (2.4-10, 11, 13, 14) may be shown to be particular configurations of two scalar integrals which are based upon a parametric description of the surface of integration.

In general, the set of equations

$$x = f_1(\xi,\zeta) \qquad y = f_2(\xi,\zeta) \qquad z = f_3(\xi,\zeta)$$

represents a surface, the parameters ξ and ζ being known as surface curvilinear coordinates. Variation of ξ, with ζ held constant, corresponds to movement over the surface along a ξ curve, while variation of ζ, with ξ held constant, relates to motion along a ζ curve. These coordinate curves may or may not be orthogonal.

If $\hat{\overline{\xi}}$ and $\hat{\overline{\zeta}}$ are unit vectors, tangential to the ξ and ζ curves at each point, then

$$\frac{\partial \overline{r}}{\partial \xi} = h_\xi \hat{\overline{\xi}} \quad ; \quad \frac{\partial \overline{r}}{\partial \zeta} = h_\zeta \hat{\overline{\zeta}}$$

where

$$h_\xi = \frac{ds_\xi}{d\xi} \quad \text{and} \quad h_\zeta = \frac{ds_\zeta}{d\zeta}$$

The vector element of area bounded by such curves is given by

$$d\overline{S} = \left(\frac{\partial \overline{r}}{\partial \xi} \times \frac{\partial \overline{r}}{\partial \zeta}\right) |d\xi \; d\zeta| = (\hat{\overline{\xi}} \times \hat{\overline{\zeta}}) \; h_\xi h_\zeta \; |d\xi \; d\zeta|$$

where $\hat{\overline{\xi}}$, $\hat{\overline{\zeta}}$ and $\hat{\overline{n}}$ form a right-handed set, $\hat{\overline{n}}$ being the unit positive normal at the surface element, whence the surface integral of V and the normal surface integral of $\overline{F}$ are seen to be

$$\iint_S V \left|\frac{\partial \overline{r}}{\partial \xi} \times \frac{\partial \overline{r}}{\partial \zeta}\right| |d\xi \; d\zeta| \quad \text{and} \quad \iint_S \overline{F} \cdot \left(\frac{\partial \overline{r}}{\partial \xi} \times \frac{\partial \overline{r}}{\partial \zeta}\right) |d\xi \; d\zeta| \qquad (2.4\text{-}15)$$

The connection between this treatment and that given previously in terms of, say, spherical coordinates is made clear by a consideration of the particular functional relationships:

$$x = f_1(\xi,\zeta) = f(\xi,\zeta) \sin\xi \cos\zeta$$

$$y = f_2(\xi,\zeta) = f(\xi,\zeta) \sin\xi \sin\zeta$$

$$z = f_3(\xi,\zeta) = f(\xi,\zeta) \cos\xi$$

In this case

$$\cos\xi = \frac{z}{(x^2+y^2+z^2)^{\frac{1}{2}}} \quad ; \quad \tan\zeta = \frac{y}{x} \quad ; \quad f(\xi,\zeta) = (x^2+y^2+z^2)^{\frac{1}{2}}$$

so that ξ and ζ correspond with the θ and Φ coordinates of the point which they specify, and $f(\xi,\zeta) = r$.

It follows that the spherical coordinate analysis is identical with that based upon surface curvilinear coordinates when the Cartesian coordinates of the surface are related to these parameters by the equations set out above and $0 < \xi < \pi$, $0 \le \zeta < 2\pi$.

2.4c Volume integration in cylindrical and spherical coordinates

The magnitude of an element of volume, which is bounded by coordinate surfaces, may be found from the substitution of (2.2-4) and (2.3-4) in (2.1-10).

For cylindrical coordinates

$$d\tau = \rho \, |d\rho \, d\Phi \, dz|$$

and for spherical coordinates

$$d\tau = r^2 \sin\theta \, |dr \, d\theta \, d\Phi|$$

The process of volume integration does not require that the shape of the volume element be matched with that of the bounding surface of the of the integration region [10], hence the complications of the previous section, associated with the introduction of non-coordinate surfaces, do not arise. The scalar integrals are simply

$$\int_\tau V \, d\tau = \iiint_\tau V \rho \, |d\rho \, d\Phi \, dz| \qquad (2.4\text{-}16)$$

10. See footnote to p. 50.

and

$$\int_\tau V\, d\tau = \iiint_\tau V\, r^2 \sin\theta\, |dr\, d\theta\, d\Phi| \qquad (2.4\text{-}17)$$

In the vector integrals, $\overline{F}$ replaces V.

EXERCISES

2-1. Confirm equation (2.1-5) for a plane or warped coordinate surface. (See Ex.1-27., p. 24).

2-2. Express the acceleration of a point in cylindrical coordinates.

$$\text{Ans:}\quad \hat{\rho}\left\{\frac{d^2\rho}{dt^2} - \rho\left(\frac{d\Phi}{dt}\right)^2\right\} + \hat{\Phi}\left\{\rho\,\frac{d^2\Phi}{dt^2} + 2\,\frac{d\rho}{dt}\,\frac{d\Phi}{dt}\right\} + \hat{z}\,\frac{d^2z}{dt^2}$$

2-3. A conical helix is defined in cylindrical coordinates by

$$\rho = az \quad , \quad \Phi = bz$$

where a and b are constants.

Find the length of the helix between z = 0 and z = z.

$$\text{Ans:}\quad \frac{ab}{2}\left\{z\left(\frac{a^2+1}{a^2b^2} + z^2\right)^{\frac{1}{2}} + \frac{a^2+1}{a^2b^2}\sinh^{-1}\frac{ab}{(a^2+1)^{\frac{1}{2}}}\, z\right\}$$

2-4. Compute the value of the closed tangential line integral of $\overline{F} = \hat{\Phi}\rho$ around the cardioid $\rho = a(1-\cos\Phi)$ by direct integration in plane polar coordinates. Confirm this by applying Stokes's theorem.

Ans: $3\pi a^2$

2-5. Determine the surface area of an ellipsoid of revolution defined by

$$\frac{\rho^2}{a^2} + \frac{z^2}{b^2} = 1$$

where a and b are constants.

$$\text{Ans:}\quad 2\pi a^2 + \frac{2\pi ab}{\left(1 - \frac{a^2}{b^2}\right)^{\frac{1}{2}}}\sin^{-1}\left(1 - \frac{a^2}{b^2}\right)^{\frac{1}{2}} \quad \text{for } b > a$$

$$2\pi a^2 + \frac{2\pi ab}{\left(\frac{a^2}{b^2} - 1\right)^{\frac{1}{2}}}\sinh^{-1}\left(\frac{a^2}{b^2} - 1\right)^{\frac{1}{2}} \quad \text{for } a > b$$

2-6. The axis of a cylinder of height h and radius a coincides with the z axis of coordinates, while the lower edge of the cylinder lies in the xy plane through the origin. A vector point function $\overline{F}$ is defined in spherical coordinates at points beyond the origin by

$$F_r = \frac{2\cos\theta}{r^3}, \quad F_\theta = \frac{\sin\theta}{r^3}, \quad F_\phi = 0$$

Use (2.4-14) to determine the outward flux of $\overline{F}$ through the cylindrical surface.

Ans: $\dfrac{2\pi}{a}\left\{1 - \dfrac{a^3}{(a^2+h^2)^{3/2}}\right\}$

2-7. Given that $\overline{F} = \overline{r}/r^3$ where $\overline{r} = \overline{i}x + \overline{j}y + \overline{k}z$, show that the flux of $\overline{F}$ through the surface of the ellipsoid of revolution

$$\frac{x^2}{a^2} + \frac{y^2}{a^2} + \frac{z^2}{b^2} = 1$$

is equal to 4π. Carry this out both in cylindrical and spherical coordinates using (2.4-11) and (2.4-14), and notice how the choice of an appropriate coordinate system simplifies the calculation.

It should be appreciated that for this particular vector field the surface integral is independent of the shape of the enclosing surface - See Sec. 1.14.

2-8. A surface defined by $\Phi = f(r,\theta)$ is intercepted by the coordinate surfaces $r = r$, $r = r + dr$, $\theta = \theta$, $\theta = \theta + d\theta$. Derive an expression for the element of area so intercepted when the positive sense of the normal at the surface makes an angle of less than 90° with $\hat{\overline{\Phi}}$, and determine the values of the projections of this element upon planes normal to $\hat{\overline{r}}$, $\hat{\overline{\theta}}$ and $\hat{\overline{\Phi}}$. Confirm these values by working directly from a diagram, and show why dS_r and dS_θ are negative when $\frac{\partial\Phi}{\partial r}$ and $\frac{\partial\Phi}{\partial\theta}$ are positive.

Ans: $dS_r = -r^2 \sin\theta \frac{\partial\Phi}{\partial r}\,|dr\,d\theta|$; $dS_\theta = -r\sin\theta\frac{\partial\Phi}{\partial\theta}\,|dr\,d\theta|$; $dS_\Phi = r\,|dr\,d\theta|$

2-9. A sphere of radius a is centred upon a point whose Cartesian coordinates are (l,m,n). Write down the equation of the sphere in terms of two parameters, ξ and ζ, determine the magnitude of the associated element of area, and integrate this to obtain the surface area of the sphere.

Ans: $x = a \sin \xi \cos \zeta + l$

$y = a \sin \xi \sin \zeta + m$

$z = a \cos \xi + n$

$dS = a^2 \sin \xi \, |d\xi \, d\zeta|$

$\int dS = 4\pi a^2$

2-10. A torus is generated by rotating a circle of radius a about the z axis in such a way that its plane contains the axis at all times. The circle initially lies in the xz plane with its centre at the point $x = R$, $z = Z$, where $R > a$.

Set up parametric equations for the surface of the torus, determine the magnitude of the element of area, and integrate to find the total surface area.

Ans: $x = (R + a \cos \xi) \cos \zeta$

$y = (R + a \cos \xi) \sin \zeta$

$z = a \sin \xi + Z$

$dS = a(R + a \cos \xi) \, |d\xi \, d\zeta|$

$\int dS = 4\pi^2 \, a \, R$

2-11. Find the area enclosed by the ellipse $\frac{x^2}{a^2} + \frac{y^2}{b^2} = 1$ by making use of the parametric equations[11]

$$x = a \cos \zeta \quad ; \quad y = b \sin \zeta$$

---

11. The use of common symbols for the parameters of the different exercises does not imply that they are related.

[Hint: The equations $x = \alpha\, a \cos\zeta$ and $y = \alpha\, b \sin\zeta$, where $0 < \alpha \leq 1$, define a system of concentric ellipses of which the outermost is the ellipse under consideration. Bearing in mind that $\tan\zeta = \frac{a}{b}\tan\Phi$, where Φ is the angle of plane polar coordinates, show that $\frac{\partial\bar{r}}{\partial\zeta}$ is tangential to the typical ellipse while $\frac{\partial\bar{r}}{\partial\alpha}$ is collinear with $\bar{r}$. Determine the value of the element of area defined by $\left|\left(\frac{\partial\bar{r}}{\partial\zeta} \times \frac{\partial\bar{r}}{\partial\alpha}\right) d\alpha\, d\zeta\right|$ and show that the area enclosed by the ellipse is given by $\int_0^{2\pi}\int_0^1 \left|\frac{\partial\bar{r}}{\partial\zeta} \times \frac{\partial\bar{r}}{\partial\alpha}\right| d\alpha\, d\zeta$. Evaluate this.]

Ans: πab

2-12. The parametric equations of the ellipsoid $\frac{x^2}{a^2} + \frac{y^2}{b^2} + \frac{z^2}{c^2} = 1$ are

$$x = a\sin\xi\cos\zeta \quad ; \quad y = b\sin\xi\sin\zeta \quad ; \quad z = c\cos\xi$$

Find the volume enclosed by the ellipsoid by utilising a three-dimensional equivalent of the procedure employed in the previous exercise.

Ans: $\frac{4}{3}\pi abc$

2.5 Grad V, Curl $\bar{F}$, Div $\bar{F}$ and $\nabla^2 V$ in Orthogonal Curvilinear Coordinates

Let $\Delta\bar{r}$ represent a small displacement at a point where u, v and w have continuous first derivatives. Then

$$\Delta\bar{r}.\nabla u \approx \Delta u \quad ; \quad \Delta\bar{r}.\nabla v \approx \Delta v \quad ; \quad \Delta\bar{r}.\nabla w \approx \Delta w$$

where Δu, Δv and Δw are the increments of u, v and w corresponding to the displacement $\Delta\bar{r}$.

It follows that

$$\frac{\Delta\bar{r}}{\Delta u}.\nabla u \approx 1 \quad ; \quad \frac{\Delta\bar{r}}{\Delta v}.\nabla v \approx 1 \quad ; \quad \frac{\Delta\bar{r}}{\Delta w}.\nabla w \approx 1$$

provided that $\Delta u \neq 0 \quad ; \quad \Delta v \neq 0 \quad ; \quad \Delta w \neq 0$

In particular, if $\Delta\bar{r}$ is directed along the u, v and w coordinate axes in turn and limits are taken as $\Delta\bar{r}\to 0$, the above approximations reduce to

$$\frac{\partial\bar{r}}{\partial u}.\nabla u = 1 \quad ; \quad \frac{\partial\bar{r}}{\partial v}.\nabla v = 1 \quad ; \quad \frac{\partial\bar{r}}{\partial w}.\nabla w = 1 \qquad (2.5\text{-}1)$$

ie $\quad h_1\hat{u} \cdot \nabla u = 1 \quad ; \quad h_2\hat{v} \cdot \nabla v = 1 \quad ; \quad h_3\hat{w} \cdot \nabla w = 1 \qquad (2.5\text{-}2)$

These results hold whether or not the coordinate system is orthogonal. Some consideration will show that, for the particular case of orthogonality, $\frac{\partial \bar{r}}{\partial u}$ and ∇u, $\frac{\partial \bar{r}}{\partial v}$ and ∇v, $\frac{\partial \bar{r}}{\partial w}$ and ∇w have the same directions (and senses) taken in pairs, whence it follows from (2.5-2) that

$$\nabla u = \frac{\hat{u}}{h_1} \quad ; \quad \nabla v = \frac{\hat{v}}{h_2} \quad ; \quad \nabla w = \frac{\hat{w}}{h_3} \qquad (2.5\text{-}3)$$

2.5a Gradient in orthogonal curvilinear coordinates

Let V be a scalar point function having continuous first derivatives throughout a region of space which includes the curvilinear volume element shown in Fig. 2.2. Then it follows that

$$V(Q) - V(P) = V(u_o+\Delta u, v_o+\Delta v, w_o+\Delta w) - V(u_o, v_o, w_o)$$

or

$$\begin{aligned} \Delta V = \; & V(u_o+\Delta u, v_o+\Delta v, w_o+\Delta w) - V(u_o, v_o+\Delta v, w_o+\Delta w) \\ & + V(u_o, v_o+\Delta v, w_o+\Delta w) - V(u_o, v_o, w_o+\Delta w) \\ & + V(u_o, v_o, w_o+\Delta w) - V(u_o, v_o, w_o) \end{aligned}$$

If we relate the component parts of this expression to motion along the u curve from S to Q, along the v curve from R to S and along the w curve from P to R, then we may replace the expression by

$$\Delta V = \left(\frac{\partial V}{\partial u}\right)_{\substack{u' \\ v_o+\Delta v \\ w_o+\Delta}} \Delta u + \left(\frac{\partial V}{\partial v}\right)_{\substack{u_o \\ v' \\ w_o+\Delta w}} \Delta v + \left(\frac{\partial V}{\partial w}\right)_{\substack{u_o \\ v_o \\ w'}} \Delta w \qquad (2.5\text{-}4)$$

where $u_o < u' < u_o + \Delta u \quad ; \quad v_o < v' < v_o + \Delta v \quad ; \quad w_o < w' < w_o + \Delta w$

The anlaysis to this point is virtually identical with that presented in Sec. 1.2 for rectangular Cartesian coordinates, although in the present instance the partial derivatives represent limits associated with non-rectilinear motion.

Suppose now that Q has been so chosen as to have the same y and z coordinates as P, and let $x(Q) - x(P) = \Delta x$. On dividing (2.5-4) by Δx and taking limits as $\Delta x \to 0$ we obtain

$$\left(\frac{\partial V}{\partial x}\right)_{-} = \left(\frac{\partial V}{\partial u}\frac{\partial u}{\partial x} + \frac{\partial V}{\partial v}\frac{\partial v}{\partial x} + \frac{\partial V}{\partial w}\frac{\partial w}{\partial x}\right)_{-}$$

whence, in general,

$$\left.\begin{aligned} \frac{\partial V}{\partial x} &= \frac{\partial V}{\partial u}\frac{\partial u}{\partial x} + \frac{\partial V}{\partial v}\frac{\partial v}{\partial x} + \frac{\partial V}{\partial w}\frac{\partial w}{\partial x} \\ \text{Similarly} \qquad \frac{\partial V}{\partial y} &= \frac{\partial V}{\partial u}\frac{\partial u}{\partial y} + \frac{\partial V}{\partial v}\frac{\partial v}{\partial y} + \frac{\partial V}{\partial w}\frac{\partial w}{\partial y} \\ \text{and} \qquad \frac{\partial V}{\partial z} &= \frac{\partial V}{\partial u}\frac{\partial u}{\partial z} + \frac{\partial V}{\partial v}\frac{\partial v}{\partial z} + \frac{\partial V}{\partial w}\frac{\partial w}{\partial z} \end{aligned}\right\} \qquad (2.5\text{-}5)$$

On multiplying these equations respectively by $\bar{i}$, $\bar{j}$ and $\bar{k}$, and adding, we find that

$$\text{grad } V = \frac{\partial V}{\partial u}\nabla u + \frac{\partial V}{\partial v}\nabla v + \frac{\partial V}{\partial w}\nabla w \qquad (2.5\text{-}6)$$

This is the general curvilinear expression for gradient.

For orthogonal coordinates it follows from (2.5-3) and (2.5-6) that

$$\text{grad } V = \frac{\hat{u}}{h_1}\frac{\partial V}{\partial u} + \frac{\hat{v}}{h_2}\frac{\partial V}{\partial v} + \frac{\hat{w}}{h_3}\frac{\partial V}{\partial w} \qquad (2.5\text{-}7)$$

2.5b Curl in orthogonal curvilinear coordinates

Since

$$\bar{F} = \hat{u}\, F_u + \hat{v}\, F_v + \hat{w}\, F_w$$

it follows from (1.16-1) and (1.16-3) that

$$\left.\begin{aligned} \text{curl } \bar{F} = {} & F_u \text{ curl } \hat{u} + \text{grad } F_u \times \hat{u} \\ & + F_v \text{ curl } \hat{v} + \text{grad } F_v \times \hat{v} \\ & + F_w \text{ curl } \hat{w} + \text{grad } F_w \times \hat{w} \end{aligned}\right\} \qquad (2.5\text{-}8)$$

To evaluate curl $\hat{u}$ for orthogonal coordinates we apply (2.5-3) to the identity curl grad u ≡ 0.

Since curl grad u = curl $\dfrac{\hat{u}}{h_1}$

$$\frac{1}{h_1}\text{ curl } \hat{u} + \text{grad } \frac{1}{h_1} \times \hat{u} = 0$$

or

$$\text{curl}\,\hat{u} = -h_1 \,\text{grad}\,\frac{1}{h_1} \times \hat{u}$$

But from (2.5-7)

$$\text{grad}\,\frac{1}{h_1} = -\frac{1}{h_1^2}\left\{\frac{\hat{u}}{h_1}\frac{\partial h_1}{\partial u} + \frac{\hat{v}}{h_2}\frac{\partial h_1}{\partial v} + \frac{\hat{w}}{h_3}\frac{\partial h_1}{\partial w}\right\}$$

whence

$$\left.\begin{aligned} &\text{curl}\,\hat{u} = \frac{\hat{v}}{h_3h_1}\frac{\partial h_1}{\partial w} - \frac{\hat{w}}{h_1h_2}\frac{\partial h_1}{\partial v} \\ \text{Similarly}\quad &\text{curl}\,\hat{v} = \frac{\hat{w}}{h_1h_2}\frac{\partial h_2}{\partial u} - \frac{\hat{u}}{h_2h_3}\frac{\partial h_2}{\partial w} \\ \text{and}\quad &\text{curl}\,\hat{w} = \frac{\hat{u}}{h_2h_3}\frac{\partial h_3}{\partial v} - \frac{\hat{v}}{h_3h_1}\frac{\partial h_3}{\partial u} \end{aligned}\right\} \tag{2.5-9}$$

On substituting (2.5-9) in (2.5-8), expanding grad F_u etc in accordance with (2.5-7) and collecting terms, we get

$$\left.\begin{aligned} \text{curl}\,\overline{F} = &\frac{\hat{u}}{h_2h_3}\left\{\frac{\partial}{\partial v}(h_3F_w) - \frac{\partial}{\partial w}(h_2F_v)\right\} \\ &+ \frac{\hat{v}}{h_3h_1}\left\{\frac{\partial}{\partial w}(h_1F_u) - \frac{\partial}{\partial u}(h_3F_w)\right\} \\ &+ \frac{\hat{w}}{h_1h_2}\left\{\frac{\partial}{\partial u}(h_2F_v) - \frac{\partial}{\partial v}(h_1F_u)\right\} \end{aligned}\right\} \tag{2.5-10}$$

This may be written in determinantal form:

$$\text{curl}\,\overline{F} = \begin{vmatrix} \dfrac{\hat{u}}{h_2h_3} & \dfrac{\hat{v}}{h_3h_1} & \dfrac{\hat{w}}{h_1h_2} \\ \dfrac{\partial}{\partial u} & \dfrac{\partial}{\partial v} & \dfrac{\partial}{\partial w} \\ h_1F_u & h_2F_v & h_3F_w \end{vmatrix} \tag{2.5-11}$$

2.5c Divergence in orthogonal curvilinear coordinates

From (1.16-2) and (1.16-4)

$$\left.\begin{aligned} \operatorname{div} \bar{F} &= \operatorname{div}(\hat{\bar{u}} F_u + \hat{\bar{v}} F_v + \hat{\bar{w}} F_w) \\ &= F_u \operatorname{div} \hat{\bar{u}} + \operatorname{grad} F_u \cdot \hat{\bar{u}} \\ &\quad + F_v \operatorname{div} \hat{\bar{v}} + \operatorname{grad} F_v \cdot \hat{\bar{v}} \\ &\quad + F_w \operatorname{div} \hat{\bar{w}} + \operatorname{grad} F_w \cdot \hat{\bar{w}} \end{aligned}\right\} \tag{2.5-12}$$

Since the coordinates are orthogonal

$$\hat{\bar{u}} = \hat{\bar{v}} \times \hat{\bar{w}}$$

hence from (1.16-7)

$$\operatorname{div} \hat{\bar{u}} = \hat{\bar{w}} \cdot \operatorname{curl} \hat{\bar{v}} - \hat{\bar{v}} \cdot \operatorname{curl} \hat{\bar{w}}$$

whence, by (2.5-9),

$$\operatorname{div} \hat{\bar{u}} = \frac{1}{h_1 h_2} \frac{\partial h_2}{\partial u} + \frac{1}{h_3 h_1} \frac{\partial h_3}{\partial u}$$

ie $\operatorname{div} \hat{\bar{u}} = \dfrac{1}{h_1 h_2 h_3} \dfrac{\partial}{\partial u} (h_2 h_3)$

Similarly $\operatorname{div} \hat{\bar{v}} = \dfrac{1}{h_1 h_2 h_3} \dfrac{\partial}{\partial v} (h_3 h_1)$ (2.5-13)

and $\operatorname{div} \hat{\bar{w}} = \dfrac{1}{h_1 h_2 h_3} \dfrac{\partial}{\partial w} (h_1 h_2)$

On substituting (2.5-13) in (2.5-12) and expanding grad F_u etc in accordance with (2.5-6) and collecting terms, we get

$$\operatorname{div} \bar{F} = \frac{1}{h_1 h_2 h_3} \left\{\frac{\partial}{\partial u} (h_2 h_3 F_u) + \frac{\partial}{\partial v} (h_3 h_1 F_v) + \frac{\partial}{\partial w} (h_1 h_2 F_w)\right\} \tag{2.5-14}$$

2.5d $\nabla^2 V$ in orthogonal curvilinear coordinates

Since $\nabla^2 V$ = div grad V it follows from (2.5-7) and (2.5-14) that

$$\nabla^2 V = \frac{1}{h_1 h_2 h_3}\left\{\frac{\partial}{\partial u}\left(\frac{h_2 h_3}{h_1}\frac{\partial V}{\partial u}\right) + \frac{\partial}{\partial v}\left(\frac{h_3 h_1}{h_2}\frac{\partial V}{\partial v}\right) + \frac{\partial}{\partial w}\left(\frac{h_1 h_2}{h_3}\frac{\partial V}{\partial w}\right)\right\} \tag{2.5-15}$$

2.6 Grad V, Curl $\overline{F}$, Div $\overline{F}$ and $\nabla^2 V$ in Cylindrical and Spherical Coordinates

These expressions are readily obtained by substitution of the appropriate values of the metrical coefficients in the formulae established above.

For cylindrical coordinates

$$u = \rho \quad ; \quad v = \Phi \quad ; \quad w = z$$

$$h_1 = 1 \quad ; \quad h_2 = \rho \quad ; \quad h_3 = 1$$

hence

$$\text{grad } V = \hat{\rho}\frac{\partial}{\partial \rho} + \hat{\Phi}\frac{1}{\rho}\frac{\partial}{\partial \Phi} + \hat{z}\frac{\partial}{\partial z} \tag{2.6-1}$$

$$\text{curl } \overline{F} = \hat{\rho}\left\{\frac{1}{\rho}\frac{\partial F_z}{\partial \Phi} - \frac{\partial F_\Phi}{\partial z}\right\} + \hat{\Phi}\left\{\frac{\partial F_\rho}{\partial z} - \frac{\partial F_z}{\partial \rho}\right\} + \hat{z}\left\{\frac{1}{\rho}\left(\frac{\partial}{\partial \rho}(\rho F_\Phi) - \frac{\partial F_\rho}{\partial \Phi}\right)\right\} \tag{2.6-2}$$

$$\text{div } \overline{F} = \frac{1}{\rho}\frac{\partial}{\partial \rho}(\rho F_\rho) + \frac{1}{\rho}\frac{\partial F_\Phi}{\partial \Phi} + \frac{\partial F_z}{\partial z} \tag{2.6-3}$$

$$\nabla^2 V = \frac{1}{\rho}\frac{\partial}{\partial \rho}\left(\rho\frac{\partial V}{\partial \rho}\right) + \frac{1}{\rho^2}\frac{\partial^2 V}{\partial \Phi^2} + \frac{\partial^2 V}{\partial z^2} \tag{2.6-4}$$

For spherical coordinates

$$u = r \quad ; \quad v = \theta \quad ; \quad w = \Phi$$

$$h_1 = 1 \quad ; \quad h_2 = r \quad ; \quad h_3 = r\sin\theta$$

hence

$$\text{grad } V = \hat{r}\frac{\partial V}{\partial r} + \hat{\theta}\frac{1}{r}\frac{\partial V}{\partial \theta} + \hat{\Phi}\frac{1}{r\sin\theta}\frac{\partial V}{\partial \Phi} \tag{2.6-5}$$

$$\left.\begin{aligned}\text{curl } \overline{F} &= \hat{r}\,\frac{1}{r \sin\theta}\left\{\frac{\partial}{\partial\theta}(F_\phi \sin\theta) - \frac{\partial F_\theta}{\partial\phi}\right\} \\ &\quad + \hat{\theta}\,\frac{1}{r}\left\{\frac{1}{\sin\theta}\frac{\partial F_r}{\partial\phi} - \frac{\partial}{\partial r}(rF_\phi)\right\} \\ &\quad + \hat{\phi}\,\frac{1}{r}\left\{\frac{\partial}{\partial r}(rF_\theta) - \frac{\partial F_r}{\partial\theta}\right\}\end{aligned}\right\} \qquad (2.6\text{-}6)$$

$$\text{div } \overline{F} = \frac{1}{r^2}\frac{\partial}{\partial r}(r^2 F_r) + \frac{1}{r \sin\theta}\frac{\partial}{\partial\theta}(F_\theta \sin\theta) + \frac{1}{r\sin\theta}\frac{\partial F_\phi}{\partial\phi} \qquad (2.6\text{-}7)$$

$$\nabla^2 V = \frac{1}{r^2}\frac{\partial}{\partial r}\left(r^2\frac{\partial V}{\partial r}\right) + \frac{1}{r^2\sin\theta}\frac{\partial}{\partial\theta}\left(\frac{\partial V}{\partial\theta}\sin\theta\right) + \frac{1}{r^2\sin^2\theta}\frac{\partial^2 V}{\partial\phi^2} \qquad (2.6\text{-}8)$$

2.7 Derivation of Grad V, Div $\overline{F}$ and Curl $\overline{F}$ in Cylindrical Coordinates by Transformation of Axes

Although the substitution of appropriate metrical coefficients in the general curvilinear expressions leads rapidly to the formulation of grad, div and curl in specific coordinate systems, there are several other methods whereby the formulae may be derived and which provide a deeper insight into the relationships existing within a given system. Two such methods will now be considered.

(1) Reference to the equations comprising (2.5-5) of which the first is

$$\frac{\partial V}{\partial x} = \frac{\partial V}{\partial u}\frac{\partial u}{\partial x} + \frac{\partial V}{\partial v}\frac{\partial v}{\partial x} + \frac{\partial V}{\partial w}\frac{\partial w}{\partial x}$$

reveals that when the analytical relationships between u, v, w and x, y, z are known, it is possible to express the partial derivatives $\frac{\partial V}{\partial x}$, $\frac{\partial V}{\partial y}$ and $\frac{\partial V}{\partial z}$ wholly in terms of curvilinear coordinates. If, in addition, the unit vectors $\overline{i}$, $\overline{j}$, $\overline{k}$ can be expressed in terms of $\hat{u}$, $\hat{v}$, $\hat{w}$, it is seen that the curvilinear form of grad V may be developed. Further, since F_x, F_y and F_z may be substituted for V, the expression of each of these rectangular components as the sum of curvilinear components enables us to derive the curvilinear forms of div $\overline{F}$ and curl $\overline{F}$ as well.

In the case of a cylindrical system

$$\rho = (x^2+y^2)^{\frac{1}{2}} \quad ; \quad \phi = \tan^{-1}\frac{y}{x} \quad ; \quad z = z$$

hence

$$\left.\begin{aligned} \frac{\partial\rho}{\partial x} &= \frac{x}{\rho} = \cos\Phi \quad ; \quad \frac{\partial\rho}{\partial y} = \frac{y}{\rho} = \sin\Phi \quad ; \quad \frac{\partial\rho}{\partial z} = 0 \\ \frac{\partial\Phi}{\partial x} &= -\frac{1}{\rho}\sin\Phi \quad ; \quad \frac{\partial\Phi}{\partial y} = \frac{1}{\rho}\cos\Phi \quad ; \quad \frac{\partial\Phi}{\partial z} = 0 \end{aligned}\right\} \quad (2.7\text{-}1)$$

Further, from Fig. 2.3

$$\left.\begin{aligned} F_x &= F_\rho \cos\Phi - F_\Phi \sin\Phi \\ F_y &= F_\rho \sin\Phi + F_\Phi \cos\Phi \end{aligned}\right\} \quad (2.7\text{-}2)$$

Then from (2.5-5)

$$\left.\begin{aligned} \frac{\partial V}{\partial x} &= \frac{\partial V}{\partial\rho}\cos\Phi - \frac{\partial V}{\partial\Phi}\frac{1}{\rho}\sin\Phi \\ \frac{\partial V}{\partial y} &= \frac{\partial V}{\partial\rho}\sin\Phi + \frac{\partial V}{\partial\Phi}\frac{1}{\rho}\cos\Phi \\ \frac{\partial V}{\partial z} &= \frac{\partial V}{\partial z} \end{aligned}\right\} \quad (2.7\text{-}3)$$

hence

$$\operatorname{grad} V = \frac{\partial V}{\partial\rho}(\bar{i}\cos\Phi + \bar{j}\sin\Phi) + \frac{1}{\rho}\frac{\partial V}{\partial\Phi}(-\bar{i}\sin\Phi + \bar{j}\cos\Phi) + \bar{k}\frac{\partial V}{\partial z}$$

whence, from (2.2-5),

$$\operatorname{grad} V = \hat{\rho}\frac{\partial V}{\partial\rho} + \hat{\Phi}\frac{1}{\rho}\frac{\partial V}{\partial\Phi} + \hat{z}\frac{\partial V}{\partial z}$$

Upon substituting F_x for V in the first line of (2.7-3) we get

$$\frac{\partial F_x}{\partial x} = \frac{\partial F_x}{\partial\rho}\cos\Phi - \frac{\partial F_x}{\partial\Phi}\frac{1}{\rho}\sin\Phi$$

whence, from (2.7-2),

$$\frac{\partial F_x}{\partial x} = \left(\frac{\partial F_\rho}{\partial \rho}\cos\Phi - \frac{\partial F_\Phi}{\partial \rho}\sin\Phi\right)\cos\Phi$$

$$- \left(\frac{\partial F_\rho}{\partial \Phi}\cos\Phi - F_\rho \sin\Phi - \frac{\partial F_\Phi}{\partial \Phi}\sin\Phi - F_\Phi\cos\Phi\right)\frac{1}{\rho}\sin\Phi$$

Similarly,

$$\frac{\partial F_y}{\partial y} = \left(\frac{\partial F_\rho}{\partial \rho}\sin\Phi + \frac{\partial F_\Phi}{\partial \rho}\cos\Phi\right)\sin\Phi$$

$$+ \left(\frac{\partial F_\rho}{\partial \Phi}\sin\Phi + F_\rho \cos\Phi + \frac{\partial F_\Phi}{\partial \Phi}\cos\Phi - F_\Phi\sin\Phi\right)\frac{1}{\rho}\cos\Phi$$

so that

$$\text{div}\,\overline{F} = \frac{1}{\rho}\frac{\partial}{\partial \rho}(\rho F_\rho) + \frac{1}{\rho}\frac{\partial F_\Phi}{\partial \Phi} + \frac{\partial F_z}{\partial z}$$

In like manner it may be shown that

$$\frac{\partial F_z}{\partial y} - \frac{\partial F_y}{\partial z} = \left\{\left(\frac{\partial F_z}{\partial \rho} - \frac{\partial F_\rho}{\partial z}\right)\sin\Phi + \left(\frac{1}{\rho}\frac{\partial F_z}{\partial \Phi} - \frac{\partial F_\Phi}{\partial z}\right)\cos\Phi\right\}$$

Similarly

$$\frac{\partial F_x}{\partial z} - \frac{\partial F_z}{\partial x} = \left\{\left(\frac{1}{\rho}\frac{\partial F_z}{\partial \Phi} - \frac{\partial F_\Phi}{\partial z}\right)\sin\Phi + \left(\frac{\partial F_\rho}{\partial z} - \frac{\partial F_z}{\partial \rho}\right)\cos\Phi\right\}$$

$$\frac{\partial F_y}{\partial x} - \frac{\partial F_x}{\partial y} = \left\{\frac{\partial F_\Phi}{\partial \rho} + \frac{F_\Phi}{\rho} - \frac{1}{\rho}\frac{\partial F_\rho}{\partial \Phi}\right\}$$

whence

$$\text{curl}\,\overline{F} = (\overline{i}\cos\Phi + \overline{j}\sin\Phi)\left(\frac{1}{\rho}\frac{\partial F_z}{\partial \Phi} - \frac{\partial F_\Phi}{\partial z}\right)$$

$$+ (-\overline{i}\sin\Phi + \overline{j}\cos\Phi)\left(\frac{\partial F_\rho}{\partial z} - \frac{\partial F_z}{\partial \rho}\right)$$

$$+ \overline{k}\left(\frac{\partial F_\Phi}{\partial \rho} + \frac{F_\Phi}{\rho} - \frac{1}{\rho}\frac{\partial F_\rho}{\partial \Phi}\right)$$

or, from (2.2-5),

$$\text{curl}\ \overline{F} = \hat{\rho}\left(\frac{1}{\rho}\frac{\partial F_z}{\partial \Phi} - \frac{\partial F_\Phi}{\partial z}\right) + \hat{\Phi}\left(\frac{\partial F_\rho}{\partial z} - \frac{\partial F_z}{\partial \rho}\right) + \hat{z}\,\frac{1}{\rho}\left(\frac{\partial}{\partial \rho}(\rho F_\Phi) - \frac{\partial F_\rho}{\partial \Phi}\right)$$

(2) In view of the invariance of grad, div and curl with respect to choice of rectangular axes, we may write

$$\left.\begin{aligned} \text{grad}\ V &= \overline{i}'\frac{\partial V}{\partial x'} + \overline{j}'\frac{\partial V}{\partial y'} + \overline{k}'\frac{\partial V}{\partial z'} \\ \text{div}\ \overline{F} &= \frac{\partial F_{x'}}{\partial x'} + \frac{\partial F_{y'}}{\partial y'} + \frac{\partial F_{z'}}{\partial z'} \\ \text{curl}\ \overline{F} &= \overline{i}'\left(\frac{\partial F_{z'}}{\partial y'} - \frac{\partial F_{y'}}{\partial z'}\right) + \overline{j}'\left(\frac{\partial F_{x'}}{\partial z'} - \frac{\partial F_{z'}}{\partial x'}\right) + \overline{k}'\left(\frac{\partial F_{y'}}{\partial x'} - \frac{\partial F_{x'}}{\partial y'}\right) \end{aligned}\right\} \quad (2.7\text{-}4)$$

where $\overline{i}'$, $\overline{j}'$, $\overline{k}'$ and x', y', z' refer to a right-handed system of <u>fixed</u> rectangular axes which coincide with $\hat{\rho}$, $\hat{\Phi}$, $\hat{z}$ at the point where the transformation is to be effected.

It will be seen from Fig. 2.3 that if the systems are coincident at P, then movement in the $\overline{i}'$ or $\overline{k}'$ directions at P leaves the orientation of $\hat{\rho}$, $\hat{\Phi}$ and $\hat{z}$ unaltered; in addition, $\Delta x' = \Delta\rho$ and $\Delta z' = \Delta z$.

Hence

$$\left.\begin{aligned} &\frac{\partial V}{\partial x'} = \frac{\partial V}{\partial \rho}\ ;\quad \frac{\partial F_{x'}}{\partial x'} = \frac{\partial F_\rho}{\partial \rho}\ ;\quad \frac{\partial F_{y'}}{\partial x'} = \frac{\partial F_\Phi}{\partial \rho}\ ;\quad \frac{\partial F_{z'}}{\partial x'} = \frac{\partial F_z}{\partial \rho} \\ &\frac{\partial V}{\partial z'} = \frac{\partial V}{\partial z}\ ;\quad \frac{\partial F_{x'}}{\partial z'} = \frac{\partial F_\rho}{\partial z}\ ;\quad \frac{\partial F_{y'}}{\partial z'} = \frac{\partial F_\Phi}{\partial z}\ ;\quad \frac{\partial F_{z'}}{\partial z'} = \frac{\partial F_z}{\partial z} \end{aligned}\right\} \quad (2.7\text{-}5)$$

Fig. 2.6 is drawn in a plane through P normal to the z axis. The cylindrical coordinates of P are ρ_1, Φ_1, z_1 and those of Q, the typical point on the y' axis through P, are ρ, Φ, z_1. $PQ = y'$. Because both ρ and Φ vary with y'

$$\frac{\partial V}{\partial y'} = \frac{\partial V}{\partial \rho}\frac{\partial \rho}{\partial y'} + \frac{\partial V}{\partial \Phi}\frac{\partial \Phi}{\partial y'}$$

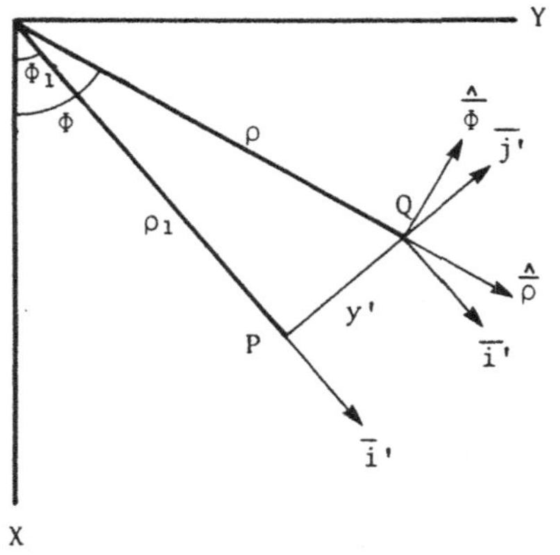

Fig. 2.6

But $\rho^2 = \rho_1^2 + y'^2$, whence $\dfrac{\partial\rho}{\partial y'} = \dfrac{y'}{\rho} = \sin(\Phi-\Phi_1)$

and $\Phi - \Phi_1 = \tan^{-1}\dfrac{y'}{\rho_1}$, whence $\dfrac{\partial\Phi}{\partial y'} = \dfrac{\cos^2(\Phi-\Phi_1)}{\rho_1}$

so that for any point on the y' axis through P

$$\frac{\partial V}{\partial y'} = \frac{\partial V}{\partial \rho}\sin(\Phi-\Phi_1) + \frac{\partial V}{\partial \Phi}\frac{\cos^2(\Phi-\Phi_1)}{\rho_1} \tag{2.7-6}$$

Since $\Phi - \Phi_1 = 0$ at P it follows[12] that

$$\left(\frac{\partial V}{\partial y'}\right)_P = \frac{1}{\rho_1}\left(\frac{\partial V}{\partial \Phi}\right)_P \tag{2.7-7}$$

Further,

$$\left.\begin{aligned} F_{x'} &= F_\rho \cos(\Phi-\Phi_1) - F_\Phi \sin(\Phi-\Phi_1) \\ F_{y'} &= F_\rho \sin(\Phi-\Phi_1) + F_\Phi \cos(\Phi-\Phi_1) \end{aligned}\right\} \tag{2.7-8}$$

12. The reader who considers that unnecessary effort has been expended in establishing a result which is intuitively obvious should exercise caution when approaching higher-order derivatives. (See equation (2.10-4).)

On substituting $F_{x'}$ and $F_{y'}$ for V in (2.7-6) and putting $(\Phi-\Phi_1) = 0$ after differentiation, we find that

$$\left(\frac{\partial F_{x'}}{\partial y'}\right)_P = \frac{1}{\rho_1}\left(\frac{\partial F_\rho}{\partial \Phi} - F_\Phi\right)_P \quad ; \quad \left(\frac{\partial F_{y'}}{\partial y'}\right)_P = \frac{1}{\rho_1}\left(F_\rho + \frac{\partial F_\Phi}{\partial \Phi}\right)_P \quad ; \quad \left(\frac{\partial F_{z'}}{\partial y'}\right)_P = \frac{1}{\rho_1}\left(\frac{\partial F_z}{\partial \Phi}\right)_P \tag{2.7-9}$$

Equations (2.7-7) and (2.7-9) with subscripts deleted, together with (2.7-5), furnish the curvilinear equivalents of all rectangular partial derivatives. Substitution in (2.7-4), with $\bar{i}', \bar{j}', \bar{k}'$ replaced by $\hat{\rho}, \hat{\Phi}, \hat{z}$, yields the required cylindrical forms of grad, div and curl.

2.8 Derivation of Grad V, Div $\bar{F}$ and Curl $\bar{F}$ in Spherical Coordinates by Transformation of Axes

(1) The background relationships required for the first type of transformation are derived from (2.3-1) and (2.3-2).

Since

$$x = r \sin\theta \cos\Phi \quad ; \quad y = r \sin\theta \sin\Phi \quad ; \quad z = r\cos\theta \tag{2.3-1}$$

$$r = (x^2+y^2+z^2)^{\frac{1}{2}} \quad ; \quad \theta = \cos^{-1}\frac{z}{(x^2+y^2+z^2)^{\frac{1}{2}}} \quad ; \quad \Phi = \tan^{-1}\frac{y}{x} \tag{2.3-2}$$

It follows that

$$\left.\begin{aligned}
&\frac{\partial r}{\partial x} = \sin\theta\cos\Phi \quad ; \quad \frac{\partial r}{\partial y} = \sin\theta\sin\Phi \quad ; \quad \frac{\partial r}{\partial z} = \cos\theta \\
&\frac{\partial \theta}{\partial x} = \frac{1}{r}\cos\theta\cos\Phi \quad ; \quad \frac{\partial \theta}{\partial y} = \frac{1}{r}\cos\theta\sin\Phi \quad ; \quad \frac{\partial \theta}{\partial z} = -\frac{1}{r}\sin\theta \\
&\frac{\partial \Phi}{\partial x} = -\frac{1}{r}\frac{\sin\Phi}{\sin\theta} \quad ; \quad \frac{\partial \Phi}{\partial y} = \frac{1}{r}\frac{\cos\Phi}{\sin\theta} \quad ; \quad \frac{\partial \Phi}{\partial z} = 0
\end{aligned}\right\} \tag{2.8-1}$$

From Fig. 2.4

$$\left.\begin{aligned}
F_x &= F_r \sin\theta\cos\Phi + F_\theta \cos\theta\cos\Phi - F_\Phi \sin\Phi \\
F_y &= F_r \sin\theta\sin\Phi + F_\theta \cos\theta\sin\Phi + F_\Phi \cos\Phi \\
F_z &= F_r \cos\theta - F_\theta \sin\theta
\end{aligned}\right\} \tag{2.8-2}$$

The relationships between $\hat{r}, \hat{\theta}, \hat{\Phi}$ and $\bar{i}, \bar{j}, \bar{k}$ have already been expressed in (2.3-5).

The detailed treatment will not be pursued here. Is is formally analogous to that developed for cylindrical coordinates and is straightforward, but extremely tedious.

(2) The second type of transformation, involving a set of fixed rectangular axes $\bar{i}'$, $\bar{j}'$, $\bar{k}'$ coincident with $\hat{r}$, $\hat{\theta}$, $\hat{\phi}$ at P, is carried out as follows.

It will be seen from Fig. 2.4 that the $\hat{r}$, $\hat{\theta}$ and $\hat{\phi}$ axes maintain their orientation in space with movement in the $\bar{i}'$ direction at P, so that the following relationships may be written down immediately.

$$\frac{\partial V}{\partial x'} = \frac{\partial V}{\partial r} \quad ; \quad \frac{\partial F_{x'}}{\partial x'} = \frac{\partial F_r}{\partial r} \quad ; \quad \frac{\partial F_{y'}}{\partial x'} = \frac{\partial F_\theta}{\partial r} \quad ; \quad \frac{\partial F_{z'}}{\partial x'} = \frac{\partial F_\phi}{\partial r} \qquad (2.8\text{-}3)$$

Movement in the $\bar{j}'$ direction at P, ie along the y' axis, involves variation of both r and θ. This is shown in Fig. 2.7 which is drawn in the plane $\phi = \phi_1$, the coordinates of P being r_1, θ_1, ϕ_1 and those of Q - the typical point on the y' axis through P - being r, θ, ϕ_1. PQ = y'.

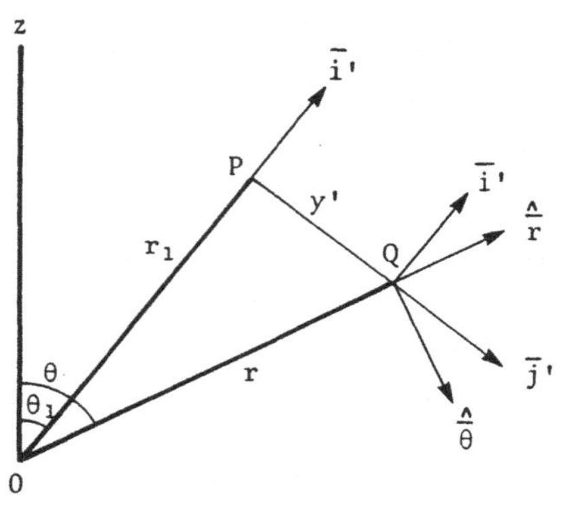

Fig. 2.7

Proceeding as in the case of cylindrical coordinates we may show that

$$\frac{\partial V}{\partial y'} = \frac{\partial V}{\partial r} \sin(\theta - \theta_1) + \frac{\partial V}{\partial \theta} \frac{\cos^2(\theta - \theta_1)}{r_1}$$

whence

$$\left(\frac{\partial V}{\partial y'}\right)_P = \frac{1}{r_1}\left(\frac{\partial V}{\partial \theta}\right)_P \tag{2.8-4}$$

Also

$$F_{x'} = F_r \cos(\theta-\theta_1) - F_\theta \sin(\theta-\theta_1)$$

$$F_{y'} = F_r \sin(\theta-\theta_1) + F_\theta \cos(\theta-\theta_1)$$

$$F_{z'} = F_\Phi$$

whence

$$\left(\frac{\partial F_{x'}}{\partial y'}\right)_P = \frac{1}{r_1}\left(\frac{\partial F_r}{\partial \theta} - F_\theta\right)_P \; ; \; \left(\frac{\partial F_{y'}}{\partial y'}\right)_P = \frac{1}{r_1}\left(F_r + \frac{\partial F_\theta}{\partial \theta}\right)_P \; ; \; \left(\frac{\partial F_{z'}}{\partial y'}\right)_P = \frac{1}{r_1}\left(\frac{\partial F_\Phi}{\partial \theta}\right)_P \tag{2.8-5}$$

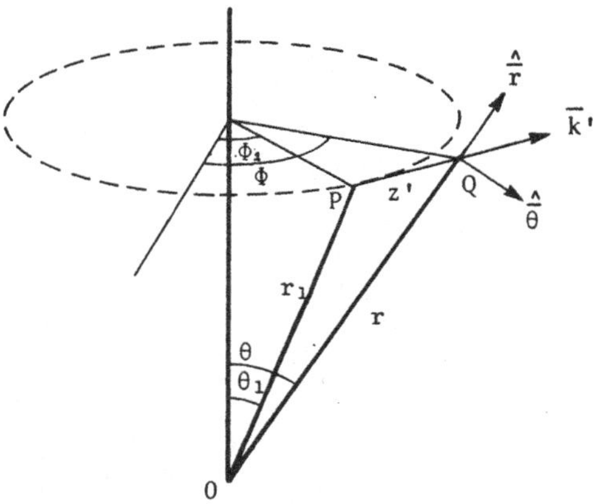

Fig. 2.8

It will be seen from Fig. 2.8 that movement in $\bar{k}'$ direction at P, ie along the z' axis, is accompanied by variation of each curvilinear coordinate, hence we must write

$$\frac{\partial V}{\partial z'} = \frac{\partial V}{\partial r}\frac{\partial r}{\partial z'} + \frac{\partial V}{\partial \theta}\frac{\partial \theta}{\partial z'} + \frac{\partial V}{\partial \Phi}\frac{\partial \Phi}{\partial z'}$$

On taking the coordinates of Q as r, θ, Φ and putting PQ = z' it follows from the trigonometry of the figure that

$$\frac{\partial r}{\partial z'} = \frac{z'}{r} \quad ; \quad \frac{\partial \theta}{\partial z'} = \frac{z' r_1 \cos \theta_1}{r^3 \sin \theta} \quad ; \quad \frac{\partial \Phi}{\partial z'} = \frac{\cos^2(\Phi - \Phi_1)}{r_1 \sin \theta_1}$$

whence

$$\left(\frac{\partial V}{\partial z'}\right)_P = \frac{1}{r_1 \sin \theta_1} \left(\frac{\partial V}{\partial \Phi}\right)_P \qquad (2.8\text{-}6)$$

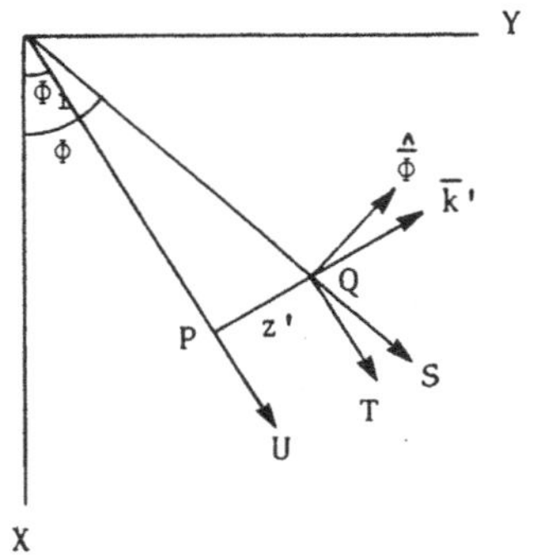

Fig. 2.9

Fig. 2.9 is drawn in the plane through P normal to the z axis. It contains the unit vectors $\hat{\Phi}$ and $\bar{k}'$ but no other. The remaining arrows of the figure show the directions of the vector projections of $\hat{r}$, $\bar{i}'$, $\hat{\theta}$ and $\bar{j}'$ in this plane.

To derive curvilinear expressions for $F_{x'}$, $F_{y'}$ and $F_{z'}$ at Q we may resolve F_r and F_θ into a vertical component (perpendicular to the plane of the figure) and a horizontal component along QS. The latter component together with F_Φ may be replaced by horizontal components in the $\bar{k}'$ and QT directions, where QT is parallel to PU.

The $\bar{i}'$ vector at P makes an angle $\frac{\pi}{2} - \theta_1$ with $\overrightarrow{PU}$, hence this is the angle between $\overrightarrow{QT}$ and $\bar{i}'$. The component along QT, together with the vertical component at Q, may be resolved accordingly along $\bar{i}'$ and $\bar{j}'$. We find that

$$F_{x'} = (F_r \sin\theta + F_\theta \cos\theta)\cos(\Phi-\Phi_1)\sin\theta_1 - F_\Phi \sin(\Phi-\Phi_1)\sin\theta_1 + (F_r\cos\theta - F_\theta\sin\theta)\cos\theta_1$$

$$F_{y'} = (F_r \sin\theta + F_\theta \cos\theta)\cos(\Phi-\Phi_1)\cos\theta_1 - F_\Phi \sin(\Phi-\Phi_1)\cos\theta_1 - (F_r\cos\theta - F_\theta\sin\theta)\sin\theta_1$$

$$F_{z'} = (F_r \sin\theta + F_\theta \cos\theta)\sin(\Phi-\Phi_1) + F_\Phi \cos(\Phi-\Phi_1)$$

Substitution of these spherical forms of $F_{x'}$, $F_{y'}$ and $F_{z'}$ for V in (2.8-6) yields

$$\left.\begin{aligned}
\left(\frac{\partial F_{x'}}{\partial z'}\right)_P &= \left\{\frac{1}{r\sin\theta}\left(\frac{\partial F_r}{\partial\Phi} - F_\Phi\sin\theta\right)\right\}_P \\
\left(\frac{\partial F_{y'}}{\partial z'}\right)_P &= \left\{\frac{1}{r\sin\theta}\left(\frac{\partial F_\theta}{\partial\Phi} - F_\Phi\cos\theta\right)\right\}_P \\
\left(\frac{\partial F_{z'}}{\partial z'}\right)_P &= \left\{\frac{1}{r\sin\theta}\left(\frac{\partial F_\Phi}{\partial\Phi} + F_r\sin\theta + F_\theta\cos\theta\right)\right\}_P
\end{aligned}\right\} \quad (2.8\text{-}7)$$

The required spherical forms of grad, div and curl are then obtained by deleting the subscripts and substituting (2.8-3) to (2.8-7) in (2.7-4) with $\bar{i}'$, $\bar{j}'$, $\bar{k}'$ replaced by $\hat{\bar{r}}$, $\hat{\bar{\theta}}$, $\hat{\bar{\Phi}}$.

EXERCISES

2-13. If u, v, w are general curvilinear coordinates, show that for all points at which the component factors are defined

$$\frac{\partial\bar{r}}{\partial u}\cdot\nabla v = \frac{\partial\bar{r}}{\partial u}\cdot\nabla w = \frac{\partial\bar{r}}{\partial v}\cdot\nabla u = \frac{\partial\bar{r}}{\partial v}\cdot\nabla w = \frac{\partial\bar{r}}{\partial w}\cdot\nabla u = \frac{\partial\bar{r}}{\partial w}\cdot\nabla v = 0$$

2-14. Derive $\nabla^2 V = \frac{\partial^2 V}{\partial x^2} + \frac{\partial^2 V}{\partial y^2} + \frac{\partial^2 V}{\partial z^2}$ in cylindrical coordinates by substituting (2.7-1) in (2.2-5) with $u = \rho$, $v = \Phi$, $w = z$, and by differentiating a second time with respect to x, y and z.

2-15. By substituting F_x, F_y and F_z for V in (2.5-5), show that in general curvilinear coordinates

$$\text{div}\,\bar{F} = \nabla u.\frac{\partial\bar{F}}{\partial u} + \nabla v.\frac{\partial\bar{F}}{\partial v} + \nabla w.\frac{\partial\bar{F}}{\partial w}$$

$$\text{curl}\,\bar{F} = \nabla u\times\frac{\partial\bar{F}}{\partial u} + \nabla v\times\frac{\partial\bar{F}}{\partial v} + \nabla w\times\frac{\partial\bar{F}}{\partial w}$$

2-16. It follows from (2.5-6) that, in general curvilinear coordinates,

$$\nabla V = \nabla u \frac{\partial V}{\partial u} + \nabla v \frac{\partial V}{\partial v} + \nabla w \frac{\partial V}{\partial w}$$

where $\nabla V = \bar{i}\frac{\partial V}{\partial x} + \bar{j}\frac{\partial V}{\partial y} + \bar{k}\frac{\partial V}{\partial z} \qquad (V = V(u,v,w))$

hence, for operation upon a scalar field,

$$\nabla u \frac{\partial}{\partial u} + \nabla v \frac{\partial}{\partial v} + \nabla w \frac{\partial}{\partial w} \equiv \bar{i}\frac{\partial}{\partial x} + \bar{j}\frac{\partial}{\partial y} + \bar{k}\frac{\partial}{\partial z}$$

Show from the results of the previous exercise that this continues to hold for operations upon a vector field.

2-17. Use the results of Ex.2-15. to show that, in the case of orthogonal coordinates, div $\bar{F}$ and curl $\bar{F}$ may be expressed as three-line expansions whose first lines are respectively

$$\frac{1}{h_1}\frac{\partial F_u}{\partial u} + \frac{\hat{\bar{u}}}{h_1} \cdot \left\{\frac{\partial \hat{\bar{u}}}{\partial u} F_u + \frac{\partial \hat{\bar{v}}}{\partial u} F_v + \frac{\partial \hat{\bar{w}}}{\partial u} F_w\right\}$$

and

$$\frac{\hat{\bar{u}}}{h_1} \times \left\{\frac{\partial \hat{\bar{u}}}{\partial u} F_u + \frac{\partial \hat{\bar{v}}}{\partial u} F_v + \frac{\partial \hat{\bar{w}}}{\partial u} F_w\right\} + \frac{\hat{\bar{w}}}{h_1}\frac{\partial F_v}{\partial u} - \frac{\hat{\bar{v}}}{h_1}\frac{\partial F_w}{\partial u}$$

Determine the values of the partial derivatives of the unit vectors for cylindrical and spherical coordinates by an appeal to appropriate diagrams or analytical functions (eg (2.3-5)), and so express divergence and curl in these coordinates.

Ans: In cylindrical coordinates $u = \rho$, $v = \Phi$, $w = z$. All partial derivatives are zero except

$$\frac{\partial \hat{\bar{\rho}}}{\partial \Phi} = \hat{\bar{\Phi}} \quad ; \quad \frac{\partial \hat{\bar{\Phi}}}{\partial \Phi} = -\hat{\bar{\rho}}$$

In spherical coordinates $u = r$, $v = \theta$, $w = \Phi$. All partial derivatives are zero except

$$\frac{\partial \hat{\bar{r}}}{\partial \theta} = \hat{\bar{\theta}} \quad ; \quad \frac{\partial \hat{\bar{\theta}}}{\partial \theta} = -\hat{\bar{r}}$$

$$\frac{\partial \hat{\bar{r}}}{\partial \Phi} = \hat{\bar{\Phi}} \sin\theta \quad ; \quad \frac{\partial \hat{\bar{\theta}}}{\partial \Phi} = \hat{\bar{\Phi}} \cos\theta \quad ; \quad \frac{\partial \hat{\bar{\Phi}}}{\partial \Phi} = -(\hat{\bar{r}} \sin\theta + \hat{\bar{\theta}} \cos\theta)$$

2-18. Substitution in (2.6-4) reveals that, for cylindrical coordinates and finite values of ρ,

$$\nabla^2\Phi = \nabla^2 z = \nabla^2 \ln \rho = 0$$

It follows that grad Φ, grad z and grad ln ρ may be expressed at points beyond the z axis as the curls of certain vector point functions.

Determine some of these functions by expressing the gradients in cylindrical coordinates, and by making an appropriate choice of F_ρ, F_Φ and F_z in (2.6-2).

Ans:
$$\text{grad } \Phi = \text{curl } \hat{\rho}\, z/\rho = - \text{curl } \hat{z} \ln \rho$$
$$\text{grad } z = \text{curl } \hat{\Phi}\, \rho/2 = - \text{curl } \hat{\rho}\, \rho\Phi$$
$$\text{grad } \ln \rho = \text{curl } \hat{z}\, \Phi = - \text{curl } \hat{\Phi}\, z/\rho$$

2-19. Given that $\overline{G} = (\overline{k}\times\overline{r})/r^3$ where $\overline{k}$ is the unit vector in the positive z direction, $\overline{r}$ is the position vector from the origin of coordinates and $r = |\overline{r}|$, express $\overline{G}$ in cylindrical coordinates and determine two expressions for $\overline{F}$ which satisfy the relationship $\overline{G} = \text{curl } \overline{F}$. Compare the result of Ex.1-48, p. 64.

Ans: $$\overline{G} = \hat{\Phi} \frac{\rho}{(\rho^2+z^2)^{3/2}} \quad ; \quad \overline{F} = \frac{\hat{\rho} z}{\rho(\rho^2+z^2)^{\frac{1}{2}}} \quad \text{or} \quad \overline{F} = \frac{\hat{z}}{(\rho^2+z^2)^{\frac{1}{2}}} = \frac{\overline{k}}{r}$$

2-20. It follows from the expression for $\nabla^2 V$ in spherical coordinates (2.6-8) that div grad $\frac{1}{r}$ is zero at all points other than the origin, where it is undefined. ($\nabla^2 \frac{1}{r}$ is also undefined along the axis, $\sin \theta = 0$, but this axis may be chosen arbitrarily for a given origin.) Express $\text{grad}\frac{1}{r} = -\frac{\hat{r}}{r^2}$ as curl $\overline{F}$, where $\overline{F}$ has a $\hat{\Phi}$ component only, and by transforming this into rectangular coordinates derive an alternative solution of Ex.1-45., p. 64. Notice the greater ease of interpretion of the polar expression.

Ans: $$\overline{F} = \hat{\Phi} \frac{\cot \theta}{r} = -\overline{i} \frac{zy}{(x^2+y^2)r} + \overline{j} \frac{zx}{(x^2+y^2)r} + \overline{k}\, 0$$ beyond the z axis. ($\sin \theta = 0$)

2-21. If

$$\overline{a} = \frac{\hat{v} \times \hat{w}}{[\hat{u}\ \hat{v}\ \hat{w}]} \qquad \overline{b} = \frac{\hat{w} \times \hat{u}}{[\hat{u}\ \hat{v}\ \hat{w}]} \qquad \overline{c} = \frac{\hat{u} \times \hat{v}}{[\hat{u}\ \hat{v}\ \hat{w}]}$$

show that

$$\bar{a}.d\bar{r} = h_1 du \qquad \bar{b}.d\bar{r} = h_2 dv \qquad \bar{c}.d\bar{r} = h_3 dw$$

and hence derive the general curvilinear forms of grad, div and curl as shown below.

$$\text{grad } V = \frac{\bar{a}}{h_1}\frac{\partial V}{\partial u} + \frac{\bar{b}}{h_2}\frac{\partial V}{\partial v} + \frac{\bar{c}}{h_3}\frac{\partial V}{\partial w}$$

$$\text{div } \bar{F} = \frac{\bar{a}}{h_1}.\frac{\partial \bar{F}}{\partial u} + \frac{\bar{b}}{h_2}.\frac{\partial \bar{F}}{\partial v} + \frac{\bar{c}}{h_3}.\frac{\partial \bar{F}}{\partial w}$$

$$\text{curl } \bar{F} = \frac{\bar{a}}{h_1}\times\frac{\partial \bar{F}}{\partial u} + \frac{\bar{b}}{h_2}\times\frac{\partial \bar{F}}{\partial v} + \frac{\bar{c}}{h_3}\times\frac{\partial \bar{F}}{\partial w}$$

2-22. Derive the general relationship

$$\frac{\partial}{\partial u}(h_2\hat{v}) = \frac{\partial}{\partial v}(h_1\hat{u})$$

by applying the identity $\oint d\bar{r} \equiv \bar{0}$ to the closed curve lying within a w coordinate surface and bounded by the coordinate lines u, u + du, v, v + dv.

Note the corresponding equalities

$$\frac{\partial}{\partial v}(h_3\hat{w}) = \frac{\partial}{\partial w}(h_2\hat{v}) \quad ; \quad \frac{\partial}{\partial w}(h_1\hat{u}) = \frac{\partial}{\partial u}(h_3\hat{w})$$

2-23. For orthogonal curvilinear coordinates in Ex.2-21., $\bar{a} = \hat{u}$, $\bar{b} = \hat{v}$, $\bar{c} = \hat{w}$. Transform the resulting expressions for div $\bar{F}$ and curl $\bar{F}$, into (2.5-14) and (2.5-10) respectively.

[Hint: Make use of the relationships derived by expanding the equalities in Ex.2-22. and forming scalar products with $\hat{u}$, $\hat{v}$ or $\hat{w}$ as required.]

2-24. By writing $(\overline{F}.\nabla)$ and $\overline{G}$ in spherical coordinates show that $(\overline{F}.\nabla)\overline{G}$

$$= \hat{r}\left\{F_r \frac{\partial G_r}{\partial r} + \frac{F_\theta}{r}\frac{\partial G_r}{\partial \theta} - \frac{F_\theta G_\theta}{r} + \frac{F_\Phi}{r \sin\theta}\frac{\partial G_r}{\partial \Phi} - \frac{F_\Phi G_\Phi}{r}\right\}$$

$$+ \hat{\theta}\left\{F_r \frac{\partial G_\theta}{\partial r} + \frac{F_\theta G_r}{r} + \frac{F_\theta}{r}\frac{\partial G_\theta}{\partial \theta} + \frac{F_\Phi}{r \sin\theta}\frac{\partial G_\theta}{\partial \Phi} - \frac{F_\Phi G_\Phi \cos\theta}{r \sin\theta}\right\}$$

$$+ \hat{\Phi}\left\{F_r \frac{\partial G_\Phi}{\partial r} + \frac{F_\theta}{r}\frac{\partial G_\Phi}{\partial \theta} + \frac{F_\Phi G_r}{r} + \frac{F_\Phi}{r \sin\theta}\frac{\partial G_\Phi}{\partial \Phi} + \frac{F_\Phi G_\theta \cos\theta}{r \sin\theta}\right\}$$

2.9 Derivation of Curl $\overline{F}$ and Div $\overline{F}$ in Orthogonal Curvilinear Coordinates via Line and Surface Integration

2.9a Derivation of curl $\overline{F}$

Fig. 2.10 represents a curvilinear parallelepiped[13] formed by the coordinate surfaces $u = u_o$, $u = u_1$, $v = v_o$, $v = v_1$, $w = w_o$, $w = w_1$.

$\overline{F}$ is a vector point function which, together with its first derivatives, is continuous throughout a region of space which includes the parallelepiped.

To determine the tangential line integral of $\overline{F}$ around the closed contour PCDEP, we note that, in virtue of the orthogonal nature of the coordinates, $\hat{u}$ and $\hat{w}$ are normal to PC and DE at every point of these curves, and that $\hat{u}$ and $\hat{v}$ and similarly normal to CD and EP. Consequently,

$$\oint_{PCDEP} \overline{F}.d\overline{r} = \int_{PC} F_v dr_v + \int_{CD} F_w dr_w + \int_{DE} F_v dr_v + \int_{EP} F_w dr_w$$

Suppose that the parallelepiped is divided by closely-spaced u, v and w surfaces. The v and w surfaces intersect the face PCDE in a set of curvilinear quadrilaterals. Those quadrilaterals lying between the j-1th and jth v surfaces (which cut PCDE in aa' and bb') are shown in the figure. The typical quadrilateral $\alpha\beta\gamma\delta$ lies between the k-1th and kth w surfaces (which cut PCDE in cc' and dd'). The coordinate spacing between aa' and bb' is Δv_j; that between cc' and dd' is Δw_k.

13. Perhaps 'curvilinear hexahedron' would be a more appropriate description, since it is supposed that opposite faces may be unequal. However, the meaning is clear.

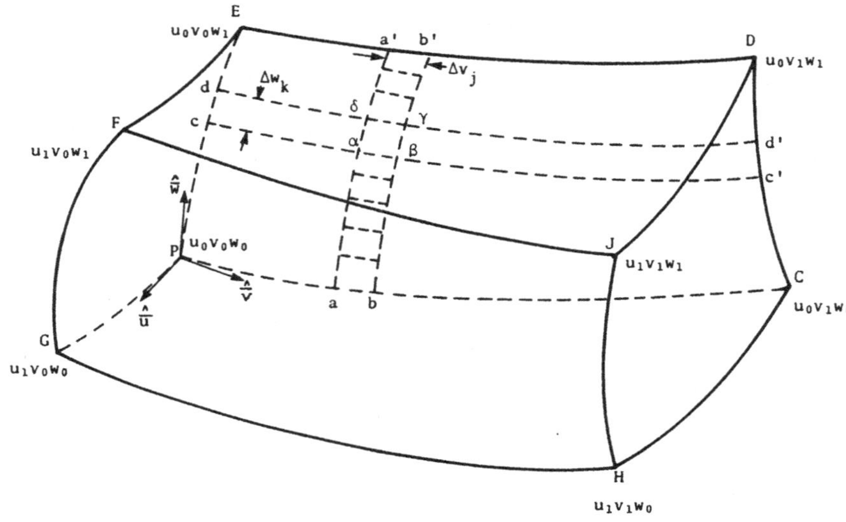

Fig. 2.10

The contribution of the line elements ab and a'b' to the integral over PC and DE is given approximately by

$$(h_2F_v)_b\ \Delta v_j - (h_2F_v)_{b'}\ \Delta v_j$$

A negative sign precedes the second term because Δv_j is taken as intrinsically positive, but the associated vector element, $\Delta\bar{r}$, points in the negative $\hat{v}$ direction, ie towards decreasing values of v.

The mean-value theorem permits of the replacement of this expression by

$$-\Delta v_j \sum_{k=1}^{p} \left(\frac{\partial}{\partial w}\ (h_2F_v)\right)_{jk'}\ \Delta w_k$$

where the subscript jk' indicates that the derivative is to be evaluated at a point between β and γ in the element shown. In this notation the zeroth w surface is PGHC and the pth is EFJD.

The corresponding approximation for the sum of the line integrals over PC and DE therefore becomes

$$- \sum_{j=1}^{m} \Delta v_j \sum_{k=1}^{p} \left(\frac{\partial}{\partial w} (h_2 F_v) \right)_{jk'} \Delta w_k$$

where the zeroth v surface is PEFG and the mth is CDJH.

If ΔS_{jk} represents the scalar area of the typical surface element and $(h_2 h_3)_{jk}$ denotes the value of the product $h_2 h_3$ at some particular point of the element, then it follows from the considerations of Sec. 2.1 that

$$(h_2 h_3)_{jk} \Delta v_j \Delta w_k = \Delta S_{jk} + \varepsilon_{jk}$$

where $(\varepsilon_{jk}/\Delta S_{jk}) \to 0$ as Δv_j, $\Delta w_k \to 0$

The above summation may therefore be written in the form

$$- \sum_{j=1}^{m} \sum_{k=1}^{p} \left(\frac{1}{h_2 h_3} \right)_{jk} \left(\frac{\partial}{\partial w} (h_2 F_v) \right)_{jk'} (\Delta S_{jk} + \varepsilon_{jk})$$

If, now, limits are taken as Δv_j, $\Delta w_k \to 0$ and m, $p \to \infty$, we immediately arrive at the relationship

$$\int_{PC} F_v dr_v + \int_{DE} F_v dr_v = - \int_{PCDE} \frac{1}{h_2 h_3} \frac{\partial}{\partial w} (h_2 F_v) \, dS$$

A similar analysis shows that

$$\int_{CD} F_w dr_w + \int_{EP} F_w dr_w = + \int_{PCDE} \frac{1}{h_2 h_3} \frac{\partial}{\partial v} (h_3 F_w) \, dS$$

On combining these equations we get

$$\oint_{PCDEP} \bar{F} . d\bar{r} = \int_{PCDE} \frac{1}{h_2 h_3} \left\{ \frac{\partial}{\partial v} (h_3 F_w) - \frac{\partial}{\partial w} (h_2 F_v) \right\} dS$$

This is the orthogonal curvilinear equivalent of Stokes's theorem in two dimensions for a coordinate surface[14].

Since

$$\oint_{PCDEP} \overline{F}.d\overline{r} = \int_{PCDE} (\text{curl}\ \overline{F})_n\ dS$$

then

$$\int_{PCDE} (\text{curl}\ \overline{F})_n\ dS = \int_{PCDE} \frac{1}{h_2h_3}\left\{\frac{\partial}{\partial v}(h_3F_w) - \frac{\partial}{\partial w}(h_2F_v)\right\} dS$$

Now let $u_1 \to u_o$, $v_1 \to v_o$, $w_1 \to w_o$. In view of the continuity of the integrands, it follows from the mean-value theorem for integrals that

$$((\text{curl}\ \overline{F})_n)_P = ((\text{curl}\ \overline{F})_u)_P = \left\{\frac{1}{h_2h_3}\left(\frac{\partial}{\partial v}(h_3F_w) - \frac{\partial}{\partial w}(h_2F_v)\right)\right\}_P$$

This is in agreement with equation (2.5-10).

The remaining components of curl $\overline{F}$ are found by equivalent subdivision of the v and w surfaces through P.

2.9b <u>Derivation of div $\overline{F}$</u>

Fig. 2.11 is a further view of the curvilinear parallelepiped of Fig. 2.10. As in the previous analysis, $\overline{F}$ and its first derivatives are continuous throughout a region of space which includes the parallelepiped.

---

14. For a surface described by the orthogonal surface curvilinear coordinates ξ and ζ, the corresponding relationship is

$$\oint_{\Gamma} \overline{F}.d\overline{r} = \int_S \frac{1}{h_\xi h_\zeta}\left\{\frac{\partial}{\partial \xi}(h_\zeta F_\zeta) - \frac{\partial}{\partial \zeta}(h_\xi F_\xi)\right\} dS$$

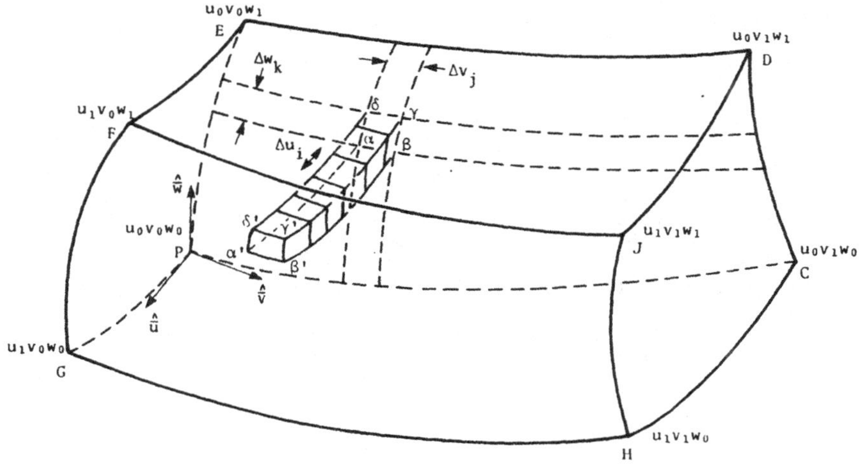

Fig. 2.11

To determine the surface integral of $\overline{F}$ over the faces PCDE and GHJF we note that $\hat{u}$ is normal to these surfaces at every point of them, while $\hat{v}$ and $\hat{w}$ are tangential.

Hence

$$\int_{PCDE} \overline{F}.d\overline{S} + \int_{GHJF} \overline{F}.d\overline{S} = \int_{PCDE} F_u dS_u + \int_{GHJF} F_u dS_u$$

The enclosure formed by the j-1th and jth v surfaces together with the k-1th and kth w surfaces is shown in the figure. This cuts the face GHJF in the curvilinear quadrilateral $\alpha'\beta'\gamma'\delta'$. The enclosure is intersected by u coordinate surfaces, the typical volume element so formed lying between the i-1th and ith u surfaces and being designated $\Delta\tau_{ijk}$. The coordinate spacing between the i-1th and ith u surfaces is Δu_i.

The contribution of the elementary quadrilaterals $\alpha\beta\gamma\delta$ and $\alpha'\beta'\gamma'\delta'$ to the normal surface integral of $\overline{F}$ over the faces PCDE and GHJF is given approximately by

$$(h_2h_3F_u)_{\gamma'} \Delta v_j\Delta w_k - (h_2h_3F_u)_\gamma \Delta v_j\Delta w_k$$

A negative sign precedes the second term because the positive normal over PCDE, being directed outwards from the parallelepiped, points in the negative $\hat{u}$ direction, and so makes dS_u negative.

This expression may be replaced by

$$\Delta v_j \Delta w_k \sum_{i=1}^{n} \left(\frac{\partial}{\partial u}(h_2 h_3 F_u)\right)_{i'jk} \Delta u_i$$

where the subscript i'jk indicates that the derivative is to be evaluated on $\gamma\gamma'$ between the i-1th and ith u surface.

In this notation the zeroth u surface is PCDE and the nth is GHJF.

The corresponding approximation for the sum of the surface integrals over PCDE and GHJF therefore becomes

$$\sum_{j=1}^{m} \sum_{k=1}^{p} \Delta v_j \Delta w_k \sum_{i=1}^{n} \left(\frac{\partial}{\partial u}(h_2 h_3 F_u)\right)_{i'jk} \Delta u_i$$

If $(h_1 h_2 h_3)_{ijk}$ denotes the value of the product $h_1 h_2 h_3$ at some particular point of the element $\Delta\tau_{ijk}$, and if the volume of this element is also represented by $\Delta\tau_{ijk}$, then

$$(h_1 h_2 h_3)_{ijk} \Delta u_i \Delta v_j \Delta w_k = \Delta\tau_{ijk} + \varepsilon_{ijk}$$

where $(\varepsilon_{ijk}/\Delta\tau_{ijk}) \to 0$ as Δu_i, Δv_j, $\Delta w_k \to 0$.

The above summation may therefore be written as

$$\sum_{i=1}^{n} \sum_{j=1}^{m} \sum_{k=1}^{p} \left(\frac{1}{h_1 h_2 h_3}\right)_{ijk} \left(\frac{\partial}{\partial u}(h_2 h_3 F_u)\right)_{i'jk} (\Delta\tau_{ijk} + \varepsilon_{ijk})$$

If, now, limits are taken as Δu_i, Δv_j, $\Delta w_k \to 0$ and n, m, $p \to \infty$, we see at once that

$$\int_{PCDE} F_u dS_u + \int_{GHJF} F_u dS_u = \int_{\tau} \frac{1}{h_1 h_2 h_3} \frac{\partial}{\partial u}(h_2 h_3 F_u)\, d\tau$$

The same system of subdivision may be used to determine the remaining surface integrals. It is found that the total surface integral is given by

$$\oint_S \overline{F}.d\overline{S} = \int_\tau \frac{1}{h_1h_2h_3}\left\{\frac{\partial}{\partial u}(h_2h_3F_u) + \frac{\partial}{\partial v}(h_3h_1F_v) + \frac{\partial}{\partial w}(h_1h_2F_w)\right\} d\tau$$

This is the orthogonal curvilinear form of the divergence theorem.

Since

$$\oint_S \overline{F}.d\overline{S} = \int_\tau \text{div } \overline{F}\, d\tau$$

then

$$\int_\tau \text{div } \overline{F}\, d\tau = \int_\tau \frac{1}{h_1h_2h_3}\left\{\frac{\partial}{\partial u}(h_2h_3F_u) + \frac{\partial}{\partial v}(h_3h_1F_v) + \frac{\partial}{\partial w}(h_1h_2F_w)\right\} d\tau$$

By invoking the mean-value theorem for integrals and taking limits as $u_1 \to u_o$, $v_1 \to v_o$, $w_1 \to w_o$ we see that

$$(\text{div } \overline{F})_P = \left(\frac{1}{h_1h_2h_3}\left\{\frac{\partial}{\partial u}(h_2h_3F_u) + \frac{\partial}{\partial v}(h_3h_1F_v) + \frac{\partial}{\partial w}(h_1h_2F_w)\right\}\right)_P$$

This is in agreement with (2.5-14).

2.10 $\nabla^2\overline{F}$ in General Orthogonal, Cylindrical and Spherical Coordinates

It was pointed out in Sec. 1.18 that it is possible to derive general or specific curvilinear expressions for $\nabla^2\overline{F}$ by substitution of appropriate curvilinear forms of grad, curl and div in the relationship

$$\nabla^2\overline{F} = \text{grad div } \overline{F} - \text{curl curl } \overline{F} \qquad (1.18\text{-}5)$$

Before proceeding with this, however, an example will be given of a direct transformation from the basic Cartesian form to the curvilinear. This will be carried out for the particular case of cylindrical coordinates.

Since ∇^2 is invariant with respect to choice of rectangular axes, equation (1.18-3) may be replaced by

$$\nabla^2\overline{F} = \overline{i}'\nabla'^2F_{x'} + \overline{j}'\nabla'^2F_{y'} + \overline{k}'\nabla'^2F_{z'} \qquad (2.10\text{-}1)$$

where $\overline{i}'$, $\overline{j}'$, $\overline{k}'$ are fixed rectangular unit axes coincident with $\hat{\overline{\rho}}$, $\hat{\overline{\Phi}}$, $\hat{\overline{z}}$ at the point $P(\rho_1,\Phi_1,z_1)$ where the transformation is to be made.

Then

$$\left.\begin{aligned}(\nabla^2\overline{F})_P &= \hat{\rho}_1\left(\frac{\partial^2 F_{x'}}{\partial x'^2}+\frac{\partial^2 F_{x'}}{\partial y'^2}+\frac{\partial^2 F_{x'}}{\partial z'^2}\right)_P\\ &+\hat{\Phi}_1\left(\frac{\partial^2 F_{y'}}{\partial x'^2}+\frac{\partial^2 F_{y'}}{\partial y'^2}+\frac{\partial^2 F_{y'}}{\partial z'^2}\right)_P\\ &+\hat{z}_1\left(\frac{\partial^2 F_{z'}}{\partial x'^2}+\frac{\partial^2 F_{z'}}{\partial y'^2}+\frac{\partial^2 F_{z'}}{\partial z'^2}\right)_P\end{aligned}\right\} \qquad (2.10\text{-}2)$$

The considerations leading to equation (2.7-5) allow us, in addition, to write down

$$\left.\begin{aligned}\frac{\partial^2 F_{x'}}{\partial x'^2} = \frac{\partial^2 F_\rho}{\partial \rho^2} \;;\quad \frac{\partial^2 F_{y'}}{\partial x'^2} = \frac{\partial^2 F_\Phi}{\partial \rho^2} \;;\quad \frac{\partial^2 F_{z'}}{\partial x'^2} = \frac{\partial^2 F_z}{\partial \rho^2}\\ \frac{\partial^2 F_{x'}}{\partial z'^2} = \frac{\partial^2 F_\rho}{\partial z^2} \;;\quad \frac{\partial^2 F_{y'}}{\partial z'^2} = \frac{\partial^2 F_\Phi}{\partial z^2} \;;\quad \frac{\partial^2 F_{z'}}{\partial z'^2} = \frac{\partial^2 F_z}{\partial z^2}\end{aligned}\right\} \qquad (2.10\text{-}3)$$

The derivatives with respect to y' remain to be transformed.

It follows from equation (2.7-6) that

$$\frac{\partial}{\partial y'}\left(\frac{\partial V}{\partial y'}\right) = \left\{\frac{\partial}{\partial \rho}\left(\frac{\partial V}{\partial y'}\right)\right\}\sin(\Phi-\Phi_1) + \left\{\frac{\partial}{\partial \Phi}\left(\frac{\partial V}{\partial y'}\right)\right\}\frac{\cos^2(\Phi-\Phi_1)}{\rho_1}$$

so that

$$\left(\frac{\partial^2 V}{\partial y'^2}\right)_P = \frac{1}{\rho_1}\left\{\frac{\partial}{\partial \Phi}\left(\frac{\partial V}{\partial y'}\right)\right\}_P$$

But

$$\frac{\partial}{\partial \Phi}\left(\frac{\partial V}{\partial y'}\right) = \frac{\partial}{\partial \Phi}\left(\frac{\partial V}{\partial \rho}\sin(\Phi-\Phi_1)\right) + \frac{\partial}{\partial \Phi}\left(\frac{\partial V}{\partial \Phi}\,\frac{\cos^2(\Phi-\Phi_1)}{\rho_1}\right)$$

whence

$$\left(\frac{\partial^2 V}{\partial y'^2}\right)_P = \frac{1}{\rho_1}\left(\frac{\partial V}{\partial \rho} + \frac{1}{\rho_1}\,\frac{\partial^2 V}{\partial \Phi^2}\right)_P \qquad (2.10\text{-}4)$$

Upon substituting the cylindrical forms $F_{x'}$, $F_{y'}$ and $F_{z'}$ for V in equation (2.10-4) in accordance with equation (2.7-8) it is found that

$$\left(\frac{\partial^2 F_{x'}}{\partial y'^2}\right)_P = \frac{1}{\rho_1}\left\{\frac{\partial F_\rho}{\partial \rho} + \frac{1}{\rho_1}\left(\frac{\partial^2 F_\rho}{\partial \Phi^2} - F_\rho - 2\frac{\partial F_\Phi}{\partial \Phi}\right)\right\}_P \tag{2.10-5}$$

$$\left(\frac{\partial^2 F_{y'}}{\partial y'^2}\right)_P = \frac{1}{\rho_1}\left\{\frac{\partial F_\Phi}{\partial \rho} + \frac{1}{\rho_1}\left(\frac{\partial^2 F_\Phi}{\partial \Phi^2} - F_\Phi + 2\frac{\partial F_\rho}{\partial \Phi}\right)\right\}_P \tag{2.10-6}$$

$$\left(\frac{\partial^2 F_{z'}}{\partial y'^2}\right)_P = \frac{1}{\rho_1}\left\{\frac{\partial F_z}{\partial \rho} + \frac{1}{\rho_1}\frac{\partial^2 F_z}{\partial \Phi^2}\right\}_P \tag{2.10-7}$$

Substitution of equations (2.10-5) to (2.10-7) (with subscripts deleted) and equation (2.10-3) in equation (2.10-2) then yields

$$\left.\begin{aligned} \nabla^2\overline{F} = {} & \hat{\rho}\left\{\frac{\partial^2 F_\rho}{\partial \rho^2} + \frac{1}{\rho}\frac{\partial F_\rho}{\partial \rho} + \frac{1}{\rho^2}\frac{\partial^2 F_\rho}{\partial \Phi^2} + \frac{\partial^2 F_\rho}{\partial z^2} - \frac{F_\rho}{\rho^2} - \frac{2}{\rho^2}\frac{\partial F_\Phi}{\partial \Phi}\right\} \\ & + \hat{\Phi}\left\{\frac{\partial^2 F_\Phi}{\partial \rho^2} + \frac{1}{\rho}\frac{\partial F_\Phi}{\partial \rho} + \frac{1}{\rho^2}\frac{\partial^2 F_\Phi}{\partial \Phi^2} + \frac{\partial^2 F_\Phi}{\partial z^2} - \frac{F_\Phi}{\rho^2} + \frac{2}{\rho^2}\frac{\partial F_\rho}{\partial \Phi}\right\} \\ & + \hat{z}\left\{\frac{\partial^2 F_z}{\partial \rho^2} + \frac{1}{\rho}\frac{\partial F_z}{\partial \rho} + \frac{1}{\rho^2}\frac{\partial^2 F_z}{\partial \Phi^2} + \frac{\partial^2 F_z}{\partial z^2}\right\} \end{aligned}\right\} \tag{2.10-8}$$

On replacing V in (2.6-4) by F_ρ, F_Φ and F_z in turn, it will be seen that (2.10-8) is equivalent to

$$\nabla^2\overline{F} = \hat{\rho}\left\{\nabla^2(F_\rho) - \frac{F_\rho}{\rho^2} - \frac{2}{\rho^2}\frac{\partial F_\Phi}{\partial \Phi}\right\} + \hat{\Phi}\left\{\nabla^2(F_\Phi) - \frac{F_\Phi}{\rho^2} + \frac{2}{\rho^2}\frac{\partial F_\rho}{\partial \Phi}\right\} + \hat{z}\,\nabla^2(F_z) \tag{2.10-9}$$

Comparison of equations (2.10-9) and (2.10-1) reveals that the simplicity of the basic Cartesian expansion is retained only for the z component of cylindrical coordinates, ie

$$(\nabla^2\overline{F})_\rho \neq \nabla^2(F_\rho) \quad ; \quad (\nabla^2\overline{F})_\Phi \neq \nabla^2(F_\Phi) \quad ; \quad (\nabla^2\overline{F})_z = \nabla^2(F_z) \tag{2.10-10}$$

To derive the general orthogonal curvilinear form of $\nabla^2\overline{F}$ we will adopt the method cited at the beginning of this section. This requires the substitution of equations (2.5-7), (2.5-10) and (2.5-14) in equation (1.18-5). The working is straightforward and leads to the following expressions.

$$(\nabla^2\overline{F})_u = \left.\begin{array}{l} \dfrac{1}{h_1}\dfrac{\partial}{\partial u}\left\{\dfrac{1}{h_1h_2h_3}\left(\dfrac{\partial}{\partial u}(h_2h_3F_u) + \dfrac{\partial}{\partial v}(h_3h_1F_v) + \dfrac{\partial}{\partial w}(h_1h_2F_w)\right)\right\} \\ - \dfrac{1}{h_2h_3}\left[\dfrac{\partial}{\partial v}\left\{\dfrac{h_3}{h_1h_2}\left(\dfrac{\partial}{\partial u}(h_2F_v) - \dfrac{\partial}{\partial v}(h_1F_u)\right)\right\}\right. \\ \left. - \dfrac{\partial}{\partial w}\left\{\dfrac{h_2}{h_3h_1}\left(\dfrac{\partial}{\partial w}(h_1F_u) - \dfrac{\partial}{\partial u}(h_3F_w)\right)\right\}\right] \end{array}\right\} \quad (2.10\text{-}11)$$

$$(\nabla^2\overline{F})_v = \left.\begin{array}{l} \dfrac{1}{h_2}\dfrac{\partial}{\partial v}\left\{\dfrac{1}{h_1h_2h_3}\left(\dfrac{\partial}{\partial u}(h_2h_3F_u) + \dfrac{\partial}{\partial v}(h_3h_1F_v) + \dfrac{\partial}{\partial w}(h_1h_2F_w)\right)\right\} \\ - \dfrac{1}{h_3h_1}\left[\dfrac{\partial}{\partial w}\left\{\dfrac{h_1}{h_2h_3}\left(\dfrac{\partial}{\partial v}(h_3F_w) - \dfrac{\partial}{\partial w}(h_2F_v)\right)\right\}\right. \\ \left. - \dfrac{\partial}{\partial u}\left\{\dfrac{h_3}{h_1h_2}\left(\dfrac{\partial}{\partial u}(h_2F_v) - \dfrac{\partial}{\partial v}(h_1F_u)\right)\right\}\right] \end{array}\right\} \quad (2.10\text{-}12)$$

$$(\nabla^2\overline{F})_w = \left.\begin{array}{l} \dfrac{1}{h_3}\dfrac{\partial}{\partial w}\left\{\dfrac{1}{h_1h_2h_3}\left(\dfrac{\partial}{\partial u}(h_2h_3F_u) + \dfrac{\partial}{\partial v}(h_3h_1F_v) + \dfrac{\partial}{\partial w}(h_1h_2F_w)\right)\right\} \\ - \dfrac{1}{h_1h_2}\left[\dfrac{\partial}{\partial u}\left\{\dfrac{h_2}{h_3h_1}\left(\dfrac{\partial}{\partial w}(h_1F_u) - \dfrac{\partial}{\partial u}(h_3F_w)\right)\right\}\right. \\ \left. - \dfrac{\partial}{\partial v}\left\{\dfrac{h_1}{h_2h_3}\left(\dfrac{\partial}{\partial v}(h_3F_w) - \dfrac{\partial}{\partial w}(h_2F_v)\right)\right\}\right] \end{array}\right\} \quad (2.10\text{-}13)$$

The spherical coordinate form of $\nabla^2\overline{F}$ then follows from substitution for u, v, w, h_1, h_2, h_3 in the above expression.

It is found that

$$(\nabla^2\overline{F})_r = \frac{\partial^2 F_r}{\partial r^2} + \frac{2}{r}\frac{\partial F_r}{\partial r} + \frac{1}{r^2}\frac{\partial^2 F_r}{\partial \theta^2} + \frac{\cot\theta}{r^2}\frac{\partial F_r}{\partial \theta} + \frac{1}{r^2\sin^2\theta}\frac{\partial^2 F_r}{\partial \phi^2}$$

$$- \frac{2}{r^2}F_r - \frac{2\cot\theta}{r^2}F_\theta - \frac{2}{r^2}\frac{\partial F_\theta}{\partial \theta} - \frac{2}{r^2\sin\theta}\frac{\partial F_\phi}{\partial \phi}$$

or

$$(\nabla^2\overline{F})_r = \nabla^2(F_r) - \frac{2}{r^2} F_r - \frac{2}{r^2 \sin\theta}\frac{\partial}{\partial\theta}(F_\theta \sin\theta) - \frac{2}{r^2 \sin\theta}\frac{\partial F_\Phi}{\partial\Phi} \tag{2.10-14}$$

$$(\nabla^2\overline{F})_\theta = \frac{\partial^2 F_\theta}{\partial r^2} + \frac{2}{r}\frac{\partial F_\theta}{\partial r} + \frac{1}{r^2}\frac{\partial^2 F_\theta}{\partial\theta^2} + \frac{\cot\theta}{r^2}\frac{\partial F_\theta}{\partial\theta} + \frac{1}{r^2 \sin^2\theta}\frac{\partial^2 F_\theta}{\partial\Phi^2}$$

$$+ \frac{2}{r^2}\frac{\partial F_r}{\partial\theta} - \frac{2\cos\theta}{r^2 \sin^2\theta}\frac{\partial F_\Phi}{\partial\Phi} - \frac{1}{r^2 \sin^2\theta} F_\theta$$

or

$$(\nabla^2\overline{F})_\theta = \nabla^2(F_\theta) + \frac{2}{r^2}\frac{\partial F_r}{\partial\theta} - \frac{2\cos\theta}{r^2 \sin^2\theta}\frac{\partial F_\Phi}{\partial\Phi} - \frac{1}{r^2 \sin^2\theta} F_\theta \tag{2.10-15}$$

$$(\nabla^2\overline{F})_\Phi = \frac{\partial^2 F_\Phi}{\partial r^2} + \frac{2}{r}\frac{\partial F_\Phi}{\partial r} + \frac{1}{r^2}\frac{\partial^2 F_\Phi}{\partial\theta^2} + \frac{\cot\theta}{r^2}\frac{\partial F_\Phi}{\partial\theta} + \frac{1}{r^2 \sin^2\theta}\frac{\partial^2 F_\Phi}{\partial\Phi^2}$$

$$- \frac{1}{r^2 \sin^2\theta} F_\Phi + \frac{2\cos\theta}{r^2 \sin^2\theta}\frac{\partial F_\theta}{\partial\Phi} + \frac{2}{r^2 \sin\theta}\frac{\partial F_r}{\partial\Phi}$$

or

$$(\nabla^2\overline{F})_\Phi = \nabla^2(F_\Phi) - \frac{1}{r^2 \sin^2\theta} F_\Phi + \frac{2\cos\theta}{r^2 \sin^2\theta}\frac{\partial F_\theta}{\partial\Phi} + \frac{2}{r^2 \sin\theta}\frac{\partial F_r}{\partial\Phi} \tag{2.10-16}$$

EXERCISES

2-25. Derive the cylindrical forms of curl and divergence by applying Stokes's theorem and the divergence theorem to the appropriate coordinate surfaces and enclosure.

Repeat this for spherical coordinates.

2-26. Derive an expression for $\oint_S V\, d\overline{S}$ where V is a well-behaved scalar point function and S is the surface of an orthogonal curvilinear parallelepiped. Then employ equation (1.17-6) to show that

$$\text{grad } V = \frac{\hat{\overline{u}}}{h_1}\frac{\partial V}{\partial u} + \frac{\hat{\overline{v}}}{h_2}\frac{\partial V}{\partial v} + \frac{\hat{\overline{w}}}{h_3}\frac{\partial V}{\partial w}$$

$$+ \frac{V}{h_1 h_2 h_3}\left\{\frac{\partial}{\partial u}(\hat{\overline{u}} h_2 h_3) + \frac{\partial}{\partial v}(\hat{\overline{v}} h_3 h_1) + \frac{\partial}{\partial w}(\hat{\overline{w}} h_1 h_2)\right\}$$

The second line of this expression must be zero since grad V cannot involve the absolute value of V. Prove this independently by means of the identity $\oint d\overline{S} \equiv \overline{0}$.

2-27. If $\overline{r} = \overline{i}x + \overline{j}y + \overline{k}z$ and $r = |\overline{r}|$ prove that

$$\nabla^2(r^n\,\overline{r}) = n(n+3)\, r^{n-2}\,\overline{r}$$

by (a) expansion in rectangular coordinates

(b) substitution in equations (2.10-14, 15, 16).

2-28. Derive $\nabla^2\overline{F} = \overline{i}\,\nabla^2 F_x + \overline{j}\,\nabla^2 F_y + \overline{k}\,\nabla^2 F_z$ in cylindrical coordinates by substituting the curvilinear forms of the Cartesian components of $\overline{F}$, equation (2.7-2), in the cylindrical expression for $\nabla^2 V$, equation (2.6-4), and by subsequently employing equation (2.2-5) to relate the rectilinear to the curvilinear unit vectors.

2-29. Derive $\nabla^2\overline{F}$ in cylindrical coordinates by substituting the cylindrical forms of grad, div and curl in the equation

$$\nabla^2\overline{F} = \text{grad div }\overline{F} - \text{curl curl }\overline{F}$$

2-30. Use the divergence theorem to demonstrate that the volume defined by a closed surface is equal to

(a) $\frac{1}{2}\oint_S (\text{grad } R^2).d\overline{S}$

where R is the distance from a fixed plane to the surface element dS.

(b) $\frac{1}{4}\oint_S (\text{grad } \rho^2).d\overline{S}$

where ρ is the radial distance from a fixed external line to the surface element dS.

(c) $^1/_6 \oint_S (\text{grad } r^2).d\bar{S}$

where r is the radial distance from a fixed external point to the surface element dS.

2.11 Change of Volume Resulting from Transformation of Coordinate Values

If the coordinates (x,y,z) of any point of a region of space are transformed to (x',y',z') in the same rectangular system, the volume of an element defined by a set of such points will be altered accordingly. Thus consider a parallelepiped initially defined by the vector edges $\overrightarrow{OP} = \bar{i}\Delta x$, $\overrightarrow{OQ} = \bar{j}\Delta y$, $\overrightarrow{OR} = \bar{k}\Delta z$. Let O move to O', P to P', etc, and let

$$\overrightarrow{O'P'} = \bar{i}\Delta x_1' + \bar{j}\Delta y_1' + \bar{k}\Delta z_1'$$

$$\overrightarrow{O'Q'} = \bar{i}\Delta x_2' + \bar{j}\Delta y_2' + \bar{k}\Delta z_2'$$

$$\overrightarrow{O'R'} = \bar{i}\Delta x_3' + \bar{j}\Delta y_3' + \bar{k}\Delta z_3'$$

Then in the limit as Δx, Δy, $\Delta z \to 0$

$$\Delta x_1' = \frac{\partial x'}{\partial x}\Delta x \qquad \Delta y_1' = \frac{\partial y'}{\partial x}\Delta x \qquad \Delta z_1' = \frac{\partial z'}{\partial x}\Delta x$$

$$\Delta x_2' = \frac{\partial x'}{\partial y}\Delta y \qquad \Delta y_2' = \frac{\partial y'}{\partial y}\Delta y \qquad \Delta z_2' = \frac{\partial z'}{\partial y}\Delta y$$

$$\Delta x_3' = \frac{\partial x'}{\partial z}\Delta z \qquad \Delta y_3' = \frac{\partial y'}{\partial z}\Delta z \qquad \Delta z_3' = \frac{\partial z'}{\partial z}\Delta z$$

where the derivatives are evaluated at O and are supposed to be continuous in a neighbourhood of O.

It follows that

$$\overrightarrow{O'P'} = \left(\bar{i}\,\frac{\partial x'}{\partial x} + \bar{j}\,\frac{\partial y'}{\partial x} + \bar{k}\,\frac{\partial z'}{\partial x}\right)\Delta x$$

$$\overrightarrow{O'Q'} = \left(\bar{i}\,\frac{\partial x'}{\partial y} + \bar{j}\,\frac{\partial y'}{\partial y} + \bar{k}\,\frac{\partial z'}{\partial y}\right)\Delta y$$

$$\overrightarrow{O'R'} = \left(\bar{i}\,\frac{\partial x'}{\partial z} + \bar{j}\,\frac{\partial y'}{\partial z} + \bar{k}\,\frac{\partial z'}{\partial z}\right)\Delta z$$

whence the volume of the transformed parallelepiped is given by

$$\left|\left(\bar{i}\frac{\partial x'}{\partial x}+\bar{j}\frac{\partial y'}{\partial x}+\bar{k}\frac{\partial z'}{\partial x}\right)\times\left(\bar{i}\frac{\partial x'}{\partial y}+\bar{j}\frac{\partial y'}{\partial y}+\bar{k}\frac{\partial z'}{\partial y}\right).\left(\bar{i}\frac{\partial x'}{\partial z}+\bar{j}\frac{\partial y'}{\partial z}+\bar{k}\frac{\partial z'}{\partial z}\right)\Delta x\Delta y\Delta z\right|$$

or

$$\Delta\tau' = \begin{vmatrix} \frac{\partial x'}{\partial x} & \frac{\partial x'}{\partial y} & \frac{\partial x'}{\partial z} \\ \frac{\partial y'}{\partial x} & \frac{\partial y'}{\partial y} & \frac{\partial y'}{\partial z} \\ \frac{\partial z'}{\partial x} & \frac{\partial z'}{\partial y} & \frac{\partial z'}{\partial z} \end{vmatrix} \Delta\tau \equiv J\left(\frac{x',y',z'}{x,y,z}\right)\Delta\tau \qquad (2.11\text{-}1)$$

2.12 Surface Relationships

2.12a Surface gradient, divergence and curl

Let ξ and ζ be orthogonal surface coordinates and let $\hat{\bar{\xi}}, \hat{\bar{\zeta}}, \hat{\bar{n}}$ form a right-handed set where $\hat{\bar{n}}$ is a unit normal to the surface (Sec. 2.4b). Then if P and Q are closely spaced points of the surface and V is a scalar point function having continuous derivatives upon the surface we may write

$$V_Q - V_P \approx \left(\frac{\partial V}{\partial\xi}\right)_P \Delta\xi + \left(\frac{\partial V}{\partial\zeta}\right)_P \Delta\zeta$$

$$\approx \left(\frac{\hat{\bar{\xi}}}{h_\xi}\frac{\partial V}{\partial\xi}+\frac{\hat{\bar{\zeta}}}{h_\zeta}\frac{\partial V}{\partial\zeta}\right)_P \cdot (\hat{\bar{\xi}}h_\xi\Delta\xi+\hat{\bar{\zeta}}h_\zeta\Delta\zeta)$$

$$\approx \left(\frac{\hat{\bar{\xi}}}{h_\xi}\frac{\partial V}{\partial\xi}+\frac{\hat{\bar{\zeta}}}{h_\zeta}\frac{\partial V}{\partial\zeta}\right)_P \cdot \Delta\bar{r}$$

where $\Delta\bar{r} = \overrightarrow{PQ}$

Hence at all points of the surface where $\frac{\partial V}{\partial\xi}$ and $\frac{\partial V}{\partial\zeta}$ are continuous

$$\frac{dV}{ds} = \hat{\bar{s}} \cdot \left(\frac{\hat{\bar{\xi}}}{h_\xi}\frac{\partial V}{\partial\xi}+\frac{\hat{\bar{\zeta}}}{h_\zeta}\frac{\partial V}{\partial\zeta}\right) \qquad (2.12\text{-}1)$$

where $\hat{\bar{s}}$ is the unit tangent to the surface in the direction of motion.

For obvious reasons we define the second factor of the scalar product as the surface gradient of V and write

$$\text{grads } V = \frac{\hat{\bar{\xi}}}{h_\xi}\frac{\partial V}{\partial\xi}+\frac{\hat{\bar{\zeta}}}{h_\zeta}\frac{\partial V}{\partial\zeta} \qquad (2.12\text{-}2)$$

Since $\frac{dV}{ds}$ is independent of coordinates, the surface gradient as defined above will remain invariant if the surface be defined in terms of alternative orthogonal coordinates. V may or may not be defined outside the surface; in any case grads V is directed tangentially. It follows from (2.12-2) that

$$\text{grads}\ \xi = \frac{\hat{\xi}}{h_\xi} \qquad \text{grads}\ \zeta = \frac{\hat{\zeta}}{h_\zeta} \tag{2.12-3}$$

The surface divergence and curl are defined in terms of orthogonal surface coordinates as follows[15]

$$\text{divs}\ \overline{F} = \frac{\hat{\xi}}{h_\xi}\cdot\frac{\partial\overline{F}}{\partial\xi} + \frac{\hat{\zeta}}{h_\zeta}\cdot\frac{\partial\overline{F}}{\partial\zeta} \tag{2.12-4}$$

$$\text{curls}\ \overline{F} = \frac{\hat{\xi}}{h_\xi}\times\frac{\partial\overline{F}}{\partial\xi} + \frac{\hat{\zeta}}{h_\zeta}\times\frac{\partial\overline{F}}{\partial\zeta} \tag{2.12-5}$$

It is easily shown by substitution in (2.12-2), (2.12-4) and (2.12-5) that the various expansions developed in Sec. 1.16 for the gradient, divergence and curl of scalar and vector products have their direct counterparts in surface relationships. Thus if V, U, $\overline{F}$ and $\overline{G}$ are scalar and vector point functions defined upon the surface but not necessarily outside it, and with or without normal components, then

$$\text{grads}\ V\,U = V\ \text{grads}\ U + U\ \text{grads}\ V \tag{2.12-6}$$

$$\text{divs}\ V\,\overline{F} = V\ \text{divs}\ \overline{F} + \text{grads}\ V.\overline{F} \tag{2.12-7}$$

$$\text{curls}\ V\,\overline{F} = V\ \text{curls}\ \overline{F} + \text{grads}\ V\times\overline{F} \tag{2.12-8}$$

$$\text{grads}(\overline{F}.\overline{G}) = (\overline{F}.\text{grads})\overline{G} + (\overline{G}.\text{grads})\overline{F} + \overline{F}\times\text{curls}\ \overline{G} + \overline{G}\times\text{curls}\ \overline{F} \tag{2.12-9}$$

where

$$(\overline{F}.\text{grads})\overline{G} \equiv \frac{F_\xi}{h_\xi}\frac{\partial\overline{G}}{\partial\xi} + \frac{F_\zeta}{h_\zeta}\frac{\partial\overline{G}}{\partial\zeta}$$

---

15. Some writers employ the terms 'surface divergence' and 'surface curl' with completely different connotations.

$$\text{curls}(\overline{F}\times\overline{G}) = (\overline{G}.\text{grads})\overline{F} - (\overline{F}.\text{grads})\overline{G} + \overline{F}\ \text{divs}\ \overline{G} - \overline{G}\ \text{divs}\ \overline{F} \tag{2.12-10}$$

$$\text{divs}(\overline{F}\times\overline{G}) = \overline{G}.\text{curls}\ \overline{F} - \overline{F}.\text{curls}\ \overline{G} \tag{2.12-11}$$

When V has continuous derivatives both upon and beyond the surface we may relate grad V and grads V at a point of the surface in the following way. Let P be a point of the surface S and let $\overrightarrow{PQ} = \overline{i}\Delta x$. In general, Q will lie beyond S. The normal from Q to S meets S in R. Let the coordinates of R be $\xi_P + \Delta\xi$, $\zeta_P + \Delta\zeta$. Then

$$V_Q - V_P \approx \left(\frac{\partial V}{\partial \xi}\right)_P \Delta\xi + \left(\frac{\partial V}{\partial \zeta}\right)_P \Delta\zeta + \left(\frac{\partial V}{\partial n}\right)_P \Delta n$$

$$\approx \left(\frac{\hat{\overline{\xi}}}{h_\xi}\frac{\partial V}{\partial \xi} + \frac{\hat{\overline{\zeta}}}{h_\zeta}\frac{\partial V}{\partial \zeta} + \hat{n}\frac{\partial V}{\partial n}\right)_P . (\hat{\overline{\xi}}h_\xi\Delta\xi + \hat{\overline{\zeta}}h_\zeta\Delta\zeta + \hat{\overline{n}}\Delta n)$$

$$\approx \overline{i}\Delta x . \left(\frac{\hat{\overline{\xi}}}{h_\xi}\frac{\partial V}{\partial \xi} + \frac{\hat{\overline{\zeta}}}{h_\zeta}\frac{\partial V}{\partial \zeta} + \hat{n}\frac{\partial V}{\partial n}\right)_P$$

Then at all points of the surface

$$\frac{\partial V}{\partial x} = \left(\frac{\hat{\overline{\xi}}}{h_\xi}\frac{\partial V}{\partial \xi} + \frac{\hat{\overline{\zeta}}}{h_\zeta}\frac{\partial V}{\partial \zeta} + \hat{n}\frac{\partial V}{\partial n}\right)_x$$

whence

$$\text{grad } V = \frac{\hat{\overline{\xi}}}{h_\xi}\frac{\partial V}{\partial \xi} + \frac{\hat{\overline{\zeta}}}{h_\zeta}\frac{\partial V}{\partial \zeta} + \hat{n}\frac{\partial V}{\partial n}$$

or

$$\text{grad } V = \text{grads } V + \hat{n}\frac{\partial V}{\partial n} \tag{2.12-12}$$

Substitution of F_x for V in the above analysis leads to

$$\frac{\partial F_x}{\partial x} = \frac{\hat{\xi}_x}{h_\xi}\frac{\partial F_x}{\partial \xi} + \frac{\hat{\zeta}_x}{h_\zeta}\frac{\partial F_x}{\partial \zeta} + \hat{n}_x\frac{\partial F_x}{\partial n}$$

whence

$$\text{div } \overline{F} = \frac{\hat{\overline{\xi}}}{h_\xi}.\frac{\partial \overline{F}}{\partial \xi} + \frac{\hat{\overline{\zeta}}}{h_\zeta}.\frac{\partial \overline{F}}{\partial \zeta} + \hat{\overline{n}}.\frac{\partial \overline{F}}{\partial n}$$

since $\overline{i}$, $\overline{j}$, $\overline{k}$ are independent of ξ, ζ, n

hence

$$\text{div } \overline{F} = \text{divs } \overline{F} + \hat{\overline{n}}.\frac{\partial \overline{F}}{\partial n} \tag{2.12-13}$$

In like manner it may be shown that

$$\text{curl } \overline{F} = \text{curls } \overline{F} + \hat{\overline{n}} \times \frac{\partial \overline{F}}{\partial n} \tag{2.12-14}$$

2.12b <u>Integral transformations</u>

From equation (2.12-5)

$$\hat{\overline{n}}.\text{curls } \overline{F} = \frac{\hat{\overline{\zeta}}}{h_\xi}.\frac{\partial \overline{F}}{\partial \xi} - \frac{\hat{\overline{\xi}}}{h_\zeta}.\frac{\partial \overline{F}}{\partial \zeta}$$

$$= \frac{\hat{\overline{\zeta}}}{h_\xi}.\frac{\partial}{\partial \xi}(\hat{\overline{\xi}}F_\xi + \hat{\overline{\zeta}}F_\zeta + \hat{\overline{n}}F_n) - \frac{\hat{\overline{\xi}}}{h_\zeta}.\frac{\partial}{\partial \zeta}(\hat{\overline{\xi}}F_\xi + \hat{\overline{\zeta}}F_\zeta + \hat{\overline{n}}F_n)$$

On expansion this yields

$$\hat{\overline{n}}.\text{curls } \overline{F} = \frac{\hat{\overline{\zeta}}}{h_\xi}.\frac{\partial \hat{\overline{\xi}}}{\partial \xi} F_\xi - \frac{\hat{\overline{\xi}}}{h_\zeta}.\frac{\partial \hat{\overline{\zeta}}}{\partial \zeta} F_\zeta + \frac{1}{h_\xi}\frac{\partial F_\zeta}{\partial \xi} - \frac{1}{h_\zeta}\frac{\partial F_\xi}{\partial \zeta}$$
$$+ \left\{\frac{\hat{\overline{\zeta}}}{h_\xi}.\frac{\partial \hat{\overline{n}}}{\partial \xi} - \frac{\hat{\overline{\xi}}}{h_\zeta}.\frac{\partial \hat{\overline{n}}}{\partial \zeta}\right\} F_n \tag{2.12-15}$$

It is easily shown (see Ex.2-22., p. 146) that

$$\frac{\partial}{\partial \xi}(\hat{\overline{\zeta}}h_\zeta) = \frac{\partial}{\partial \zeta}(\hat{\overline{\xi}}h_\xi)$$

or

$$h_\zeta \frac{\partial \hat{\overline{\zeta}}}{\partial \xi} + \hat{\overline{\zeta}}\frac{\partial h_\zeta}{\partial \xi} = h_\xi \frac{\partial \hat{\overline{\xi}}}{\partial \zeta} + \hat{\overline{\xi}}\frac{\partial h_\xi}{\partial \zeta} \tag{2.12-16}$$

whence, upon scalar multiplication by $\hat{n}$, we obtain

$$\frac{\hat{n}}{h_\xi}\cdot\frac{\partial\hat{\zeta}}{\partial\xi} = \frac{\hat{n}}{h_\zeta}\cdot\frac{\partial\hat{\xi}}{\partial\zeta} \quad (h_\xi\neq 0, h_\zeta\neq 0) \tag{2.12-17}$$

On writing $\hat{n} = \hat{\xi}\times\hat{\zeta}$ and expanding the last term of (2.12-15) accordingly we obtain

$$\left\{\left(\frac{\hat{\zeta}}{h_\xi}\cdot\hat{\xi}\times\frac{\partial\hat{\zeta}}{\partial\xi}\right)-\left(\frac{\hat{\xi}}{h_\zeta}\cdot\frac{\partial\hat{\xi}}{\partial\zeta}\times\hat{\zeta}\right)\right\}F_n = \left\{\frac{\hat{n}}{h_\zeta}\cdot\frac{\partial\hat{\xi}}{\partial\zeta} - \frac{\hat{n}}{h_\xi}\cdot\frac{\partial\hat{\zeta}}{\partial\xi}\right\}F_n = 0 \quad \text{by (2.12-17)}$$

whence

$$\hat{n}.\text{curl}_s\,\bar{F} = \frac{\hat{\zeta}}{h_\zeta}\cdot\frac{\partial\hat{\xi}}{\partial\xi}\,F_\xi - \frac{\hat{\xi}}{h_\zeta}\cdot\frac{\partial\hat{\zeta}}{\partial\zeta}\,F_\zeta + \frac{1}{h_\xi}\frac{\partial F_\zeta}{\partial\xi} - \frac{1}{h_\zeta}\frac{\partial F_\xi}{\partial\zeta} \tag{2.12-18}$$

Scalar multiplication of (2.12-16) by $\hat{\xi}$ and $\hat{\zeta}$ in turn yields

$$\frac{\hat{\xi}}{h_\xi}\cdot\frac{\partial\hat{\zeta}}{\partial\xi} = \frac{1}{h_\xi h_\zeta}\frac{\partial h_\xi}{\partial\zeta} \tag{2.12-19}$$

and

$$\frac{\hat{\zeta}}{h_\zeta}\cdot\frac{\partial\hat{\xi}}{\partial\zeta} = \frac{1}{h_\xi h_\zeta}\frac{\partial h_\zeta}{\partial\xi} \tag{2.12-20}$$

Since $\frac{\partial}{\partial\xi}(\hat{\xi}.\hat{\zeta}) = 0$ and $\frac{\partial}{\partial\zeta}(\hat{\xi}.\hat{\zeta}) = 0$ we have

$$\frac{\partial\hat{\xi}}{\partial\xi}\cdot\hat{\zeta} + \hat{\xi}\cdot\frac{\partial\hat{\zeta}}{\partial\xi} = 0 \quad\text{and}\quad \frac{\partial\hat{\xi}}{\partial\zeta}\cdot\hat{\zeta} + \hat{\xi}\cdot\frac{\partial\hat{\zeta}}{\partial\zeta} = 0$$

hence (2.12-19) and (2.12-20) may be transformed into

$$-\frac{\hat{\zeta}}{h_\xi}\cdot\frac{\partial\hat{\xi}}{\partial\xi} = \frac{1}{h_\xi h_\zeta}\frac{\partial h_\xi}{\partial\zeta} \tag{2.12-21}$$

and

$$-\frac{\hat{\xi}}{h_\zeta}\cdot\frac{\partial\hat{\zeta}}{\partial\zeta} = \frac{1}{h_\xi h_\zeta}\frac{\partial h_\zeta}{\partial\xi} \tag{2.12-22}$$

Substitution of (2.12-21) and (2.12-22) in (2.12-18) then yields

$$\hat{n}.\text{curls }\bar{F} = \frac{1}{h_\xi}\frac{\partial F_\zeta}{\partial \xi} - \frac{1}{h_\zeta}\frac{\partial F_\xi}{\partial \zeta} - \frac{F_\xi}{h_\xi h_\zeta}\frac{\partial h_\xi}{\partial \zeta} + \frac{F_\zeta}{h_\xi h_\zeta}\frac{\partial h_\zeta}{\partial \xi}$$

or

$$\hat{n}.\text{curls }\bar{F} = \frac{1}{h_\xi h_\zeta}\left\{\frac{\partial}{\partial \xi}(h_\zeta F_\zeta) - \frac{\partial}{\partial \zeta}(h_\xi F_\xi)\right\} \tag{2.12-23}$$

An analysis formally identical with that given in Sec. 2.9a then leads to Stokes's theorem for curvilinear surface coordinates, viz

$$\int_S (\text{curls }\bar{F}).d\bar{S} = \oint_\Gamma \bar{F}.d\bar{r} \tag{2.12-24}$$

This is seen to hold whether or not $\bar{F}$ is tangential to S.

The expression for surface divergence (2.12-4) may be transformed in a similar manner. We obtain

$$\text{divs }\bar{F} = \frac{1}{h_\xi h_\zeta}\left\{\frac{\partial}{\partial \xi}(h_\zeta F_\xi) + \frac{\partial}{\partial \zeta}(h_\xi F_\zeta)\right\} + \left\{\frac{\hat{\xi}}{h_\xi}.\frac{\partial \hat{n}}{\partial \xi} + \frac{\hat{\zeta}}{h_\zeta}.\frac{\partial \hat{n}}{\partial \zeta}\right\}F_n \tag{2.12-25}$$

It may also be shown that

$$\int_S \frac{1}{h_\xi h_\zeta}\left\{\frac{\partial}{\partial \xi}(h_\zeta F_\xi) + \frac{\partial}{\partial \zeta}(h_\xi F_\zeta)\right\} dS = \oint_\Gamma \bar{F}.\hat{n}'\, ds \tag{2.12-26}$$

where $\hat{n}'$ is the outward normal to Γ tangential to the surface, hence the curvilinear surface form of the divergence theorem becomes

$$\int_S \text{divs }\bar{F}\, dS = \oint_\Gamma \bar{F}.\hat{n}'\, ds + \int_S (\text{divs }\hat{n})\,\bar{F}.d\bar{S} \tag{2.12-27}$$

When the surface is planar, or $\bar{F}$ is everywhere tangential to it, we have

$$\int_S \text{divs }\bar{F}\, dS = \oint_\Gamma \bar{F}.\hat{n}'\, ds \tag{2.12-27a}$$

The integral transformations expressed by equations (1.17-1), (1.17-3) and (1.17-5) have the following surface counterparts:

$$\int_S d\bar{S} \times \text{grads } V = \oint_\Gamma V\, d\bar{r} \tag{2.12-28}$$

$$\int_S \text{curls } \bar{F}\, dS = \oint_\Gamma (\hat{\bar{n}}' \times \bar{F})\, ds + \int_S \text{divs } \hat{\bar{n}}\, (d\bar{S} \times \bar{F}) \tag{2.12-29}$$

$$\int_S \text{grads } V\, dS = \oint_\Gamma V\, \hat{\bar{n}}'\, ds + \int_S (\text{divs } \hat{\bar{n}})\, V\, d\bar{S} \tag{2.12-30}$$

Equation (2.12-28) is most easily verified by expanding curls V $\bar{a}$ in accordance with (2.12-8), where $\bar{a}$ is a constant vector, and applying (2.12-24). (c.f. the alternative proof of (1.17-1)).

Equation (2.12-29) represents the surface form of Ostrogradsky's theorem. It is readily demonstrated by the expansion of divs($\bar{a}\times\bar{F}$) in accordance with (2.12-11) and a subsequent application of (2.12-27).

To obtain (2.12-30) expand divs V $\bar{a}$ in accordance with (2.12-7) and apply (2.12-27).

EXERCISES

2-31. Prove (2.12-10) and (2.12-14).

2-32. Show that curls $\hat{\bar{n}} = 0$ by considering the $\hat{\bar{\xi}}$, $\hat{\bar{\zeta}}$ and $\hat{\bar{n}}$ components in turn.

2-33. Develop a proof of

$$\oint_\Gamma V\, d\bar{r} = \int_S d\bar{S} \times \text{grads } V$$

independently of that suggested in the text by showing that

$$\oint_\Gamma V\, d\bar{r} = \iint_S \left\{ \frac{\partial}{\partial \xi} (V h_\zeta \hat{\bar{\zeta}} d\zeta)\, d\xi - \frac{\partial}{\partial \zeta} (V h_\xi \hat{\bar{\xi}} d\xi)\, d\zeta \right\}$$

and noting subsequently that

$$\int_S \frac{V}{h_\xi h_\zeta} \left\{ \frac{\partial}{\partial \xi} (\hat{\bar{\zeta}} h_\zeta) - \frac{\partial}{\partial \zeta} (\hat{\bar{\xi}} h_\xi) \right\} dS = 0$$

2-34. Show that

$$\text{divs}\,\overline{F} = \frac{1}{h_\xi h_\zeta}\left\{\frac{\partial}{\partial\xi}(h_\zeta F_\xi) + \frac{\partial}{\partial\zeta}(h_\xi F_\zeta)\right\} + (\text{divs}\,\hat{\overline{n}})\,F_n$$

2-35. Show that

$$\int_S \frac{1}{h_\xi h_\zeta}\left\{\frac{\partial}{\partial\xi}(h_\zeta F_\xi) + \frac{\partial}{\partial\zeta}(h_\xi F_\zeta)\right\} dS = \oint_\Gamma \overline{F}.\hat{\overline{n}}'\, ds$$

2-36. Develop a proof of (2.12-30) in the following way.

Show by direct integration that

$$\int_S \text{grads}\, V\, dS = \oint_\Gamma V\,\hat{\overline{n}}'\, ds - \int_S \frac{1}{h_\xi h_\zeta}\left\{\frac{\partial}{\partial\xi}(\hat{\overline{\xi}} h_\zeta) + \frac{\partial}{\partial\zeta}(\hat{\overline{\zeta}} h_\xi)\right\} V\, dS$$

and by consideration of the $\hat{\overline{\xi}}$, $\hat{\overline{\zeta}}$ and $\hat{\overline{n}}$ components in turn demonstrate that

$$\frac{1}{h_\xi h_\zeta}\left\{\frac{\partial}{\partial\xi}(\hat{\overline{\xi}} h_\zeta) + \frac{\partial}{\partial\zeta}(\hat{\overline{\zeta}} h_\xi)\right\} = -\hat{\overline{n}}\,\text{divs}\,\hat{\overline{n}}$$

2-37. Employ (2.12-30) to show that divs $\hat{\overline{n}}$ is invariant with respect to change of orthogonal coordinates and apply this result to (2.12-27) and (2.12-29) to show that divs $\overline{F}$ and curls $\overline{F}$ are likewise invariant.

2-38. Derive Green's theorem for surface coordinates, viz

$$\int_S (V\,\nabla^2_S U - U\,\nabla^2_S V)\, dS = \oint_\Gamma (V\,\nabla_S U - U\,\nabla_S V).\hat{\overline{n}}'\, ds$$

where ∇_S = grads and ∇^2_S = divs grads

2-39. If V is well-behaved upon a closed surface show that

$$\oint_S \nabla^2_S V\, dS = 0$$

CHAPTER 3

GREEN'S THEOREM AND ALLIED TOPICS

3.1 Green's Theorem

The symmetrical and asymmetrical forms of Green's theorem have been stated and proved in Sec. 1.17. They are re-introduced here in rather more general terms as follows.

Let S_1, S_2 --- S_n be closed regular surfaces which neither intersect nor enclose one another, and let Σ be a closed regular surface which encloses all of these (Fig. 3.1). The simply connected region of space bounded by the surfaces is designated R (or τ, for the purpose of volume integration).

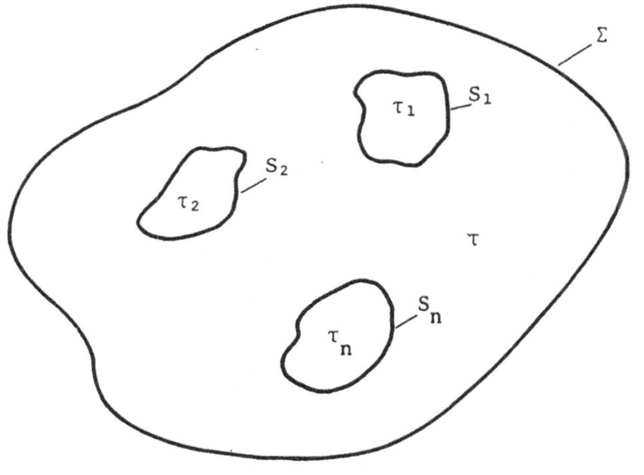

Fig. 3.1

If V and U are single-valued scalar point functions with continuous second derivatives in R, then

$$\oint_{S_{1..n}\Sigma} V \,\mathrm{grad}\, U.d\bar{S} = \int_\tau V \nabla^2 U \, d\tau + \int_\tau \mathrm{grad}\, V.\mathrm{grad}\, U \, d\tau \qquad \left.\begin{matrix}1.17\text{-}10\\3.1\text{-}1\end{matrix}\right\}$$

and

$$\oint_{S_{1..n}\Sigma} (V \text{ grad } U - U \text{ grad } V).d\bar{S} = \int_\tau (V \nabla^2 U - U \nabla^2 V)\, d\tau \qquad \begin{matrix}1.17\text{-}11\\ 3.1\text{-}2\end{matrix}$$

Substitution of grad V for $\bar{F}$ in (1.17-9) yields

$$\oint_{S_{1..n}\Sigma} V \text{ grad } V.d\bar{S} = \int_\tau V \nabla^2 V\, d\tau + \int_\tau \text{grad } V.\text{grad } V\, d\tau$$

or

$$\oint_{S_{1..n}\Sigma} V \frac{\partial V}{\partial n}\, dS = \int_\tau V \nabla^2 V\, d\tau + \int_\tau |\text{grad } V|^2\, d\tau \qquad (3.1\text{-}3)$$

where n is distance measured along the outward normal[1].

Since (V-U) is single-valued with continuous second derivatives in R, it may replace V in equation (3.1-3), in which case

$$\oint_{S_{1..n}\Sigma} (V-U) \frac{\partial}{\partial n}(V-U)\, dS = \int_\tau (V-U)\, \nabla^2 (V-U)\, d\tau + \int_\tau |\text{grad}(V-U)|^2\, d\tau \qquad (3.1\text{-}4)$$

We will have occasion to use both of these modified forms of Green's theorem, equations (3.1-3) and (3.1-4), in the next section.

Because V and U are required to be well-behaved functions within R it is necessary to exclude discontinuities by means of closed surfaces. Such surfaces, however, may be invoked in the absence of discontinuities, and the related surface integrals then replace the volume integrals taken over the regions enclosed by these surfaces. Thus, if S_1 represents such a surface, Green's theorem may be applied to the region R_1 (τ_1) enclosed by S_1. Upon adding the associated equation to that obtaining for the region R the two surface integrals over S_1 cancel because of the reversal of the positive sense of the normal on passing through the surface, so that we are left with surface integrals over $S_{2..n}\Sigma$ and an integration volume $\tau + \tau_1$. This expression could, of course, have been written down directly.

---

1. It is usual to write the derivative in partial form. There are arguments for and against this.

It is often necessary to apply Green's theorem to an unbounded region[2] comprising all space outside the set of local surfaces $S_{1..n}$. In this case the Σ surface is moved out to infinity in all directions. The value of the surface integral over Σ then depends upon the manner in which the scalar integrand behaves at infinite distance. If R is distance measured from a local origin, then the function V is said to be regular at infinity if

(a) RV is bounded as $R\to\infty$

(b) $R^2\nabla V$ is bounded as $R\to\infty$

ie V is regular at infinity if V vanishes at least as $\frac{1}{R}$ and grad V vanishes at least as $\frac{1}{R^2}$ for sufficiently large values of R.

When V and U are both regular at infinity it is seen that the surface integrals over Σ vanish at infinite distance (surface area increases only as R^2), so that equations (3.1-1) to (3.1-4) hold with Σ deleted, while τ includes all space outside $S_{1..n}$.

3.2 The Harmonic Function

A scalar point function V is said to be harmonic at a point if it has continuous second derivatives and satisfies Laplace's equation ($\nabla^2V=0$) throughout some neighbourhood[3] of that point. Correspondingly, a function is harmonic in an open region if it is harmonic at every point of that region. When the region is closed we will suppose, in addition, that the second derivatives of the function are continuous at all points of the boundary[4]. Any function which is stated to be harmonic in an unbounded region is, by implication, regular at infinity.

The following theorems relate to functions which are (a) harmonic in the closed region $\underline{R}$ bounded by the surfaces $S_{1..n}\Sigma$, where all surfaces are at finite distance, or (b) harmonic everywhere outside the surfaces $S_{1..n}$, all of which are at finite distance.

Theorem 3.2-1

If V is harmonic in the closed region bounded by the surfaces $S_{1..n}\Sigma$, then the total surface integral of the normal derivative of V is zero.

---

2. A region is bounded or finite if all of its points lie within a sphere of finite radius; otherwise it is unbounded or infinite. It is seen that the term 'unbounded' does not necessarily imply 'devoid of all boundaries' but rather 'devoid of a closed exterior boundary'.

3. The word 'neighbourhood' has been used in a colloquial sense in earlier pages to refer to the vicinity of a point or line. Strictly, a neighbourhood of a point is any open region which contains that point. The word will be used in this sense from now on.

4. This limitation may be unnecessarily restrictive. See Kellogg, pp. 211-2.

Proof: Interchange V and U in equation (3.1-1) and put U = 1. Since $\nabla^2 V = 0$ and grad U = 0 it follows that

$$\oint_{S_{1..n}\Sigma} \frac{\partial V}{\partial n}\, dS = 0$$

Alternatively, we have from the divergence theorem

$$\oint_{S_{1..n}\Sigma} \frac{\partial V}{\partial n}\, dS = \oint_{S_{1..n}\Sigma} \text{grad } V.d\bar{S} = \int_{\tau} \nabla^2 V\, d\tau = 0$$

[When V is harmonic everywhere outside $S_{1..n}$ the surface integral over Σ remains equal and opposite to that over $S_{1..n}$ as Σ recedes to infinity. This does not preclude the possibility that V is regular at infinity because $\frac{\partial V}{\partial n}$ may decrease only as $\frac{1}{R^2}$ while the area of surface integration increases as R^2. However, should $\frac{\partial V}{\partial n}$ decrease at a greater rate, eg as $\frac{1}{R^3}$, the surface integral over Σ will vanish at infinity and the integral over $S_{1..n}$ must then also vanish.]

Theorem 3.2-2

If V is harmonic in the closed region $\underline{R}$ bounded by $S_{1..n}\Sigma$ and has the same value at all points of the surfaces, then V is constant throughout $\underline{R}$ and equal to its value on the surfaces.

Proof: If V = V' upon the surfaces, then from equation (3.1-3)

$$V' \oint_{S_{1..n}\Sigma} \frac{\partial V}{\partial n}\, dS = \int_{\tau} |\text{grad } V|^2\, d\tau$$

But $\oint_{S_{1..n}\Sigma} \frac{\partial V}{\partial n}\, dS = 0$ from Theorem 3.2-1

hence $\int_{\tau} |\text{grad } V|^2\, d\tau = 0$

Since $|\text{grad } V|^2$ is positive or zero for any volume element, it must be zero for all volume elements if the volume integral is zero. Hence grad V = 0 throughout $\underline{R}$, and V is constant throughout $\underline{R}$ and equal to its value on the surfaces.

<u>Theorem 3.2-3</u>

If V is harmonic in the closed region $\underline{R}$ bounded by the surfaces $S_{1..n}\Sigma$, and its normal derivative is zero at all points of the surfaces, then V is constant within $\underline{R}$.

Proof: Since $\frac{\partial V}{\partial n}$ is zero at all points of the surfaces, it follows from equation (3.1.3) that $\int_\tau |\text{grad } V|^2 \, d\tau = 0$, whence, for reasons given above, V is constant as stated.

<u>Theorem 3.2-3a</u>

If V is harmonic everywhere outside the surfaces $S_{1..n}$, and if its normal derivative is zero at all points of these surfaces, then V is zero upon and outside the surfaces.

Proof: If Σ is a closed surface surrounding $S_{1..n}$ then

$$\oint_{S_{1..n}} V \frac{\partial V}{\partial n} \, dS + \oint_{\Sigma} V \frac{\partial V}{\partial n} \, dS = \int_\tau |\text{grad } V|^2 \, d\tau$$

The surface integral over $S_{1..n}$ is zero because $\frac{\partial V}{\partial n} = 0$ at each point. If Σ recedes to infinity the associated surface integral disappears, because $V \frac{\partial V}{\partial n}$ vanishes at least as $\frac{1}{R^3}$ while the surface area increases as R^2.

Hence

$$\int_\tau |\text{grad } V|^2 \, d\tau = 0$$

where τ is the space external to $S_{1..n}$.

V is consequently constant within this region and, being zero at infinity, is zero throughout.

<u>Theorem 3.2-4</u>

If V is harmonic in the closed region $\underline{R}$ bounded by the surfaces $S_{1..n}\Sigma$, and its value is specified at each point of the surfaces, then V is uniquely determined at all points of $\underline{R}$[5].

---

5. It is assumed in these theorems that the function V exists. See Sec. 3.8.

Proof: Let U be a scalar point function, harmonic in R, which has the same value as V upon the surfaces $S_{1..n}\Sigma$. Then $V - U = 0$ at each point of the surfaces and $\nabla^2(V-U) = \nabla^2 V - \nabla^2 U = 0$ in R.

Hence equation (3.1-4) becomes

$$\int_\tau |\text{grad}(V-U)|^2 \, d\tau = 0$$

ie

$$\int_\tau |\text{grad } V - \text{grad } U|^2 \, d\tau = 0$$

It follows that grad V = grad U at all points of R so that V and U can differ only by a constant. But V = U upon the surfaces so that V and U are equal throughout R, hence V is uniquely determined.

Theorem 3.2-4a

If V is harmonic everywhere outside the surfaces $S_{1..n}$ and its value is specified at each point of these surfaces, then V is determined uniquely at all points outside $S_{1..n}$.

Proof: If the point function U is harmonic outside $S_{1..n}$ and has the same value as V at all points of these surfaces, and if Σ is a closed surface surrounding $S_{1..n}$, then from equation (3.1-4)

$$\oint_{S_{1..n}} (V-U) \frac{\partial}{\partial n}(V-U) dS + \oint_\Sigma (V-U) \frac{\partial}{\partial n}(V-U) dS = \int_\tau |\text{grad}(V-U)|^2 \, d\tau$$

The surface integral over $S_{1..n}$ is zero because $V - U = 0$ at each point. As Σ recedes to infinity the associated surface integral disappears because $(V-U) \frac{\partial}{\partial n}(V-U)$ vanishes at least as $\frac{1}{R^3}$ while the surface area increases as R^2.

Hence

$$\int_\tau |\text{grad}(V-U)|^2 \, d\tau = 0$$

where τ includes all space beyond $S_{1..n}$.

The arguments of the previous theorem then apply.

Theorem 3.2-5

If V is harmonic in the closed region R bounded by the surfaces $S_{1..n}\Sigma$ and is constant over each of these surfaces in turn, and if the surface integral of the normal derivative of V is specified for each surface, then V is determined throughout R to within an additive constant.

Proof: If the point function U also satisfies the given conditions, then from equation (3.1-4)

$$(V-U)_{S_1} \oint_{S_1} \frac{\partial}{\partial n}(V-U)dS \text{ ------ } + (V-U)_{\Sigma} \oint_{\Sigma} \frac{\partial}{\partial n}(V-U)dS = \int_{\tau} |\text{grad}(V-U)|^2 \, d\tau$$

In this case V-U is an unknown constant for each surface. The equality of $\oint \frac{\partial V}{\partial n} dS$ and $\oint \frac{\partial U}{\partial n} dS$ for each surface is sufficient to reduce the total surface integral to zero, so that

$$\int_{\tau} |\text{grad}(V-U)|^2 \, d\tau = 0$$

Previous arguments then shown that V and U can differ only by a constant within R.

It should be noted that the theorem continues to hold when the surface integral of the normal derivative is specified for all but one of the surfaces; this is sufficient to define its value over the remaining surface in accordance with Theorem 3.2-1.

Theorem 3.2-5a

If V is harmonic everywhere outside the surfaces $S_{1..n}$ and is constant over each of the surfaces in turn, and if the surface integral of the normal derivative of V is specified for each surface, then V is uniquely determined on and outside the surfaces.

Proof: Suppose that U also satisfies the given conditions and that Σ is a closed surface surrounding $S_{1..n}$. Then from equation (3.1-4)

$$(V-U)_{S_1} \oint_{S_1} \frac{\partial}{\partial n}(V-U)dS \text{ ---- } + \oint_{\Sigma} (V-U) \frac{\partial}{\partial n}(V-U)dS = \int_{\tau} |\text{grad}(V-U)|^2 \, d\tau$$

As in the previous theorem the terms associated with the surfaces $S_{1..n}$ vanish. As Σ recedes to infinity the associated surface integral approaches zero (see Theorem 3.2-4a), so that

$$\int_{\tau} |\text{grad}(V-U)|^2 \, d\tau = 0$$

where τ includes all space beyond $S_{1..n}$.

It follows that V and U can differ only by a constant on and outside $S_{1..n}$, but this constant is zero since both V and U vanish at infinity.

Theorem 3.2-6

If V is harmonic in the closed region $\underline{R}$ bounded by the surfaces $S_{1..n}\Sigma$ and its normal derivative on the surfaces is a specified function of position, then V is determined throughout $\underline{R}$ and upon the surfaces to within an additive constant.

Theorem 3.2-6a

If V is harmonic everywhere outside the surfaces $S_{1..n}$ and its normal derivative on the surfaces is a specified function of position, then V is determined uniquely on and outside the surfaces.

Theorems 3.2-6 and 3.2-6a follows from equation (3.1-4) in much the same way as the other uniqueness theorems.

Theorem 3.2-7

If V is harmonic in a closed region $\underline{R}$, then V cannot be a maximum or a minimum at any interior point of $\underline{R}$.

Proof: Suppose that V is a maximum at some interior point 0 of $\underline{R}$. If S' is a spherical surface of radius δ centred upon 0 then, for sufficiently small values of δ, $\frac{\partial V}{\partial r}$ must be negative at all points of the surface, where r is distance measured from 0. The surface integral of $\frac{\partial V}{\partial r}$ over S' must likewise be negative and is given by

$$\oint_{S'} \frac{\partial V}{\partial r} \, dS = \oint_{S'} \frac{\partial V}{\partial n} \, dS = \oint_{S'} \text{grad } V.d\overline{S}$$

But

$$\oint_{S'} \text{grad } V.d\overline{S} = \int_{\tau'} \nabla^2 V \, d\tau = 0$$

where τ' is the volume enclosed by the sphere, hence the assumption that V is a maximum at 0 involves a contradiction. A similar contradiction appears when it is supposed that V is a minimum at 0. It follows that V can be neither a maximum nor a minimum at an interior point of $\underline{R}$.

3.3 Green's Formula

Let O (x_o,y_o,z_o) be a point exterior to the closed region R in Fig. 3.1, ie let O lie outside Σ or within the region enclosed by S_1 or S_2, etc. If the point function U in equation (3.1-2) is identified with $\frac{1}{r}$, where r is distance measured from O, then U will be harmonic in R. This is readily demonstrated by substitution in equation (2.6-8)[6]. In this case equation (3.1-2) reduces to

$$0 = \oint_{S_{1..n}\Sigma} \left\{\frac{1}{r}\frac{\partial V}{\partial n} - V\frac{\partial}{\partial n}\left(\frac{1}{r}\right)\right\} dS - \int_{\tau} \frac{\nabla^2 V}{r}\, d\tau \tag{3.3-1}$$

Now let O be an interior point of R. Since $\nabla^2 \frac{1}{r}$ is undefined at the origin or r, the point O must be excluded from the region of integration before equation (3.1-2) is applied. Suppose that a small sphere of radius δ is centred upon O. Then if $\tau-\tau_\delta$ represents the modified integration volume and S_δ denotes the surface of the sphere, equation (3.1-2) becomes

$$\oint_{S_{1..n}\Sigma} \left\{V\frac{\partial}{\partial n}\left(\frac{1}{r}\right) - \frac{1}{r}\frac{\partial V}{\partial n}\right\} dS + \oint_{S_\delta} \left\{V\frac{\partial}{\partial n}\left(\frac{1}{r}\right) - \frac{1}{r}\frac{\partial V}{\partial n}\right\} dS = -\int_{\tau-\tau_\delta} \frac{\nabla^2 V}{r}\, d\tau \tag{3.3-2}$$

Since the positive normal at the surface of the sphere is directed radially inwards, the surface integral over the sphere is given by

$$\oint_{S_\delta} \left\{-V\frac{\partial}{\partial r}\left(\frac{1}{r}\right) + \frac{1}{r}\frac{\partial V}{\partial r}\right\} dS = \frac{1}{\delta^2}\oint_{S_\delta} V\, dS + \frac{1}{\delta}\oint_{S_\delta} \frac{\partial V}{\partial r}\, dS$$

Now as $\delta\to 0$, $V\to V_o$ at all points of S_δ, so that the first term of the right-hand side of the above equation approaches the value $\frac{1}{\delta^2} V_o\, 4\pi\delta^2 = 4\pi V_o$. The second term approaches zero because $\frac{\partial V}{\partial r}$ is bounded over S_δ and $\frac{1}{\delta}\oint_{S_\delta} dS \to 0$ as $\delta\to 0$. It then follows from equation (3.3-2) that the volume integral over $\tau-\tau_\delta$ must likewise approach a limit as $\delta\to 0$. This may be demonstrated independently by integration over τ_δ (see Ex.3-1, p. 180).

6. For this purpose the origin of spherical coordinates is temporarily located at O. Note, however, that O is no longer identified with the origin of a fixed coordinate system.

If we write[7]

$$\lim_{\delta\to 0} \int_{\tau-\tau_\delta} \frac{\nabla^2 V}{r}\, d\tau \equiv \int_\tau \frac{\nabla^2 V}{r}\, d\tau$$

then

$$4\pi V_o = \oint_{S_{1..n}\Sigma} \left\{ \frac{1}{r}\frac{\partial V}{\partial n} - V\frac{\partial}{\partial n}\left(\frac{1}{r}\right) \right\} dS - \int_\tau \frac{\nabla^2 V}{r}\, d\tau \qquad (3.3\text{-}3)$$

Equations (3.3-1) and (3.3-3) are known as Green's formula. It should be noted that equation (3.3-1) is not a particular case of equation (3.3-3) where $V_o = 0$; the considerations leading to equation (3.3-1) take no account of the value of V_o, since 0 lies beyond the region of integration.

We see from equation (3.3-3) that when V is harmonic within $S_{1..n}\Sigma$ it may be expressed as a surface integral over these surfaces. If V should be harmonic everywhere beyond $S_{1..n}$ then only these surfaces need appear in an expression for V since the integral over Σ vanishes at infinity.

On the other hand, if V is well-behaved throughout all space and vanishes at infinity in such a way as to render the surface integral over Σ zero at infinite distance, then V may be expressed solely as a volume integral[8]. But in this case V cannot be harmonic everywhere since it must then vanish at all points.

The extension of Green's formula to a vector point function with the requisite degree of continuity is straightforward. By substituting F_x, F_y and F_z for V in equations (3.3-1) and (3.3-3), multiplying by $\bar{i}$, $\bar{j}$, $\bar{k}$ and adding, we obtain

---

7. It follows from a modification of the above analysis that the same limit obtains when S_δ is replaced by any regular closed surface which shrinks about 0 in such a way that its maximum chord approaches zero. When this is the case the integral is said to be convergent. The matter is discussed in greater detail in Sec. 4.4.

8. It is assumed here that the point of evaluation of V is located at finite distance from $S_{1..n}$ or from the region (supposed bounded) over which $\nabla^2 V$ is non-zero. However, this restriction may be shown to be unnecessary. (See Ex.4-43., p. 272 and Ex.4-58., p. 288.)

$$\left.\begin{matrix}\overline{0} \\ 4\pi\overline{F}_o\end{matrix}\right\} = \oint_{S_{1..n}\Sigma}\left\{\frac{1}{r}\frac{\partial\overline{F}}{\partial n} - \overline{F}\frac{\partial}{\partial n}\left(\frac{1}{r}\right)\right\} dS - \int_\tau \frac{\nabla^2\overline{F}}{r}\, d\tau \qquad (3.3\text{-}4)$$

3.4 Gauss's Integral

Solid Angle

On putting V = 1 in equations (3.3-1) and (3.3-3) we obtain

$$0 = -\oint_{S_{1..n}\Sigma}\frac{\partial}{\partial n}\left(\frac{1}{r}\right) dS \qquad \text{(where the origin of r is a point exterior to } \underline{R}.) \qquad (3.4\text{-}1)$$

$$4\pi = -\oint_{S_{1..n}\Sigma}\frac{\partial}{\partial n}\left(\frac{1}{r}\right) dS \qquad \text{(where the origin of r is an interior point of } \underline{R}.) \qquad (3.4\text{-}2)$$

For the case of a single closed surface, S, these equations reduce to

$$0 = -\oint_S\frac{\partial}{\partial n}\left(\frac{1}{r}\right) dS \qquad \text{(exterior origin)} \qquad (3.4\text{-}1a)$$

$$4\pi = -\oint_S\frac{\partial}{\partial n}\left(\frac{1}{r}\right) dS \qquad \text{(interior origin)} \qquad (3.4\text{-}2a)$$

The integral is known as Gauss's integral.

The solid angle subtended by a regular surface S at a point 0 (not lying upon the surface) is defined by

$$\Omega = \int_S \frac{\overline{r}}{r^3}\cdot d\overline{S} \qquad (3.4\text{-}3)$$

where $\overline{r}$ is the position vector of the element dS relative to 0 and r is distance measured from 0.

Since $\operatorname{grad}\frac{1}{r} = \hat{r}\frac{d}{dr}\left(\frac{1}{r}\right) = -\frac{\overline{r}}{r^3}$ at points other than 0,

$$\Omega = -\int_S \operatorname{grad}\frac{1}{r}\cdot d\overline{S} = -\int_S\frac{\partial}{\partial n}\left(\frac{1}{r}\right) dS \qquad (3.4\text{-}4)$$

hence from equations (3.4-1a) and (3.4-2a) the solid angle subtended at a point by a closed surface is zero or 4π according as the point lies beyond or within the enclosure.

It further follows that all open surfaces bounded by the same regular curve subtend equal solid angles at a point, provided that they lie on the same side of the point, when the positive sense of the normal at the surfaces is taken to correspond to a particular currency around the bounding curve (Ex.1-46., p. 64).

EXERCISES

3-1. Suppose that in the anaylsis leading to equation (3.3-3) S_δ is enclosed by a regular surface S. By applying (3.1-2) with $U = \frac{1}{r}$ to the region bounded by S and S_δ, shows that as S shrinks uniformly about 0

$$\oint_S \left\{V \frac{\partial}{\partial n}\left(\frac{1}{r}\right) - \frac{1}{r}\frac{\partial V}{\partial n}\right\} dS + \oint_{S_\delta} \left\{V \frac{\partial}{\partial n}\left(\frac{1}{r}\right) - \frac{1}{r}\frac{\partial V}{\partial n}\right\} dS \to 4\pi V_o$$

(where the positive normal is directed towards 0) and so demonstrate that $\int_\tau \frac{\nabla^2 V}{r}\, d\tau$ is convergent.

3-2. Use Green's formula to show that when V is harmonic in the closed region bounded by a spherical surface, its value at the centre of the sphere is equal to (a) its average value over the surface (b) its average value throughout the sphere.

[(a) is known as Gauss's average-value theorem.]

3-3. A square of side 2h and centre P is orientated parallel to the x and y axes of coordinates. If the square lies within a region throughout which the scalar point function V is well-behaved, show, by expansion in a Taylor series, that

$$\frac{1}{h^2}\{V(1) + V(2) + V(3) + V(4) - 4\,V(P)\} = \left(\frac{\partial^2 V}{\partial x^2} + \frac{\partial^2 V}{\partial y^2}\right)_P$$

correct to the third order of smallness, where 1, 2, 3, 4 are the centre points of the sides of the square.

Extend this to the three-dimensional case to show that

$$\frac{1}{h^2}\{V(1) + V(2) \text{ ---- } + V(5) + V(6) - 6\,V(P)\} = \left(\frac{\partial^2 V}{\partial x^2} + \frac{\partial^2 V}{\partial y^2} + \frac{\partial^2 V}{\partial z^2}\right)_P$$

to the same order of accuracy, where 1---6 are the centre points of the faces of a cube of edge 2h and centre P.

Show further that if the points 1 to 8 are the vertices of the cube, then

$$\frac{1}{4h^2}\{V(1) + V(2) --- + V(7) + V(8) - 8\,V(P)\} = \left(\frac{\partial^2 V}{\partial x^2} + \frac{\partial^2 V}{\partial y^2} + \frac{\partial^2 V}{\partial z^2}\right)_P$$

[When V is harmonic and two-dimensional or planar in the xy plane, we see that

$$V(1) + V(2) + V(3) + V(4) = 4\,V(P)$$

For the harmonic, three-dimensional case

$$V(1) + V(2) --- + V(5) + V(6) = 6\,V(P)$$

and

$$V(1) + V(2) --- + V(7) + V(8) = 8\,V(P)$$

These relationships constitute finite-difference forms of Gauss's average-value theorem. They are the bases of the numerical solution of harmonic fields with given boundary values, corresponding to square and cubic lattice subdivision of the region under consideration.]

3-4. Apply Green's theorem in the form (3.1-1) with V = 1 to a z-aligned cylindrical enclosure in a two-dimensional xy field (the section of the cylinder comprising any regular closed curve) and so show that for both two-dimensional and planar xy fields

$$\oint_\Gamma \left(\frac{\partial U}{\partial x}\,dr_y - \frac{\partial U}{\partial y}\,dr_x\right) = \int_{S_z} \left(\frac{\partial^2 U}{\partial x^2} + \frac{\partial^2 U}{\partial y^2}\right) dS_z$$

Note that appropriate replacement of symbols in Ex.1-43., p. 63 also leads to the above relationship.

3-5. Let R be the region bounded by a closed surface S_1 and a closed surface S_2 which surrounds S_1. If V and U are scalar point functions, harmonic in R and having different constant values over S_1 and S_2 in turn, show that at all points of R

$$\text{grad } V = \left\{\frac{V_{S_1} - V_{S_2}}{U_{S_1} - U_{S_2}}\right\} \text{grad } U$$

and

$$\frac{V_{S_1} - V_{S_2}}{\oint\limits_{S_1} \frac{\partial V}{\partial n}\,dS} = \frac{U_{S_1} - U_{S_2}}{\oint\limits_{S_1} \frac{\partial U}{\partial n}\,dS}$$

[Hint: Substitute $(V - \alpha U)$ for V in equation (3.1-3) and choose α to eliminate the surface integrals.]

3-6. If V is harmonic in the closed region R bounded by the surfaces $S_{1..n}\Sigma$, show that V is uniquely determined by R provided that (a) V is specified at every point of one or more surfaces (b) $\frac{\partial V}{\partial n}$ is specified at every point of certain of the remaining surfaces (c) V is constant over the rest of the surfaces and $\oint \frac{\partial V}{\partial n}\,dS$ is specified for each of these.

3-7. If V is harmonic in the closed region R bounded by the surfaces $S_{1..n}\Sigma$, show that V is uniquely determined in R provided that for all points of the surfaces $\frac{\partial V}{\partial n} + \alpha V = \beta$, where α and β are specified continuous functions of position and α is everywhere positive.

Extend this theorem to the unbounded region outside $S_{1..n}$.

3-8. If V is harmonic within the closed region R bounded by the surfaces $S_{1..n}\Sigma$, show that V is determined in R to within an additive constant provided that the vector tangential component of grad V (and hence the derivative of V in any tangential direction) is specified at each point of the surfaces, and $\oint \frac{\partial V}{\partial n}\,dS$ is specified for each surface in turn.

Extend this theorem to the unbounded region outside $S_{1..n}$.

[Theorems 3.2-5 and 3.2-5a represent particular cases of the above in which the tangential derivatives of V are zero at all points of the surfaces.]

3-9. If V is harmonic within the closed region R bounded (externally) by a single surface S, show that V is determined in R to within an additive constant provided that the tangential derivatives of V are specified at each point of S.

3-10. Derive equations (3.4-1) and (3.4-2) from equations (3.4-1a) and (3.4-2a) by superposition, remembering that the positive sense of the normal at any one surface may be required to reverse with change of position of the origin of r.

3-11. Show that if the position vectors drawn from a point 0 to all elements of a surface S cut a sphere of unit radius centred upon 0 not more than once at any point of the sphere, then the scalar area so defined upon the sphere is equal to the magnitude of the solid angle subtended at 0 by the surface.

3-12. Derive an expression for the solid angle subtended by a plane disc of radius a at a point on its axis at a distance d. Hence show that when an exterior point O approaches any interior point of a bounded plane surface along the normal on the positive side, the limiting value of the solid angle subtended at O by the surface is -2π. (cf Ex.1-47., p. 64)

$$\text{Ans:}\quad \Omega = \pm\, 2\pi \left\{1 - \frac{d}{(d^2 + a^2)^{\frac{1}{2}}}\right\}$$

3-13. Show that when a point O passes through an open regular surface S at some interior point of S, the solid angle subtended at O by S changes numerically by 4π.

[Hint: Close the system with a second surface and make use of equations (3.4-1a) and (3.4-2a).]

3-14. A line drawn normally from a point to the edge of a half-plane makes an angle θ with the half-plane. Use the result of Ex.3-11. to determine the magnitude of the solid angle subtended by the half-plane at the point.

Ans: $2(\pi - \theta)$

3-15. By substituting $\frac{1}{r}e^{\gamma r}$ for U in equation (3.1-2), where γ is a real constant, and proceeding as in the development of Green's formula, prove that

$$\left.\begin{matrix}4\pi V_o\\ 0\end{matrix}\right\} = \oint_{S_{1..n}\Sigma}\left\{\frac{1}{r}e^{\gamma r}\frac{\partial V}{\partial n} - V\frac{\partial}{\partial n}\left(\frac{1}{r}e^{\gamma r}\right)\right\}dS - \int_\tau \frac{1}{r}e^{\gamma r}\,(\nabla^2-\gamma^2)V\,d\tau$$

according as the origin of r lies within or without the integration space.

3-16. Let V be a scalar point function which is well-behaved throughout the region $\underline{R}$ bounded by surfaces $S_{1..n}\Sigma$, and let $(\nabla^2-\gamma^2)V$ be a specified function of position in $\underline{R}$, where γ is a real constant. Show that V is uniquely determined in $\underline{R}$ provided that V or $\frac{\partial V}{\partial n}$ is specified at all points of the surfaces, or provided that the boundary conditions of Ex.3-7. or 3-8. are satisfied.

3-17. By substituting $\frac{1}{r}\cos\alpha r$ and $\frac{1}{r}\sin\alpha r$ in turn for U in equation (3.1-2), where α is a constant, show that

$$\left.\begin{matrix}4\pi V_o\\ 0\end{matrix}\right\} = \oint_{S_{1..n}\Sigma}\left\{\frac{1}{r}\cos\alpha r\,\frac{\partial V}{\partial n} - V\frac{\partial}{\partial n}\left(\frac{1}{r}\cos\alpha r\right)\right\}dS - \int_\tau \frac{1}{r}\cos\alpha r\,(\nabla^2+\alpha^2)V\,d\tau$$

$$0 = \oint_{S_{1..n}\Sigma}\left\{\frac{1}{r}\sin\alpha r\,\frac{\partial V}{\partial n} - V\frac{\partial}{\partial n}\left(\frac{1}{r}\sin\alpha r\right)\right\}dS - \int_\tau \frac{1}{r}\sin\alpha r\,(\nabla^2+\alpha^2)V\,d\tau$$

3-18. If V is harmonic throughout a region R and V and $\frac{\partial V}{\partial n}$ are zero upon a regular surface element which bounds R in part or lies within it, then V is zero throughout R.

Prove this in the following way:

Let a sphere of radius a be so drawn as to project a short distance through the surface element into R, the intercepted portion of the element being designated S_1, the spherical cap S_2 and the enclosed region τ. If r is distance measured from the centre of the sphere, show, by identification of $\left(\frac{1}{r} - \frac{1}{a}\right)$ with U in equation (3.1-2) that

$$\frac{1}{a^2}\int_{S_2} V\, dS = \int_{\tau} \left(\frac{1}{r} - \frac{1}{a}\right) \nabla^2 V\, d\tau = 0$$

Assuming that τ is sufficiently small to ensure that V does not change sign within it, show that V must be zero throughout τ, and hence, by iteration, throughout R.

3.5 Treatment of Surface and Point Discontinuities in Scalar Fields

Suppose that the scalar point function V and/or its derivatives are discontinuous upon an open surface S which lies within the region R of Fig.3.1. We may exclude S from the integration space by surrounding it with a tightly-fitting envelope which comprises essentially the two surfaces S' and S" (Fig.3.2). The additional surface integral which then appears in equations (3.3-1) and (3.3-3) is

$$\int_{S'} \frac{1}{r}\frac{\partial V}{\partial n}\, dS - \int_{S'} V \frac{\partial}{\partial n}\left(\frac{1}{r}\right) dS + \int_{S''} \frac{1}{r}\frac{\partial V}{\partial n}\, dS - \int_{S''} V \frac{\partial}{\partial n}\left(\frac{1}{r}\right) dS$$

where the positive normal is directed in each case towards S[9].

There is a 1:1 correspondence of surface elements in S' and S" since they are indefinitely close together; in particular, the same value of r may be assigned to corresponding elements relative to any point outside S, and the positive normals are oppositely directed. The above integral expression may therefore by brought into the form

$$\int_{S'} \frac{1}{r}\left\{\left(\frac{\partial V}{\partial n}\right)_{S'} - \left(\frac{\partial V}{\partial n}\right)_{S''}\right\} dS - \int_{S'} (V_{S'} - V_{S''}) \frac{\partial}{\partial n}\left(\frac{1}{r}\right) dS$$

where the positive sense of the normal for both S' and S" is that defined by S',

---

9. It is assumed that the contribution to the surface integral from the strip which joins S' and S" at the boundary of S approaches zero as the envelope shrinks about S.

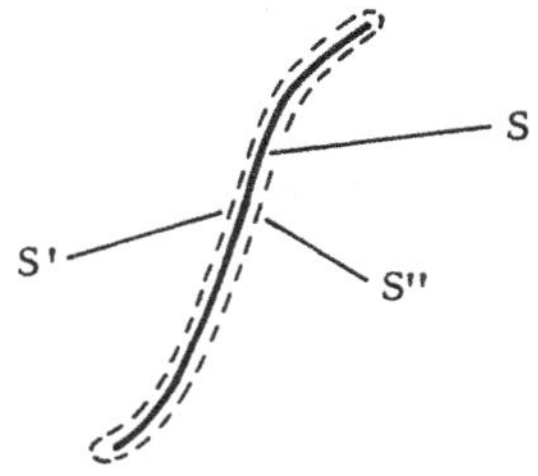

Fig. 3.2

or, alternatively,

$$\int_{S''} \frac{1}{r} \left\{ \left(\frac{\partial V}{\partial n}\right)_{S''} - \left(\frac{\partial V}{\partial n}\right)_{S'} \right\} dS - \int_{S''} (V_{S''} - V_{S'}) \frac{\partial}{\partial n} \left(\frac{1}{r}\right) dS$$

where the positive sense of the normal for both S' and S" is that defined by S".

Each of these two expressions is equivalent to

$$- \int_S \frac{1}{r} \Delta \left(\frac{\partial V}{\partial n}\right) dS + \int_S \Delta V \frac{\partial}{\partial n} \left(\frac{1}{r}\right) dS$$

where ΔV and $\Delta \left(\frac{\partial V}{\partial n}\right)$ are the increments of V and $\frac{\partial V}{\partial n}$ corresponding to positive motion through S when the same arbitrarily-defined positive sense of the normal is assigned to both sides of the surface, and $\frac{\partial}{\partial n}\left(\frac{1}{r}\right)$ is the associated normal derivative of $\frac{1}{r}$.

The modified form of Green's formula then becomes

$$\left.\begin{matrix} 0 \\ 4\pi V_o \end{matrix}\right\} = \oint_{S_{1..n}\Sigma} \left\{ \frac{1}{r} \frac{\partial V}{\partial n} - V \frac{\partial}{\partial n} \left(\frac{1}{r}\right) \right\} dS + \int_S \left\{ \Delta V \frac{\partial}{\partial n} \left(\frac{1}{r}\right) - \frac{1}{r} \Delta \left(\frac{\partial V}{\partial n}\right) \right\} dS - \int_\tau \frac{\nabla^2 V}{r} d\tau \tag{3.5-1}$$

where $\int_\tau$ is now employed to denote the limiting value approached by the volume integral as the S'S" surface shrinks about S (and the δ sphere shrinks about 0).

This equation continues to apply when S is closed, so long as τ includes the region embraced by S. As before, the right-hand side is zero when 0 lies inside S_1 or S_2 etc, or outside Σ. Alternatively, we may write

$$\left.\begin{matrix}0\\4\pi V_o\end{matrix}\right\} = \oint\limits_{S_{1..n}\Sigma,S}\left\{\frac{1}{r}\frac{\partial V}{\partial n} - V\frac{\partial}{\partial n}\left(\frac{1}{r}\right)\right\}dS - \int\limits_{\tau-\tau_S}\frac{\nabla^2 V}{r}\,d\tau \tag{3.5-2}$$

where τ_S is the region enclosed by S.

In this case the right-hand side is zero when 0 lies inside S, S_1 or S_2 etc, or outside Σ. Since S represents a surface of discontinuity, V and $\frac{\partial V}{\partial n}$ are evaluated at points lying just within the region of integration.

Now suppose that points of discontinuity[10] (singular points) are present within R. Let a sphere of radius ε be centred upon some such point P. If, within this ε sphere, V takes the form $f(r') + V'$, where r' is distance measured from P and V' is a well-behaved point function, then the surface integral over the sphere is given by

$$\oint\limits_{S_\varepsilon}\left\{\frac{1}{r}\frac{\partial V}{\partial n} - V\frac{\partial}{\partial n}\left(\frac{1}{r}\right)\right\}dS$$

$$= \oint\limits_{S_\varepsilon}\left\{-\frac{1}{r}\frac{\partial}{\partial r'}(f(r')+V') - (f(r')+V')\frac{\partial}{\partial n}\left(\frac{1}{r}\right)\right\}dS$$

$$= -f'(\varepsilon)\oint\limits_{S_\varepsilon}\frac{1}{r}\,dS - \oint\limits_{S_\varepsilon}\frac{1}{r}\frac{\partial V'}{\partial r'}\,dS - f(\varepsilon)\oint\limits_{S_\varepsilon}\frac{\partial}{\partial n}\left(\frac{1}{r}\right)dS - \oint\limits_{S_\varepsilon}V'\frac{\partial}{\partial n}\left(\frac{1}{r}\right)dS$$

Since the origin of r lies beyond the ε sphere, the third term of this expression is zero. The first term may be shown[11] to be equal to $-f'(\varepsilon)\,\frac{4\pi\varepsilon^2}{r_p}$ for all values of $\varepsilon < r_p$ so long as the ε sphere does not cut other bounding surfaces; in any case, it clearly reduces to this as ε approaches zero. Furthermore, as $\varepsilon\to 0$, the remaining terms approach

$$\left(\frac{1}{r}\,\text{grad}\,V'\right)_P\cdot\oint\limits_{S_\varepsilon}d\bar{S} - V'_P\oint\limits_{S_\varepsilon}\frac{\partial}{\partial n}\left(\frac{1}{r}\right)dS$$

---

10. ie discontinuity of V or its first or second derivatives.

11. See Ex.3-19., p. 188.

and each of these is zero, hence

$$\lim_{\varepsilon\to 0} \oint_{S_\varepsilon} \left\{ \frac{1}{r}\frac{\partial V}{\partial n} - V\frac{\partial}{\partial n}\left(\frac{1}{r}\right) \right\} dS = -f'(\varepsilon)\,\frac{4\pi\varepsilon^2}{r_p}$$

Cases of particular interest are those for which $f(r') = \frac{a}{r'^n}$ where a is a constant. The limiting value of the surface integral over S_ε is then $\frac{4\pi a}{r_p} n\varepsilon^{1-n}$. If $n > 1$ this expression approaches infinity as $\varepsilon\to 0$; the associated volume integral consequently approaches infinity in the same way since the difference between the two integrals remains constant[12]. When $n = 1$ the limiting value is $\frac{4\pi a}{r_p}$ and when $n < 1$ the limiting value is zero. Hence for $n = 1$ and $n < 1$ Green's formula becomes respectively

$$\left.\begin{matrix} 0 \\ 4\pi V_o \end{matrix}\right\} = \oint_{S_{1..n}\Sigma} \left\{ \frac{1}{r}\frac{\partial V}{\partial n} - V\frac{\partial}{\partial n}\left(\frac{1}{r}\right) \right\} dS + \frac{4\pi a}{r_p} - \int_\tau \frac{\nabla^2 V}{r}\, d\tau \tag{3.5-3}$$

and

$$\left.\begin{matrix} 0 \\ 4\pi V_o \end{matrix}\right\} = \oint_{S_{1..n}\Sigma} \left\{ \frac{1}{r}\frac{\partial V}{\partial n} - V\frac{\partial}{\partial n}\left(\frac{1}{r}\right) \right\} dS - \int_\tau \frac{\nabla^2 V}{r}\, d\tau \tag{3.5-4}$$

where $\int_\tau$ denotes the limiting value of the volume integral as the ε sphere shrinks about P and the δ sphere shrinks about O.

These formulae may be extended directly to cover multiple point singularities.

Before leaving the subject it should be remarked that while Green's formula and its variants may be applied to any scalar field having the requisite degree of continuity, no useful purpose is served by its application to a field whose value is arbitrarily assigned from point to point (within the limitations imposed by requirements of continuity). In this case its value at the point O is uncorrelated with its value outside a neighbourhood of O, so that the ability to express V_o in terms of V and its derivatives elsewhere is largely illusory. Indeed, it is clear from

---

12. This, of course, is true in all cases where the alteration of a bounding surface changes both the associated surface and volume integrals, so long as discontinuities continue to be excluded from the integration space.

equation (3.3-1), when one of the surfaces $S_{1..n}$ is identified with S_δ, that the contribution to V_o of the component of the field outside a neighbourhood of 0 is necessarily zero. However, when the V field is derived by some mathematical process from a primary field, and in such a way that the value of V at any point is dependent upon the value of the primary field at all points, then V is determinate and inter-related everywhere and the interpretation of Green's formula is no longer trivial. It is in this context that Green's formula is applied in subsequent pages.

EXERCISES

3-19. Let the scalar point function V be defined by $V = \frac{1}{r'}$, where r' is distance measured from some point P. If 0 is a point not coincident with P and r is distance measured from 0, use Green's formula (3.3-3) to express V_o in terms of a surface integral over a spherical surface S of radius R centred upon P for the case $R < r_p$, by showing that both the surface integral at infinity and the volume integral outside S are zero. Hence prove that

$$\oint_S \frac{dS}{r} = \frac{1}{r_p} \oint_S dS = \frac{4\pi R^2}{r_p}$$

Confirm this by applying equation (3.3-1) to the region bounded externally by S and internally by a small auxiliary sphere centred upon P, or, alternatively, by applying (3.5-3) to the region bounded externally by S.

Now suppose that $R > r_p$. Make use of equation (3.3-3) or (3.5-3) to show that in this case

$$\oint_S \frac{dS}{r} = \frac{1}{R} \oint_S dS = 4\pi R$$

These results have important consequences in potential theory and its applications.

3-20. Make use of equation (2.6-8) to obtain an approximation for the volume integral of $\frac{\nabla^2 V}{r}$ over a spherical shell centred upon the point P, given that $V = a/r'^n$ where r' is distance measured from P and a is a constant, ε and ε' are the internal and external radii of the shell and r is distance measured from some external point 0. Show that for small values of ε' and $n > 1$ the integral is approximately equal to $4\pi\, an(1/\varepsilon^{n-1} - 1/\varepsilon'^{n-1})/r_p$; for $n = 1$ it is zero for all non-zero ε; and for $n < 1$ it is given approximately by $4\pi\, an(\varepsilon^{1-n} - \varepsilon'^{1-n})/r_p$. Confirm that these results are consistent in respect of limiting behaviour with the results obtained for the surface integral over the ε

3-21. A scalar point function V is well-behaved throughout the region bounded by $S_{1--n}\Sigma$. Its value at an interior point 0 is consequently given by equation (3.3-3).

The magnitude of V is now altered at an interior point P by the addition of a component which takes the form $\frac{a}{r'}$ between $r' = 0$ and $r' = r'_1$, where r' is distance measured from P. Between $r' = r'_1$ and $r' = r'_2$ the additional component tapers smoothly to zero. The value of V at 0 is obviously unaffected by the modification at P, provided that $r_p > r'_2$, but equation (3.5-3) is now applicable.

Since $\nabla^2 V$ has the same value as for the unmodified field in the regions defined by $r' < r'_1$ $(r' \neq 0)$ and $r' > r'_2$, and since the transitional region may be made very small, how does one dispose of the unwanted term $\frac{4\pi a}{r_p}$?

Ans: The volume integral of $-\frac{\nabla^2 V}{r}$ for the additional component, when taken over the transitional region, just cancels the unwanted term, irrespective of the size of this region. (Prove this by evaluation of the integral for the particular case in which the additional component takes the form f(r') between $r' = r'_1$ and $r' = r'_2$, bearing in mind that $f'(r'_2) = 0$ and $f'(r'_1) = -\frac{a}{r'^2_1}$.)

3.6 Uniqueness Theorem for Scalar Fields

The uniqueness theorems set out above for harmonic fields (Theorems 3.2-4 to 3.2-6a) are valid for non-harmonic fields containing surface and point discontinuities so long as certain additional requirements are met[13].

Suppose that the region R of Fig.3.1 contains an open or closed surface of discontinuity S and an isolated point of discontinuity P. Let ΔV and $\Delta \frac{\partial V}{\partial n}$ be specified functions of position over S and let V take the form $f(r') + V'$ in a neighbourhood of P as discussed in the previous section. Let $\nabla^2 V$ be a specified function of position at all points of R not coincident with S or P. Then the additional surface specification laid down in any of the above theorems is sufficient to ensure that V (or, at worst, grad V) is uniquely determined at all points of R beyond S and P. This may be demonstrated as follows.

13. This also applies to the theorems presented in Ex.3-6. to 3-9., p. 182.

Suppose that the scalar point function U has the same specifications as V. From equation (3.1-4)

$$\oint_{S_{1..n}\Sigma} (V-U)\,\frac{\partial}{\partial n}(V-U)\,dS + \int_{S'S''} (V-U)\,\frac{\partial}{\partial n}(V-U)\,dS + \oint_{S_\varepsilon} (V-U)\,\frac{\partial}{\partial n}(V-U)\,dS$$
$$= \int_{\tau'} (V-U)\,\nabla^2(V-U)\,d\tau + \int_{\tau'} |\mathrm{grad}(V-U)|^2\,d\tau \qquad (3.6\text{-}1)$$

where S' and S" are paired surfaces closely fitting S, and S_ε is a small spherical surface of radius ε centred upon P. The integration volume τ' comprises R minus the regions enclosed by S'S" and S_ε.

If S should be closed, the above equation is formed by the addition of two equations, one referring to the region within S and the other to that outside S and within $S_{1..n}\Sigma$.

The specifications stated in one or other of the uniqueness theorems 3.2-4 to 3.2-6a are sufficient to ensure the vanishing of the first surface integral of equation (3.6-1). The first volume integral is zero because $\nabla^2 V = \nabla^2 U$ throughout the integration region. If the remaining surface integrals can be shown to be zero, then, for reasons discussed previously, grad V = grad U throughout the region under consideration.

The surface integral associated with S' and S" may be written as

$$\int_S (V-U)_{S'}\,\frac{\partial}{\partial n_{S'}}(V-U)_{S'}\,dS - \int_S (V-U)_{S''}\,\frac{\partial}{\partial n_{S'}}(V-U)_{S''}\,dS$$

where $\frac{\partial}{\partial n_{S'}}$ denotes differentiation along the positive normal as defined by S'

Now

$$V_{S'} - V_{S''} = U_{S'} - U_{S''}$$

or

$$(V-U)_{S'} = (V-U)_{S''}$$

and

$$\frac{\partial V_{S'}}{\partial n_{S'}} - \frac{\partial V_{S''}}{\partial n_{S'}} = \frac{\partial U_{S'}}{\partial n_{S'}} - \frac{\partial U_{S''}}{\partial n_{S'}}$$

or

$$\frac{\partial}{\partial n_{S'}} (V-U)_{S'} = \frac{\partial}{\partial n_{S'}} (V-U)_{S''}$$

hence the surface integrals over S' and S" cancel.

On choosing ε to bring S_ε into that neighbourhood of P where $V = f(r') + V'$ and $U = f(r') + U'$ the surface integral over S_ε becomes

$$\oint_{S_\varepsilon} (V'-U') \frac{\partial}{\partial n} (V'-U') \, dS$$

In the limit as $\varepsilon \to 0$ this reduces to

$$((V'-U') \operatorname{grad}(V'-U'))_P \cdot \oint_{S_\varepsilon} d\bar{S} = 0$$

It follows that all surface integrals are zero and that grad V = grad U everywhere in $\underline{R}$ beyond S and P. Further, if V be specified at any point of $S_{1..n}\Sigma$, then V is uniquely determined throughout $\underline{R}$.

The analysis continues to hold in the absence of the bounding surface Σ, so long as V is regular at infinity. The extension to multiple surface and point discontinuities is straightforward.

An important generalisation of the above treatment leads to the following theorem.

<u>Theorem 3.6-1</u>

Let the region $\underline{R}$, bounded by the surfaces $S_{1..n}\Sigma$, be divided into a set of sub-regions with interfaces S_a, S_b --- such that a given point function g and its first derivatives are continuous at interior points of the sub-regions but are discontinuous upon the interfaces, and let g, where defined, be everywhere positive or everywhere negative. Further, let V be a point function with continuous second derivatives in $\underline{R}$ except upon S_a, S_b ---, and upon certain surfaces S_α, S_β ---- and points P_1, P_2 --- lying within the sub-regions.

Then V is rendered piecewise determinate in $\underline{R}$, at least to within the same additive constant, by the specification of all of the following factors:

(1) div(g grad V) at all interior points of the sub-regions excluding S_α, S_β ---, P_1 P_2 ---.

(2) V or $\frac{\partial V}{\partial n}$ at each point of $S_{1..n}\Sigma$, or $\oint g \frac{\partial V}{\partial n} dS$ for any bounding surface together with the tangential derivatives of V at each point of the surface, or, in the case of a single bounding surface, the tangential derivatives alone. (If $S_{1..n}\Sigma$ comprise surfaces of discontinuity the expressions should be evaluated just within the integration space.)

(3) ΔV and $\Delta\left(g \frac{\partial V}{\partial n}\right)$ through all points of the surfaces S_a, S_b --- S_α, S_β ---.

(4) $V = V' + V''$ in a neighbourhood of each of the points P_1, P_2 --- where V' is an individually specified discontinuous point function and V" is any continuous point function.

The proof of this theorem, which involves the expansion of the function div((V-U) g grad(V-U)) and its subsequent integration (cf Ex.1-68., p. 79) is left as an exercise for the reader, who should extend it to the (externally) unbounded case. When $g = 1$ at all points, the problem reduces to that treated above.

The planar equivalent is set out in Ex.3-46, p. 215.

3.7 Theorems Relating to Vector Fields
Vector Analogue of Green's Theorem

Theorem 3.7-1

If a vector point function $\overline{F}$ has continuous first derivatives and specified values of curl and divergence throughout the closed, simply connected region R bounded by the surfaces $S_{1..n}\Sigma$, then $\overline{F}$ is uniquely determined in R provided that either

(1) the normal component of $\overline{F}$ is specified at each point of the surfaces, or

(2) $\oint \overline{F}.d\overline{S}$ is specified for each surface in turn and $\hat{n} \times \overline{F}$ is specified at each point of the surfaces, where $\hat{n}$ is the unit normal to the surface. (The specification of $\hat{n} \times \overline{F}$ is equivalent, for known values of $\hat{n}$, to the specification of $(\hat{n} \times \overline{F}) \times \hat{n}$ and this represents the vector tangential component of $\overline{F}$ upon the surface.)

In the event that the region R is bounded (externally) by a single surface S, then (2) above reduces to the single requirement that $\hat{n} \times \overline{F}$ be specified at each point of S.

Both conditions (1) and (2) are met by the more severe requirement that $\overline{F}$ be specified at each point of the surfaces.

Proof: Let $\overline{F}_1$ and $\overline{F}_2$ be vector point functions having the specified values of curl and divergence in $\underline{R}$. Then $\text{curl}(\overline{F}_1-\overline{F}_2) = \overline{0}$ and $\text{div}(\overline{F}_1-\overline{F}_2) = 0$, whence $\overline{F}_1 - \overline{F}_2 = \text{grad } V$ and $\text{div grad } V = \nabla^2 V = 0$ in $\underline{R}$. If F_n is specified over $S_{1..n}\Sigma$, $(\overline{F}_1-\overline{F}_2).\hat{n} = 0$ or $\frac{\partial V}{\partial n} = 0$ at each point of the surfaces. It then follows from Theorem 3.2-3 that V is constant and $\overline{F}_1 = \overline{F}_2$ in $\underline{R}$.

If $\oint \overline{F}.d\overline{S}$ is specified for each surface in turn, $\oint (\overline{F}_1-\overline{F}_2).d\overline{S} = 0$ or $\oint \frac{\partial V}{\partial n}\, dS = 0$ for each surface. If, in addition, $\hat{n} \times \overline{F}$ is specified at each point of the surfaces, $\hat{n} \times(\overline{F}_1-\overline{F}_2) = \overline{0}$, whence the tangential component of grad V is zero and V is constant over each surface in turn. It then follows from equation (3.1-3) that V is constant and $\overline{F}_1 = \overline{F}_2$ in $\underline{R}$.

For the case of a single enclosing surface, the constancy of V over this surface and its harmonic nature within the enclosure are sufficient to ensure that V is constant and $\overline{F}_1 = \overline{F}_2$ in $\underline{R}$ (Theorem 3.2-2).

Theorem 3.7-2

If a vector point function $\overline{F}$ has continuous first derivatives in the closed, simply connected region $\underline{R}$ bounded by the surfaces $S_{1..n}\Sigma$, and if curl $\overline{F}$ and div $\overline{F}$ are zero in $\underline{R}$, then $\overline{F}$ is zero in $\underline{R}$ provided that either

(1) the normal component of $\overline{F}$ is zero at each point of the surfaces

or

(2) $\oint \overline{F}.d\overline{S}$ is zero for each surface and $\hat{n} \times \overline{F}$ is zero at each point of the surfaces.

In the event that the region $\underline{R}$ is bounded (externally) by a single surface S, then (2) above reduces to the single requirement that the tangential component of $\overline{F}$ be zero at each point of S.

Both (1) and (2) are met by the requirement that $\overline{F}$ be zero at all points of the surfaces.

The proof is similar to that of Theorem 3.7-1.

Theorem 3.7-3

If $\overline{F}$ has continuous second derivatives in the closed region $\underline{R}$ and $\nabla^2\overline{F}$ is a specified function of position in $\underline{R}$ while $\overline{F}$ is specified at each point of the surfaces, then $\overline{F}$ is uniquely determined in $\underline{R}$. If $\frac{\partial \overline{F}}{\partial n}$, rather than $\overline{F}$, is specified at each point of the surfaces, then $\overline{F}$ is determined in $\underline{R}$ to within an additive constant vector.

Theorem 3.7-4

If $\overline{F}$ has continuous second derivatives in the closed region R and $\nabla^2\overline{F} = \overline{0}$ in R, while $\overline{F}$ has the same value at all points of the bounding surfaces, then $\overline{F}$ is constant in R.

Theorems 3.7-3 and 3.7-4 are easily proved by the resolution of $\overline{F}$ into rectangular components.

Theorem 3.7-5

If the vector point functions $\overline{F}$ and $\overline{G}$ are continuously differentiable throughout all space and curl $\overline{F}$ and div $\overline{G}$ are everywhere zero while $R^2|\overline{F}|$ and $R^2|\overline{G}|$ are bounded as $R\to\infty$, then $\int \overline{F}.\overline{G}\, d\tau$ is zero when taken over all space.

Proof: Since $\overline{F}$ is irrotational we may write $\overline{F} = \text{grad } V$.

But

$$\text{div } V\overline{G} = V \text{ div } \overline{G} + \text{grad } V.\overline{G} = \overline{F}.\overline{G}$$

hence

$$\oint_\Sigma V\, \overline{G}.d\overline{S} = \int_\tau \overline{F}.\overline{G}\, d\tau$$

where τ is the region bounded by some closed surface Σ.

The value of the surface integral must, of course, be independent of any constant component of V since the datum for the latter is arbitrarily chosen, and this is confirmed by the relationship

$$\oint_\Sigma V\, \overline{G}.d\overline{S} = V_o \oint_\Sigma \overline{G}.d\overline{S} + \oint_\Sigma \Delta V\, \overline{G}.d\overline{S}$$

$$= \oint_\Sigma \Delta V\, \overline{G}.d\overline{S} \qquad \text{(since } \overline{G} \text{ is solenoidal)}$$

where V_o may be taken as the value of V at some point of Σ.

If, now, Σ recedes to infinity, the increment of V over Σ will fall off at least as $\frac{1}{R}$ since ΔV is proportional to $R|\text{grad } V|$, and $R^2|\text{grad } V|$ is bounded. If, in addition, $R^2|\overline{G}|$ is bounded, the surface integral over Σ vanishes and the volume integral over all space is zero.

This result continues to hold when a finite portion of space is divided into a system of sub-regions throughout which $\overline{F}$ and $\overline{G}$ are piecewise differentiable, provided that the vector tangential component of $\overline{F}$ and the normal component of $\overline{G}$ are continuous through the interfaces. Since V may be chosen arbitrarily for any one point of a given sub-region, it is permissible to choose equal values for V at two adjacent surface points in each contiguous pair of sub-regions, and so ensure that V is everywhere continuous through the interfaces. The surface integrals over the interfaces then cancel in the sum leaving only the integral over the exterior bounding surface Σ, and this approaches zero as Σ recedes to infinity.

A vector analogue of Green's theorem can be derived for vector point functions which have continuous first and second derivatives in a closed region.

If $\overline{F}$ and $\overline{C}$ are such functions then from equation (1.16-7)

$$\text{div}(\overline{F} \times \text{curl}\ \overline{C}) = \text{curl}\ \overline{C}.\text{curl}\ \overline{F} - \overline{F}.\text{curl curl}\ \overline{C}$$

hence, from the divergence theorem,

$$\oint_{S_{1..n}\Sigma} (\overline{F} \times \text{curl}\ \overline{C}).d\overline{S} = -\int_\tau \overline{F}.\text{curl curl}\ \overline{C}\ d\tau + \int_\tau \text{curl}\ \overline{F}.\text{curl}\ \overline{C}\ d\tau \tag{3.7-1}$$

The symmetrical form of this equation is seen to be

$$\oint_{S_{1..n}\Sigma} (\overline{F} \times \text{curl}\ \overline{C} - \overline{C} \times \text{curl}\ \overline{F}).d\overline{S} = -\int_\tau (\overline{F}.\text{curl curl}\ \overline{C} - \overline{C}.\text{curl curl}\ \overline{F})\ d\tau \tag{3.7-2}$$

When $\overline{C} = \overline{F}$, equation (3-7.1) becomes

$$\oint_{S_{1..n}\Sigma} (\overline{F} \times \text{curl}\ \overline{F}).d\overline{S} = -\int_\tau \overline{F}.\text{curl curl}\ \overline{F}\ d\tau + \int_\tau |\text{curl}\ \overline{F}|^2\ d\tau \tag{3.7-3}$$

These equations may be considered to be vector equivalents of equations (3.1-1) to (3.1-3).

When grad div $\overline{F}$ = grad div $\overline{C} = \overline{0}$, curl curl may be replaced by $-\nabla^2$ in accordance with equation (1.18-5) and the formal analogy becomes complete. Otherwise, (3.7-2) is equivalent to

$$\oint_{S_{1..n}\Sigma} (\overline{F} \times \text{curl } \overline{C} - \overline{C} \times \text{curl } \overline{F}).d\overline{S} = -\int_\tau (\overline{F}.\text{grad div } \overline{C} - \overline{C}.\text{grad div } \overline{F})\, d\tau + \int_\tau (\overline{F}.\nabla^2\overline{C} - \overline{C}.\nabla^2\overline{F})\, d\tau \tag{3.7-4}$$

But

$$\text{div}(\overline{F} \text{ div } \overline{C}) = \text{div } \overline{F} \text{ div } \overline{C} + \overline{F}.\text{grad div } \overline{C}$$

whence

$$\oint_{S_{1..n}\Sigma} (\overline{F} \text{ div } \overline{C}).d\overline{S} = \int_\tau \text{div } \overline{F} \text{ div } \overline{C}\, d\tau + \int_\tau \overline{F}.\text{grad div } \overline{C}\, d\tau$$

and

$$\oint_{S_{1..n}\Sigma} (\overline{F} \text{ div } \overline{C} - \overline{C} \text{ div } \overline{F}).d\overline{S} = \int_\tau (\overline{F}.\text{grad div } \overline{C} - \overline{C}.\text{grad div } \overline{F})\, d\tau$$

hence equation (3.7-4) may be replaced by

$$\oint_{S_{1..n}\Sigma} (\overline{F} \times \text{curl } \overline{C} - \overline{C} \times \text{curl } \overline{F} + \overline{F} \text{ div } \overline{C} - \overline{C} \text{ div } \overline{F}).d\overline{S} = \int_\tau (\overline{F}.\nabla^2\overline{C} - \overline{C}.\nabla^2\overline{F})\, d\tau \tag{3.7-5}$$

Similarly equation (3.7-3) may be replaced by

$$\oint_{S_{1..n}\Sigma} (\overline{F} \times \text{curl } \overline{F} + \overline{F} \text{ div } \overline{F}).d\overline{S} = \int_\tau \{(\text{div } \overline{F})^2 + |\text{curl } \overline{F}|^2\}\, d\tau + \int_\tau \overline{F}.\nabla^2\overline{F}\, d\tau \tag{3.7-6}$$

EXERCISES

Prove the following propositions.

3-22. If a vector point function $\overline{F}$ has continuous first derivatives throughout the infinite region $\underline{R}$ bounded locally by the surfaces $S_{1..n}$ and has specified values of curl and divergence in $\underline{R}$, and if $R^2|\overline{F}|$ is bounded as $R\to\infty$, then $\overline{F}$ is uniquely determined in $\underline{R}$, provided that either

(1) the normal component of $\overline{F}$ is specified at each point of $S_{1..n}$

or

(2) $\oint \overline{F}.d\overline{S}$ is specified for each surface in turn and $\hat{\overline{n}} \times \overline{F}$ is specified at each point of the surfaces.

[Hint: If $\overline{F}_1$ and $\overline{F}_2$ satisfy the specifications, then $\overline{F}_1 - \overline{F}_2 = \text{grad } V$, and $\nabla^2 V = 0$ outside $S_{1..n}$. Write down equation (3.1-3) with V replaced by $V - V_o$, where V_o is the value of V at some point of Σ. Eliminate the surface integral over Σ by showing that as Σ recedes to infinity $\oint_\Sigma (V-V_o) \frac{\partial V}{\partial n} dS\to 0$.]

3-23. If a vector point function $\overline{F}$ has continuous first derivatives and zero curl and divergence throughout the infinite region $\underline{R}$ bounded locally by the surfaces $S_{1..n}$, and if $R^2|\overline{F}|$ is bounded as $R\to\infty$, then $\overline{F}$ is zero in $\underline{R}$, provided that either

(1) the normal component of $\overline{F}$ is zero at all points of the surfaces

or

(2) $\oint \overline{F}.d\overline{S}$ is zero for each surface and $\hat{\overline{n}} \times \overline{F}$ is zero at all points of the surfaces.

3-24. If a vector point function $\overline{F}$ has continuous second derivatives in the infinite region $\underline{R}$ bounded locally by the surfaces $S_{1..n}$, and if $\nabla^2\overline{F}$ is a specified function of position throughout $\underline{R}$ then

(1) $\overline{F}$ is uniquely determined in $\underline{R}$ if $\overline{F}$ is specified at each point of $S_{1..n}$ and if $R|\overline{F}|$ and $R^2|\nabla F_x|$ etc are bounded as $R\to\infty$

(2) $\overline{F}$ is determined to within an additive constant vector if $\frac{\partial\overline{F}}{\partial n}$ is specified at each point of the surfaces and $R^2|\nabla F_x|$ etc are bounded as $R\to\infty$.

3-25. If a vector point function $\overline{F}$ has continuous second derivatives in the infinite region $\underline{R}$ bounded locally by the surfaces $S_{1..n}$, and if $\nabla^2\overline{F} = \overline{0}$ in $\underline{R}$, then

(1) $\overline{F}$ is zero in $\underline{R}$ if $\overline{F}$ is zero over the surfaces and if $R|\overline{F}|$ and $R^2|\nabla F_x|$ etc are bounded as $R\to\infty$

(2) $\overline{F}$ is a constant vector in $\underline{R}$ if $\frac{\partial \overline{F}}{\partial n} = \overline{0}$ at each point of the surfaces and $R^2|\nabla F_x|$ etc are bounded as $R\to\infty$.

-oOOo-

Make use of equation (3.7-3) to verify the following propositions.

3-26. If $\overline{F}$ has continuous second derivatives in the region $\underline{R}$ bounded by the surfaces $S_{1..n}\Sigma$ and curl curl $\overline{F}$ is a specified function of position, then curl $\overline{F}$ is uniquely determined in $\underline{R}$ by the specification, at all points of the surfaces, of any one of the following factors:

(1) $\overline{F}$ (to within an additive constant vector)

(2) curl $\overline{F}$

(3) $\hat{\underline{n}} \times \overline{F}$

(4) $\hat{\underline{n}} \times$ curl $\overline{F}$

A further sufficient condition is the specification of $\hat{\underline{n}}$.curl $\overline{F}$ over $S_{1..n}\Sigma$. Prove this by showing that curl$(\overline{F}_1-\overline{F}_2)$ may be written as grad V, where $\overline{F}_1$ and $\overline{F}_2$ are possible fields, and by demonstrating that V is constant within $\underline{R}$. (It is supposed that $\underline{R}$ is simply connected.)

3-27. If, in Ex.3-26., curl curl $\overline{F}$ is everywhere zero, then curl $\overline{F} = \overline{0}$ in $\underline{R}$ provided that $\overline{F}$ or curl $\overline{F}$ is everywhere normal to the surfaces, or curl $\overline{F}$ is zero upon the surfaces, or $\overline{F}$ is constant upon the surfaces. Further, if both curl curl $\overline{F}$ and div $\overline{F}$ are zero in $\underline{R}$ and $\overline{F} = \overline{0}$ upon the surfaces, then $\overline{F}$ is zero in $\underline{R}$. (The latter proposition is a particular case of Theorem 3.7-4.)

3.8 Green's Function
The Dirichlet and Neumann Problems

Let U be harmonic and let V have continuous second derivatives in the closed region $\underline{R}$ bounded by the surfaces $S_{1..n}\Sigma$. If a small spherical surface of radius δ is centred upon some interior point 0 of $\underline{R}$, then it follows from equation (3.1-2) that

$$\oint_{S_{1..n}\Sigma} \left\{V \frac{\partial U}{\partial n} - U \frac{\partial V}{\partial n}\right\} dS + \oint_{S_\delta} \left\{V \frac{\partial U}{\partial n} - U \frac{\partial V}{\partial n}\right\} dS + \int_{\tau-\tau_\delta} U\, \nabla^2 V\, d\tau = 0$$

Since U and V are well-behaved throughout a neighbourhood of 0 the above equation leads to

$$\oint_{S_{1..n}\Sigma} \left\{ V \frac{\partial U}{\partial n} - U \frac{\partial V}{\partial n} \right\} dS + \int_{\tau} U \nabla^2 V \, d\tau = 0 \qquad (3.8\text{-}1)$$

where $\int_\tau$ represents the limiting value of the volume integral as the δ sphere shrinks about 0.

It has already been shown that if r is distance measured from 0, then

$$4\pi V_o = \oint_{S_{1..n}\Sigma} \left\{ \frac{1}{r} \frac{\partial V}{\partial n} - V \frac{\partial}{\partial n} \left(\frac{1}{r}\right) \right\} dS - \int_{\tau} \frac{\nabla^2 V}{r} \, d\tau \qquad (3.3\text{-}3)$$

where the volume integral has the above limiting significance[14].

On subtracting equation (3.8-1) from (3.3.3) we obtain

$$4\pi V_o = \oint_{S_{1..n}\Sigma} \left\{ \left(U + \frac{1}{r}\right) \frac{\partial V}{\partial n} - V \frac{\partial}{\partial n} \left(U + \frac{1}{r}\right) \right\} dS - \int_{\tau} \left(U + \frac{1}{r}\right) \nabla^2 V \, d\tau \qquad (3.8\text{-}2)$$

If a point function U_1 can be found which is harmonic in $\underline{R}$ and assumes the value $-\frac{1}{r}$ at each point of the surfaces $S_{1..n}\Sigma$, and if $G_1 = U_1 + \frac{1}{r}$, then equation (3.8-2) reduces to

$$4\pi V_o = -\oint_{S_{1..n}\Sigma} V \frac{\partial G_1}{\partial n} \, dS - \int_{\tau} G_1 \nabla^2 V \, d\tau \qquad (3.8\text{-}3)$$

There are physical grounds for supposing that U_1 must exist for most configurations of surfaces and positions of 0[15], and, this being so, it follows from Theorem 3.2-4 that U_1 will be unique in each situation. Correspondingly, G_1 is uniquely defined and harmonic except at the pole 0, where it becomes infinite as $\frac{1}{r}$; it is known as a Green's function for Laplace's equation. It is seen from (3.8-3) that by means of this function it has been possible to express the independent point function V

14. The δ sphere may be replaced by any regular region of which 0 is an interior point, as mentioned in the footnote to p. 178.

15. See Ex.3-28. to 3-30., pp. 209-10 and Kellogg, Ch.9

at any interior point of $\underline{R}$ in terms of $\nabla^2 V$ throughout $\underline{R}$ and V upon the bounding surfaces - the normal derivative of V which appeared in (3.3-3) has been eliminated. This result is not unexpected since the considerations of Sec. 3.6 show that V is uniquely determined in $\underline{R}$ when $\nabla^2 V$ is specified throughout $\underline{R}$ and V is specified at all points of the surfaces. It will be appreciated that a different Green's function is required for the evaluation of V at each point of $\underline{R}$.

Now suppose that a harmonic function U_2 can be found such that $\frac{\partial U_2}{\partial n} = -\frac{\partial}{\partial n}\left(\frac{1}{r}\right) + C$ at all points of $S_{1..n}\Sigma$, where C is a constant.

It is clear from Theorem 3.2-1 and equation (3.4-2) that $C = -4\pi/$(total area of $S_{1..n}\Sigma$). Then if $G_2 = U_2 + \frac{1}{r}$, equation (3.8-2) reduces to

$$4\pi V_o = \oint_{S_{1..n}\Sigma} G_2 \frac{\partial V}{\partial n}\, dS - C \oint_{S_{1..n}\Sigma} V\, dS - \int_\tau G_2\, \nabla^2 V\, d\tau \qquad (3.8\text{-}4)$$

so that V_o is now expressed in terms of $\nabla^2 V$ throughout $\underline{R}$ and $\frac{\partial V}{\partial n}$ upon the bounding surfaces, to within an additive constant.

It may be shown that such a representation is generally possible[16], and this might have been predicted on the basis of Sec. 3.6. The functions U_2 and G_2 are not unique since only $\frac{\partial U_2}{\partial n}$ has been specified over $S_{1..n}\Sigma$. If required, they can be made unique by the specification $\oint_{S_{1..n}\Sigma} U_2\, dS = 0$. A separate function is required for the evaluation of V at each point of $\underline{R}$. Like G_1, G_2 is known as a Green's function for Laplace's equation; it is sometimes called a Green's function of the second kind or a Neumann function.

The above considerations may be extended to the case where Σ recedes to infinity and V, which is taken to be regular at infinity, is evaluated at some point 0 outside the local surfaces $S_{1..n}$.

Since it is still possible to find an appropriate point function G_1, harmonic outside 0, zero upon $S_{1..n}$ and regular at infinity, the surface integral over Σ may be eliminated in equation (3.8-2), while (3.8-3) is replaced by

16. See Ex.3-31., p. 210.

$$4\pi V_o = -\oint_{S_{1..n}} V \frac{\partial G_1}{\partial n} dS - \int_\tau G_1 \nabla^2 V d\tau \qquad (3.8\text{-}3(a))$$

where τ includes all space outside $S_{1..n}$.

Equation (3.8-4) is replaced, in similar circumstances, by

$$4\pi V_o = \oint_{S_{1..n}} G_2 \frac{\partial V}{\partial n} dS - \int_\tau G_2 \nabla^2 V d\tau \qquad (3.8\text{-}4(a))$$

The term which involves C is missing in equation (3.8-4(a)) because $C \to 0$ and $C \oint_\Sigma dS \to -4\pi$ as Σ recedes to infinity; as a consequence, both $C \oint_{S_{1..n}} V\, dS$ and $C \oint_\Sigma V\, dS$ approach zero.

When $\nabla^2 V = 0$ in R, equation (3.8-3) reduces to

$$4\pi V_o = -\oint_{S_{1..n}\Sigma} V \frac{\partial G_1}{\partial n} dS \qquad (3.8\text{-}3(b))$$

This equation shows how a point function V, which is known to be harmonic in R, may be computed from its boundary values alone (given G_1), as would be expected from Theorem 3.2-4. We now enquire whether, corresponding to some arbitrary smooth distribution of surface values, there is a matching point function which is harmonic in R. This particular existence problem is known as Dirichlet's problem or the first boundary problem of potential theory[17]. It may, in fact, be shown that for most configurations of surfaces the required function does exist[18], hence its value may be computed via equation (3.8-3(b)). In particular: If f denotes a point function which varies smoothly over each of the surfaces $S_{1..n}\Sigma$, and if $G_1(0)$ is a Green's function for Laplace's equation with a pole at 0 and vanishing on the surfaces $S_{1..n}\Sigma$, then the formula

17. The specification of boundary values is known as the Dirichlet condition. When the boundary value is everywhere zero the condition is said to be homogeneous; otherwise it is non-homogeneous.

18. See Kellogg, Ch.11.

$$4\pi u_o = -\oint_{S_{1..n}\Sigma} f \frac{\partial}{\partial n} G_1(0)\, dS \qquad (3.8\text{-}5)$$

defines u at any point 0 of R in such a way as to make u harmonic in R and take on the value f at every point of the bounding surfaces.

The allied problem, associated with the harmonic form of (3.8-4), may be formulated as follows: "To find a point function v, harmonic in R and taking on prescribed values g of $\frac{\partial V}{\partial n}$ upon $S_{1..n}\Sigma$, given that g is continuous upon the surfaces and $\oint_{S_{1..n}\Sigma} g\, dS = 0$."

This is the Neumann problem or the second boundary problem of potential theory[19]. The required function may, in general, be shown to exist, in which case the solution is

$$4\pi v_o = \oint_{S_{1..n}\Sigma} G_2(0)\, g\, dS + \text{const.} \qquad (3.8\text{-}6)$$

The Dirichlet and Neumann problems continue to be soluble when R comprises all space outside the local surfaces $S_{1..n}$, the solutions being represented by equations (3.8-5) and (3.8-6) with the surface integrals over Σ and the constant deleted. No restriction is placed upon the value of $\oint_{S_{1..n}} g\, dS$.

3.9 Scalar Fields in Plane Regions

The earlier sections of this chapter have been devoted exclusively to those scalar point functions which are defined throughout a region of space. Similar analyses may be developed for point functions which are defined only in a plane. A number of planar formulae may be derived as in Ex.3-4., p. 181, by an application of the three-dimensional formulae to a two-dimensional (xy) field in space with appropriate bounding surfaces, and a subsequent cancellation of the common z factor; alternatively, the formulae may be developed for a region in the plane ab initio, as in Ex.1-31., p. 45.

---

19. The specification of the normal derivative upon the boundary is known as the Neumann condition.

The planar forms of Green's theorem, (3.1-1) and (3.1-2), are

$$\oint_{\Gamma_{1..n}\Gamma'} V \frac{\partial U}{\partial n'} \, ds = \int_S V \nabla^2 U \, dS + \int_S \text{grad } V.\text{grad } U \, dS \tag{3.9-1}$$

and

$$\oint_{\Gamma_{1..n}\Gamma'} \left\{ V \frac{\partial U}{\partial n'} - U \frac{\partial V}{\partial n'} \right\} ds = \int_S (V \nabla^2 U - U \nabla^2 V) \, dS \tag{3.9-2}$$

where $\Gamma_{1..n}$ are regular closed curves, none of which encloses or has a point in common with any other, and all of which are enclosed by the regular curve Γ'. S denotes the surface of the multiply connected region $\underline{R}$ of the plane bounded externally by Γ' and internally by $\Gamma_{1..n}$. Differentiation along the outward normal to the contour element ds is denoted by $\frac{\partial}{\partial n'}$; grad $V \equiv \bar{i} \frac{\partial V}{\partial x} + \bar{j} \frac{\partial V}{\partial y}$ and $\nabla^2 V \equiv \frac{\partial^2 V}{\partial x^2} + \frac{\partial^2 V}{\partial y^2}$

The theorems relating to harmonic functions in bounded regions of space have their counterparts for a region in the plane. The correspondence continues to hold for unbounded regions so long as the line integral around Γ' vanishes as Γ' recedes to infinity. We can ensure the disappearance of an integral of the type $\oint_{\Gamma'} V \frac{\partial V}{\partial n'} ds$ in this circumstance by stipulating that both V and R^2 grad V should be bounded in absolute value as $R \to \infty$, where R denotes distance from a local origin. This requirement is somewhat lighter than that associated with three-dimensional regularity at infinity because $\oint_{\Gamma'} ds$ increases linearly with R while $\oint_{\Sigma} dS$ increases as the square.

The same requirement also serves to eliminate the line integral $\oint_{\Gamma'} (V-U) \frac{\partial}{\partial n'}(V-U) \, ds$ which appears in the development of planar uniqueness theorems, but it is unnecessarily restrictive in this context since we are concerned with differences rather than absolute values. Indeed, a point function may increase without limit as $R \to \infty$ and still submit to a demonstration of uniqueness. Thus a planar harmonic function V may meet the specification

$$\left(V - \alpha \ln \frac{\beta}{R} \right) \to 0 \quad \text{like} \quad \frac{1}{R} \quad \text{as} \quad R \to \infty$$

$$\frac{\partial}{\partial R}\left(V-\alpha \ln \frac{\beta}{R}\right) \to 0 \quad \text{like } \frac{1}{R^2} \quad \text{as} \quad R\to\infty$$

where α and β are constants.

If a function U satisfies the same conditions then it is clear that both $R(V-U)$ and $R^2 \frac{\partial}{\partial R}(V-U)$ are bounded at infinity, so that

$$\oint_{\Gamma'} (V-U) \frac{\partial}{\partial n'}(V-U)\, ds \to 0 \quad \text{like } \frac{1}{R^2} \quad \text{as} \quad R\to\infty$$

despite the fact that both functions tend to a logarithmic infinity as $R\to\infty$.

The planar form of Green's formula follows from a substitution of $\ln \frac{1}{\rho}$ for U in equation (3.9-2) where ρ is distance measured from the point O at which V is to be determined[20]. When O lies within R the surface integration is carried out over $S-S_\delta$ where S_δ is the surface enclosed by a small circle[21] centred upon O. On taking limits as the circle shrinks about O it is found that

$$2\pi V_o = \oint_{\Gamma_{1..n}\Gamma'} \left\{\left(\ln \frac{1}{\rho}\right) \frac{\partial V}{\partial n'} - V \frac{\partial}{\partial n'}\left(\ln \frac{1}{\rho}\right)\right\} ds - \int_S \left(\ln \frac{1}{\rho}\right) \nabla^2 V \, dS \qquad (3.9\text{-}3)$$

When O lies outside R, $\ln \frac{1}{\rho}$ has no singularity within R and the surface integration is carried out over S in its entirety. The right-hand side of equation (3.9-3) is then found to be zero.

---

20. It follows from equation (2.6-4) that $\ln \frac{1}{\rho}$ is harmonic in the plane for non-zero values of ρ.

21. The circle may be replaced by any regular closed curve; the surface integral in equation (3.9-3) is consequently convergent.

A substitution of unity for V leads directly to Gauss's integral for the plane:

$$\left.\begin{matrix}2\pi\\0\end{matrix}\right\} = -\oint_{\Gamma_{1..n}\Gamma'} \frac{\partial}{\partial n'}\left(\ln\frac{1}{\rho}\right) ds \qquad (3.9\text{-}4)$$

according as the origin of ρ lies within or without $\underline{R}$.

When U is harmonic throughout $\underline{R}$ the addition of equations (3.9-2) and (3.9-3) yields the planar equivalent of equation (3.8-2), viz

$$2\pi V_o = \oint_{\Gamma_{1..n}\Gamma'} \left\{\left(U + \ln\frac{1}{\rho}\right)\frac{\partial V}{\partial n'} - V\frac{\partial}{\partial n'}\left(U + \ln\frac{1}{\rho}\right)\right\} ds - \int_S \left(U + \ln\frac{1}{\rho}\right)\nabla^2 V\, dS \qquad (3.9\text{-}5)$$

If U can be so chosen as to make $U = -\ln\frac{1}{\rho}$ at all point of $\Gamma_{1..n}\Gamma'$, then $U + \ln\frac{1}{\rho}$ defines a Green's function of the first kind in the plane with a pole at 0 for the particular set of contours involved, and (3.9-5) reduces to

$$2\pi V_o = -\oint_{\Gamma_{1..n}\Gamma'} V\frac{\partial G_1}{\partial n'} ds - \int_S G_1\nabla^2 V\, dS \qquad (3.9\text{-}6)$$

When V is harmonic everywhere outside $\Gamma_{1..n}$ and V and $\frac{\partial V}{\partial n'}$ assume appropriate limiting forms at infinite distance, it may be possible to express V_o in terms of local line integrals alone, although G_1 itself does not vanish at infinity. (See Ex.3-42. and 3-43. p. 213.)

It was shown in Sec. 3.5 that when V takes the form $\frac{a}{r'} + V'$ throughout a neighbourhood of some point P in a three-dimensional region (where r' is distance from P and V' is a well-behaved point function), the singularity at P gives rise to the term $\frac{a}{r_P}$ in the expression for V_o (equation (3.5-3)). In the planar case a singularity taking the parallel form $a\ln\frac{1}{\rho'} + w'$ gives rise to the corresponding term $a\ln\frac{1}{\rho_P}$.

3.10 Minimal Theorems

Theorem 3.10-1

Let V_1 and V_2 be scalar point functions, well-behaved in the closed region $\underline{R}$ bounded by the surfaces $S_{1..n}\Sigma$. If $V_1 = V_2$ at all points of the surfaces and V_1 is harmonic in $\underline{R}$ while V_2 is not, then

$$\int_\tau |\text{grad } V_1|^2 \, d\tau < \int_\tau |\text{grad } V_2|^2 \, d\tau \qquad (3.10\text{-}1)$$

Proof: Let $V_1 - V_2 = V'$. Then from equation (3.1-1)

$$\oint_{S_{1..n}\Sigma} V' \frac{\partial V_1}{\partial n} \, dS = \int_\tau V' \, \nabla^2 V_1 \, d\tau + \int_\tau \text{grad } V' . \text{grad } V_1 \, d\tau$$

Since $V' = 0$ on $S_{1..n}\Sigma$ and $\nabla^2 V_1 = 0$ in $\underline{R}$ it follows that

$$\int_\tau \text{grad } V' . \text{grad } V_1 \, d\tau = 0$$

or

$$\int_\tau |\text{grad } V_1|^2 \, d\tau = \int_\tau \text{grad } V_1 . \text{grad } V_2 \, d\tau$$

Now

$$\text{grad } V' . \text{grad } V' = (\text{grad } V_1 - \text{grad } V_2).(\text{grad } V_1 - \text{grad } V_2)$$

$$= |\text{grad } V_1|^2 + |\text{grad } V_2|^2 - 2 \, \text{grad } V_1 . \text{grad } V_2$$

hence

$$\int_\tau |\text{grad } V'|^2 \, d\tau = \int_\tau |\text{grad } V_1|^2 \, d\tau + \int_\tau |\text{grad } V_2|^2 \, d\tau - 2 \int_\tau |\text{grad } V_1|^2 \, d\tau$$

or

$$\int_\tau |\text{grad } V_1|^2 \, d\tau + \int_\tau |\text{grad } V'|^2 \, d\tau = \int_\tau |\text{grad } V_2|^2 \, d\tau \qquad (3.10\text{-}1(a))$$

Since V_2 is not harmonic throughout $\underline{R}$ there will be at least some subregion of $\underline{R}$ in which $\nabla^2(V_1-V_2) \neq 0$, ie div grad$(V_1-V_2) \neq 0$. Grad(V_1-V_2) cannot be continuously zero in this subregion hence its contribution to $\int_\tau |\text{grad } V'|^2 \, d\tau$ will be positive. The contribution of the remainder of $\underline{R}$ to the integral must be positive or zero hence the total integral is positive. This being so, equation (3.10-1(a)) leads directly to the required inequality.

The result may also be expressed in the following way: If V has prescribed values upon the surfaces $S_{1..n}\Sigma$ then

$$\int_\tau \left\{\left(\frac{\partial V}{\partial x}\right)^2 + \left(\frac{\partial V}{\partial y}\right)^2 + \left(\frac{\partial V}{\partial z}\right)^2\right\} d\tau$$

is a minimum when V is the solution of the associated Dirichlet problem.

Theorem 3.10-2

Let $\overline{F}_1$ and $\overline{F}_2$ be vector point functions with continuous first derivatives in the closed region $\underline{R}$ bounded by the surfaces $S_{1..n}\Sigma$, and let div $\overline{F}_1$ = div $\overline{F}_2$ in $\underline{R}$ and $(F_1)_n = (F_2)_n$ at all points of the surfaces. If $\overline{F}_1$ can be represented by $W(\text{grad } V + \overline{u})$ where W, V and $\overline{u}$ are well-behaved point functions, W being everywhere positive, and if $\overline{F}_1 \neq \overline{F}_2$ throughout some portion of $\underline{R}$, then

$$\int_\tau \frac{F_1^2}{W} d\tau - 2\int_\tau \overline{F}_1.\overline{u} \, d\tau < \int_\tau \frac{F_2^2}{W} d\tau - 2\int_\tau \overline{F}_2.\overline{u} \, d\tau \qquad (3.10\text{-}2)$$

where $F_1^2 = |\overline{F}_1|^2$ etc.

Proof: Since $(\overline{F}_1-\overline{F}_2).(\overline{F}_1-\overline{F}_2) = F_1^2 + F_2^2 - 2\overline{F}_1.\overline{F}_2$

$$\int_\tau \frac{|\overline{F}_1-\overline{F}_2|^2}{W} d\tau = \int_\tau \frac{F_1^2}{W} d\tau + \int_\tau \frac{F_2^2}{W} d\tau - 2\int_\tau \overline{F}_2.(\text{grad } V + \overline{u}) \, d\tau$$

Now

$$\text{div } V(\overline{F}_1-\overline{F}_2) = V \text{ div}(\overline{F}_1-\overline{F}_2) + (\overline{F}_1-\overline{F}_2).\text{grad } V$$

hence

$$\oint_{S_{1..n}\Sigma} V(\overline{F}_1-\overline{F}_2).d\overline{S} = \int_\tau V\,\mathrm{div}(\overline{F}_1-\overline{F}_2)\,d\tau + \int_\tau (\overline{F}_1-\overline{F}_2).\mathrm{grad}\,V\,d\tau$$

But $(F_1)_n - (F_2)_n = 0$ over $S_{1..n}\Sigma$ and $\mathrm{div}(\overline{F}_1-\overline{F}_2) = 0$ in $\underline{R}$, so that

$$\int_\tau (\overline{F}_1-\overline{F}_2).\mathrm{grad}\,V\,d\tau = 0$$

or

$$\int_\tau \overline{F}_1.\mathrm{grad}\,V\,d\tau = \int_\tau \overline{F}_2.\mathrm{grad}\,V\,d\tau$$

It then follows that

$$\int_\tau \frac{|\overline{F}_1-\overline{F}_2|^2}{W}\,d\tau = \int_\tau \frac{F_1{}^2}{W}\,d\tau + \int_\tau \frac{F_2{}^2}{W}\,d\tau - 2\int_\tau \overline{F}_2.\overline{u}\,d\tau - 2\int_\tau \overline{F}_1.\mathrm{grad}\,V\,d\tau$$

or

$$\int_\tau \frac{F_1{}^2}{W}\,d\tau - 2\int_\tau \overline{F}_1.\overline{u}\,d\tau + \int_\tau \frac{|\overline{F}_1-\overline{F}_2|^2}{W}\,d\tau = \int_\tau \frac{F_2{}^2}{W}\,d\tau - 2\int_\tau \overline{F}_2.\overline{u}\,d\tau$$

(3.10-2(a))

But $(\overline{F}_1-\overline{F}_2) \neq \overline{0}$ over some portion $\underline{R}$, hence $\int_\tau \frac{|\overline{F}_1-\overline{F}_2|^2}{W}\,d\tau$ is positive; equation (3.10-2) then follows from (3.10-2(a))

On putting $W = 1$ and $\overline{u} = \overline{0}$, equation (3.10-2) reduces to

$$\int_\tau F_1{}^2\,d\tau < \int_\tau F_2{}^2\,d\tau$$

From this inequality we deduce that if $\overline{F}$ has specified values of divergence in $\underline{R}$ and of normal component upon $S_{1..n}\Sigma$, then the volume integral of F^2 over $\underline{R}$ is a minimum if $\overline{F}$ can be expressed as the gradient of a scalar point function. This form of expression is certainly possible when $\mathrm{div}\,\overline{F} = 0$ since the required scalar function is then the solution of the Neumann problem for the particular set of surfaces involved and the value of the normal derivative corresponding to F_n.

EXERCISES

3.28. When heat flows through an isotropic medium of thermal conductivity k, the rate of heat transfer in the positive sense through a geometrical surface S within the medium is given by

$$\frac{dQ}{dt} = -\int_S k \text{ grad } T.d\bar{S}$$

where T is the (scalar) temperature.

If q is a point function representing the rate of heat generation per unit volume in the medium and ρ and c denote density and specific heat respectively, show that at interior points of the medium

$$\text{div}(k \text{ grad } T) + q = \rho c \frac{\partial T}{\partial t}$$

Hence show that in a homogeneous medium free from sources (and sinks) of heat[22], and under steady state conditions (ie temperature everywhere independent of time), the temperature field satisfies Laplace's equation, viz

$$\nabla^2 T = 0$$

3-29. A point source of heat of strength k[23] is located at the point P in a homogeneous isotropic medium of thermal conductivity k. If the medium were devoid of bounding surfaces the heat flow would exhibit radial symmetry; in the present instance, however, it is supposed that the medium is physically terminated in the surfaces $S_{1..n}\Sigma$, and that heat is extracted through these surfaces at an overall rate which balances the heat input.

Assuming that the bounding of the system produces only a finite perturbation of the radial flow field in some neighbourhood of P (it being supposed that P does not lie in any surface), make use of the arguments leading to equation (3.5-3) to demonstrate that the steady-state temperature field corresponding to some fixed surface distribution may be expressed as

$$T = \frac{1}{r'} + U'$$

where r' is distance measured from P and U' is some harmonic point function.

---

22. ie regions of generation (and extraction) of heat.

23. A source of unit strength is one which generates 4π units of heat in unit time.

3-30. Let the point source of heat in the previous exercise be now placed at some interior point 0. The steady-state temperature distribution becomes

$$T = \frac{1}{r} + U$$

where r is distance measured from 0 and U is harmonic throughout the medium.

Show that for the particular case in which all points of the surfaces are held at the constant temperature T', the point function T - T' is the Green's function G_1 appropriate to the surfaces $S_{1..n}\Sigma$ and the pole 0.

3-31. A point source of incompressible fluid of unit strength is located at an interior point 0 of a simply connected region $\underline{R}$ bounded by the surfaces $S_{1..n}\Sigma$. On the assumption that curl $\bar{v} = \bar{0}$ in $\underline{R}$, where $\bar{v}$ is the vector velocity field of flow, we may write $\bar{v} = -\text{grad}\,\psi$, where ψ is a point scalar known as the velocity potential. If fluid is extracted through the surfaces at the rate of $4\pi/(\text{total area of } S_{1..n}\Sigma)$ per unit area per unit time, show that the velocity potential of the resulting fluid flow is a Green's function G_2, to within an additive constant, for the surfaces $S_{1..n}\Sigma$ and the pole 0.

[Note that this continues to hold when the Σ surface recedes to infinity, in which case the local surfaces are required to be impervious to the flow.]

3-32. A point source of incompressible fluid of strength a_1 is located at an interior point of $\underline{R}$. At each point of the bounding surfaces $S_{1..n}\Sigma$ there is a prescribed rate of fluid extraction per unit area per unit time represented by the continuous scalar point function σ_1, the overall rate of extraction corresponding to that of generation. The corresponding velocity field, which is assumed to be irrotational in $\underline{R}$, is designated $\bar{v}_1$.

This is repeated for source strengths a_2, a_3 --- located at various interior points of $\underline{R}$ with extraction rates σ_2, σ_3 --- and velocity distributions $\bar{v}_2$, $\bar{v}_3$ ---. Show that when all sources are present simultaneously, and the rate of fluid extraction corresponds with $\sigma_1 + \sigma_2 + \sigma_3$ ---, the velocity field is given by $\bar{v}_1 + \bar{v}_2 + \bar{v}_3 +$ ---.

3-33. Let S be the surface of a sphere of radius a and centre T. If the point 0 lies within the sphere at a distance b from T, and if 0' is the point inverse to 0 (ie if TOO' is a straight line and TO.TO' = a^2) and if r is distance measured from 0 while r' is distance measured from 0', show that $U = -\frac{a}{br'}$ is harmonic throughout the sphere and has the value $-\frac{1}{r}$ upon its surface. Hence show by means of equation (3.8-3(b)) that if the point function V is harmonic within the sphere,

$$V_o = \frac{a^2 - b^2}{4\pi a} \oint_S \frac{V}{r^3}\, dS \quad \text{(Poisson's integral)}$$

Show further that if 0 lies outside the sphere and V is harmonic outside the sphere, then

$$V_o = \frac{b^2 - a^2}{4\pi a} \oint_S \frac{V}{r^3}\, dS$$

whence

$$V_o = \frac{a}{b} V_a \quad \text{when } V = V_a = \text{constant over S.}$$

3-34. Find the Green's function G_1 appropriate to a pole at 0 in a half-space[24] at a distance d from the bounding plane S. Hence show on the basis of equation (3.8-3(a)), where $S_{1..n}$ is replaced by S, that if V is a point function with continuous second derivatives throughout the half-space and regular at infinity, its value at 0 is probably given by

$$V_o = \frac{d}{2\pi} \int_S \frac{V}{r^3}\, dS - \frac{1}{4\pi} \int_\tau \left\{\frac{1}{r} - \frac{1}{r'}\right\} \nabla^2 V\, d\tau$$

where r is distance from 0 and r' is distance from the image of 0 in S, and where the region of volume integration extends throughout the half-space.

Confirm the validity of this expression by evaluating the surface integrals of equation (3.8-2) over a central circular disc in S and a matching hemispherical cap which assumes a constantly increasing radius. Hence show that the requirement that V be regular at infinity may be replaced by the lighter requirement that V be bounded and $|\text{grad } V| \to 0$ as $R \to \infty$.

Show further that when V is harmonic the above expression represents the limiting value of the expressions derived for V in the previous exercise as a, $b \to \infty$ and $|a-b| = d$ (constant).

3-35. Find the Green's function G_2 for a point in a half-space, and so obtain an alternative expression for V_o to that of the previous exercise, assuming that V is regular at infinity.

---

24. ie an unbounded region comprising all points upon and to one side of an infinite plane.

$$\text{Ans:}\quad V_o = \frac{1}{2\pi}\int_S \frac{1}{r}\frac{\partial V}{\partial n}\,dS - \frac{1}{4\pi}\int_\tau \left\{\frac{1}{r} + \frac{1}{r'}\right\} \nabla^2 V\, d\tau$$

[It should be noted that the results of this and the previous exercise may be derived by a direct application of equations (3.3-1) and (3.3-3) to the point 0 and its image, bearing in mind that these points lie respectively within and without the integration region.]

3-36. In equation (3.8-2), and in those equations which derive from it, it has been supposed that V is continuous throughout the region under consideration. Show that when a singularity occurs at the point P such that $V = \frac{a}{r'} + U'$ in a neighbourhood of P (where r' is distance from P and U' is a continuous point function) equations (3.8-3) and (3.8-4) are replaced by

$$4\pi V_o = -\oint_{S_{1..n}\Sigma} V \frac{\partial G_1}{\partial n}\, dS + [G_1]_P\, 4\pi a - \int_\tau G_1\, \nabla^2 V\, d\tau$$

$$4\pi V_o = \oint_{S_{1..n}\Sigma} G_2 \frac{\partial V}{\partial n}\, dS - C\oint_{S_{1..n}\Sigma} V\, dS + [G_2]_P\, 4\pi a - \int_\tau G_2\, \nabla^2 V\, d\tau$$

where the volume integral has the same limiting significance as in equation (3.5-3).

3-37. Since the value of Green's function G_1 depends, for a given set of surfaces, both upon the point of evaluation, say Q, and the position of the pole 0, it is often written as $G_1(Q,0)$. With this notation, show that $G_1(0',0) = G_1(0,0')$ ie show that the value of G_1 at the point 0' for a pole at 0 and a given set of bounding surfaces (which need not include Σ if G_1 is regular at infinity) is equal to its value at 0 when the pole is at 0'.

[Hint: Substitute $G_1(Q,0)$ and $G_1(Q,0')$ for V and U in equation (3.1-2) and integrate over the region bounded by the given surfaces and two small spherical surfaces centred upon 0 and 0', noting that G_1 is harmonic in the integration region. The required relationship is obtained when the evaluation of the spherical surface integrals is carried out for the limiting case in which these surfaces shrink indefinitely about 0 and 0'.]

3-38. Derive equation (3.9-1) from (3.1-1) by an appropriate choice of bounding surfaces in a two-dimensional field. Obtain the same result by working directly from the planar form of the divergence theorem.

3-39. State and prove the planar counterparts of Theorems 3.2-1 to 3.2-7.

3-40. Prove equation (3.9-3). Extend the analysis to cover the case in which the excluding circle is replaced by a regular closed curve by applying equation (3.9-2), with $U = \ln \frac{1}{\rho}$, to the region of the surface bounded by this curve and an interior circle centred upon 0, and showing that the surface integral approaches zero as the closed curve shrinks uniformly about 0.

3-41. Let a circle of radius a enclose a plane region $\underline{R}$, and let 0 be an interior point of $\underline{R}$ at a distance b from its centre. If 0' is the point inverse to 0 with respect to the circle and ρ and ρ' denote distance from 0 and 0' respectively, show that $\ln \frac{b\rho'}{a\rho}$ vanishes upon the circle. Show further that $\ln \frac{b\rho'}{a\rho}$ is harmonic in $\underline{R}$ except at 0 by successively locating the origin of plane polar coordinates at 0 and 0' and making use of equation (2.6-4) to evaluate $\nabla^2 \ln \rho' - \nabla^2 \ln \rho$.

It is seen that $\ln \frac{b\rho'}{a\rho}$ is a Green's function of the first kind, vanishing upon the circle and with a pole at 0. Hence deduce from equation (3.9-6) that if V is harmonic in $\underline{R}$

$$V_o = \frac{a^2 - b^2}{2\pi a} \oint_\Gamma \frac{V}{\rho^2}\, ds$$

where Γ represents the circle.

3-42. Let the point 0 of the previous exercise now lie outside $\underline{R}$. Show that $\ln \frac{b\rho'}{a\rho}$ continues to vanish upon Γ and that it is harmonic outside $\underline{R}$ except at 0, where it becomes logarithmically infinite. If V is harmonic everywhere outside $\underline{R}$ derive an expression for V_o in terms of line integrals around Γ and Γ', where Γ' is a closed curve which embraces both 0 and Γ.

What limiting behaviour of V and grad V ensures that the line integral around Γ' vanishes at infinity for finite values of b?

Ans: $$V_o = \frac{b^2 - a^2}{2\pi a} \oint_\Gamma \frac{V}{\rho^2}\, ds + \frac{1}{2\pi} \oint_{\Gamma'} \left\{\left(\ln \frac{b\rho'}{a\rho}\right) \frac{\partial V}{\partial n'} - V \frac{\partial}{\partial n'} \left(\ln \frac{b\rho'}{a\rho}\right)\right\} ds$$

$\oint_\infty$ vanishes when $\frac{V}{R} \to 0$ and $R \frac{\partial V}{\partial R} \to 0$ as $R \to \infty$.

These requirements are satisfied if V and $R^2|\text{grad } V|$ are bounded at infinity.

3-43. Derive an alternative expression for V_o to that of Ex.3-42. by replacing $\ln \frac{b\rho'}{a\rho}$ with $\ln \frac{b\rho'}{\rho''\rho}$, where ρ'' is distance measured from the centre of $\underline{R}$.

Under what conditions does the integral around Γ' now vanish at infinity for finite values of b?

Ans:

$$V_o = \frac{b^2-a^2}{2\pi a}\oint_\Gamma \frac{V}{\rho^2}\,ds - \frac{1}{2\pi a}\oint_\Gamma V\,ds + \frac{1}{2\pi}\oint_{\Gamma'}\left\{\left(\ln\frac{b\rho'}{\rho''\rho}\right)\frac{\partial V}{\partial n'} - V\frac{\partial}{\partial n'}\left(\ln\frac{b\rho'}{\rho''\rho}\right)\right\}ds$$

$\oint_\infty$ vanishes when $V\to 0$ and $(R \ln R)\,\frac{\partial V}{\partial R}\to 0$ as $R\to\infty$.

These requirements are satisfied if R V and $R^2|\text{grad } V|$ are bounded at infinity.

3-44. If the more severe of the limiting conditions in Ex.3-42. and 3-43. are imposed upon V, viz that $V\to 0$ and $(R \ln R)\,\frac{\partial V}{\partial R}\to 0$ as $R\to\infty$, it follows that V_o may be expressed either as

$$\frac{b^2-a^2}{2\pi a}\oint_\Gamma \frac{V}{\rho^2}\,ds \quad \text{or} \quad \frac{b^2-a^2}{2\pi a}\oint_\Gamma \frac{V}{\rho^2}\,ds - \frac{1}{2\pi a}\oint_\Gamma V\,ds$$

In these circumstances $\oint_\Gamma V\,ds$ must be zero.

Give an independent proof of this and show that $\oint_\Gamma \frac{\partial V}{\partial n'}\,ds = 0.$

[Hint: Evaluate the integrals of equation (3.9-3) for the particular case in which 0 is located at the centre of R (and therefore outside the region of integration), bearing in mind that $\oint_\Gamma \frac{\partial V}{\partial n'}\,ds = -\oint_{\Gamma'} \frac{\partial V}{\partial n'}\,ds.$]

3-45. Let the planar point function V be defined by $V = \ln\frac{1}{\rho'}$ where ρ' is distance measured from some point P. If 0 is a point not coincident with P and ρ is distance measured from 0 use equation (3.9-3) to express V_o in terms of a line integral around a circle Γ of radius a centred upon P for the case $a < \rho_P$, by showing that both the line integral at infinity and the surface integral outside the circle are zero.

Hence prove that

$$\oint_\Gamma \ln\frac{1}{\rho}\,ds = \ln\frac{1}{\rho_P}\oint_\Gamma ds = 2\pi a \ln\frac{1}{\rho_P}$$

Now suppose that $a > \rho_P$. By taking account of the singularity of V which now lies within the region of integration show that

$$\oint_\Gamma \ln \frac{1}{\rho}\, ds = \ln \frac{1}{a} \oint_\Gamma ds = 2\pi a \ln \frac{1}{a}$$

(cf Ex.3-19., p. 188).

3-46. Let the region $\underline{R}$ of the plane, bounded by the closed curves $\Gamma_{1..n}\Gamma'$, be divided into a set of sub-regions by internal boundaries Γ_a, Γ_b --- such that a given point function g and its first derivatives are continuous at interior points of the sub-regions but are discontinuous upon Γ_a, Γ_b --- and let g, where defined, be everywhere positive or everywhere negative. Further, let V be a point function having continuous second derivatives in $\underline{R}$ except upon Γ_a, Γ_b ---, and upon certain lines Γ_α, Γ_β --- and points P_1, P_2, --- lying within the sub-regions.

Show that V is rendered piecewise determinate, at least to within the same additive constant, by the specification of all of the following factors:

(1) div(g grad V) at all interior points of the sub-regions excluding Γ_α, Γ_β --- P_1, P_2 ---.

(2) V or $\frac{\partial V}{\partial n'}$ at each point of $\Gamma_{1..n}\Gamma'$, or $\oint g \frac{\partial V}{\partial n'} ds$ for any of these contours together with the tangential derivative of V at each point of the contour, or, in the case of a single bounding curve, the tangential derivative alone. (If $\Gamma_{1..n}\Gamma'$ comprise lines of discontinuity, the expressions should be evaluated just within $\underline{R}$.)

(3) ΔV and $\Delta\left(g \frac{\partial V}{\partial n'}\right)$ through all points of Γ_a, Γ_b --- Γ_α, Γ_β ---.

(4) $V = V' + V''$ in a neighbourhood of P_1, P_2 ---, where V' is an individually specified discontinuous point function and V'' is any continuous point function.

This theorem also covers the case of a two-dimensional field in space, whose bounding surfaces and surfaces of discontinuity are closed and open cylinders with parallel axes, which, in combination with lines of discontinuity, cut a transverse plane in the above contours and points; for in these circumstances the conditions for uniqueness in the plane necessarily lead to uniqueness in space. However, it should be noted that the operators grad and div grad in the above context are of an essentially planar nature, and that in those cases where g and V are defined outside the plane with other than two-dimensional symmetry, it is unlikely that the planar and spatial operations will lead to the same result. (See Ex.4-29, p. 257.)

3-47. Let the scalar point functions, W, V_1 and V_2 be well-behaved in the region $\underline{R}$ bounded by the surfaces $S_{1..n}\Sigma$, with W everywhere positive, and let $\overline{F}_1 = W \text{ grad } V_1$ and $\overline{F}_2 = W \text{ grad } V_2$.

If $\text{div } \overline{F}_1 = \text{div } \overline{F}_2$ at interior points of $\underline{R}$ and $\oint \overline{F}_1.d\overline{S} = \oint \overline{F}_2.d\overline{S}$ over each of the surfaces in turn, show that

$$\int_\tau W|\text{grad } V_1|^2 \, d\tau < \int_\tau W|\text{grad } V_2|^2 \, d\tau$$

when V_1 is constant over each of the surfaces in turn and V_2 is not.

3-48. Extend the result of Ex.3-47. to the case where W is piecewise continuous in $\underline{R}$, V_1 is continuous through the interfaces of the sub-regions, and the normal components of $\overline{F}_1$ and $\overline{F}_2$ are either continuous or equally discontinuous through these interfaces.

Show that when the Σ surface is deleted the above inequality continues to hold for integration over all space outside $S_{1..n}$, provided that V_1 and V_2 are regular at infinity and W is continuous everywhere outside a sphere of finite radius centred upon some local origin.

3-49. Show by means of Theorem 3.10-2, or otherwise, that the kinetic energy of a mass of incompressible fluid bounded by the surfaces $S_{1..n}\Sigma$, and having prescribed values of normal velocity upon these surfaces, is a minimum when the flow is irrotational. (Kelvin's minimum energy theorem.)

3-50. Let V be harmonic in the region of the plane outside Γ_{1--n} and let V or $\frac{\partial V}{\partial n}$, be zero upon Γ_{1--n}. Show that if, in addition, $V/\ln R$ is bounded and $R\left(\frac{\partial V}{\partial R} \ln R + \frac{V}{R}\right) \to 0$ uniformly in all directions as $R \to \infty$, then V is zero everywhere outside Γ_{1--n}.

CHAPTER 4

UNRETARDED POTENTIAL THEORY

4.1 The Scalar Potential of Point Sources

Let the scalar magnitudes a_1--a_i--a_n be allocated respectively to the fixed points P_1--P_i--P_n. Then with each position of a variable point O, not coincident with P_1, P_2--, we may associate a scalar potential ϕ_o, defined by[1]

$$\phi_o = \sum_{i=1}^{n} \frac{a_i}{OP_i} \qquad (4.1\text{-}1)$$

It is seen that ϕ represents a single-valued scalar point function having continuous derivatives of all orders wherever it is defined. The points P_1, P_2-- are known as the sources of the potential, and a_1, a_2-- are said to be the strengths of the sources.

If r is chosen to denote distance measured from O, as in previous sections, then $\frac{1}{r}$ continues to define a unique scalar field for any given position of O, and we may write

---

1. The particular potential function here defined for point sources, and later extended to distributed sources, is known as a Newtonian potential. There is another type of potential which is defined only in the plane and is known as a logarithmic potential. For point sources this is given by

$$\phi_o = \sum_{i=1}^{n} a_i \ln \frac{1}{OP_i}$$

It is shown in Sec. 4.2 that the logarithmic potential of point sources is related to the Newtonian potential of certain rectilinear source systems of infinite extent.

As used in this document the word 'potential', when unqualified, implies 'Newtonian potential'.

$$\phi_o = \sum_{i=1}^{n} a_i \left(\frac{1}{r}\right)_{P_i} \tag{4.1-1(a)}$$

Again, we may define a set of scalar point functions typified by $\frac{1}{r'_i}$, where r'_i is distance measured from P_i, so that

$$\phi_o = \sum_{i=1}^{n} a_i \left(\frac{1}{r'_i}\right)_o \tag{4.1-1(b)}$$

The fields defined by $\frac{1}{r}$ and $\frac{1}{r'_i}$ are quite distinct, but have the common value $\frac{1}{OP_i}$ when evaluated at P_i and 0 respectively.

The following related notation will also be adopted.

$$(\bar{r})_{P_i} = \overrightarrow{OP_i} \quad ; \quad (\bar{r}'_i)_o = \overrightarrow{P_iO}$$

A single point source is called a monopole or singlet. Certain combinations of monopoles are known as multipoles; of these the simplest is the dipole.

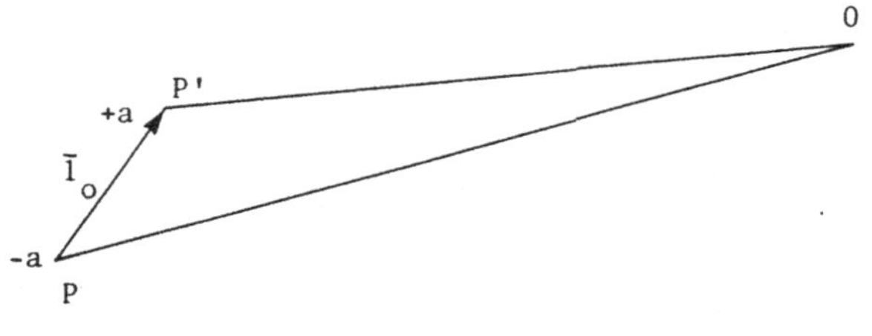

Fig. 4.1

In Fig. 4.1 point sources of strengths -a and +a are located at P and P' respectively, where $\overrightarrow{PP'} = \bar{1}_o = \hat{\bar{1}}_o 1_o$. If P' is allowed to approach P along the fixed direction $-\hat{\bar{1}}_o$, and if the source strengths are adjusted to maintain al_o constant during the operation, then the limiting configuration as $l_o \to 0$ and $a \to \infty$ is said to constitute a dipole[2] at P. The product $al_o = p^{(1)}$ is known as the scalar moment of the dipole and $a\bar{l}_o = \bar{p}^{(1)}$ the vector moment.

From equation (4.1-1(a)) the potential at 0 of the point sources at P and P' is given by

$$-a\left(\frac{1}{r}\right)_P + a\left(\frac{1}{r}\right)_{P'}$$

and this may be expressed as

$$-a\left(\frac{1}{r}\right)_P + a\left\{\left(\frac{1}{r}\right)_P + l_o\left(\frac{d}{dl_o}\left(\frac{1}{r}\right)\right)_P + \frac{1}{2!}\,l^2{}_o\left(\frac{d^2}{dl^2{}_o}\left(\frac{1}{r}\right)\right)_P ---\right\}$$

$$= al_o\left(\frac{d}{dl_o}\left(\frac{1}{r}\right)\right)_P + \frac{1}{2!}\,al_o{}^2\left(\frac{d^2}{dl^2{}_o}\left(\frac{1}{r}\right)\right)_P + ---$$

where $\frac{d}{dl_o}$ = spatial rate of change in the direction $\hat{l}_o$.

The dipole potential at 0 is consequently given by

$$\phi_o{}^{(1)} = p^{(1)}\left(\frac{d}{dl_o}\left(\frac{1}{r}\right)\right)_P = p^{(1)}\left((\hat{l}_o.\nabla)\,\frac{1}{r}\right)_P = \bar{p}^{(1)}.\left(\nabla\left(\frac{1}{r}\right)\right)_P \qquad (4.1\text{-}2)$$

If the strengths of the point sources comprising the dipole were to be reversed in sign the potential at 0 would likewise be reversed in sign; if the dipole were moved without change of orientation to some point P" where $\overrightarrow{PP''} = \bar{l}_1 = \hat{l}_1\, l_1$ its potential at 0 would become

$$p^{(1)}\left(\frac{d}{dl_o}\left(\frac{1}{r}\right)\right)_{P''}$$

It follows that a pair of dipoles, one with reversed sign at P and the other translated to P" without reversal, give rise to a potential at 0 of

2. This is sometimes referred to as a 'point dipole' or 'doublet', the term 'dipole' being reserved for the non-limiting configuration. For the sake of uniformity of nomenclature 'dipole' will be used in this section with the connotation 'point dipole'; in the later sections it will be replaced by 'doublet'.

$$-p^{(1)}\left(\frac{d}{dl_o}\left(\frac{1}{r}\right)\right)_P + p^{(1)}\left\{\left(\frac{d}{dl_o}\left(\frac{1}{r}\right)\right)_P + l_1\left(\frac{d}{dl_1}\frac{d}{dl_o}\left(\frac{1}{r}\right)\right)_P + \frac{1}{2!}l_1{}^2\left(\frac{d^2}{dl_1{}^2}\frac{d}{dl_o}\left(\frac{1}{r}\right)\right)_P + ---\right\}$$

$$= p^{(1)}\,l_1\left(\frac{d}{dl_1}\frac{d}{dl_o}\left(\frac{1}{r}\right)\right)_P + \frac{1}{2!}p^{(1)}\,l_1{}^2\left(\frac{d^2}{dl_1{}^2}\frac{d}{dl_o}\left(\frac{1}{r}\right)\right)_P + ---$$

where $\frac{d}{dl_1}$ = spatial rate of change in the direction $\hat{l}_1$.

If $p^{(1)}l_1$ is maintained constant while $l_1 \to 0$ and $p^{(1)} \to \infty$ the limiting source configuration is said to comprise a quadrupole at P, and the potential at 0 is then

$$\phi_o{}^{(2)} = \frac{1}{2!}\,p^{(2)}\left(\frac{d^2}{dl_1\,dl_o}\left(\frac{1}{r}\right)\right)_P \tag{4.1-3}$$

where $p^{(2)} = 2\,p^{(1)}l_1$

When $\hat{l}_1$ is normal to $\hat{l}_o$ the quadrupole is two-dimensional and rectangular; in general it is two-dimensional and oblique.

We can build up higher order multipoles at P by a process of iteration. The potential at 0 of the 2^m pole is seen to be

$$\phi_o{}^{(m)} = \frac{1}{m!}\,p^{(m)}\left(\frac{d^m}{dl_{m-1}\;---\;dl_o}\left(\frac{1}{r}\right)\right)_P \tag{4.1-4}$$

where

$$p^{(m)} = mp^{(m-1)}\;l_{m-1} \tag{4.1-5}$$

When $\hat{l}_o = \hat{l}_1 = \hat{l}_2$ --- the resulting quadrupole, octupole and higher order multipoles are said to be axial or linear. They are clearly one-dimensional. In this case we may simplify the associated analysis by supposing that the positive sense of the z axis of Cartesian coordinates is aligned with $\hat{l}_o$. Then equation (4.1-4) reduces to

$$\phi_{o_{\text{axial}}}^{(m)} = \frac{1}{m!}\,p^{(m)}\left(\frac{\partial^m}{\partial z^m}\left(\frac{1}{r}\right)\right)_P \tag{4.1-6}$$

If (x_o, y_o, z_o) and (x', y', z') are the coordinates of 0 and P respectively, and if $\overrightarrow{PO} = \bar{R}$ and θ is the angle between $\hat{z}$ and $\bar{R}$ (Fig. 4.2), then (4.1-6) becomes

$$\phi_{o_{\text{axial}}}^{(m)} = \frac{1}{m!}\,p^{(m)}\left(\frac{\partial^m}{\partial z^m}\{(x-x_o)^2 + (y-y_o)^2 + (z-z_o)^2\}^{-\frac{1}{2}}\right)$$

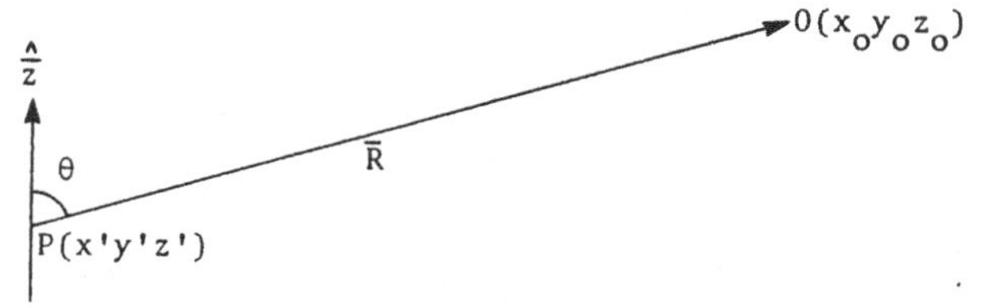

Fig. 4.2

This yields a dipole potential at O of

$$\left.\begin{aligned} \phi_o^{(1)} &= p^{(1)} \frac{(z_o - z')}{R^3} = \frac{\bar{p}^{(1)}.\bar{R}}{R^3} = \frac{p^{(1)} \cos\theta}{R^2} \\ \text{or} \quad & \\ \phi_o^{(1)} &= \bar{p}^{(1)}.\left(\frac{\bar{r}'}{(r')^3}\right)_o = -\bar{p}^{(1)}.\left(\frac{\bar{r}}{r^3}\right)_P \end{aligned}\right\} \quad (4.1\text{-}7)$$

where r' is distance measured from P,

and an axial quadrupole potential of

$$\phi_{o_{\text{axial}}}^{(2)} = \frac{1}{2!} p^{(2)} \left\{ \frac{3(z_o - z')^2}{R^5} - \frac{1}{R^3} \right\} = p^{(2)} \frac{(3\cos^2\theta - 1)}{2R^3}$$

It is seen that R and θ are spherical coordinates of O in a system centred on P with $\hat{\theta}$ and $\hat{z}$ aligned. The angular coordinate Φ is missing because the potentials are axially symmetrical about $\hat{z}$[3].

By treating the monopole as a multipole of order zero and identifying $p^{(0)}$ with the strength of the monopole, the sequence $\phi^{(0)}$, $\phi^{(1)}$, $\phi^{(2)}_{\text{axial}}$, $\phi^{(3)}_{\text{axial}}$ --- is brought into the form

$$\frac{p^{(0)}}{R},\ p^{(1)} \frac{\cos\theta}{R^2},\ p^{(2)} \frac{(3\cos^2\theta - 1)}{2R^3},\ p^{(3)} \frac{(5\cos^3\theta - 3\cos\theta)}{2R^4} \text{ ----} \quad (4.1\text{-}8)$$

The student of Legendre functions will recognise the general term of the sequence as

---

3. Since r currently denotes distance from O we will henceforth represent spherical coordinates by R, θ, Φ.

$$\phi_o^{(m)} = p^{(m)} \frac{P_m(\cos\theta)}{R^{m+1}} \qquad (4.1\text{-}9)$$

where $P_m(\cos\theta)$ is the mth degree Legendre polynomial in $\cos\theta$.

These polynomials appear in the solution of Laplace's equation in spherical coordinates and usually defined by

$$P_m(\cos\theta) = \frac{1}{2^m m!} \frac{d^m}{d(\cos\theta)^m} (\cos^2\theta - 1)^m \qquad (4.1\text{-}10)$$

but the above treatment leads to an alternative definition:

$$P_m(\cos\theta) = \frac{1}{m!} \left(r^{m+1} \frac{\partial^m}{\partial z^m} \left(\frac{1}{r}\right)\right)_P \qquad (4.1\text{-}11)$$

where r is distance measured from an arbitrary point 0 and P is any point such that the angle between $\overrightarrow{PO}$ and $\hat{z}$ is θ.

$|P_m(\cos\theta)|$ cannot exceed unity. This is evident for the first few orders when written out in the following way.

$$\begin{aligned} P_o(\cos\theta) &= 1 \\ P_1(\cos\theta) &= \cos\theta \\ P_2(\cos\theta) &= \tfrac{1}{2}(3\cos^2\theta - 1) = \tfrac{1}{4}(3\cos 2\theta + 1) \\ P_3(\cos\theta) &= \tfrac{1}{2}(5\cos^3\theta - 3\cos\theta) = \tfrac{1}{8}(5\cos 3\theta + 3\cos\theta) \end{aligned} \qquad (4.1\text{-}12)$$

The subject of Legendre polynomials will not be pursued further within this context. For the general expansions of non-axial multipoles the interested reader is referred to Stratton[4].

We now proceed to determine the potential at 0 of a distribution of point sources in a neighbourhood of P. Let $\bar{s}_i$ be the position vector of P_i relative to P and let θ_i be the angle between $\bar{s}_i$ and $\bar{R}$ (Fig. 4.3).

---

4. "Electromagnetic Theory", Sec. 3.12

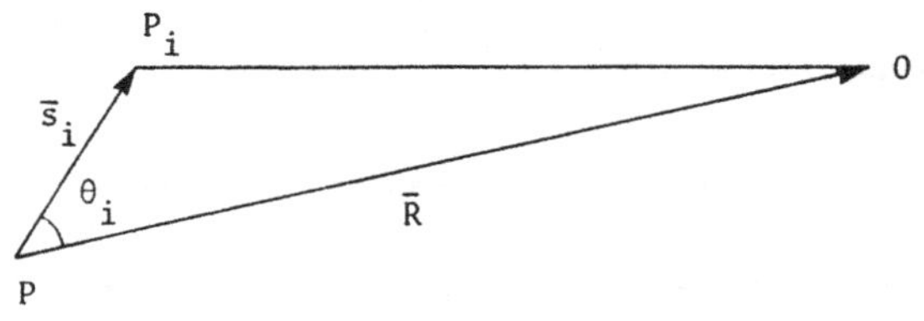

Fig. 4.3

As before, r is identified with distance measured from 0. Then from equation (1.2-9), with V replaced by $\frac{1}{r}$ and $\Delta\bar{s}$ replaced by $\bar{s}_i$, we have

$$\left(\frac{1}{r}\right)_{P_i} = \left(\frac{1}{r}\right)_P + \left((\bar{s}_i\cdot\nabla)\,\frac{1}{r}\right)_P + \frac{1}{2!}\left((\bar{s}_i\cdot\nabla)^2\,\frac{1}{r}\right)_P + ---- \qquad (4.1\text{-}13)$$

or

$$\begin{aligned}\left(\frac{1}{r}\right)_{P_i} = \frac{1}{R} &+ \left\{ s_{i_x}\left(\frac{\partial}{\partial x}\left(\frac{1}{r}\right)\right)_P + s_{i_y}\left(\frac{\partial}{\partial y}\left(\frac{1}{r}\right)\right)_P + s_{i_z}\left(\frac{\partial}{\partial z}\left(\frac{1}{r}\right)\right)_P \right\} \\ &+ \left\{ \frac{1}{2}\,s_{i_x}^2\left(\frac{\partial^2}{\partial x^2}\left(\frac{1}{r}\right)\right)_P + \frac{1}{2}\,s_{i_y}^2\left(\frac{\partial^2}{\partial y^2}\left(\frac{1}{r}\right)\right)_P + \frac{1}{2}\,s_{i_z}^2\left(\frac{\partial^2}{\partial z^2}\left(\frac{1}{r}\right)\right)_P \right. \\ &\left. + s_{i_x}s_{i_y}\left(\frac{\partial^2}{\partial x\partial y}\left(\frac{1}{r}\right)\right)_P + s_{i_x}s_{i_z}\left(\frac{\partial^2}{\partial x\partial z}\left(\frac{1}{r}\right)\right)_P + s_{i_y}s_{i_z}\left(\frac{\partial^2}{\partial y\partial z}\left(\frac{1}{r}\right)\right)_P \right\} \\ &+ ---\end{aligned} \qquad (4.1\text{-}14)$$

and

$$\begin{aligned}\phi_o = \frac{1}{R}\sum_{i=1}^{n} a_i &+ \left\{ \sum_{i=1}^{n} a_i\,s_{i_x}\left(\frac{\partial}{\partial x}\left(\frac{1}{r}\right)\right)_P + --- \right\} \\ &+ \left\{ \frac{1}{2}\sum_{i=1}^{n} a_i\,s_{i_x}^2\left(\frac{\partial^2}{\partial x^2}\left(\frac{1}{r}\right)\right)_P + ------ + \sum_{i=1}^{n} a_i\,s_{i_x}s_{i_y}\left(\frac{\partial^2}{\partial x\partial y}\left(\frac{1}{r}\right)\right)_P + ---- \right\} \\ &+ ---\end{aligned} \qquad (4.1\text{-}15)$$

The first term of (4.1-15) is seen to be identical with the potential at 0 of a monopole of strength $\sum_{i=1}^{n} a_i$ at P. The three terms within the first pair of brackets correspond to the potentials of x, y and z - aligned dipoles at P of scalar moments $\sum_{i=1}^{n} a_i s_{i_x}$, $\sum_{i=1}^{n} a_i s_{i_y}$, $\sum_{i=1}^{n} a_i s_{i_z}$. Since the three terms may be replaced by $\left\{\sum_{i=1}^{n} a_i \bar{s}_i\right\}.\left(\nabla \frac{1}{r}\right)_P$ it follows from equation (4.1-2) that the individual dipoles may be replaced by a single dipole of vector moment $\sum_{i=1}^{n} a_i \bar{s}_i$. This vector quantity is said to be the dipole moment or polarisation of the complex relative to P.

The second derivative terms within the second pair of brackets in equation (4.1-15) correspond to the potentials of axial quadrupoles of x, y and z alignment at P and of moments $\sum_{i=1}^{n} a_i s_{i_x}^2$, $\sum_{i=1}^{n} a_i s_{i_y}^2$ and $\sum_{i=1}^{n} a_i s_{i_z}^2$. The cross derivative terms correspond to rectangular quadrupoles in the xy, zy and yz planes having moments

$$2\sum_{i=1}^{n} a_i s_{i_x} s_{i_y} \quad ; \quad 2\sum_{i=1}^{n} a_i s_{i_x} s_{i_z} \quad \text{and} \quad 2\sum_{i=1}^{n} a_i s_{i_y} s_{i_z}$$

It is clear that further terms in the expansion can be related to higher-order multipoles at P, each of which will derive from a pair of multipoles one order lower and displaced along one or other coordinate axis through P.

Expansion of equation (4.1-14) yields

$$\left(\frac{1}{r}\right)_{P_i} = \frac{1}{R} + \frac{\bar{s}_i.\bar{R}}{R^3} + \frac{1}{2R^3}\left\{3\,\frac{(\bar{s}_i.\bar{R})^2}{R^2} - \bar{s}_i.\bar{s}_i\right\} + ---$$

or

$$\left(\frac{1}{r}\right)_{P_i} = \frac{1}{R} + \frac{s_i \cos\theta_i}{R^2} + \frac{s_i^2}{2R^3}(3\cos^2\theta_i - 1) + ---$$

ie

$$\frac{1}{OP_i} = \frac{1}{R}\sum_{m=o}^{\infty}\left(\frac{s_i}{R}\right)^m P_m(\cos\theta_i) \tag{4.1-16}$$

The series is convergent for $s_i<R$. It follows that the associated series for ϕ_o, viz

$$\phi_o = \frac{1}{R}\sum_{i=1}^{n} a_i + \frac{1}{R^2}\sum_{i=1}^{n} a_i s_i \cos\theta_i + \frac{1}{2R^3}\sum_{i=1}^{n} a_i s_i^2 (3\cos^2\theta_i - 1) + --- \tag{4.1-17}$$

is likewise convergent provided that 0 is an exterior point of some sphere which is centred upon P and contains all of the sources.

The relative magnitudes of the various terms depends upon both the distance of 0 from the source complex and the nature of the complex itself. When all source strengths are of the same sign the first term predominates and relative to this term the following terms fall off at least as $(s_{i_{max}}/R)$, $(s_{i_{max}}^2/R^2)$ --- where $s_{i_{max}}$ is the greatest value of s_i. When the total source strength is zero the first term vanishes and the second term may or may not predominate. In this case the value of the second term may be shown to be independent of the position of the datum point P from which s_i is measured. When the total source strength is not zero the second term may be made to vanish by locating P at the source centre (see Ex.4-5. p. 231).

4.2 The Scalar Potential of Line Sources

Let λ be a finite, piecewise continuous scalar function of length of arc s measured from one end of a regular curve Γ (or from some specified point of the curve if Γ be closed). Then the associated scalar potential at any point 0 outside the curve is defined by

$$\phi_o = \int_\Gamma \frac{\lambda}{r}\, ds \qquad (4.2\text{-}0)$$

where r is the distance of ds from O[5].

Γ is said to comprise a line source of potential, whose local density or strength per unit length is λ.

The simplest form of line source is rectilinear and of constant density. Consider such a source AB of density λ and length 2c, lying in the z axis through the origin of coordinates and bisected by the origin (Fig. 4.4).

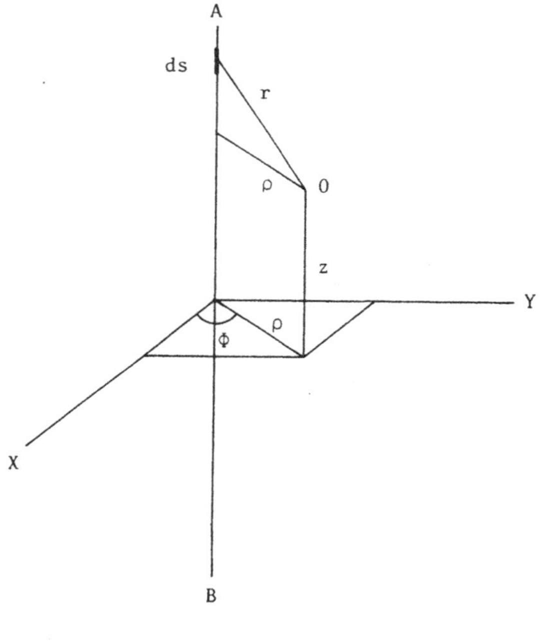

Fig. 4.4

5. The integral is, of course, shorthand for

$$\lim_{\substack{n\to\infty \\ \Delta s_i \to 0}} \sum_{i=1}^{n} \left(\frac{\lambda}{r}\right)_{P_i} \Delta s_i$$

where P_i is some point of that element of the curve whose chord is of length Δs_i. (See Sec. 1.5).

If the z coordinate of a source element is represented by ζ and the cylindrical coordinates of O are ρ, Φ, z, then

$$\phi_o = \lambda \int_{-c}^{+c} \frac{1}{\{\rho^2+(\zeta-z)^2\}^{\frac{1}{2}}} d\zeta$$

$$= \lambda \left[\ln \{(\zeta-z)+((\zeta-z)^2+\rho^2)^{\frac{1}{2}}\}\right]_{-c}^{+c}$$

or

$$\phi_o = \lambda \ln \frac{c - z + R_1}{-c - z + R_2} \tag{4.2-1}$$

where

$$R_1 = OA \quad \text{and} \quad R_2 = OB$$

Now

$$R_1{}^2 = \rho^2 + (c-z)^2 \quad \text{and} \quad R_2{}^2 = \rho^2 + (c+z)^2$$

hence

$$R_2{}^2 - R_1{}^2 = 4\,cz$$

Substitution for z in equation (4.2-1) yields

$$\phi_o = \lambda \ln \frac{R_1 + R_2 + 2c}{R_1 + R_2 - 2c} \tag{4.2-2}$$

As O approaches any point of the source, $R_1 + R_2 \rightarrow 2c$ and the potential at O approaches a logarithmic infinity. This is also seen to be true for the general case of a non-rectilinear source of variable density so long as Γ is smooth in a neighbourhood of O, for it is possible to divide the curve into several portions, one of which is a small element which is virtually straight and of constant density, intersected by the normal through O and approached continuously by O. By the above analysis the potential associated with this segment is logarithmically infinite at the segment; this must also apply to the potential of the complete source.

An immediate consequence of equation (4.2-2) is that the equipotential surfaces of a uniform rectilinear source comprise confocal ellipsoids of revolution, since for such surfaces

$$\frac{R_1 + R_2 + 2c}{R_1 + R_2 - 2c} = k \text{ (const)}$$

whence

$$R_1 + R_2 = 2c\,\frac{(k + 1)}{(k - 1)} \tag{4.2-3}$$

A binominal expansion of equations (4.2-1) or (4.2-2) for the particular case $|z|<c$ and $\rho<c-|z|$ yields

$$\phi_o = \lambda \ln\left\{\frac{4(c^2-z^2)}{\rho^2} + 1 + \frac{c^2 + 3z^2}{c^2 - z^2} - \frac{\rho^2}{2c^2}\,\frac{\left(1 + \frac{8z^2}{c^2} - \frac{z^4}{c^4}\right)}{\left(1 - \frac{z^2}{c^2}\right)^3} + ---\right\}$$

As $\rho\to 0$,

$$\phi_o \to \lambda \ln 4\,\frac{(c^2-z^2)}{\rho^2} \tag{4.2-4}$$

and as $c\to\infty$ with finite values of ρ and z

$$\phi_o \to \lambda \ln \frac{4c^2}{\rho^2} = 2\lambda(\ln 2c - \ln \rho) \tag{4.2-5}$$

It follows that for an infinite uniform rectilinear source the potential is logarithmically infinite at all finite distances. Again, for $z^2 \ll c^2$ and $\rho^2 \ll c^2$

$$\left(\frac{\partial\phi}{\partial z}\right)_o = \frac{-2\lambda z}{c^2} \quad \text{and} \quad \left(\frac{\partial^2\phi}{\partial z^2}\right)_o = \frac{-2\lambda}{c^2} \tag{4.2-6}$$

hence as $c\to\infty$, the first and second axial derivatives of ϕ approach zero for any given finite values of z and ρ. The scalar potential field is therefore two-dimensional at finite distance from the origin of coordinates and is, of course, symmetrical about the axis defined by the source.

If several parallel rectilinear sources of equal lengths and densities λ_1, λ_2 -- are present, each being bisected by the xy plane through the origin of coordinates, the limiting value of ϕ_o as $c\to\infty$ becomes

$$\phi_o = 2\left\{\sum_{i=1}^{n} \lambda_i\right\} \ln 2c - 2\sum_{i=1}^{n} \lambda_i \ln \rho_i \tag{4.2-7}$$

where ρ_i is the normal distance of 0 from the ith line source.

When $\sum_{i=1}^{n} \lambda_i = 0$ the logarithmically infinite component of the potential vanishes and

$$\phi_o = -2\lambda_1 \ln \rho_1 - 2\lambda_2 \ln \rho_2 \text{ ---} \tag{4.2-8}$$

The same result would have obtained for any plane region normal to the line sources had the points of intersection of the line sources and the plane been treated as point sources of logarithmic potential of strengths $2\lambda_1$, $2\lambda_2$ --- as referred to in the footnote to p. 217.

However, in the latter case equation (4.2-8) would continue to represent the potential at infinite distance whereas in the three-dimensional case the expression holds only at finite distance from the sources. Unless the total source strength is zero the logarithmic potential in the plane does not vanish at infinity but becomes logarithmically infinite.

A planar uniform rectilinear doublet is the limiting configuration of two parallel uniform rectilinear sources of equal lengths and equal and opposite line densities, $\pm\lambda$, whose centres are displaced transversely by a distance d, when $d \to o$ and $\lambda \to \infty$ in such a way as to maintain λd constant.

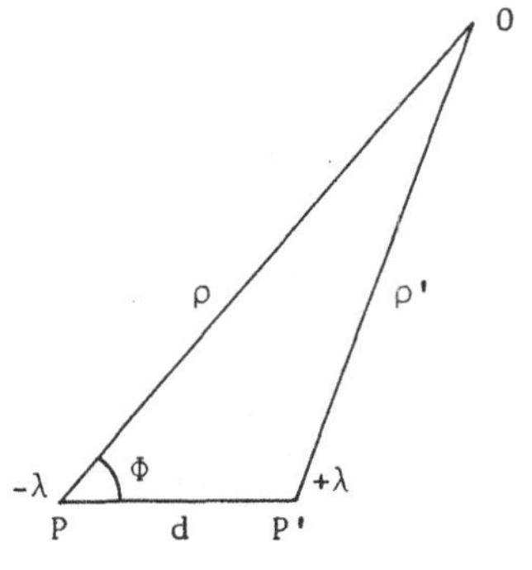

Fig. 4.5

In Fig. 4.5 uniform, parallel line sources of strengths $-\lambda$ and $+\lambda$ are bisected normally by the plane of the paper in P and P'. If the length of each source is 2c and $\rho^2 << c^2$ then

$$\phi_o \approx -2\lambda \ln 2c + 2\lambda \ln \rho + 2\lambda \ln 2c - 2\lambda \ln \rho'$$

$$\approx 2\lambda \ln \frac{\rho}{\rho'}$$

As $d \to 0$,

$$\rho' \to \rho \left\{ 1 - \frac{d \cos \Phi}{\rho} \right\}$$

and

$$\phi_o \to 2\lambda \ln \left\{ 1 + \frac{d \cos \Phi}{\rho} \right\}$$

whence

$$\phi_o \to \frac{2\lambda d \cos \Phi}{\rho}$$

Hence at non-zero and finite distance from an infinite uniform rectilinear doublet (or at all non-zero distances in the plane from a doublet pair of logarithmic point sources of moment $2\lambda d$)

$$\phi_o = \frac{2\lambda d \cos \Phi}{\rho} \qquad (4.2\text{-}9)$$

The equipotential surfaces defined by $(\cos \Phi)/\rho$ = const comprise circular cylinders which contain the source and are bisected by the half-plane $\Phi = 0$ or $\Phi = \pi$.

The rectilinear line doublet under consideration may also be treated as the limiting case of a set of co-planar dipoles aligned normally to their line of centres where the number of dipoles per unit length increases without limit while the individual moments approach zero in such a way as to maintain the total scalar moment per unit length constant and equal to λd[6]. This approach can clearly be extended to the more general type of line doublet which may be curved, non-planar and of variable density; indeed, such a source is best defined in terms of a linear distribution of normally aligned point doublets. It therefore follows from (4.1-2) and (4.1-7) that the potential at 0 of any regular line doublet which does not contain 0 may be expressed as

$$\phi_o = \int_\Gamma L \frac{\partial}{\partial n'} \left(\frac{1}{r}\right) ds \qquad (4.2\text{-}10)$$

where $\overline{L} = L\, \hat{\overline{n}}'$ is the transverse vector dipole moment per unit length of contour.

6. Similarly, the simple line source, characterised by the piecewise - continuous density λ, is equivalent, at exterior points, to a limiting configuration of point sources upon Γ of strength λ per unit length. For this reason we may refer to it as a line singlet.

EXERCISES

4-1. The potential at 0 of a dipole of vector moment $\bar{p}^{(1)}$ at P is given in equation (4.1-2) as

$$\phi_o^{(1)} = \bar{p}^{(1)}.\left(\nabla\left(\frac{1}{r}\right)\right)_P$$

where r is distance measured from 0.

If r' is distance measured from P show that

$$\phi_o^{(1)} = -\bar{p}^{(1)}.\left(\nabla\left(\frac{1}{r'}\right)\right)_0$$

[In this connection see Sec. 4.6.]

4-2. Prove the equivalent relationships:

$$\left(\frac{\partial^m}{\partial z^m}\left(\frac{1}{r}\right)\right)_P = (-1)^m\left(\frac{\partial^m}{\partial z^m}\left(\frac{1}{r'}\right)\right)_0$$

$$\frac{\partial^m}{\partial z'^m}\{(x_o-x')^2+(y_o-y')^2+(z_o-z')^2\}^{-\frac{1}{2}} = (-1)^m\frac{\partial^m}{\partial z_o^m}\{(x_o-x')^2+(y_o-y')^2+(z_o-z')^2\}^{-\frac{1}{2}}$$

4-3. Derive equation (4.1-16) by expressing OP_i in terms of s_i, R and $\cos\theta_i$ (Fig. 4.3), and expanding $\frac{1}{OP_i}$ in a binomial series on the assumption that R is sufficiently large to make the series absolutely convergent.

4-4. Expand $\left(\frac{1}{r_i'}\right)_0$ about $\left(\frac{1}{r_i'}\right)_P$ in Fig. 4.3, where r_i' is distance measured from P_i. Hence show that when $R < s_i$

$$\left(\frac{1}{r_i'}\right)_0 = \frac{1}{OP_i} = \frac{1}{s_i}\sum_{m=0}^{\infty}\left(\frac{R}{s_i}\right)^m P_m(\cos\theta_i)$$

4-5. So long as the total strength of a system of point sources is zero the polarisation (dipole moment) of the complex is independent of the datum point from which it is computed. Prove this, and show further that when the total source strength is not zero the position vector

$$\bar{s}_G = \sum_{i=1}^{n} a_i \bar{s}_i \Big/ \sum_{i=1}^{n} a_i$$

relative to the (arbitrary) point P defines a point G which is fixed in relation to the complex and about which the polarisation is zero. [G is known as the source centre. It corresponds with the centre of mass in mechanics except insofar as charge magnitude may be positive or negative.]

4-6. If the point O lies at a distance R from the source centre G of a point source system, show that

$$\phi_O = \frac{1}{R} \sum_{i=1}^{n} a_i + \frac{1}{2R^3} (A+B+C-3I) + ---$$

where A, B and C are the second moments of the source strengths about three perpendicular axes through G, and I is the second moment of the source strengths about the axis OG. (MacCullagh's formula.)

4-7. Show that the terms within the second pair of brackets in equation (4.1-15) are equal to the potential at O of three oblique quadrupoles located at P and lying in the xy, yz and zx planes, with constituent dipoles orientated parallel to the x, y and z axes respectively.

4-8. Use equation (4.1-17) to prove that the potential of a collinear set of sources of strengths $-a$, $+3a$, $-3a$, $+a$ and equal spacing d becomes identical with that of an axial octupole of moment $p^{(3)} = 6ad^3$ when ad^3 is held constant as $a \to \infty$ and $d \to 0$.

This demonstrates that an axial multipole - or its equivalent - may be generated by a single limiting process rather than by a series of such processes. Show that this is also true of a rectangular quadrupole by expanding the potential of an appropriate four-source system in a Taylor series.

4-9. The potential of a system of logarithmic point sources in the plane may be expanded about an arbitrary origin in a manner similar to that developed for the Newtonian potential in the corresponding three-dimensional problem. Thus if Fig. 4.3 now describes the relative positions of the typical logarithmic source P_i of strength $2\lambda_i$, the point of observation O and some point P in the vicinity of P_i, where $PP_i < PO$, show that

$$\ln OP_i = \ln R - \frac{s_i}{R} \cos \theta_i - \frac{1}{2} \frac{s_i^2}{R^2} \cos 2\theta_i - \frac{1}{3} \frac{s_i^3}{R^3} \cos 3\theta_i ---$$

and

$$\phi_o = -\left\{\sum_{i=1}^{n} 2\lambda_i\right\}\ln R + \frac{1}{R}\sum_{i=1}^{n} 2\lambda_i\, s_i \cos\theta_i + \frac{1}{2R^2}\sum_{i=1}^{n} 2\lambda_i\, s_i^2 \cos 2\theta_i ---$$

[Hint: The factor $\left\{1 - \frac{2s_i}{R}\cos\theta_i + \frac{s_i^2}{R^2}\right\}$ may be written as $\left\{1 - \frac{s_i}{R}e^{+j\theta_i}\right\}$ $\left\{1 - \frac{s_i}{R}e^{-j\theta_i}\right\}$ where $j = \sqrt{-1}$ so that $\ln\left\{1 - \frac{2s_i}{R}\cos\theta_i + \frac{s_i^2}{R^2}\right\}^{\frac{1}{2}}$ may be expressed as the sum of two convergent series provided that $s_i < R$.]

4-10. If a system of parallel, uniform, rectilinear sources of equal length 2c and densities λ_1, λ_2 --- are bisected normally in the points P_1, P_2 --- by the plane of the previous exercise, show that ϕ_o is given by the previous expression increased by $2\left\{\sum_{i=1}^{n}\lambda_i\right\}\ln 2c$ when $c\to\infty$, provided that PO is finite.

4-11. Derive equation (4.2-9) from (4.1-7) by treating the line doublet as an infinite set of transversely orientated point doublets of moment λd per unit length.

4.3 The Scalar Potential of Surface Sources

Let σ be a finite, piecewise continuous function of position upon a surface S. Then the associated scalar potential at any point 0 is defined by

$$\phi_o = \int_S \frac{\sigma}{r}\, dS \qquad (4.3\text{-}1)$$

where r is the distance of dS from 0.

S is said to comprise a surface source of potential, of local surface density (or strength per unit area) σ.

ϕ is seen to be finite and continuous at all points outside a bounded surface because r is non-zero for each surface element, but it is not immediately apparent that ϕ will be finite upon the surface itself because the integrand is infinite at 0 and the integral is consequently improper. It can, nevertheless, be shown that the potential integral is both convergent and continuous at all points of a smooth surface, including those points where σ is discontinuous[7]; the treatment, however, is best postponed until similar proofs have been presented in connection with the potential of a volume source. (See Ex. 4-16. and 4-17., pp. 242-3.)

Meanwhile we may show, by plausible argument, that the potential is convergent upon the surface and continuous through it at those points inside the boundary where the surface is smooth and σ is continuous. For this purpose we first examine the potential distribution along the axis of a circular disc of constant density σ, centred upon the origin of coordinates and lying in the xy plane. If the radius of the disc is a and the coordinates of O are (0,0,z) then

$$\phi_o = \int_0^a \frac{\sigma 2\pi\rho d\rho}{(\rho^2+z^2)^{\frac{1}{2}}} \qquad (\rho^2=x^2+y^2)$$

or

$$\phi_o = 2\pi\sigma \{(a^2+z^2)^{\frac{1}{2}}-(z^2)^{\frac{1}{2}}\} \tag{4.3-2}$$

As z approaches zero through positive or negative values, ϕ_o approaches $2\pi\sigma a$, and this value of the potential obtains upon the surface itself, the integral being no longer improper for $z = 0$ when dS is written as $2\pi\rho d\rho$. The potential is therefore continuous through the disc along the central axis.

Now suppose that O lies close to a smooth, curved surface source of variable density. If the normal to the surface through O meets it in some interior point P and a circle is drawn on the surface about P, it will, if sufficiently small, define a sensibly plane disc of constant density. The potential at O is the sum of that associated with the disc and that associated with the remainder of the surface. The latter is continuous at O since the distance of each source element exceeds some non-zero value; the former is continuous through the surface at P as proved above. The total potential is therefore continuous through the surface at interior points, and, being continuous tangentially at points just outside the surface, is likewise continuous upon the surface itself.

The surface source presently under consideration is said to comprise a simple layer[8]. As a source of potential it is seen to be equivalent, at least at points off the surface, to the limiting configuration of a smooth distribution of singlets (point sources) whose number per unit

---

7. The result is even more general than this. For the continuity of potential at a conical point, a cusp and a point of intersection of a finite number of surfaces, see H. and B.S. Jeffreys, loc. cit., Ch.6.

8. Or 'single layer', as distinct from the 'double layer' introduced below.

area of the surface is increased without limit, and whose individual magnitudes approach zero in such a way as to maintain the total source strength per unit area at any point equal to σ[9].

There is a second type of surface source known as a surface doublet, doublet shell or double layer. This comprises the limiting configuration of a smooth distribution of point doublets orientated normally to a surface, when the number of doublets per unit area of the surface is increased without limit while the individual doublet moments approach zero in such a way as to make the total scalar moment per unit area, μ, some specified, piecewise continuous function of position on the surface. μ is known as the surface density, or, if constant, the strength of the shell[10]. The term 'double layer' derives from the equivalence of a plane doublet shell and two plane simple layers of densities $\pm\sigma$ and spacing d, when the product σd is maintained equal to μ while $\sigma\to\infty$ and $d\to 0$. Although the equivalence fails when the surface is curved[11], we will continue to use the terms interchangeably.

The potential at an exterior point O of a doublet shell S may be written down directly by substitution of $\mu d\bar{S}$ for $\bar{p}^{(1)}$ in equations (4.1-2) or (4.1-7). We obtain

$$\phi_O = \int_S \mu\nabla\left(\frac{1}{r}\right).d\bar{S} = \int_S \mu\,\frac{\partial}{\partial n}\left(\frac{1}{r}\right) dS \tag{4.3-3}$$

or

$$\phi_O = -\int_S \mu\,\frac{\bar{r}}{r^3}.d\bar{S} = -\int \mu d\Omega \tag{4.3-4}$$

where $d\Omega$ is the element of solid angle subtended at O by dS.

It is seen that μ must be taken as positive or negative according as the positive sense of motion through the surface corresponds with that of the doublet orientation or not.

When the shell is uniform (ie of constant density) these equations reduce to

---

9. For this reason we may also refer to the source as a surface singlet.

10. The term 'strength' is more appropriately reserved for simple sources, where it signifies $\sum a_i$, $\int \lambda\, ds$, $\int \sigma\, dS$, etc.

11. See Ex.4-12. and 4-13., p. 241.

$$\phi_o = \mu \int_S \frac{\partial}{\partial n}\left(\frac{1}{r}\right) dS = -\mu\Omega \qquad (4.3\text{-}5)$$

It then follows from equations (3.4-1) and (3.4-2) that for a closed uniform doublet shell the potential at 0 is zero or $-4\pi\mu$ according as 0 is an exterior or interior point of the closed region of space bounded by S. There is consequently a discontinuity of potential of $4\pi\mu$ on passing through the surface at any point in direction of the positive normal.

The same expression holds for the discontinuity of potential which obtains at non-boundary points of any smooth surface doublet of continuously variable density if μ is identified with its local value. This follows from a consideration of the contributions of a surface element to the potential at adjacent points on either side of the surface. Suppose that these points lie upon the normal to the surface through an interior point of the element. Since the element, if sufficiently small, may be treated as planar and of constant density, the associated partial potentials are seen from equation (4.3-5) to approach $\pm 2\pi\mu$. That component deriving from the remainder of the surface must be continuous through the surface along the normal since all elements are removed by more than some non-zero distance; the limiting values of the potential, as the surface is approached along the normal on either side, accordingly take the form $\phi' - 2\pi\mu$ and $\phi' + 2\pi\mu$, where μ is the local surface density.

4.4 The Scalar Potential of a Volume Source

Let ρ be a scalar point function[12], finite and piecewise continuous throughout a finite region of space τ. Then the associated scalar potential at any point 0 is defined by

$$\phi_o = \int_\tau \frac{\rho}{r}\, d\tau \qquad (4.4\text{-}1)$$

where r is the distance of $d\tau$ from 0.

τ is said to comprise a volume source of potential whose local density, or strength per unit volume, is ρ.

The potential is clearly finite and continuous at exterior points of τ, but at interior and boundary points the integral becomes improper because the integrand is infinite at 0. In this case the value of the integral is taken to be

12. There will be no cause to confuse this with the cylindrical coordinate ρ.

$$\lim_{d \to 0} \int_{\tau - \tau_1} \frac{\rho}{r} \, d\tau \qquad (4.4\text{-}2)$$

where τ_1 is a regular region of which 0 is an interior point and d its maximum chord.

It is required that the limit be independent of the shape of τ_1. To show that such a limit exists we make use of Cauchy's criterion[13], which, for the present purpose, may be stated in the following form: Let τ_δ denote a sphere of radius δ centred upon 0, and let τ_1 and τ_2 be any regular regions contained by τ_δ and having 0 as an interior point. Then the integral is convergent if, for any positive number ε, a value of Δ can be found such that

$$\left| \int_{\tau - \tau_1} \frac{\rho}{r} \, d\tau - \int_{\tau - \tau_2} \frac{\rho}{r} \, d\tau \right| < \varepsilon \quad \text{for} \quad 0 < \delta < \Delta \qquad (4.4\text{-}3)$$

Since the region of integration outside τ_δ is common to both integrals, equation (4.4-3) may be replaced by

$$\left| \int_{\tau_\delta - \tau_1} \frac{\rho}{r} \, d\tau - \int_{\tau_\delta - \tau_2} \frac{\rho}{r} \, d\tau \right| < \varepsilon \quad \text{for} \quad 0 < \delta < \Delta \qquad (4.4\text{-}3(a))$$

The existence of Δ may be demonstrated in the following way.

Because ρ is finite throughout the closed region τ

$$\left| \int_{\tau_\delta - \tau_1} \frac{\rho}{r} \, d\tau \right| \leq |\rho|_{max} \int_{\tau_\delta - \tau_1} \frac{1}{r} \, d\tau$$

where $|\rho|_{max}$ is the greatest numerical value of ρ in τ.

If, now, $d\tau$ is written in spherical coordinates as $r \sin\theta \, d\Phi \, r \, d\theta \, dr$ the integral on the right hand side of the inequality ceases to be improper, so that we have

$$\int_{\tau_\delta - \tau_1} \frac{1}{r} \, d\tau < \int_{\tau_\delta} \frac{1}{r} \, d\tau = \int_0^\delta \int_0^\pi \int_0^{2\pi} r \sin\theta \, d\Phi \, d\theta \, dr = 2\pi\delta^2$$

---

13. Kellogg, Ch. 6.

whence

$$\left| \int_{\tau_\delta - \tau_1} \frac{\rho}{r}\, d\tau \right| < 2\pi\delta^2 \, |\rho|_{max}$$

The same relationship holds for integration over $\tau_\delta - \tau_2$ hence

$$\left| \int_{\tau_\delta - \tau_1} \frac{\rho}{r}\, d\tau - \int_{\tau_\delta - \tau_2} \frac{\rho}{r}\, d\tau \right| < 4\pi\delta^2 \, |\rho|_{max}$$

in which case equation (4.4-3(a)) is satisfied when $\Delta = \left\{ \dfrac{\varepsilon}{4\pi |\rho|_{max}} \right\}^{\frac{1}{2}}$

The integral $\int_\tau \frac{\rho}{r}\, d\tau$ is therefore convergent.

Since it has not been required that ρ be continuous, the analysis may also be applied to boundary points of τ by treating these as interior points of an extended system where ρ passes discontinuously to zero.

We now proceed to show that ϕ is everywhere continuous.

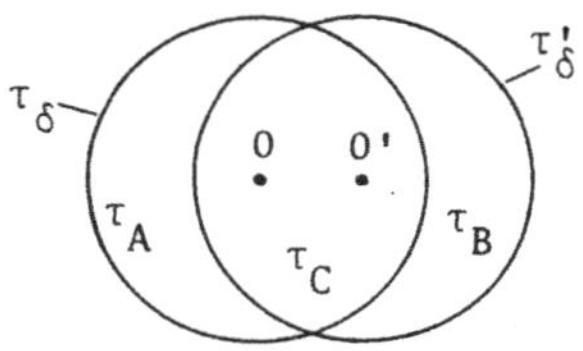

Fig. 4.6

Let 0 and 0' be interior points of τ of spacing less than δ and let spheres of radius δ be centred upon each (Fig. 4.6). If the regions enclosed by the spheres are denoted by τ_δ and τ'_δ then $\tau_\delta = \tau_C + \tau_A$, $\tau'_\delta = \tau_C + \tau_B$ where τ_C is the region common to both, and τ_A and τ_B are respectively subregions of τ_δ and τ'_δ alone. Let r be distance measured from 0 and r' distance measured from 0'.

Then

$$\int_{\tau_\delta} \frac{1}{r}\, d\tau - \int_{\tau'_\delta} \frac{1}{r}\, d\tau = \int_{\tau_A} \frac{1}{r}\, d\tau - \int_{\tau_B} \frac{1}{r}\, d\tau$$

At interior points of τ_A, $\frac{1}{r} > \frac{1}{\delta}$, and at interior points of τ_B, $\frac{1}{r} < \frac{1}{\delta}$, hence

$$\int_{\tau_\delta} \frac{1}{r}\, d\tau - \int_{\tau'_\delta} \frac{1}{r}\, d\tau > 0$$

since $\tau_A = \tau_B$.

But from symmetry

$$\int_{\tau'_\delta} \frac{1}{r}\, d\tau = \int_{\tau_\delta} \frac{1}{r'}\, d\tau$$

so that

$$\int_{\tau_\delta} \frac{1}{r}\, d\tau > \int_{\tau_\delta} \frac{1}{r'}\, d\tau \tag{4.4-4}$$

If $(\phi_o)_{\tau_\delta}$ and $(\phi_o')_{\tau_\delta}$ represent the potentials at 0 and 0' associated with source elements within τ_δ, then

$$|(\phi_o)_{\tau_\delta}| = \left|\int_{\tau_\delta} \frac{\rho}{r}\, d\tau\right| \le |\rho|_{max} \int_{\tau_\delta} \frac{1}{r}\, d\tau$$

and

$$|(\phi_o')_{\tau_\delta}| = \left|\int_{\tau_\delta} \frac{\rho}{r'}\, d\tau\right| \le |\rho|_{max} \int_{\tau_\delta} \frac{1}{r'}\, d\tau < |\rho|_{max} \int_{\tau_\delta} \frac{1}{r}\, d\tau$$

Hence

$$|(\phi_o')_{\tau_\delta} - (\phi_o)_{\tau_\delta}| < 2\, |\rho|_{max} \int_{\tau_\delta} \frac{1}{r}\, d\tau$$

in which case

$$|(\phi_o')_{\tau_\delta} - (\phi_o)_{\tau_\delta}| < \tfrac{2}{3}\varepsilon \quad \text{when} \quad 0 < \delta \le \left\{\frac{\varepsilon}{6\pi|\rho|_{max}}\right\}^{\frac{1}{2}} \tag{4.4-5}$$

If $(\phi_0)_{\tau-\tau_\delta}$ and $(\phi_0')_{\tau-\tau_\delta}$ represent the potentials at 0 and 0' associated with source elements outside τ_δ, then

$$(\phi_{0'})_{\tau-\tau_\delta} - (\phi_0)_{\tau-\tau_\delta} = \int_{\tau-\tau_\delta} \frac{\rho}{r'}\,d\tau - \int_{\tau-\tau_\delta} \frac{\rho}{r}\,d\tau = \int_{\tau-\tau_\delta} \rho\left(\frac{1}{r'} - \frac{1}{r}\right) d\tau$$

hence

$$|(\phi_{0'})_{\tau-\tau_\delta} - (\phi_0)_{\tau-\tau_\delta}| < |\rho|_{max} \left|\frac{1}{r'} - \frac{1}{r}\right|_{max} \int_{\tau-\tau_\delta} d\tau$$

Now $\left|\frac{1}{r'} - \frac{1}{r}\right|$ assumes its greatest value when the source element lies just outside τ_δ and is in line with 00', in which case it is equal to $d/\delta(\delta-d)$, where $d = 00'$, so that

$$|(\phi_{0'})_{\tau-\tau_\delta} - (\phi_0)_{\tau-\tau_\delta}| < |\rho|_{max} \frac{d}{\delta(\delta-d)} (\tau-\tau_\delta)$$

whence

$$|(\phi_{0'})_{\tau-\tau_\delta} - (\phi_0)_{\tau-\tau_\delta}| < \frac{\varepsilon}{3} \quad \text{for} \quad 0 < d \le \frac{\delta}{1 + 3\,|\rho|_{max}\,(\tau-\tau_\delta)/\varepsilon\delta} \qquad (4.4\text{-}6)$$

If, then, some given δ satisfies equation (4.4-5) and, for this value of δ, d satisfies equation (4.4-6), the potentials at 0 and 0' of all source elements are such that

$$|(\phi_{0'})_\tau - (\phi_0)_\tau| < \varepsilon$$

It follows that ϕ is continuous at 0.

For reasons discussed previously the analysis remains valid at boundary points of τ.

At exterior points the result follows immediately from the inequality

$$|(\phi_{0'})_\tau - (\phi_0)_\tau| = \left|\int_\tau \frac{\rho}{r'}\,d\tau - \int_\tau \frac{\rho}{r}\,d\tau\right| < |\rho|_{max} \left|\frac{1}{r'} - \frac{1}{r}\right|_{max} \tau$$

since $\frac{1}{r}$ and $\frac{1}{r'}$ are finite for all source elements and $\left|\frac{1}{r'} - \frac{1}{r}\right|_{max} \to 0$ as $00' \to 0$.

It has been supposed in the foregoing treatment that the region of integration is bounded. This does not imply that ρ is necessarily zero outside τ but that our concern is with that potential function which

exterior points is consequently irrelevant and must be equated to zero if it is necessary to extend the region of integration while maintaining the original source system (as in the above treatment of boundary points). We may equally well deal with unbounded source systems so long as the inclusion of all source elements does not lead to infinite values of ϕ at all finite distances from some local origin as would be the case, for example, if ρ were everywhere constant. It may be shown that if R is distance measured from some local origin, and R' is finite, then ϕ is everywhere finite provided that, for some value of $n > 2$, $|\rho|R^n$ is bounded for all $R > R'$. While this is a sufficient condition it is not a necessary one, and a patchy distribution of source density may admit of a lower value of the exponent.

A singular point at which ρ becomes infinite has only a finite effect on the potential at points outside a neighbourhood of the singularity provided that the rate at which ρ approaches infinity is suitably restricted. Thus if r' is distance measured from a point singularity P, then the total source strength associated with a neighbourhood of P remains finite if $|\rho|r'^n$ is bounded throughout that neighbourhood for some value of $n < 3$. For a line (or surface) of finite extent upon which ρ becomes infinite the corresponding requirement is that $|\rho|r'^n$ be bounded for some value of $n < 2$ (or <1) where r' is distance measured normally from the line (or surface). Moreover, an additional restriction upon the rate of growth of ρ may allow ϕ to be both finite and continuous at the singularity itself. (See Ex.4-20., p. 244).

EXERCISES

4-12. Let S be an open or closed doublet shell of variable density μ whose surface coincides with a part or the whole of a sphere of radius R, and whose vector dipole moment is directed out of the sphere. Show that, in respect of potential, the shell cannot be replaced by the limiting configuration of simple layers on spherical surfaces of radii R and $R + \delta R$ having equal and opposite densities along any radius and such that

$$\lim_{\substack{\sigma\to\infty \\ \delta R\to 0}} \sigma\,\delta R = \mu$$

What potential is associated with such a system?

$$\text{Ans:}\quad \phi_o = \int_S \mu \frac{\partial}{\partial n}\left(\frac{1}{r}\right) dS + \frac{2}{R}\int_S \frac{\mu dS}{r}$$

4-13. Show that the values of surface density required for the equivalence of the double layer and doublet shell in the above exercise are

$$\sigma_R = -\frac{\mu}{\delta R} \quad \text{and} \quad \sigma_{R+\delta R} = +\frac{\mu}{\delta R} - \frac{2\mu}{R} \quad \text{where} \quad \delta R\to 0.$$

4-14. Make use of equation (4.3-2) to establish the value of the potential along the axis of a disc-shaped double layer of radius a and uniform surface density μ, centred upon the origin of coordinates and lying within the xy plane with the dipole moment directed along $\hat{z}$. Confirm that $\phi \to +2\pi\mu$ or $-2\pi\mu$ according as the positive or negative side of the double layer is approached along the axis.

$$\text{Ans:}\quad \phi = 2\pi\mu\left\{\frac{z}{(z^2)^{\frac{1}{2}}} - \frac{z}{(a^2+z^2)^{\frac{1}{2}}}\right\}$$

4-15. If f and g are point functions which become infinite at some point 0 of τ, but are piecewise continuous with $|f| \leq g$ in any region not containing 0, then $\int_\tau f\, d\tau$ is convergent when $\int_\tau g\, d\tau$ is convergent.

Prove this.

[Hint: Construct spheres of radius δ and 2δ about 0 and let τ_1 and τ_2 be regular regions contained by the inner sphere. The region τ_3 has 0 for an interior point and is contained by both τ_1 and τ_2.

First show that

$$\int_{\tau_\delta - \tau_3} g\, d\tau < \frac{\varepsilon}{2} \quad \text{for} \quad 0 < 2\delta < 2\Delta$$

and then proceed to the inequality

$$\left| \int_{\tau_\delta - \tau_1} f\, d\tau - \int_{\tau_\delta - \tau_2} f\, d\tau \right| < \varepsilon \quad \text{for} \quad 0 < \delta < \Delta \;]$$

4-16. A tangent plane is drawn through a point 0 of a smooth surface S. A cylinder of radius a whose axis coincides with the normal to S through 0 cuts S in the region S_1 and the tangent plane in the region S_1'. If a is sufficiently small the normal at any point of S_1 makes an angle θ of less than 90° with the normal through 0. Let σ' be a point function, defined upon S_1', and equal to $\sigma/\cos\theta$ where σ is the source density at the corresponding point (by axial projection) on S_1. Form the integrals

$$\int_{S_1} \frac{\sigma}{r}\, dS \quad \text{and} \quad \int_{S_1'} \frac{\sigma'}{r'}\, dS'$$

where r and r' are respectively the distances of dS and dS' from 0.

By modifying the result of the previous exercise to apply to surface integrals show that $\int_{S_1} \frac{\sigma}{r}\, dS$, and consequently $\int_{S} \frac{\sigma}{r}\, dS$, is convergent at 0.

4-17. Develop the planar analogue of the analysis which establishes the continuity of the potential function within a volume source, and so demonstrate that the potential of a plane surface source of piecewise continuous density is continuous at all interior and boundary points of the surface.

4-18. Show (a) by direct integration

(b) by utilising the results of Ex.3-19., p. 188, that the potential of a uniform spherical shell of radius a, thickness Δt and source density ρ is given by

(1) $\frac{4\pi a^2 \rho \Delta t}{R}$ at points outside the shell

(2) $4\pi a \rho \Delta t$ at points inside the shell

where R denotes distance from the centre.

Hence show that the potential of a uniform spherical source is the same at exterior points as that of a central point source whose strength is equal to the product of the volume of the sphere and its density. Show further that if the radius of the sphere is a and its density ρ the potential at interior points is given by $2\pi\rho\left(a^2 - \frac{1}{3}R^2\right)$.

Confirm that the potential is continuous through the surface of the sphere, and sketch its magnitude as a function of distance from the source centre.

4-19. If the density ρ of a volume source of potential is everywhere bounded, and if $\rho = \alpha/R^n$ for $R > R'$ where α is a constant, R is distance measured from some local origin and R' is finite, show that the potential is everywhere finite when $n > 2$ and is regular at infinity when $n > 3$.

[If d is the distance of 0 from the local origin and ϕ_o' is the contribution to the potential at 0 of sources within a radius R' of the origin then

(a) for $d \le R'$

$$\phi_o = \phi_o' + \frac{4\pi\alpha}{n-2}\frac{1}{R'^{\,n-2}} \quad \text{for} \quad n > 2$$

(b) for $d \ge R'$

$$\phi_o = \phi_o' + \frac{4\pi\alpha}{n-3}\left\{\frac{d^{2-n}}{2-n} + \frac{R'^{\,3-n}}{d}\right\} \quad \text{for} \quad n > 2,\ n \ne 3$$

$$\phi_o = \phi_o' + \frac{4\pi\alpha}{d}(1+\ln d-\ln R') \quad \text{for} \quad n = 3]$$

4-20. Within a spherical region of radius R about a fixed point P the density of a volume source of potential is given by $\rho = \alpha/r'^n$ where α is a constant and r' is distance from P. If O is a variable point show that the potential at O deriving from sources within the region is finite when O coincides with P, provided that $n < 2$. Show further that, for this condition, ϕ_o is continuous at P.

$$[\phi_P = \frac{4\pi\alpha}{2-n} R^{2-n} \qquad \text{for} \quad n < 2$$

For $0 < d \leq R$ where $d = OP$

$$\phi_o = \frac{4\pi\alpha}{2-n}\left\{R^{2-n} - \frac{d^{2-n}}{3-n}\right\} \qquad \text{for} \quad n < 3, \; n \neq 2$$

$$\phi_o = 4\pi\alpha\,(1+\ln R - \ln d) \qquad \text{for} \quad n = 2]$$

4.5 The Representation of a Scalar Point Function as the Combined Potentials of Surface and Volume Sources

If the scalar point function V is well-behaved in the region of space τ, bounded by the surfaces $S_{1..n}\Sigma$, then its value at any interior point O of τ, as expressed by Green's formula, is

$$V_o = \frac{1}{4\pi}\oint_{S_{1..n}\Sigma}\left\{\frac{1}{r}\frac{\partial V}{\partial n} - V\frac{\partial}{\partial n}\left(\frac{1}{r}\right)\right\} dS - \frac{1}{4\pi}\int_\tau \frac{\nabla^2 V}{r}\, d\tau \tag{4.5-1}$$

where r is distance measured from O.

The potential at O of simple and double layer surface sources of densities σ and μ on $S_{1..n}\Sigma$, and of volume sources of density ρ in τ, is given by

$$\phi_o = \oint_{S_{1..n}\Sigma}\left\{\frac{\sigma}{r} + \mu\frac{\partial}{\partial n}\left(\frac{1}{r}\right)\right\} dS + \int_\tau \frac{\rho}{r}\, d\tau \tag{4.5-2}$$

where $\hat{n}$ is directed out of τ, as in equation (4.5-1), and μ is positive for a corresponding doublet orientation.

Hence if we put $\sigma = \frac{1}{4\pi}\frac{\partial V}{\partial n}$, $\mu = -\frac{V}{4\pi}$ and $\rho = -\frac{\nabla^2 V}{4\pi}$ the potential of these sources (whose densities are independent of the position of O) will be equal at any interior point of τ to the value of V obtaining there.

This representation is not unique as we now proceed to show.

Let U_1 be a scalar point function which is well-behaved in the region τ_1 bounded externally by S_1. Then the arguments which led to

$$0 = \frac{1}{4\pi}\oint_{S_1}\left\{\frac{1}{r}\frac{\partial U_1}{\partial n'} - U_1\frac{\partial}{\partial n'}\left(\frac{1}{r}\right)\right\}dS - \frac{1}{4\pi}\int_{\tau_1}\frac{\nabla^2 U_1}{r}\,d\tau \qquad (4.5\text{-}3)$$

where r continues to be measured from 0 in τ and the positive sense of the normal n' is directed into τ.

Similar relationships hold for each of the surfaces $S_{2..n}$ and the associated regions $\tau_{2..n}$. The functions U_1, U_2 .... will, in general, be unrelated.

If, in addition, U_Σ is well-behaved outside Σ and is regular at infinity, then

$$0 = \frac{1}{4\pi}\oint_{\Sigma}\left\{\frac{1}{r}\frac{\partial U_\Sigma}{\partial n'} - U_\Sigma\frac{\partial}{\partial n'}\left(\frac{1}{r}\right)\right\}dS - \frac{1}{4\pi}\int_{\tau_\Sigma}\frac{\nabla^2 U_\Sigma}{r}\,d\tau \qquad (4.5\text{-}4)$$

where τ_Σ represents all space outside Σ.

On adding equations (4.5-1), (4.5-4) and the set of equations typified by equation (4.5-3), we obtain

$$V_o = \frac{1}{4\pi}\oint_{S_{1..n}\Sigma}\left\{\frac{1}{r}\left(\frac{\partial V}{\partial n} + \frac{\partial U}{\partial n'}\right) - (V-U)\frac{\partial}{\partial n}\left(\frac{1}{r}\right)\right\}dS - \frac{1}{4\pi}\int_{\tau}\frac{\nabla^2 V}{r}\,d\tau - \frac{1}{4\pi}\int_{\tau_{1..n}\tau_\Sigma}\frac{\nabla^2 U}{r}\,d\tau \qquad (4.5\text{-}5)$$

This equation postulates a set of surface sources quite different to those deriving from equation (4.5-1) and involves additional volume sources. Since the choice of U is unrestricted, apart from the requirement of continuity within individual regions and, in the case of U_Σ, behaviour at infinity, there are an infinite number of possible source configurations whose potential equates V at interior points of τ. This remains true when U is restricted to harmonic functions, in which case

$$V_o = \frac{1}{4\pi}\oint_{S_{1..n}\Sigma}\left\{\frac{1}{r}\left(\frac{\partial V}{\partial n} + \frac{\partial U}{\partial n'}\right) - (V-U)\frac{\partial}{\partial n}\left(\frac{1}{r}\right)\right\}dS - \frac{1}{4\pi}\int_{\tau}\frac{\nabla^2 V}{r}\,d\tau \qquad (4.5\text{-}6)$$

There are, however, two unique forms of equation (4.5-6). The first emerges when U is equated to V (or to its limiting value) at all points of $S_{1..n}\Sigma$, making U the solutions of interior Dirichlet problems in $\tau_{1..n}$ and an exterior Dirichlet problem in τ_Σ. This does not impose any restriction on V and leads to the relationship

$$V_o = \frac{1}{4\pi} \oint_{S_{1..n}\Sigma} \frac{1}{r} \left(\frac{\partial V}{\partial n} + \frac{\partial U}{\partial n'} \right) dS - \frac{1}{4\pi} \int_\tau \frac{\nabla^2 V}{r} d\tau \qquad (4.5\text{-}7)$$

so that V_o is expressed as the potential of a single layer source on $S_{1..n}\Sigma$ and a volume source within τ.

The second unique solution requires that $\frac{\partial U}{\partial n'} = -\frac{\partial V}{\partial n}$ over $S_{1..n}\Sigma$ and consequently identifies U with the solutions of interior Neumann problems in $\tau_{1..n}$ and an exterior Neumann problem in τ_Σ, in which case

$$V_o = -\frac{1}{4\pi} \oint_{S_{1..n}\Sigma} (V-U) \frac{\partial}{\partial n} \left(\frac{1}{r} \right) dS - \frac{1}{4\pi} \int_\tau \frac{\nabla^2 V}{r} d\tau \qquad (4.5\text{-}8)$$

so that V_o is expressed as the potential of a double layer source on $S_{1..n}\Sigma$ and a volume within τ. This representation, however, imposes restrictions on V because the interior Neumann problems are soluble only if $\oint \frac{\partial V}{\partial n} dS = 0$ for each of the surfaces $S_{1..n}$ in turn. When the surfaces $S_{1..n}$ do not exist, ie when τ is a region bounded by Σ alone, the difficulty does not arise and equation (4.5-8) is applicable to all V.

We have confined our attention in the above analysis to the description of V as a potential function in a bounded portion of space only. V may or may not be defined outside this region. If it is, then V and/or $\frac{\partial V}{\partial n}$ may or may not be continuous across the bounding surface or surfaces. It is clear that the potential source system deriving from equation (4.5-1) and matching V at interior points of τ can give rise only to zero potential outside τ because the right hand side of equation (4.5-1) is zero when the origin of r is exterior to τ. Thus unless V is zero outside τ it cannot be matched everywhere by this source system.

Suppose now that V is defined everywhere, and that it is well-behaved in the regions $\tau_{1..n}$, τ_Σ as well as in τ. For convenience let V be designated V_1, V_2 .... V_Σ in τ_1, τ_2 .... τ_Σ and V in τ. Then on putting $U_1 = V_1$, $U_2 = V_2$ etc in the equations leading to (4.5-5) and on pairing both V and $\frac{\partial V}{\partial n}$ on opposite sides of $S_{1..n}\Sigma$ as in Sec. 3.5, we find that for interior points of τ

$$V_o = \frac{1}{4\pi} \oint_{S_{1..n}\Sigma} \left\{ -\frac{1}{r} \Delta \frac{\partial V}{\partial n} + \Delta V \frac{\partial}{\partial n} \left(\frac{1}{r} \right) \right\} dS - \frac{1}{4\pi} \int_\infty \frac{\nabla^2 V}{r} d\tau \qquad (4.5\text{-}9)$$

where ΔV and $\Delta \frac{\partial V}{\partial n}$ are the increments of V and $\frac{\partial V}{\partial n}$ corresponding to positive motion through the surface when the same arbitrarily defined positive sense of the normal is assigned to each side of the surface. The volume integral is taken over all space and it is assumed that V is regular at infinity.

Suppose now that the origin of r is transferred to τ_1. Then

$$(V_1)_o = \frac{1}{4\pi}\oint_{S_1}\left\{\frac{1}{r}\frac{\partial V_1}{\partial n_1} - V_1\frac{\partial}{\partial n_1}\left(\frac{1}{r}\right)\right\} dS - \frac{1}{4\pi}\int_{\tau_1}\frac{\nabla^2 V_1}{r}\,d\tau \tag{4.5-10}$$

where the positive sense of the normal is directed out of τ_1.

We have also

$$0 = \frac{1}{4\pi}\oint_{S_{1..n}\Sigma}\left\{\frac{1}{r}\frac{\partial V}{\partial n} - V\frac{\partial}{\partial n}\left(\frac{1}{r}\right)\right\} dS - \frac{1}{4\pi}\int_{\tau}\frac{\nabla^2 V}{r}\,d\tau \tag{4.5-11}$$

$$0 = \frac{1}{4\pi}\oint_{S_2}\left\{\frac{1}{r}\frac{\partial V_2}{\partial n_2} - V_2\frac{\partial}{\partial n_2}\left(\frac{1}{r}\right)\right\} dS - \frac{1}{4\pi}\int_{\tau_2}\frac{\nabla^2 V_2}{r}\,d\tau \tag{4.5-12}$$

with similar relationships for $S_{3..n}\Sigma$.

Upon adding equations (4.5-10), (4.5-11) and (4.5-12) etc, and pairing as before, we obtain

$$(V_1)_o = \frac{1}{4\pi}\oint_{S_{1..n}\Sigma}\left\{-\frac{1}{r}\Delta\frac{\partial V}{\partial n} + \Delta V\frac{\partial}{\partial n}\left(\frac{1}{r}\right)\right\} dS - \frac{1}{4\pi}\int_{\infty}\frac{\nabla^2 V}{r}\,d\tau \tag{4.5-13}$$

The right hand side of equation (4.5-13) is identical with that of equation (4.5-9) hence V is represented both in τ and τ_1 by the same expression. Similar arguments show that this holds also for the remaining regions. It follows that any point function V, which is well-behaved at points not coincident with a set of surfaces and is regular at infinity, may be represented at all points removed from such surfaces by the potential of simple and double layer sources upon these surfaces and of volume sources throughout space, the respective source densities being $-\frac{1}{4\pi}\Delta\frac{\partial V}{\partial n}$, $\frac{\Delta V}{4\pi}$ and $\frac{-\nabla^2 V}{4\pi}$. This distribution may be shown to be unique.

The result is not restricted to functions whose surfaces of discontinuity comprise the set of closed surfaces $S_{1..n}\Sigma$ as shown in Fig. 3.1. It is clear that the surfaces may be open or that a number of surfaces may enclose each other in turn. When point or line discontinuities are present their exclusion from the region of integration gives rise to the usual unpaired surface integrals which, of course, are required to approach a limit as the excluding surface shrinks about the discontinuity, and which lead, for non-zero limits, to the introduction of point and line sources in the equivalent potential source system.

4.6 The Gradient and Laplacian of the Scalar Potential of Point Sources

Gauss's Law

4.6a Gradient of the potential of singlet sources

The potential at 0 of singlet sources of strengths a_1--a_i--a_n at P_1--P_i--P_n is given by

$$\phi_o = \sum_{i=1}^{n} \frac{a_i}{OP_i} \qquad (4.1\text{-}1)$$

If (x_o,y_o,z_o) and (x_i,y_i,z_i) are the Cartesian coordinates of 0 and P_i respectively, then

$$OP_i = \{(x_i-x_o)^2+(y_i-y_o)^2+(z_i-z_o)^2\}^{\frac{1}{2}}$$

and

$$\frac{\partial}{\partial x_o}\left(\frac{1}{OP_i}\right) = \frac{(x_i-x_o)}{OP_i^3} \qquad (4.6\text{-}1(a))$$

whence

$$\left(\text{grad}\,\frac{1}{r_i'}\right)_o = \left(\frac{\bar{r}}{r^3}\right)_{P_i} = -\left(\frac{\bar{r}_i'}{(r_i')^3}\right)_o$$

where r represents distance measured from 0 and r_i' represents distance measured from P_i.

In addition,

$$\frac{\partial}{\partial x_i}\left(\frac{1}{OP_i}\right) = -\frac{(x_i-x_o)}{OP_i^3} \qquad (4.6\text{-}1(b))$$

whence

$$\left(\text{grad}\,\frac{1}{r}\right)_{P_i} = -\left(\frac{\bar{r}}{r^3}\right)_{P_i} = \left(\frac{\bar{r}_i'}{(r_i')^3}\right)_o$$

so that

$$(\mathrm{grad}\ \phi)_o = \sum_{i=1}^{n} a_i \left(\mathrm{grad}\ \frac{1}{r'_i}\right)_o = -\sum_{i=1}^{n} a_i \left(\frac{\bar{r}'_i}{(r'_i)^3}\right)_o \qquad (4.6\text{-}2)$$

$$= -\sum_{i=1}^{n} a_i \left(\mathrm{grad}\ \frac{1}{r}\right)_{P_i} = \sum_{i=1}^{n} a_i \left(\frac{\bar{r}}{r^3}\right)_{P_i} \qquad (4.6\text{-}3)$$

The expressions $\left(\mathrm{grad}\ \frac{1}{r}\right)_{P_i}$ and $\left(\mathrm{grad}\ \frac{1}{r'_i}\right)_o$ are commonly replaced in the literature by $\mathrm{grad}_{P_i} \frac{1}{r}$ and $\mathrm{grad}_o \frac{1}{r}$. Since the gradient operator can act only upon a scalar point function and since no such function is adequately represented by the inverse of a distance between points unless an origin of the distance is specified, the common factor $\frac{1}{r}$ in the second pair of expressions must be identified with either $\frac{1}{r}$ or $\frac{1}{r'_i}$ of the previous pair. The use of a common symbol to denote two distinct point functions is confusing and for this reason we will adhere to the notation which identifies r with distance measured from the point of evaluation of the scalar or vector field, viz 0, and r'_i with distance measured from the source location P_i.

The above results have been obtained by working within a single system of Cartesian coordinates. The same results follow from an application of the distributive law

$$\mathrm{grad}(\phi_1+\phi_2---) = \mathrm{grad}\ \phi_1 + \mathrm{grad}\ \phi_2 --$$

where ϕ_1, ϕ_2 are the partial potential fields associated with each of the point sources in turn, combined with the known invariance of the analytical formulation of grad ϕ with respect to choice of axes (and origin). This invariance has been demonstrated for systems of Cartesian axes and must also hold for cylindrical and spherical coordinate systems since these duplicate the Cartesian systems in evaluating the components of grad ϕ in three mutually perpendicular directions. Hence, if $(\mathrm{grad}\ \phi_1)_o$, $(\mathrm{grad}\ \phi_2)_o$ --- are evaluated in turn by locating the origin of spherical coordinates at P_1, P_2 --- and applying equation (2.6-5) to the associated spherically symmetrical fields, we may form $[\mathrm{grad}\ \phi]_o$ by addition of the resulting vectors.

Thus, in terms of the current notation,

$$(\mathrm{grad}\ \phi_i)_o = a_i\left(\hat{r}'_i \frac{d}{dr'_i}\left(\frac{1}{r'_i}\right)\right)_o = -a_i\left(\frac{\hat{r}'_i}{(r'_i)^2}\right)_o$$

whence

$$(\text{grad } \phi)_o = -\sum_{i=1}^{n} a_i \left(\frac{\bar{r}_i'}{(r_i')^3}\right)_o$$

and this is identical with equation (4.6-2).

Grad ϕ has continuous derivatives of all orders at points not coincident with the sources. It is irrotational for all closed curves which do not pass through any source, as may be demonstrated by an application of Stokes's theorem (with curl grad $\phi \equiv 0$) to a surface which spans the closed curve and contains no sources.

4.6b Gradient of the potential of a doublet source

Let a point doublet of moment $\bar{p}^{(1)}$ be located at the origin of spherical coordinates and aligned with the direction $\theta = 0$. Then according to equation (4.1-7) the potential of the doublet at the point (R,θ,Φ) is

$$\phi = \frac{p^{(1)} \cos\theta}{R^2}$$

whence from equation (2.6-5) the components of grad ϕ are

$$\left.\begin{aligned}
(\text{grad } \phi)_R &= \frac{\partial}{\partial R}\left(\frac{p^{(1)}\cos\theta}{R^2}\right) = -\frac{2p^{(1)}\cos\theta}{R^3} \\
(\text{grad } \phi)_\theta &= \frac{1}{R}\frac{\partial}{\partial\theta}\left(\frac{p^{(1)}\cos\theta}{R^2}\right) = -\frac{p^{(1)}\sin\theta}{R^3} \\
(\text{grad } \phi)_\Phi &= \frac{1}{R\sin\theta}\frac{\partial}{\partial\Phi}\left(\frac{p^{(1)}\cos\theta}{R^2}\right) = 0
\end{aligned}\right\} \qquad (4.6\text{-}4)$$

As in the case of the singlet distribution the gradient field is undefined at the source itself.

For a system of doublets grad ϕ may be computed by vector addition of the components associated with the individual doublets, these being evaluated by locating the origin of spherical coordinates at each of the doublets in turn. The magnitude of the gradient field of a single doublet is seen to fall off as the cube of distance, whereas the field of the singlet falls off as the square. It is evident that the gradient of the potential of a 2^m pole varies as $\frac{1}{R^{m+2}}$.

4.6c The Laplacian of the scalar potential of point sources

It follows from equation (4.6-1(a)) that for a system of singlet sources

$$\frac{\partial^2\phi_o}{\partial x_o^2} = \sum_{i=1}^{n} \frac{a_i}{(OP_i)^3}\left\{-1 + 3\,\frac{(x_i-x_o)^2}{(OP_i)^2}\right\}$$

whence

$$\frac{\partial^2\phi_o}{\partial x_o^2} + \frac{\partial^2\phi_o}{\partial y_o^2} + \frac{\partial^2\phi_o}{\partial z_o^2} = \left(\frac{\partial^2\phi}{\partial x^2} + \frac{\partial^2\phi}{\partial y^2} + \frac{\partial^2\phi}{\partial z^2}\right)_o = 0 \quad (r_i\neq 0)$$

ie

$$\nabla^2\phi = \text{div grad } \phi = 0 \quad \text{everywhere outside the sources.} \tag{4.6-5}$$

Alternatively, we may locate the origin of spherical coordinates at each of the sources in turn, bearing in mind that

$$\nabla^2(\phi_1+\phi_2---) = \nabla^2\,\phi_1 + \nabla^2\,\phi_2\ ---$$

and that $\nabla^2\phi$ is invariant with respect to choice of axes and origin.

Then from equation (2.6-8)

$$(\nabla^2(\phi_i))_o = \left\{\frac{1}{(r_i')^2}\frac{d}{dr_i'}\left((r_i')^2\frac{d}{dr_i'}\left(\frac{a_i}{r_i'}\right)\right)\right\}_o = \left\{\frac{1}{(r_i')^2}\frac{d}{dr_i'}(-a_i)\right\}_o$$

ie

$$(\nabla^2(\phi_i))_o = 0 \qquad (r_i' \neq 0)$$

hence $\nabla^2\ \phi = 0$ at all points outside the sources.

Equation (4.6-5) is known as Laplace's equation.

Since this result holds for any finite collection of singlets we may suppose that it will continue to hold for a multipole. This is obviously true when the limiting process which leads to the formation of the multipole is arrested at some point so as to leave the associated monopole magnitudes finite, but it may also be shown to be true for the limiting case itself. Thus for an axial $2^{(m)}$ pole situated at P the potential at 0 is given by

$$\phi_{o_{axial}}^{(m)} = \frac{1}{m!} p^{(m)} \left(\frac{\partial^m}{\partial z^m}\left(\frac{1}{r}\right)\right)_P \tag{4.1-6}$$

or, by the result of Ex.4-2., p. 231,

$$\phi_{o_{axial}}^{(m)} = \frac{1}{m!} p^{(m)} (-1)^m \left(\frac{\partial^m}{\partial z^m}\left(\frac{1}{r'}\right)\right)_o$$

Then

$$\left(\nabla^2\phi_{axial}^{(m)}\right)_o = \frac{1}{m!} p^{(m)} (-1)^m \left\{\left(\frac{\partial^2}{\partial x^2} + \frac{\partial^2}{\partial y^2} + \frac{\partial^2}{\partial z^2}\right) \frac{\partial^m}{\partial z^m}\left(\frac{1}{r'}\right)\right\}_o$$

$$= \frac{1}{m!} p^{(m)} (-1)^m \left\{\frac{\partial^m}{\partial z^m}\left(\frac{\partial^2}{\partial x^2} + \frac{\partial^2}{\partial y^2} + \frac{\partial^2}{\partial z^2}\right) \frac{1}{r'}\right\}_o$$

since all derivatives (including cross derivatives) are continuous at 0, ie

$$\left(\nabla^2\phi_{axial}^{(m)}\right)_o = \frac{1}{m!} p^{(m)} (-1)^m \left(\frac{\partial^m}{\partial z^m} \nabla^2 \left(\frac{1}{r'}\right)\right)_o$$

$$= 0$$

The argument is readily extended to non-axial multipoles.

Since div grad ϕ is zero at all points other than those occupied by sources, grad ϕ is solenoidal in source-free aperiphractic regions. It may or may not be solenoidal in periphractic (non-aperiphractic) regions. The simplest example of this is the case already treated in Sec. 1.14 where the vector $\overline{F}$ is now seen to be identical with the negative gradient of the potential of a unit source situated at the origin of coordinates. In the region beyond the δ sphere div $\overline{F} = 0$ but $\overline{F}$ is not solenoidal. It would be rendered solenoidal if the singlet at the origin were replaced by a doublet or other multipole because the flux of $\overline{F}$ through the δ sphere would then become zero. (See next sub-section.)

4.6d <u>Gauss's Law</u>

Let point sources of potential of strengths a_1--a_j be situated at the interior points P_1--P_j of a region of space $\underline{R}$ bounded by the surfaces $S_{1..n}\Sigma$, and let sources of strengths a_k--a_m be situated at the exterior points P_k--P_m. Then the outward flux of (- grad ϕ) through $S_{1..n}\Sigma$ is equal to 4π × total strength of sources enclosed,

ie

$$\oint_{S_{1..n}\Sigma} (-\operatorname{grad}\phi).d\bar{S} = 4\pi \sum_{i=1}^{j} a_i$$

Proof: The value of grad ϕ at any point 0 of the surfaces is given by equation (4.6-2) as

$$(\operatorname{grad}\phi)_o = \sum_{i=1}^{m} a_i \left(\operatorname{grad}\left(\frac{1}{r_i'}\right)\right)_o = -\sum_{i=1}^{m} a_i \left(\frac{\bar{r}_i'}{(r_i')^3}\right)_o$$

hence

$$\oint_{S_{1..n}\Sigma} (-\operatorname{grad}\phi).d\bar{S} = -\oint_{S_{1..n}\Sigma} \left\{\sum_{i=1}^{m} a_i \operatorname{grad}\left(\frac{1}{r_i'}\right)\right\}.d\bar{S} = \oint_{S_{1..n}\Sigma} \left\{\sum_{i=1}^{m} a_i \frac{\bar{r}_i'}{(r_i')^3}\right\}.d\bar{S}$$

$$= -\sum_{i=1}^{m} a_i \oint_{S_{1..n}\Sigma} \operatorname{grad}\left(\frac{1}{r_i'}\right).d\bar{S} = -\sum_{i=1}^{m} a_i \oint_{S_{1..n}\Sigma} \frac{\partial}{\partial n}\left(\frac{1}{r_i'}\right) dS$$

But from equations (3.4-1) and (3.4-2)

$$\oint_{S_{1..n}\Sigma} \frac{\partial}{\partial n}\left(\frac{1}{r_i'}\right) dS = -4\pi \qquad (i = 1\text{---}\ j)$$

$$\oint_{S_{1..n}\Sigma} \frac{\partial}{\partial n}\left(\frac{1}{r_i'}\right) dS = 0 \qquad (i = k\text{---}\ m)$$

hence

$$\oint_{S_{1..n}\Sigma} (-\,\mathrm{grad}\,\phi).d\overline{S} = \oint_{S_{1..n}\Sigma} \left\{\sum_{i=1}^{m} a_i \frac{\overline{r}_i'}{(r_i')^3}\right\}.d\overline{S} = 4\pi \sum_{i=1}^{j} a_i \qquad (4.6\text{-}6)$$

Alternatively, let small spherical surfaces, centred upon the interior sources, be designated S_{δ_1}---S_{δ_j}, and let these enclose the regions τ_{δ_1}------τ_{δ_j}. Then the potential field is harmonic at those interior points of $\underline{R}$ not occupied by the spheres, whence it follows from Theorem 3.2-$\overline{1}$ that

$$\oint_{S_{1..n}\Sigma} (-\,\mathrm{grad}\,\phi).d\overline{S} = \oint_{S_{\delta_1}..S_{\delta_j}} \mathrm{grad}\,\phi.d\overline{S}$$

Now

$$\oint_{S_{\delta_1}} \mathrm{grad}\,\phi.d\overline{S} = \oint_{S_{\delta_1}} \frac{\partial\phi_1}{\partial n}\,dS + \sum_{i=2}^{m} \oint_{S_{\delta_1}} \frac{\partial\phi_i}{\partial n}\,dS$$

where ϕ_i is the partial potential deriving from a_i, hence

$$\oint_{S_{\delta_1}} \mathrm{grad}\,\phi.d\overline{S} = -\oint_{S_{\delta_1}} \frac{\partial}{\partial r_1'}\left(\frac{a_1}{r_1'}\right) dS - \sum_{i=2}^{m} \int_{\tau_{\delta_1}} \nabla^2\,\phi_i\,d\tau$$

$$= 4\pi a_1$$

since $\nabla^2\phi_i = 0$ in τ_{δ_1} for $i = 2$--- m

whence

$$\oint_{S_{1..n}\Sigma} (-\,\mathrm{grad}\,\phi).d\overline{S} = 4\pi \sum_{i=1}^{j} a_i$$

Substitution of λds, σdS and $\rho d\tau$ for a_i in the above analysis leads to Gauss's law for distributed sources for those cases where the source neither intersects nor touches the bounding surfaces[14]. We have

$$\oint_{S_{1..n}\Sigma} (-\text{ grad } \phi).d\overline{S} = 4\pi \int \lambda ds$$

$$\text{or } 4\pi \int \sigma dS$$

$$\text{or } 4\pi \int \rho d\tau$$

where the right hand integrations are carried out over interior points of R.

For mixed sources the individual contributions are additive, since there is no mutual interference. (A point source has no length, a line source no area and a surface source no volume.)

EXERCISES

4-21. Let a vector field $\overline{F}$ be defined by

$$\overline{F} = \sum_{i=1}^{n} a_i \frac{\overline{r}'_i}{(r'_i)^{\alpha+1}} = \sum_{i=1}^{n} a_i \frac{\hat{\overline{r}}'_i}{(r'_i)^{\alpha}}$$

where a_i is a scalar magnitude associated with a fixed point P_i, $\overline{r}'_i$ is the position vector of any point relative to P_i and α is a constant.

$\overline{F}$ is seen to comprise the sum of a set of central vector fields, so called because the component fields are radially directed in relation to P_1, P_2 ---. $\overline{F}$ may be shown to be irrotational for all closed curves which do not pass through any of the points P_1, P_2 --, irrespective of the value of α. Prove this

(a) by direct integration, making use of the relationships $d\overline{r}'_1 = d\overline{r}'_2$ etc, and $\overline{r}'_i.d\overline{r}'_i = r'_i\, dr'_i$

(b) by means of equation (1.16-1) and the result of Ex.1-59., p. 78

(c) by expressing $\overline{F}$ as the gradient of a scalar field.

14. A justification of this procedure will be found in the next section.

4-22. With $\overline{F}$ defined as above, show that div $\overline{F}$ is negative, zero, or positive at all points other than P_1, P_2 --- according as α is greater than, equal to, or less than 2.

4-23. A doublet of vector moment $\overline{p}$ is situated at P. If the position vector of O relative to P is $\overline{r}'$, show, by resolution of the spherical components, that the gradient of the potential of the doublet at O may be written as

$$\text{grad } \phi = \frac{\overline{p}^{(1)}}{(r')^3} - \frac{3(\overline{p}^{(1)}.\overline{r}')\overline{r}'}{(r')^5}$$

4-24. Extend the proof of the harmonic nature of the potential of an axial multipole to an oblique quadrupole.

4-25. State and prove the planar form of Gauss's law.

Ans: The law may be stated as follows:

Let a multiply connected region $\underline{R}$ in the plane be bounded externally by the regular closed curve Γ' and internally by $\Gamma_{1..n}$. Let P_1--P_j and P_k--P_m be interior and exterior points of $\underline{R}$ at which sources of logarithmic potential of strengths a_1--a_m are situated. Then if ϕ is the logarithmic potential of the complex

$$\oint_{\Gamma_{1..n}\Gamma'} (-\text{ grad } \phi).\hat{n}' \, ds = 2\pi \sum_{i=1}^{j} a_i$$

or

$$\oint_{\Gamma_{1..n}\Gamma'} \left\{ \sum_{i=1}^{m} \frac{a_i \, \overline{\rho}_i'}{(\rho_i')^2} \right\}.\hat{n}' \, ds = 2\pi \sum_{i=1}^{j} a_i$$

where $\hat{n}'$ is the unit outward normal in the plane to the contour element ds, and $\overline{\rho}_i'$ is the position vector of ds relative to P_i.

The proof parallels that for the three dimensional case.

4-26. It follows from Gauss's average-value theorem (theorem of the arithmetic mean) Ex.3-2., p. 180, and from the harmonic nature of the potential field outside its sources that the mean value, over a spherical surface, of the potential deriving from point sources entirely outside the sphere is equal to the value of their potential at the centre of the sphere. Deduce this from one of the results of Ex.3-19., p. 188, and from the other result of this exercise deduce the 'second average value theorem', viz. 'The mean value over a spherical surface of the potential of point sources lying entirely within the sphere is equal to the sum of the source strengths divided by the radius of the sphere.' [The theorem continues to hold for distributed sources, as a simple extension of the analysis will show.]

4-27. Let the planar point function V be well-behaved in the region S of the plane bounded internally by the closed curves $\Gamma_{1..n}$ and externally by Γ'. Make use of equation (3.9-3) to show that V may be expressed at interior points of S as the logarithmic potential of line singlets and doublets coincident with $\Gamma_{1..n}\Gamma'$ and having the respective densities $\frac{1}{2\pi}\frac{\partial V}{\partial n'}$ and $\frac{-V}{2\pi}$, (where the doublet density is positive if orientation corresponds with $\hat{n}'$), together with a surface source on S of density $\frac{-1}{2\pi}\nabla^2 V$.

Following the analysis of Sec. 4.5 show that an infinite number of logarithmic source combinations can be found whose potential within S is equal to V.

4-28. Extend the analysis of the previous exercise to show that if V is well-behaved everywhere in the plane except upon $\Gamma_{1..n}\Gamma'$, it may be expressed at any point O in S', not coincident with $\Gamma_{1--n}\Gamma'\Gamma''$, as

$$V_o = \oint_{\Gamma_{1..n}\Gamma'} \left\{\left(\ln\frac{1}{\rho}\right)\left(-\frac{1}{2\pi}\,\Delta\,\frac{\partial V}{\partial n'}\right) + \frac{\Delta V}{2\pi}\frac{\partial}{\partial n'}\left(\ln\frac{1}{\rho}\right)\right\} ds + \int_{S'} \left(\ln\frac{1}{\rho}\right)\left(-\frac{1}{2\pi}\nabla^2 V\right) dS$$

$$+ \frac{1}{2\pi}\oint_{\Gamma''}\left\{\left(\ln\frac{1}{\rho}\right)\frac{\partial V}{\partial n'} - V\frac{\partial}{\partial n'}\left(\ln\frac{1}{\rho}\right)\right\} ds$$

where Γ'' is some contour which encloses $\Gamma_{1..n}\Gamma'$, and S' is the region of the plane bounded externally by Γ''. ρ denotes distance from O, and the factors $\Delta\frac{\partial V}{\partial n'}$ and ΔV represent the increments of $\frac{\partial V}{\partial n'}$ and V for positive motion through a contour when the same arbitrarily defined positive sense of the normal is assigned to each side and the positive sense of doublet alignment conforms with this.

Hence show that if $R^2\frac{\partial V}{\partial R}$ is bounded and $V \to 0$ at infinity[15], V is expressible as the combined potential of logarithmic line singlets and doublets coincident with $\Gamma_{1..n}\Gamma'$ and of logarithmic surface sources in the plane wherever $\nabla^2 V$ is non-zero.

4-29. The results of Ex.4-27. and 4-28. above are applicable to any scalar function which is well-behaved at most points of a bounded or unbounded region in a plane, whether or not the function is defined only in the plane or derives from a cut across a two or three-dimensional field in space.

---

15. or, more generally, if the line integral around Γ'' vanishes at infinity.

It should therefore be possible to express a Newtonian potential at points of a plane as the potential of a set of logarithmic sources in the plane. Carry this out for the Newtonian potential ϕ of a uniform, spherical, surface source of radius a and strength q when the plane passes through its centre T, ie express $\phi_o\left(=\frac{q}{a}\right.$ inside the circular section and $=\frac{q}{R_o}$ outside it, where R is distance measured from $\left.T\right)$ as the combined potential of a logarithmic line source around the circle and a logartihmic surface source upon the plane. Prove that this expression reduces to $\frac{q}{a}$ and $\frac{q}{R_o}$ by evaluating the line and surface integrals.

[Hint: For R > a, first show that the line integral around Γ'' in Ex.4-28. vanishes when Γ'' recedes to infinity. Then compute $\nabla^2\phi$, bearing in mind that the relevant expression is not that associated with a three-dimensional system but with a plane. To confirm the result, evaluate the line integral around the circle and the areal integral between $R = a$ and $R = R_o$ by making use of the expansion for $\ln \rho$ given in Ex.4-9., p. 232. Then derive an appropriate expansion for $\ln \rho$ when $R > R_o$ and complete the evaluation of the surface integral out to infinity.]

Ans: For R > a

$$\phi_o = \oint \frac{q}{2\pi a^2} \ln \frac{1}{\rho} \, ds - \int \frac{q}{2\pi R_o{}^3} \ln \frac{1}{\rho} \, dS$$

where the line integral is evaluated around the circle and the surface integral at all points of the plane outside the circle.]

4-30. Show from the result of Ex.3-15., p. 183 that a point function V, which is well-behaved within the region $\underline{R}$ bounded by the surfaces $S_{1..n}\Sigma$, may be expressed within $\underline{R}$ as the sum of either exponentially enhanced or exponentially attenuated potentials (γ real, and positive or negative) deriving from simple and double layer sources $S_{1..n}\Sigma$ and volume sources in $\underline{R}$, of respective densities $\frac{1}{4\pi}\frac{\partial V}{\partial n}$, $-\frac{V}{4\pi}$ and $-\frac{1}{4\pi}(\nabla^2-\gamma^2)V$.

[Note particularly that the potential associated with an element of double layer of density μ is given by $\mu \frac{\partial}{\partial n}\left(\frac{1}{r} e^{\gamma r}\right) dS$ and not by $\mu e^{\gamma r} \frac{\partial}{\partial n}\left(\frac{1}{r}\right) dS$, where $\hat{n}$ is directed out of $\underline{R}$.]

Show further that there are an infinite number of ways in which V may be expressed in the finite region $\underline{R}$, and for a given exponent, in terms of surface sources upon $S_{1..n}\Sigma$ and volume sources within τ, τ_1---τ_n, and that when V is defined everywhere outside $S_{1..n}$ the surface integral over Σ vanishes as Σ recedes to infinity, provided that the associated exponent is real and negative and V and grad V are bounded at infinity.

4-31. Show that there is a unique combination of simple and double surface sources and of volume sources whose potential is everywhere identical with an arbitrarily specified scalar point function which is well-behaved at all points removed from certain surfaces and which is regular at infinity.

4.7 The Gradient and Laplacian of the Scalar Potential of Line and Surface Sources

4.7a The gradient and Laplacian of the potential of line singlets and doublets

The gradient of the potential of a finite line singlet (simple line source) at points outside it can be found by approximating the source with a set of point singlets of strength $\lambda\Delta s$, and proceeding as in Sec. 4.6a. We have

$$\phi_o = \int_\Gamma \frac{\lambda}{r}\, ds \approx \sum_{i=1}^{n} \left(\frac{\lambda}{r}\right)_{P_i} \Delta s_i = \sum_{i=1}^{n} (\lambda)_{P_i} \left(\frac{1}{OP_i}\right) \Delta s_i$$

whence

$$\frac{\partial \phi_o}{\partial x_o} \approx \frac{\partial}{\partial x_o} \sum_{i=1}^{n} (\lambda)_{P_i} \left(\frac{1}{OP_i}\right) \Delta s_i = \sum_{i=1}^{n} (\lambda)_{P_i} \frac{\partial}{\partial x_o} \left(\frac{1}{OP_i}\right) \Delta s_i$$

since λ is independent of x_o, or

$$\frac{\partial \phi_o}{\partial x_o} \approx - \sum_{i=1}^{n} (\lambda)_{P_i} \frac{\partial}{\partial x_i} \left(\frac{1}{OP_i}\right) \Delta s_i = \sum_{i=1}^{n} (\lambda)_{P_i} \frac{(x_i - x_o)}{(OP_i)^3} \Delta s_i$$

$$\approx - \sum_{i=1}^{n} (\lambda)_{P_i} \left(\frac{\partial}{\partial x}\left(\frac{1}{r}\right)\right)_{P_i} \Delta s_i$$

hence

$$(\text{grad } \phi)_o \approx - \sum_{i=1}^{n} \left(\lambda \text{ grad } \frac{1}{r}\right)_{P_i} \Delta s_i = \sum_{i=1}^{n} \left(\lambda \frac{\bar{r}}{r^3}\right)_{P_i} \Delta s_i$$

The integral form of this relationship is

$$\text{grad}\,\phi = \text{grad}\int_\Gamma \frac{\lambda}{r}\,ds = -\int_\Gamma \lambda\,\text{grad}\,\frac{1}{r}\,ds = \int_\Gamma \lambda\,\frac{\bar{r}}{r^3}\,ds \qquad (4.7\text{-}1)$$

If it is borne in mind that ϕ and grad ϕ are always evaluated at the variable point 0, which is the origin of r, and that the components of the integrand are evaluated upon ds, then the interpretation is unambiguous.

This allows us to drop the subscript 0 from grad ϕ in equation (4.7-1) and subsequent equations[16].

The above treatment, in which the finite differentiated series is ultimately replaced by an integral, presupposes that the order of the limiting processes associated with integration and differentiation may be reversed, ie differentiation may be carried out under the integral sign. This procedure can be shown to be legitimate provided that the integrand is a continuous function of the coordinates of 0 and ds, and the region of integration is finite. Since r is non-zero for each surface element, and λ, if piecewise continuous, may be treated as the sum of continuous density functions, the required conditions are fulfilled. Higher order derivatives of the potential at exterior points ('points of free space') may be found by the same procedure[17].

The radial and axial components of grad ϕ which obtain for a uniform rectilinear source (Fig. 4.4) may be determined by differentiation of equation (4.2-1) with respect to ρ and z, or, more easily, by integrating the components of grad ϕ in accordance with equation (4.7-1). It is found that

$$(\text{grad}\,\phi)_\rho = \frac{\partial\phi}{\partial\rho} = -\frac{\lambda}{\rho}\left\{\frac{c-z}{R_1} + \frac{c+z}{R_2}\right\} \qquad (4.7\text{-}2)$$

---

16. The alternative form of grad ϕ, which parallels equation (4.6-2) rather than equation (4.6-3), viz grad $\phi = \int_\Gamma \lambda\left(\text{grad}\left(\frac{1}{r'}\right)\right)_0 ds$, where r' is distance measured from ds, does not comprise a surface integral in the same sense as equation (4.7-1), since grad $\left(\frac{1}{r'}\right)$ is not evaluated upon the the contour element. For this reason we will make no use of it.

17. In writing $\frac{\partial^n}{\partial x_o{}^n}\int_\Gamma \frac{\lambda}{r}\,ds = \int_\Gamma \lambda\,\frac{\partial^n}{\partial x_o{}^n}\left(\frac{1}{r}\right) ds$ as an intermediate step in the routine differentiation of the potential function, it is, of course, to be understood that $\frac{\partial^n}{\partial x_o{}^n}\left(\frac{1}{r}\right)$ does not stand alone but in apposition with ds, so that it is equivalent to $\frac{\partial^n}{\partial x_o{}^n}\left(\frac{1}{OP}\right)$ where P is a point of ds.

for non-zero values of ρ, and

$$(\text{grad } \phi)_z = \frac{\partial \phi}{\partial z} = -\lambda \left\{ \frac{1}{R_1} - \frac{1}{R_2} \right\} \qquad (4.7\text{-}3)$$

at points outside the source.

These expressions may be written in terms of the angles shown in Figs. 4.7a and 4.7b.

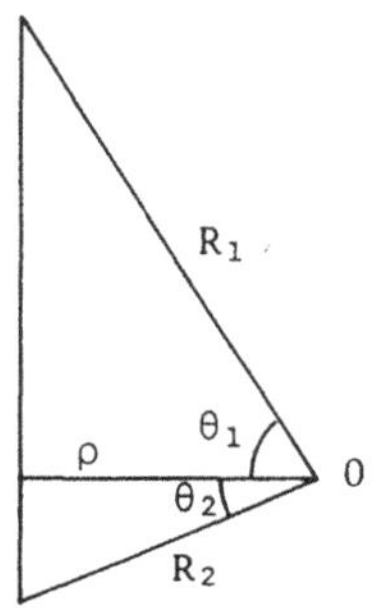

Fig. 4.7a

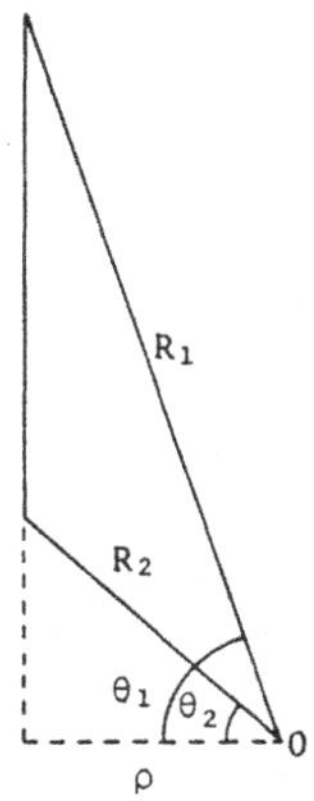

Fig. 4.7b

For non-zero values of ρ

$$\frac{\partial \phi}{\partial \rho} = -\frac{\lambda}{\rho} (\sin \theta_1 + \sin \theta_2) \qquad (\text{Fig. 4.7a})$$

or (4.7-2(a))

$$\frac{\partial \phi}{\partial \rho} = -\frac{\lambda}{\rho} (\sin \theta_1 - \sin \theta_2) \qquad (\text{Fig. 4.7b})$$

and

$$\frac{\partial \phi}{\partial z} = -\frac{\lambda}{\rho} (\cos \theta_1 - \cos \theta_2) \qquad (4.7\text{-}3(a))$$

It follows that at all finite distances from a uniform rectilinear source which extends to infinity in both directions, the radial and axial components of grad ϕ are given respectively by

$$(\text{grad } \phi)_\rho = -\frac{2\lambda}{\rho} \qquad (4.7\text{-}4)$$

$$(\text{grad } \phi)_z = 0 \qquad (4.7\text{-}5)$$

The most general expression for the gradient of the potential of a line doublet follows from equation (4.2-10) by differentiation under the integral sign.

$$\text{grad}\ \phi = -\,\text{grad}\int_\Gamma \frac{\bar{L}.\bar{r}}{r^3}\,ds = -\sum \bar{i}\,\frac{\partial}{\partial x_o}\int_\Gamma \frac{\bar{L}.\bar{r}}{r^3}\,ds$$

$$= \sum \bar{i}\int_\Gamma \left\{L_x \frac{\partial}{\partial x}\frac{(x-x_o)}{r^3} + L_y \frac{\partial}{\partial x}\frac{(y-y_o)}{r^3} + L_z \frac{\partial}{\partial x}\frac{(z-z_o)}{r^3}\right\} ds$$

where the coordinates of ds are (x,y,z)

$$= \sum \bar{i}\int_\Gamma \left\{\frac{L_x}{r^3} - \frac{3(x-x_o)\ \bar{L}.\bar{r}}{r^5}\right\} ds$$

ie

$$\text{grad}\ \phi = \int_\Gamma \left\{\frac{\bar{L}}{r^3} - \frac{3\ \bar{r}\ \bar{L}.\bar{r}}{r^5}\right\} ds \qquad (4.7\text{-}6)$$

This duplicates the result of Ex.4-23., p. 256 in integral form, the point doublet moment $\bar{p}$ being replaced by $\bar{L}ds$.

The case of greatest interest is that in which the doublet is rectilinear, infinite, uniform and planar. While it is possible to evaluate its potential gradient via equation (4.7-6), it is easier to differentiate the potential function as expressed in equation (4.2-9), or to perform a vectorial addition of the component gradients associated with a pair of infinite, rectilinear, simple sources of equal and opposite density in a limiting configuration. It is found that the magnitude of the gradient field falls off as the square of the perpendicular distance from the source and is independent of the orientation of the point of observation relative to the plane defined by the doublet moment ($\Phi=0$). The field is directed at an angle 2Φ to this plane and is consequently tangential to a circular cylinder which contains both the source and the point of observation and is bisected by the half-plane $\Phi = \frac{\pi}{2}$ or $\Phi = \frac{3\pi}{2}$.

A formally identical analysis applies to a logarithmic point doublet in the plane, in which case the cylinder is replaced by its circular section.

The Laplacian of the potential of line singlets and doublets is undefined upon the source itself since ϕ and its derivatives are undefined there. At points outside the source $\nabla^2\phi = 0$. Thus

$$\nabla^2\int_\Gamma \frac{\lambda}{r}\,ds = \int_\Gamma \lambda\,\nabla^2\left(\frac{1}{r}\right) ds = 0$$

since $$\nabla^2\left(\frac{1}{r}\right) = 0 \qquad (r\neq 0)$$

Similarly,

$$\nabla^2 \int_\Gamma \bar{L}.\nabla\left(\frac{1}{r}\right) ds = \nabla^2 \int_\Gamma \left\{L_x \frac{\partial}{\partial x}\left(\frac{1}{r}\right) + L_y \frac{\partial}{\partial y}\left(\frac{1}{r}\right) + L_z \frac{\partial}{\partial z}\left(\frac{1}{r}\right)\right\} ds$$

$$= \int_\Gamma \left\{L_x \nabla^2\left(\frac{\partial}{\partial x}\left(\frac{1}{r}\right)\right) + L_y \nabla^2\left(\frac{\partial}{\partial y}\left(\frac{1}{r}\right)\right) + L_z \nabla^2\left(\frac{\partial}{\partial z}\left(\frac{1}{r}\right)\right)\right\} ds$$

$$= \int_\Gamma \left\{L_x \frac{\partial}{\partial x}\left(\nabla^2\left(\frac{1}{r}\right)\right) + L_y \frac{\partial}{\partial y}\left(\nabla^2\left(\frac{1}{r}\right)\right) + L_z \frac{\partial}{\partial z}\left(\nabla^2\left(\frac{1}{r}\right)\right)\right\} ds$$

since all partial (and mixed) derivatives are continuous at ds,

hence

$$\nabla^2 \int_\Gamma \bar{L}.\nabla\left(\frac{1}{r}\right) ds = 0 \qquad (r\neq 0)$$

4.7b The gradient and Laplacian of the potential of surface singlets and doublets

The gradient of the potential of a surface singlet (simple layer) takes a form similar to that of the line singlet:

$$\text{grad}\,\phi = \text{grad}\int_S \frac{\sigma}{r}\,dS = -\int_S \sigma\,\text{grad}\,\frac{1}{r}\,dS = \int_S \sigma\frac{\bar{r}}{r^3}\,dS \qquad (4.7\text{-}7)$$

For the particular case of a disc-shaped surface source of constant density σ and radius a, lying in the xy plane and centred upon the origin, the potential upon the z axis is given by

$$\phi = 2\pi\sigma\,\{(a^2+z^2)^{\frac{1}{2}}-(z^2)^{\frac{1}{2}}\} \qquad (4.3\text{-}2)$$

so that

$$\frac{\partial\phi}{\partial z} = 2\pi\sigma\left\{\frac{z}{(a^2+z^2)^{\frac{1}{2}}} - \frac{z}{(z^2)^{\frac{1}{2}}}\right\} \qquad (z\neq 0) \qquad (4.7\text{-}8)$$

The normal derivative of ϕ consequently approaches $-2\pi\sigma$ or $+2\pi\sigma$ according as z approaches zero through positive or negative values. This result, taken in conjunction with the known continuity of the derivative of the

any regular surface source of piecewise continuous density the increment of $\frac{\partial\phi}{\partial n}$ in passing through the surface at an interior point in the direction of the positive normal is $-4\pi\sigma$, where σ is the local surface density (assumed to be continuous at the point in question). It may, in fact, be shown that this is the case[18], and that the limiting values of $\frac{\partial\phi}{\partial n}$ on the opposite sides of the surface are

$$\left(\frac{\partial\phi}{\partial n}\right)_+ = -2\pi\sigma + \hat{n}.\int_S \sigma \frac{\bar{r}}{r^3}\, dS$$

$$\left(\frac{\partial\phi}{\partial n}\right)_- = +2\pi\sigma + \hat{n}.\int_S \sigma \frac{\bar{r}}{r^3}\, dS \qquad (4.7\text{-}9)$$

where $\hat{n}$ is the unit positive normal to the surface, and where the integral is evaluated at that point of the surface cut by the normal.

Since ϕ is continuous on and through a simple surface source, it would appear that the derivative of ϕ in the direction of any tangent to the surface at an interior point P must have the same limiting value when P is approached along the normal through it from either side of the surface. This may be shown to be true provided that a limit exists, the required condition being that

$$|\sigma(Q)-\sigma(P)| \le Ar^{\alpha} \quad \text{for} \quad r = PQ \le c \qquad (4.7\text{-}10)$$

where Q is a point of the surface and A, α and c are positive constants.

This is known as a Hölder condition. We will meet it again in connection with volume sources.

If the disc-shaped simple layer source considered above is replaced by a surface doublet (double layer) of constant density μ, whose moment is aligned with the positive z direction, the potential upon the axis is easily shown to be given by

$$\phi = 2\pi\mu \left\{ \frac{z}{(z^2)^{\frac{1}{2}}} - \frac{z}{(a^2+z^2)^{\frac{1}{2}}} \right\} \quad (z\neq 0) \qquad (4.7\text{-}11)$$

whence

$$\frac{\partial\phi}{\partial z} = -2\pi\mu \frac{a^2}{(a^2+z^2)^{3/2}} \quad (z\neq 0) \qquad (4.7\text{-}12)$$

---

18. See Kellogg, Ch. 6 for a rigorous treatment of this and other aspects of surface sources.

Thus the derivative of ϕ upon the axis in the direction of the positive normal has the same value at points spaced equally on either side of the disc. It is possible to generalise this result as follows: If P is an interior point of a regular surface doublet of piecewise continuous density μ, and a normal to the surface is drawn through P, then the difference between the normal derivatives of ϕ at points of the normal equidistant from P approaches zero as P is approached from both sides, provided that μ is continuous at P.

Whereas the tangential derivative of ϕ is always continuous through a simple surface source at points where a Hölder condition is fulfilled, the tangential derivatives on either side of a surface doublet cannot approach a common limit as the surface is approached, unless μ is constant in a neighbourhood of the point of evaluation. This is an immediate consequence of the relationship

$$\phi_+ - \phi_- = 4\pi\mu$$

The gradient of the potential of a surface doublet may be written down at once by substitution of $\mu d\bar{S}$ for $\bar{L}ds$ in equation (4.7-6). We have

$$\text{grad } \phi = \text{grad} \int_S \mu\nabla\left(\frac{1}{r}\right).d\bar{S} = \int_S \left\{\frac{\mu d\bar{S}}{r^3} - \frac{3\,\bar{r}\,\mu d\bar{S}.\bar{r}}{r^5}\right\} \tag{4.7-13}$$

When μ is constant over the surface, grad ϕ admits of an important vector transformation which may be developed in the following way.

$$(\text{grad } \phi)_x = \frac{\partial}{\partial x_o}\,\mu \int_S \nabla\left(\frac{1}{r}\right).d\bar{S}$$

$$= -\frac{\partial}{\partial x_o}\,\mu \int_S \left\{dS_x\,\frac{(x-x_o)}{r^3} + dS_y\,\frac{(y-y_o)}{r^3} + dS_z\,\frac{(z-z_o)}{r^3}\right\}$$

$$= \mu \int_S \left\{dS_x\,\frac{\partial}{\partial x}\,\frac{(x-x_o)}{r^3} + dS_y\,\frac{\partial}{\partial y}\,\frac{(x-x_o)}{r^3} + dS_z\,\frac{\partial}{\partial z}\,\frac{(x-x_o)}{r^3}\right\}$$

since $\dfrac{\partial}{\partial x}\,\dfrac{(y-y_o)}{r^3} = \dfrac{\partial}{\partial y}\,\dfrac{(x-x_o)}{r^3}$ etc

hence

$$(\text{grad } \phi)_x = \mu \int_S d\bar{S}.\nabla \left(\frac{(x-x_o)}{r^3}\right)$$

and

$$\text{grad } \phi = \mu \int_S (d\bar{S}.\nabla) \frac{\bar{r}}{r^3} \tag{4.7-14}$$

But from equation (1.17-12), with $\frac{\bar{r}}{r^3}$ substituted for $\bar{F}$, we have

$$\int_S (d\bar{S}.\nabla) \frac{\bar{r}}{r^3} = \oint_\Gamma \frac{d\bar{r} \times \bar{r}}{r^3} + \int_S \text{div} \frac{\bar{r}}{r^3} d\bar{S} - \int_S d\bar{S} \times \text{curl} \frac{\bar{r}}{r^3}$$

where Γ is the boundary of S,

and, since

$$\text{div} \frac{\bar{r}}{r^3} = \text{curl} \frac{\bar{r}}{r^3} = 0 \qquad (r \neq 0)$$

$$\int_S (d\bar{S}.\nabla) \frac{\bar{r}}{r^3} = \oint_\Gamma \frac{d\bar{r} \times \bar{r}}{r^3} \tag{4.7-15}$$

hence

$$\text{grad } \phi = -\mu \text{ grad } \Omega = -\mu \oint_\Gamma \frac{\bar{r} \times d\bar{r}}{r^3} \tag{4.7-16}$$

We will return to this relationship and its further transformations subsequent to the treatment of vector potential.

The Laplacian of the potential of surface sources is undefined upon the surfaces themeselves because ϕ and/or its derivatives are undefined there. Outside the surfaces $\nabla^2\phi = 0$. The proofs parallel those for line sources.

EXERCISES

4-32. Let an infinite, rectilinear doublet of density $\bar{L}$ coincide with the z axis of cylindrical coordinates, and let $\bar{L}$ be directed along $\Phi = 0$. Show that at any point (ρ,Φ,z), $(\rho\neq 0)$, the vector (− grad ϕ) is directed at an angle Φ to the radius vector $\rho\hat{\rho}$ in the plane z = z and has the magnitude $\frac{2L}{\rho^2}$.

Deduce that (− grad ϕ) is tangential to a circular cylinder which contains the z axis and the point (ρ,Φ,z), and is bisected by the half-plane $\Phi = \frac{\pi}{2}$ or $\Phi = \frac{3\pi}{2}$.

Show from equation (4.2-9) that the equipotential surfaces are circular cylinders which contain the z axis and are bisected by the half-plane $\Phi = 0$ or $\Phi = \pi$, and confirm that the two systems of cylinders are orthogonal as required.

4-33. A uniform, cylindrical surface source of length 2c, radius a and density σ is aligned centrally with the z axis of cylindrical coordinates, and bisected by the $\rho\Phi$ plane through the origin. By treating the surface as a system of axial strips of linear density $\sigma a d\Phi$ and using the result of Ex.4-10., p. 233, derive an expression for the potential outside the cylinder at finite distance from the origin, for the condition $c\to\infty$.

Extend the analysis of Ex.4-9., pp. 232-3 to cover the case $PP_1 > PO$, and so derive the corresponding expression for the potential inside the cylinder.

Ans: For $\rho > a$

$$\phi = 4\pi a\sigma\left\{\ln 2c - \ln\rho + \frac{1}{2\pi}\int_0^{2\pi}\left(\frac{a}{\rho}\cos\Phi + \frac{a^2}{2\rho^2}\cos 2\Phi + -- \right)d\Phi\right\}$$

$$= 2q\,(\ln 2c - \ln\rho)$$

where q = total source strength per unit length of cylinder.

For $\rho < a$

$$\phi = 4\pi a\sigma\left\{\ln 2c - \ln a + \frac{1}{2\pi}\int_0^{2\pi}\left(\frac{\rho}{a}\cos\Phi + \frac{\rho^2}{2a^2}\cos 2\Phi + -- \right)d\Phi\right\}$$

$$= 2q\,(\ln 2c - \ln a)$$

4-34. It follows from the results of the previous exercise that $\text{grad}\,\phi = -\hat{\rho}\,2q/\rho$ for $\rho > a$, ie, at points outside the cylinder the gradient of the potential is identical with that deriving from a source of linear density q coincident with the axis. At points inside the cylinder the gradient is zero. Confirm these conclusions (a) by integration of the components of grad ϕ associated with the individual strips, and (b) by an application of Gauss's law to closed, coaxial cylindrical surfaces of appropriate radii.

[When the radius of the Gaussian cylinder exceeds that of the source its end surfaces cut the source, and Gauss's law is not directly applicable. In this case let the end surfaces lie in the planes $z = \pm z'$, and let those portions of the source lying between the planes $z = +z' \pm \Delta z$ and $z = -z' \pm \Delta z$ give rise to partial potentials designated ϕ_2 and ϕ_3 respectively. If the partial potential deriving from the remainder of the source is ϕ_1 then $\phi_1 = \phi - \phi_2 - \phi_3$. From symmetry, the flux of grad ϕ_2 through the end surface $z = +z'$ is zero, and the flux of grad ϕ_3 through the end surface $z = -z'$ is likewise zero. The required result is obtained by integrating grad ϕ_1 over the Gaussian surface, applying Gauss's law and taking limits as $\Delta z \to 0$.]

4-35. Let the source of Ex.4-32. and 4-33. above be replaced by one of surface density $\sigma_1 \cos \Phi$. Show that the potential at the point $(\rho,0,z)$ is given by

$$\phi = \frac{2\pi a^2 \sigma_1}{\rho} \qquad (\rho>a)$$

and (ρ, z finite)

$$\phi = 2\pi\sigma_1\rho \qquad (\rho<a)$$

It is evident from symmetry that a surface density of the form $\sigma_1 \sin \Phi$ makes no contribution to the potential in the half-planes $\Phi = 0$ and $\Phi = \pi$. By writing

$$\sigma_1 \cos \Phi = \sigma_1 \cos \Phi_o \cos \Phi' - \sigma_1 \sin \Phi_o \sin \Phi'$$

where

$$\Phi' = \Phi - \Phi_o$$

the original source distribution is resolved into two sinoidal distributions in space quadrature, having maximum values of $|\sigma_1 \cos \Phi_o|$ and $|\sigma_1 \sin \Phi_o|$ and neutral planes defined by $\Phi = \Phi_o \pm \frac{\pi}{2}$ and $\Phi = \Phi_o$, $\Phi = \Phi_o + \pi$. Make use of this contrivance to show that the potential at (ρ,Φ,z) is given by

$$\frac{2\pi a^2 \sigma_1 \cos\Phi}{\rho} \qquad (\rho > a)$$

$$(\rho,\ z \text{ finite})$$

$$2\pi\sigma_1 \rho \cos\Phi \qquad (\rho < a)$$

Hence prove (a) that the potential outside the cylinder is identical with that deriving from an infinite line doublet on the axis, having a moment $\pi a^2 \sigma_1$ per unit length and an orientation $\Phi = 0$, and (b) that at points inside the cylinder grad ϕ has the constant magnitude $2\pi\sigma_1$ and is directed parallel to the half-plane $\Phi = 0$.

4-36. Extend the analysis of the previous exercise to show that when

$$\sigma = \sigma_n \cos n\Phi + \sigma'_n \sin n\Phi \qquad (n=1,2,3---)$$

the potential at finite distance from the origin is given by

$$\phi = \frac{2\pi a^{n+1}}{n\rho^n} (\sigma_n \cos n\Phi + \sigma'_n \sin n\Phi) \qquad (\rho > a)$$

$$\phi = \frac{2\pi\rho^n}{na^{n-1}} (\sigma_n \cos n\Phi + \sigma'_n \sin n\Phi) \qquad (\rho < a)$$

Confirm that the tangential derivative of the potential is continuous through the surface and that ϕ is harmonic inside and outside the cylinder.

Show that the normal derivative of ϕ changes from $+2\pi\sigma$ to $-2\pi\sigma$ on passing radially outwards through the surface and hence deduce that for all density distributions which are a function of Φ alone and give rise to zero total source strength per unit length of cylinder, the normal derivatives at the surface depend only upon the local surface density.

4-37. Let the surface density of the cylindrical source of the previous exercises be single-valued and given by $\sigma = f(\Phi)$. Evaluate $\hat{n}.\int \sigma \frac{\bar{r}}{r^3}\, dS$ at a point of the surface at finite distance from the origin and use equation (4.7-9) to show that the limiting values of the outward normal derivative of ϕ, as the surface is approached from within and without, are

$$-\frac{q}{a} + 2\pi\sigma \quad \text{and} \quad -\frac{q}{a} - 2\pi\sigma$$

where q is the total source strength per unit length of the cylinder. Hence confirm the values of $\frac{\partial\phi}{\partial n}$ found in Ex.4-34. and 4-36.

4-38. The potential of a uniform, spherical surface source (or its mathematical equivalent) has been treated in Ex.4-18, p. 243. Determine the value of grad ϕ at points inside and outside the sphere by differentiation of this function. Check the result (a) by integration of the components of grad ϕ associated with circular strips of the surface normal to an axis of symmetry, and (b) by an application of Gauss's law.

What independent approach demonstrates that grad ϕ is zero inside the sphere?

$$\text{Ans:}\quad \text{grad}\,\phi = -4\pi a^2\sigma/R^2 \qquad (R>a)$$

$$\text{grad}\,\phi = \overline{0} \qquad (R<a)$$

Since $\nabla^2\phi = 0$ inside the sphere and since, by symmetry, ϕ is constant over any concentric spherical surface, it follows from Theorem 3.2-2 that ϕ is constant and grad ϕ is zero within the sphere.

4-39. Two uniform, spherical volume sources of radius a and densities $\pm\rho$ are centred upon the origin of spherical coordinates. If the positive source is now displaced a distance Δz along the axis $\theta = 0$, show that the combined sources behave like a spherical surface source of density $\sigma_1 \cos\theta$ when $\rho\to\infty$ and $\Delta z\to 0$ in such a way as to maintain $\rho\Delta z$ constant and equal to σ_1. Hence show that a spherical surface source of density $\sigma_1 \cos\theta$ gives rise to a potential outside the sphere identical with that of a point doublet of moment $\frac{4}{3}\pi a^3\sigma_1$, and orientation $\theta = 0$, at the centre of the sphere, ie

$$\phi = \frac{4}{3}\frac{\pi a^3\sigma_1 \cos\theta}{R^2} \qquad (R>a)$$

Confirm this expression for points on the axis $\theta = 0$ by integration of the partial potentials deriving from individual surface elements.

Make use of the result of Ex.4-18., p. 243 for the potential inside a uniform spherical volume source, to show that the potential inside the surface source under consideration is given by

$$\phi = \frac{4}{3}\pi\sigma_1 R\cos\theta \qquad (R<a)$$

and hence show that for $R < a$ the vector field, grad ϕ, has the constant magnitude $\frac{4}{3}\pi\sigma_1$ and is directed parallel to $\theta = 0$. Confirm the value of the potential at points on the axis by surface integration, and examine the normal and tangential derivatives of ϕ in the vicinity of $R = a$.

4-40. The potential at an exterior point 0 of a uniform double layer of density μ is equal to $-\mu\Omega$, where Ω is the solid angle subtended by the layer at 0. The change of potential associated with a displacement $\Delta\bar{s}$ from 0 to 0' is equal to that obtaining at 0 when the layer is translated by $-\Delta\bar{s}$. Determine the change of solid angle resulting from this translation in terms of a triple scalar product involving the vector area of the elements of the peripheral strip traced out by the boundary Γ, and, by interchange of dot and cross, arrive at the relationship

$$\Delta\Omega = \Delta\bar{s}.\oint_{\Gamma} \frac{\bar{r} \times d\bar{r}}{r^3}$$

irrespective of the choice of positive currency.

Hence develop an independent proof of equation (4.7-16).

4-41. Devise a further proof of equation (4.7-16), similar to that given in the text, without making use of equation (1.17-12).

[Hint: Transform

$$-\frac{\partial}{\partial x_o}\int_S \left\{dS_x \frac{(x-x_o)}{r^3} + dS_y \frac{(y-y_o)}{r^3} + dS_z \frac{(z-z_o)}{r^3}\right\}$$

into

$$\int_S dS_x \operatorname{div}\frac{\bar{r}}{r^3} - \int_S \left\{dS_x\left(\operatorname{grad}\frac{(y-y_o)}{r^3}\right)_y - dS_y\left(\operatorname{grad}\frac{(y-y_o)}{r^3}\right)_x\right\}$$

$$+ \int_S \left\{dS_z\left(\operatorname{grad}\frac{(z-z_o)}{r^3}\right)_x - dS_x\left(\operatorname{grad}\frac{(z-z_o)}{r^3}\right)_z\right\}$$

and apply equation (1.17-1).]

4-42. A uniform rectilinear singlet logarithmic source of length 2c and density λ' lies upon the y axis of coordinates in the xy plane and is bisected by the origin. Determine the value of the associated potential at any exterior point (x,y), and at the point (0,y) where $|y| \le c$, and so show that it is everywhere defined and continuous. Find the rate of change of this potential along the x axis, and hence the value, upon the x axis, of the potential of a logarithmic line doublet of vector density $\bar{L}' = \bar{i}L'$ coincident with the line singlet.

Deduce the occurrence of a discontinuity of magnitude $2\pi\lambda'$ in the normal derivative of the potential of any logarithmic line singlet and of $2\pi L'$ in the potential of any logarithmic line doublet on passing normally through these sources at points where the densities are continuous. Compare these results with those of Ex.4-28., p. 257. when V is identified with the logarithmic potential of singlet and doublet sources on $\Gamma_{1..n}\Gamma'$, showing first that the required conditions are satisfied at infinity.

Ans: Line singlet

$$\phi(x,y) = \lambda' \left\{ (c-y)\ \ln \frac{1}{(x^2+(c-y)^2)^{\frac{1}{2}}} + (c+y)\ \ln \frac{1}{(x^2+(c+y)^2)^{\frac{1}{2}}} \right.$$

$$\left. + 2c - x\left(\tan^{-1}\frac{c-y}{x} + \tan^{-1}\frac{c+y}{x}\right)\right\}$$

$$\phi(0,y) = \lambda' \left\{ (c-y)\ \ln \frac{1}{(c-y)} + (c+y)\ \ln \frac{1}{(c+y)} + 2c \right\} \qquad (|y| \le c)$$

$$\frac{\partial \phi}{\partial x}(x,0) = -\,2\lambda' \tan^{-1}\frac{c}{x} \qquad (x \ne 0)$$

Line doublet

$$\phi(x,0) = 2L' \tan^{-1}\frac{c}{x} \qquad (x \ne 0)$$

4-43. It was shown in Sec. 3.3 that if V is any function, harmonic outside the local surfaces $S_{1..n}$, then

$$4\pi V_o = \oint_{S_{1..n}} \left(\frac{1}{r}\frac{\partial V}{\partial n} - V\frac{\partial}{\partial n}\left(\frac{1}{r}\right)\right) dS$$

where $\hat{n}$ is directed into $S_{1..n}$.

This relationship holds provided that the point of evaluation of V is located at finite distance from $S_{1..n}$. Demonstrate that the latter restriction is unnecessary by proceeding in the following way:

Let ϕ be the potential of singlet and doublet surface sources upon $S_{1..n}$ having the respective densities $\frac{1}{4\pi}\frac{\partial V}{\partial n}$ and $-\frac{V}{4\pi}$ (where the doublet density is positive if doublet alignment corresponds with $\hat{n}$.)

Show that ϕ is harmonic outside $S_{1..n}$ and that $\phi = V$ upon $S_{1..n}$. Hence deduce, by means of equation (3.1-4) (with U replaced by ϕ and Σ located at infinity), that ϕ and V are identical outside $S_{1..n}$ and that V may consequently be expressed everywhere in the above form.

4.8 The Gradient of the Scalar Potential of a Volume Source

The gradient of the potential of a finite, piecewise continuous volume source of density ρ, occupying a region of space τ bounded by the surfaces $S_{1..n}\Sigma$, is given at exterior points by

$$\operatorname{grad} \phi = \operatorname{grad} \int_\tau \frac{\rho}{r}\, d\tau = -\int_\tau \rho \operatorname{grad} \frac{1}{r}\, d\tau = \int_\tau \frac{\rho \bar{r}}{r^3}\, d\tau \qquad (4.8\text{-}1)$$

This follows from arguments similar to those leading to the corresponding expressions for simple line and surface sources.

At any source element where ρ is differentiable

$$\operatorname{grad} \frac{\rho}{r} = \frac{1}{r} \operatorname{grad} \rho + \rho \operatorname{grad} \frac{1}{r} \qquad (4.8\text{-}2)$$

hence if ρ is differentiable throughout τ

$$\operatorname{grad} \phi = \int_\tau \frac{1}{r} \operatorname{grad} \rho\, d\tau - \int_\tau \operatorname{grad} \frac{\rho}{r}\, d\tau$$

or, by equation (1.17-5)

$$\operatorname{grad} \phi = \int_\tau \frac{1}{r} \operatorname{grad} \rho\, d\tau - \oint_{S_{1..n}\Sigma} \frac{\rho}{r}\, d\bar{S} \qquad (4.8\text{-}3)$$

where the surface integration is carried out just inside τ.

When ρ is piecewise continuous in τ it becomes necessary to replace equation (4.8-3) with a set of similar equations involving integration over individual subregions and their interfaces.

We now proceed to investigate the gradient of the potential at interior points of a volume source throughout which ρ has continuous first derivatives. Before commencing the analysis it is convenient to introduce two additional potential functions, viz the partial potential and the cavity potential. The partial potential is defined at interior points of the source by the integral $\int_{\tau-\tau_\delta} \frac{\rho}{r}\, d\tau$ where τ_δ is a small sphere of fixed radius δ centred upon the point of evaluation of ϕ. Thus in the evaluation of this function the δ sphere is carried from point to point. The scalar potential at any interior point is seen to be equal to the limiting value of the partial potential at that point as $\delta \to 0$. The cavity potential is defined for some fixed position of the δ sphere, and is given at all points within the sphere by $\int_{\tau-\tau_\delta} \frac{\rho}{r}\,d\tau$. In general, the cavity and partial potentials will be equal in value only at the centre of a fixed sphere.

We may write down expressions for the gradient of the cavity potential (grad cavity pot ρ) at once since the points of evaluation lie outside the region of integration. On taking account of the variation of the latter, we have from equations (4.8-1) and (4.8-3)

$$\text{grad cavity pot } \rho = -\int_{\tau-\tau_\delta} \rho \,\text{grad}\,\frac{1}{r}\,d\tau \qquad (4.8\text{-}4)$$

or

$$\text{grad cavity pot } \rho = \int_{\tau-\tau_\delta} \frac{1}{r}\,\text{grad}\,\rho\,d\tau - \oint_{S_{1..n}\Sigma,\ S_\delta} \frac{\rho}{r}\,d\overline{S} \qquad (4.8\text{-}5)$$

where S_δ is the surface of the δ sphere about 0.

To determine the gradient of the partial potential we proceed as follows:

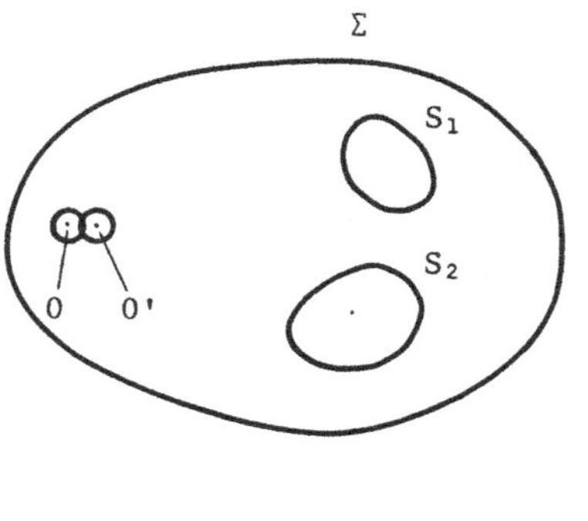

Fig. 4.8a

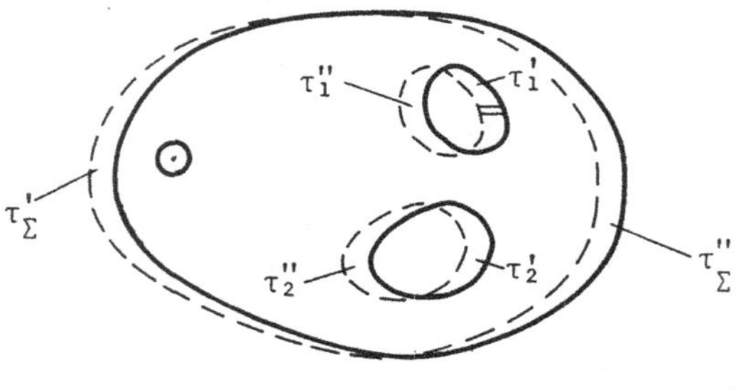

Fig. 4.8b

In Fig. 4.8a the point 0 at which the gradient is to be found lies within a region τ bounded externally by the surface Σ and internally by S_1 and S_2. The point 0' is located on the x axis through 0 so that $\overrightarrow{00'} = \overline{1}\,\Delta x$ where Δx is intrinsically positive. Equal spheres of radius δ are drawn about 0 and 0'. Since the evaluation of the partial potentials at 0 and 0' involves integrations which exclude those source elements lying within the δ spheres centred upon each point in turn, it follows that the increment of partial potential associated with movement from 0 to 0' is equal to the increment of partial potential which obtains at 0 when the scalar source field ρ is slipped a distance Δx in the negative x direction, while 0 and its δ sphere remain fixed in space.

In Fig. 4.8b the boundaries of the region of integration subsequent to this slip are shown dotted, except for the sphere around 0 which constitutes an integration boundary both before and after slip. If we suppose that individual volume elements maintain their positions relative to 0 as the source field moves through them, the increment of partial potential at 0 becomes a consequence of the variation of ρ experienced by these elements. This contrasts with the earlier approach leading to equations (4.8-1) and (4.8-4) where fixed values of ρ and variable values of r were attributed to each element.

The subregions marked τ_1', τ_2' and τ_Σ' in Fig. 4.8b are composed of volume elements for which ρ was zero prior to the slip and non-zero after it. The reverse is true of those regions marked τ_1'', τ_2'' and τ_Σ''. The bulk of the integration space, however, is common to both sets of bounding surfaces, and within this region, which will be designated τ_c, there is a smooth variation of ρ during the slipping process. If ρ and ρ' represent the initial and final values of the source density in the typical volume element, the resulting increment of partial potential at 0 is given by $\int_{\tau_c} \frac{(\rho'-\rho)}{r}\,d\tau$, and the corresponding contribution to the gradient of the partial potential at 0 becomes

$$\lim_{\Delta x\to 0} \frac{1}{\Delta x}\int_{\tau_c} \frac{(\rho'-\rho)}{r}\,d\tau = \lim_{\Delta x\to 0}\int_{\tau_c} \frac{1}{r}\,\frac{(\rho'-\rho)}{\Delta x}\,d\tau = \int_{\tau-\tau_\delta} \frac{1}{r}\,\frac{\partial\rho}{\partial x}\,d\tau$$

The subregions τ_1', τ_2', τ_Σ' and τ_1'', τ_2'', τ_Σ'' may be divided into elementary prisms of length Δx lying parallel to the x axis. A typical element of τ_1' is shown in Fig. 4.8b. The right-hand end of the prism comprises portion of the surface S_1 while the left-hand end comprises the same portion of S_1 displaced. If the end surface is represented by the vector $d\overline{S}$ then the volume of the prism is $-dS_x\,\Delta x$. The same expression holds for the typical prism in τ_2' and τ_Σ' since, in each case, the positive sense of $d\overline{S}$ is such as to render dS_x negative. The overall contribution of these elements to the potential gradient at 0 is consequently given by

$$\lim_{\Delta x\to 0} \frac{1}{\Delta x}\int_{\tau_1'\tau_2'\tau_\Sigma'} \frac{\rho}{r}\,d\tau = \frac{1}{\Delta x}\int_{S'} \frac{\rho}{r}\,dS_x\,(-\Delta x) = -\int_{S'} \frac{\rho}{r}\,dS_x$$

where S' represents those parts of S_1, S_2 and Σ contiguous with τ_1', τ_2' and τ_Σ'.

Similar reasoning shows that the component of the potential gradient arising from the elimination of the source field from the regions τ_1'', τ_2'' and τ_Σ'' is given by

$$\lim_{\Delta x\to 0} -\frac{1}{\Delta x}\int_{\tau_1''\tau_2''\tau_\Sigma''} \frac{\rho}{r}\,d\tau = -\frac{1}{\Delta x}\int_{S''} \frac{\rho}{r}\,dS_x\,\Delta x = -\int_{S''} \frac{\rho}{r}\,dS_x$$

where S" represents those parts of S_1, S_2 and Σ contiguous with τ_1'', τ_2'' and τ_Σ''.

On summing all contributions we get

$$\frac{\partial}{\partial x_o}\text{ partial pot }\rho = \int_{\tau-\tau_\delta} \frac{1}{r}\frac{\partial\rho}{\partial x}\,d\tau - \oint_{S_{1,2}\Sigma} \frac{\rho}{r}\,dS_x \tag{4.8-6}$$

whence we may proceed immediately to the generalisation

$$\text{grad partial pot }\rho = \int_{\tau-\tau_\delta} \frac{1}{r}\text{ grad }\rho\,d\tau - \oint_{S_{1..n}\Sigma} \frac{\rho}{r}\,d\bar{S} \tag{4.8-7}$$

By integrating equation (4.8-2) over the region $\tau-\tau_\delta$ and combining the result with equation (4.8-7) we obtain the alternative expression

$$\text{grad partial pot }\rho = -\int_{\tau-\tau_\delta} \rho\text{ grad }\frac{1}{r}\,d\tau + \oint_{S_\delta} \frac{\rho}{r}\,d\bar{S} \tag{4.8-8}$$

It is not immediately apparent that grad pot ρ may be found by taking the limit of equation (4.8-7) or of equation (4.8-8) as $\delta\to 0$, since this operation yields

$$\lim_{\delta\to 0}\text{ grad partial pot }\rho$$

rather than

$$\text{grad }\lim_{\delta\to 0}\text{ partial pot }\rho$$

as required. However, on pursuing the above treatment for this particular case, we see that, for a field slip Δx, the increment of that component of pot ρ arising from common volume elements (say pot' ρ) is given by

$$\Delta\text{ pot}'\,\rho = \lim_{\delta\to 0}\int_{\tau-\tau_\delta} \frac{\rho'}{r}\,d\tau - \lim_{\delta\to 0}\int_{\tau-\tau_\delta} \frac{\rho}{r}\,d\tau$$

Since each integral is convergent this may be written as

$$\Delta \text{ pot}' \rho = \lim_{\delta \to 0} \int_{\tau-\tau_\delta} \frac{\rho' - \rho}{r} \, d\tau$$

Now the first derivative of the source density has been postulated to be continuous throughout τ, hence from Taylor's theorem

$$\rho' - \rho = \boxed{\frac{\partial \rho}{\partial x}} \Delta x$$

at each point, where $\frac{\partial \rho}{\partial x}$ is evaluated somewhere upon the associated slip path.

Thus

$$\Delta \text{ pot}' \rho = \lim_{\delta \to 0} \int_{\tau-\tau_\delta} \frac{1}{r} \left\{ \boxed{\frac{\partial \rho}{\partial x}} \Delta x \right\} d\tau$$

or

$$\Delta \text{ pot}' \rho = \Delta x \lim_{\delta \to 0} \int_{\tau-\tau_\delta} \frac{1}{r} \boxed{\frac{\partial \rho}{\partial x}} \, d\tau$$

Hence

$$\frac{\partial}{\partial x_o} \text{ pot}' \rho = \lim_{\Delta x \to 0} \lim_{\delta \to 0} \int_{\tau-\tau_\delta} \frac{1}{r} \boxed{\frac{\partial \rho}{\partial x}} \, d\tau = \int_\tau \frac{1}{r} \frac{\partial \rho}{\partial x} \, d\tau$$

which confirms the relationship

$$\text{grad} \lim_{\delta\to 0} \text{partial pot } \rho = \lim_{\delta\to 0} \text{grad partial pot } \rho \qquad (4.8\text{-}11)$$

Since the surface integral in equation (4.8-8) vanishes as $\delta\to 0$, we have, in addition,

$$\text{grad pot } \rho = -\int_\tau \rho \text{ grad } \frac{1}{r} \, d\tau \qquad (4.8\text{-}12)$$

Any attempt to replace the field-slipping technique by a 'variable r' approach leads to difficulties in the limiting process. If Δx is allowed to approach zero prior to δ, the resulting expression represents the gradient of the cavity potential within a vanishingly small δ sphere; if the order of taking limits is reversed, the integrand of the potential is no longer a continuous function of the coordinates of 0 in a neighbourhood of the point of evaluation and the significance of the operation is not clear. Nevertheless, the above analysis reveals that the limiting values of grad cavity pot and grad partial pot are identical. In addition, the relationship

$$\frac{\partial}{\partial x_o} \int_\tau \frac{\rho}{r} \, d\tau = -\int_\tau \rho \frac{\partial}{\partial x}\left(\frac{1}{r}\right) d\tau$$

which was employed in the derivation of equation (4.8-1) where r was everywhere greater than some positive number, is now seen to subsist when the integral is improper, ie when 0 lies within the source. However, it should not be supposed that the latter behaviour extrapolates to higher-order derivatives. Thus, $\int_\tau \rho\nabla^2\left(\frac{1}{r}\right) d\tau \left[\text{ie} \lim_{\delta\to 0} \int_{\tau-\tau_\delta} \rho\nabla^2\left(\frac{1}{r}\right) d\tau\right]$ is always zero because $\nabla^2\left(\frac{1}{r}\right) = 0$ for all volume elements in the integral, whereas $\nabla^2 \int_\tau \frac{\rho}{r} \, d\tau$ is, in general, non-zero.

The field-slipping technique is equally effective in the determination of grad pot ρ at exterior and boundary points of τ. For exterior points, where no δ sphere need be invoked, we arrive directly at equation (4.8-10) without the associated limiting process, and this may be transformed into equation (4.8-12) by integrating equation (4.8-2) over the entire region bounded by $S_{1..n}\Sigma$. At boundary points, grad partial pot ρ continues to be given by equation (4.8-7), except insofar as the δ sphere excludes portion of the surface integral over that surface upon which 0 is located. As $\delta\to 0$ the volume and surface integrals converge and grad pot ρ is represented by equation (4.8-10) as before. Since the x component of grad pot ρ is consequently equal, at all points, to the sum of the potential of a volume source of density $\frac{\partial\rho}{\partial x}$ and that of surfaces sources of density $-\rho \frac{dS_x}{dS}$, both of which are defined and continuous everywhere, it follows that grad pot ρ is itself continuous upon and through the bounding surfaces.

When the source density has piecewise continuous first derivatives in τ and 0 lies within the source, but not upon a surface of discontinuity, a simple extension of the above analysis shows that equation (4.8-7) must be replaced by

$$\text{grad partial pot } \rho = \int_{\tau-\tau_\delta} \frac{1}{r} \text{ grad } \rho \, d\tau - \oint_{S_{1..n}\Sigma} \frac{\rho}{r} \, d\bar{S} - \int_{S_A S_B} \frac{\rho}{r} \, d\bar{S} \qquad (4.8\text{-}7(a))$$

where the last term involves integration over both sides of each interior surface of discontinuity, the postive sense of $d\bar{S}$ corresponding to motion through the surface away from the side of integration.

The latter surface integral is cancelled when equation (4.8-7(a)) is combined with the set of equations resulting from the integration of equation (4.8-2) over each subregion in turn, so that equation (4.8-8) applies as before. Hence in this circumstance grad pot ρ is given by equation (4.8-12) or by equation (4.8-7(a)) with $\tau-\tau_\delta$ replaced by τ. It will be noted that when piecewise differentiability of ρ is accompanied by continuity of ρ throughout τ the additional surface integral of equation (4.8-7(a)) vanishes. These expressions continue to hold when 0 lies upon an interior surface of discontinuity.

4.9 The Laplacian of the Scalar Potential of a Volume Source

Poisson's Equation

Extension of Gauss's Law

When the point of evaluation of the potential lies outside the source

$$\frac{\partial}{\partial x_o} \int_\tau \frac{\rho}{r} \, d\tau = - \int_\tau \rho \frac{\partial}{\partial x} \left(\frac{1}{r}\right) d\tau$$

hence

$$\frac{\partial^2}{\partial x_o^2} \int_\tau \frac{\rho}{r} \, d\tau = - \frac{\partial}{\partial x_o} \int_\tau \rho \frac{\partial}{\partial x} \left(\frac{1}{r}\right) d\tau = \int_\tau \rho \frac{\partial^2}{\partial x^2} \left(\frac{1}{r}\right) d\tau$$

and

$$\nabla^2 \int_\tau \frac{\rho}{r} \, d\tau = \int_\tau \rho \, \nabla^2 \left(\frac{1}{r}\right) d\tau = 0$$

Thus, the potential satisfies Laplace's equation at points outside the source, ie

$$\text{div grad pot } \rho = 0 \qquad (4.9\text{-}1)$$

The same analysis applies at the interior points of a fixed cavity created within the source, so that

$$\text{div grad cavity pot } \rho = 0 \qquad (4.9\text{-}2)$$

To determine the Laplacian of the potential at interior points of the source we note that the x derivative of the potential of ρ may be expressed as the potential of the x derivative of ρ together with a certain surface integral (4.8-9). Accordingly, if $\frac{\partial\rho}{\partial x}$ has continuous first derivatives in τ, then

$$\frac{\partial^2}{\partial x_o^2}\int_\tau \frac{\rho}{r}\,d\tau = \frac{\partial}{\partial x_o}\int_\tau \frac{1}{r}\frac{\partial\rho}{\partial x}\,d\tau - \frac{\partial}{\partial x_o}\oint_{S_{1..n}\Sigma}\frac{\rho}{r}\,dS_x$$

$$= \int_\tau \frac{1}{r}\frac{\partial^2\rho}{\partial x^2}\,d\tau - \oint_{S_{1..n}\Sigma}\frac{1}{r}\frac{\partial\rho}{\partial x}\,dS_x + \oint_{S_{1..n}\Sigma}\rho\,\frac{\partial}{\partial x}\left(\frac{1}{r}\right)dS_x$$

and

$$\nabla^2\int_\tau \frac{\rho}{r}\,d\tau = \int_\tau \frac{1}{r}\nabla^2\rho\,d\tau - \oint_{S_{1..n}\Sigma}\frac{1}{r}\,\text{grad }\rho.d\bar{S} + \oint_{S_{1..n}\Sigma}\rho\,\text{grad }\frac{1}{r}.d\bar{S}$$

ie

$$\nabla^2\int_\tau \frac{\rho}{r}\,d\tau = \int_\tau \frac{1}{r}\nabla^2\rho\,d\tau + \oint_{S_{1..n}\Sigma}\left\{\rho\,\frac{\partial}{\partial n}\left(\frac{1}{r}\right) - \frac{1}{r}\frac{\partial\rho}{\partial n}\right\}dS \qquad (4.9\text{-}3)$$

whence from equation (3.3-3)

$$\nabla^2\int_\tau \frac{\rho}{r}\,d\tau = -4\pi\rho$$

Hence, at interior points of the source,

$$\text{div grad pot } \rho = -4\pi\rho \qquad (4.9\text{-}4)$$

This important result is known as Poisson's equation.

The corresponding expression for the partial potential is easily shown to be

$$\text{div grad partial pot } \rho = \oint_{S_\delta} \left\{ \frac{1}{r} \frac{\partial \rho}{\partial n} - \rho \frac{\partial}{\partial n} \left(\frac{1}{r} \right) \right\} dS \qquad (4.9\text{-}5)$$

The field-slipping technique is equally applicable at points outside the source.

In this case we again obtain equation (4.9-3), but because the origin of r is exterior to the source, the right hand side of equation (4.9-3) is zero in accordance with equation (3.3-1).

Poisson's equation is seen to reduce to Laplace's equation at points outside the source, or at any interior point where ρ is zero. In this connection it should be noted that if ρ is zero at an isolated point within the source, the value of the Laplacian cannot be derived legitimately by the 'variable r' approach.

It is clear that Poisson's equation will continue to hold when ρ is piecewise continuous, provided that discontinuities of ρ and its derivatives are excluded from a neighbourhood of 0; for the 'variable r' analysis, which shows that the excluded source elements contribute nothing to the Laplacian at 0, is insensitive to a finite number of such discontinuities.

The conditions under which Poisson's equation is valid may be shown to be much less restrictive than would be suggested by the field-slipping analysis. It is, in fact, only necessary that ρ should satisfy a Hölder condition at the point in question, although this admits of the possibility that ρ is not differentiable there[19] (see Ex.4-48. and 4-49., pp. 284-5).

An application of Poisson's equation permits of an extension of Gauss's law to include the case in which the surfaces $S_{1..n}\Sigma$ of integration of grad ϕ are immersed in a volume source in which are embedded point, line and surface sources which have no point in common with $S_{1..n}\Sigma$. Upon integrating equation (4.9-4) for that component of the potential deriving from the volume source alone, we get

$$\oint_{S_{1..n}\Sigma} \text{grad pot } \rho . d\bar{S} = -4\pi \int_\tau \rho \, d\tau$$

19. Kellogg, Ch. 6. In the present context, ρ satisfies a Hölder condition at 0 if there is a neighbourhood of 0 for which $|\rho(0')-\rho(0)| \leq Ar^\alpha$, where $r = 00'$ and A and α are positive numbers. A Hölder condition implies continuity but not necessarily differentiability.

This continues to hold when ρ is bounded and satisfies a Hölder condition everywhere except upon interior surfaces of discontinuity; for if the region is divided into a set of subregions lying just within the surfaces of discontinuity, equations of the above type hold for each of the subregions in turn and the limiting values of the surfaces integrals over the interfaces cancel in pairs since grad pot ρ is continuous through them. The components of $\oint_{S_{1..n}\Sigma} \text{grad}\,\phi.d\bar{S}$ arising from the remaining sources retain the values found previously because the superposition is linear, hence for the complete source system

$$\oint_{S_{1..n}\Sigma} (-\text{ grad }\phi).d\bar{S} = 4\pi \times \text{total strength of sources enclosed} \qquad (4.9\text{-}6)$$

As a consequence of the definition of the partial potential in terms of an excluding δ sphere, the foregoing expressions for the potential and its derivatives - cavity potential apart - involve the limiting values of volume integrals as $\delta\to 0$. It is easily shown that the same limiting values obtain for all regular excluding regions. This continues to be so in most of the subsequent analyses (unless the geometrical properties of the δ sphere are invoked for the purpose of evaluation) although, in general, the δ notation will be retained. In the few cases in which volume integrals are non-convergent, attention will be drawn to the fact.

EXERCISES

4-44. The relationship

$$\text{grad}\int_\tau \frac{\rho}{r}\,d\tau = -\int_\tau \rho\ \text{grad}\,\frac{1}{r}\,d\tau = \int_\tau \rho\,\frac{\bar{r}}{r^3}\,d\tau$$

subsists at each point of a volume source in which ρ is piecewise continuous.

Demonstrate this in the following way.

Let 0 be an interior or boundary point of the source. With 0 as centre draw a sphere of fixed radius a. Then the above relationship clearly holds for source elements outside the sphere. Now draw a concentric sphere of radius δ within the first. Let 0' be a point displaced from 0 by a distance Δx ($<\delta$) along the positive x axis through 0, and let r' and r be distances measured from 0' and 0 respectively.

Make use of the inequalities $|r-r'| \le \Delta x$ and $\frac{1}{rr'} \le \frac{1}{2}\left(\frac{1}{r^2}+\frac{1}{r'^2}\right)$ to show that a positive Δ exists such that for $\delta < \Delta$

$$\left| \left\{ \underset{\tau_\delta}{\phi(0')} - \underset{\tau_\delta}{\phi(0)} \right\} \Big/ \Delta x - \int_{\tau_\delta} \frac{\rho(x-x_o)}{r^3}\, d\tau \right| < \frac{\varepsilon}{2}$$

where $\underset{\tau_\delta}{\phi(0')} = \int_{\tau_\delta} \frac{\rho}{r'}\, d\tau$ and ε has any positive value.

Then show that for this value of Δ a positive X exists such that for $\Delta x < X$ an inequality identical with the above holds for source elements lying between the spheres. For this purpose expand $\frac{1}{r'}$ in terms of $\frac{1}{r}$ as in Sec. 4.1, bearing in mind that $|P_m(\cos\theta)| \le 1$.

Hence, prove that

$$\lim_{\Delta x \to 0} \left\{ \underset{\tau}{\phi(0')} - \underset{\tau}{\phi(0)} \right\} \Big/ \Delta x = \int_{\tau} \frac{\rho(x-x_o)}{r^3}\, d\tau$$

and complete the demonstration.

4-45. It follows from equation (4.8-8) that

$$\frac{\partial}{\partial x_o} \int_{\tau-\tau_\delta} \frac{\rho}{r}\, d\tau = -\int_{\tau-\tau_\delta} \rho \frac{\partial}{\partial x}\left(\frac{1}{r}\right) d\tau + \oint_{S_\delta} \frac{\rho}{r}\, dS_x$$

hence

$$\frac{\partial^2}{\partial x_o^2} \int_{\tau-\tau_\delta} \frac{\rho}{r}\, d\tau = -\frac{\partial}{\partial x_o} \int_{\tau-\tau_\delta} \rho \frac{\partial}{\partial x}\left(\frac{1}{r}\right) d\tau + \frac{\partial}{\partial x_o} \oint_{S_\delta} \frac{\rho}{r}\, dS_x$$

Show that the field-slipping technique may be used to expand the right-hand side of the lower equation, and so derive an expression for $\nabla^2 \int_{\tau-\tau_\delta} \frac{\rho}{r}\, d\tau$ in terms of volume and surface integrals. Reduce this expression to (4.9-5) by expanding div $\left(\rho \text{ grad } \frac{1}{r}\right)$ and forming its volume integral over $\tau-\tau_\delta$.

4-46. Show that

$$\nabla^2 \int_{S_{1..n}\Sigma} V \frac{\partial}{\partial n}\left(\frac{1}{r}\right) dS = \nabla^2 \int_{S_{1..n}\Sigma} \frac{1}{r} \frac{\partial V}{\partial n}\, dS = 0$$

where V and $\frac{\partial V}{\partial n}$ are piecewise continuous upon $S_{1..n}\Sigma$ and the point of

Hence show by means of Green's formula (equation (3.3-3)) that if V is any function well-behaved in τ, then, at interior points of τ,

$$-4\pi\nabla^2 V = \nabla^2 \int_\tau \frac{\nabla^2 V}{r}\, d\tau$$

or

$$-4\pi\gamma = \nabla^2 \int_\tau \frac{\gamma}{r}\, d\tau \qquad \text{where} \quad \gamma = \nabla^2 V$$

[This does not constitute a general proof of Poisson's equation because γ is not an independent point function.]

4-47. Make use of the result of Ex.4-18., p. 243 to prove that $\nabla^2\phi = -4\pi\rho$ at all interior points of a spherical volume source of constant density ρ.

4-48. The potential at a distance R from the centre of a spherical volume source of radius a and density f(R) is given by

$$\phi = \frac{1}{R}\int_0^R f(R)\, 4\pi R^2\, dR + \int_R^a f(R)\, 4\pi R\, dR \qquad (R \leq a)$$

$$\phi = \frac{1}{R}\int_0^a f(R)\, 4\pi R^2\, dR \qquad (R \geq a)$$

in accordance with the results of Ex.4-18., p. 243.

If f(R) is everywhere continuous, show that

$$\frac{d\phi}{dR} = -\frac{1}{R^2}\int_0^R f(R)\, 4\pi R^2\, dR \quad : \quad \nabla^2\phi = -4\pi f(R) \qquad (0<R<a)$$

$$\frac{d\phi}{dR} = 0 \quad : \quad \nabla^2\phi = -4\pi f(0) \qquad (R=0)$$

$$\frac{d\phi}{dR} = -\frac{1}{R^2}\int_0^a f(R)\, 4\pi R^2\, dR \quad : \quad \nabla^2\phi = 0 \qquad (R>a)$$

4-49. Let the density of a spherical volume source be given by $\rho = R \sin \frac{1}{R}$, where R represents distance from the centre of the source. Show that the derivatives of ρ are undefined at $R = 0$ and that $\int_\tau \frac{\nabla^2\rho}{r}\, d\tau$ is non-convergent there. Hence conclude that, in this case, the field-slipping technique fails to define a value of $\nabla^2\phi$. Note, however, that on the basis of the previous exercise (or because ρ satisfies a Hölder condition) the Laplacian of ϕ does exist at $R = 0$ and is zero there.

4-50. Let σ_1, μ_1, ρ_1, and σ_2, μ_2, ρ_2 denote two sets of finite, continuous source distributions on given, non-intersecting surfaces and in space, and let ρ_1 and ρ_2 satisfy a Hölder condition at all points. If the two sets give rise to the same potential at all points outside the surfaces, show that the corresponding source densities are everywhere identical.

[Hint: Consider the behaviour of the potential field deriving from the difference distributions $\sigma_1 - \sigma_2$, $\mu_1 - \mu_2$, $\rho_1 - \rho_2$.]

4-51. It is possible to develop planar equivalents of the relationships derived in Secs. 4.8 and 4.9.

Consider the potential in the xy plane of a system of logarithmic surface sources of density σ' in the region S of the plane bounded by the closed curves $\Gamma_{1..n}\Gamma'$. If σ' is finite and piecewise continuous in S we may define

$$\text{partial pot } \sigma' = \int_{S-S_\delta} \sigma' \ln \frac{1}{\rho}\, dS$$

where S_δ is the region within a circle of radius δ centred upon the point of evaluation 0, and

$$\text{pot } \sigma' = \int_S \sigma' \ln \frac{1}{\rho}\, dS$$

Devise a procedure similar to that adopted in Sec. 4.4 to demonstrate that the integral for pot σ' is convergent, and that the potential is continuous everywhere in the plane.

Develop a planar field-slipping analysis to show that at interior points of S, where it is supposed that σ' is well-behaved,

$$\frac{\partial}{\partial x_o} \text{partial pot } \sigma' = \int_{S-S_\delta} \frac{\partial \sigma'}{\partial x} \ln \frac{1}{\rho}\, dS - \oint_{\Gamma_{1..n}\Gamma'} \sigma' \ln \frac{1}{\rho}\, \hat{n}'.\bar{i}\, ds$$

and

$$\text{grad partial pot } \sigma' = \int_{S-S_\delta} (\text{grad } \sigma') \ln \frac{1}{\rho}\, dS - \oint_{\Gamma_{1..n}\Gamma'} \sigma' \ln \frac{1}{\rho}\, \hat{\bar{n}}'\, ds$$

where $\hat{\bar{n}}'$ is the unit outward normal to the bounding curve.

Expand $\text{grad}\left(\sigma' \ln \frac{1}{\rho}\right)$ and integrate over $S-S_\delta$ by means of the planar analogue of equation (1.17-3) to obtain

$$\oint_{\Gamma_{1..n}\Gamma'\Gamma_\delta} \sigma' \ln \frac{1}{\rho}\, \hat{\bar{n}}'\, ds = \int_{S-S_\delta} \sigma' \,\text{grad} \ln \frac{1}{\rho}\, dS + \int_{S-S_\delta} \left(\ln \frac{1}{\rho}\right) \text{grad } \sigma'\, dS$$

where Γ_δ is the boundary of S_δ, and hence show that within the region S

$$\text{grad pot } \sigma' = \int_S (\text{grad } \sigma') \ln \frac{1}{\rho}\, dS - \oint_{\Gamma_{1..n}\Gamma'} \sigma' \ln \frac{1}{\rho}\, \hat{\bar{n}}'\, ds$$

$$= -\int_S \sigma' \,\text{grad} \ln \frac{1}{\rho}\, dS$$

Devise an independent proof of the latter relationship for points exterior to the source system.

Finally, show that

$$\text{grad cavity pot } \sigma' = -\int_{S-S_\delta} \sigma' \,\text{grad} \ln \frac{1}{\rho}\, dS$$

$$= \int_{S-S_\delta} (\text{grad } \sigma') \ln \frac{1}{\rho}\, dS - \oint_{\Gamma_{1..n}\Gamma'\Gamma_\delta} \sigma' \ln \frac{1}{\rho}\, \hat{\bar{n}}'\, ds$$

4-52. Extend the analysis of Ex.4-51. to show that

$$\frac{\partial^2}{\partial x_o{}^2} \text{partial pot } \sigma' = \int_{S-S_\delta} \frac{\partial^2 \sigma'}{\partial x^2} \ln \frac{1}{\rho}\, dS - \oint_{\Gamma_{1..n}\Gamma'} \frac{\partial \sigma'}{\partial x} \ln \frac{1}{\rho}\, \hat{\bar{n}}'.\bar{i}\, ds$$

$$+ \oint_{\Gamma \quad \Gamma'} \sigma' \frac{\partial}{\partial x} \left(\ln \frac{1}{\rho}\right) \hat{\bar{n}}'.\bar{i}\, ds$$

and proceed to the relationship

$$\nabla^2 \text{ pot } \sigma' = \int_S (\nabla^2\sigma') \ln\frac{1}{\rho}\, dS - \oint_{\Gamma_{1..n}\Gamma'} \frac{\partial\sigma'}{\partial n'} \ln\frac{1}{\rho}\, ds + \oint_{\Gamma_{1..n}\Gamma'} \sigma' \frac{\partial}{\partial n'}\left(\ln\frac{1}{\rho}\right) ds$$

By combining this with equation (3.9-3) derive Poisson's equation in the plane, viz

$$\text{div grad pot } \sigma' = \nabla^2 \int_S \sigma' \ln\frac{1}{\rho}\, dS = -2\pi\sigma'$$

4-53. Show that, with the exception of Poisson's equation, the formulae of Ex.4-51. and 4-52. continue to hold when the problem is converted to one of Newtonian potential in the plane, provided that $\ln\frac{1}{\rho}$ is replaced everywhere by $\frac{1}{\rho}$.

4-54. If the origin of r lies within the region bounded by the surfaces $S_{1..n}\Sigma$ and if ϕ is the potential of interior sources, show that

$$\oint_{S_{1..n}\Sigma} \left\{\frac{1}{r}\frac{\partial\phi}{\partial n} - \phi\frac{\partial}{\partial n}\left(\frac{1}{r}\right)\right\} dS = 0$$

4-55. Prove that Poisson's equation continues to hold at points where ρ has continuous second derivatives when ρ is piecewise continuous within the region of integration with open or closed surfaces of discontinuity.

4-56. Employ a field-slipping analysis to show that

$$\nabla^2 \text{ (partial)} \int_{\tau-\tau_\delta} \frac{\rho}{r}\cos\alpha r\, d\tau = \int_{\tau-\tau_\delta} (\nabla^2\rho)\frac{1}{r}\cos\alpha r\, d\tau$$

$$+ \oint_{S_{1..n}\Sigma} \left\{\rho\frac{\partial}{\partial n}\left(\frac{1}{r}\cos\alpha r\right) - \frac{1}{r}\cos\alpha r\frac{\partial\rho}{\partial n}\right\} dS$$

where α is a constant, and transform this result via a modified form of Green's theorem into

$$\nabla^2 \text{ (partial)} \int_{\tau-\tau_\delta} \frac{\rho}{r} \cos \alpha r \, d\tau = -\alpha^2 \int_{\tau-\tau_\delta} \frac{\rho}{r} \cos \alpha r \, d\tau$$

$$+ \oint_{S_\delta} \left\{ \frac{1}{r} \cos \alpha r \frac{\partial \rho}{\partial n} - \rho \frac{\partial}{\partial n} \left(\frac{1}{r} \cos \alpha r \right) \right\} dS$$

Hence demonstrate that

$$(\nabla^2+\alpha^2) \int_\tau \frac{\rho}{r} \cos \alpha r \, d\tau = -4\pi\rho$$

Show similarly that

$$(\nabla^2+\alpha^2) \int_\tau \frac{\rho}{r} \sin \alpha r \, d\tau = 0$$

and deduce that

$$(\nabla^2+\alpha^2) \int_\tau \frac{\rho}{r} e^{\pm j\alpha r} \, d\tau = -4\pi\rho$$

4-57. Proceed as in the previous exercise to show that

$$(\nabla^2-\gamma^2) \int_\tau \frac{\rho}{r} e^{\pm \gamma r} \, d\tau = -4\pi\rho$$

where γ is a real, imaginary or complex constant, and so derive

$$(\nabla^2+(\alpha^2-\beta^2)-2j\alpha\beta) \int_\tau \frac{\rho}{r} e^{\pm j\alpha r} e^{\pm \beta r} \, d\tau = -4\pi\rho$$

where α and β are real constants.

4-58. Let the point function V have continuous second derivatives throughout all space and be regular at infinity, and let $\nabla^2 V$ be zero outside a sphere of finite radius. If ϕ is the potential of a volume source of density $-\frac{1}{4\pi} \nabla^2 V$, modify the argument of Ex.4-43., p. 272 to show that V and ϕ are identical, and that V may therefore be expressed everywhere as

$$4\pi V_o = -\int \frac{\nabla^2 V}{r} \, d\tau$$

4-59. Develop the planar analogues of the previous exercise and that of Ex.4-43., p. 272.

4.10 Equivalent Layer Theorems in Scalar Potential Theory

Let the surfaces $S_{1..n}\Sigma$ divide a finite system of sources into two parts, viz that within the region $\underline{R}$ bounded by $S_{1..n}\Sigma$ and that outside it. It is supposed that no point source lies upon the bounding surfaces and that no line or surface source has a point in common with them. However, volume sources may or may not be continuous through them. Let the potential fields deriving from sources within and without $\underline{R}$ be designated ϕ_i and ϕ_e respectively, and let $\phi = \phi_i + \phi_e$. Since ϕ_e is harmonic in $\underline{R}$ it follows from Green's formula that at any interior point 0 of $\underline{R}$

$$\phi_{e_0} = \frac{1}{4\pi}\oint_{S_{1..n}\Sigma}\left\{\frac{1}{r}\frac{\partial\phi_e}{\partial n} - \phi_e\frac{\partial}{\partial n}\left(\frac{1}{r}\right)\right\}dS \qquad (4.10\text{-}1)$$

where $\hat{n}$ is directed out of τ.

Thus the potential at 0 due to sources outside $\underline{R}$ (integration volume τ_e) is equal to that deriving from simple and double layer surface sources on $S_{1..n}\Sigma$ of densities $\frac{1}{4\pi}\frac{\partial\phi_e}{\partial n}$ and $\frac{-\phi_e}{4\pi}$ respectively, where the doublet density is positive if doublet alignment corresponds with $\hat{n}$.

These surface sources are, for obvious reasons, known as equivalent layers.

It is easily shown that layers of densities $\frac{1}{4\pi}\frac{\partial\phi}{\partial n}$ and $\frac{-\phi}{4\pi}$ likewise satisfy the requirement. For the same origin of r within $\underline{R}$ and integration of ϕ_i over each of the components of τ_e in turn, Green's formula becomes

$$0 = \frac{1}{4\pi}\oint_{S_m}\left\{\frac{1}{r}\frac{\partial\phi_i}{\partial n'} - \phi_i\frac{\partial}{\partial n'}\left(\frac{1}{r}\right)\right\}dS \quad \text{for} \quad m = 1\text{---}n \qquad (4.10\text{-}2)$$

and

$$0 = \frac{1}{4\pi}\oint_{\Sigma}\left\{\frac{1}{r}\frac{\partial\phi_i}{\partial n'} - \phi_i\frac{\partial}{\partial n'}\left(\frac{1}{r}\right)\right\}dS + \frac{1}{4\pi}\oint_{\infty}\left\{\frac{1}{r}\frac{\partial\phi_i}{\partial n} - \phi_i\frac{\partial}{\partial n}\left(\frac{1}{r}\right)\right\}dS \qquad (4.10\text{-}3)$$

where $\hat{n'}$ is directed into $\underline{R}$.

The surface integral at infinity is zero because ϕ_i is harmonic outside $\underline{R}$ so that upon subtracting equations (4.10-2) and (4.10-3) from (4.10-1) we obtain

$$\phi_{e_o} = \frac{1}{4\pi} \oint_{S_{1..n}\Sigma} \left\{ \frac{1}{r}\frac{\partial\phi}{\partial n} - \phi \frac{\partial}{\partial n}\left(\frac{1}{r}\right) \right\} dS \tag{4.10-4}$$

which leads to the result stated.

ϕ and $\frac{\partial\phi}{\partial n}$ refer to values actually obtaining upon $S_{1..n}\Sigma$ since the components ϕ_i and ϕ_e, together with their normal derivatives, are continuous through and upon S, in spite of possible discontinuities of ρ.

It is evident from the considerations of Sec. 4.5 that an infinite number of equivalent layer combinations exist, having the densities

$$\frac{1}{4\pi}\left\{ \frac{\partial\phi_e}{\partial n} + \frac{\partial U}{\partial n'} \right\} \quad \text{and} \quad -\frac{1}{4\pi}(\phi_e - U)$$

where U represents a set of functions harmonic in the regions comprising τ_e, and $\frac{\partial U}{\partial n'}$ is the single-ended derivative on $S_{1..n}\Sigma$. A solution of the Dirichlet or Neumann problem for an individual region permits of the elimination of the double or simple layer upon the associated surface.

The above analysis includes the cases of interior and exterior equivalence when only one bounding surface is involved. By deleting the surfaces $S_{1..n}$ and identifying Σ with the given surface S we see that the potential within the enclosure due to sources outside it is duplicated by that of surface layers of the prescribed densities where the positive sense of the normal is directed out of the enclosure. If, on the other hand, the surfaces $S_{2..n}$ are deleted and S_1 identified with S while Σ is removed to infinity, the potential outside the enclosure due to sources within it is seen to be duplicated by that of the prescribed surface layers, where the positive sense of the normal is directed into the enclosure. Hence, if the layer densities are expressed in terms of the total potential and its normal derivative on S, the magnitudes of the equivalent layers are unchanged in passing from the problem of interior to that of exterior equivalence, for given interior and exterior sources, but the polarities are reversed.

The foregoing results are embodied in part in the following equivalent layer theorem.

Theorem 4.10-1

If sources of potential exist within and without the region bounded by the set of surfaces $S_{1..n}\Sigma$, the potential within the region resulting from exterior sources is identical with that deriving from simple and double layer surface sources on $S_{1..n}\Sigma$ of densities $\frac{1}{4\pi}\frac{\partial\phi}{\partial n}$ and $\frac{-\phi}{4\pi}$ respectively, where the positive normal is directed out of the region and ϕ represents either the potential of the exterior sources or the total potential.

Equivalent layer theory has a wider significance in certain physical applications than appears in the above considerations. It was pointed out in Ex.3-31., p. 210 that the velocity field $\bar{v}$ of an incompressible fluid undergoing irrotational motion could be expressed as the gradient of a scalar point function ψ which is everywhere harmonic within the field of flow except in a neighbourhood of a source (or sink) or upon a surface of discontinuity of $\bar{v}$. Suppose that the closed surface S bounds a region free from fluid sources and surfaces of discontinuity of $\bar{v}$. Then

$$\psi_o = \oint_S \left\{\frac{1}{r}\frac{1}{4\pi}\frac{\partial\psi}{\partial n} - \frac{\psi}{4\pi}\frac{\partial}{\partial n}\left(\frac{1}{r}\right)\right\} dS$$

Since, for any given disposition of exterior sources, the field of flow within S is modified by the introduction of exterior barriers to the flow in the form of impervious bodies, it is clear that the equivalent potential sources defined by the above integral take account not only of the effect of exterior sources but also of the modification introduced by exterior surfaces of discontinuity of $\bar{v}$ - this has no parallel in the potential theory treated above where sources alone produce discontinuities.

Similar arguments apply in the case of heat flow.

EXERCISES

4-60. Let a closed geometrical surface S be equipotential in the presence of sources within and without the enclosure. Show that the interior sources may be removed and a simple layer source of equal total strength defined upon S in such a way as to leave the exterior potential field unaltered.

Show further that the potential field within the enclosure may be everywhere reduced by an amount equal to the original potential of S by the removal of the exterior sources and the appropriate disposition upon S of a simple layer source whose total strength is equal and opposite to that of the interior sources.

Prove that in each case the required distribution of surface density is unique.

4-61. Show that Theorem 4.10-1 applies to logarithmic sources in the plane provided that contours replace surfaces, line singlets and doublets replace simple and double layer surface sources, and $\frac{1}{2\pi}\frac{\partial\phi}{\partial n'}$, $-\frac{\phi}{2\pi}$ replace $\frac{1}{4\pi}\frac{\partial\phi}{\partial n}$, $-\frac{\phi}{4\pi}$.

[Proceed from equation (3.9-3), showing that the line integral

$$\oint\left\{\left(\ln\frac{1}{\rho}\right)\frac{\partial\phi_i}{\partial\rho} - \phi_i\frac{\partial}{\partial\rho}\left(\ln\frac{1}{\rho}\right)\right\}ds$$

is always zero at infinity.]

4-62. Let A and B be points of the plane having the coordinates (+d,0) and (-d,0). If ρ_1 and ρ_2 denote distance measured from A and B respectively, show that the locus of points which maintain $\frac{\rho_2}{\rho_1} = k(\text{const})$ is a circle of centre $d\frac{(k^2+1)}{(k^2-1)},0$ and radius $\frac{2dk}{\{(k^2-1)^2\}^{\frac{1}{2}}}$. Sketch a set of these circles for various values of k, noting that those having reciprocal values of k form mirror images in the y axis.

If logarithmic point sources of strength $+\alpha'$ and $-\alpha'$ are located at A and B respectively, show that the potential field within the region bounded internally, or internally and externally, by any two circles is identical with that deriving from simple logarithmic line sources upon the circles having the density $\frac{1}{2\pi}\frac{\partial\phi}{\partial n'}$, where $\hat{n}'$ is directed out of the region. Show further that the potential gradient at any point is tangential to a circle passing through that point and having AB as a chord.

[The equipotential contours and those orthogonal to them are said to constitute a system of coaxal circles having the radical axis x = 0 and the limiting points A and B.]

4.11 The Method of Images in Scalar Potential Theory

Whereas equivalent layer theory is primarily concerned with the substitution of boundary sources for sources which lie outside the region under consideration, image theory seeks to replace pre-existing boundary sources by exterior sources. Thus if a region $\underline{R}$ is bounded by the surfaces $S_{1..n}\Sigma$, upon which are disposed simple and/or double layer sources whose potential field is denoted by ϕ_s, we are required to find a system of exterior sources of potential ϕ_e such that $\phi_e = \phi_s$ at interior points of $\underline{R}$. If ϕ_i is the potential field associated with existing interior sources then an equivalent requirement is that $\phi_e + \phi_i = \phi_s + \phi_i$ at interior points of $\underline{R}$[20].

These remarks present the problem in its most general form, but in practice the situation is rather simpler. Double layer sources are not normally encountered in image treatments and the number of discrete surfaces involved rarely exceeds two. Nor are the surface distributions arbitrarily assigned; the terms of the problem invariably require that $\phi_i + \phi_s$, or its normal derivative, satisfy stated conditions upon the surfaces.

The validity of a proposed image system, ie system of surrogate sources, is determined by means of criteria established in Ch. 3. In terms of the present problem it was there shown that ϕ_e and ϕ_s, being harmonic in $\underline{R}$, are identical in $\underline{R}$, at least to within an additive constant, provided that

(a) $\phi_e = \phi_s$ at all points of $S_{1..n}\Sigma$

or (b) $\dfrac{\partial\phi_e}{\partial n} = \dfrac{\partial\phi_s}{\partial n}$ at all points of $S_{1..n}\Sigma$

or (c) the tangential component of grad ϕ_e is identical with that of grad ϕ_s at all points of $S_{1..n}\Sigma$, and $\oint \dfrac{\partial\phi_e}{\partial n}\,dS = \oint \dfrac{\partial\phi_s}{\partial n}\,dS$ for each surface in turn.

For the particular case in which the interior bounding surfaces $S_{1..n}$ are deleted, ie when $\underline{R}$ is aperiphractic, the surface integrals of the normal derivatives of ϕ_e and ϕ_s over Σ are necessarily zero and need not be specified. When $\underline{R}$ includes all space outside $S_{1..n}$, ie when Σ recedes to infinity, ϕ_e and ϕ_s are identical in $\underline{R}$ provided that either (a) or (b) or (c) is satisfied at all points of the local surfaces. Image equivalence is often considered complete when it leads to equality of potential gradients throughout $\underline{R}$ rather than equality of potentials, in which case a constant difference of potential is of no significance.

---

20. It is not possible to equate ϕ_e and ϕ_s upon a bounding surface itself in the presence of a double layer surface source since ϕ_s is then undefined. Similarly $\dfrac{\partial\phi_s}{\partial n}$ is undefined upon a simple layer source. In subsequent references to the matching of potential functions and their derivatives over $S_{1..n}\Sigma$ we refer to the limiting values of these functions as the surfaces are approached normally from within $\underline{R}$. Alternatively, we may identify integration over $S_{1..n}\Sigma$ as integration over a corresponding

Since ϕ_i and its derivatives are supposed to be continuous upon $S_{1..n}\Sigma$, the conditions for the identity of ϕ_e and ϕ_s (or of $\phi_e + \phi_i$ and $\phi_s + \phi_i$) in R, as listed above, may be replaced by

(a') $\phi_e + \phi_i = \phi_s + \phi_i$ at all points of $S_{1..n}\Sigma$

or (b') $\frac{\partial}{\partial n}(\phi_e+\phi_i) = \frac{\partial}{\partial n}(\phi_s+\phi_i)$ at all points of $S_{1..n}\Sigma$

or (c') the tangential component of $\mathrm{grad}(\phi_e+\phi_i)$ is identical with that of $\mathrm{grad}(\phi_s+\phi_i)$ at all points of $S_{1..n}\Sigma$ and $\oint \frac{\partial}{\partial n}(\phi_e+\phi_i)\,dS = \oint \frac{\partial}{\partial n}(\phi_s+\phi_i)\,dS$ for each surface in turn.

Combinations of these conditions (and of (a), (b) and (c) above) are also admissible.

It should be noted that the uniqueness theorems developed in Sec. 3.6 for non-harmonic fields lead to (a'), (b') and (c') without dissection of the potential functions into harmonic and non-harmonic components.

The term 'image' first arose in connection with the problem of a half space upon whose boundary plane a simple layer source is so distributed as to maintain the plane at zero potential in the presence of a point source within the half space. In this case the potential of the surface source is duplicated, within the half space, by that of an equal and opposite point source located at the position of the mirror image of the original source in the plane[21]. (See Ex.4-77., p. 303.)

Image systems are not restricted to point sources. Thus in the above half space problem an extended object source would demand an extended image source since matched object/image elements, when taken in pairs, maintain the required boundary potential. This remains true when the geometry of the system requires that the object and image complexes have different shapes (Ex.4-75., p. 302). On the other hand, a single point source may require more than one point image and possibly an infinite number, depending upon the nature of the surface sources to be replaced (Ex.4-76., p. 302 and Ex.4-78., p. 304).

Image methods may be applied to the solution of physical problems in which the scalar fields under consideration are not primarily defined as Newtonian (or logarithmic[22]) potentials. While we continue to seek a set of sources whose potential equates the scalar field at interior points of a bounded region, it may no longer be possible to identify the role of the image component as that of a substitute for boundary sources since the latter may have no physical existence. Consider, for example, the irrotational flow of an incompressible fluid from a point source of unit strength in the vicinity of an infinite, plane, impervious boundary. From previous considerations the velocity field may be expressed as

---

21. In this connection it should be borne in mind that the criteria of identity established above are not directly applicable to systems which involve unclosed surfaces of infinite extent.

22. The treatment of images in logarithmic potential theory parallels that for Newtonian potentials.

$$\bar{v} = -\text{grad } \psi$$

where

(1) $\dfrac{\partial\psi}{\partial n} = 0$ upon the bounding surface

(2) $\nabla^2\psi = 0$ outside the source

(3) $\psi = \dfrac{1}{r'} + U$ within a neighbourhood of the source

where r' denotes distance from the source and U is some continuous scalar function.

Now a potential function ϕ, which derives in part from a source of unit strength coincident with the fluid source and in part from a point source of equal magnitude and sign in the mirror image position, has the following characteristics:

(1) $\dfrac{\partial\phi}{\partial n} = 0$ upon the bounding surface

(2) $\nabla^2\phi = 0$ in the half space outside the parent source

(3) $\phi = \dfrac{1}{r'} + U'$ within a neighbourhood of the parent source, where U' is some continuous scalar function

(4) ϕ is regular at infinity.

Hence for ψ regular at infinity, it follows from the arguments of Sec. 3.6 that $\phi = \psi$ throughout the operating half space. The modification of the point function ψ which attends the introduction of the barrier to the flow is seen to be paralleled in the mathematical potential model by the effect of the image source. A similar analysis is applicable to the steady state flow of heat in a half space of uniform thermal conductivity contiguous with a half space of zero conductivity.

It will be recalled that the solution of similar problems in Ch. 3, by means of Green's functions, made use of mirror image points to define one component of these functions. However, the image point was then required to mirror the point of evaluation of the field and consequently moved with it, whereas it currently mirrors the source and remains fixed in space. Nevertheless, the earlier form of solution may be transformed easily into the later, as will be evident from an examination of the solutions of Ex.3-34. and 3-35., p. 211.

The method of images may be extended, in certain cases, to the evaluation of a scalar field over contiguous regions of space when the field or its normal derivative is discontinuous through the common boundary. Thus consider the steady state flow of heat from a point source of unit strength located at P in a homogeneous half space of thermal conductivity k_1, contiguous with a half space of conductivity k_2. It is known from the results of Ex.3-28. and 3-29., p. 209 that the temperature T is harmonic at interior points of each half space except in a neighbourhood of the source, where it takes the form

$$T = \frac{1}{k_1 r'} + U$$

where r' is distance measured from the source and U is some continuous function.

The normal component of the heat flow vector, -k grad T, must be everywhere continuous through the interface for a common sense of the normal, since there can be no accumulation of heat upon it under steady state conditions, hence

$$\Delta\left(k\,\frac{\partial T}{\partial n}\right) = 0$$

We will suppose that there is no discontinuity of temperature through the interface[23] so that

$$\Delta T = 0$$

Finally, it will be assumed that T may be treated as regular at infinity.

Reference to a simplified form of Theorem 3.6-1 shows that the above set of relationships is sufficient to define T uniquely throughout all space. (For this purpose T is identified with V, and k with g; all surfaces other than S_a and Σ are deleted, S_a being identified with the interface of the half spaces and Σ with a spherical surface of infinite radius centred upon a local origin.) Now it is easily shown that the relationships in T are duplicated by a potential function ϕ which is defined in the k_1 half space as the combined potential of a point source of strength $\frac{1}{k_1}$ at P and a point source of strength $\frac{1}{k_1}\frac{(k_1-k_2)}{(k_1+k_2)}$ at the image position P', and in the k_2 half space as the potential of a point source of strength $\frac{2}{k_1+k_2}$ at P. It follows that ϕ is everywhere identical with T and that the temperature at any point 0 is given by

$$T_{o_{k_1}} = \frac{1}{k_1}\left\{\frac{1}{r_p} + \frac{(k_1-k_2)}{(k_1+k_2)}\,\frac{1}{r_{p'}}\right\}$$

$$T_{o_{k_2}} = \frac{2}{k_1+k_2}\,\frac{1}{r_p}$$

where r is distance measured from 0.

---

23. This is one of a number of possibilities. See H.S. Carslaw and J.C. Jaeger, "Conduction of Heat in Solids", p. 23, 2nd ed., Oxford University Press (1959).

It will be observed that neither image source is situated in the region whose temperature it defines; this is clearly a requirement if T is to remain harmonic at all interior points of the half-spaces beyond a neighbourhood of the source of heat.

EXERCISES

4-63. A simple layer source is distributed over a spherical surface S in such a way as to maintain its potential constant in the presence of other sources. Show that if these additional sources lie outside the sphere, the potential throughout the region enclosed by the sphere is constant and equal to its value on the surface, and that if they lie inside the sphere, the potential at exterior points and upon the surface is identical with that which would obtain if all source elements were concentrated at the centre.

4-64. A point source of strength α is located at P, a distance f from the centre T of a sphere of radius a (>f). A simple layer source is distributed over the surface in such a way as to maintain it at the constant potential ϕ_s in the presence of the point source. If the total source strength upon S is α_s deduce from the result of Ex.4-63. that

$$\phi_S = (\alpha + \alpha_S)/a$$

Show that the potential of the simple layer is duplicated at interior points of the sphere by that of an exterior image system comprising a point source of strength $-\frac{a}{f}\alpha$ at the point P', inverse to P in the sphere (see Ex.3-33., p. 210), together with a concentric spherical surface source of any radius d (>a) and uniform density $\phi_S/4\pi d$, so that the potential of the original system at an interior point O is given by

$$\phi_o = \frac{\alpha}{r_P} - \frac{a}{f}\frac{\alpha}{r_{P'}} + \phi_S$$

where r is distance measured from O.

Use this expression and the result of Ex.4-63. to prove that the surface density $\sigma \left(= -\frac{1}{4\pi}\Delta\frac{\partial\phi}{\partial n}\right)$ at any point Q of the surface is given by

$$\sigma_Q = -\frac{\alpha(a^2-f^2)}{4\pi a\, PQ^3} + \frac{\alpha + \alpha_S}{4\pi a^2}$$

and confirm that the surface integral is equal to α_S.

4-65. Repeat Ex.4-64. for the case f > a.

Ans: $\phi_S = \frac{\alpha_S}{a} + \frac{\alpha}{f}$

The image system comprises a point source of strength $-\frac{a}{f}\alpha$ at the inverse point P' together with a point source of strength $\alpha_S + \frac{a}{f}\alpha$ at T, so that at an interior point O

$$\phi_O = \frac{\alpha}{r_P} - \frac{a}{f}\frac{\alpha}{r_{P'}} + \frac{a\phi_S}{r_T}$$

At any point Q of the surface

$$\sigma_Q = -\frac{\alpha(f^2-a^2)}{4\pi a\ PQ^3} + \frac{\alpha_S}{4\pi a^2} + \frac{\alpha}{4\pi af}$$

4-66. A simple logarithmic line source is distributed around a circle Γ in such a way as to maintain its potential constant in the presence of other logarithmic sources in the plane. Show that if these additional sources lie outside the circle, the potential throughout the region of the plane bounded by the circle is constant and equal to its value on the boundary, and that if they are enclosed by the circle, the potential at exterior points of the plane and upon the circle is identical with that which would obtain if all source elements were concentrated at the centre.

[In the second case, make use of the expansion for ln ρ given in Ex.4-9, pp. 232-3, to show that if ϕ is the potential of the actual configuration of sources and ψ is the potential field when all sources are concentrated at the centre, $\phi - \psi \to 0$ as $\frac{1}{R}$ and $\frac{\partial}{\partial R}(\phi-\psi)\to 0$ as $\frac{1}{R^2}$ when $R\to\infty$.

This represents a particular case of the boundary conditions mentioned on pp. 203-4.]

4-67. A logarithmic point source of strength α' is located at P, a distance f from the centre T of a circle Γ of radius a (>f), and in the plane of the circle. A simple logarithmic line source is distributed around the circle in such a way as to maintain it at the constant potential ϕ_Γ in the presence of the point source. If the total source strength around Γ is α'_Γ deduce from the result of Ex.4-66. that

$$\phi_\Gamma = -(\alpha'+\alpha'_\Gamma)\ \ln a$$

Show that the potential of the circular source is duplicated at points of the plane within it by that of a coplanar image system comprising a logarithmic point source of strength $-\alpha'$ at the point P', inverse to P in the circle, together with a concentric circular source of any radius d (>a) and of uniform density $\left(\alpha' \ln\frac{a}{f} - \phi_\Gamma\right)\Big/ 2\pi d \ln d$, so that the potential of the original system at an interior point O is given by

$$\phi_O = \alpha' \ln\frac{f\rho_{P'}}{a\rho_P} + \phi_\Gamma$$

Use this expression and the result of Ex.4-66. to demonstrate that the line density $\lambda' \left(= \frac{-1}{2\pi} \Delta \frac{\partial\phi}{\partial n'}\right)$ at any point Q of the circle is given by

$$\lambda'_Q = -\frac{\alpha'(a^2-f^2)}{2\pi a\ PQ^2} + \frac{\alpha' + \alpha'_\Gamma}{2\pi a}$$

and confirm that the line integral is equal to α'_Γ.

[Hint:

$$\int \left\{1 - \frac{2f}{a}\cos\theta + \frac{f^2}{a^2}\right\}^{-1} d\theta = \frac{a^2}{a^2 - f^2}\int\left(1 + \frac{2f}{a}\cos\theta + \frac{2f^2}{a^2}\cos 2\theta\text{---}\right) d\theta$$

for $f < a$]

4-68. Repeat Ex.4-67. for the case $f > a$.

Ans: $\phi_\Gamma = -\alpha'_\Gamma \ln a - \alpha' \ln f$

The image system comprises a logarithmic point source of strength $-\alpha'$ at the inverse point P', together with a logarithmic point source of strength $\alpha' + \alpha'_\Gamma$ at T, so that at an exterior point O

$$\phi_O = \alpha' \ln \frac{\rho_{P'}}{\rho_P} - (\alpha'+\alpha'_\Gamma) \ln \rho_T$$

At any point Q of the circle

$$\lambda'_Q = -\frac{\alpha'(f^2-a^2)}{2\pi a\ PQ^2} + \frac{\alpha' + \alpha'_\Gamma}{2\pi a}$$

-- oOOo --

Exercises 4-69. to 4-71. deal with source systems which may be decomposed into sets of parallel, uniform, rectilinear elements of equal length, bisected by a common transverse plane. The analysis of each system is to be confined to that region about the central plane where the potential field is sensibly two-dimensional and the non-constant component of magnitude varies logarithmically with radial distance from the associated line element (equation (4.2-5)). Planar theorems may consequently be invoked in proofs of uniqueness, but it should be borne in mind that the demonstration of exterior uniqueness requires that the radius of an outer bounding contour be made large compared with the displacement of the line elements from a central axis; in this circumstance equation (4.2-5) will continue to hold upon the boundary only if the length of the system be maintained large compared with the contour radius itself.

4-69. A simple layer source, whose density is a function of angular position only, is distributed over the surface of a cylinder in such a way as to maintain its potential constant in the presence of other uniform axial sources. Show that if these additional sources are exterior to the cylinder the potential within the cylinder is constant and equal to its value on the surface, and that if they lie within the cylinder the potential at exterior points and upon the surface is identical with that which would obtain if all source elements were concentrated upon the axis.

4-70. The axis of a simple cylindrical surface source of radius a and length 2c cuts a transverse plane in the point T, while a parallel rectilinear source of constant density λ and equal length cuts it in P, where PT = f(<a). The surface density on the cylinder is a function of angular position only, and is such as to maintain the potential of the cylinder constant and equal to ϕ_c in the presence of the line source.

If the strength per unit length of the cylindrical source is λ_c deduce from the result of Ex.4-69. that

$$\phi_c = 2(\lambda+\lambda_c) \ln \frac{2c}{a}$$

Show that the potential of the simple layer is duplicated at interior points of the cylinder by that of an exterior image system comprising an axial line source of density $-\lambda$ through the point P', inverse to P in the circular cross-section, together with a coaxial cylindrical source of any radius d (>a) and length 2c having the uniform surface density $\left(\phi_c - 2\lambda \ln \frac{a}{f}\right)\Big/ 4\pi d \ln \frac{2c}{d}$, so that the potential of the original system at an interior point O in the transverse plane is given by

$$\phi_o = 2\lambda \ln \frac{f\rho_{P'}}{a\rho_P} + \phi_c$$

where ρ is distance measured from O.

Use this expression and the result of Ex.4-69. to demonstrate that the surface density $\sigma \left(= -\frac{1}{4\pi}\,\Delta\,\frac{\partial\phi}{\partial n}\right)$ at any point Q on the periphery of the cylinder in the transverse plane is given by

$$\sigma_Q = -\frac{\lambda}{2\pi a}\,\frac{(a^2-f^2)}{PQ^2} + \frac{\lambda+\lambda_c}{2\pi a}$$

4-71. Repeat Ex.4-70. for the case f > a and the region exterior to the cylinder.

Ans: $\phi_c = 2\lambda \ln \frac{2c}{f} + 2\lambda_c \ln \frac{2c}{a}$

The image system comprises a uniform axial line source of density $-\lambda$ through the inverse point P', together with a uniform axial line source of density $\lambda + \lambda_c$ through T. At an exterior point O in the transverse plane

$$\phi_o = 2\lambda \ln \frac{\rho_{P'}}{\rho_P} + 2(\lambda+\lambda_c) \ln \frac{2c}{\rho_T}$$

At any point Q on the periphery of the cylinder in the transverse plane

$$\sigma_Q = -\frac{\lambda}{2\pi a} \frac{(f^2-a^2)}{PQ^2} + \frac{\lambda + \lambda_c}{2\pi a}$$

4-72. A point source of incompressible fluid of unit strength is located at P, a distance f from the centre of an impervious sphere of radius a. Show that for irrotational flow the velocity field may be expressed as $\bar{v} = -\operatorname{grad}\phi$, where ϕ is the combined potential of a unit point source at P and image sources within the sphere comprising (1) a point source of strength $\frac{a}{f}$ at P', inverse to P in the sphere, and (2) a uniform rectilinear source of density $-\frac{1}{a}$ whose end points are P' and the centre of the sphere. (ϕ represents the Neumann function for the sphere and exterior pole P.)

4-73. Two simple, circular, logarithmic line sources of radii a_1 and a_2 have a centre to centre spacing of g. The source strengths are respectively $+\alpha'$ and $-\alpha'$ and the densities are so arranged as to make each circle equipotential. Use the result of Ex.4-62., p. 292 to find the difference between the potentials of the circles (a) when each is exterior to the other, (b) when the circle of radius a_1 is interior to that of radius a_2.

[Hint: Combine the data to obtain

(a) $g = \dfrac{a_1}{k_1} + \dfrac{a_2}{k_2} \qquad g = k_1 a_1 + k_2 a_2$

(b) $g = \dfrac{a_2}{k_2} - \dfrac{a_1}{k_1} \qquad g = k_2 a_2 - k_1 a_1$

and solve for k_1 and k_2.]

Ans:

(a) $$\Delta\phi = \alpha' \ln \left\{\frac{g^2 - a_1^2 - a_2^2 + \sqrt{C}}{2\, a_1\, a_2}\right\}$$

(b) $$\Delta\phi = \alpha' \ln \left\{\frac{a_2^2 + a_1^2 - g^2 + \sqrt{C}}{2\, a_1\, a_2}\right\}$$

where $C = g^4 - 2g^2 (a_1^2+a_2^2) + (a_2^2-a_1^2)^2$

These results may be expressed more elegantly as

$$\text{(a)} \quad \Delta\phi = \alpha' \cosh^{-1}\left\{\frac{g^2 - a_1^2 - a_2^2}{2\, a_1\, a_2}\right\}$$

$$\text{(b)} \quad \Delta\phi = \alpha' \cosh^{-1} - \left\{\frac{g^2 - a_1^2 - a_2^2}{2\, a_1\, a_2}\right\}$$

4-74. A uniform rectilinear source of length 2c and density λ lies in the z axis of Cartesian coordinates and is bisected by the origin. Make use of equations (4.2-2) and (4.2-3) to show that equipotential surfaces take the form of prolate ellipsoids of revolution having the end points of the source as common foci, and such that the ellipsoid of potential $\lambda \ln k$ has an elliptical section in the xz plane defined by

$$\frac{z^2}{(k+1)^2} + \frac{x^2}{4k} = \frac{c^2}{(k-1)^2}$$

Hence, show that if a simple surface source of strength α is distributed upon a prolate ellipsoid of revolution of semi major and minor axes a and b respectively, in such a way as to maintain the surface equipotential, then the potential of the surface is given by

$$\frac{\alpha}{(a^2-b^2)^{\frac{1}{2}}} \ln \frac{a + (a^2-b^2)^{\frac{1}{2}}}{b}$$

while the potential of any exterior confocal ellipsoidal surface of semi axes a' and b' is given by

$$\frac{\alpha}{(a^2-b^2)^{\frac{1}{2}}} \ln \frac{a' + (a^2-b^2)^{\frac{1}{2}}}{b'}$$

4-75. A uniform rectilinear source of length 2c and density λ lies in the z axis of Cartesian coordinates. A simple layer source is distributed over a spherical surface of centre (d,0,0) and radius $a(<d)$ in such a way as to maintain the surface equipotential in the presence of the line source. If the total surface strength is zero show that the image system, which duplicates the potential of the surface source at points outside the sphere, comprises a point source at the centre (or a uniform concentric spherical source of equal strength and radius less than a) together with a curved line source which lies in the xz plane and has the form of a circular arc of radius $\frac{a^2}{2d}$ and centre $\left\{d - \frac{a^2}{2d}, 0, 0\right\}$. Determine the potential of the spherical surface.

Ans: $2\lambda \sinh^{-1} \frac{c}{d}$

4-76. Two spherical, concentric surface sources of radii a and b are equipotential in the presence of each other and a point source of strength α, located at a distance d from their centre, where $a < d < b$. If the potentials of the surfaces are ϕ_a and ϕ_b find their respective source strengths.

[Hint: Introduce interior and exterior images p_1 and q_1 such that p_1, in combination with α, renders the inner surface S_a zero-potential, and q_1, in combination with α, renders the outer S_b zero-potential. The presence of p_1 upsets the equipotential status of S_b while q_1 upsets that of S_a, but these effects may be cancelled by the importation of additional images q_2 and p_2. However, the latter images necessitate, in turn, the introduction of a further pair, and so on. Show that this procedure leads to two systems of images whose sums are convergent series. Set up additional sources to bring the surfaces to the required potentials, and apply Gauss's Law and the result of Ex.4-63. above.]

Ans: Strength of $S_a = \dfrac{ab}{b-a}\left\{\phi_a-\phi_b+\dfrac{\alpha}{b}-\dfrac{\alpha}{d}\right\}$

Strength of $S_b = \dfrac{ab}{b-a}\left\{\dfrac{b}{a}\,\phi_b-\phi_a+\dfrac{\alpha}{d}-\dfrac{\alpha}{a}\right\}$

4-77. A point source of magnitude a is located at P, a distance d from an infinite plane surface upon which a simple surface source is so distributed as to maintain the plane at zero potential in the presence of the point source. P' is the image of P in the plane and PP' is bisected by the plane in the point Q. A hemisphere with Q as centre is described in the half space R which contains P. If ϕ denotes the potential of the complete surface source and ψ that of an image source of magnitude -a, show that the surface integral $\int (\phi-\psi)\,\frac{\partial}{\partial n}\,(\phi-\psi)\,dS$, taken over the hemisphere, vanishes as the radius of the hemisphere approaches infinity. Hence, demonstrate that ϕ and ψ are identical in R.

Show similarly that the combined potential of the surface source and point source at P is everywhere zero in the half space containing P'.

[Hint: Let Q be the origin of spherical coordinates and let $\vec{QP}$ be the axis from which θ is measured. Then ϕ is everywhere finite upon $\theta = 0$ and approaches zero at $\theta = \frac{\pi}{2}$ as $r\to\infty$; further, the symmetry of the surface density is such as to make ϕ independent of Φ. The general solution of Laplace's equation which is compatible with these restrictions, and which consequently includes the expression for ϕ, may be shown to be given, for $r > d$, by

$$\frac{A_o}{r}+\frac{A_1}{r^2}\,P_1(\cos\theta)+\frac{A_2}{r^3}\,P_2(\cos\theta)+\frac{A_3}{r^4}\,P_3(\cos\theta)\ ---$$

where A_o, A_1, A_2 are constants and $P_m(\cos\theta)$ in the mth degree Legendre polynomial in $\cos\theta$ (Sec. 4.1).

Express ψ in the above form with known constants for $r > d$ by means of equation (4.1-17), and by equating ϕ and ψ for $\theta = \frac{\pi}{2}$ show that

$$A_o = -a,\quad A_2 = -ad^2,\quad A_4 = -ad^4\ ----$$

so that

$$\phi - \psi = \frac{(A_1-ad)}{r^2} P_1(\cos\theta) + \frac{(A_3-ad^3)}{r^4} P_3(\cos\theta) ----$$

Hence, show that the surface integral reduces to a series of the form

$$\frac{B_1}{R^3} + \frac{B_2}{R^5} + \frac{B_3}{R^7} ----$$

where B_1, B_2 --- are constants and R is the radius of the hemisphere.]

4-78. Two half planes, which meet to form a wedge of dihedral angle θ, carry a distribution of simple surface sources which maintain them at zero potential in the presence of a point source at some point P within the wedge. Show that if $\theta = \frac{\pi}{n}$, with n integral, the potential of the surface sources may be duplicated within the wedge by that of 2n - 1 image sources disposed around a circle centred normally upon the common edge and passing through P. Assume that the potential of the surface sources is regular at infinity.

Why cannot this treatment be extended to a wedge for which $\theta = \frac{2\pi}{m}$, where m is odd?

Ans: A finite number of image sources, in conjuction with the parent source, will reduce the potential of the planes to zero, but not all images will lie outside the wedge, hence it is not possible to meet the requirement that the potential be harmonic at all points of the wedge beyond a neighbourhood of P.

4.12 The Vector Potential of Line, Surface and Volume Sources

4.12a The vector potential of simple and double line sources

Let $\bar{I}$ be a bounded and piecewise continuous function of length of arc s measured from one end of a regular curve Γ, or from a specified point of it if the curve is closed. Then the Newtonian vector potential at 0 of the line source so defined is given by

$$\bar{A}_o = \int_\Gamma \frac{\bar{I}}{r}\, ds \qquad (4.12\text{-}1)$$

where r is the distance of ds from 0.

$\bar{I}$ is known as the linear source density.

When $\bar{I}$ is everywhere tangential to Γ

$$\bar{A}_o = \int_\Gamma \frac{I}{r}\, d\bar{r} \qquad (4.12\text{-}2)$$

where I is positive or negative according as $\bar{I}$ has the same sense as $d\bar{r}$ or not.

A uniform rectilinear source of density $\bar{I}$ gives rise to a vector potential whose magnitude is equal to the scalar potential of an identical source of density I, for all orientations of $\bar{I}$ with respect to Γ. A similar equivalence holds for parallel combinations of such sources where the individual vector densities are linearly related, so that the analyses of Sec. 4.2 are directly applicable to systems of this type.

The vector potential is logarithmically infinite at points of a line source where $\bar{I}$ is non-zero, as in the scalar case.

When the source is closed and of uniform density, and $\bar{I}$ is everywhere tangential to Γ, equation (4.12-2) may be replaced by a surface integral in accordance with equation (1.17-1). We have

$$\bar{A}_o = I \oint_\Gamma \frac{1}{r}\, d\bar{r} = I \int_S d\bar{S} \times \operatorname{grad} \frac{1}{r} \qquad (4.12\text{-}3)$$

where S is any regular surface spanning Γ, and the positive sense of the normal at the surface is right handedly related to the sense of integration around Γ.

It follows that a line source of this type, whose dimensions approach zero and whose density approaches infinity in such a way as to maintain $I\bar{S}$ constant[24] and equal to, say $\bar{m}$, gives rise to a vector potential

$$\bar{A}_o = \bar{m} \times \operatorname{grad} \frac{1}{r} \qquad (4.12\text{-}4)$$

for all positions of 0 outside it. This expression is the vector analogue of the scalar potential of a point doublet, viz $\bar{p}^{(1)}.\operatorname{grad} \frac{1}{r}$ (p. 219)[25].

---

24. A source of this type will be referred to subsequently as a 'whirl'.

25. Since we will not be concerned subsequently with multipoles of order higher than unity, we will henceforth drop the superscript (1) and express dipole potential as $\bar{p}.\operatorname{grad} \frac{1}{r}$.

Two vector line sources may be combined to form a double source. A uniform rectilinear double source comprises the limiting configuration of two parallel uniform rectilinear sources of equal and opposite vectorial line density $\pm\bar{I}$ and spacing d, where $d\to 0$ and $\bar{I}d$ remains constant and equal to, say, $\bar{L}_1$. Here $\bar{I}$ is the density of one of the line pair, arbitrarily chosen and designated 'positive'. The potential at any point 0, not coincident with the source, is then given by

$$\bar{A}_o = \int_\Gamma \bar{L}_1 \frac{\partial}{\partial n'}\left(\frac{1}{r}\right) ds \qquad (4.12\text{-}5)$$

where $\frac{\partial}{\partial n'}$ denotes differentiation along the normal in the plane of the line pair from the 'negative' to the 'positive' element.

The same expression continues to hold for the general form of line doublet where $\bar{L}_1$ is a function of distance along Γ, and Γ may be curved and twisted, provided that equal and opposite source strengths ($\bar{I}$ ds) are intercepted in the parent filaments by adjacent normals (cf the requirement for scalar double layers as considered in Ex.4-12. and 4-13., p. 241 and extended below to vector double layers).

4.12b The vector potential of simple and double surface sources

Let $\bar{K}$ be a bounded and piecewise continuous function of position upon a regular surface S. The associated vector potential at any point 0 is given by

$$\bar{A}_o = \int_S \frac{\bar{K}}{r}\, dS \qquad (4.12\text{-}6)$$

where r is the distance of dS from 0.

The density function $\bar{K}$, which may or may not be tangential to S, can be resolved at all points where it is defined into scalar components, and these are piecewise continuous upon S. Since

$$\int_S \frac{\bar{K}}{r}\, dS = \bar{i}\int_S \frac{K_x}{r}\, dS + \bar{j}\int_S \frac{K_y}{r}\, dS + \bar{k}\int_S \frac{K_z}{r}\, dS \qquad (4.12\text{-}7)$$

and since the component scalar potentials are everywhere finite and continuous (Sec. 4.3) it follows that $\bar{A}$ is finite and continuous at all interior, boundary and exterior points of S.

Two plane, parallel surface sources of uniform densities $\pm\bar{K}$ (where $+\bar{K}$ is the density of the arbitrarily chosen 'positive' surface) reduce to a uniform double layer source when $\bar{K}d$ is maintained constant and equal to, say, $\bar{\mu}_1$ as $d\to 0$. The potential at any point 0 outside the source is then

$$\bar{A}_o = \int_S \bar{\mu}_1 \frac{\partial}{\partial n}\left(\frac{1}{r}\right) dS = -\int_S \bar{\mu}_1 \, d\Omega \tag{4.12-8}$$

where $\frac{\partial}{\partial n}$ denotes differentiation along the normal from the 'negative' to the 'positive' surface.

This continues to hold when μ_1 is a function of position and the composite surface is curved, provided that in the latter case the density of the 'negative' side is so adjusted that if a small closed curve is drawn anywhere on the 'positive' surface and projected normally onto the 'negative' surface, equal and opposite values of $\bar{K}$ dS obtain for the two elements so delineated.

By resolving the vector potential into scalar components and applying the arguments of Sec. 4.3, we see that $\bar{A}$ increases by $4\pi\bar{\mu}_1$ with normal movement through the surface from the 'negative' to the 'positive' side at any interior point of S where $\bar{\mu}_1$ is continuous.

4.12c <u>The vector potential of a volume source</u>

Let $\bar{J}$ be a bounded and piecewise continuous function of position within a bounded region of space τ. Then the vector potential associated with this volume source is given at any point 0 by

$$\bar{A}_o = \int_\tau \frac{\bar{J}}{r} d\tau = \bar{i}\int_\tau \frac{J_x}{r} d\tau + \bar{j}\int_\tau \frac{J_y}{r} d\tau + \bar{k}\int_\tau \frac{J_z}{r} d\tau \tag{4.12-9}$$

where r is the distance of $d\tau$ from 0.

It follows from the considerations of Sec. 4.4 that $\bar{A}$ is finite and continuous at interior, boundary and exterior points of τ. This remains true when τ is unbounded externally, provided that $\bar{A}$ continues to be finite everywhere.

The definitions of partial and cavity vector potential parallel those for the scalar case.

4.13 Reciprocal Relationships in Scalar and Vector Potential Theory

Consider two systems of point sources of magnitudes $a_1, a_2 \ldots a_i \ldots a_n$ and $a_1', a_2' \ldots a_i' \ldots a_m'$ such that no two sources coincide. Then

$$\sum_{i=1}^{n} a_i \phi_i' = \sum_{i=1}^{m} a_i' \phi_i \tag{4.13-1}$$

where ϕ_i' is the potential at a_i of the sources $a_1' \ldots a_m'$, and ϕ_i is the potential at a_i' of the sources $a_1 \ldots a_n$.

This becomes evident when equation (4.13-1) is expressed in the form of the double summation

$$\sum_{i=1}^{n} a_i \sum_{j=1}^{m} a_j'/r_{a_i a_j'} = \sum_{i=1}^{m} a_i' \sum_{j=1}^{n} a_j/r_{a_i' a_j} \qquad (4.13\text{-}1(a))$$

and subsequently expanded.

If a pair of sources from the two sets coincide in space, then the above equations continue to hold provided that the infinite components of ϕ and ϕ' are deleted, eg if a_p and a_q' coincide, then ϕ_p' should be calculated ignoring a_q' and ϕ_q ignoring a_p. Similar arguments lead to the related proposition: If two sets of point sources of magnitudes $a_1 \ldots a_n$ and $a_1' \ldots a_n'$ occupy the fixed positions $P_1 \ldots P_n$ successively, then

$$\sum_{i=1}^{n} a_i \, \phi_i' = \sum_{i=1}^{n} a_i' \, \phi_i \qquad (4.13\text{-}2)$$

where a_i' is deleted in the computation of ϕ_i' and a_i is deleted in the computation of ϕ_i.

Since equations (4.13-1) and (4.13-2) are consequences of the symmetrical nature of the associated double summations they are not restricted to inverse distance (Newtonian) potential functions but are equally valid in logarithmic potential theory.

It is intuitively evident that these relationships may be extended to continuous source systems. Thus for non-intersecting, open or closed line and surface sources in space, we have

$$\int_{\Gamma_{1..n}} \lambda \, \phi' \, ds = \int_{\Gamma'_{1..m}} \lambda' \, \phi \, ds' \qquad (4.13\text{-}3)$$

$$\int_{S_{1..n}} \sigma \, \phi' \, dS = \int_{S'_{1..m}} \sigma' \, \phi \, dS' \qquad (4.13\text{-}4)$$

and for volume sources

$$\int_{\tau_{1..n}} \rho \, \phi' \, d\tau = \int_{\tau'_{1..m}} \rho' \, \phi \, d\tau' \qquad (4.13\text{-}5)$$

The corresponding expressions for point, line and surface doublets are found by pairing elements in equations (4.13-1), (4.13-3) and (4.13-4), which then take the form

$$\sum_{i=1}^{n} \bar{p}_i \cdot \text{grad}\, \phi'_i = \sum_{i=1}^{m} \bar{p}'_i \cdot \text{grad}\, \phi_i \tag{4.13-6}$$

$$\int_{\Gamma_{1..n}} \bar{L} \cdot \text{grad}\, \phi'\, ds = \int_{\Gamma'_{1..m}} \bar{L}' \cdot \text{grad}\, \phi\, ds' \tag{4.13-7}$$

$$\int_{S_{1..n}} \mu\, d\bar{S} \cdot \text{grad}\, \phi' = \int_{S'_{1..m}} \mu'\, d\bar{S}' \cdot \text{grad}\, \phi \tag{4.13-8}$$

It follows from equation (4.13-8) that for two non-intersecting, uniform, double layer sources of equal density, the flux through either surface of the gradient of the potential of the other has the same value.

Since the potentials deriving from piecewise continuous simple surface and volume sources are everywhere finite, the requirement of non-intersection of component source regions in equations (4.13-4) and (4.13-5) is unnecessary, and it becomes possible to formulate the following proposition: If volume sources of density ρ occupy a region of space τ containing surface sources $S_{1..n}$ of density σ and ϕ is the associated potential function, and if alternative densities ρ' and σ' give rise to ϕ', then

$$\int_{S_{1..n}} \sigma\, \phi'\, dS + \int_{\tau} \rho\, \phi'\, d\tau = \int_{S_{1..n}} \sigma'\, \phi\, dS + \int_{\tau} \rho'\, \phi\, d\tau \tag{4.13-9}$$

Since the potential of a logarithmic line source is finite within the source, the corresponding relationship in logarithmic potential theory is

$$\int_{\Gamma_{1..n}} \lambda\, \phi'\, ds + \int_{S} \sigma\, \phi'\, dS = \int_{\Gamma_{1..n}} \lambda'\, \phi\, ds + \int_{S} \sigma'\, \phi\, dS \tag{4.13-10}$$

where λ and σ now denote logarithmic source densities.

Similar reciprocal relationships occur in vector potential theory. Thus for two systems of non-intersecting vector line sources of densities $\bar{I}$ and $\bar{I}'$ and potentials $\bar{A}$ and $\bar{A}'$, we have

$$\int_{\Gamma_{1..n}} \bar{I}.\bar{A}'\, ds = \int_{\Gamma'_{1..m}} \bar{I}'.\bar{A}\, ds' \qquad (4.13\text{-}11)$$

since this is equivalent to

$$\int_{\Gamma_{1..n}} \bar{I} \,.\int_{\Gamma'_{1..m}} \frac{\bar{I}'}{r}\, ds'\, ds = \int_{\Gamma'_{1..m}} \bar{I}' \,.\int_{\Gamma_{1..n}} \frac{\bar{I}}{r}\, ds\, ds'$$

where r is the distance between ds and ds'.

It is not necessary that the contours be closed or that the densities be uniform and directed along the contours. However, in the case of two closed, uniform, vector line sources whose densities are of equal magnitude and directed along the contours, equation (4.13-11) reduces to

$$\oint_{\Gamma} \bar{A}'.d\bar{r} = \oint_{\Gamma'} \bar{A}.d\bar{r}' \qquad (4.13\text{-}12)$$

whence

$$\int_{S} (\text{curl } \bar{A}').d\bar{S} = \int_{S'} (\text{curl } \bar{A}).dS' \qquad (4.13\text{-}13)$$

where S and S' are regular surfaces spanning Γ and Γ'. Under these circumstances the flux through S of the curl of the vector potential of Γ' is equal to the flux through S' of the curl of the vector potential of Γ.

Relations similar to equation (4.13-11) may be written down both for surface and volume sources. Since the vector potentials are everywhere finite for finite, piecewise continuous sources, these equalities hold, in addition, for alternative source distributions upon given surfaces and in given regions.

EXERCISES

4-79. Extend the planar form of Green's formula (3.9-3) to vector fields, and repeat Ex.4-27., p. 257 in this context.

4-80. Show that the vectorial form of Gauss's law in space is

$$-\oint_{S_{1..n}\Sigma} \frac{\partial \bar{A}}{\partial n}\, dS = 4\pi \left\{ \int \bar{I}\, ds + \int \bar{K}\, dS + \int_{\tau} \bar{J}\, d\tau \right\}$$

4-81. Show that Gauss's law for logarithmic vector potentials in the plane is

$$-\oint_{\Gamma_{1..n}\Gamma'} \frac{\partial \bar{A}'}{\partial n'}\, ds = 2\pi \left\{ \int \bar{I}'\, ds + \int_S \bar{K}'\, dS \right\}$$

where

$$\bar{A}' = \int \bar{I}' \ln \frac{1}{\rho}\, ds + \int_S \bar{K}' \ln \frac{1}{\rho}\, dS$$

and the line source of the right hand side is contained within S and has no point in common with $\Gamma_{1..n}\Gamma'$.

4-82. The scalar potential of a long cylindrical volume source of constant density is identical, at points of the bounding surface near the mid plane, with that obtaining when the source elements are concentrated upon the axis of the cylinder. The potential upon the inner surface of a similar hollow cylindrical source is equal to that obtaining upon the axis.

Prove the above assertions by means of an analysis similar to that undertaken for Ex.4-69., p. 300. Hence derive, via equation (4.2-5), an expression for the scalar potential at an interior point of a solid, uniform, cylindrical source of radius a and length 2c as $c \to \infty$, and write down the corresponding value of the vector potential when the source density $\bar{J}$ is everywhere constant.

$$\text{Ans:}\quad \bar{A} = \pi\bar{J}\left\{2a^2 \ln \frac{2c}{a} + a^2 - \rho^2\right\}$$

where ρ is distance measured from the axis.

4-83. By expressing each integral as a double summation, show that

$$\int_\tau \rho \operatorname{grad} \phi'\, d\tau = -\int_\tau \rho' \operatorname{grad} \phi\, d\tau$$

where ϕ is the potential associated with a volume source of density ρ in the bounded region of space τ, and ϕ' is the potential of an alternative distribution ρ'.

Confirm this result by putting $\bar{F} = \operatorname{grad} \phi$ and $\bar{G} = \operatorname{grad} \phi'$ in equation (1.17-17), transposing the prime, and adding the resulting equations. Transform the unwanted volume integral into a surface integral by making use of equation (1.16-5) integrated in accordance with equation (1.17-5). Then show that the surface integrals vanish at infinity while the component volume integrals are unchanged beyond τ. [When ρ and ρ' are discontinuous upon the boundary of τ the procedure should be carried out both for τ and for all space beyond τ. (On addition, the surface integrals over the boundary of τ cancel because grad pot ρ remains continuous at discontinuities of ρ (p. 278).)]

4.14 The Divergence, Curl and Laplacian of the Vector Potential of Simple Line and Surface Sources

4.14a Line sources

It follows from the argument of Sec. 4.7a that at any point 0 outside the source

$$\operatorname{div}\bar{A} = \operatorname{div}\int_\Gamma \frac{\bar{I}}{r}\,ds = \sum \frac{\partial}{\partial x_o}\int_\Gamma \frac{I_x}{r}\,ds = -\sum\int_\Gamma I_x \frac{\partial}{\partial x}\frac{1}{r}\,ds$$

$$= -\int_\Gamma \bar{I}\,.\,\operatorname{grad}\frac{1}{r}\,ds = \int_\Gamma \bar{I}\,.\,\frac{\bar{r}}{r^3}\,ds \qquad (4.14\text{-}1)$$

where $\bar{r}$ is directed from 0 to ds.

Similarly

$$\operatorname{curl}\bar{A} = \operatorname{curl}\int_\Gamma \frac{\bar{I}}{r}\,ds = \sum\left\{\frac{\partial}{\partial y_o}\int_\Gamma \frac{I_z}{r}\,ds - \frac{\partial}{\partial z_o}\int_\Gamma \frac{I_y}{r}\,ds\right\}$$

$$= \sum\int_\Gamma \left\{I_y\frac{\partial}{\partial z}\left(\frac{1}{r}\right) - I_z\frac{\partial}{\partial y}\left(\frac{1}{r}\right)\right\}ds$$

$$= \int_\Gamma \bar{I}\times\operatorname{grad}\frac{1}{r}\,ds = -\int_\Gamma \bar{I}\times\frac{\bar{r}}{r^3}\,ds \qquad (4.14\text{-}2)$$

When the line source is closed and $\bar{I}$ is everywhere tangential to it and of constant magnitude

$$\operatorname{div}\bar{A} = \operatorname{div} I\oint_\Gamma \frac{d\bar{r}}{r} = -I\oint_\Gamma\left(\operatorname{grad}\frac{1}{r}\right).d\bar{r} = -I\int_S\left(\operatorname{curl}\operatorname{grad}\frac{1}{r}\right).d\bar{S}$$

where S spans Γ but does not cut 0, hence

$$\operatorname{div}\bar{A} = \operatorname{div} I\oint_\Gamma \frac{d\bar{r}}{r} = 0 \qquad (4.14\text{-}1(a))$$

In the same circumstances

$$\text{curl } \bar{A} = \text{curl } I \oint_{\Gamma} \frac{d\bar{r}}{r} = I \oint_{\Gamma} \frac{\bar{r}}{r^3} \times d\bar{r} \qquad (4.14\text{-}2(a))$$

whence from equation (4.7-16)

$$\text{curl } I \oint_{\Gamma} \frac{d\bar{r}}{r} = -\text{ grad } \phi = I \text{ grad } \Omega \qquad (4.14\text{-}2(b))$$

where ϕ is the scalar potential of a uniform surface doublet of density I which spans Γ but does not cut 0, and which subtends the solid angle Ω at 0.

The usual right-handed relationship holds between the positive sense of integration around Γ and the orientation of the surface doublets.

At points outside Γ the Laplacian of $\bar{A}$ is zero because

$$\nabla^2 \bar{A} = \nabla^2 \int_{\Gamma} \frac{\bar{I}}{r} \, ds = \int_{\Gamma} \bar{I} \, \nabla^2 \left(\frac{1}{r}\right) ds = \bar{0} \qquad (4.14\text{-}3)$$

The Laplacian is undefined upon the contour itself.

4.14b Surface sources

Since $\bar{A}$ is continuous for normal movement through a simple surface source, it follows that any tangential derivative of the vector potential is likewise continuous at each interior point of the surface where it is defined. The normal derivative, however, is discontinuous. Proceeding as for the normal derivative of a simple scalar surface source (Sec. 4.7b) we find that

$$\Delta \frac{\partial \bar{A}}{\partial n} = \Delta \frac{\partial}{\partial n} \int_{S} \frac{\bar{K}}{r} \, dS = -4\pi \bar{K} \qquad (4.14\text{-}4)$$

where $\bar{K}$ is the local (continuous) surface density, and $\Delta \frac{\partial \bar{A}}{\partial n}$ is the increment of $\frac{\partial \bar{A}}{\partial n}$ for positive motion through the surface, with a common positive sense of the normal on both sides.

At points outside the surface, div $\bar{A}$ and curl $\bar{A}$ may be written down by analogy with equations (4.14-1) and (4.14-2).

$$\text{div } \bar{A} = \text{div} \int_{S} \frac{\bar{K}}{r} \, dS = -\int_{S} \bar{K} \cdot \text{grad } \frac{1}{r} \, dS \qquad (4.14\text{-}5)$$

$$\text{curl } \bar{A} = \text{curl} \int_S \frac{\bar{K}}{r} dS = \int_S \bar{K} \times \text{grad} \frac{1}{r} dS \qquad (4.14\text{-}6)$$

For the particular case in which $\bar{K}$ is everywhere normal to the surface and of constant magnitude

$$\text{div } \bar{A} = \text{div } K \int_S \frac{d\bar{S}}{r} = -K \int_S \left(\text{grad } \frac{1}{r}\right) . d\bar{S} = K \int_S \frac{\bar{r}}{r^3} . d\bar{S} = K\,\Omega \qquad (4.14\text{-}5(a))$$

$$\text{curl } \bar{A} = \text{curl } K \int_S \frac{d\bar{S}}{r} = K \int_S d\bar{S} \times \text{grad } \frac{1}{r} = K \oint_\Gamma \frac{d\bar{r}}{r} \qquad (4.14\text{-}6(a))$$

The latter expression is the vector potential at 0 of a peripheral line source whose density is everywhere tangential to the contour and of magnitude K.

If, in addition, the surface is closed, then

$$\text{div } \bar{A} = \text{div } K \oint_S \frac{d\bar{S}}{r} = 4\pi K \quad \text{or} \quad 0 \qquad (4.14\text{-}5(b))$$

according as 0 lies within or without the enclosure, and

$$\text{curl } \bar{A} = \text{curl } K \oint_S \frac{d\bar{S}}{r} = K \oint_S d\bar{S} \times \text{grad } \frac{1}{r} = \bar{0} \qquad (4.14\text{-}6(b))$$

at all points not coincident with S (from equation (1.17-2)).

Surface sources in which the source density is tangential to the surface are of considerable importance. For such sources it may be shown that the divergence of the vector potential is zero at all exterior points whenever the surface divergence of $\bar{K}$ (divs $\bar{K}$) is zero at interior points of the surface and $\bar{K}$ has no component normal to the boundary when the surface is unclosed. The proof of this proposition is the subject of Ex.4-88., p. 317.

Both div $\bar{A}$ and curl $\bar{A}$ are discontinuous for normal movement through the surface at an interior point where $\bar{K}$ is continuous but neither normal nor tangential to the local surface. It may be shown, in terms of the notation employed above, that

$$\Delta \operatorname{div} \overline{A} = \Delta \operatorname{div} \int_S \frac{\overline{K}}{r}\, dS = -4\pi\, (\hat{n}.\overline{K}) \qquad (4.14\text{-}7)$$

$$\Delta \operatorname{curl} \overline{A} = \Delta \operatorname{curl} \int_S \frac{\overline{K}}{r}\, dS = -4\pi\, (\hat{n}\times\overline{K}) \qquad (4.14\text{-}8)$$

As in the case of the line source,

$$\nabla^2 \overline{A} = \int_S \overline{K}\, \nabla^2 \left(\frac{1}{r}\right) dS = \overline{0} \qquad (4.14\text{-}9)$$

at points outside the surface.
It is undefined upon the surface itself.

EXERCISES

4-84. Devise an alternative proof of equation (4.14-1(a)) by expanding $\frac{d}{ds}\left(\frac{I}{r}\right)$ and integrating around Γ.

4-85. Show that for a disc-shaped source of radius a and constant density $\overline{K}$, the values of div $\overline{A}$ and curl $\overline{A}$ at a distance d from the disc upon the axis are given by

$$\operatorname{div} \overline{A} = 2\pi\, (\hat{n}.\overline{K}) \left\{1 - \frac{d}{(a^2+d^2)^{\frac{1}{2}}}\right\} \qquad (d \neq 0)$$

$$\operatorname{curl} \overline{A} = 2\pi\, (\hat{n}\times\overline{K}) \left\{1 - \frac{d}{(a^2+d^2)^{\frac{1}{2}}}\right\} \qquad (d \neq 0)$$

where $\hat{n}$ is directed towards the disc.

Use these results

(a) to deduce equations (4.14-7) and (4.14-8)

(b) to demonstrate that for a plane tangential surface source

$$\operatorname{curl} \overline{A} \to 2\pi\, (\hat{n}\times\overline{K})$$

as the surface is approached from one side or another at an interior point, where $\overline{K}$ is the (continuous) local surface density.

4-86. By differentiation of the derivatives of $\frac{1}{r}$, show that all points outside the sources

$$\operatorname{curl} \int_{\Gamma} \bar{I} \times \operatorname{grad} \frac{1}{r} \, ds = - \operatorname{grad} \int_{\Gamma} \bar{I} \cdot \operatorname{grad} \frac{1}{r} \, ds$$

and

$$\operatorname{curl} \int_{S} \bar{K} \times \operatorname{grad} \frac{1}{r} \, dS = - \operatorname{grad} \int_{S} \bar{K} \cdot \operatorname{grad} \frac{1}{r} \, dS$$

Γ and S may or may not be closed, and the relationships continue to hold for individual source elements.

Confirm the above equations by applying (1.18-5) to (4.14-1), (4.14-2) and (4.14-3) etc.

Rewrite the second equation in a form appropriate to an open surface source where $\bar{K}$ is everywhere normal to the surface and of constant magnitude, and so derive (4.14-2(b)) by substituting $\bar{I}$ for $\bar{K}$.

Show also that

$$\operatorname{curl} \int_{S} V \, d\bar{S} \times \operatorname{grad} \frac{1}{r} = - \operatorname{grad} \int_{S} V \, d\bar{S} \cdot \operatorname{grad} \frac{1}{r}$$

where V is any piecewise continuous scalar point function which is defined upon S.

4-87. If S is a closed surface source having a surface density $\bar{K}$ which is everywhere normal to the surface and of constant magnitude, express the vector potential of $\bar{K}$ at points outside the enclosed region τ as a volume integral over τ, by means of equation (1.17-5) with $V = \frac{1}{r}$. Then show by differentiation of the derivatives of $\frac{1}{r}$ that div $\bar{A}$ = curl $\bar{A}$ = 0 outside τ.

Show further that when 0 is an interior point of τ

$$\bar{A} = K \int_{\tau-\tau_\delta} \operatorname{grad} \frac{1}{r} \, d\tau$$

where τ_δ is a δ sphere centred upon 0, whence

$$\operatorname{div} \bar{A} = - \sum \frac{\partial}{\partial x_o} K \int_{\tau-\tau_\delta} \frac{(x-x_o)}{r^3} \, d\tau$$

Now develop a field-slipping analysis, as in Sec. 4.8, in which $\frac{1}{r}$ is replaced by $\frac{(x-x_o)}{r^3}$, and, by putting $\rho = 1$, derive the relationship

$$\frac{\partial}{\partial x_o}\int_{\tau-\tau_\delta}\frac{(x-x_o)}{r^3}\,d\tau = -\oint_S \frac{(x-x_o)}{r^3}\,dS_x$$

Hence prove that div $\overline{A} = 4\pi K$ within τ.

Show similarly that curl $\overline{A} = \overline{0}$ within τ.

4-88. It follows from Ex.2-34. and 2-35., p. 167 that if $\overline{F}$ is a well-behaved vector point function which is defined upon the open surface S (but not necessarily outside it) and is everywhere tangential to S, then

$$\oint_\Gamma \overline{F}.\hat{n}'\,|d\overline{r}| = \int_S \text{divs}\,\overline{F}\,dS = \int_S \frac{1}{h_\xi h_\zeta}\left\{\frac{\partial}{\partial\xi}(h_\zeta F_\xi) + \frac{\partial}{\partial\zeta}(h_\xi F_\zeta)\right\} dS$$

Let $\overline{F} = \frac{\overline{K}}{r}$, where $\overline{K}$ denotes surface source density (assumed to be tangential to the surface) and r is distance measured from an exterior origin. Expand the resulting equation to show that

$$\oint_\Gamma \frac{\overline{K}}{r}\,.\,\hat{n}'\,ds = \int_S \frac{\text{divs}\,\overline{K}}{r}\,dS + \int_S \left\{\frac{K_\xi}{h_\xi}\frac{\partial}{\partial\xi}\left(\frac{1}{r}\right) + \frac{K_\zeta}{h_\zeta}\frac{\partial}{\partial\zeta}\left(\frac{1}{r}\right)\right\} dS$$

Reduce this to

$$\oint_\Gamma \frac{\overline{K}}{r}\,.\,\hat{n}'\,ds = \int_S \frac{\text{divs}\,\overline{K}}{r}\,dS + \int_S \overline{K}\,.\,\text{grad}\,\frac{1}{r}\,dS$$

by showing that

$$\text{grad}\,\frac{1}{r} = \frac{\hat{\xi}}{h_\xi}\frac{\partial}{\partial\xi}\left(\frac{1}{r}\right) + \frac{\hat{\zeta}}{h_\zeta}\frac{\partial}{\partial\zeta}\left(\frac{1}{r}\right) + \hat{n}\frac{\partial}{\partial n}\left(\frac{1}{r}\right)$$

where $\hat{\xi}$, $\hat{\zeta}$, $\hat{n}$ form a right-handed set.

Hence show that the divergence of the vector potential of a well-behaved tangential surface source of density $\overline{K}$ is zero at exterior points if divs $\overline{K}$ is everywhere zero and the surface is (a) closed or (b) open, with $\overline{K}.\hat{n}' = 0$ at all points of Γ (ie with no component of $\overline{K}$ normal to the periphery).

Extend this result to a surface source which may be divided into subregions over which $\text{divs}\,\bar{K} = 0$ and for which the normal component of $\bar{K}$ is continuous through internal boundaries.

4-89. A rectilinear source of constant tangential density $\bar{I}$ subtends an angle 2θ at the point O upon a perpendicular bisector of the source through the mid point P.

Show that

(1) $(\text{curl}\,\bar{A})_o = (\hat{\bar{r}}\times\bar{I})\,\dfrac{2\sin\theta}{d}$ where $\vec{OP} = \hat{\bar{r}}\,d$

(2) the right-handed tangential line integral of curl $\bar{A}$ around a concentric circle in the mid plane approaches $4\pi I$ as the circle shrinks about the source.

Show further that in the latter circumstance the contribution to the integral from source elements outside a neighbourhood of P approaches zero, and so demonstrate that the above value for the integral continues to hold at interior points of a curvilinear source having a continuously turning tangent, where I is the magnitude of the local (continuous) source density.

Now let the line source be closed and let I be constant. Prove that $\text{curl curl}\,\bar{A} = \bar{0}$ outside the source and hence show that the line integral of curl $\bar{A}$ around any regular closed curve which threads the source contour once is equal to $4\pi I$. Devise a simple geometrical construction to demonstrate that if the curve of integration threads the source contour n times the corresponding line integral is $4\pi nI$.

Arrive at the same result by working from equation (4.14-2(b)), bearing in mind that curl $\bar{A}$ is continuous everywhere outside the source.

4-90. A tangential surface source takes the form of a closed strip having $\text{divs}\,\bar{K} = 0$ at interior points and $\bar{K}$ parallel to the edge along the edges. Let a simple closed curve which threads the source once contract about the strip in such a way as to give rise to matching contour elements on the two sides, and let P and Q be the points of the curve which approach the edges of the strip. Make use of equation (4.14-8) to show that the line integral of curl $\bar{A}$ around the curve in the $\vec{PQ}$ direction is given by

$$-\int_P^Q 4\pi\,(\hat{\bar{n}}\times\bar{K}).d\bar{r}$$

where the integral is taken over one side only and $\hat{\bar{n}}$ is the corresponding outward normal.

Prove that this value of line integral obtains for all closed curves which thread the source once, and that it is equal in magnitude to $4\pi \times$ flux of $\bar{K}$ through any simple curve lying within the strip and joining the opposite edges.

4-91. A torus of arbitrary section, whose axis coincides with the z axis of cylindrical coordinates, carries a tangential surface density $\overline{K}$ which has no Φ component at any point. The magnitude K is a function of ρ only, and is such as to maintain divs $\overline{K}$ zero everywhere upon the surface. By expanding in cylindrical coordinates, show from symmetry that $(\text{curl } \overline{A})_\rho = (\text{curl } \overline{A})_z = \overline{0}$ both within and without the torus. Invoke Stokes's theorem and the relationship curl curl $\overline{A} = \overline{0}$ to prove that $(\text{curl } \overline{A})_\Phi$ is zero at all exterior points; by taking account of equation (4.14-8) in addition, and employing a doubly-bounded surface of integration show that, for interior points,

$$|(\text{curl } \overline{A})_\Phi| = \frac{4\pi K_o \rho_o}{\rho} = \frac{\text{const}}{\rho}$$

where K_o is the magnitude of $\overline{K}$ along a circle of arbitrary radius ρ_o drawn upon the surface.

4-92. The vector density of a cylindrical surface source of length 2c and radius a is given by $\overline{K} = \hat{\Phi}K$, where K is a constant. Show that $(\text{curl } \overline{A})_\Phi$ is everywhere zero, and that at points on the axis within the cylinder

$$(\text{curl } \overline{A})_z = 2\pi K (2 - \sin\theta_1 - \sin\theta_2)$$

where θ_1 and θ_2 are the angles subtended by the end radii.

Prove that $(\text{curl } \overline{A})_\rho \to 0$ within a finite distance of the source centre as $c \to \infty$, and, by expansion of $(\text{curl curl } \overline{A})_\Phi$, demonstrate that under these conditions $\frac{\partial}{\partial\rho}(\text{curl } \overline{A})_z \to 0$. Hence show that for an infinite, uniform, circumferential source (infinite solenoid)

$$(\text{curl } \overline{A})_z = 4\pi K \qquad (\rho<a)$$

$$(\text{curl } \overline{A})_z = 0 \qquad (\rho>a)$$

By integrating curl $\overline{A}$ over a transverse section, derive

$$\overline{A} = \hat{\Phi}\, 2\pi K\rho \qquad (\rho \leq a)$$

$$\overline{A} = \hat{\Phi}\, \frac{2\pi K a^2}{\rho} \qquad (\rho \geq a)$$

4-93. Consider a system of non-intersecting cylindrical surface sources of arbitrary cross-section and infinite length, aligned with the z axis of coordinates. These carry (1) a scalar density σ which is restricted only by the requirement that $\frac{\partial\sigma}{\partial z} = 0$ everywhere, and (2) a vector density given by $\overline{K} = \hat{z}\sigma$. Show that grad ϕ and curl $\overline{A}$ are orthogonal and of equal magnitude at all points exterior to the sources and at finite distance from a local origin. Show also that the flux of curl $\overline{A}$ through an axial strip of unit axial width, whose cross-section is a simple curve which does not cut any source, is equal to the difference of scalar potential between the end points.

[Note that when σ is such as to maintain each surface equipotential in the presence of the others, the flux of curl $\overline{A}$ through an axial strip joining any two cylinders is independent of the position of the ends of the strip on the cylinders.]

4-94. A cylindrical surface source of radius a and length 2c is centred upon the z axis of coordinates. The source density is single-valued and given everywhere by $\overline{K} = \hat{z}\, f(\Phi)$. Show that if $c\to\infty$ and the surface is approached along a normal from within and without at finite distance from the source centre, the limiting values of $(\text{curl}\,\overline{A})_\Phi$ are given respectively by

$$\frac{1}{a}\oint K\, ds - 2\pi K_o \qquad \text{and} \qquad \frac{1}{a}\oint K\, ds + 2\pi K_o$$

where the line integral is taken around the clyinder and K_o is the magnitude of the local surface density. (cf Ex.4-37., p. 269)

[Since $\oint \overline{K}\, ds = \int \overline{K}\, dS$ taken over unit axial length of surface, we may refer to $\oint \overline{K}\, ds = \hat{z}\oint K\, ds$ as the strength of the source per unit length.]

4.15 The Divergence, Curl and Laplacian of the Vector Potential of a Volume Source

It follows from Sec. 4.8 that both at interior and exterior points of τ

$$\frac{\partial}{\partial x_o}\int_\tau \frac{\rho}{r}\, d\tau = \int_\tau \frac{1}{r}\frac{\partial\rho}{\partial x}\, d\tau - \oint_{S_{1..n}\Sigma} \frac{\rho}{r}\, dS_x = -\int_\tau \rho\,\frac{\partial}{\partial x}\left(\frac{1}{r}\right) d\tau \qquad (4.15\text{-}1)$$

provided that ρ is well-behaved throughout τ.

From Sec. 4.9

$$\nabla^2 \int_\tau \frac{\rho}{r}\, d\tau = 0 \qquad \text{at exterior points of } \tau \qquad (4.15\text{-}2)$$

and

$$\nabla^2 \int_\tau \frac{\rho}{r}\, d\tau = -4\pi\rho \qquad \text{at interior points of } \tau \tag{4.15-3}$$

Expressions for div pot $\bar{J}$, curl pot $\bar{J}$ and ∇^2 pot $\bar{J}$ are readily developed by substitution of the scalar components of $\bar{J}$ for ρ in the above equations. Thus

$$\operatorname{div} \int_\tau \frac{\bar{J}}{r}\, d\tau = \sum \frac{\partial}{\partial x_o} \int_\tau \frac{J_x}{r}\, d\tau = \sum \int_\tau \frac{1}{r} \frac{\partial J_x}{\partial x}\, d\tau - \sum \oint_{S_{1..n}\Sigma} \frac{J_x}{r}\, dS_x$$

$$= -\sum \int_\tau J_x \frac{\partial}{\partial x}\left(\frac{1}{r}\right) d\tau$$

hence, at interior and exterior points of τ,

$$\operatorname{div} \operatorname{pot} \bar{J} = \int_\tau \frac{\operatorname{div} \bar{J}}{r}\, d\tau - \oint_{S_{1..n}\Sigma} \frac{\bar{J}}{r} \cdot d\bar{S} \tag{4.15-4}$$

or

$$\operatorname{div} \operatorname{pot} \bar{J} = -\int_\tau \bar{J} \cdot \operatorname{grad} \frac{1}{r}\, d\tau \tag{4.15-5}$$

Similarly

$$\operatorname{curl} \int_\tau \frac{\bar{J}}{r}\, d\tau = \sum \bar{i} \left\{ \frac{\partial}{\partial y_o} \int_\tau \frac{J_z}{r}\, d\tau - \frac{\partial}{\partial z_o} \int_\tau \frac{J_y}{r}\, d\tau \right\}$$

$$= \sum \bar{i} \left\{ \int_\tau \frac{1}{r} \left(\frac{\partial J_z}{\partial y} - \frac{\partial J_y}{\partial z} \right) d\tau - \oint_{S_{1..n}\Sigma} \frac{1}{r} \left(J_z dS_y - J_y dS_z \right) \right\}$$

$$= -\sum \bar{i} \int_\tau \left\{ J_z \frac{\partial}{\partial y}\left(\frac{1}{r}\right) - J_y \frac{\partial}{\partial z}\left(\frac{1}{r}\right) \right\} d\tau$$

hence, at interior and exterior points of τ,

$$\operatorname{curl} \operatorname{pot} \bar{J} = \int_\tau \frac{\operatorname{curl} \bar{J}}{r} d\tau - \oint_{S_{1..n}\Sigma} d\bar{S} \times \frac{\bar{J}}{r} \tag{4.15-6}$$

or

$$\operatorname{curl} \operatorname{pot} \bar{J} = \int_\tau \bar{J} \times \operatorname{grad} \frac{1}{r} d\tau \tag{4.15-7}$$

Div pot $\bar{J}$ and curl pot $\bar{J}$ are continuous through $S_{1..n}\Sigma$ since each may be expressed as the sum of the scalar potentials of volume and surface sources multiplied, if necessary, by constant vectors.

Finally,

$$\nabla^2 \int_\tau \frac{\bar{J}}{r} d\tau = \sum \bar{i} \, \nabla^2 \int_\tau \frac{J_x}{r} d\tau$$

whence

$$\nabla^2 \operatorname{pot} \bar{J} = 0 \qquad \text{at exterior points of } \tau \tag{4.15-8}$$

and

$$\nabla^2 \operatorname{pot} \bar{J} = -4\pi\bar{J} \qquad \text{at interior points of } \tau \tag{4.15-9}$$

provided that each component of $\bar{J}$ satisfies a Hölder condition throughout τ.

Expressions for the divergence and curl of the cavity potential and partial potential may be developed from the scalar components of equations (4.8-4), (4.8-5), (4.8-7) and (4.8-8), while ∇^2 cavity pot and ∇^2 partial pot follows from (4.9-2) and (4.9-5). Grad div and curl curl partial pot are subjects of Ex.4-100. and 4-102., pp. 326-7.

The more important formulae relating to the scalar and vector volume sources are listed in Tables 1 and 2, pp. 329-333.

EXERCISES

4-95. By treating a closed line source and an open surface source as limiting configurations of a volume source, make use of equation (4.15-4) to plausibly substantiate the requirements for source density, as set down in Sec. 4.14a and Ex.4-88., p. 317, in order that the divergence of the vector potential shall be zero at exterior points.

4-96. If mixed volume and surface sources are contained within a finite region of space and are such that the divergence of the associated vector potential is everywhere zero, show that the vector potential may be expressed at interior and exterior points as

$$\bar{A} = \frac{1}{4\pi}\text{ curl pot curl } \bar{A}$$

4-97. The vector density of a solid cylindrical source of length 2c and radius a is given by $\bar{J} = \hat{z}f(\rho)$ where $\hat{z}$ is axial and ρ denotes distance from the axis. If the transverse mid plane cuts the source at $z = 0$, show from equation (4.15-4) that for $\rho^2 << c^2$ and $z^2 << c^2$

$$\text{grad div } \bar{A} \approx \frac{-2I}{c^2}\hat{z}$$

where

$$I = \int_0^a 2\pi\rho\ f(\rho)\ d\rho$$

whence

$$\text{curl curl } \bar{A} \approx 4\pi\bar{J} - \frac{2I}{c^2}\hat{z} \qquad \text{at interior points}$$

and

$$\text{curl curl } \bar{A} \approx \frac{-2I}{c^2}\hat{z} \qquad \text{at exterior points}$$

It is clear from equation (4.15-7) that curl $\bar{A}$ has no radial or axial component. By applying Stokes's theorem to curl $\bar{A}$ for the region of a transverse plane bounded by a circle centred upon the axis, show, from arguments of symmetry, that

$$(\text{curl A})_\Phi \approx \frac{2I}{\rho} - \frac{I\rho}{c^2} \approx \frac{2I}{\rho}(1-\cos^2\theta)^{\frac{1}{2}} = \frac{2I}{\rho}\sin\theta \qquad \text{at exterior points}$$

where 2θ is the angle subtended by the axis of the source at the point of evaluation of $(\text{curl } \bar{A})_\Phi$

and

$$(\text{curl } \bar{A})_\Phi \approx \frac{2I'}{\rho} - \frac{I\rho}{c^2} \qquad \text{at interior points}$$

where

$$I' = \int_0^{\rho} 2\pi\rho\ f(\rho)\ d\rho$$

Note that $(\text{curl}\ \bar{A})_\phi$ is unchanged at exterior points when the source elements are concentrated upon the axis, and that $(\text{curl}\ \bar{A})_\phi \to 0$ in the region bounded externally by a hollow cylindrical source of the above type as $c \to \infty$. [For such a source it is clear from equation (4.15-7) and considerations of symmetry that $\text{curl}\ \bar{A} = \bar{0}$ upon the axis for all values of c.]

4-98. Use the result of the previous exercise to show that

$$\frac{-\partial A_z}{\partial \rho} = 2\pi\rho J$$

at interior points of a cylindrical source of uniform axial density $\bar{J}$ when the half length $c \to \infty$, and so confirm the value of the variable component of $\bar{A}$ as derived for this case in Ex.4-82., p. 311.

4-99. A vector volume source is bounded by spherical surfaces of radii a_1 and $a_2 (a_2 > a_1)$ centred upon the origin of spherical coordinates. The source density is everywhere radial and given by $\bar{J} = \hat{R}\ f(R)$, where R is distance measured from the origin. Integrate by parts to determine the contribution of a spherical shell to the vector potential at points of greater and lesser radius, and proceed to evaluate $\bar{A}$ for $R < a_1$; $a_1 \le R \le a_2$; $R > a_2$. Find div $\bar{A}$ and grad div $\bar{A}$ at corresponding points by application of equations (2.6-7) and (2.6-5).

Confirm the results for grad div $\bar{A}$ by showing that curl curl $\bar{A} = \bar{0}$ at interior and exterior points of the source (see Ex.1-59., p. 78) and utilising equations (4.15-8) and (4.15-9).

Ans:

$$\bar{A} = \hat{R}\,\frac{4\pi}{3}\,R \int_{a_1}^{a_2} f(R)\ dR \qquad (R \le a_1)$$

$$\bar{A} = \hat{R}\,\frac{4\pi}{3} \left\{ \frac{1}{R^2} \int_{a_1}^{R} R^3\ f(R)\ dR + R \int_{R}^{a_2} f(R)\ dR \right\} \qquad (a_1 \le R \le a_2)$$

$$\overline{A} = \hat{R}\,\frac{4\pi}{3}\,\frac{1}{R^2}\int_{a_1}^{a_2} R^3\, f(R)\, dR \qquad (R \geq a_2)$$

$$\text{div}\,\overline{A} = 4\pi \int_{a_1}^{a_2} f(R)\, dR \qquad (R \leq a_1)$$

$$\text{div}\,\overline{A} = 4\pi \int_{R}^{a_2} f(R)\, dR \qquad (a_1 \leq R \leq a_2)$$

$$\text{div}\,\overline{A} = 0 \qquad (R \geq a_2)$$

$$\text{grad div}\,\overline{A} = \overline{0} \qquad (R < a_1 \text{ and } R > a_2)$$

$$\text{grad div}\,\overline{A} = -\hat{R}\, 4\pi\, f(R) = -4\pi\, \overline{J} \qquad (a_1 < R < a_2)$$

4-100. Show that

$$\text{grad div (partial)} \int_{\tau-\tau_\delta} \frac{\overline{J}}{r}\, d\tau$$

$$= \int_{\tau-\tau_\delta} \frac{1}{r}\,\text{grad div}\,\overline{J}\, d\tau - \oint_{S_{1..n}^{\Sigma}} \frac{1}{r}\,\text{div}\,\overline{J}\, d\overline{S} - \text{grad} \oint_{S_{1..n}^{\Sigma}} \frac{\overline{J}}{r} \cdot d\overline{S}$$

$$= -\int_{\tau-\tau_\delta} \text{div}\,\overline{J}\,\text{grad}\,\frac{1}{r}\, d\tau + \oint_{S_\delta} \frac{1}{r}\,\text{div}\,\overline{J}\, d\overline{S} + \oint_{S_{1..n}^{\Sigma}} \overline{J}.d\overline{S}\,\text{grad}\,\frac{1}{r}$$

By taking the x component of the above expression and making use of the volume integral of the expansion of div $\overline{J}\,\frac{\partial}{\partial x}\left(\frac{1}{r}\right)$, or by working directly from equation (1.17-15) with $\overline{F} = \overline{J}$ and $\overline{G} = \text{grad}\,\frac{1}{r}$ derive the relationship

$$\text{grad div (partial)} \int_{\tau-\tau_\delta} \frac{\bar{J}}{r}\, d\tau$$

$$= \int_{\tau-\tau_\delta} (\bar{J}.\nabla)\ \text{grad}\ \frac{1}{r}\, d\tau + \oint_{S_\delta} \left\{\frac{1}{r}\ \text{div}\ \bar{J}\ d\bar{S} - \bar{J}.d\bar{S}\ \text{grad}\ \frac{1}{r}\right\}$$

4-101. The results of the previous exercise continue to hold when τ_δ and S_δ are replaced by τ' and S', where these symbols refer to any regular region containing 0.

Evaluate $\lim\limits_{S'\to 0} \oint_{S'} \bar{J}.d\bar{S}\ \text{grad}\ \frac{1}{r}$ when S' comprises

(a) a spherical surface centred upon 0

(b) a cylindrical surface of radius a and length 2d aligned with $\bar{J}_o$ and centred upon 0

which shrinks about 0 without change of shape.

Hence conclude that the volume integral $\int_\tau (\bar{J}.\nabla)\ \text{grad}\ \frac{1}{r}\, d\tau$ is non-convergent, and that whereas we may write

$$\text{grad div} \int_\tau \frac{\bar{J}}{r}\, d\tau = -\int_\tau \text{div}\ \bar{J}\ \text{grad}\ \frac{1}{r}\, d\tau + \oint_{S_{1..n}\Sigma} \bar{J}.d\bar{S}\ \text{grad}\ \frac{1}{r}$$

at interior points of τ, it is necessary to express the alternative form as

$$\text{grad div} \int_\tau \frac{\bar{J}}{r}\, d\tau = \lim_{\tau'\to 0} \int_{\tau-\tau'} (\bar{J}.\nabla)\ \text{grad}\ \frac{1}{r}\, d\tau - \lim_{S'\to 0} \oint_{S'} \bar{J}.d\bar{S}\ \text{grad}\ \frac{1}{r}$$

Ans: $\lim\limits_{S'\to 0} \oint_{S'} \bar{J}.d\bar{S}\ \text{grad}\ \frac{1}{r} = \frac{4}{3}\pi\,\bar{J}_o$ for sphere.

$= 4\pi\,\bar{J}_o \left\{1 - \frac{d}{(a^2+d^2)^{\frac{1}{2}}}\right\}$ for cylinder.

4-102. Show that

$$\text{curl curl (partial)} \int_{\tau-\tau_\delta} \frac{\bar{J}}{r}\, d\tau$$

$$= \int_{\tau-\tau_\delta} \frac{1}{r}\,\text{curl curl}\,\bar{J}\, d\tau - \oint_{S_{1..n}\Sigma} \frac{1}{r}\, d\bar{S} \times \text{curl}\,\bar{J} - \text{curl} \oint_{S_{1..n}\Sigma} d\bar{S} \times \frac{\bar{J}}{r}$$

$$= -\int_{\tau-\tau_\delta} \left(\text{grad}\,\frac{1}{r} \times \text{curl}\,\bar{J}\right) d\tau + \oint_{S_\delta} \frac{1}{r}\,(d\bar{S} \times \text{curl}\,\bar{J}) + \oint_{S_{1..n}\Sigma} \text{grad}\,\frac{1}{r} \times (d\bar{S}\times\bar{J})$$

Make use of equation (1.17-17) with $\bar{F} = \bar{J}$ and $\bar{G} = \text{grad}\,\frac{1}{r}$ to transform the equation into

$$\text{curl curl (partial)} \int_{\tau-\tau_\delta} \frac{\bar{J}}{r}\, d\tau$$

$$= \int_{\tau-\tau_\delta} (\bar{J}.\nabla)\,\text{grad}\,\frac{1}{r}\, d\tau + \oint_{S_\delta} \left\{\frac{1}{r}\,(d\bar{S} \times \text{curl}\,\bar{J}) - \text{grad}\,\frac{1}{r} \times (d\bar{S}\times\bar{J})\right\}$$

4-103. Evaluate $\lim_{S'\to 0} \oint_{S'} \text{grad}\,\frac{1}{r} \times (d\bar{S}\times\bar{J})$ when S' comprises

(a) a spherical surface centred upon 0

(b) a cylindrical surface of radius a and length 2d aligned with $\bar{J}_o$ and centred upon 0

which shrinks about 0 without change of shape.

Hence conclude from the results of the previous exercise that

$$\text{curl curl} \int_\tau \frac{\bar{J}}{r}\, d\tau = -\int_\tau \text{grad}\,\frac{1}{r} \times \text{curl}\,\bar{J}\, d\tau + \oint_{S_{1..n}\Sigma} \text{grad}\,\frac{1}{r} \times (d\bar{S}\times\bar{J})$$

at interior points of τ, and that

$$\text{curl curl} \int \frac{\bar{J}}{r}\, d\tau = \lim_{\tau'\to 0} \int (\bar{J}.\nabla)\,\text{grad}\,\frac{1}{r}\, d\tau - \lim_{S'\to 0} \oint \text{grad}\,\frac{1}{r} \times (d\bar{S}\times\bar{J})$$

Ans: $$\lim_{S'\to 0} \oint_{S'} \text{grad}\,\frac{1}{r} \times (d\bar{S}\times\bar{J}) = -\frac{8}{3}\pi\bar{J}_o \quad \text{for sphere.}$$

$$= -4\pi\bar{J}_o\,\frac{d}{(a^2+d^2)^{\frac{1}{2}}} \quad \text{for cylinder.}$$

4-104. Combine the results of Ex.4-100. and 4-101. to show that

$$\nabla^2 \text{ (partial)} \int_{\tau-\tau_\delta} \frac{\bar{J}}{r}\,d\tau = \oint_{S_\delta} \left\{ \text{div}\,\frac{\bar{J}}{r}\,d\bar{S} - d\bar{S}\times\text{curl}\,\frac{\bar{J}}{r} - 2\bar{J}\,d\bar{S}.\text{grad}\,\frac{1}{r} \right\}$$

and make use of equation (1.17-13) to reduce this to

$$\nabla^2 \text{ (partial)} \int_{\tau-\tau_\delta} \frac{\bar{J}}{r}\,d\tau = \oint_{S_\delta} \left\{ (d\bar{S}.\nabla)\,\frac{\bar{J}}{r} - 2\bar{J}\,d\bar{S}.\text{grad}\,\frac{1}{r} \right\}$$

$$= \oint_{S_\delta} \left\{ \frac{1}{r}\frac{\partial\bar{J}}{\partial n} - \bar{J}\frac{\partial}{\partial n}\left(\frac{1}{r}\right) \right\} dS$$

4-105. Transform $(\bar{J}.\nabla)\,\text{grad}\,\frac{1}{r}$ into $\text{grad}\left(\bar{J}.\nabla\left(\frac{1}{r}\right)\right)$ and make use of (1.17-5) to show that the value of

$$\int_{\tau-\tau'} (\bar{J}.\nabla)\,\text{grad}\,\frac{1}{r}\,d\tau - \oint_{S'} \bar{J}.\text{grad}\,\frac{1}{r}\,d\bar{S}$$

is independent of the shape and size of S'. Hence demonstrate that if the shape and orientation of S' and its configuration with respect to 0 are maintained constant during the limiting process, then the value of

$$\lim_{\tau'\to 0} \int_{\tau-\tau'} (\bar{J}.\nabla)\,\text{grad}\,\frac{1}{r}\,d\tau - \lim_{S'\to 0} \oint_{S'} \bar{J}.\text{grad}\,\frac{1}{r}\,d\bar{S}$$

will be independent of the values of the individual limits.

Show further that this continues to hold when $\bar{J}.\text{grad}\,\frac{1}{r}\,d\bar{S}$ is replaced by $\bar{J}.d\bar{S}\,\text{grad}\,\frac{1}{r}$ or $d\bar{S}\times\left(\text{grad}\,\frac{1}{r}\times\bar{J}\right)$ or $(\bar{J}\times d\bar{S})\times\text{grad}\,\frac{1}{r}$.

TABLE 1

The Scalar Potential Function $\int \frac{\rho}{r}\, d\tau$ and its Derivatives

(1)

$$\text{pot}\ \rho = \int_{\tau} \frac{\rho}{r}\, d\tau$$ (interior and exterior points of τ)

(2)

$$\text{partial pot}\ \rho = \int_{\tau-\tau_\delta} \frac{\rho}{r}\, d\tau$$ (evaluated at centre of moving δ sphere within τ)

(3)

$$\text{cavity pot}\ \rho = \int_{\tau-\tau_\delta} \frac{\rho}{r}\, d\tau$$ (defined throughout fixed δ sphere within τ)

(4)

$$\text{grad pot}\ \rho = \int_{\tau} \frac{\text{grad}\ \rho}{r}\, d\tau - \oint_{S_{1..n}{}^{\Sigma}} \frac{\rho}{r}\, d\bar{S} = -\int_{\tau} \rho\ \text{grad}\ \frac{1}{r}\, d\tau$$ (interior and exterior points of τ)

(5)

$$\text{grad partial pot}\ \rho = \int_{\tau-\tau_\delta} \frac{\text{grad}\ \rho}{r}\, d\tau - \oint_{S_{1..n}{}^{\Sigma}} \frac{\rho}{r}\, d\bar{S} = -\int_{\tau-\tau_\delta} \rho\ \text{grad}\ \frac{1}{r}\, d\tau + \oint_{S_\delta} \frac{\rho}{r}\, d\bar{S}$$

(6)

$$\text{grad cavity pot}\ \rho = \int_{\tau-\tau_\delta} \frac{\text{grad}\ \rho}{r}\, d\tau - \oint_{S_{1..n}{}^{\Sigma},\ S_\delta} \frac{\rho}{r}\, d\bar{S} = -\int_{\tau-\tau_\delta} \rho\ \text{grad}\ \frac{1}{r}\, d\tau$$

(7)

$$\nabla^2\ \text{pot}\ \rho = 0$$ at exterior points of τ

(8)

$$\nabla^2\ \text{pot}\ \rho = -4\pi\rho$$ at interior points of τ

(9)

$$\nabla^2\ \text{pot}\ \rho = \oint_{S_{1..n}{}^{\Sigma}} \left\{ \rho \frac{\partial}{\partial n}\left(\frac{1}{r}\right) - \frac{1}{r}\frac{\partial \rho}{\partial n} \right\} dS + \text{pot}\ \nabla^2 \rho$$ (interior and exterior points of τ)

TABLE 1(CONTD.)

(10)

$$\nabla^2 \text{ partial pot } \rho = \oint_{S_{1..n}\Sigma} \left\{ \rho \frac{\partial}{\partial n}\left(\frac{1}{r}\right) - \frac{1}{r}\frac{\partial \rho}{\partial n} \right\} dS + \text{partial pot } \nabla^2 \rho$$

$$= \oint_{S_\delta} \left\{ \frac{1}{r}\frac{\partial \rho}{\partial n} - \rho \frac{\partial}{\partial n}\left(\frac{1}{r}\right) \right\} dS$$

(11)

$$\nabla^2 \text{ cavity pot } \rho = 0$$

TABLE 2

The Vector Potential Function $\int \frac{\overline{J}}{r}\, d\tau$ and its Derivatives

(1)

$$\text{pot } \overline{J} = \int_\tau \frac{\overline{J}}{r}\, d\tau \quad \text{(interior and exterior points of } \tau)$$

(2)

$$\text{partial pot } \overline{J} = \int_{\tau-\tau_\delta} \frac{\overline{J}}{r}\, d\tau \quad \text{(evaluated at centre of moving } \delta \text{ sphere within } \tau)$$

(3)

$$\text{cavity pot } \overline{J} = \int_{\tau-\tau_\delta} \frac{\overline{J}}{r}\, d\tau \quad \text{(defined throughout fixed } \delta \text{ sphere within } \tau)$$

(4)

$$\text{div pot } \overline{J} = \int_\tau \frac{\text{div } \overline{J}}{r}\, d\tau - \oint_{S_{1..n}\Sigma} \frac{\overline{J}}{r} \cdot d\overline{S} = -\int_\tau \overline{J} \cdot \text{grad } \frac{1}{r}\, d\tau \quad \text{(interior and exterior points of } \tau)$$

(5)

$$\text{div partial pot } \overline{J} = \int_{\tau-\tau_\delta} \frac{\text{div } \overline{J}}{r}\, d\tau - \oint_{S_{1..n}\Sigma} \frac{\overline{J}}{r} \cdot d\overline{S} = -\int_{\tau-\tau_\delta} \overline{J} \cdot \text{grad } \frac{1}{r}\, d\tau + \oint_{S_\delta} \frac{\overline{J}}{r} \cdot d\overline{S}$$

(6)

$$\text{div cavity pot } \overline{J} = \int_{\tau-\tau_\delta} \frac{\text{div } \overline{J}}{r}\, d\tau - \oint_{S_{1..n}\Sigma,\ S_\delta} \frac{\overline{J}}{r} \cdot d\overline{S} = -\int_{\tau-\tau_\delta} \overline{J} \cdot \text{grad } \frac{1}{r}\, d\tau$$

TABLE 2(CONTD.)

(7)

$$\text{curl pot } \bar{J} = \int_{\tau} \frac{\text{curl } \bar{J}}{r} d\tau - \oint_{S_{1..n}\Sigma} d\bar{S} \times \frac{\bar{J}}{r} = \int_{\tau} \bar{J} \times \text{grad } \frac{1}{r} d\tau$$ (interior and exterior points of τ)

(8)

$$\text{curl partial pot } \bar{J} = \int_{\tau-\tau_\delta} \frac{\text{curl } \bar{J}}{r} d\tau - \oint_{S_{1..n}\Sigma} d\bar{S} \times \frac{\bar{J}}{r} = \int_{\tau-\tau_\delta} \bar{J} \times \text{grad } \frac{1}{r} d\tau + \oint_{S_\delta} d\bar{S} \times \frac{\bar{J}}{r}$$

(9)

$$\text{curl cavity pot } \bar{J} = \int_{\tau-\tau_\delta} \frac{\text{curl } \bar{J}}{r} d\tau - \oint_{S_{1..n}\Sigma,\ S_\delta} d\bar{S} \times \frac{\bar{J}}{r} = \int_{\tau-\tau_\delta} \bar{J} \times \text{grad } \frac{1}{r} d\tau$$

(10)

∇^2 pot $\bar{J}$ = 0 at exterior points of τ

(11)

∇^2 pot $\bar{J}$ = $-4\pi\bar{J}$ at interior points of τ

(12)

$$\nabla^2 \text{ pot } \bar{J} = \oint_{S_{1..n}\Sigma} \left\{ \bar{J} \frac{\partial}{\partial n}\left(\frac{1}{r}\right) - \frac{1}{r}\frac{\partial \bar{J}}{\partial n} \right\} dS + \text{pot } \nabla^2 \bar{J}$$ (interior and exterior points of τ)

(13)

$$\nabla^2 \text{ partial pot } \bar{J} = \oint_{S_{1..n}\Sigma} \left\{ \bar{J} \frac{\partial}{\partial n}\left(\frac{1}{r}\right) - \frac{1}{r}\frac{\partial \bar{J}}{\partial n} \right\} dS + \text{partial pot } \nabla^2 \bar{J}$$

$$= \oint_{S_\delta} \left\{ \frac{1}{r}\frac{\partial \bar{J}}{\partial n} - \bar{J} \frac{\partial}{\partial n}\left(\frac{1}{r}\right) \right\} dS$$

(14)

∇^2 cavity pot $\bar{J}$ = 0

(15)

$$\text{grad div partial pot } \bar{J} = \int_{\tau-\tau_\delta} \frac{\bar{r}}{r^3} \text{ div } \bar{J}\, d\tau + \oint_{S_\delta} \frac{1}{r} \text{ div } \bar{J}\, d\bar{S} - \oint_{S_{1..n}\Sigma} \frac{\bar{r}}{r^3} \bar{J}.d\bar{S}$$

TABLE 2(CONTD.)

(16)

$$\text{grad div partial pot } \bar{J} = \int_{\tau-\tau_\delta} (\bar{J}.\nabla)\,\text{grad}\,\frac{1}{r}\,d\tau + \oint_{S_\delta} \left\{\frac{1}{r}\,\text{div}\,\bar{J}\,d\bar{S} + \frac{\bar{r}}{r^3}\,\bar{J}.d\bar{S}\right\}$$

(17)

$$\text{grad div pot } \bar{J} = \int_{\tau} \frac{\bar{r}}{r^3}\,\text{div}\,\bar{J}\,d\tau - \oint_{S_{1..n}^{\Sigma}} \frac{\bar{r}}{r^3}\,\bar{J}.d\bar{S} \quad \text{(interior and exterior points of } \tau)$$

(18)

$$\text{grad div pot } \bar{J} = \int_{\tau} (\bar{J}.\nabla)\,\text{grad}\,\frac{1}{r}\,d\tau \quad \text{at exterior points of } \tau$$

(19)

$$\text{grad div pot } \bar{J} = \lim_{\tau'\to 0} \int_{\tau-\tau'} (\bar{J}.\nabla)\,\text{grad}\,\frac{1}{r}\,d\tau + \lim_{S'\to 0} \oint_{S'} \frac{\bar{r}}{r^3}\,\bar{J}.d\bar{S} \quad \text{at interior points of } \tau$$

(20)

$$\text{curl curl partial pot } \bar{J} = \int_{\tau-\tau_\delta} \left(\frac{\bar{r}}{r^3} \times \text{curl}\,\bar{J}\right) d\tau + \oint_{S_\delta} \frac{1}{r}\,(d\bar{S} \times \text{curl}\,\bar{J}) - \oint_{S_{1..n}^{\Sigma}} \frac{\bar{r}}{r^3} \times (d\bar{S}\times\bar{J})$$

(21)

$$\text{curl curl partial pot } \bar{J} = \int_{\tau-\tau_\delta} (\bar{J}.\nabla)\,\text{grad}\,\frac{1}{r}\,d\tau + \oint_{S_\delta} \left\{\frac{1}{r}\,(d\bar{S} \times \text{curl}\,\bar{J}) + \frac{\bar{r}}{r^3} \times (d\bar{S}\times\bar{J})\right\}$$

(22)

$$\text{curl curl pot } \bar{J} = \int_{\tau} \left(\frac{\bar{r}}{r^3} \times \text{curl}\,\bar{J}\right) d\tau - \oint_{S_{1..n}^{\Sigma}} \frac{\bar{r}}{r^3} \times (d\bar{S}\times\bar{J}) \quad \text{(interior and exterior points of } \tau)$$

(23)

$$\text{curl curl pot } \bar{J} = \int_{\tau} (\bar{J}.\nabla)\,\text{grad}\,\frac{1}{r}\,d\tau \quad \text{at exterior points of } \tau$$

TABLE 2(CONTD.)

(24)

$$\text{curl curl pot } \bar{J} = \lim_{\tau'\to 0} \int_{\tau-\tau'} (\bar{J}.\nabla)\,\text{grad}\,\frac{1}{r}\,d\tau + \lim_{S'\to 0} \oint_{S'} \frac{\bar{r}}{r^3} \times (d\bar{S}\times\bar{J}) \quad \text{at interior points of } \tau.$$

(It is, of course, assumed in expressions of this type that the dimensions of S' are reduced in such a manner that the limit exists.)

4.16 Equivalent Layers and Image Systems in Vector Potential Theory

The analyses of Sec. 4.5 may be readily extended, via equation (3.3-4), to the representation of vector point functions as the vector potentials of simple and double layer surface sources and of volume sources, both for bounded and unbounded regions of space. This leads directly to a theory of equivalent layers for those functions defined primarily as vector potentials. It should be noted that the surface densities of such equivalent layers are not, in general, directed parallel to the surfaces.

Surface sources are often considered to be equivalent to exterior sources if the curls of the associated vector potentials (rather than the potentials themselves) are identical within the bounded region. In this connection let $\bar{V}$ be the curl of the vector potential $\bar{A}$ deriving from sources exterior to the region $\underline{R}$ bounded by the surfaces $S_{1..n}\Sigma$. Suppose that $\text{div}\,\bar{A} = 0$ beyond the sources[26]. Then $\bar{V}$ is solenoidal in $\underline{R}$ because $\oint (\text{curl}\,\bar{A}).d\bar{S} = 0$ for all interior surfaces, and $\text{curl}\,\bar{V} = \text{grad div}\,\bar{A} - \nabla^2\bar{A} = \bar{0}$. Now it is shown in Sec. 4.17[27] that in these circumstances

$$4\pi\bar{V}_o = -\,\text{curl} \oint_{S_1} \frac{(\hat{\bar{n}}\times\bar{V})+(\hat{\bar{n}}_1\times\bar{V}_1)}{r}\,dS \;----\; -\,\text{curl} \oint_{\Sigma} \frac{(\hat{\bar{n}}\times\bar{V})+(\hat{\bar{n}}_\Sigma\times\bar{V}_\Sigma)}{r}\,dS \tag{4.16-1}$$

where the auxiliary functions $\bar{V}_1 --- \bar{V}_\Sigma$ are the gradient fields of certain scalar point functions $U_1 --- U_\Sigma$ defined within $\tau_1 --- \tau_\Sigma$, and $\hat{\bar{n}}_1 --- \hat{\bar{n}}_\Sigma$ are the corresponding outward normals.

Equation (4.16-1) represents $\bar{V}$ as the curl of the vector potential of surface sources on $S_{1..n}\Sigma$; such sources consequently comprise equivalent layers in the extended sense of the term. It is apparent that the surface densities are tangential, irrespective of the orientation of $\bar{V}$. Their surface divergence is zero, as we now proceed to show.

---

26. See Sec. 4.19.

27. See also Ex.4-115., p. 341.

Let $\overline{K}_1 = (\hat{n}\times\overline{V}) + (\hat{n}_1\times\overline{V}_1)$. The outward flux of $\overline{K}_1$ through a small closed curve Γ_1 drawn upon S_1 is given by $\oint_{\Gamma_1} \overline{K}_1.\hat{n}_1' \, ds$, where $\hat{n}_1'$ is tangential to S_1 and outwardly normal to Γ_1. This is equal to

$$\oint_{\Gamma_1} \hat{n} \times (\overline{V}-\overline{V}_1).\hat{n}_1' \, ds = \oint_{\Gamma_1} (\hat{n}_1'\times\hat{n}).(\overline{V}-\overline{V}_1) \, ds = \oint_{\Gamma_1} (\overline{V}_1-\overline{V}).d\overline{r}$$

$$= \int (\text{curl } \overline{V}_1).d\overline{S} - \int (\text{curl } \overline{V}).d\overline{S}$$

where the surface integrations are carried out just inside τ_1 and τ respectively, $\overline{V}_1$ and $\overline{V}$ having continuous derivatives up to the common boundary.

But curl $\overline{V}$ is zero in τ and curl $\overline{V}_1$ = curl grad $U_1 = \overline{0}$ in τ_1, hence the surface integrals vanish and the flux of $\overline{K}_1$ through Γ_1 is zero. Since $\overline{K}_1$ is well-behaved it follows that divs $\overline{K}_1 = 0$ on S_1. Identical arguments apply for the other surfaces.

The concept of equivalence is sometimes carried still further by postulating that scalar surface sources are equivalent to exterior vector sources when the negative[28] gradient of the scalar potential is identical with the curl of the vector potential at interior points of the bounded region.

If curl $\overline{A} = \overline{V}$ then, with div $\overline{A} = 0$, curl $\overline{V} = \overline{0}$ in $\underline{R}$, whence for $\underline{R}$ simply connected, $\overline{V} = -$ grad ψ, where

$$\psi_Q = -\int_P^Q \overline{V}.d\overline{r}, \qquad P \text{ being fixed in } \underline{R}.$$

But div $\overline{V}$ = div curl $\overline{A} = 0$ in $\underline{R}$, hence $\nabla^2\psi = 0$, and

$$\psi_o = \oint_{S_{1..n}\Sigma} \left\{\frac{1}{4\pi}\,\frac{1}{r}\,\frac{\partial\psi}{\partial n} - \frac{\psi}{4\pi}\,\frac{\partial}{\partial n}\left(\frac{1}{r}\right)\right\} dS$$

28. The reason for this choice of sign, which is clearly arbitrary in the present context, need not concern us here.

whence

$$\overline{V}_o = -\text{grad}\oint_{S_{1..n}\Sigma} \frac{1}{4\pi}\frac{1}{r}\frac{\partial\psi}{\partial n}\,dS - \text{grad}\oint_{S_{1..n}\Sigma} \frac{-\psi}{4\pi}\frac{\partial}{\partial n}\left(\frac{1}{r}\right)dS \tag{4.16-2}$$

Further, $\overline{V}$ is solenoidal in $\underline{R}$ since $\oint (\text{curl}\,\overline{A}).d\overline{S} = 0$ for all closed surfaces, hence $\oint \frac{\partial\psi}{\partial n}\,dS = 0$ over S_1---S_n in turn. The Neumann problem may consequently be solved for each of the regions τ_1---τ_n, and curl $\overline{A}$ may be expressed as the negative gradient of the potential of simple and/or double layer sources of $S_{1..n}\Sigma$.

It is clear that this result holds for any point function which is irrotational and solenoidal in $\underline{R}$.

It will be recalled that the method of images in scalar potential theory is concerned with the determination of an exterior scalar source system which gives rise, at interior points of a bounded region $\underline{R}$, to a potential (or potential gradient) identical with that deriving from specified surface sources, or their equivalents, on the boundary of $\underline{R}$. The concept of image equivalence in vector potential theory is rather wider since mixed sources may be involved. Consideration will be restricted, in the present instance, to the determination of an exterior vector source system which gives rise, at interior points of $\underline{R}$, to

(1) a vector potential which is identical with that deriving from specified boundary sources

or (2) a vector potential whose curl is identical with the curl of the potential deriving from specified boundary sources

or (3) a vector potential whose curl is identical with the negative gradient of the scalar potential of specified boundary sources.

It will be supposed that the divergence of each component vector potential is zero outside the source.

As in the case of scalar image problems, the surface densities are specified indirectly by the requirement that the combined potentials of these and other sources, or their derivatives, satisfy certain boundary conditions[29].

The relevant uniqueness criteria have been established in Sec. 3.7. Let $\overline{A}_e$ and $\overline{A}_s$ represent respectively the vector potentials of exterior (image) and surface sources, and let ϕ_s be the scalar potential of surface sources.

29. Some of these conditions have their origin in the physical behaviour of polarised systems and can best be appreciated in that context. (See Ex.4-109. to 4-113., pp. 339-40.)

(1) Since $\nabla^2\overline{A}_e = \nabla^2\overline{A}_s = \overline{0}$ at interior points of $\underline{R}$, it follows from Theorem 3.7-3 that $\overline{A}_e = \overline{A}_s$ in $\underline{R}$ if $\overline{A}_e = \overline{A}_s$ at all points of $S_{1..n}\Sigma$. This is true whether or not $\underline{R}$ is simply connected. When $\underline{R}$ is bounded internally but not externally the equality will continue to hold if $R|\overline{A}_e-\overline{A}_s|$ and $R^2 \left|\frac{\partial}{\partial R}(\overline{A}_e-\overline{A}_s)\right|$ are bounded at infinity.

It may be noted in this connection that when the sources comprise parallel systems of cylinders and lines of axially-directed density, the magnitude of the vector potential is everywhere equal to the scalar potential deriving from sources of equal magnitude. As a result, the vector equivalents of Ex.4-69. to 4-71., p. 300 may be propounded and solved by substitution of $\overline{A}$ for ϕ, $\overline{I}$ for λ and $\overline{K}$ for σ, etc, without recourse to a uniqueness theorem for vector fields.

(2) When it is required that curl $\overline{A}_e$ = curl $\overline{A}_s$ in $\underline{R}$, rather than $\overline{A}_e = \overline{A}_s$, a relaxation of the boundary requirement is possible. On writing curl $\overline{A}_e = \overline{V}_e$ and curl $\overline{A}_s = \overline{V}_s$ we have

$$\left.\begin{aligned} \text{curl } \overline{V}_e &= \text{curl } \overline{V}_s = \overline{0} \\ \text{div } \overline{V}_e &= \text{div } \overline{V}_s = 0 \end{aligned}\right\} \text{ in } \underline{R}$$

It then follows from Theorem 3.7-1 (with $\overline{V}$ substituted for $\overline{F}$) that

curl $\overline{A}_e$ = curl $\overline{A}_s$ in a simply connected region $\underline{R}$

provided that either

$$\text{(a)} \quad \hat{n} \cdot \text{curl } \overline{A}_e = \hat{n} \cdot \text{curl } \overline{A}_s \text{ on } S_{1..n}\Sigma \tag{4.16-3}$$

or

$$\text{(b)} \quad \hat{n} \times \text{curl } \overline{A}_e = \hat{n} \times \text{curl } \overline{A}_s \text{ on } S_{1..n}\Sigma \tag{4.16-4}$$

it being understood that curl $\overline{A}_s$ is to be evaluated just inside $\underline{R}$. The additional requirement postulated by Theorem 3.7-1 in connection with (b), viz $\oint (\text{curl } \overline{A}_e).d\overline{S} = \oint (\text{curl } \overline{A}_s).d\overline{S}$ for each surface in turn, is automatically satisfied since each integral is zero.

In addition, it is easily shown that

$$\text{curl}\ \overline{A}_e = \text{curl}\ \overline{A}_s \quad \text{in a multiply connected region } \underline{R}$$

provided that either

(c) $\text{curl}\ (\overline{A}_e - \overline{A}_s)$ is irrotational in $\underline{R}$
and $\hat{n} \cdot \text{curl}\ \overline{A}_e = \hat{n} \cdot \text{curl}\ \overline{A}_s$ on $S_{1..n}\Sigma$ (4.16-5)

or

(d) $\hat{n} \times \text{curl}\ \overline{A}_e = \hat{n} \times \text{curl}\ \overline{A}_s$ on $S_{1..n}\Sigma$ (4.16-6)

(3) Let $\text{curl}\ \overline{A}_e = \overline{V}_e$ and $-\text{grad}\ \phi_s = \overline{V}_s$. Then

$$\left.\begin{aligned} \text{curl}\ \overline{V}_e &= \text{curl}\ \overline{V}_s = \overline{0} \\ \text{div}\ \overline{V}_e &= \text{div}\ \overline{V}_s = 0 \end{aligned}\right\} \quad \text{in } \underline{R}$$

whence, from Theorem 3.7-1,

$$\text{curl}\ \overline{A}_e = -\text{grad}\ \phi_s \quad \text{in a simply connected region } \underline{R}$$

provided that either

(a) $\hat{n} \cdot \text{curl}\ \overline{A}_e = -\hat{n} \cdot \text{grad}\ \phi_s$ on $S_{1..n}\Sigma$ (4.16-7)

or

(b) $\hat{n} \times \text{curl}\ \overline{A}_e = -\hat{n} \times \text{grad}\ \phi_s$ on $S_{1..n}\Sigma$ (4.16-8)
and $\oint \text{grad}\ \phi_s . d\overline{S} = 0$ for each surface in turn.

[The latter condition stems from the requirement that $\oint \text{curl}\ \overline{A}_e . d\overline{S} = \oint -\text{grad}\ \phi_s . d\overline{S}$ for each surface.]

Further,

$$\text{curl}\ \overline{A}_e = -\text{grad}\ \phi_s \quad \text{in a multiply connected region } \underline{R}$$

provided that either

(c) $\text{curl}\ \overline{A}_e$ is irrotational in $\underline{R}$
and $\hat{n} \cdot \text{curl}\ \overline{A}_e = -\hat{n} \cdot \text{grad}\ \phi_s$ on $S_{1..n}\Sigma$ (4.16-9)

or

$$\text{(d)} \quad \hat{\bar{n}} \times \text{curl } \bar{A}_e = - \hat{\bar{n}} \times \text{grad } \phi_s \text{ on } S_{1..n}\Sigma \qquad (4.16\text{-}10)$$

$$\text{and } \oint \text{grad } \phi_s . d\bar{S} = 0 \quad \text{for each surface in turn.}$$

In the event that $\underline{R}$ is unbounded externally in cases (1) and (2) above, the conditions for equality are maintained provided that $R^2 \; |\text{curl}(\bar{A}_e - \bar{A}_s)|$ and $R^2 \; |\text{curl } \bar{A}_e + \text{grad } \phi_s|$ are bounded at infinity.

EXERCISES

4-106. Extend the analyses of Sec. 4.5 to the representation of a vector point function as the vector potential of simple and double surface sources and of volume sources, both for bounded and unbounded regions of space, and rework Sec. 4.10 in terms of vector point functions.

4-107. Two cylindrical vector surface sources of length 2c, radii a_1 and a_2, and centre-to-centre spacing g ($>a_1+a_2$) are bisected by the same transverse plane. The surface density $\bar{K}$ is everywhere directed axially, without axial variation, and is such as to maintain each surface equipotential in the presence of the other, within finite distance of the source centre, as $c\to\infty$. The centres of the circles in which the cylinders cut the transverse plane are P and Q, corresponding to radii a_1 and a_2, and the strengths of the cylinders per unit length (ie $\oint \bar{K}\, ds$ taken around each periphery) are $+\hat{\bar{z}}\, I$ and $-\hat{\bar{z}}\, I$ respectively.

Show that the value of the vector potential, at points outside the cylinders, is identical with that deriving from axial line sources of density $+\hat{\bar{z}}\, I$ and $-\hat{\bar{z}}\, I$ displaced inwards from P and Q by

$$\frac{g^2 + a_1^2 - a_2^2 - \sqrt{C}}{2g} \quad \text{and} \quad \frac{g^2 - a_1^2 + a_2^2 - \sqrt{C}}{2g}$$

where $C = g^4 - 2g^2\,(a_1^2+a_2^2) + (a_2^2-a_1^2)^2$

Make use of the result of Ex.4-93., p. 320 and of Ex.4-73., p. 301, appropriately transformed for a two-dimensional field in space, to determine the magnitude of the flux of curl $\bar{A}$ between the cylinders per unit axial length.

[Hint: Indentify the cross-section in the central plane with the circles in Ex.4-62., p. 292, taking P adjacent to A and Q adjacent to B. Show that $OP^2 = d^2 + a_1^2$ and $OQ^2 = d^2 + a_2^2$. Hence derive $d = \sqrt{C}/2g$.]

Ans: Flux magnitude per unit axial length $= 2I \cosh^{-1} \dfrac{(g^2-a_1^2-a_2^2)}{2a_1 a_2}$

4-108. Confirm the requirements for image equivalence in multiply connected regions as set down in equations (4.16-5/6) and (4.16-9/10). Show that identical requirements follow from the results of Ex.3-26., p. 198, and that a further sufficient condition for the equality of curl $\bar{A}$ and

4-109. A rectilinear source Γ of uniform tangential density $\bar{I}$ lies parallel to and outside a cylindrical surface source of radius a, centred upon the z axis of cylindrical coordinates. The surface density $\bar{K}$ is axially directed and a function of Φ alone. It is so distributed that $(\text{curl}(\bar{A}_\Gamma+\bar{A}_s))_\Phi$ increases α times with normal movement through the surface into the cylinder at any point, where $\bar{A}_\Gamma$ and $\bar{A}_s$ denote the vector potentials of the line and surface sources and α is a constant greater than unity.

Show that the divergence of $\bar{A}$ for the individual sources approaches zero at all points outside them, and within finite distance of a local origin, as the source lengths extend to infinity in both directions.

Use the result of Ex.4-94., p. 320, to show that the boundary requirement stated above leads to the relationship

$$2\pi K = -\frac{\alpha - 1}{\alpha + 1}\left\{(\text{curl } A_\Gamma)_\Phi + \frac{1}{a}\oint K\, ds\right\}$$

By taking the tangential line integral of both sides of this equation around the periphery and integrating the normal component of curl curl $\bar{A}_\Gamma$ over a surface which spans the contour, deduce that $\oint K\, ds = 0$, ie the source strength per unit length of cylinder is zero. Hence show that the tangential component of the curl of the vector potential of the surface source is duplicated just inside the cylinder by that of a line source coincident with Γ and of density $\frac{\alpha - 1}{\alpha + 1}\bar{I}$.

Develop a uniqueness criterion for the two-dimensional fields under consideration and so prove that

$$\text{curl } (\bar{A}_\Gamma+\bar{A}_s) = \frac{2\alpha}{1 + \alpha}\text{ curl } \bar{A}_\Gamma$$

at interior points of the cylinder.

4-110. Show that the conditions laid down in equations (4.16-5/6) for the equality of curl $\bar{A}_e$ and curl $\bar{A}_s$ in a multiply connected region bounded by a finite source system remain sufficient for the region bounded internally by the infinite cylindrical source system of the previous exercise if supplemented by the requirement that $\rho^n|\text{curl}(\bar{A}_e-\bar{A}_s)|$ be bounded for large values of ρ when $n > 1$. Hence prove that the image complex whose curl pot duplicates that of the cylindrical surface source at $\rho > a$ comprises a line source inverse to Γ in the cylinder and of density $\frac{\alpha - 1}{\alpha + 1}\bar{I}$, together with an equal and opposite line source upon the axis.

[Hint: Evaluate $(\text{curl } A)_\Phi = \frac{-\partial A_z}{\partial\rho}$ for object and image systems at $\rho = a$.]

4-111. Suppose that in Ex.4-109. and 4-110. the vector surface source is replaced by a scalar source whose density σ is a function of Φ alone, and is so distributed that the normal component of curl $\bar{A}_\Gamma$ - grad ϕ_s is reduced by α times with normal movement through the surface into the cylinder at any point.

Use the result of Ex.4-37., p. 269 to show that

$$2\pi\sigma = \frac{\alpha - 1}{\alpha + 1}\left\{\frac{q}{a} + \hat{n} \cdot \text{curl } \bar{A}_\Gamma\right\}$$

where q is the strength/unit length of the cylinder and $\hat{n}$ is the outward normal.

Prove that q is zero by integration of both sides of the above equation around the cylinder and by an application of the divergence theorem to a closed cylindrical surface of unit length. Then show that the negative gradient of the surface potential is matched at interior points by the curl of the vector potential of a line source of density $-\frac{\alpha - 1}{\alpha + 1}\bar{I}$ coincident with Γ, and conclude that

$$\text{curl } \bar{A}_\Gamma - \text{grad } \phi_s = \frac{2}{\alpha + 1}\text{ curl } \bar{A}_\Gamma$$

at points within the cylinder.

Now use the relationship $\hat{n} \cdot \text{curl } \bar{A} = \frac{1}{\rho}\frac{\partial A_z}{\partial \Phi}$ to show that the negative gradient of the surface potential is duplicated just outside the cylinder by the curl of the vector potential of a line source of density $\frac{\alpha - 1}{\alpha + 1}\bar{I}$, inverse to Γ in the cylinder, and confirm that the image system for the exterior region is completed by the addition of a line source of density $-\frac{\alpha - 1}{\alpha + 1}\bar{I}$ upon the axis.

4-112. A uniform, closed, tangential line source Γ of magnitude I is situated to one side of an infinite, plane, tangential surface source whose density $\bar{K}$ is such that the tangential component of curl$(\bar{A}_\Gamma + \bar{A}_s)$ is increased α times with normal movement through the surface from the half space $\underline{R}$ containing Γ to the half space $\underline{R}'$. Use a result of Ex.4-85., p. 315 to show that the surface divergence of $\bar{K}$ is everywhere zero.

4-113. On the assumption that $R^2|\text{curl}(\bar{A}_\Gamma - \bar{A}_s)|$ is bounded upon the surface of a hemisphere of local origin as its radius approaches infinity, show that the curl of the vector potential of the surface source in the above exercise is duplicated in $\underline{R}'$ by that of a line source of magnitude $\frac{\alpha - 1}{\alpha + 1}$ I coincident with Γ and having the same sense, and in $\underline{R}$ by that of an optical image source in $\underline{R}'$, of magnitude $\frac{\alpha - 1}{\alpha + 1}$ I and so directed that when object and image sources are resolved into components parallel to and normal to the plane the former have the same sense and the latter opposite senses.

4-114. If the surface density in Ex.4-112. is so distributed that curl$(\overline{A}_T+\overline{A}_s)$ is reduced to zero just inside $\underline{R}'$, rather than increased α times, show that divs $\overline{K}$ continues to be zero everywhere but that the image source located in $\underline{R}'$ is now of magnitude I and of opposite sense to the image in Ex.4-113.

4-115. Let $\overline{V}$ be a vector point function, irrotational and solenoidal within the region $\underline{R}$ bounded by the surfaces $S_{1--n}\Sigma$. Show that $\overline{V}$ may be expressed within $\underline{R}$ as

$$4\pi\overline{V}_o = -\text{grad}\oint_{S_1}(\psi-U_1)\frac{\partial}{\partial n}\left(\frac{1}{r}\right)dS \text{ ----- } -\text{grad}\oint_{\Sigma}(\psi-U_\Sigma)\frac{\partial}{\partial n}\left(\frac{1}{r}\right)dS$$

where

$$\psi_Q = \int_P^Q \overline{V}.d\overline{r} \quad \text{(P being a fixed and Q a variable point of } \underline{R})$$

and U_1---U_Σ are the solutions of Neumann problems in the regions τ_1---τ_Σ.

Now use a result of Ex.4-86., p. 316 to show that

$$4\pi\overline{V}_o = \text{curl}\oint_{S_1} d\overline{S}\times(\psi-U_1)\,\text{grad}\,\frac{1}{r} \text{ ------- } +\text{curl}\oint_{\Sigma} d\overline{S}\times(\psi-U_\Sigma)\,\text{grad}\,\frac{1}{r}$$

and transform this into

$$4\pi\overline{V}_o = -\text{curl}\oint_{S_1}\frac{(\hat{n}\times\overline{V}-\hat{n}\times\overline{V}_1)}{r}dS \text{ ------ } -\text{curl}\oint_{\Sigma}\frac{(\hat{n}\times\overline{V}-\hat{n}\times\overline{V}_\Sigma)}{r}dS$$

where $\overline{V}_1 = \text{grad}\,U_1$ etc.

4.17 The Grad-Curl Theorem[30]

Let $\overline{F}$ be any vector point function, well-behaved throughout the region τ bounded by the surfaces $S_{1..n}\Sigma$; $\overline{F}$ may or may not be defined outside τ. Let $\overline{A}$ be the vector potential of τ when treated as a volume source of density $\overline{F}$. Since

$$\nabla^2\overline{A} = \text{grad div } \overline{A} - \text{curl curl } \overline{A}$$

wherever $\overline{A}$ is well-behaved, it follows that

$$\nabla^2 \int_\tau \frac{\overline{F}}{r}\, d\tau = \text{grad div} \int_\tau \frac{\overline{F}}{r}\, d\tau - \text{curl curl} \int_\tau \frac{\overline{F}}{r}\, d\tau$$

both for interior and exterior origins of r.

Hence

$$\left.\begin{matrix} 4\pi\overline{F}_o \\ \overline{0} \end{matrix}\right\} = -\text{grad div} \int_\tau \frac{\overline{F}}{r}\, d\tau + \text{curl curl} \int_\tau \frac{\overline{F}}{r}\, d\tau \qquad (4.17\text{-}1)$$

where the left hand side is $4\pi\overline{F}$ at interior points of τ and zero at exterior points.

Since equations (4.15-4) to (4.15-7) hold within and without the source we have also

$$\left.\begin{matrix} 4\pi\overline{F}_o \\ \overline{0} \end{matrix}\right\} = -\text{grad} \int_\tau \frac{\text{div } \overline{F}}{r}\, d\tau + \text{grad} \oint_{S_{1..n}\Sigma} \frac{\overline{F}}{r} \cdot d\overline{S}$$

$$+ \text{curl} \int_\tau \frac{\text{curl } \overline{F}}{r}\, d\tau - \text{curl} \oint_{S_{1..n}\Sigma} d\overline{S} \times \frac{\overline{F}}{r} \qquad (4.17\text{-}2)$$

30. So called by A. O'Rahilly "Electromagnetic Theory", p. 14, Dover, New York (1965). It is generally expressed as equation (4.17-2) without the surface integrals and known as Helmholtz's theorem.

and

$$\left.\begin{matrix} 4\pi\overline{F}_o \\ \overline{0} \end{matrix}\right\} = \operatorname{grad}\int_\tau \overline{F}\,.\,\operatorname{grad}\frac{1}{r}\,d\tau + \operatorname{curl}\int_\tau \overline{F}\times\operatorname{grad}\frac{1}{r}\cdot d\tau \qquad (4.17\text{-}3)$$

or, from equation (4.17-2),

$$\left.\begin{matrix} 4\pi\overline{F}_o \\ \overline{0} \end{matrix}\right\} = -\int_\tau (\operatorname{div}\overline{F})\,\frac{\overline{r}}{r^3}\,d\tau + \oint_{S_{1..n}\Sigma} \overline{F}.d\overline{S}\,\frac{\overline{r}}{r^3}$$

$$- \int_\tau (\operatorname{curl}\overline{F})\times\frac{\overline{r}}{r^3}\,d\tau + \oint_{S_{1..n}\Sigma} (d\overline{S}\times\overline{F})\times\frac{\overline{r}}{r^3} \qquad (4.17\text{-}4)$$

Equations (4.17-1) to (4.17-4) represent various forms of the grad-curl theorem.

Now suppose that some point function $\overline{F}_1$ is well-behaved throughout the region τ_1, bounded externally by S_1. Then for an origin of r within τ we have

$$\overline{0} = -\operatorname{grad}\int_{\tau_1}\frac{\operatorname{div}\overline{F}_1}{r}\,d\tau + \operatorname{grad}\oint_{S_1}\frac{\overline{F}_1}{r}\,.\,\hat{n}_1\,dS$$

$$+ \operatorname{curl}\int_{\tau_1}\frac{\operatorname{curl}\overline{F}_1}{r}\,d\tau - \operatorname{curl}\oint_{S_1}\hat{n}_1\times\frac{\overline{F}_1}{r}\,dS \qquad (4.17\text{-}5)$$

where $\hat{n}_1$ is the outward normal from τ_1.

On adding equations (4.17-5) and (4.17-2) we obtain an alternative expression for $4\pi\overline{F}_o$ within τ which involves additional volume and surface integrals over τ_1 and S_1 in terms of an arbitrary function $\overline{F}_1$. Since the same treatment can be extended to τ_1--τ_n (and to the region outside Σ so long as the choice of auxiliary function is such as to maintain the volume integrals finite and reduce the surface integrals at infinity to zero) it is clear that any well-behaved function $\overline{F}$ may be expressed within the bounded region τ as $-\operatorname{grad}\psi + \operatorname{curl}\overline{Q}$ where ψ and $\overline{Q}$ may assume an infinite number of forms. This behaviour parallels that for scalar point functions as described in Sec. 4.5.

We are now in a position to demonstrate that a vector point function $\overline{V}$ which is solenoidal throughout a region τ may be expressed within τ as the curl of a point function (See p. 62).

Since $\oint \overline{V}.d\overline{S} = 0$ for any closed surface which can be drawn within τ, $\text{div}\,\overline{V} = 0$ at interior points of τ and $\oint \overline{V}.d\overline{S} = 0$ for each of the surfaces $S_{1..n}\Sigma$ in turn. Let U_1 be a scalar point function, harmonic in τ_1, and such that the outward normal derivative $\dfrac{\partial U_1}{\partial n_1}$ is equal to $-\overline{V}.\hat{n}$ at all points of S_1[31]. Put $\overline{V}_1 = \text{grad}\,U_1$. Then

$$\text{div}\,\overline{V}_1 = \nabla^2 U_1 = 0 \quad \text{in } \tau_1$$

$$\text{curl}\,\overline{V}_1 = \text{curl grad}\,U_1 = \overline{0} \quad \text{in } \tau_1$$

$$\overline{V}_1.\hat{n}_1 = -\overline{V}.\hat{n} \quad \text{on } S_1$$

Under these conditions equation (4.17-5) becomes

$$\overline{0} = -\,\text{grad}\oint_{S_1} \frac{\overline{V}}{r}\,.\,\hat{n}\,dS - \text{curl}\oint_{S_1} \hat{n}_1 \times \frac{\overline{V}_1}{r}\,dS$$

and on adding this to equation (4.17-2), where $\text{div}\,\overline{V}$ is already zero, there is a cancellation of $\text{grad}\oint_{S_1} \dfrac{\overline{V}}{r}\,.\,d\overline{S}$.

Proceeding in the same way for the regions τ_2--τ_n, we eliminate $\text{grad}\oint_{S_{2..n}} \dfrac{\overline{V}}{r}\,.\,d\overline{S}$ from equation (4.17-2) at the expense of an increased number of curl terms. Finally, let U_Σ be harmonic in the region τ_Σ outside Σ, regular at infinity, and such that $\overline{V}_\Sigma.\hat{n}_\Sigma = -\overline{V}.\hat{n}$ on Σ. Then for this component equation (4.17-5) becomes

---

31. U_1 exists, being a solution of an interior Neumann problem.

$$\overline{0} = -\operatorname{grad}\oint_{\Sigma}\frac{\overline{V}}{r}\cdot d\overline{S} + \operatorname{grad}\oint_{\infty}\frac{\overline{V}_{\Sigma}}{r}\cdot\hat{n}_{\Sigma}\,dS - \operatorname{curl}\oint_{\Sigma}\hat{n}_{\Sigma}\times\frac{\overline{V}_{\Sigma}}{r}\,dS$$

$$-\operatorname{curl}\oint_{\infty}\hat{n}_{\Sigma}\times\frac{\overline{V}_{\Sigma}}{r}\,dS$$

But U_{Σ} is regular at infinity, hence $R^2|\overline{V}_{\Sigma}|$ is bounded at infinity and the surface integrals at infinity consequently vanish. Upon summing the resulting equation with that developed above, $\operatorname{grad}\oint_{\Sigma}\frac{\overline{V}}{r}\cdot d\overline{S}$ is eliminated and $\overline{V}$ is expressed entirely as curl functions:

$$4\pi\overline{V}_o = \operatorname{curl}\int_{\tau}\frac{\operatorname{curl}\overline{V}}{r}\,d\tau - \operatorname{curl}\oint_{S_1}\frac{\{(\hat{n}\times\overline{V})+(\hat{n}_1\times\overline{V}_1)\}}{r}\,dS \;---$$

$$-\operatorname{curl}\oint_{\Sigma}\frac{\{(\hat{n}\times\overline{V})+(\hat{n}_{\Sigma}\times\overline{V}_{\Sigma})\}}{r}\,dS \qquad (4.17\text{-}6)$$

Reverting to a general case, suppose that $\overline{F}$ is defined everywhere within Σ but is discontinuous or has discontinuous derivatives upon interior surfaces which we identify with $S_{1..n}$. Following the procedure adopted in Sec. 4.5 for scalar functions, we find that $\overline{F}$ is given at interior points of the enclosure, not coincident with $S_{1..n}$, by

$$4\pi\overline{F}_o = -\operatorname{grad}\int_{\tau'}\frac{\operatorname{div}\overline{F}}{r}\,d\tau + \operatorname{grad}\oint_{S_{1..n}}\frac{-\Delta(\overline{F}.\hat{n})}{r}\,dS + \operatorname{grad}\oint_{\Sigma}\frac{\overline{F}.\hat{n}}{r}\,dS$$

$$+\operatorname{curl}\int_{\tau'}\frac{\operatorname{curl}\overline{F}}{r}\,d\tau - \operatorname{curl}\oint_{S_{1..n}}\frac{-\Delta(\hat{n}\times\overline{F})}{r}\,dS - \operatorname{curl}\oint_{\Sigma}\frac{\hat{n}\times\overline{F}}{r}\,dS \qquad (4.17\text{-}7)$$

where τ' is the set of regions enclosed by Σ, and $\Delta(\ )$ indicates the increment in the bracketed term corresponding to positive motion through the surface when the same arbitrarily-defined sense of the normal is assigned to both sides of the surface[32].

If there are no discontinuities of $\overline{F}$ and its derivatives outside Σ, and if Σ is allowed to expand to infinity, equation (4.17-7) becomes

$$4\pi\overline{F}_o = -\,\text{grad}\int_{\infty} \frac{\text{div}\,\overline{F}}{r}\,d\tau + \text{grad}\oint_{S_{1..n}} \frac{-\Delta(\overline{F}.\hat{n})}{r}\,dS + \text{curl}\int_{\infty} \frac{\text{curl}\,\overline{F}}{r}\,d\tau - \text{curl}\oint_{S_{1..n}} \frac{-\Delta(\hat{n}\times\overline{F})}{r}\,dS \tag{4.17-8}$$

provided that the volume integrals converge and the terms involving Σ vanish. This will certainly be the case if $|\overline{F}|R^n$ is bounded for $n > 1$ and R^n div $\overline{F}$ and $R^n|\text{curl}\,\overline{F}|$ are bounded for $n > 2$.

By proceeding from equation (4.17-4) rather than (4.17-2) we find that

$$4\pi\overline{F}_o = -\int_{\infty} (\text{div}\,\overline{F})\,\frac{\overline{r}}{r^3}\,d\tau + \oint_{S_{1..n}} -\Delta(\overline{F}.\hat{n})\,\frac{\overline{r}}{r^3}\,dS - \int_{\infty} (\text{curl}\,\overline{F})\times\frac{\overline{r}}{r^3}\,d\tau + \oint_{S_{1..n}} -\Delta(\hat{n}\times\overline{F})\times\frac{\overline{r}}{r^3}\,dS \tag{4.17-9}$$

provided that R^n div $\overline{F}$ and $R^n|\text{curl}\,\overline{F}|$ are bounded for $n > 1$ and $|\overline{F}|\to 0$ as $R\to\infty$.

Since the vectorial content of $\overline{F}$ and curl $\overline{F}$ has been ignored in arriving at the above conditions for validity, the latter should be treated as sufficient but not necessary.

---

32. The surfaces $S_{1..n}$ are not necessarily closed.

4.18 The Gradient and Laplacian of the Scalar Point Function $\int \overline{P}.\text{grad}\,\frac{1}{r}\,d\tau$

Let $\overline{P}$ be a vector point function, well-behaved throughout the region τ bounded by the surfaces $S_{1..n}\Sigma$, and let r be the distance of $d\tau$ from the point of evaluation 0. Then $\int_\tau \overline{P}.\text{grad}\,\frac{1}{r}\,d\tau$ defines a scalar point function which is identical with the sum of two Newtonian potential functions and is everywhere convergent and continuous. This may be demonstrated as follows:

When 0 is exterior to τ we have, at interior points of τ,

$$\text{div}\,\frac{\overline{P}}{r} = \frac{1}{r}\,\text{div}\,\overline{P} + \overline{P}.\text{grad}\,\frac{1}{r}$$

whence

$$\int_\tau \text{div}\,\frac{\overline{P}}{r}\,d\tau = \oint_{S_{1..n}\Sigma} \frac{\overline{P}}{r}.d\overline{S} = \int_\tau \frac{\text{div}\,\overline{P}}{r}\,d\tau + \int_\tau \overline{P}.\text{grad}\,\frac{1}{r}\,d\tau$$

or

$$\int_\tau \overline{P}.\text{grad}\,\frac{1}{r}\,d\tau = \int_\tau \frac{(-\text{div}\,\overline{P})}{r}\,d\tau + \oint_{S_{1..n}\Sigma} \frac{\overline{P}}{r}.\hat{n}\,dS \qquad (4.18\text{-}1)$$

Hence at exterior points of τ the integral under consideration is equal to the combined potentials of the volume source τ of density $(-\text{div}\,\overline{P})$ and the surface sources $S_{1..n}\Sigma$ of density $P_n = \overline{P}.\hat{n}$ where $\hat{n}$ is the outward unit normal from τ.

When 0 lies within τ the integral $\int_\tau \overline{P}.\text{grad}\,\frac{1}{r}\,d\tau$ is improper. Let 0 be an interior point of the region τ' bounded by the closed regular surface S'. Then corresponding to equation (4.18-1) we now have

$$\int_{\tau-\tau'} \overline{P}.\text{grad}\,\frac{1}{r}\,d\tau = \int_{\tau-\tau'} \frac{(-\text{div}\,\overline{P})}{r}\,d\tau + \oint_{S_{1..n}\Sigma} \frac{\overline{P}}{r}.d\overline{S} + \oint_{S'} \frac{\overline{P}}{r}.d\overline{S} \qquad (4.18\text{-}2)$$

As S' shrinks about 0, $\int_{\tau-\tau'} \frac{(-\text{div}\,\overline{P})}{r}\,d\tau$ converges for reasons discussed in Sec. 4.4 and $\oint_{S'} \frac{\overline{P}}{r}.d\overline{S} \to 0$ (see Ex.4-117., p. 354).

Hence at interior points of τ

$$\lim_{\tau'\to 0}\int_{\tau-\tau'} \overline{P}.\text{grad}\,\frac{1}{r}\,d\tau \equiv \int_{\tau} \overline{P}.\text{grad}\,\frac{1}{r}\,d\tau = \int_{\tau} \frac{(-\text{div}\,\overline{P})}{r}\,d\tau + \oint_{S_{1..n}\Sigma} \frac{\overline{P}}{r}.d\overline{S} \qquad (4.18\text{-}3)$$

This relationship may be shown to subsist upon $S_{1..n}\Sigma$, so that $\int_{\tau} \overline{P}.\text{grad}\,\frac{1}{r}\,d\tau$ is everywhere equal to the sum of the potential functions described above. We will consequently refer to it as a potential function[33]. It is everywhere continuous, but its normal derivative is discontinuous upon $S_{1..n}\Sigma$ by $-4\pi P_n$ for a common positive sense of the normal on both sides.

4.18a <u>Gradient of $\int_{\tau} \overline{P}.\text{grad}\,\frac{1}{r}\,d\tau$ at interior and exterior points of τ</u>

It follows from equations (4.18-1) and (4.18-3) that at interior and exterior points of τ

$$\text{grad}\int_{\tau} \overline{P}.\text{grad}\,\frac{1}{r}\,d\tau = \text{grad}\int_{\tau} \frac{(-\text{div}\,\overline{P})}{r}\,d\tau + \text{grad}\oint_{S_{1..n}\Sigma} \frac{\overline{P}}{r}.d\overline{S}$$

whence from equation (4.7-7) with $\overline{P}.\hat{n}$ substituted for σ, and from (4.8-1) and (4.8-12) with $(-\text{div}\,\overline{P})$ substituted for ρ

$$\text{grad}\int_{\tau} \overline{P}.\text{grad}\,\frac{1}{r}\,d\tau = \int_{\tau} (-\text{div}\,\overline{P})\,\frac{\overline{r}}{r^3}\,d\tau + \oint_{S_{1..n}\Sigma} \overline{P}.d\overline{S}\,\frac{\overline{r}}{r^3} \qquad (4.18\text{-}4)$$

---

33. At exterior points of τ, $\int_{\tau} \overline{P}.\text{grad}\,\frac{1}{r}\,d\tau$ is the potential of a limiting configuration of scalar point doublets within τ, and it is in this context that the function usually arises (see Sec. 4.20a). At interior points of τ, however, the potential of such a complex is indeterminate, but we may continue to manipulate the integral independently of any such interpretation.

Alternatively, since equation (4.8-10) applies at interior and exterior points of τ,

$$\operatorname{grad}\int_\tau \overline{P}.\operatorname{grad}\frac{1}{r}\,d\tau = -\int_\tau \frac{\operatorname{grad}\operatorname{div}\overline{P}}{r}\,d\tau + \oint_{S_{1..n}\Sigma}\frac{\operatorname{div}\overline{P}}{r}\,d\overline{S} + \oint_{S_{1..n}\Sigma}\overline{P}.d\overline{S}\,\frac{\overline{r}}{r^3} \tag{4.18-5}$$

Again, by combining equation (4.18-4) with (4.17-4) we obtain

$$\operatorname{grad}\int_\tau \overline{P}.\operatorname{grad}\frac{1}{r}\,d\tau = \int_\tau (\operatorname{curl}\overline{P})\times\frac{\overline{r}}{r^3}\,d\tau - \oint_{S_{1..n}\Sigma}(d\overline{S}\times\overline{P})\times\frac{\overline{r}}{r^3} \quad \text{at exterior points} \tag{4.18-6}$$

$$= \int_\tau (\operatorname{curl}\overline{P})\times\frac{\overline{r}}{r^3}\,d\tau - \oint_{S_{1..n}\Sigma}(d\overline{S}\times\overline{P})\times\frac{\overline{r}}{r^3} + 4\pi\overline{P} \quad \text{at interior points} \tag{4.18-7}$$

Finally, from equation (4.17-3),

$$\operatorname{grad}\int_\tau \overline{P}.\operatorname{grad}\frac{1}{r}\,d\tau = -\operatorname{curl}\int_\tau \overline{P}\times\operatorname{grad}\frac{1}{r}\,d\tau \quad \text{at exterior points} \tag{4.18-8}$$

$$= -\operatorname{curl}\int_\tau \overline{P}\times\operatorname{grad}\frac{1}{r}\,d\tau + 4\pi\overline{P} \quad \text{at interior points} \tag{4.18-9}$$

4.18b <u>Gradient of cavity potential</u> $\int_{\tau-\tau_\delta}\overline{P}.\operatorname{grad}\frac{1}{r}\,d\tau$

As in Sec. 4.8 the cavity potential is defined at all points of a fixed δ sphere within τ, so that in accordance with equation (4.8-2) the source system now reduces to a volume source $\tau-\tau_\delta$ of density $(-\operatorname{div}\overline{P})$ and surface sources $S_{1..n}\Sigma$, S_δ of density $\overline{P}.\hat{\overline{n}}$, the point of evaluation being exterior to the region of integration. Various expressions for grad (cavity)$\int_{\tau-\tau_\delta} \overline{P} \cdot \text{grad}\,\frac{1}{r}\,d\tau$ may be obtained by appropriate amendment of the results for grad $\int_{\tau} \overline{P} \cdot \text{grad}\,\frac{1}{r}\,d\tau$ at exterior points of τ and these expressions will hold at all interior points of the δ sphere. We get

$$\text{grad (cavity)} \int_{\tau-\tau_\delta} \overline{P} \cdot \text{grad}\,\frac{1}{r}\,d\tau$$

$$= \int_{\tau-\tau_\delta} (-\text{div}\,\overline{P})\,\frac{\overline{r}}{r^3}\,d\tau + \oint_{S_{1..n}\Sigma,\ S_\delta} \overline{P}.d\overline{S}\,\frac{\overline{r}}{r^3} \qquad (4.18\text{-}10)$$

$$= -\int_{\tau-\tau_\delta} \frac{\text{grad div}\,\overline{P}}{r}\,d\tau + \oint_{S_{1..n}\Sigma,\ S_\delta} \frac{\text{div}\,\overline{P}}{r}\,d\overline{S} + \oint_{S_{1..n}\Sigma,\ S_\delta} \overline{P}.d\overline{S}\,\frac{\overline{r}}{r^3} \qquad (4.18\text{-}11)$$

$$= \int_{\tau-\tau_\delta} (\text{curl}\,\overline{P}) \times \frac{\overline{r}}{r^3}\,d\tau - \oint_{S_{1..n}\Sigma,\ S_\delta} (d\overline{S}\times\overline{P}) \times \frac{\overline{r}}{r^3} \qquad (4.18\text{-}12)$$

$$= -\,\text{curl} \int_{\tau-\tau_\delta} \overline{P} \times \text{grad}\,\frac{1}{r}\,d\tau \qquad (4.18\text{-}13)$$

It follows from equation (4.18-10) that when the gradient of the cavity potential is evaluated at the centre of the δ sphere

$$\lim_{\delta\to 0} \text{grad (cavity)} \int_{\tau-\tau_\delta} \overline{P} \cdot \text{grad}\,\frac{1}{r}\,d\tau$$

$$= \int_{\tau} (-\text{div}\,\overline{P})\,\frac{\overline{r}}{r^3}\,d\tau + \oint_{S_{1..n}\Sigma} \overline{P}.d\overline{S}\,\frac{\overline{r}}{r^3} + \lim_{\delta\to 0} \oint_{S_\delta} \overline{P}.d\overline{S}\,\frac{\overline{r}}{r^3}$$

$$= \text{grad} \int_{\tau} \overline{P} \cdot \text{grad}\,\frac{1}{r}\,d\tau - \frac{4}{3}\pi\overline{P}_o \qquad (4.18\text{-}14)$$

This result, which is of considerable importance in the theory of polarised systems, contrasts with that found in Sec. 4.8 for a simple volume source, viz.

$$\lim_{\delta\to 0} \text{grad (cavity)} \int_{\tau-\tau_\delta} \frac{\rho}{r}\, d\tau = \text{grad} \int_{\tau} \frac{\rho}{r}\, d\tau .$$

4.18c <u>Laplacian of</u> $\int_\tau \overline{P} \cdot \text{grad}\,\frac{1}{r}\, d\tau$ <u>at interior and exterior points of τ</u>

From equations (4.18-1) and (4.18-3)

$$\nabla^2 \int_\tau \overline{P} \cdot \text{grad}\,\frac{1}{r}\, d\tau = \nabla^2 \int_\tau \frac{(-\text{div}\,\overline{P})}{r}\, d\tau + \nabla^2 \oint_{S_{1..n}\Sigma} \frac{\overline{P}}{r} \cdot d\overline{S}$$

But

$$\nabla^2 \oint_{S_{1..n}\Sigma} \frac{\overline{P}}{r} \cdot d\overline{S} = \oint_{S_{1..n}\Sigma} \overline{P}.d\overline{S}\; \nabla^2 \left(\frac{1}{r}\right) = 0$$

hence, from (4.9-1) and (4.9-4) with ($-$div $\overline{P}$) substituted for ρ,

$$\nabla^2 \int_\tau \overline{P} \cdot \text{grad}\,\frac{1}{r}\, d\tau = 0 \quad \text{at exterior points} \qquad (4.18\text{-}15)$$

$$\nabla^2 \int_\tau \overline{P} \cdot \text{grad}\,\frac{1}{r}\, d\tau = 4\pi\, \text{div}\,\overline{P} \quad \text{at interior points} \qquad (4.18\text{-}16)$$

4.18d <u>Laplacian of cavity potential</u> $\int_{\tau-\tau_\delta} \overline{P} \cdot \text{grad}\,\frac{1}{r}\, d\tau$

From equation (4.18-2)

$$\nabla^2 \text{ (cavity)} \int_{\tau-\tau_\delta} \overline{P} \cdot \text{grad}\,\frac{1}{r}\, d\tau = \nabla^2 \text{ (cavity)} \int_{\tau-\tau_\delta} \frac{(-\text{div}\,\overline{P})}{r}\, d\tau + \nabla^2 \oint_{S_{1..n}\Sigma,\ S_\delta} \frac{\overline{P}}{r} \cdot d\overline{S}$$

But

$$\nabla^2 \oint_{S_{1..n}\Sigma,\ S_\delta} \frac{\overline{P}}{r} \cdot d\overline{S} = \oint_{S_{1..n}\Sigma,\ S_\delta} \overline{P}.d\overline{S}\ \nabla^2 \left(\frac{1}{r}\right) = 0$$

since the δ sphere is fixed in space,

hence from equation (4.9-2)

$$\nabla^2 \text{ (cavity)} \int_{\tau-\tau_\delta} \overline{P} \cdot \text{grad} \frac{1}{r}\, d\tau = 0 \qquad (4.18\text{-}17)$$

4.18e Discontinuity of grad $\int_\tau \overline{P} \cdot \text{grad} \frac{1}{r}\, d\tau$ at the bounding surfaces

We have

$$\text{grad} \int_\tau \overline{P} \cdot \text{grad} \frac{1}{r}\, d\tau = \text{grad} \int_\tau \frac{(-\text{div}\ \overline{P})}{r}\, d\tau + \text{grad} \oint_{S_{1..n}\Sigma} \frac{\overline{P}.\hat{n}}{r}\, dS$$

where $\hat{n}$ is the outward normal from τ.

The first term of the right hand side is continuous through $S_{1..n}\Sigma$ since it comprises the gradient of the potential of a well-behaved volume source of density (-div $\overline{P}$). The second term is discontinuous since it comprises the gradient of the potential of a surface source of density $\overline{P}.\hat{n}$. In accordance with the findings of Sec. 4.7b this term suffers a discontinuity of $-4\pi(\overline{P}.\hat{n})\hat{n}$ with movement through the surface out of τ.

4.18f The partial potential and its derivatives

Use of the transformation (4.18-3), coupled with the results of Sec. 4.8, has rendered it unnecessary to invoke a partial potential or its derivatives in the determination of grad $\int_\tau \overline{P} \cdot \text{grad} \frac{1}{r}\, d\tau$ and $\nabla^2 \int_\tau \overline{P} \cdot \text{grad} \frac{1}{r}\, d\tau$ at interior points of τ. Such point functions may, of course, be employed, and the manipulations involved in their derivation afford useful practice in handling this branch of potential theory. Accordingly, a number of exercises at the end of the section are devoted to the topic, and are set out in some detail to assist the reader.

EXERCISES

4-116. By means of an analysis similar to that employed in Sec. 4.4 in connection with the improper integral $\int_\tau \frac{\rho}{r}\,d\tau$, show that $\int_\tau P_x \frac{(x-x_o)}{r^3}\,d\tau$ (and consequently $\int_\tau \overline{P} \cdot \text{grad}\,\frac{1}{r}\,d\tau$) is convergent at interior points of τ.

4-117. Show that, for any closed regular surface S, the integral $\oint_S \frac{d\overline{S}}{r}$ approaches zero as S shrinks about the origin of r, and that consequently integrals of the type $\oint_S \frac{\overline{P}}{r} \cdot d\overline{S}$ and $\oint_S d\overline{S} \times \frac{\overline{P}}{r}$ likewise approach zero, provided that $\overline{P}$ is finite and continuous at the origin. [Hint: Express $\oint_S \frac{d\overline{S}}{r}$ as the volume integral of grad $\frac{1}{r}$ between S and an included spherical surface centred upon 0, and show that its component scalar magnitudes approach zero as these surfaces shrink about 0.]

4-118. Demonstrate that (4.18-3) continues to hold at boundary points of τ.

4-119. Prove that

$$\sum \overline{i}\,\frac{\partial}{\partial x_o} \int_\tau \overline{P} \cdot \text{grad}\,\frac{1}{r}\,d\tau = \sum \overline{i} \int_\tau \overline{P} \cdot \text{grad}\,\frac{(x-x_o)}{r^3}\,d\tau$$

at exterior points of τ.

Expand div $\frac{(x-x_o)}{r^3}\,\overline{P}$ and integrate over τ to obtain

$$\int_\tau \overline{P} \cdot \text{grad}\,\frac{(x-x_o)}{r^3}\,d\tau = \oint_{S_{1..n}\Sigma} \frac{(x-x_o)}{r^3}\,\overline{P}.d\overline{S} - \int_\tau \frac{(x-x_o)}{r^3} \cdot \text{div}\,\overline{P}\,d\tau$$

and hence obtain grad $\int_\tau \overline{P} \cdot \text{grad}\,\frac{1}{r}\,d\tau$ in the form (4.18-4).

4-120. By writing

$$\text{grad (partial)} \int_{\tau-\tau_\delta} \overline{P} \cdot \text{grad}\,\frac{1}{r}\,d\tau$$

$$= \text{grad (partial)} \int_{\tau-\tau_\delta} \frac{(-\text{div}\,\overline{P})}{r}\,d\tau + \text{grad} \oint_{S_1 \ldots \Sigma} \frac{\overline{P}}{r} \cdot d\overline{S} + \text{grad} \oint_{S_\delta} \frac{\overline{P}}{r} \cdot d\overline{S}$$

show that

$$\text{grad (partial)} \int_{\tau-\tau_\delta} \overline{P} \cdot \text{grad} \frac{1}{r} \, d\tau$$

$$= \int_{\tau-\tau_\delta} (-\text{div } \overline{P}) \frac{\overline{r}}{r^3} \, d\tau - \oint_{S_\delta} \frac{\text{div } \overline{P}}{r} \, d\overline{S} + \oint_{S_{1..n}\Sigma} \overline{P}.d\overline{S} \frac{\overline{r}}{r^3} + \text{grad} \oint_{S_\delta} \frac{\overline{P}}{r} \cdot d\overline{S}$$

Bearing in mind that the δ sphere moves with 0, derive the relationship

$$\left\{ \text{grad} \oint_{S_\delta} \frac{\overline{P}}{r} \cdot d\overline{S} \right\}_x = \oint_{S_\delta} \frac{1}{r} \left\{ \frac{\partial P_x}{\partial x} dS_x + \frac{\partial P_y}{\partial x} dS_y + \frac{\partial P_z}{\partial x} dS_z \right\}$$

and combine this with the x component of the remaining surface integral over S_δ to obtain

$$\oint_{S_\delta} \frac{1}{r} \{ (d\overline{S} \times \text{grad } P_z)_y - (d\overline{S} \times \text{grad } P_y)_z \}$$

Make use of equation (1.17-2) to show that this is zero, and hence obtain

$$\text{grad (partial)} \int_{\tau-\tau_\delta} \overline{P} \cdot \text{grad} \frac{1}{r} \, d\tau = \int_{\tau-\tau_\delta} (-\text{div } \overline{P}) \frac{\overline{r}}{r^3} \, d\tau + \oint_{S_{1..n}\Sigma} \overline{P}.d\overline{S} \frac{\overline{r}}{r^3}$$

4-121. It is possible to derive the result of the previous exercise without making use of the initial transformation (4.18-2).

Write

$$\text{grad (partial)} \int_{\tau-\tau_\delta} \overline{P} \cdot \text{grad} \frac{1}{r} \, d\tau$$

$$= - \text{grad (partial)} \int_{\tau-\tau_\delta} \left\{ P_x \frac{(x-x_o)}{r^3} + P_y \frac{(y-y_o)}{r^3} + P_z \frac{(z-z_o)}{r^3} \right\} d\tau$$

and show that the field-slipping technique developed in Sec. 4.8 for the evaluation of grad partial pot ρ can be applied to the evaluation of grad (partial) $\int_{\tau-\tau_\delta} P_x \frac{(x-x_o)}{r^3}\, d\tau$, since $\frac{x-x_o}{r^3}$ remains constant for each volume element during the slip. Hence derive

$$\text{grad (partial)} \int_{\tau-\tau_\delta} \overline{P} \cdot \text{grad}\,\frac{1}{r}\, d\tau$$

$$= -\int_{\tau-\tau_\delta} \left\{ \frac{(x-x_o)}{r^3}\,\text{grad}\,P_x + \frac{(y-y_o)}{r^3}\,\text{grad}\,P_y + \frac{(z-z_o)}{r^3}\,\text{grad}\,P_z \right\} d\tau$$

$$+ \oint_{S_{1..n}\Sigma} \left\{ P_x \frac{(x-x_o)}{r^3} + P_y \frac{(y-y_o)}{r^3} + P_z \frac{(z-z_o)}{r^3} \right\} d\overline{S}$$

Integrate grad $\frac{(x-x_o)}{r^3} P_x$ and its expansion over $\tau-\tau_\delta$, combining the resulting equation and similar equations in $(y-y_o)$ and $(z-z_o)$ with that above to get

$$\text{grad (partial)} \int_{\tau-\tau_\delta} \overline{P} \cdot \text{grad}\,\frac{1}{r}\, d\tau$$

$$= \int_{\tau-\tau_\delta} \left\{ P_x\,\text{grad}\,\frac{(x-x_o)}{r^3} + P_y\,\text{grad}\,\frac{(y-y_o)}{r^3} + P_z\,\text{grad}\,\frac{z-z_o}{r^3} \right\} d\tau$$

$$- \oint_{S_\delta} \left\{ P_x \frac{(x-x_o)}{r^3} + P_y \frac{(y-y_o)}{r^3} + P_z \frac{(z-z_o)}{r^3} \right\} d\overline{S}$$

By noting that $\frac{\partial}{\partial x}\frac{(y-y_o)}{r^3} = \frac{\partial}{\partial y}\frac{(x-x_o)}{r^3}$ etc. show that the x component of the above expression may be written as

$$\int_{\tau-\tau_\delta} \overline{P} \cdot \text{grad}\,\frac{(x-x_o)}{r^3}\, d\tau + \oint_{S_\delta} \overline{P} \cdot \text{grad}\,\frac{1}{r}\, dS_x$$

Utilise the integral of div $\frac{(x-x_O)}{r^3}\,\overline{P}$ and its expansion over $\tau-\tau_\delta$ to replace this expression by

$$-\int_{\tau-\tau_\delta}\frac{(x-x_O)}{r^3}\,\text{div}\,\overline{P}\,d\tau+\oint_{S_{1..n}\Sigma}\frac{(x-x_O)}{r^3}\,\overline{P}.d\overline{S}+\oint_{S_\delta}\left\{\frac{(x-x_O)}{r^3}\,\overline{P}.d\overline{S}+\overline{P}\,.\,\text{grad}\,\frac{1}{r}\,dS_x\right\}$$

Hence show that

$$\text{grad (partial)}\int_{\tau-\tau_\delta}\overline{P}\,.\,\text{grad}\,\frac{1}{r}\,d\tau$$

$$=\int_{\tau-\tau_\delta}(-\text{div}\,\overline{P})\,\frac{\overline{r}}{r^3}\,d\tau+\oint_{S_{1..n}\Sigma}\overline{P}.d\overline{S}\,\frac{\overline{r}}{r^3}+\oint_{S_\delta}\overline{P}\times\left(d\overline{S}\times\text{grad}\,\frac{1}{r}\right)$$

$$=\int_{\tau-\tau_\delta}(-\text{div}\,\overline{P})\,\frac{\overline{r}}{r^3}\,d\tau+\oint_{S_{1..n}\Sigma}\overline{P}.d\overline{S}\,\frac{\overline{r}}{r^3}$$

since $d\overline{S}\times\text{grad}\,\frac{1}{r}=\overline{0}$ on S_δ

Make use of equation (4.17-4), modified for an integration region $\tau-\tau_\delta$ and surfaces $S_{1..n}\Sigma$, S_δ, to derive from the above result

$$\text{grad (partial)}\int_{\tau-\tau_\delta}\overline{P}\,.\,\text{grad}\,\frac{1}{r}\,d\tau$$

$$=\int_{\tau-\tau_\delta}(\text{curl}\,\overline{P})\times\frac{\overline{r}}{r^3}\,d\tau-\oint_{S_{1..n}\Sigma}(d\overline{S}\times\overline{P})\times\frac{\overline{r}}{r^3}-\oint_{S_\delta}\overline{P}\,d\overline{S}\,.\,\frac{\overline{r}}{r^3}$$

4-122. The major portions of the analyses of Ex.4-120. and 4-121. remain valid when the excluded region τ_δ is neither spherical nor centred upon O.

It is only in the last lines of the analyses that the properties of the sphere are invoked in order to eliminate certain terms. On this basis it would be expected that

$$\sum i\oint_S\frac{1}{r}\,\{(d\overline{S}\times\text{grad}\,P_z)_y-(d\overline{S}\times\text{grad}\,P_y)_z\}=\oint_S\overline{P}\times\left(d\overline{S}\times\text{grad}\,\frac{1}{r}\right)$$

where S is any regular closed surface and the origin of r is not a point of S.

Devise an independent proof of this.

4-123. Show that $\operatorname{div} \operatorname{grad} \int_\tau \bar{P} \cdot \operatorname{grad} \frac{1}{r} \, d\tau = 0$ at exterior points of τ by evaluating $\operatorname{div} \int_\tau (-\operatorname{div} \bar{P}) \frac{\bar{r}}{r^3} \, d\tau + \operatorname{div} \oint_{S_{1..n}\Sigma} \bar{P} . d\bar{S} \frac{\bar{r}}{r^3}$ at exterior points.

4-124. From equation (4.18-2)

$$\nabla^2 \text{ (partial)} \int_{\tau-\tau_\delta} \bar{P} \cdot \operatorname{grad} \frac{1}{r} \, d\tau$$

$$= \nabla^2 \text{ (partial)} \int_{\tau-\tau_\delta} \frac{(-\operatorname{div} \bar{P})}{r} \, d\tau + \nabla^2 \oint_{S_{1..n}\Sigma} \frac{\bar{P}}{r} \cdot d\bar{S} + \nabla^2 \oint_{S_\delta} \frac{\bar{P}}{r} \cdot d\bar{S}$$

Show that

(1)

$$\nabla^2 \text{ (partial)} \int_{\tau-\tau_\delta} \frac{(-\operatorname{div} P)}{r} \, d\tau$$

$$= - \int_{\tau-\tau_\delta} \frac{\nabla^2(\operatorname{div} \bar{P})}{r} \, d\tau + \oint_{S_{1..n}\Sigma} \left\{ \frac{1}{r} \operatorname{grad} \operatorname{div} \bar{P} - (\operatorname{div} \bar{P}) \operatorname{grad} \frac{1}{r} \right\} \cdot d\bar{S}$$

$$= \oint_{S_\delta} \operatorname{div} \bar{P} \left(\operatorname{grad} \frac{1}{r} \right) \cdot d\bar{S} - \oint_{S_\delta} \frac{1}{r} (\operatorname{grad} \operatorname{div} \bar{P}) . d\bar{S}$$

(2)

$$\nabla^2 \oint_{S_{1..n}\Sigma} \frac{\bar{P}}{r} \cdot d\bar{S} = 0$$

(3)

$$\nabla^2 \oint_{S_\delta} \frac{\overline{P}}{r} \,.\, d\overline{S} = \oint_{S_\delta} \frac{1}{r} (\nabla^2 \overline{P}).d\overline{S}$$

$$= \oint_{S_\delta} \frac{1}{r} (\text{grad div } \overline{P}).d\overline{S} - \oint_{S_\delta} \frac{1}{r} (\text{curl curl } \overline{P}).d\overline{S}$$

By integrating curl $\frac{(\text{curl } \overline{P})}{r}$ and its expansion over S_δ, show that

$$- \oint_{S_\delta} \frac{1}{r} (\text{curl curl } \overline{P}).d\overline{S} = \oint_{S_\delta} (\text{curl } \overline{P}).\left(d\overline{S} \times \text{grad } \frac{1}{r}\right)$$

Hence derive

$$\nabla^2 \text{ (partial)} \int_{\tau-\tau_\delta} \overline{P} \,.\, \text{grad } \frac{1}{r} \, d\tau = \oint_{S_\delta} (-\text{div } \overline{P}) \frac{\overline{r}}{r^3} \,.\, d\overline{S} + \oint_{S_\delta} (\text{curl } \overline{P}).\left(d\overline{S} \times \text{grad } \frac{1}{r}\right)$$

$$= \oint_{S_\delta} (-\text{div } \overline{P}) \frac{\overline{r}}{r^3} \,.\, d\overline{S}$$

4-125. Derive the more general result of the previous exercise (which does not require that S_δ represent a spherical region) by working from the general form of grad (partial) $\int_{\tau-\tau_\delta} \overline{P} \,.\, \text{grad } \frac{1}{r} \, d\tau$ as developed in Ex.4-121., ie by expanding

$$\text{div (partial)} \int_{\tau-\tau_\delta} (-\text{div } \overline{P}) \frac{\overline{r}}{r^3} \, d\tau + \text{div} \oint_{S_{1..n}\Sigma} \overline{P}.d\overline{S} \frac{\overline{r}}{r^3} + \text{div} \oint_{S_\delta} \overline{P} \times \left(d\overline{S} \times \text{grad } \frac{1}{r}\right)$$

Firstly, transform div (partial) $\int_{\tau-\tau_\delta} (-\text{div } \overline{P}) \frac{\overline{r}}{r^3} \, d\tau$ by means of the field-slipping technique into

$$\int_{\tau-\tau_\delta} (\text{grad div } \overline{P}) \,.\, \text{grad } \frac{1}{r} \, d\tau - \oint_{S_{1..n}\Sigma} \text{div } \overline{P} \;\; \text{grad } \frac{1}{r} \,.d\overline{S}$$

and by integrating div $\left(\text{div}\,\overline{P}\,\text{grad}\,\frac{1}{r}\right)$ and its expansion over $\tau-\tau_\delta$ transform this in turn into $\oint_{S_\delta} (-\text{div}\,P)\,\frac{\overline{r}}{r^3}\cdot d\overline{S}$

Secondly, show that

$$\text{div}\oint_{S_{1..n}\Sigma} \overline{P}.d\overline{S}\,\frac{\overline{r}}{r^3} = 0$$

Thirdly, bearing in mind that $\overline{P}$ alone is a function of x_o, develop the relationship

$$\text{div}\oint_{S_\delta} \overline{P}\times\left(d\overline{S}\times\text{grad}\,\frac{1}{r}\right) = \sum\oint_{S_\delta} dS_x\left\{\frac{\partial P_x}{\partial x}\frac{\partial}{\partial x}\left(\frac{1}{r}\right)+\frac{\partial P_y}{\partial x}\frac{\partial}{\partial y}\left(\frac{1}{r}\right)+\frac{\partial P_z}{\partial x}\frac{\partial}{\partial z}\left(\frac{1}{r}\right)\right\}$$

$$-\sum\oint_{S_\delta}\frac{\partial}{\partial x}\left(\frac{1}{r}\right)\left\{\frac{\partial P_x}{\partial x}\,dS_x+\frac{\partial P_y}{\partial x}\,dS_y+\frac{\partial P_z}{\partial x}\,dS_z\right\}$$

and transform this into

$$\oint_{S_\delta}\sum(\text{curl}\,\overline{P})_x\left(d\overline{S}\times\text{grad}\,\frac{1}{r}\right)_x = \oint_{S_\delta}(\text{curl}\,\overline{P})\cdot\left(d\overline{S}\times\text{grad}\,\frac{1}{r}\right)$$

4-126. In Sec. 4.18b it was stated that

$$\lim_{\delta\to 0}\oint_{S_\delta}\overline{P}.d\overline{S}\,\frac{\overline{r}}{r^3} = -\frac{4}{3}\pi\overline{P}_o$$

where 0 is the centre of the δ sphere.

Prove this.

4.19 The Divergence, Curl and Laplacian of the Vector Point Function[34]

$$\int \overline{M}\times\text{grad}\,\frac{1}{r}\,d\tau$$

---

34. The reader will notice the parallel nature of the treatments in Secs. 4.18 and 4.19, even to the extent, in part, of a common wording.

Let $\overline{M}$ be a vector point function, well-behaved throughout the region τ bounded by the surfaces $S_{1..n}\Sigma$, and let r be the distance of $d\tau$ from the point of evaluation 0. Then $\int_\tau \overline{M} \times \text{grad}\,\frac{1}{r}\,d\tau$ defines a vector point function which is identical with the sum of two Newtonian potential functions and is everywhere convergent and continuous. This may be demonstrated as follows:

When 0 is exterior to τ we have, at interior points of τ,

$$\text{curl}\,\frac{\overline{M}}{r} = \frac{1}{r}\,\text{curl}\,\overline{M} + \left(\text{grad}\,\frac{1}{r}\right) \times \overline{M}$$

whence

$$\int_\tau \text{curl}\,\frac{\overline{M}}{r}\,d\tau = \oint_{S_{1..n}\Sigma} d\overline{S} \times \frac{\overline{M}}{r} = \int_\tau \frac{\text{curl}\,\overline{M}}{r}\,d\tau + \int_\tau \left(\text{grad}\,\frac{1}{r}\right) \times \overline{M}\,d\tau$$

or

$$\int_\tau \overline{M} \times \text{grad}\,\frac{1}{r}\,d\tau = \int_\tau \frac{\text{curl}\,\overline{M}}{r}\,d\tau + \oint_{S_{1..n}\Sigma} \frac{\overline{M}}{r} \times \hat{n}\,dS \qquad (4.19\text{-}1)$$

Hence at exterior points of τ the integral under consideration is equal to the combined potentials of the volume source τ of density curl $\overline{M}$ and the surface sources $S_{1..n}\Sigma$ of density $\overline{M} \times \hat{n}$, where $\hat{n}$ is the outward unit normal from τ.

When 0 lies within τ the integral $\int_\tau \overline{M} \times \text{grad}\,\frac{1}{r}\,d\tau$ is improper. Let 0 be an interior point of the region τ' bounded by the closed regular surface S'. Then corresponding to equation (4.19-1) we now have

$$\int_{\tau-\tau'} \overline{M} \times \text{grad}\,\frac{1}{r}\,d\tau = \int_{\tau-\tau'} \frac{\text{curl}\,\overline{M}}{r}\,d\tau - \oint_{S_{1..n}\Sigma} d\overline{S} \times \frac{\overline{M}}{r} - \oint_{S'} d\overline{S} \times \frac{\overline{M}}{r} \qquad (4.19\text{-}2)$$

As S' shrinks about 0, $\int_{\tau-\tau'} \frac{\text{curl}\,\overline{M}}{r}\,d\tau$ converges and $\oint_{S'} d\overline{S} \times \frac{\overline{M}}{r} \to 0$ for reasons discussed previously (Ex.4-117., p. 353). Hence at interior points of τ

$$\lim_{\tau'\to 0}\int_{\tau-\tau'} \overline{M}\times \operatorname{grad}\frac{1}{r}\,d\tau \equiv \int_{\tau} \overline{M}\times \operatorname{grad}\frac{1}{r}\,d\tau = \int_{\tau}\frac{\operatorname{curl}\overline{M}}{r}\,d\tau - \oint_{S_{1..n}\Sigma} d\overline{S}\times\frac{\overline{M}}{r} \qquad (4.19\text{-}3)$$

This relationship may be shown to subsist upon $S_{1..n}\Sigma$, so that $\int_{\tau} \overline{M}\times \operatorname{grad}\frac{1}{r}\,d\tau$ is everywhere equal to the sum of the potential functions described above. We will consequently refer to it as a (vector) potential function[35]. It is everywhere continuous, but its normal derivative is discontinuous upon $S_{1..n}\Sigma$ by $-4\pi(\overline{M}\times\hat{n})$ for a common positive sense of the normal on both sides.

4.19a Divergence of $\int_{\tau} \overline{M}\times \operatorname{grad}\frac{1}{r}\,d\tau$ at interior and exterior points of τ

From equation (4.15-7) with $\overline{M}$ substituted for $\overline{J}$

$$\operatorname{curl}\int_{\tau}\frac{\overline{M}}{r}\,d\tau = \int_{\tau}\overline{M}\times \operatorname{grad}\frac{1}{r}\,d\tau$$

at interior and exterior points of τ,

hence

$$\operatorname{div}\int_{\tau}\overline{M}\times \operatorname{grad}\frac{1}{r}\,d\tau = \operatorname{div}\operatorname{curl}\int_{\tau}\frac{\overline{M}}{r}\,d\tau = 0 \qquad (4.19\text{-}4)$$

wherever it is defined.

4.19b Divergence of cavity potential $\int_{\tau-\tau_\delta}\overline{M}\times \operatorname{grad}\frac{1}{r}\,d\tau$

Since the δ sphere is fixed in space, the point of evaluation is an exterior point of the region of integration in the sense of the preceding sub-section, and the same result applies, viz.

---

35. At exterior points of τ, $\int_{\tau}\overline{M}\times \operatorname{grad}\frac{1}{r}\,d\tau$ is the potential of a limiting configuration of whirls within τ (p. 305) and it is in this context that the function usually arises (see Sec. 4.20b). At interior points of τ, however, the potential of such a complex is indeterminate, but we may continue to manipulate the integral indpendently of any such interpretation.

$$\text{div (cavity)} \int_{\tau-\tau_\delta} \overline{M} \times \text{grad}\,\frac{1}{r}\,d\tau = 0 \tag{4.19-5}$$

4.19c <u>Curl of</u> $\int_\tau \overline{M} \times \text{grad}\,\frac{1}{r}\,d\tau$ <u>at interior and exterior points of τ</u>

It follows from equations (4.19-1) and (4.19-3) that at interior and exterior points of τ

$$\text{curl} \int_\tau \overline{M} \times \text{grad}\,\frac{1}{r}\,d\tau = \text{curl} \int_\tau \frac{\text{curl}\,\overline{M}}{r}\,d\tau + \text{curl} \oint_{S_{1..n}\Sigma} \frac{\overline{M}}{r} \times \hat{n}\,dS$$

whence from equation (4.14-6) with $\overline{M} \times \hat{n}$ substituted for $\overline{K}$ and from (4.15-7) with curl $\overline{M}$ substituted for $\overline{J}$

$$\text{curl} \int_\tau \overline{M} \times \text{grad}\,\frac{1}{r}\,d\tau = -\int_\tau (\text{curl}\,\overline{M}) \times \frac{\overline{r}}{r^3}\,d\tau + \oint_{S_{1..n}\Sigma} (d\overline{S}\times\overline{M}) \times \frac{\overline{r}}{r^3} \tag{4.19-6}$$

Alternatively, from equation (4.15-6)

$$\text{curl} \int_\tau \overline{M} \times \text{grad}\,\frac{1}{r}\,d\tau = \int_\tau \frac{\text{curl curl}\,\overline{M}}{r}\,d\tau - \oint_{S_{1..n}\Sigma} d\overline{S} \times \frac{\text{curl}\,\overline{M}}{r} + \oint_{S_{1..n}\Sigma} (d\overline{S}\times\overline{M}) \times \frac{\overline{r}}{r^3} \tag{4.19-7}$$

Again, by combining (4.19-6) and (4.17-4), with $\overline{M}$ substituted for $\overline{F}$, we get

$$\text{curl} \int_\tau \overline{M} \times \text{grad}\,\frac{1}{r}\,d\tau = \int_\tau (\text{div}\,\overline{M})\,\frac{\overline{r}}{r^3}\,d\tau - \oint_{S_{1..n}\Sigma} \overline{M}.d\overline{S}\,\frac{\overline{r}}{r^3} \quad \text{at exterior points} \tag{4.19-8}$$

$$= \int_\tau (\text{div}\,\overline{M})\,\frac{\overline{r}}{r^3}\,d\tau - \oint_{S_{1..n}\Sigma} \overline{M}.d\overline{S}\,\frac{\overline{r}}{r^3} + 4\pi\overline{M} \quad \text{at interior points} \tag{4.19-9}$$

Finally, from equation (4.17-3),

$$\text{curl}\int_\tau \overline{M} \times \text{grad}\,\frac{1}{r}\,d\tau = -\,\text{grad}\int_\tau \overline{M}\,.\,\text{grad}\,\frac{1}{r}\,d\tau \quad \text{at exterior points} \tag{4.19-10}$$

$$= -\,\text{grad}\int_\tau \overline{M}\,.\,\text{grad}\,\frac{1}{r}\,d\tau + 4\pi\overline{M} \quad \text{at interior points} \tag{4.19-11}$$

4.19d <u>Curl of cavity potential</u> $\int_{\tau-\tau_\delta} \overline{M} \times \text{grad}\,\frac{1}{r}\,d\tau$

The appropriate expressions may be derived directly from (4.19-9) to (4.19-11) by taking account of the alteration of integration volume, the introduction of the additional S_δ and the exterior nature of the point of observation. We have

$$\text{curl (cavity)}\int_{\tau-\tau_\delta} \overline{M} \times \text{grad}\,\frac{1}{r}\,d\tau = -\int_{\tau-\tau_\delta} (\text{curl}\,\overline{M}) \times \frac{\overline{r}}{r^3}\,d\tau + \oint_{S_{1..n}\Sigma,\ S_\delta} (d\overline{S}\times\overline{M}) \times \frac{\overline{r}}{r^3} \tag{4.19-12}$$

$$= \int_{\tau-\tau_\delta} \frac{(\text{curl curl}\,\overline{M})}{r}\,d\tau - \oint_{S_{1..n}\Sigma,\ S_\delta} d\overline{S} \times \frac{\text{curl}\,\overline{M}}{r} + \oint_{S_{1..n}\Sigma,\ S_\delta} (d\overline{S}\times\overline{M}) \times \frac{\overline{r}}{r^3} \tag{4.19-13}$$

$$= \int_{\tau-\tau_\delta} (\text{div}\,\overline{M})\,\frac{\overline{r}}{r^3}\,d\tau - \oint_{S_{1..n}\Sigma,\ S_\delta} \overline{M}.d\overline{S}\,\frac{\overline{r}}{r^3} \tag{4.19-14}$$

$$= -\,\text{grad}\int_{\tau-\tau_\delta} \overline{M}\,.\,\text{grad}\,\frac{1}{r}\,d\tau \tag{4.19-15}$$

It follows from equation (4.19-12) that when the curl of the cavity potential is evaluated at the centre of the δ sphere

$$\lim_{\delta\to 0} \text{curl (cavity)} \int_{\tau-\tau_\delta} \overline{M} \times \text{grad}\,\frac{1}{r}\,d\tau$$

$$= -\int_\tau (\text{curl}\,\overline{M}) \times \frac{\overline{r}}{r^3}\,d\tau + \oint_{S_{1..n}\Sigma} (d\overline{S}\times\overline{M}) \times \frac{\overline{r}}{r^3} + \lim_{\delta\to 0} \oint_{S_\delta} (d\overline{S}\times\overline{M}) \times \frac{\overline{r}}{r^3}$$

$$= \text{curl} \int_\tau \overline{M} \times \text{grad}\,\frac{1}{r}\,d\tau - \frac{8}{3}\,\pi\overline{M}_o \qquad (4.19\text{-}16)$$

Like equation (4.18-14), this result is of considerable importance in the theory of polarised systems. It may be contrasted with the result found in Sec. 4.15 for a simple vector volume source, viz.

$$\lim_{\delta\to 0} \text{curl (cavity)} \int_{\tau-\tau_\delta} \frac{\overline{J}}{r}\,d\tau = \text{curl} \int_\tau \frac{\overline{J}}{r}\,d\tau$$

4.19e <u>Laplacian of</u> $\int_\tau \overline{M} \times \text{grad}\,\frac{1}{r}\,d\tau$ <u>at interior and exterior points of</u> τ

From equations (4.19-1) and (4.19-3)

$$\nabla^2 \int_\tau \overline{M} \times \text{grad}\,\frac{1}{r}\,d\tau = \nabla^2 \int_\tau \frac{\text{curl}\,\overline{M}}{r}\,d\tau - \nabla^2 \oint_{S_{1..n}\Sigma} d\overline{S} \times \frac{\overline{M}}{r}$$

But

$$\nabla^2 \oint_{S_{1..n}\Sigma} d\overline{S} \times \frac{\overline{M}}{r} = \oint_{S_{1..n}\Sigma} (d\overline{S}\times\overline{M})\,\nabla^2\left(\frac{1}{r}\right) = \overline{0}$$

hence from equations (4.15-8) and (4.15-9) with curl $\overline{M}$ substituted for $\overline{J}$

$$\nabla^2 \int_\tau \overline{M} \times \text{grad}\,\frac{1}{r}\,d\tau = \overline{0} \quad \text{at exterior points} \qquad (4.19\text{-}17)$$

$$\nabla^2 \int_\tau \overline{M} \times \text{grad}\,\frac{1}{r}\,d\tau = -4\pi\,\text{curl}\,\overline{M} \quad \text{at interior points} \qquad (4.19\text{-}18)$$

4.19f <u>Laplacian of cavity potential</u> $\int_{\tau-\tau_\delta} \overline{M} \times \text{grad}\,\frac{1}{r}\,d\tau$

From equation (4.19-2)

$$\nabla^2 \text{ (cavity)} \int_{\tau-\tau_\delta} \overline{M} \times \text{grad}\,\frac{1}{r}\,d\tau = \nabla^2 \text{ (cavity)} \int_{\tau-\tau_\delta} \frac{\text{curl}\,\overline{M}}{r}\,d\tau - \nabla^2 \cdot \oint_{S_{1..n}\Sigma,\ S_\delta} d\overline{S} \times \frac{\overline{M}}{r}$$

But

$$\nabla^2 \oint_{S_{1..n}\Sigma,\ S_\delta} d\overline{S} \times \frac{\overline{M}}{r} = \oint_{S_{1..n}\Sigma,\ S_\delta} (d\overline{S}\times\overline{M})\ \nabla^2\left(\frac{1}{r}\right) = \overline{0}$$

since the δ sphere is fixed in space. Further, each of the scalar components of ∇^2 (cavity) $\int_{\tau-\tau_\delta} \frac{\text{curl}\,\overline{M}}{r}\,d\tau$ is zero in accordance with (4.9-2), hence

$$\nabla^2 \text{ (cavity)} \int_{\tau-\tau_\delta} \overline{M} \times \text{grad}\,\frac{1}{r}\,d\tau = \overline{0} \qquad (4.19\text{-}19)$$

4.19g <u>Discontinuity of curl</u> $\int_\tau \overline{M} \times \text{grad}\,\frac{1}{r}\,d\tau$ <u>at the bounding surfaces</u>

We have

$$\text{curl} \int_\tau \overline{M} \times \text{curl}\,\frac{1}{r}\,d\tau = \text{curl} \int_\tau \frac{\text{curl}\,\overline{M}}{r}\,d\tau + \text{curl} \oint_{S_{1..n}\Sigma} \frac{\overline{M} \times \hat{n}}{r}\,dS$$

where $\hat{n}$ is the outward normal from τ.

The first term of the right hand side is continuous through $S_{1..n}\Sigma$ since it comprises the curl of the potential of a well-behaved volume source of density curl $\overline{M}$. The second term is discontinuous since it comprises the curl of the potential of a surface source of density $\overline{M} \times \hat{n}$. According to equation (4.14-8) it suffers a discontinuity of $-4\pi(\hat{n}\times(\overline{M}\times\hat{n}))$ with outward movement through $S_{1..n}\Sigma$, and this may be written as $-4\pi\overline{M}_t$, where $\overline{M}_t$ is the vector tangential component of $\overline{M}$.

4.19h The partial potential and its derivatives

Exercises involving the derivatives of the partial potential are introduced in the following pages. In these exercises, terms which include the factor $d\bar{S} \times \text{grad}\,\frac{1}{r}$ have been left intact (despite their being zero at each point of S_δ) in order to confirm, via equation (1.17-2), the convergent nature of the associated volume integrals at all points where $\bar{M}$ is well behaved.

Formulae developed in this and the preceding section are brought together in Tables 3 and 4, pp. 373-8.

EXERCISES

4-127. Prove that $\text{div} \int_\tau \bar{M} \times \text{grad}\,\frac{1}{r}\, d\tau = 0$ at interior and exterior points of τ by taking the divergence of equation (4.19-1) and utilising (4.15-4), together with the integral over $S_{1..n}\Sigma$ of curl $\frac{\bar{M}}{r}$ and its expansion.

4-128. Expand

$$\text{div} \int_\tau \bar{M} \times \text{grad}\,\frac{1}{r}\, d\tau = \sum \frac{\partial}{\partial x_o} \int_\tau \left\{ M_y \frac{\partial}{\partial z}\left(\frac{1}{r}\right) - M_z \frac{\partial}{\partial y}\left(\frac{1}{r}\right) \right\} d\tau$$

at points outside τ by differentiation of the derivatives of $\frac{1}{r}$, and so show that $\text{div} \int_\tau \bar{M} \times \text{grad}\,\frac{1}{r} = 0$ at exterior points.

4-129. Expand

$$\text{div (partial)} \int_{\tau-\tau_\delta} \bar{M} \times \text{grad}\,\frac{1}{r}\, d\tau = \sum \frac{\partial}{\partial x_o} \text{(partial)} \int_{\tau-\tau_\delta} \left\{ M_y \frac{\partial}{\partial z}\left(\frac{1}{r}\right) - M_z \frac{\partial}{\partial y}\left(\frac{1}{r}\right) \right\} d\tau$$

by the field-slipping technique to obtain

$$\int_{\tau-\tau_\delta} \left(\text{grad}\,\frac{1}{r}\right) . (\text{curl}\,\bar{M})\, d\tau - \oint_{S_{1..n}\Sigma} \left(\bar{M} \times \text{grad}\,\frac{1}{r}\right) . d\bar{S}$$

By expansion and integration of $\text{div}\,\frac{(\text{curl}\,\bar{M})}{r}$ over $\tau-\tau_\delta$ and curl $\frac{\bar{M}}{r}$ over $S_{1..n}\Sigma$, reduce this to

$$\oint_{S_\delta} \frac{(\text{curl}\,\bar{M})}{r} . d\bar{S} = \oint_{S_\delta} \bar{M} . \left(\text{grad}\,\frac{1}{r} \times d\bar{S}\right)$$

4-130. Obtain the result of the previous exercise by writing

$$\text{div (partial)} \int_{\tau-\tau_\delta} \bar{M} \times \text{grad}\,\frac{1}{r}\,d\tau$$

$$= \text{div (partial)} \int_{\tau-\tau_\delta} \frac{\text{curl}\,\bar{M}}{r}\,d\tau - \text{div} \oint_{S_{1..n}\Sigma} d\bar{S} \times \frac{\bar{M}}{r} - \text{div} \oint_{S_\delta} d\bar{S} \times \frac{\bar{M}}{r}$$

in accordance with equation (4.19-2). Bearing in mind that the δ sphere moves with 0, show that the above expressions may be brought into the form

$$-\oint_{S_{1..n}\Sigma} \frac{(\text{curl}\,\bar{M})}{r} \cdot d\bar{S} + \oint_{S_{1..n}\Sigma} (d\bar{S}\times\bar{M}) \cdot \text{grad}\,\frac{1}{r} - \sum \oint_{S_\delta} \frac{1}{r} \left\{dS_y \frac{\partial M_z}{\partial x} - dS_z \frac{\partial M_y}{\partial x}\right\}$$

Integrate curl $\frac{\bar{M}}{r}$ and its expansion over $S_{1..n}\Sigma$ to demonstrate the mutual cancellation of the first two terms, and show that the third term may be expressed as

$$\oint_{S_\delta} \frac{(\text{curl}\,\bar{M})}{r} \cdot d\bar{S} = \oint_{S_\delta} \bar{M} \cdot \left(\text{grad}\,\frac{1}{r} \times d\bar{S}\right)$$

4-131. Substitution of $\bar{M}$ for $\bar{J}$ in the expression for curl partial pot $\bar{J}$ in Table 2, p. 331 yields

$$\text{curl (partial)} \int_{\tau-\tau_\delta} \frac{\bar{M}}{r}\,d\tau = \int_{\tau-\tau_\delta} \bar{M} \times \text{grad}\,\frac{1}{r}\,d\tau + \oint_{S_\delta} d\bar{S} \times \frac{\bar{M}}{r}$$

Obtain the result of the previous two exercises by taking the divergence of both sides of this equation, bearing in mind that the δ sphere moves with 0.

4-132. Expand

$$\text{curl} \int_\tau \bar{M} \times \text{grad}\,\frac{1}{r}\,d\tau = \sum \bar{i} \left\{\frac{\partial}{\partial y_o} \int_\tau \left(\bar{M}\times \text{grad}\,\frac{1}{r}\right)_z d\tau - \frac{\partial}{\partial z_o} \int_\tau \left(\bar{M}\times \text{grad}\,\frac{1}{r}\right)_y d\tau\right\}$$

at exterior point of τ by differentiation of the derivatives of $\frac{1}{r}$ to obtain

$$\text{curl} \int_\tau \overline{M} \times \text{grad} \frac{1}{r} \, d\tau = - \text{grad} \int_\tau \overline{M} \cdot \text{grad} \frac{1}{r} \, d\tau$$

4-133. By writing

$$\text{curl (partial)} \int_{\tau-\tau_\delta} \overline{M} \times \text{grad} \frac{1}{r} \, d\tau$$

$$= \text{curl (partial)} \int_{\tau-\tau_\delta} \frac{\text{curl } \overline{M}}{r} \, d\tau - \text{curl} \oint_{S_{1..n}\Sigma} d\overline{S} \times \frac{\overline{M}}{r} - \text{curl} \oint_{S_\delta} d\overline{S} \times \frac{\overline{M}}{r}$$

and making use of the volume integral of curl $\frac{(\text{curl } \overline{M})}{r}$ and its expansion over $\tau-\tau_\delta$, show that

$$\text{curl (partial)} \int_{\tau-\tau_\delta} \overline{M} \times \text{grad} \frac{1}{r} \, d\tau$$

$$= \oint_{S_\delta} d\overline{S} \times \frac{\text{curl } \overline{M}}{r} - \int_{\tau-\tau_\delta} (\text{curl } \overline{M}) \times \frac{\overline{r}}{r^3} \, d\tau + \oint_{S_{1..n}\Sigma} (d\overline{S} \times \overline{M}) \times \frac{\overline{r}}{r^3} - \text{curl} \oint_{S_\delta} d\overline{S} \times \frac{\overline{M}}{r}$$

Expand the last term (moving δ sphere) and combine with the first to reduce the above expression to

$$- \int_{\tau-\tau_\delta} (\text{curl } \overline{M}) \times \frac{\overline{r}}{r^3} \, d\tau + \oint_{S_{1..n}\Sigma} (d\overline{S} \times \overline{M}) \times \frac{\overline{r}}{r^3} + \sum \overline{i} \oint_{S_\delta} \frac{1}{r} \{(d\overline{S} \times \text{grad } M_z)_y - (d\overline{S} \times \text{grad } M_y)_z\}$$

Show that the final term may be replaced by

$$\sum \overline{i} \oint_{S_\delta} \left\{ \left(d\overline{S} \times M_y \, \text{grad} \frac{1}{r} \right)_z - \left(d\overline{S} \times M_z \, \text{grad} \frac{1}{r} \right)_y \right\} = \oint_{S_\delta} \overline{M} \times \left(d\overline{S} \times \text{grad} \frac{1}{r} \right)$$

4-134. Make use of the field-slipping technique to expand

$$\text{curl (partial)} \int_{\tau-\tau_\delta} \overline{M} \times \text{grad}\,\frac{1}{r}\,d\tau$$

$$= \sum \bar{i} \left\{ \frac{\partial}{\partial y_o} \int_{\tau-\tau_\delta} \left(M_x \frac{\partial}{\partial y}\left(\frac{1}{r}\right) - M_y \frac{\partial}{\partial x}\left(\frac{1}{r}\right) \right) d\tau - \frac{\partial}{\partial z_o} \int_{\tau-\tau_\delta} \left(M_z \frac{\partial}{\partial x}\left(\frac{1}{r}\right) - M_x \frac{\partial}{\partial z}\left(\frac{1}{r}\right) \right) d\tau \right\}$$

and show that the result can be brought into the form

$$\int_{\tau-\tau_\delta} (\text{div}\,\overline{M})\,\frac{\overline{r}}{r^3}\,d\tau - \oint_{S_{1..n}\Sigma} \overline{M}.d\overline{S}\,\frac{\overline{r}}{r^3} + \oint_{S_\delta} \overline{M}\,\text{grad}\,\frac{1}{r}\,.\,d\overline{S}$$

Transform this via equation (4.17-4), modified for an integration region $\tau-\tau_\delta$ and surfaces $S_{1..n}\Sigma$, S_δ, into

$$-\int_{\tau-\tau_\delta} (\text{curl}\,\overline{M}) \times \frac{\overline{r}}{r^3}\,d\tau + \oint_{S_{1..n}\Sigma} (d\overline{S}\times\overline{M}) \times \frac{\overline{r}}{r^3} + \oint_{S_\delta} \left\{ \overline{M}.d\overline{S}\,\frac{\overline{r}}{r^3} + (d\overline{S}\times\overline{M}) \times \frac{\overline{r}}{r^3} - \overline{M}\,\frac{\overline{r}}{r^3}\,.\,d\overline{S} \right\}$$

Show that the last term may be reduced to

$$\oint_{S_\delta} \overline{M} \times \left(d\overline{S} \times \text{grad}\,\frac{1}{r} \right)$$

whence obtain the result of the previous exercise.

Show that the first form of curl (partial) $\int_{\tau-\tau_\delta} \overline{M} \times \text{grad}\,\frac{1}{r}\,d\tau$ developed above may be replaced by

$$-\,\text{grad (partial)} \int_{\tau-\tau_\delta} \overline{M}\,.\,\text{grad}\,\frac{1}{r}\,d\tau + \oint_{S_\delta} \left\{ \overline{M}\,\text{grad}\,\frac{1}{r}\,.\,d\overline{S} + \overline{M} \times \left(d\overline{S} \times \text{grad}\,\frac{1}{r} \right) \right\}$$

4-135. In Sec. 4.19d it was stated that

$$\lim_{\delta\to 0} \oint_{S_\delta} (d\bar{S}\times\bar{M}) \times \frac{\bar{r}}{r^3} = -\frac{8}{3}\pi\bar{M}_o$$

where 0 is the centre of the δ sphere.

Prove this.

4-136. Derive the values of the Laplacian of $\int_\tau \bar{M}\times \text{grad}\,\frac{1}{r}\,d\tau$ at interior and exterior points of τ by writing

$$\nabla^2 \int_\tau \bar{M}\times \text{grad}\,\frac{1}{r}\,d\tau = \text{grad div} \int_\tau \bar{M}\times \text{grad}\,\frac{1}{r}\,d\tau - \text{curl curl} \int_\tau \bar{M}\times \text{grad}\,\frac{1}{r}\,d\tau$$

and (a) applying (4.19-4), (4.19-10) and (4.19-11).

(b) putting $\int_\tau \bar{M}\times \text{grad}\,\frac{1}{r}\,d\tau = \text{curl} \int_\tau \frac{\bar{M}}{r}\,d\tau$ and writing

curl curl curl as curl (grad div $-\nabla^2$).

4-137. By writing

$$\nabla^2 \text{ (partial)} \int_{\tau-\tau_\delta} \bar{M}\times \text{grad}\,\frac{1}{r}\,d\tau$$

$$= \text{grad div (partial)} \int_{\tau-\tau_\delta} \bar{M}\times \text{grad}\,\frac{1}{r}\,d\tau - \text{curl curl (partial)} \int_{\tau-\tau_\delta} \bar{M}\times \text{grad}\,\frac{1}{r}\,d\tau$$

and applying the results of Ex.4-129. and 4-134., show that

$$\nabla^2 \text{ (partial)} \int_{\tau-\tau_\delta} \bar{M}\times \text{grad}\,\frac{1}{r}\,d\tau = -\oint_{S_\delta} \left\{\left(d\bar{S}\times\frac{\bar{r}}{r^3}\right) \text{div}\,\bar{M} + \left(d\bar{S}\times\frac{\bar{r}}{r^3}\right)\times \text{curl}\,\bar{M}\right\}$$

$$+ 2\oint_{S_\delta} \left\{\sum\left(d\bar{S}\times\frac{\bar{r}}{r^3}\right)_x \text{grad}\,M_x + \frac{\bar{r}}{r^3}.d\bar{S}\ \text{curl}\,\bar{M}\right\}$$

4-138. By writing

$$\nabla^2 \text{ (partial)} \int_{\tau-\tau_\delta} \overline{M} \times \text{grad } \frac{1}{r} \, d\tau = \nabla^2 \text{ (partial)} \int_{\tau-\tau_\delta} \frac{\text{curl } \overline{M}}{r} \, d\tau - \nabla^2 \oint_{S_{1..n}\Sigma,\ S_\delta} d\overline{S} \times \frac{\overline{M}}{r}$$

and demonstrating that

$$\nabla^2 \oint_{S_{1..n}\Sigma} d\overline{S} \times \frac{\overline{M}}{r} = \overline{0}$$

and

$$\nabla^2 \oint_{S_\delta} d\overline{S} \times \frac{\overline{M}}{r} = \oint_{S_\delta} \left\{ - \left(d\overline{S} \times \frac{\overline{r}}{r^3} \right) \times \text{curl } \overline{M} - 2 \sum \overline{i} \left(d\overline{S} \times \frac{\overline{r}}{r^3} \right) . \text{ grad } M_x \right.$$

$$\left. + \text{ div } \overline{M} \left(d\overline{S} \times \frac{\overline{r}}{r^3} \right) + \frac{1}{r} \frac{\partial}{\partial n} \text{ (curl } \overline{M}) \text{ } dS \right\}$$

arrive at

$$\nabla^2 \text{ (partial)} \int_{\tau-\tau_\delta} \overline{M} \times \text{grad } \frac{1}{r} \, d\tau$$

$$= \oint_{S_\delta} \left\{ \left(d\overline{S} \, . \, \frac{\overline{r}}{r^3} \right) \text{curl } \overline{M} + \left(d\overline{S} \times \frac{\overline{r}}{r^3} \right) \times \text{curl } \overline{M} - \left(d\overline{S} \times \frac{\overline{r}}{r^3} \right) \text{div } \overline{M} \right.$$

$$\left. + 2 \sum \overline{i} \left(d\overline{S} \times \frac{\overline{r}}{r^3} \right) . \text{ grad } \overline{M}_x \right\}$$

and show that this expression is identical with that derived in the previous exercise.

4-139. Since

$$\text{curl (partial)} \int_{\tau-\tau_\delta} \frac{\overline{M}}{r} \, d\tau = \int_{\tau-\tau_\delta} \overline{M} \times \text{grad } \frac{1}{r} \, d\tau + \oint_{S_\delta} d\overline{S} \times \frac{\overline{M}}{r}$$

we may write

$$\text{curl curl (partial)} \int_{\tau-\tau_\delta} \frac{\overline{M}}{r}\, d\tau = \text{curl (partial)} \int_{\tau-\tau_\delta} \overline{M} \times \text{grad}\, \frac{1}{r}\, d\tau + \text{curl} \oint_{S_\delta} d\overline{S} \times \frac{\overline{M}}{r}$$

whence, from a result of Ex.4-134. above,

$$\text{curl curl (partial)} \int_{\tau-\tau_\delta} \frac{\overline{M}}{r}\, d\tau$$

$$= - \int_{\tau-\tau_\delta} \text{div}\, \overline{M}\, \text{grad}\, \frac{1}{r}\, d\tau + \oint_{S_{1..n}\Sigma} \overline{M}.d\overline{S}\, \text{grad}\, \frac{1}{r} + \oint_{S_\delta} \overline{M}\, \text{grad}\, \frac{1}{r} \,.\, d\overline{S} + \text{curl} \oint_{S_\delta} d\overline{S} \times \frac{\overline{M}}{r}$$

Expand the final term, bearing in mind that in this case $\overline{M}$ alone is variable, and by making use of an expansion for grad div partial pot $\overline{J}$ in Table 2, p. 331, arrive at

$$\text{curl curl partial pot}\, \overline{M} = \text{grad div partial pot}\, \overline{M} + \oint_{S_\delta} \left\{ \overline{M} \frac{\partial}{\partial n}\left(\frac{1}{r}\right) - \frac{1}{r} \frac{\partial \overline{M}}{\partial n} \right\} dS$$

4-140. Simplify the expressions in Ex.4-137. and 4-138. by demonstrating that

$$\oint_{S_\delta} \sum \left(d\overline{S} \times \frac{\overline{r}}{r^3}\right)_x \text{grad}\, M_x = \oint_{S_\delta} \left(d\overline{S} \times \frac{\overline{r}}{r^3}\right) \times \text{curl}\, \overline{M}$$

and

$$\oint_{S_\delta} \sum \overline{i} \left(d\overline{S} \times \frac{\overline{r}}{r^3}\right) .\, \text{grad}\, M_x = \overline{0}$$

TABLE 3

The Scalar Potential Function $\int \overline{P} \cdot \text{grad}\,\frac{1}{r}\,d\tau$ and its Derivatives

(1)

$$\int_{\tau} \overline{P} \cdot \text{grad}\,\frac{1}{r}\,d\tau = \int_{\tau} \frac{(-\text{div}\,\overline{P})}{r}\,d\tau + \oint_{S_{1..n}\Sigma} \frac{P}{r} \cdot d\overline{S}$$

(interior and exterior points of τ)

(2)

$$\text{partial} \int_{\tau-\tau_{\delta}} \overline{P} \cdot \text{grad}\,\frac{1}{r}\,d\tau = \int_{\tau-\tau_{\delta}} \frac{(-\text{div}\,\overline{P})}{r}\,d\tau + \oint_{S_{1..n}\Sigma,\ S_{\delta}} \frac{\overline{P}}{r} \cdot d\overline{S}$$

(evaluated at centre of moving δ sphere within τ)

(3)

$$\text{cavity} \int_{\tau-\tau_{\delta}} \overline{P} \cdot \text{grad}\,\frac{1}{r}\,d\tau = \int_{\tau-\tau_{\delta}} \frac{(-\text{div}\,\overline{P})}{r}\,d\tau + \oint_{S_{1..n}\Sigma,\ S_{\delta}} \frac{\overline{P}}{r} \cdot d\overline{S}$$

(defined throughout fixed δ sphere within τ)

(4)

$$\text{grad} \int_{\tau} \overline{P} \cdot \text{grad}\,\frac{1}{r}\,d\tau = \int_{\tau} (-\text{div}\,\overline{P})\,\frac{\overline{r}}{r^3}\,d\tau + \oint_{S_{1..n}\Sigma} \overline{P}.d\overline{S}\,\frac{\overline{r}}{r^3}$$

(interior and exterior points of τ)

$$= \int_{\tau} (\text{curl}\,\overline{P}) \times \frac{\overline{r}}{r^3}\,d\tau - \oint_{S_{1..n}\Sigma} (d\overline{S}\times\overline{P}) \times \frac{\overline{r}}{r^3}$$

$$= -\,\text{curl} \int_{\tau} \overline{P} \times \text{grad}\,\frac{1}{r}\,d\tau$$

(exterior points of τ)

$$= \int_{\tau} (\text{curl}\,\overline{P}) \times \frac{\overline{r}}{r^3}\,d\tau - \oint_{S_{1..n}\Sigma} (d\overline{S}\times\overline{P}) \times \frac{\overline{r}}{r^3} + 4\pi\overline{P}$$

$$= -\,\text{curl} \int_{\tau} \overline{P} \times \text{grad}\,\frac{1}{r}\,d\tau + 4\pi\overline{P}$$

(interior points of τ)

TABLE 3(CONTD.)

(5)

$$\text{grad (cavity)} \int_{\tau-\tau_\delta} \overline{P} \cdot \text{grad} \frac{1}{r} \, d\tau = \int_{\tau-\tau_\delta} (-\text{div}\, \overline{P}) \frac{\overline{r}}{r^3} \, d\tau + \oint_{S_{1..n}\Sigma,\, S_\delta} \overline{P} . d\overline{S} \frac{\overline{r}}{r^3}$$

$$= \int_{\tau-\tau_\delta} (\text{curl}\, \overline{P}) \times \frac{\overline{r}}{r^3} \, d\tau - \oint_{S_{1..n}\Sigma,\, S_\delta} (d\overline{S} \times \overline{P}) \times \frac{\overline{r}}{r^3}$$

$$= - \text{curl} \int_{\tau-\tau_\delta} \overline{P} \times \text{grad} \frac{1}{r} \, d\tau$$

(6)

$$\lim_{\delta\to 0} \text{grad (cavity)} \int_{\tau-\tau_\delta} \overline{P} \cdot \text{grad} \frac{1}{r} \, d\tau = \text{grad} \int_{\tau} \overline{P} \cdot \text{grad} \frac{1}{r} \, d\tau - \frac{4}{3} \pi \overline{P}$$

for evaluation at the centre of the δ sphere.

(7)

$$\text{grad (partial)} \int_{\tau-\tau_\delta} \overline{P} \cdot \text{grad} \frac{1}{r} \, d\tau$$

$$= \int_{\tau-\tau_\delta} (-\text{div}\, \overline{P}) \frac{\overline{r}}{r^3} \, d\tau + \oint_{S_{1..n}\Sigma} \overline{P} . d\overline{S} \frac{\overline{r}}{r^3} - \oint_{S_\delta} \overline{P} \times \left(d\overline{S} \times \frac{\overline{r}}{r^3} \right)$$

$$= \int_{\tau-\tau_\delta} (\text{curl}\, \overline{P}) \times \frac{\overline{r}}{r^3} \, d\tau - \oint_{S_{1..n}\Sigma} (d\overline{S} \times \overline{P}) \times \frac{\overline{r}}{r^3} + \oint_{S_\delta} \overline{P} \, \text{grad} \frac{1}{r} \cdot d\overline{S}$$

$$= - \text{curl (partial)} \int_{\tau-\tau_\delta} \overline{P} \times \text{grad} \frac{1}{r} \, d\tau - \oint_{S_\delta} \left\{ \overline{P} \, d\overline{S} \cdot \frac{\overline{r}}{r^3} + \overline{P} \times \left(d\overline{S} \times \frac{\overline{r}}{r^3} \right) \right\}$$

(8)

$$\nabla^2 \int_{\tau} \overline{P} \cdot \text{grad} \frac{1}{r} \, d\tau = 0 \quad \text{at exterior points of } \tau$$

TABLE 3(CONTD.)

(9)

$$\nabla^2 \int_\tau \overline{P} \cdot \text{grad} \frac{1}{r} \, d\tau = 4\pi \, \text{div} \, \overline{P} \quad \text{at interior points of } \tau$$

(10)

$$\nabla^2 \text{ (cavity)} \int_{\tau-\tau_\delta} \overline{P} \cdot \text{grad} \frac{1}{r} \, d\tau = 0$$

(11)

$$\nabla^2 \text{ (partial)} \int_{\tau-\tau_\delta} \overline{P} \cdot \text{grad} \frac{1}{r} \, d\tau = - \oint_{S_\delta} \left\{ \text{div} \, \overline{P} \, d\overline{S} \cdot \frac{\overline{r}}{r^3} + (\text{curl} \, \overline{P}) \cdot \left(d\overline{S} \times \frac{\overline{r}}{r^3} \right) \right\}$$

TABLE 4

The Vector Potential Function $\int \overline{M} \times \text{grad} \frac{1}{r} \, d\tau$ and its Derivatives

(1)

$$\int_\tau \overline{M} \times \text{grad} \frac{1}{r} \, d\tau = \int_\tau \frac{\text{curl} \, \overline{M}}{r} \, d\tau - \oint_{S_{1..n}\Sigma} d\overline{S} \times \frac{\overline{M}}{r} \quad \text{(interior and exterior points of } \tau)$$

(2)

$$\text{partial} \int_{\tau-\tau_\delta} \overline{M} \times \text{grad} \frac{1}{r} \, d\tau = \int_{\tau-\tau_\delta} \frac{\text{curl} \, \overline{M}}{r} \, d\tau - \oint_{S_{1..n}\Sigma, \, S_\delta} d\overline{S} \times \frac{\overline{M}}{r} \quad \text{(evaluated at centre of moving } \delta \text{ sphere within } \tau)$$

(3)

$$\text{cavity} \int_{\tau-\tau_\delta} \overline{M} \times \text{grad} \frac{1}{r} \, d\tau = \int_{\tau-\tau_\delta} \frac{\text{curl} \, \overline{M}}{r} \, d\tau - \oint_{S_{1..n}\Sigma, \, S_\delta} d\overline{S} \times \frac{\overline{M}}{r} \quad \text{(defined throughout fixed } \delta \text{ sphere within } \tau)$$

(4)

$$\text{div} \int_\tau \overline{M} \times \text{grad} \frac{1}{r} \, d\tau = 0 \quad \text{(interior and exterior points of } \tau)$$

(5)

$$\text{div (cavity)} \int_{\tau-\tau_\delta} \overline{M} \times \text{grad} \frac{1}{r} \, d\tau = 0$$

TABLE 4(CONTD.)

(6)

$$\text{div (partial)} \int_{\tau-\tau_\delta} \overline{M} \times \text{grad}\,\frac{1}{r}\,d\tau = \oint_{S_\delta} \overline{M} \cdot \left(d\overline{S} \times \frac{\overline{r}}{r^3}\right)$$

(7)

$$\text{curl} \int_\tau \overline{M} \times \text{grad}\,\frac{1}{r}\,d\tau = -\int_\tau (\text{curl}\,\overline{M}) \times \frac{\overline{r}}{r^3}\,d\tau + \oint_{S_{1..n}\Sigma} (d\overline{S}\times\overline{M}) \times \frac{\overline{r}}{r^3} \quad \text{(interior and exterior points of } \tau)$$

$$\left.\begin{aligned} &= \int_\tau (\text{div}\,\overline{M})\,\frac{\overline{r}}{r^3}\,d\tau - \oint_{S_{1..n}\Sigma} \overline{M}.d\overline{S}\,\frac{\overline{r}}{r^3} \\ &= -\,\text{grad} \int_\tau \overline{M}\,.\,\text{grad}\,\frac{1}{r}\,d\tau \end{aligned}\right\} \quad \text{(exterior points of } \tau)$$

$$\left.\begin{aligned} &= \int_\tau (\text{div}\,\overline{M})\,\frac{\overline{r}}{r^3}\,d\tau - \oint_{S_{1..n}\Sigma} \overline{M}.d\overline{S}\,\frac{\overline{r}}{r^3} + 4\pi\overline{M} \\ &= -\,\text{grad} \int_\tau \overline{M}\,.\,\text{grad}\,\frac{1}{r}\,d\tau + 4\pi\overline{M} \end{aligned}\right\} \quad \text{(interior points of } \tau)$$

(8)

$$\text{curl (cavity)} \int_{\tau-\tau_\delta} \overline{M} \times \text{grad}\,\frac{1}{r}\,d\tau = -\int_{\tau-\tau_\delta} (\text{curl}\,\overline{M}) \times \frac{\overline{r}}{r^3}\,d\tau + \oint_{S_{1..n}\Sigma,\ S_\delta} (d\overline{S}\times\overline{M}) \times \frac{\overline{r}}{r^3}$$

$$= \int_{\tau-\tau_\delta} (\text{div}\,\overline{M})\,\frac{\overline{r}}{r^3}\,d\tau - \oint_{S_{1..n}\Sigma,\ S_\delta} \overline{M}.d\overline{S}\,\frac{\overline{r}}{r^3}$$

$$= -\,\text{grad} \int_{\tau-\tau_\delta} \overline{M}\,.\,\text{grad}\,\frac{1}{r}\,d\tau$$

TABLE 4 (CONTD.)

(9)

$$\lim_{\delta\to 0} \text{curl (cavity)} \int_{\tau-\tau_\delta} \bar{M} \times \text{grad} \frac{1}{r}\, d\tau = \text{curl} \int_{\tau} \bar{M} \times \text{grad} \frac{1}{r}\, d\tau - \frac{8}{3}\pi\bar{M}$$

for evaluation at the centre of the δ sphere.

(10)

$$\text{curl (partial)} \int_{\tau-\tau_\delta} \bar{M} \times \text{grad} \frac{1}{r}\, d\tau$$

$$= -\int_{\tau-\tau_\delta} (\text{curl } \bar{M}) \times \frac{\bar{r}}{r^3}\, d\tau + \oint_{S_{1..n}\Sigma} (d\bar{S}\times\bar{M}) \times \frac{\bar{r}}{r^3} - \oint_{S_\delta} \bar{M} \times \left(d\bar{S} \times \frac{\bar{r}}{r^3}\right)$$

$$= \int_{\tau-\tau_\delta} (\text{div } \bar{M}) \frac{\bar{r}}{r^3}\, d\tau - \oint_{S_{1..n}\Sigma} \bar{M}.d\bar{S}\, \frac{\bar{r}}{r^3} - \oint_{S_\delta} \bar{M}\, d\bar{S} \,.\, \frac{\bar{r}}{r^3}$$

$$= -\text{grad (partial)} \int_{\tau-\tau_\delta} \bar{M} \,.\, \text{grad} \frac{1}{r}\, d\tau - \oint_{S_\delta} \left\{\bar{M}\, d\bar{S} \,.\, \frac{\bar{r}}{r^3} + \bar{M} \times \left(d\bar{S} \times \frac{\bar{r}}{r^3}\right)\right\}$$

(11)

$$\nabla^2 \int_{\tau} \bar{M} \times \text{grad} \frac{1}{r}\, d\tau = \bar{0} \quad \text{at exterior points of } \tau$$

(12)

$$\nabla^2 \int_{\tau} \bar{M} \times \text{grad} \frac{1}{r}\, d\tau = -4\pi \text{ curl } \bar{M} \quad \text{at interior points of } \tau$$

(13)

$$\nabla^2 \text{ (cavity)} \int_{\tau-\tau_\delta} \bar{M} \times \text{grad} \frac{1}{r}\, d\tau = \bar{0}$$

(14)

$$\text{grad div} \int_{\tau} \bar{M} \times \text{grad} \frac{1}{r}\, d\tau = \bar{0} \quad \text{(interior and exterior points of } \tau)$$

TABLE 4(CONTD.)

(15)

$$\text{curl curl}\int_\tau \overline{M}\times \text{grad}\,\frac{1}{r}\,d\tau = \overline{0} \quad \text{at exterior points of } \tau$$

(16)

$$\text{curl curl}\int_\tau \overline{M}\times \text{grad}\,\frac{1}{r}\,d\tau = 4\pi\,\text{curl}\,\overline{M} \quad \text{at interior points of } \tau$$

(17)

$$\nabla^2\ (\text{partial})\int_{\tau-\tau_\delta} \overline{M}\times \text{grad}\,\frac{1}{r}\,d\tau$$

$$= \oint_{S_\delta}\left\{d\overline{S}\,.\,\frac{\overline{r}}{r^3}\,\text{curl}\,\overline{M} - \left(d\overline{S}\times\frac{\overline{r}}{r^3}\right)\text{div}\,\overline{M} + \left(d\overline{S}\times\frac{\overline{r}}{r^3}\right)\times\text{curl}\,\overline{M}\right\}$$

(18)

$$\text{grad div (partial)}\int_{\tau-\tau_\delta} \overline{M}\times \text{grad}\,\frac{1}{r}\,d\tau = \oint_{S_\delta}\left\{\left(d\overline{S}\times\frac{\overline{r}}{r^3}\right)\times\text{curl}\,\overline{M}\right\}$$

(19)

$$\text{curl curl (partial)}\int_{\tau-\tau_\delta} \overline{M}\times \text{grad}\,\frac{1}{r}\,d\tau = \oint_{S_\delta}\left\{\left(d\overline{S}\times\frac{\overline{r}}{r^3}\right)\text{div}\,\overline{M} - d\overline{S}\,.\,\frac{\overline{r}}{r^3}\,\text{curl}\,\overline{M}\right\}$$

4.20 Introduction to the Macroscopic Potentials[36]

4.20a The macroscopic scalar potential

Suppose that a system of point singlets and doublets occupies a region τ of space. Then provided that the distribution exhibits a sufficient degree of statistical regularity, its potential may be approximated at exterior points of τ by the expression

$$\int_\tau \frac{\rho}{r}\,d\tau + \int_\tau \overline{P}\,.\,\text{grad}\,\frac{1}{r}\,d\tau \qquad (4.20\text{-}1)$$

36. This subject is treated in more general form and in greater detail in Sec. 5.18.

where ρ and $\overline{P}$ are continuous point functions which are related to the source strength/unit volume, and the source polarisation/unit volume relative to a local origin[37], in a neighbourhood of the point of evaluation.

For the most general type of mixed distribution the second term of equation (4.20-1) may be shown to derive in part from the singlet component, although the contribution from this will usually be small; when doublets alone are present, only the second term survives. This may be demonstrated plausibly in the following way.

It is shown in Sec. 4.1 that the potential at 0 of a point doublet located at P is given by

$$\phi_O = -\overline{p} \cdot \left(\frac{\overline{r}}{r^3}\right)_P = \overline{p} \cdot \left(\text{grad } \frac{1}{r}\right)_P \qquad (4.20\text{-}2)$$

where $\overline{p}$ is the vector moment of the doublet and $\overline{r} = \overrightarrow{OP}$.

A set of such doublets, when occupying the subregion $\Delta\tau$ whose distance from 0 is large compared with its dimensions, gives rise to the potential

$$\phi_O = \sum_{\Delta\tau} \left\{\overline{p} \cdot \left(\text{grad } \frac{1}{r}\right)_P\right\} \approx \left(\frac{\Sigma\overline{p}}{\Delta\tau}\right) \cdot \left(\text{grad } \frac{1}{r}\right)_{P'} \Delta\tau$$

where P' is a point of the subregion.

The potential of a set of juxtaposed subregions comprising the region τ is consequently given approximately at exterior points by

$$\sum_{\tau} \left\{\left(\frac{\Sigma\overline{p}}{\Delta\tau}\right) \cdot \left(\text{grad } \frac{1}{r}\right)_{P'} \Delta\tau\right\}$$

If the dimensions of the individual subregions, when restricted to values small compared with their distance from 0, are nevertheless sufficiently large in relation to the 'grain size' of the structure for conditions of statistical regularity to prevail within them, with $\frac{\Sigma\overline{p}}{\Delta\tau}$ varying gradually from one subregion to the next, it may be assumed that the above summation can be replaced by the integral

$$\int_{\tau} \overline{P} \cdot \text{grad } \frac{1}{r} \, d\tau$$

37. See p. 224.

where $\bar{P}$ is smoothly interpolated between the individual values of $\frac{\Sigma \bar{p}}{\Delta\tau}$ assigned, say, to the centre of each subregion.

Equation (4.20-1) serves to define the 'macroscopic' potential of the mixed source system. This point function is well-behaved, not only at exterior points of τ, where it approximates the 'true' or 'microscopic' potential, $\sum \frac{a_i}{r_i}$, but at interior and boundary points as well, where the microscopic potential may be undefined. The more finely-grained the distribution, the more closely does the macroscopic potential approximate the microscopic right up to the source boundary.

Macroscopic potentials may also be defined for point singlet and doublet line and surface sources. In this connection it will be observed that equations (4.2-10) and (4.3-3) may be looked upon as the macroscopic potentials of normally-orientated line and surface doublets in a limiting, finely-grained configuration, or as the potentials of paired, continuous line and surface sources. (Where continuous sources are concerned, the terms 'macroscopic' and 'microscopic' are superfluous.)

For present purposes we will prescind from singlet distributions altogether, and suppose that potentials of the form $\int_\tau \frac{\rho}{r}\, d\tau$ and $\int_S \frac{\sigma}{r}\, dS$ invariably derive from continuous volume and surface sources. Correspondingly, we will identify $\int_\tau \bar{P} \cdot \operatorname{grad} \frac{1}{r}\, d\tau$ (or its surface equivalent) as the macroscopic potential of a statistically-regular volume (or surface) distribution of point doublets.

4.20b The macroscopic vector potential

It has been shown in Sec. 4.12a that the vector potential of a closed, uniform, tangential line source of magnitude I and vector area $\bar{S}$, which shrinks about a point P while maintaining $I\bar{S}$ constant, is given at an exterior point O by

$$\bar{A}_O = \bar{m} \times \left(\operatorname{grad} \frac{1}{r}\right)_P \qquad (4.20\text{-}3)$$

where

$$\bar{m} = I\bar{S}$$

The limiting configuration of this source is known as a whirl or vector doublet, and $\overline{m}$ is said to be the moment of the whirl, by analogy with $\overline{p}$ in equation (4.20-2)[38].

It is clear that by proceeding from (4.20-3) as from (4.20-2), we can demonstrate with equal plausibility that the vector potential at exterior points of a statistically-regular volume distribution of whirls is approximated by

$$\int_\tau \overline{M} \times \text{grad}\,\frac{1}{r}\,d\tau$$

where $\overline{M}$ is a continuous point function which is smoothly interpolated between individual values of $\frac{\Sigma\overline{m}}{\Delta\tau}$ assigned, say, to the centre of each subregion.

This expression serves to define the macroscopic potential of such a distribution. Unlike the microscopic potential, $\sum \oint I\,\frac{d\overline{r}}{r}$, which is undefined at the source points, the macroscopic potential is defined and continuous everywhere.

For a mixed volume distribution of vector singlets and doublets the macroscopic potential takes the form

$$\int_\tau \frac{\overline{J}}{r}\,d\tau + \int_\tau \overline{M} \times \text{grad}\,\frac{1}{r}\,d\tau \qquad (4.20\text{-}4)$$

where $\overline{J}$ is evaluated statistically in terms of the singlet distribution. This is entirely analogous to equation (4.20-1). We have not yet described the nature of the vector singlet, nor need it concern us here. For present purposes it will be supposed that integrals of the form $\int_\tau \frac{\overline{J}}{r}\,d\tau$ derive from continuous density distributions and that the function $\int_\tau \overline{M} \times \text{grad}\,\frac{1}{r}\,d\tau$ has its sole origin in a statistically-regular volume distribution of whirls.

---

38. This involves an inconsistency since it might reasonably be expected that the moment of the whirl about some exterior point O' would be defined as $\oint \overline{r}' \times I\,d\overline{r}'$. However, the result of Ex.1-61., p. 78 identifies this integral with $2\overline{m}$ rather than $\overline{m}$. As in the case of the scalar doublet, the moment is independent of the position of O'.

A statistically-regular open surface distribution of whirls gives rise to the macroscopic potential $\int_S \overline{M} \times \text{grad}\,\frac{1}{r}\,dS$, where $\overline{M}$ is now related to the vector moment/unit area in a neighbourhood of the point of evaluation.

When the whirls are aligned normally, ie when the whirl moments are everywhere normal to the surface, the potential is continuous through the surface at interior points where $\overline{M}$ is continuous.

This may be seen by writing

$$\int_S \overline{M} \times \text{grad}\,\frac{1}{r}\,dS = \text{curl}\int_S \frac{\overline{M}}{r}\,dS$$

and making use of the relationship (4.14-8).

If $\overline{M} = \hat{n}\,M$ and M is constant over the surface

$$\int_S \overline{M} \times \text{grad}\,\frac{1}{r}\,dS = M\int_S d\overline{S} \times \text{grad}\,\frac{1}{r} = M\oint_\Gamma \frac{d\overline{r}}{r}$$

where the contour integration is carried out right-handedly in relation to the positive sense of the normal at the surface. It follows that the macroscopic potential of this particular surface distribution of whirls is identical with the vector potential of a tangential line source of constant magnitude M coincident with the periphery.

4.21 Inverse-Square Vector Fields and their Relationship to the Potential Functions

4.21a Inverse-square fields deriving from scalar sources

The inverse-square vector point function $\overline{U}$ deriving from Newtonian point sources is defined at any point 0 not coincident with a source by

$$\overline{U}_o = -\sum_{i=1}^{n} a_i \frac{\overline{r}_i}{r_i^3} = -\sum_{i=1}^{n} a_i \frac{\hat{\overline{r}}_i}{r_i^2} \qquad (4.21\text{-}1)$$

where $\hat{\overline{r}}_i$ is a unit vector directed from 0 to the ith source element.

It is seen that each source element of positive magnitude is supposed to give rise at 0 to a vector contribution directed radially away from the element and diminishing as the square of the distance from it.

For simple line, surface and volume sources of piecewise continuous density we have

$$\overline{U}_o = -\int_\Gamma \lambda \frac{\overline{r}}{r^3}\, ds \quad ; \quad \overline{U}_o = -\int_S \sigma \frac{\overline{r}}{r^3}\, dS \quad ; \quad \overline{U}_o = -\int_\tau \rho \frac{\overline{r}}{r^3}\, d\tau \qquad (4.21\text{-}2)$$

Reference to (4.6-3), (4.7-1), (4.7-7) and (4.8-1) reveals that in each case

$$\overline{U} = -\operatorname{grad} \phi \qquad (4.21\text{-}3)$$

at exterior points of the source, where ϕ is the associated scalar potential.

It is clear that equation (4.21-3) will continue to hold for all combinations of source elements and must therefore be valid for point, line and surface doublets and for combinations of these.

The volume integral in equation (4.21-2) is convergent at interior points of a piecewise continuous source (including those points at which ρ is discontinuous), and (4.21-3) continues to apply there. An outline of the proof appears in Ex.4-44., p. 282. The relationship has been derived for a continuous source in Sec. 4.8.

It follows from equation (4.21-1) that in the case of a volume distribution of point doublets, $\overline{U}$ must be interpreted as the negative gradient of the microscopic potential. At exterior points of the distribution, where the microscopic potential is adequately represented by the macroscopic, we may consequently write

$$\overline{U} = -\operatorname{grad} \int_\tau \overline{P} \cdot \operatorname{grad} \frac{1}{r}\, d\tau \qquad (4.21\text{-}4)$$

$$= \operatorname{curl} \int_\tau \overline{P} \times \operatorname{grad} \frac{1}{r}\, d\tau \quad \text{from (4.18-8)} \qquad (4.21\text{-}5)$$

This fails at interior points.

Let a cavity be created within the distribution. Then provided that the dimensions of the cavity are sufficiently large, and the point of evaluation sufficiently removed from the walls, the inverse-square field at O deriving from sources outside the cavity will be given by

$$\overline{U}_o \text{ (cavity)} = -\operatorname{grad} \text{ (cavity)} \int_{\tau-\tau'} \overline{P} \cdot \operatorname{grad} \frac{1}{r}\, d\tau \qquad (4.21\text{-}6)$$

(The cavity is denoted by τ' rather than τ_δ since the analysis will hold for any regular region.)

By combining equations (4.18-4) and (4.18-10) we find that

$$\overline{U}_o \text{ (cavity)} = -\text{grad}\int_\tau \overline{P}\,.\,\text{grad}\,\frac{1}{r}\,d\tau + \int_{\tau'} (-\text{div}\,\overline{P})\,\frac{\overline{r}}{r^3}\,d\tau - \oint_{S'} \overline{P}.d\overline{S}\,\frac{\overline{r}}{r^3} \tag{4.21-7}$$

where S' is the bounding surface of the cavity and the volume integrals relate to the undisturbed distribution.

Now suppose that the cavity exhibits point symmetry about 0. Then so long as div $\overline{P}$ is sensibly constant over the excised region, the volume integral over τ' vanishes and

$$\overline{U}_o \text{ (cavity)} = -\text{grad}\int_\tau \overline{P}\,.\,\text{grad}\,\frac{1}{r}\,d\tau - \oint_{S'} \overline{P}.d\overline{S}\,\frac{\overline{r}}{r^3} \tag{4.21-8}$$

We will consider explicitly three types of cavity:

(1) a cylinder of large length/diameter ratio whose axis is aligned with the direction of $\overline{P}$ at 0 (needle-shaped)

(2) a cylinder of small length/diameter ratio of the above alignment (disc-shaped)

(3) a sphere.

The overall dimensions of these cavities are supposed to be the minimum consistent with the requirement that the microscopic potential of the distribution in a neighbourhood of 0 is closely matched by the macroscopic. Then if $\overline{P}$ is constant over the excised region[39], an evaluation of the surface integral shows that

$$\overline{U}_o \text{ (needle-shaped cavity)} = -\text{grad}\int_\tau \overline{P}\,.\,\text{grad}\,\frac{1}{r}\,d\tau \tag{4.21-9}$$

$$\overline{U}_o \text{ (disc-shaped cavity)} = -\text{grad}\int_\tau \overline{P}\,.\,\text{grad}\,\frac{1}{r}\,d\tau + 4\pi\overline{P}_o \tag{4.21-10}$$

$$\overline{U}_o \text{ (spherical cavity)} = -\text{grad}\int_\tau \overline{P}\,.\,\text{grad}\,\frac{1}{r}\,d\tau + \frac{4}{3}\pi\overline{P}_o \tag{4.21-11}$$

---

39. This requirement is more severe than necessary. See Ex.4-145. and 4-146., p. 391.

By combining equations (4.21-9) to (4.21-11) with (4.18-9) we obtain the alternative forms

$$\overline{U}_o \text{ (needle-shaped cavity)} = \text{curl} \int_\tau \overline{P} \times \text{grad} \frac{1}{r} \, d\tau - 4\pi \overline{P}_o \qquad (4.21\text{-}12)$$

$$\overline{U}_o \text{ (disc-shaped cavity)} = \text{curl} \int_\tau \overline{P} \times \text{grad} \frac{1}{r} \, d\tau \qquad (4.21\text{-}13)$$

$$\overline{U}_o \text{ (spherical cavity)} = \text{curl} \int_\tau \overline{P} \times \text{grad} \frac{1}{r} \, d\tau - \frac{8}{3} \pi \overline{P}_o \qquad (4.21\text{-}14)$$

It should be noted that the absolute dimensions of the cavity are not contained in any expression.

The surface integral which appears in equation (4.21-8), and is evaluated in subsequent equations, stems from the restriction of the integration region to $\tau - \tau'$. The presence of the cavity, as such, has no bearing upon the situation[40]. Thus, in the determination of the value of $\overline{U}_o$ for a complete source, the contribution from doublets enclosed by S' is added to the external contribution as expressed above. Since the value of $\overline{U}_o$ must be independent of the mode of division, as determined by the shape and size of S' within the restriction already imposed, it is apparent that the interior contribution is primarily dependent upon the shape of τ'. In a number of cases of practical importance the mathematical model postulates a configuration of doublets of such symmetry that when the point 0 coincides with any one doublet the inverse-square field at 0 deriving from all other doublets in τ' is zero, provided that S' is spherical[41]. This result lends a special importance to (4.21-11) and (4.21-14), since it allows us to express an essentially microscopic effect in terms of macroscopic functions. The subject is taken up again in Sec. 5.20 where it is developed against a background of retarded potentials.

Equation (4.21-9) plays a significant part in an alternative approach to the definition of scalar potential. Suppose, in the first instance, that we are concerned with continuous volume sources of bounded density and finite extent. Then $\overline{U} = -\text{grad}\,\phi$ everywhere, so that if we define a scalar point function ϕ' by

40. It is, of course, assumed in the present context that the removal of part of the distribution has no effect on the remainder.

41. Or cubical. See Ex.4-148., p. 392.

$$\phi'_O - \phi'_P = -\int_P^O \overline{U}.d\overline{r} \qquad (4.21\text{-}15)$$

we have

$$\phi'_O - \phi'_P = \int_P^O \text{grad}\ \phi.d\overline{r} = \phi_O - \phi_P$$

If ϕ'_P is postulated to be zero when P is some point of an infinite spherical surface centred upon a local origin, then ϕ' will be zero at all points of the surface, since $|\overline{U}|$ decreases as the square of the distance from individual source elements and the circumference of the sphere increases as the first power of the radius. Because ϕ also is zero everywhere on the sphere we may write

$$\phi'_O = -\int_\infty^O \overline{U}.d\overline{r} = \phi_O \qquad (4.21\text{-}16)$$

for all positions of O.

A difficulty arises when closed, continuous surface sources are introduced. To reach interior points of any enclosure the contour of integration must pass through the bounding surface. $\overline{U}$ is undefined upon this surface and it is not possible to determine whether or not a discontinuity of ϕ' should be assigned to the point of intersection. Thus, the presence of a uniform double layer with its attendant discontinuity of potential cannot be detected by observation of the inverse-square field or its derivatives on either side of the surface.

Equation (4.21-16) remains valid for discrete distributions provided that the integration contour avoids the sources and ϕ and ϕ' are understood to refer to microscopic potentials. However, in order to derive a macroscopic potential at interior points of a doublet distribution some non-microscopic form of $\overline{U}$ must be employed and it is in this context that equation (4.21-9) becomes significant, for it will be seen that by defining a 'macroscopic $\overline{U}$', identical with the inverse-square field at the centre of a needle-shaped cavity, the associated potential will approximate $\int_\tau \overline{P}\ .\ \text{grad}\ \frac{1}{r}\ d\tau$. This approach to the macroscopic potential via a cavity-defined gradient function is of historical interest only and will not be pursued[42].

---

42. Likewise the substitution of the inverse-square field at the centre of a disc-shaped cavity for the analytical expression on the right-hand side of equation (4.21-10).

4.21b Inverse-square fields deriving from vector sources

The inverse-square vector point function $\overline{W}$ deriving from piecewise continous simple line or surface sources is defined at an exterior point 0 by

$$\overline{W}_o = -\int_\Gamma \overline{I} \times \frac{\overline{r}}{r^3}\, ds \quad ; \quad \overline{W}_o = -\int_S \overline{K} \times \frac{\overline{r}}{r^3}\, dS \qquad (4.21\text{-}17)$$

where $\overline{r}$ is directed from 0 to the source element.

Each source element is seen to give rise at 0 to a vector contribution which diminishes as the square of the distance from the element, and is directed normally to the plane containing $\overline{r}$ and $\overline{I}$ or $\overline{r}$ and $\overline{K}$. Reference to equations (4.14-2) and (4.14-6) reveals that in each case

$$\overline{W} = \text{curl}\ \overline{A} \qquad (4.21\text{-}18)$$

where $\overline{A}$ is the associated vector potential.

It is clear that this relationship will continue to hold for all combinations of source elements, including line and surface doublets.

The inverse-square field deriving from a piecewise continuous volume source of density $\overline{J}$ is given by

$$\overline{W}_o = -\int_\tau \overline{J} \times \frac{\overline{r}}{r^3}\, d\tau \qquad (4.21\text{-}19)$$

The integral is everywhere convergent and the relationship $\overline{W} = \text{curl}\ \overline{A}$ may be shown to hold at all points, including points of discontinuity of $\overline{J}$. This may be proved by an obvious extension of the analysis of Ex.4-44., p. 282. It has already been demonstrated for points at which $\overline{J}$ is continuous (Sec. 4.15).

It follows from equation (4.21-18) that in the case of a volume distribution of whirls, $\overline{W}$ is identical with the curl of the microscopic potential. At exterior points of the distribution, where the macroscopic and microscopic potentials are sensibly equal, we have

$$\overline{W} = \text{curl} \int_\tau \overline{M} \times \text{grad}\,\frac{1}{r}\, d\tau \qquad (4.21\text{-}20)$$

$$= -\,\text{grad} \int_\tau \overline{M}\,.\,\text{grad}\,\frac{1}{r}\, d\tau \quad \text{from (4.19-10)} \qquad (4.21\text{-}21)$$

This fails at interior points.

Proceeding as in Sec. 4.21a, let a cavity be created within the distribution. Then, with the earlier nomenclature,

$$\overline{W}_o \text{ (cavity)} = \text{curl (cavity)} \int_{\tau-\tau'} \overline{M} \times \text{grad} \frac{1}{r} \, d\tau \qquad (4.21\text{-}22)$$

whence, on combining equations (4.19-6) and (4.19-12), we find that

$$\overline{W}_o \text{ (cavity)} = \text{curl} \int_\tau \overline{M} \times \text{grad} \frac{1}{r} \, d\tau + \int_{\tau'} (\text{curl } \overline{M}) \times \frac{\overline{r}}{r^3} \, d\tau + \oint_{S'} (d\overline{S} \times \overline{M}) \times \frac{\overline{r}}{r^3} \qquad (4.21\text{-}23)$$

The volume integral over τ' vanishes when the cavity exhibits point symmetry about 0, provided that curl $\overline{M}$ is sensibly constant across it, in which case equation (4.21-23) reduces to

$$\overline{W}_o \text{ (cavity)} = \text{curl} \int_\tau \overline{M} \times \text{grad} \frac{1}{r} \, d\tau + \oint_{S'} (d\overline{S} \times \overline{M}) \times \frac{\overline{r}}{r^3} \qquad (4.21\text{-}24)$$

On evaluating the surface integral for needle and disc-shaped cavities whose axes are aligned with the local direction of $\overline{M}$ (assumed constant over τ'), and for a sphere, we get

$$\overline{W}_o \text{ (needle-shaped cavity)} = \text{curl} \int_\tau \overline{M} \times \text{grad} \frac{1}{r} \, d\tau - 4\pi \overline{M}_o \qquad (4.21\text{-}25)$$

$$\overline{W}_o \text{ (disc-shaped cavity)} = \text{curl} \int_\tau \overline{M} \times \text{grad} \frac{1}{r} \, d\tau \qquad (4.21\text{-}26)$$

$$\overline{W}_o \text{ (spherical cavity)} = \text{curl} \int_\tau \overline{M} \times \text{grad} \frac{1}{r} \, d\tau - \frac{8}{3} \pi \overline{M}_o \qquad (4.21\text{-}27)$$

Equation (4.21-27) represents an approximation of (4.19-6), viz

$$\lim_{\delta \to 0} \text{curl (cavity)} \int_{\tau-\tau_\delta} \overline{M} \times \text{grad} \frac{1}{r} \, d\tau = \text{curl} \int_\tau \overline{M} \times \text{grad} \frac{1}{r} \, d\tau - \frac{8}{3} \pi \overline{M}_o$$

By combining equations (4.21-25) to (4.21-27) with (4.19-11) we obtain

$$\overline{W}_o \text{ (needle-shaped cavity)} = - \operatorname{grad} \int_\tau \overline{M} \cdot \operatorname{grad} \frac{1}{r} \, d\tau \qquad (4.21\text{-}28)$$

$$\overline{W}_o \text{ (disc-shaped cavity)} = - \operatorname{grad} \int_\tau \overline{M} \cdot \operatorname{grad} \frac{1}{r} \, d\tau + 4\pi\overline{M}_o \qquad (4.21\text{-}29)$$

$$\overline{W}_o \text{ (spherical cavity)} = - \operatorname{grad} \int_\tau \overline{M} \cdot \operatorname{grad} \frac{1}{r} \, d\tau + \frac{4}{3}\pi\overline{M}_o \qquad (4.21\text{-}30)$$

4.21c The representation of a vector point function as the combined inverse-square fields of scalar and vector sources

It follows from equation (4.17-4) that a vector point function $\overline{F}$, which is well-behaved within the region τ bounded by the surfaces $S_{1..n}\Sigma$, may be represented within τ as the sum of inverse-square fields deriving from volume sources of densities $\frac{1}{4\pi}\operatorname{div}\overline{F}$ and $\frac{1}{4\pi}\operatorname{curl}\overline{F}$ throughout τ, and surface sources on $S_{1..n}\Sigma$ of densities $\frac{-\hat{n}.\overline{F}}{4\pi}$ and $\frac{-(\hat{n}\times\overline{F})}{4\pi}$, where $\hat{n}$ is the outward normal.

This form of expression is not unique. There are, in fact, an infinite number of possible combinations of sources covering the surfaces $S_{1..n}\Sigma$ and the space within and without τ which produce the required values of $\overline{F}$ at interior points, as seen from the considerations of Sec. 4.17. However, when $\overline{F}$ is defined and well-behaved at all points of space outside given surfaces and vanishes at infinity, the associated source system is unique and corresponds to volume sources of densities $\frac{1}{4\pi}\operatorname{div}\overline{F}$ and $\frac{1}{4\pi}\operatorname{curl}\overline{F}$ taken over all space, together with sources upon the surfaces of discontinuity having the densities $\frac{1}{4\pi}\Delta(\hat{n}.\overline{F})$ and $\frac{1}{4\pi}\Delta(\hat{n}\times\overline{F})$ in the notation of Sec. 4.5.

4.21d Alternative definitions of scalar and vector potential

The terms 'scalar potential' and 'vector potential' are frequently defined without reference to scalar and vector sources.

Thus ψ is said to be the scalar potential of the point function $\overline{F}$ in the region τ if $\overline{F} = \operatorname{grad}\psi$ (or $-\operatorname{grad}\psi$) in τ[43]. This relationship requires that curl $\overline{F}$ be zero, and that $\overline{F}$ be irrotational in a multiply connected region if ψ is to be single-valued (see Ex.1-38., p. 47). In any case, ψ is determinate only to within an additive constant.

43.
ψ is, of course, a scalar potential, as previously defined, whether $\overline{F}$ is the inverse-square field of scalar sources or not, because any scalar point function can be expressed as the sum of potential functions.

Similarly, $\bar{c}$ is said to be the vector potential of $\bar{F}$ in τ if $\bar{F} = \text{curl}\ \bar{c}$ in τ. $\bar{F}$ must be solenoidal for this to be possible (p. 62). While $\bar{c}$, like any other vector field, must be expressible as a vector potential in the general form given by equation (3.3-4), it is specifically related to volume sources in τ and tangential surface sources upon its boundaries by equation (4.17-6). $\bar{c}$ is determinate only to within an additive gradient function since curl grad is identically zero.

It will be observed that with the above definitions of scalar and vector potential it is no longer possible to predict the values of the discontinuities obtaining at the interfaces of juxtaposed regions[44].

EXERCISES

4-141. A closed surface S is immersed in a statistically-regular volume distribution of (a) point doublets and (b) whirls, in such a way that it does not intersect any source element. If the subscripts 1 and 2 refer respectively to microscopic and macroscopic functions, show that

(a.1) $\oint_S \bar{U}.d\bar{S} = \oint_S (-\text{ grad } \phi_1).d\bar{S} = 0$

(a.2) $-\text{ div grad } \phi_1 = \text{div } \bar{U} = 0$ outside each doublet

(a.3) $-\text{ div grad } \phi_2 = -4\pi \text{ div } \bar{P}$ throughout the distribution

(b.1) $\oint_S \bar{W}.d\bar{S} = \oint_S (\text{curl } \bar{A}_1).d\bar{S} = 0$

(b.2) $\text{div curl } \bar{A}_1 = 0$ outside each whirl

(b.3) $\text{div curl } \bar{A}_2 = 0$ throughout the distribution

(b.4) $\text{curl curl } \bar{A}_1 = \bar{0}$ outside each whirl

(b.5) $\text{curl curl } \bar{A}_2 = 4\pi \text{ curl } \bar{M}$ throughout the distribution

4-142. A volume distribution of point doublets occupies an aperiphractic region τ bounded by the surface S. Two points P and Q, exterior to τ, are at such distance from S that the macroscopic potential of the distribution is sensibly equal to the microscopic potential at each point. If a regular curve is drawn between P and Q, passing into and out of the distribution but not cutting any doublet, show that the tangential line integral of the inverse-square field along PQ is equal to the tangential line integral of the gradient of the macroscopic potential. (Note that the normal derivative of the macroscopic potential is discontinuous through S.)

---

44. The author is convinced that these definitions (which appear in practically every text book) are the primary cause of the confusion which continues to permeate the subject of electrical fundamentals.

4-143. A volume distribution of whirls occupies an aperiphractic region τ bounded by the surface S. Γ is a closed curve exterior to τ and at such distance from S that the macroscopic and microscopic vector potentials are sensibly equal at each point of it. A regular surface S' spans Γ and cuts τ but does not pass through any whirl. Show that the flux of the inverse-square field through S' is equal to the flux of the curl of the macroscopic potential. (Note that the tangential component of the curl of the macroscopic potential is discontinuous through S.)

4-144. Show that

$$\text{curl} \int_\tau \frac{\bar{J}}{r}\, d\tau = -\int_\tau \bar{J} \times \frac{\bar{r}}{r^3}\, d\tau$$

at points of discontinuity of $\bar{J}$.

4-145. Evaluate the surface integrals in equations (4.21-8) and (4.21-24) for the case of a cylindrical enclosure of length l and diameter d, given that

$$\bar{P} = (P_o + \alpha x + \beta y + \gamma z)\, \hat{z}$$

and

$$\bar{M} = (M_o + \alpha x + \beta y + \gamma z)\, \hat{z}$$

where $\hat{z}$ is axial, the origin of coordinates is the centre of the cylinder, and α, β, γ are constants.

[Note that under these conditions div $\bar{P}$ and curl $\bar{M}$ are constant throughout the enclosure.]

Ans:

$$4\pi\bar{P}_o \left\{1 - \frac{1}{(1+\alpha^2)^{\frac{1}{2}}}\right\} \quad ; \quad -4\pi\bar{M}_o \frac{1}{(1+\alpha^2)^{\frac{1}{2}}}$$

where $\alpha = d/l$

4-146. Show that for the values of $\bar{P}$ and $\bar{M}$ given in Ex.4-145, the surface integrals, when taken over a spherical enclosure, continue to be given by $\frac{4}{3}\pi\bar{P}_o$ and $-\frac{8}{3}\pi\bar{M}_o$.

4-147. The potential of a plane rectangular surface source of constant density σ, at a point O of the normal to the surface through one corner A, is given by

$$\phi = \sigma\left\{c \ln \frac{b + d_3}{d_2} + b \ln \frac{c + d_3}{d_1} - z \tan^{-1} \frac{bc}{zd_3}\right\}$$

where B and C are corners adjacent to A and distant b and c from it, D is the diagonally opposite corner, and

$$z = AO \quad , \quad d_1 = OB \quad , \quad d_2 = OC \quad , \quad d_3 = OD$$

Make use of this to evaluate the surface integral of equation (4.21-8) for a rectangular enclosure of dimensions $2b \times 2c \times 2d$ where the side of length 2d is parallel to $\overline{P}$ which is assumed to be constant over the enclosure. Hence show that the result for a cubical enclosure is identical with that for a sphere, viz $\frac{4}{3}\pi\overline{P}$.

Ans: $8\overline{P}\tan^{-1}\dfrac{bc}{d(b^2+c^2+d^2)^{\frac{1}{2}}}$

4-148. A cube is so orientated in space that one set of edges is vertical. Equal scalar point doublets, directed vertically, are located

(a) at the centre of each face

(b) at each vertex

(c) at the centre of each edge.

Use the result of Ex.4-23., p. 256 to show that the inverse-square field deriving from each set of doublets is zero at the centre of the cube. Invoke an appropriate transformation to demonstrate, without further analysis, that the inverse-square fields deriving from sets of vector point doublets (whirls) of identical orientation and disposition are likewise zero at the centre.

4-149. Develop a field-slipping analysis to show that an inverse nth power vector field $\overline{F}$, deriving from a well-behaved volume source of density ρ, may be expressed both at interior and exterior points of the source as

$$\overline{F} = -\int_\tau \rho\frac{\hat{r}}{r^n}\,d\tau = -\operatorname{grad} U$$

where

$$U = \frac{1}{(n-1)}\int_\tau \frac{\rho}{r^{n-1}}\,d\tau \qquad (n < 3, n \neq 1)$$

$$U = \int_\tau \rho\ln\frac{1}{r}\,d\tau \qquad (n = 1)$$

Note, however, that $\overline{F}$ does not converge at infinity unless $n \geq 0$, and U does not converge unless $n > 1$.

Show likewise that at exterior points

$$\operatorname{div} \overline{F} = (2-n) \int_{\tau} \frac{\rho}{r^{n+1}} \, d\tau$$

and at interior points

$$\operatorname{div} \overline{F} = (2-n) \int_{\tau} \frac{\rho}{r^{n+1}} \, d\tau \qquad (n < 2)$$

$$\operatorname{div} \overline{F} = 4\pi\rho \qquad (n = 2)$$

Observe that only for $n = 2$ is $\operatorname{div} \overline{F}$ directly related to the ambient source density. Note also that $\operatorname{curl} \overline{F} = 0$ for all values of n which allow $\overline{F}$ to be expressed as $-\operatorname{grad} U$.

4-150. An inverse nth power radial field, deriving from a point source at P, is given by

$$\overline{F} = \hat{\overline{R}} \, \alpha \, R^{-n}$$

where α is a constant and R denotes distance from P.

A region τ is bounded by inner and outer spherical surfaces S_1 and S_2, of radii a_1 and a_2, centred on P.

Express $\overline{F}$ at any interior point of τ as the sum of inverse-square vector fields deriving from volume sources of density $\frac{(2-n)}{4\pi} \alpha R^{-(n+1)}$ throughout τ, and surface sources on S_1 and S_2 of densities $\frac{\alpha}{4\pi} a_1^{-n}$ and $\frac{-\alpha}{4\pi} a_2^{-n}$.

Confirm this result by an application of equation (4.9-6), noting that source elements of greater distance from P than the point of observation yield no net contribution to the inverse-square component.

4-151. If $\overline{F}$ is irrotational in the simply connected region τ bounded by the surfaces $S_{1..n}\Sigma$, show that it may be expressed within τ, in the notation of Sec. 4.17, as

$$4\pi\overline{F}_o = -\operatorname{grad} \int_{\tau} \frac{\operatorname{div} \overline{F}}{r} \, d\tau + \operatorname{grad} \oint_{S_1} \frac{(\hat{\overline{n}}.\overline{F} + \hat{\overline{n}}_1.\overline{F}_1)}{r} \, dS + \text{-----}$$

$$\text{-----} + \operatorname{grad} \oint \frac{(\hat{\overline{n}}.\overline{F} + \hat{\overline{n}}_\Sigma.\overline{F}_\Sigma)}{r} \, dS$$

where $\overline{F}_1$----$\overline{F}_\Sigma$ are defined as follows:

Let the scalar function U be defined at any point Q of τ by

$$U_Q = \int_P^Q \overline{F}.d\overline{r}$$

where P is a fixed point of τ.

Let U_1---U_n be the solutions of the interior Dirichlet problems in τ_1---τ_n when U_1---U_n are equated to U over S_1---S_n. Similarly, let U_Σ be the solution of the exterior Dirichlet problem in τ_Σ when U_Σ is equated to U over Σ. Then $\overline{F}_1$ = grad U_1 in τ_1, etc and $\overline{F}_\Sigma$ = grad U_Σ in τ_Σ.

CHAPTER 5

RETARDED POTENTIAL THEORY

5.1 Retarded Scalar and Vector Fields

Let V be any continuous function of space and time within the region $\underline{R}$ bounded by the surfaces $S_{1..n}\Sigma$, over some particular time interval. It is clear that each of the relationships derived previously for time-invariant fields will apply to the V field under consideration at any instant of time within the specified interval.

Now consider a point function [V] which is derived from V by associating with each point of $\underline{R}$ at any time t the value of V which obtained there at the time $t - \frac{r'}{c}$, where r' denotes distance from a fixed point Q within or without $\underline{R}$, and c is a constant[1].

When c is positive, [V] is seen to be a 'retarded' field (except at the point Q when this lies within $\underline{R}$) and when c is negative it becomes an 'advanced' field. To simplify nomenclature we will refer to [V] as a retarded field but will bear in mind that c may be negative. We may generate from the parent V field as many [V] fields as we please by choice of the position of Q and the value of c. Each [V] field will be well-behaved throughout $\underline{R}$ for each point of time within some particular interval, provided that V is well-behaved in space and time over an appropriate interval. Under these conditions many of the relationships derived for static fields in earlier sections remain valid for retarded fields.

Thus, inter alia, we have

$$\int_\tau \text{grad}\,[V]\,d\tau = \oint_{S_{1..n}\Sigma} [V]\,d\bar{S}$$

$$\text{grad}\,\frac{[V]}{r} = \frac{1}{r}\,\text{grad}\,[V] + [V]\,\text{grad}\,\frac{1}{r}$$

$$\left.\begin{matrix} 4\pi[V]_0 \\ 0 \end{matrix}\right\} = \oint_{S_{1..n}\Sigma}\left\{\frac{1}{r}\frac{\partial[V]}{\partial n} - [V]\frac{\partial}{\partial n}\left(\frac{1}{r}\right)\right\} dS - \int_\tau \frac{1}{r}\nabla^2[V]\,d\tau \qquad (5.1\text{-}1)$$

---

1. It will be seen that c has the dimensions of velocity.

In equation (5.1-1) the origin of r may or may not coincide with the origin of retardation (Q). If it does, then

$$4\pi\,[V]_o = 4\pi V_o$$

In like manner we may derive a retarded vector field $[\overline{F}]$ from any parent field $\overline{F}$ which is appropriately space and time-dependent in $\underline{R}$.

Then

$$\int_\tau \operatorname{div}\,[\overline{F}]\,d\tau = \oint_{S_{1..n}\Sigma} [\overline{F}].d\overline{S}$$

$$\operatorname{curl}\operatorname{curl}\,[\overline{F}] = \operatorname{grad}\operatorname{div}\,[\overline{F}] - \nabla^2[\overline{F}]$$

$$\operatorname{div}\frac{[\overline{F}]}{r} = \frac{1}{r}\operatorname{div}\,[\overline{F}] + \operatorname{grad}\frac{1}{r}\,.\,[\overline{F}]$$

$$\left.\begin{matrix}4\pi[\overline{F}]_o\\ \overline{0}\end{matrix}\right\} = \oint_{S_{1..n}\Sigma}\left\{\frac{1}{r}\frac{\partial[\overline{F}]}{\partial n} - [\overline{F}]\frac{\partial}{\partial n}\left(\frac{1}{r}\right)\right\}dS - \int_\tau \frac{1}{r}\nabla^2[\overline{F}]\,d\tau \qquad (5.1\text{-}2)$$

$$\left.\begin{matrix}4\pi[\overline{F}]_o\\ \overline{0}\end{matrix}\right\} = \int_\tau (-\operatorname{div}[\overline{F}])\frac{\overline{r}}{r^3}\,d\tau + \oint_{S_{1..n}\Sigma}[\overline{F}].d\overline{S}\,\frac{\overline{r}}{r^3}$$

$$- \int_\tau (\operatorname{curl}[\overline{F}])\times\frac{\overline{r}}{r^3}\,d\tau + \oint_{S_{1..n}\Sigma}(d\overline{S}\times[\overline{F}])\times\frac{\overline{r}}{r^3} \qquad (5.1\text{-}3)$$

where the origin of r may or may not coincide with Q.

It will be observed that in the above equations the origin of retardation (and of r) is supposed to remain fixed in space while the various field operations are carried out. When this is not the case - as in the differentiation of potential functions, where the origin of retardation is identified with the origin of r - it is not possible to transform non-retarded into retarded relationships by the simple substitution of retarded for unretarded quantities. This matter is treated in detail in subsequent sections.

5.2 Expansion of Grad[V], Div[$\bar{F}$] and Curl[$\bar{F}$]

Let the coordinates of adjacent points P_1 and P be (x_1,y_1,z_1) and $(x_1+\Delta x,y_1,z_1)$ respectively, and let their distances from the origin of retardation, Q, be r_1' and $r_1' + \Delta r'$. Then at the time t

$$[V]_P - [V]_{P_1} = V\left(x_1+\Delta x,y_1,z_1,t - \frac{(r_1'+\Delta r')}{c}\right) - V\left(x_1,y_1,z_1,t - \frac{r_1'}{c}\right)$$

$$= V\left(x_1+\Delta x,y_1,z_1,t - \frac{(r_1'+\Delta r')}{c}\right) - V\left(x_1,y_1,z_1,t - \frac{(r_1'+\Delta r')}{c}\right)$$

$$+ V\left(x_1,y_1,z_1,t - \frac{(r_1'+\Delta r')}{c}\right) - V\left(x_1,y_1,z_1,t - \frac{r_1'}{c}\right)$$

whence, by the mean-value theorem,

$$[V]_P - [V]_{P_1} = \left.\frac{\partial V}{\partial x}\right|_{\substack{x' \\ y_1 \\ z_1 \\ t - \frac{(r_1'+\Delta r')}{c}}} \Delta x + \left.\frac{\partial V}{\partial t}\right|_{\substack{x_1 \\ y_1 \\ z_1 \\ t - \frac{(r_1'+\Delta r'')}{c}}} \left(-\frac{\Delta r'}{c}\right)$$

where

$$x_1 < x' < x_1 + \Delta x$$

$$0 < \Delta r'' < \Delta r'$$

On dividing by Δx and taking limits we get

$$\left(\frac{\partial [V]}{\partial x}\right)_{P_1} = \left[\frac{\partial V}{\partial x}\right]_{P_1} - \frac{1}{c}\left(\frac{\partial r'}{\partial x}\right)_{P_1}\left[\frac{\partial V}{\partial t}\right]_{P_1} \qquad (5.2\text{-}1)$$

whence, in general,

$$\text{grad } [V] = [\text{grad } V] - \frac{1}{c}\frac{\bar{r}'}{r'}\left[\frac{\partial V}{\partial t}\right] \qquad (5.2\text{-}2)$$

where $\bar{r}'$ is the radius vector directed from the origin of retardation, and the brackets around grad V and $\frac{\partial V}{\partial t}$ imply that these terms are to be evaluated at the time $t - \frac{r'}{c}$.

It also follows from equation (5.2-1) that for any well-behaved vector point function $\bar{F}$

$$\operatorname{div}[\bar{F}] = [\operatorname{div} \bar{F}] - \frac{1}{c}\frac{\bar{r}'}{r'} \cdot \left[\frac{\partial \bar{F}}{\partial t}\right] \qquad (5.2\text{-}3)$$

$$\operatorname{curl}[\bar{F}] = [\operatorname{curl} \bar{F}] - \frac{1}{c}\frac{\bar{r}'}{r'} \times \left[\frac{\partial \bar{F}}{\partial t}\right] \qquad (5.2\text{-}4)$$

It should be noted that

$$\frac{\partial [V]}{\partial t} = \left[\frac{\partial V}{\partial t}\right] \quad \text{and} \quad \frac{\partial [\bar{F}]}{\partial t} = \left[\frac{\partial \bar{F}}{\partial t}\right] \qquad (5.2\text{-}5)$$

so that

$$\frac{\partial}{\partial t}[\operatorname{div} \bar{F}] = \left[\frac{\partial}{\partial t}\operatorname{div} \bar{F}\right] = \left[\operatorname{div} \frac{\partial \bar{F}}{\partial t}\right] \qquad (5.2\text{-}6)$$

$$\frac{\partial}{\partial t}[\operatorname{grad} V] = \left[\frac{\partial}{\partial t}\operatorname{grad} V\right] = \left[\operatorname{grad} \frac{\partial V}{\partial t}\right] \qquad (5.2\text{-}7)$$

$$\frac{\partial}{\partial t}[\operatorname{curl} \bar{F}] = \left[\frac{\partial}{\partial t}\operatorname{curl} \bar{F}\right] = \left[\operatorname{curl} \frac{\partial \bar{F}}{\partial t}\right] \qquad (5.2\text{-}8)$$

It then follows that

$$\frac{\partial}{\partial t}\operatorname{div}[\bar{F}] = \operatorname{div}\left[\frac{\partial \bar{F}}{\partial t}\right] \qquad (5.2\text{-}9)$$

$$\frac{\partial}{\partial t}\operatorname{grad}[V] = \operatorname{grad}\left[\frac{\partial V}{\partial t}\right] \qquad (5.2\text{-}10)$$

$$\frac{\partial}{\partial t}\operatorname{curl}[\bar{F}] = \operatorname{curl}\left[\frac{\partial \bar{F}}{\partial t}\right] \qquad (5.2\text{-}11)$$

but

$$\frac{\partial}{\partial t}[\operatorname{div} \bar{F}] \neq \operatorname{div}\left[\frac{\partial \bar{F}}{\partial t}\right]$$

etc.

Now consider the differentiation of a retarded function [V] with respect to the coordinates of the origin of retardation - say (x_o', y_o', z_o'). In expanding $\frac{\partial}{\partial x_o'}[V]$ we note that [V] is a function of x_o' in virtue of retardation alone, so that at the point (x,y,z) we have

$$\frac{\partial [V]}{\partial x_o'} = -\frac{1}{c}\frac{\partial r'}{\partial x_o'}\left[\frac{\partial V}{\partial t}\right] = \frac{(x-x_o')}{cr'}\left[\frac{\partial V}{\partial t}\right] \qquad (5.2\text{-}12)$$

The local space-variation component which was evident in equation (5.2-1) is now missing.

On the other hand

$$\frac{\partial}{\partial x_o'}\left(\frac{\partial [V]}{\partial x}\right) \neq -\frac{1}{c}\frac{\partial r'}{\partial x_o'}\frac{\partial}{\partial t}\left(\frac{\partial [V]}{\partial x}\right) = -\frac{1}{c}\frac{\partial r'}{\partial x_o'}\left[\frac{\partial^2 V}{\partial t \partial x}\right] + \frac{1}{c^2}\frac{\partial r'}{\partial x_o'}\frac{\partial r'}{\partial x}\left[\frac{\partial^2 V}{\partial t^2}\right]$$

since $\frac{\partial [V]}{\partial x}$ is a function of x_o' both through space and time-dependence. It is necessary to expand $\frac{\partial [V]}{\partial x}$ before differentiating with respect to x_o'. We then obtain

$$\frac{\partial}{\partial x_o'}\left(\frac{\partial [V]}{\partial x}\right) = \frac{\partial}{\partial x_o'}\left\{\left[\frac{\partial V}{\partial x}\right] - \frac{1}{c}\frac{\partial r'}{\partial x}\left[\frac{\partial V}{\partial t}\right]\right\}$$

$$= -\frac{1}{c}\frac{\partial r'}{\partial x_o'}\left[\frac{\partial^2 V}{\partial t \partial x}\right] - \frac{1}{c}\left\{\frac{\partial}{\partial x_o'}\left(\frac{\partial r'}{\partial x}\right)\right\}\left[\frac{\partial V}{\partial t}\right] + \frac{1}{c^2}\frac{\partial r'}{\partial x_o'}\frac{\partial r'}{\partial x}\left[\frac{\partial^2 V}{\partial t^2}\right] \qquad (5.2\text{-}13)$$

5.3 Dynamical Extension of Green's Formula

The theorem to be derived extends Green's formula (3.3-1) and (3.3-3) to the case in which V is both space and time-dependent.

Let V be any point function, well-behaved in space and time within the region $\underline{R}$ bounded by the surfaces $S_{1..n}\Sigma$ over the time interval required by the analysis, and let the origin of retardation coincide with an interior point 0. Then r' = r where r is distance measured from 0. It follows from equations (5.2-2) and (5.2-3) that at points within $\underline{R}$

$$\text{div grad}[V] = \text{div}[\text{grad } V] - \text{div}\,\frac{\bar{r}}{cr}\left[\frac{\partial V}{\partial t}\right]$$

$$= [\text{div grad } V] - \frac{\bar{r}}{cr}\cdot\frac{\partial}{\partial t}[\text{grad } V] - \text{div}\,\frac{\bar{r}}{cr}\left[\frac{\partial V}{\partial t}\right]$$

$$= [\nabla^2 V] - \frac{\bar{r}}{cr}\cdot\left[\text{grad}\,\frac{\partial V}{\partial t}\right] - \left[\frac{\partial V}{\partial t}\right]\text{div}\,\frac{\bar{r}}{cr} - \frac{\bar{r}}{cr}\cdot\text{grad}\left[\frac{\partial V}{\partial t}\right]$$

or

$$\nabla^2[V] = [\nabla^2 V] - 2\,\frac{\bar{r}}{cr}.\mathrm{grad}\left[\frac{\partial V}{\partial t}\right] - \frac{1}{c^2}\left[\frac{\partial^2 V}{\partial t^2}\right] - \frac{2}{cr}\left[\frac{\partial V}{\partial t}\right] \qquad (5.3\text{-}1)$$

On surrounding 0 with a δ sphere and proceeding as in Sec. 3.3 with [V] replacing V, we obtain

$$\oint\limits_{S_{1..n}\Sigma,\ S_\delta} \left\{[V]\frac{\partial}{\partial n}\left(\frac{1}{r}\right) - \frac{1}{r}\frac{\partial}{\partial n}[V]\right\} dS = -\int\limits_{\tau-\tau_\delta} \frac{1}{r}\left[\nabla^2 V - \frac{1}{c^2}\frac{\partial^2 V}{\partial t^2}\right] d\tau$$

$$+\ 2\int\limits_{\tau-\tau_\delta} \left\{\frac{\bar{r}}{cr^2}.\mathrm{grad}\left[\frac{\partial V}{\partial t}\right] + \frac{1}{cr^2}\left[\frac{\partial V}{\partial t}\right]\right\} d\tau \qquad (5.3\text{-}2)$$

But

$$\mathrm{div}\,\frac{\bar{r}}{cr^2}\left[\frac{\partial V}{\partial t}\right] = \frac{\bar{r}}{cr^2}.\mathrm{grad}\left[\frac{\partial V}{\partial t}\right] + \frac{1}{cr^2}\left[\frac{\partial V}{\partial t}\right]$$

hence the second volume integral in equation (5.3-2) may be replaced by the surface integral

$$2\oint\limits_{S_{1..n}\Sigma,\ S_\delta} \left[\frac{\partial V}{\partial t}\right]\frac{\bar{r}}{cr^2}.d\bar{S} = 2\oint\limits_{S_{1..n}\Sigma,\ S_\delta} \frac{1}{cr}\frac{\partial r}{\partial n}\left[\frac{\partial V}{\partial t}\right] dS$$

whence

$$\oint\limits_{S_{1..n}\Sigma,\ S_\delta} \left\{[V]\frac{\partial}{\partial n}\left(\frac{1}{r}\right) - \frac{1}{r}\frac{\partial [V]}{\partial n} - \frac{2}{cr}\frac{\partial r}{\partial n}\left[\frac{\partial V}{\partial t}\right]\right\} dS = -\int\limits_{\tau-\tau_\delta} \frac{1}{r}\left[\nabla^2 V - \frac{1}{c^2}\frac{\partial^2 V}{\partial t^2}\right] d\tau$$

or

$$\oint\limits_{S_{1..n}\Sigma,\ S_\delta} \left\{[V]\frac{\partial}{\partial n}\left(\frac{1}{r}\right) - \frac{1}{r}\left[\frac{\partial V}{\partial n}\right] - \frac{1}{cr}\frac{\partial r}{\partial n}\left[\frac{\partial V}{\partial t}\right]\right\} dS = -\int\limits_{\tau-\tau_\delta} \frac{1}{r}\left[\nabla^2 V - \frac{1}{c^2}\frac{\partial^2 V}{\partial t^2}\right] d\tau$$

On taking limits as the δ sphere shrinks about 0[2] and noting that $[V]_o = V_o$ we get[3]

$$4\pi V_o = \oint_{S_{1..n}\Sigma} \left\{ \frac{1}{r}\left[\frac{\partial V}{\partial n}\right] - [V]\frac{\partial}{\partial n}\left(\frac{1}{r}\right) + \frac{1}{cr}\frac{\partial r}{\partial n}\left[\frac{\partial V}{\partial t}\right] \right\} dS - \int_\tau \frac{1}{r}\left[\nabla^2 V - \frac{1}{c^2}\frac{\partial^2 V}{\partial t^2}\right] d\tau \tag{5.3-3}$$

When 0 is exterior to $\underline{R}$ no limiting process is involved and the left hand side of equation (5.3-3) is zero.

It will be seen that equation (5.3-3) reduces to (3.3-3) when $c = \pm\infty$ or when V is invariant with respect to time over the maximum retardation interval involved.

If U_1 is a point function which is well-behaved within τ_1, ie the region bounded by S_1, then for an origin of r within τ_1 we have

$$0 = \oint_{S_1} \left\{ \frac{1}{r}\left[\frac{\partial U_1}{\partial n'}\right] - [U_1]\frac{\partial}{\partial n'}\left(\frac{1}{r}\right) + \frac{1}{cr}\frac{\partial r}{\partial n'}\left[\frac{\partial U_1}{\partial t}\right] \right\} dS - \int_{\tau_1} \frac{1}{r}\left[\nabla^2 U_1 - \frac{1}{c^2}\frac{\partial^2 U_1}{\partial t^2}\right] d\tau$$

where the positive sense of n' is directed into τ.

By combining this with similar equations for the regions $\tau_2 \ldots . \tau_n$ and with equation (5.3-3) we get

$$\begin{aligned} 4\pi V_o = & \oint_{S_{1..n}} \left\{ \frac{1}{r}\left[\frac{\partial V}{\partial n} - \frac{\partial U}{\partial n}\right] - [V-U]\frac{\partial}{\partial n}\left(\frac{1}{r}\right) + \frac{1}{cr}\frac{\partial r}{\partial n}\left[\frac{\partial V}{\partial t} - \frac{\partial U}{\partial t}\right] \right\} dS \\ & - \int_\tau \frac{1}{r}\left[\nabla^2 V - \frac{1}{c^2}\frac{\partial^2 V}{\partial t^2}\right] d\tau - \int_{\tau_{1..n}} \frac{1}{r}\left[\nabla^2 U - \frac{1}{c^2}\frac{\partial^2 U}{\partial t^2}\right] d\tau \\ & + \oint_\Sigma \left\{ \frac{1}{r}\left[\frac{\partial V}{\partial n}\right] - [V]\frac{\partial}{\partial n}\left(\frac{1}{r}\right) + \frac{1}{cr}\frac{\partial r}{\partial n}\left[\frac{\partial V}{\partial t}\right] \right\} dS \end{aligned} \tag{5.3-4}$$

---

2. It is clear that the same limit obtains when S_δ is replaced by any regular closed surface.

3. Equation (5.3-3), with the surface integral equated to zero, was proved by Lorenz in 1861; with the volume integral equated to zero by Kirchhoff in 1882, and in general by Beltrami in 1895.

Since U_1, U_2... are arbitrary functions it is clear that V_o may be expressed in terms of surface and volume integrals over $S_{1..n}\Sigma$ and τ, $\tau_{1..n}$ in an infinite number of ways for any given value of c. When V is defined throughout all space and is well-behaved everywhere except upon the surfaces $S_{1..n}$ then, provided that the surface integral over Σ vanishes as Σ recedes to infinity, V is uniquely represented at all points not coincident with $S_{1..n}$ by

$$4\pi V_o = \oint_{S_{1..n}} \left\{ -\frac{1}{r}\,\Delta \left[\frac{\partial V}{\partial n}\right] + \Delta[V]\,\frac{\partial}{\partial n}\left(\frac{1}{r}\right) - \frac{1}{cr}\,\frac{\partial r}{\partial n}\,\Delta \left[\frac{\partial V}{\partial t}\right] \right\} dS - \int_{\infty} \frac{1}{r}\left[\nabla^2 V - \frac{1}{c^2}\frac{\partial^2 V}{\partial t^2}\right] d\tau \tag{5.3-5}$$

where the Δ notation is that used previously (p. 246).

It is a sufficient condition for the disappearance of the surface integral at infinity that $R\left[\frac{\partial V}{\partial R}\right]$, $[V]$ and $R\left[\frac{\partial V}{\partial t}\right]$ vanish as $R\to\infty$ for non-zero values of c, where R represents distance from a local origin. This is not a necessary condition. Thus, if V assume the asymptotic form $\frac{1}{R}\, f\left(t - \frac{R}{c}\right)$, the surface integral vanishes although $R\left[\frac{\partial V}{\partial R}\right]$ and $R\left[\frac{\partial V}{\partial t}\right]$ do not vanish individually.

It should be noted that equation (5.3-5) continues to hold when different values of c are assigned to different regions, provided, of course, that the volume integral and the components of the surface integral are interpreted accordingly.

The results are applicable to vector fields having the required degree of continuity since the scalar field may be identified with each of the Cartesian components of the vector field in turn. Multiplication by the unit vectors and subsequent addition leads to formulae in which $\overline{F}$ replaces V. In particular,

$$\left.\begin{matrix} 4\pi\overline{F}_o \\ \overline{0} \end{matrix}\right\} = \oint_{S_{1..n}\Sigma} \left\{ \frac{1}{r}\left[\frac{\partial \overline{F}}{\partial n}\right] - [\overline{F}]\,\frac{\partial}{\partial n}\left(\frac{1}{r}\right) + \frac{1}{cr}\,\frac{\partial r}{\partial n}\left[\frac{\partial \overline{F}}{\partial t}\right] \right\} dS - \int_{\tau} \frac{1}{r}\left[\nabla^2 \overline{F} - \frac{1}{c^2}\frac{\partial^2 \overline{F}}{\partial t^2}\right] d\tau \tag{5.3-6}$$

for an interior or exterior origin of r.

The operator $\left(\nabla^2 - \frac{1}{c^2}\frac{\partial^2}{\partial t^2}\right)$ is known as the d'Alembertian, and is variously written as $\Box$, $\Box^2$ and dal (and sometimes as $-\Box^2$).

EXERCISES

5-1. Equation (5.1-3) has been derived from one form of the unretarded grad-curl theorem (4.17-4) by substitution of $[\overline{F}]$ for $\overline{F}$. By expanding in accordance with equation (5.2-3) and (5.2-4), and taking the origin of retardation to coincide with the origin of r, derive one form of the retarded grad-curl theorem, viz

$$\left.\begin{matrix}4\pi\overline{F}_o \\ \overline{0}\end{matrix}\right\} = -\int_\tau [\mathrm{div}\ \overline{F}]\frac{\overline{r}}{r^3}\,d\tau + \oint_{S_{1..n}\Sigma} [\overline{F}].d\overline{S}\,\frac{\overline{r}}{r^3}$$

$$- \int_\tau [\mathrm{curl}\ \overline{F}]\times\frac{\overline{r}}{r^3}\,d\tau + \oint_{S_{1..n}\Sigma} (d\overline{S}\times[\overline{F}])\times\frac{\overline{r}}{r^3} + \int_\tau \frac{1}{cr^2}\left[\frac{\partial\overline{F}}{\partial t}\right] d\tau$$

5-2. If $\overline{F}$ is a vector point function having continuous first derivatives in space and time and if (x_o', y_o', z_o') are the coordinates of the origin of retardation, prove, by expansion after the manner of equation (5.2-13), that

$$\frac{\partial}{\partial x_o'}\mathrm{div}[\overline{F}] = -\frac{1}{c}\,\mathrm{div}\left\{\frac{\partial r'}{\partial x_o'}\left[\frac{\partial\overline{F}}{\partial t}\right]\right\} = \mathrm{div}\,\frac{\partial[\overline{F}]}{\partial x_o'}$$

5-3. By replacing V with [V] and substituting $\frac{1}{r}e^{\gamma r}$ for U in (3.1-2), with γ a real constant and the origin of retardation identified with the origin of r, derive the relationship

$$\oint_{S_{1..n}\Sigma,\ S_\delta}\left\{[V]\frac{\partial}{\partial n}\left(\frac{1}{r}e^{\gamma r}\right) - \frac{1}{r}e^{\gamma r}\frac{\partial[V]}{\partial n}\right\} dS = \int_{\tau-\tau_\delta}\left\{[V]\nabla^2\left(\frac{1}{r}e^{\gamma r}\right) - \frac{1}{r}e^{\gamma r}\nabla^2[V]\right\} d\tau$$

Make use of equation (5.3-1) and the expansion and subsequent volume integration of $\mathrm{div}\left\{\frac{\overline{r}\,e^{\gamma r}}{cr^2}\left[\frac{\partial V}{\partial t}\right]\right\}$ to arrive at

$$\oint_{S_{1..n}\Sigma,\ S_\delta}\left\{[V]\frac{\partial}{\partial n}\left(\frac{1}{r}e^{\gamma r}\right) - \frac{1}{r}e^{\gamma r}\left[\frac{\partial V}{\partial n}\right] - \frac{1}{cr}e^{\gamma r}\frac{\partial r}{\partial n}\left[\frac{\partial V}{\partial t}\right]\right\} dS$$

$$= -\int_{\tau-\tau_\delta}\frac{1}{r}e^{\gamma r}\left[\nabla^2V - \gamma^2V + \frac{2\gamma}{c}\frac{\partial V}{\partial t} - \frac{1}{c^2}\frac{\partial^2 V}{\partial t^2}\right] d\tau$$

and so obtain

$$\left.\begin{matrix}4\pi V_o\\0\end{matrix}\right\} = \oint_{S_{1..n}\Sigma}\left\{\frac{1}{r}\,e^{\gamma r}\left[\frac{\partial V}{\partial n}\right] - [V]\frac{\partial}{\partial n}\left(\frac{1}{r}\,e^{\gamma r}\right) + \frac{1}{cr}\,e^{\gamma r}\,\frac{\partial r}{\partial n}\left[\frac{\partial V}{\partial t}\right]\right\} dS$$

$$-\int_\tau \frac{1}{r}\,e^{\gamma r}\left[\nabla^2 V - \gamma^2 V + \frac{2\gamma}{c}\frac{\partial V}{\partial t} - \frac{1}{c^2}\frac{\partial^2 V}{\partial t^2}\right] d\tau$$

according as 0 lies within or without τ.

5-4. Let the scalar point function V take the form $\frac{1}{r'}\,f_1(r',t) + f_2(x,y,z,t)$ in a neighbourhood of the point P, where r' is distance measured from P and f_1 and f_2 are well-behaved functions of space and time. Show that to take account of this singularity within the integration space, the final equation of the previous exercise must be modified by the addition to the right-hand side of the term $\frac{4\pi e^{\gamma r_P}}{r_P}\,f_1\left(0,t - \frac{r_P}{c}\right)$

5-5. Substitute [V] and [U] for V and U in equation (3.1-1) and expand in accordance with equations (5.2-2) and (5.3-1). Then apply the divergence theorem to the expansion of $\mathrm{div}[V]\left\{\frac{\overline{r}}{cr}\left[\frac{\partial U}{\partial t}\right]\right\}$ and proceed to obtain

$$\oint_{S_{1..n}\Sigma}\left[V\frac{\partial U}{\partial n} - U\frac{\partial V}{\partial n}\right] dS = \int_\tau \left\{[V\,\nabla^2 U - U\,\nabla^2 V] - \frac{\overline{r}}{cr}\cdot\frac{\partial}{\partial t}\,[V\ \mathrm{grad}\ U - U\ \mathrm{grad}\ V]\right\} d\tau$$

and

$$\oint_{S_{1..n}\Sigma}\left[V\frac{\partial V}{\partial n}\right] dS = \int_\tau \left\{[V\,\nabla^2 V] + |[\mathrm{grad}\ V]|^2 - \frac{\overline{r}}{cr}\cdot\left[\tfrac{1}{2}\,\mathrm{grad}\,\frac{\partial}{\partial t}\,V^2\right]\right\} d\tau$$

5-6. Use the results of Sec. 2.12a to show that the curvilinear surface counterparts of equations (5.2-2), (5.2-3) and (5.2-4) are

$$\mathrm{grads}[V] = [\mathrm{grads}\ V] - \frac{\overline{r'}}{cr'}\left[\frac{\partial V}{\partial t}\right] + \frac{\hat{\overline{n}}}{c}\frac{\partial r'}{\partial n}\left[\frac{\partial V}{\partial t}\right]$$

$$\mathrm{divs}[\overline{F}] = [\mathrm{divs}\ \overline{F}] - \frac{\overline{r'}}{cr'}\cdot\left[\frac{\partial \overline{F}}{\partial t}\right] + \frac{\hat{\overline{n}}}{c}\frac{\partial r'}{\partial n}\cdot\left[\frac{\partial \overline{F}}{\partial t}\right]$$

$$\text{curls}[\bar{F}] = [\text{curls } \bar{F}] - \frac{\bar{r}'}{cr'} \times \boxed{\frac{\partial \bar{F}}{\partial t}} + \frac{\hat{\bar{n}}}{c}\frac{\partial r'}{\partial n} \times \boxed{\frac{\partial \bar{F}}{\partial t}}$$

5.4 Uniqueness Theorems for Time-Dependent Fields[4]

5.4a Uniqueness of the scalar field

Let the scalar point function V' be well-behaved in space and time in the closed region R bounded by the surfaces $S_{1..n}\Sigma$. Then

$$\begin{aligned} \text{div}\left(\frac{\partial V'}{\partial t}\text{ grad } V'\right) &= \frac{\partial V'}{\partial t}\nabla^2 V' + \text{grad }\frac{\partial V'}{\partial t}\cdot\text{grad } V' \\ &= \frac{\partial V'}{\partial t}\left\{\nabla^2 V' - pV' - q\frac{\partial V'}{\partial t} - r\frac{\partial^2 V'}{\partial t^2}\right\} \\ &\quad + \frac{\partial}{\partial t}\left\{\tfrac{1}{2}pV'^2 + \tfrac{1}{2}r\left(\frac{\partial V'}{\partial t}\right)^2 + \tfrac{1}{2}|\text{grad } V'|^2\right\} + q\left(\frac{\partial V'}{\partial t}\right)^2 \end{aligned}$$

where p, q and r are functions of position (or constants).

Hence

$$\oint_{S_{1..n}\Sigma}\frac{\partial V'}{\partial t}\frac{\partial V'}{\partial n}\,dS = \int_\tau \frac{\partial V'}{\partial t}\left\{\nabla^2 V' - pV' - q\frac{\partial V'}{\partial t} - r\frac{\partial^2 V'}{\partial t^2}\right\}d\tau$$

$$+\frac{\partial}{\partial t}\int_\tau\left\{\tfrac{1}{2}pV'^2 + \tfrac{1}{2}r\left(\frac{\partial V'}{\partial t}\right)^2 + \tfrac{1}{2}|\text{grad } V'|^2\right\}d\tau + \int_\tau q\left(\frac{\partial V'}{\partial t}\right)^2 d\tau$$

Let V_1 and V_2 be well-behaved functions of space and time which satisfy the following conditions:

(1) $\nabla^2 V - pV - q\frac{\partial V}{\partial t} - r\frac{\partial^2 V}{\partial t^2}$ is a specified function of position in R for $t \geq 0$

(2) V and $\frac{\partial V}{\partial t}$ are specified throughout R at $t = 0$

(3) either $\frac{\partial V}{\partial n}$ or $\frac{\partial V}{\partial t}$ is specified upon $S_{1..n}\Sigma$ for $t > 0$.

---

4. We are concerned here primarily with fields which are not periodic in time, although the analysis is quite general. Alternative boundary conditions apply to periodic fields and these are most easily determined by a complex exponential treatment (Ch.6).

On writing $V_1 - V_2 = V'$ we then obtain

$$0 = \frac{\partial}{\partial t}\int_\tau \left\{\tfrac{1}{2}pV'^2+\tfrac{1}{2}r\left(\frac{\partial V'}{\partial t}\right)^2 +\tfrac{1}{2}|\text{grad } V'|^2\right\} d\tau + \int_\tau q\left(\frac{\partial V'}{\partial t}\right)^2 d\tau \quad \text{for} \quad t \geq 0$$

Then provided that q is nowhere negative, the second volume integral must be zero or positive for $t \geq 0$ and the first must consequently have a constant or decreasing value. But the latter integral is zero at $t = 0$ and can never become negative so long as p and r are nowhere negative. Hence in these circumstances both integrals remain zero for $t > 0$. It then follows that if conditions (1) to (3) are satisfied and p, q, r are nowhere negative, grad $V' = \overline{0}$ and grad V is unique at all points of $\underline{R}$ for $t > 0$. Further, $V' = 0$ and V is unique if, in addition, either p or q or r is positive throughout some subregion of $\underline{R}$. When r is everywhere zero it is unnecessary to specify $\frac{\partial V}{\partial t}$ throughout $\underline{R}$ at $t = 0$.

5.4b Uniqueness of the vector field

The results of the scalar investigation above are directly applicable to vector fields. Specification of $\nabla^2\overline{F} - p\overline{F} - q\frac{\partial \overline{F}}{\partial t} - r\frac{\partial^2\overline{F}}{\partial t^2}$ throughout $\underline{R}$ for $t \geq 0$, together with the specification of $\overline{F}$ and $\frac{\partial \overline{F}}{\partial t}$ throughout $\underline{R}$ at $t = 0$ and of $\frac{\partial \overline{F}}{\partial t}$ or $\frac{\partial \overline{F}}{\partial n}$ on $S_{1..n}\Sigma$ thereafter, determines the three Cartesian components of $\overline{F}$ in $\underline{R}$ for $t > 0$ and consequently $\overline{F}$ itself (for appropriate values of p, q, r).

A vector analogue of the scalar treatment may also be developed as follows:

Let the vector point function $\overline{F}'$ be well-behaved in space and time in the closed region $\underline{R}$ bounded by the surfaces $S_{1..n}\Sigma$. Then

$$\begin{aligned}\text{div}\left(\frac{\partial \overline{F}'}{\partial t} \times \text{curl } \overline{F}'\right) &= \text{curl } \overline{F}'.\text{curl } \frac{\partial \overline{F}'}{\partial t} - \frac{\partial \overline{F}'}{\partial t}.\text{curl curl } \overline{F}' \\ &= \text{curl } \overline{F}'.\text{curl } \frac{\partial \overline{F}'}{\partial t} - \frac{\partial \overline{F}'}{\partial t}.\text{grad div } \overline{F}' + \frac{\partial \overline{F}'}{\partial t}\, .\, \nabla^2\overline{F}' \\ &= \frac{\partial \overline{F}'}{\partial t}.\left\{\nabla^2\overline{F}'-p\overline{F}'-q\frac{\partial \overline{F}'}{\partial t} - r\frac{\partial^2\overline{F}'}{\partial t^2}\right\} - \frac{\partial \overline{F}'}{\partial t}.\text{grad div } \overline{F}' \\ &\quad + \frac{\partial}{\partial t}\left\{\tfrac{1}{2}p|\overline{F}'|^2+\tfrac{1}{2}r\left|\frac{\partial \overline{F}'}{\partial t}\right|^2+\tfrac{1}{2}|\text{curl } \overline{F}'|^2\right\} + q\left|\frac{\partial \overline{F}'}{\partial t}\right|^2\end{aligned}$$

where p, q and r are functions of position (or constants).

Hence

$$\oint_{S_{1..n}\Sigma} \left(\frac{\partial \bar{F}'}{\partial t} \times \mathrm{curl}\ \bar{F}'\right).d\bar{S} = \int_\tau \frac{\partial \bar{F}'}{\partial t}.\left\{\nabla^2\bar{F}' - p\bar{F}' - q\frac{\partial \bar{F}'}{\partial t} - r\frac{\partial^2 \bar{F}'}{\partial t^2}\right\} d\tau - \int_\tau \frac{\partial \bar{F}'}{\partial t}.\mathrm{grad}\ \mathrm{div}\ \bar{F}'\ d\tau$$

$$+ \frac{\partial}{\partial t}\int_\tau \left\{\tfrac{1}{2}p|\bar{F}'|^2 + \tfrac{1}{2}r\left|\frac{\partial \bar{F}'}{\partial t}\right|^2 + \tfrac{1}{2}|\mathrm{curl}\ \bar{F}'|^2\right\} d\tau + \int_\tau q\left|\frac{\partial \bar{F}'}{\partial t}\right|^2 d\tau$$

Let $\bar{F}_1$ and $\bar{F}_2$ be well-behaved functions of space and time which satisfy the following conditions:

(1) $\nabla^2\bar{F} - p\bar{F} - q\dfrac{\partial \bar{F}}{\partial t} - r\dfrac{\partial^2 \bar{F}}{\partial t^2}$ is a specified function of position in $\underline{R}$ for $t \geq 0$

(2) div $\bar{F}$ is a specified function of position in $\underline{R}$ for $t \geq 0$

(3) $\bar{F}$ and $\dfrac{\partial \bar{F}}{\partial t}$ are specified throughout $\underline{R}$ at $t = 0$

(4) either $\hat{n} \times \dfrac{\partial \bar{F}}{\partial t}$ or $\hat{n} \times \mathrm{curl}\ \bar{F}$ is specified upon $S_{1..n}\Sigma$ for $t > 0$

On writing $\bar{F}_1 - \bar{F}_2 = \bar{F}'$ we then obtain

$$0 = \frac{\partial}{\partial t}\int_\tau \left\{\tfrac{1}{2}p|\bar{F}'|^2 + \tfrac{1}{2}r\left|\frac{\partial \bar{F}'}{\partial t}\right|^2 + \tfrac{1}{2}|\mathrm{curl}\ \bar{F}'|^2\right\} d\tau + \int_\tau q\left|\frac{\partial \bar{F}'}{\partial t}\right|^2 d\tau \qquad \text{for} \quad t \geq 0$$

Then provided that q is nowhere negative, the second volume integral must be zero or positive for $t \geq 0$ and the first must consequently have a constant or decreasing value. But the latter integral is zero at $t = 0$ and can never become negative so long as p and r are nowhere negative. Hence in these circumstances both integrals remain zero for $t > 0$. It then follows that if conditions (1) to (4) are satisfied and p, q, r are nowhere negative, curl $\bar{F}' = \bar{0}$ and curl $\bar{F}$ is unique at all points of $\underline{R}$ for $t > 0$. Further, $\bar{F}' = \bar{0}$ and $\bar{F}$ is unique if, in addition, $\underline{R}$ can be divided into a set of subregions such that throughout each subregion one or more of the factors p, q, r is positive. When r is everywhere zero it is unnecessary to specify $\dfrac{\partial \bar{F}}{\partial t}$ throughout $\underline{R}$ at $t = 0$.

EXERCISES

5-7. Given that grad $V' = \bar{0}$ throughout $\underline{R}$ for $t \geq 0$ under the conditions stated in Sec. 5.4a, show that V' is likewise zero provided that p or q or r is positive throughout some subregion of $\underline{R}$.

Given that curl $\overline{F}' = \overline{0}$ throughout $\underline{R}$ for $t \geq 0$ under the conditions stated in Sec. 5.4b, show that $\overline{F}'$ is likewise zero provided that $\underline{R}$ can be divided into subregions throughout each of which one or more of the factors p, q, r is positive. Show further that $\overline{F}' = \overline{0}$ throughout $\underline{R}$ for $t > 0$ provided that

(a) conditions (1) to (3) are satisified

(b) p, q, r are nowhere negative

and either

(c) $\frac{\partial \overline{F}}{\partial t}$ is specified over $S_{1..n}\Sigma$ for $t \geq 0$

or in the event of a single bounding surface S,

(d) $\hat{n} \times \frac{\partial \overline{F}}{\partial t}$ is specified over S for $t \geq 0$.

5-8. The thermal conductivity, density and specific heat of a conducting region $\underline{R}$ bounded by geometrical or physical surfaces $S_{1..n}\Sigma$ are well-behaved positive functions of position. The region is free from sources and sinks of heat. Use the result of Ex.3-28., p. 208 and an application of the divergence theorem to the expansion of div T(k grad T), where k represents conductivity, to show that the temperature T at each point of $\underline{R}$ is uniquely determined as a function of time t provided that the temperature at all points of $\underline{R}$ is a given continuous function of position at $t = 0$ and the temperature at all points of $S_{1..n}\Sigma$ is a given function of position for $t > 0$. [Note that the result of Sec. 5.4a is applicable to the present problem only when k is constant throughout $\underline{R}$.]

5.5 The Retarded Potentials of Scalar and Vector Sources

Retarded potential functions are derived from their unretarded counterparts by time retardation of the source density functions, the origin of retardation being identified with the point of evaluation. Thus the particular source field involved in the evaluation of the retarded potential ϕ at a point 0 at a given instant will, in general, depend upon the position of 0.

For a volume source we have

$$\phi_o = \int_\tau \frac{[\rho]}{r}\, d\tau \quad \text{where} \quad [\rho] = (\rho)_{t - \frac{r}{c}} \tag{5.5-1}$$

The integration region τ must include all points where the retarded density is non-zero; it may consequently extend beyond the parent unretarded field as seen at the moment of evaluation since the latter may, in virtue of its time dependence, appear or disappear in certain subregions[5].

ϕ will be everywhere convergent and continuous, for bounded values of τ, provided that $[\rho]$ is piecewise continuous and everywhere finite. This will be so if ρ satisfies the same conditions and is everywhere piecewise continuous in time over a finite time interval, with c assumed to be non-zero. In these circumstances ϕ will also be continuous in time at all points, in spite of possible time-discontinuities of ρ. This behaviour is best appreciated with the assistance of the following constructions:

(a) Let a spherical surface move outwards from 0 at $t = t_o$ with velocity c through the unretarded source field reversed in time from $t = t_o$. Then the value of ρ at any point of the spherical surface is the value to be assigned to $[\rho]$ in the evaluation of the retarded potential at 0 at $t = t_o$.

[If the advanced density is required the field is not reversed in time.]

(b) Let the region occupied by the source field be divided into a set of volume elements fixed in position relative to 0. From each element let a continuous stream of particles be emitted with velocity c in the direction of 0. The magnitude of each particle is equated to the unretarded source strength associated with the volume element at the time of emission, divided by its distance from 0. The sum of the magnitudes arriving at 0 at any instant is equal to the value of the retarded potential at 0 at that instant. [It does not seem to be possible to provide a satisfactory construction of this type for the advanced potential.]

It will be apparent from (b) that a time discontinuity of ρ which produces a jump in the value of $\int_\tau \frac{\rho}{r}\,d\tau$ cannot be felt all at once at 0. The initial effect derives from a point, or a line, or a surface, but not from an element of volume, consequently $\int_\tau \frac{[\rho]}{r}\,d\tau$ remains continuous in time. Its first derivative, however, may be discontinuous (Ex.5-9., p. 415).

---

5. A particular case is that of a moving source of fixed density distribution. Such sources are given special treatment later, but are included, nevertheless, in general terms in equation (5.5-1) when τ is adjusted accordingly.

The partial and cavity retarded potentials are defined in the same way as their unretarded counterparts, and are required to perform the same functions in subsequent analyses.

In writing

$$\phi_o = \int_S \frac{[\sigma]}{r} \, dS \qquad (5.5\text{-}2)$$

we restrict our consideration to surfaces at rest relative to 0 and having bounding curves which enclose all points where σ may be non-zero. The integral is then adequate to describe the retarded potential function because $[\sigma]$ is zero everywhere outside S.

ϕ is everywhere convergent and continuous in space and time provided that σ satisfies the conditions laid down above for ρ.

The retarded potential of a line source is given, correspondingly, by

$$\phi_o = \int_\Gamma \frac{[\lambda]}{r} \, ds \qquad (5.5\text{-}3)$$

where Γ is at rest relative at 0 and the end points of Γ enclose all elements where λ may be non-zero.

The potential is logarithmically infinite at those points of Γ where λ is non-zero.

In writing the retarded potential of a set of point sources in the form

$$\phi_o = \sum_{i=1}^{n} \frac{[a_i]}{r} \qquad (5.5\text{-}4)$$

it must be supposed that the sources are stationary with respect to 0, since the notation cannot adequately describe the integration region associated with movement. The sources are also required to be non-conservative since conservation, in the case of a point source system, implies the constancy of individual source strengths with respect to time. [The latter restriction does not apply to the piecewise continuous source systems considered previously because temporal variation of density from point to point is not inconsistent with the constancy of the spatial density integral].

We may define a time-dependent double layer source to have the same limiting geometrical configuration as the time-invariant surface doublet described in Sec. 4.3 with the additional requirement that the unretarded simple surface densities have equal and opposite instantaneous values (apart from correction for surface curvature) at corresponding points of the component surfaces. Thus if we designate the 'negative' surface by the subscript 1 and the 'positive' by 2 the contribution to the retarded potential of a matched pair of surface elements is given by

$$d\phi_o = -\frac{[\sigma_1 dS_1]_1}{r_1} + \frac{[\sigma_2 dS_2]_2}{r_2}$$

$$= -\frac{[\sigma dS]_1}{r_1} + \frac{[\sigma dS]_2}{r_2} \qquad \text{since } \sigma_1 dS_1 = \sigma_2 dS_2 \equiv \sigma dS$$

$$= \frac{[\sigma dS]_1}{r_2} - \frac{[\sigma dS]_1}{r_1} + \frac{[\sigma dS]_2}{r_2} - \frac{[\sigma dS]_1}{r_2}$$

$$= [\sigma dS]_1 \left(\frac{1}{r_2} - \frac{1}{r_1}\right) + \frac{1}{r_2}\left([\sigma dS]_2 - [\sigma dS]_1\right)$$

Then on making use of mean-value theorem expansions, on the assumption that the first time derivative of σ is continuous throughout the interval defined by $[\sigma]_1$ and $[\sigma]_2$, and on taking limits as surface 2 approaches surface 1, we obtain for the complete source

$$\phi_o = \int_S \left\{[\mu]\frac{\partial}{\partial n}\left(\frac{1}{r}\right) - \frac{1}{cr}\frac{\partial r}{\partial n}\left[\frac{\partial \mu}{\partial t}\right]\right\} dS \tag{5.5-5}$$

where $\hat{n}$ is the unit normal directed from the 'negative' to the 'positive' surface and μ is the unretarded doublet density (moment per unit area).

It is supposed, of course, that the surface is at rest relative to O. As in the unretarded, time-invariant case there is a discontinuity of potential of $+4\pi\mu$ in passing through the surface in a positive sense at an interior point where the unretarded density is continuous in space and time and its instantaneous value is μ.

Corresponding expressions for the line and point doublet may be written down by inspection. In the notation employed for the unretarded cases we have

$$\phi_o = \int_\Gamma \left\{[L]\frac{\partial}{\partial n'}\left(\frac{1}{r}\right) - \frac{1}{cr}\frac{\partial r}{\partial n'}\left[\frac{\partial L}{\partial t}\right]\right\} ds \tag{5.5-6}$$

and

$$\phi_o = [p]\frac{d}{dl}\left(\frac{1}{r}\right) - \frac{1}{cr}\frac{dr}{dl}\left[\frac{dp}{dt}\right] \tag{5.5-7}$$

$$= [\bar{p}].\text{grad}\,\frac{1}{r} - \frac{\bar{r}}{cr^2}.\left[\frac{d\bar{p}}{dt}\right] \qquad (5.5\text{-}7(a))$$

Retarded vector potentials derive from vectorial volume, surface and line sources in the same way as their scalar equivalents derive from the corresponding scalar sources. The same behaviour with respect to convergence and continuity will therefore obtain, provided that the scalar components of the source densities satisfy the conditions laid down above for scalar sources. In terms of the notation employed for unretarded potentials we have

$$\bar{A}_o = \int_\tau \frac{[\bar{J}]}{r}\,d\tau \qquad \text{volume source} \qquad (5.5\text{-}8)$$

$$\bar{A}_o = \int_S \frac{[\bar{K}]}{r}\,dS \qquad \text{simple source source} \qquad (5.5\text{-}9)$$

$$\bar{A}_o = \int_\Gamma \frac{[\bar{I}]}{r}\,ds \qquad \text{simple line source} \qquad (5.5\text{-}10)$$

$$\bar{A}_o = \int_S \left\{[\bar{\mu}_1]\,\frac{\partial}{\partial n}\left(\frac{1}{r}\right) - \frac{1}{cr}\,\frac{\partial r}{\partial n}\left[\frac{\partial\bar{\mu}_1}{\partial t}\right]\right\} dS \qquad \text{surface doublet} \qquad (5.5\text{-}11)$$

$$\bar{A}_o = \int_\Gamma \left\{[\bar{L}_1]\,\frac{\partial}{\partial n'}\left(\frac{1}{r}\right) - \frac{1}{cr}\,\frac{\partial r}{\partial n'}\left[\frac{\partial\bar{L}_1}{\partial t}\right]\right\} ds \qquad \text{line doublet} \qquad (5.5\text{-}12)$$

The association of a vector magnitude with a stationary point, and the subsequent definition of an unretarded point singlet vector potential, has little practical significance and was consequently omitted from the considerations of Sec. 4.12. On the other hand, the association of a retarded vector potential with a moving point has considerable practical application and merits a detailed examination. This will be found in Sec. 5.10. Correspondingly, the unretarded source whose potential is the vector analogue of the scalar point doublet, as discussed on p. 305, does not comprise a vector point doublet in the conventional sense - although the potential of the latter would take the form $\bar{p}_1 \frac{d}{dl}\left(\frac{1}{r}\right)$ and could justly be called the vector equivalent of the scalar doublet potential

$p\,\frac{d}{dl}\left(\frac{1}{r}\right)$ - but a whirl, ie the limiting configuration of a closed tangential line source whose potential, $\bar{m} \times \text{grad}\,\frac{1}{r}$, is treated as the vector analogue of the scalar doublet potential when expressed in the form $\bar{p}.\text{grad}\,\frac{1}{r}$.

We now proceed to show that this equivalence continues to hold in the retarded case, where the potential of the whirl may be expressed as

$$[\bar{m}] \times \text{grad}\,\frac{1}{r} - \boxed{\frac{d\bar{m}}{dt}} \times \frac{\bar{r}}{cr^2}$$

and consequently treated as the vector analogue of the retarded scalar doublet potential when written in the form

$$[\bar{p}].\text{grad}\,\frac{1}{r} - \boxed{\frac{d\bar{p}}{\partial t}} \cdot \frac{\bar{r}}{cr^2}$$

Prior to limiting, the potential of the whirl at an exterior point 0 is given by

$$\bar{A}_o = \oint_\Gamma \frac{[\bar{I}]}{r}\,ds = \oint_\Gamma [I]\,\frac{d\bar{r}}{r}$$

Let Q be a point in the vicinity of Γ, distant r' from 0. Then provided that $\frac{dI}{dt}$ is continuous throughout the interval defined by $t - \frac{r}{c}$ and $t - \frac{r'}{c}$ we have

$$[I] = (I)_{t-\frac{r}{c}} = (I)_{t-\frac{r'}{c}} - \frac{1}{c}\,(r-r')\left(\frac{dI}{dt}\right)_{t-\frac{r''}{c}}$$

where $t - \frac{r''}{c}$ is a point of the interval.

Hence

$$\bar{A}_o = (I)_{t-\frac{r'}{c}} \oint_\Gamma \frac{d\bar{r}}{r} - \frac{1}{c}\left(\frac{dI}{dt}\right)_{t-\frac{r''}{c}} \oint_\Gamma \frac{(r-r')}{r}\, d\bar{r}$$

$$= (I)_{t-\frac{r'}{c}} \oint_\Gamma \frac{d\bar{r}}{r} + \frac{r'}{c}\left(\frac{dI}{dt}\right)_{t-\frac{r''}{c}} \oint_\Gamma \frac{d\bar{r}}{r} \qquad \text{since} \quad \oint_\Gamma d\bar{r} = \bar{0}$$

$$= (I)_{t-\frac{r'}{c}} \int_S d\bar{S} \times \text{grad}\,\frac{1}{r} - \frac{r'}{c}\left(\frac{dI}{dt}\right)_{t-\frac{r''}{c}} \int_S d\bar{S} \times \frac{\bar{r}}{r^3}$$

where S spans Γ.

Then in the limit as Γ shrinks about Q

$$\bar{A}_o = (I)_{t-\frac{r'}{c}}\, d\bar{S} \times \left(\text{grad}\,\frac{1}{r}\right)_Q - \left(\frac{dI}{dt}\right)_{t-\frac{r'}{c}} d\bar{S} \times \frac{\bar{r'}}{cr'^2}$$

so that in general

$$\bar{A} = [\bar{m}] \times \text{grad}\,\frac{1}{r} - \left[\frac{d\bar{m}}{dt}\right] \times \frac{\bar{r}}{cr^2} \qquad (5.5\text{-}13)$$

where $\bar{m}$ is the limiting value of $Id\bar{S}$.

It is now evident that we may interpret equation (5.3-3) in the light of the retarded potentials described by equations (5.5-1), (5.5-2) and (5.5-5). We see that any scalar point function V, with continuous second-order space and time derivatives in a bounded region of space $\underline{R}$, may be expressed within $\underline{R}$ as the sum of the retarded potentials associated with a volume source throughout $\underline{R}$ of density $-\frac{1}{4\pi}$ dal V and simple and double surface sources on the boundary of respective densities $\frac{1}{4\pi}\frac{\partial V}{\partial n}$ and $-\frac{V}{4\pi}$. A well-behaved vector point function may be described similarly in terms of vector potentials.

It will also be apparent from the result of Ex.5-3., p. 403 that a scalar (or vector) point function with appropriate degrees of continuity may be represented as the sum of retarded potentials which are exponentially enhanced or attenuated with distance from the source, ie potentials based upon the factor $\frac{1}{r}e^{\gamma r}$ rather than $\frac{1}{r}$. In the case of an unbounded integration region the surface integral at infinity vanishes for negative values of γ, provided that R grad V, RV and $\frac{R}{c}\frac{\partial V}{\partial t}$ are bounded at infinity in space and time. The corresponding representation for the non-retarded case was considered in Ex.3-15., p. 183 and Ex.4-30., p. 258.

EXERCISES

5-9. Let ϕ be the retarded potential at 0 of a bounded volume source which is everywhere finite and piecewise continuous in space, and constant in time until $t = t_o$. At $t = t_o$ the density jumps by the common value $\Delta\rho$ over a subregion of the source which may or may not embrace 0. Show that ϕ is continuous in time for $t \geq 0$. Show also that the first time derivative of ϕ may or may not be continuous for $t > 0$, depending upon the shape and orientation of the subregion, but that if the subregion embrace 0 the first time derivative of ϕ is discontinuous at $t = t_o$.

5-10. Let $\overline{E}$ be a well-behaved point function which satisfies the equation

$$\nabla^2\overline{E} - \alpha\frac{\partial\overline{E}}{\partial t} - \beta\frac{\partial^2\overline{E}}{\partial t^2} = \overline{0}$$

throughout the region $\underline{R}$ bounded by the surfaces $S_{1..n}\Sigma$, where α and β are positive constants.

Show that $\overline{E}$ may be expressed at any interior point of $\underline{R}$ as the sum of exponentially-modified, retarded (or advanced) vector potentials of surface and volume sources, the volume density being given by $\frac{\gamma^2}{4\pi}\overline{E}$ and the constants γ and c by

$$\gamma = \mp\frac{\alpha}{2\beta^{\frac{1}{2}}} \qquad c = \pm\frac{1}{\beta^{\frac{1}{2}}}$$

Show further that when, in addition, $\overline{E}$ varies sinusoidally in time at all points of $\underline{R}$ with angular velocity ω and fixed direction at each point, the volume integral may be eliminated by putting

$$\gamma^2 = \frac{\omega^2\beta}{2}\left\{-1 + \left(1 + \frac{\alpha^2}{\omega^2\beta^2}\right)^{\frac{1}{2}}\right\}$$

$$\frac{1}{c^2} = \frac{\beta}{2}\left\{1 + \left(1 + \frac{\alpha^2}{\omega^2\beta^2}\right)^{\frac{1}{2}}\right\}$$

where γ and c have opposite signs.

5.6 The Gradient, Divergence and Curl of Retarded Potentials

5.6a Gradient of the retarded scalar potential beyond the source

Consider first the case of a volume source of integration region τ bounded by the surfaces $S_{1..n}\Sigma$. Let the first space and time derivatives of the density ρ be continuous. Then if 0 is an exterior point of τ

$$\text{grad} \int_\tau \frac{[\rho]}{r}\, d\tau = \sum \bar{i} \frac{\partial}{\partial x_o} \int_\tau \frac{[\rho]}{r}\, d\tau = \sum \bar{i} \int_\tau \left\{ [\rho] \frac{\partial}{\partial x_o}\left(\frac{1}{r}\right) - \frac{1}{cr}\frac{\partial r}{\partial x_o} \boxed{\frac{\partial \rho}{\partial t}} \right\} d\tau$$

or

$$\text{grad} \int_\tau \frac{[\rho]}{r}\, d\tau = -\int_\tau \left\{ [\rho]\, \text{grad}\, \frac{1}{r} - \boxed{\frac{\partial \rho}{\partial t}} \frac{\bar{r}}{cr^2} \right\} d\tau \qquad (5.6\text{-}1)$$

But

$$\text{grad}\, \frac{[\rho]}{r} = \frac{1}{r}\, \text{grad}\, [\rho] + [\rho]\, \text{grad}\, \frac{1}{r} \qquad (5.6\text{-}2)$$

$$= \frac{1}{r}\, [\text{grad}\, \rho] - \boxed{\frac{\partial \rho}{\partial t}} \frac{\bar{r}}{cr^2} + [\rho]\, \text{grad}\, \frac{1}{r} \qquad (5.6\text{-}3)$$

whence by substitution in equation (5.6-1) and application of (1.17-5) we get

$$\text{grad} \int_\tau \frac{[\rho]}{r}\, d\tau = \int_\tau \frac{1}{r}\, [\text{grad}\, \rho]\, d\tau - \oint_{S_{1..n}\Sigma} \frac{[\rho]}{r}\, d\bar{S} \qquad (5.6\text{-}4)$$

and

$$\text{grad} \int_\tau \frac{[\rho]}{r}\, d\tau = \int_\tau \left\{ \frac{1}{r}\, \text{grad}\, [\rho] + \boxed{\frac{\partial \rho}{\partial t}} \frac{\bar{r}}{cr^2} \right\} d\tau - \oint_{S_{1..n}\Sigma} \frac{[\rho]}{r}\, d\bar{S} \qquad (5.6\text{-}5)$$

The gradient of the retarded potential is continuous through the boundary of a volume source provided that ρ and its space derivatives are finite and piecewise continuous in space and time. This follows from arguments similar to those adduced in connection with the gradient of the unretarded potential (p. 278).

The results for simple surface, line and point sources may be written down by analogy with equation (5.6-1).

$$\text{grad}\int_S \frac{[\sigma]}{r}\,dS = -\int_S \left\{[\sigma]\,\text{grad}\,\frac{1}{r} - \boxed{\frac{\partial\sigma}{\partial t}}\,\frac{\bar{r}}{cr^2}\right\} dS \tag{5.6-6}$$

$$\text{grad}\int_\Gamma \frac{[\lambda]}{r}\,ds = -\int_\Gamma \left\{[\lambda]\,\text{grad}\,\frac{1}{r} - \boxed{\frac{\partial\lambda}{\partial t}}\,\frac{\bar{r}}{cr^2}\right\} ds \tag{5.6-7}$$

$$\text{grad}\,\frac{[a]}{r} = -\left\{[a]\,\text{grad}\,\frac{1}{r} - \boxed{\frac{da}{dt}}\,\frac{\bar{r}}{cr^2}\right\} \tag{5.6-8}$$

The gradient operation on the right hand side of equation (5.6-8) is, of course, carried out at the source while that on the left is carried out at 0.

It is clear that the retarded potential of a simple surface source is continuous for movement through the source; as a consequence the tangential component of the potential gradient is likewise continuous. The normal component, however, is discontinuous to the same extent as in the non-retarded case, ie

$$\Delta\ \text{grad}\int_S \frac{[\sigma]}{r}\,dS = -\,4\pi\sigma\,\hat{n} \tag{5.6-9}$$

where $\hat{n}$ indicates the sense of movement and σ denotes the local surface density at the instant of evaluation (see Ex.5-11., p. 423).

5.6b Gradient of the retarded scalar potential within a volume source

We proceed as in Sec. 4.8 for the unretarded case by first determining the variation of partial potential deriving from movement of the point of evaluation. Following the notation of that section and Fig. 4.8b it is observed that when the unretarded field is slipped backwards, the increment at 0 of the retarded potential associated with the typical volume element $\Delta\tau$ is $\frac{([\rho']-[\rho])}{r}\,\Delta\tau$, where the time delay for both ρ' and ρ is $\frac{r}{c}$. Some reflection will then reveal that $([\rho']-[\rho])$ may be replaced by $\frac{\partial\rho}{\partial x}\,\Delta x$, where the derivative is evaluated at some point of the slip path. It will be apparent that it is the derivative of the unretarded field at the approximate time $t - \frac{r}{c}$ with which we are concerned. On following the procedure in Sec. 4.8 we arrive at

$$\frac{\partial}{\partial x_0}\,(\text{partial})\int_{\tau-\tau_\delta} \frac{[\rho]}{r}\,d\tau = \int_{\tau-\tau_\delta} \frac{1}{r}\boxed{\frac{\partial\rho}{\partial x}}\,d\tau - \oint_{S_{1..n}\Sigma} \frac{[\rho]}{r}\,dS_x \tag{5.6-10}$$

whence

$$\text{grad (partial)} \int_{\tau-\tau_\delta} \frac{[\rho]}{r}\, d\tau = \int_{\tau-\tau_\delta} \frac{1}{r}\,[\text{grad}\ \rho]\, d\tau - \oint_{S_{1..n}\Sigma} \frac{[\rho]}{r}\, d\bar{S} \tag{5.6-11}$$

On substituting equations (5.6-2) and (5.6-3) in (5.6-11) we get

$$\text{grad (partial)} \int_{\tau-\tau_\delta} \frac{[\rho]}{r}\, d\tau = \int_{\tau-\tau_\delta} \left\{\frac{1}{r}\,\text{grad}\,[\rho] + \left[\frac{\partial\rho}{\partial t}\right] \frac{\bar{r}}{cr^2}\right\} d\tau - \oint_{S_{1..n}\Sigma} \frac{[\rho]}{r}\, d\bar{S} \tag{5.6-12}$$

and

$$\text{grad (partial)} \int_{\tau-\tau_\delta} \frac{[\rho]}{r}\, d\tau = -\int_{\tau-\tau_\delta} \left\{[\rho]\,\text{grad}\,\frac{1}{r} - \left[\frac{\partial\rho}{\partial t}\right] \frac{\bar{r}}{cr^2}\right\} d\tau + \oint_{S_\delta} \frac{[\rho]}{r}\, d\bar{S} \tag{5.6-13}$$

The limiting forms of the above expressions as $\delta \to 0$ are identical with equations (5.6-4), (5.6-5) and (5.6-1), hence the latter expressions are valid for points of evaluation within or without the source.

The gradient of the cavity potential corresponds with that obtaining for an exterior source but with the additional fixed bounding surface S_δ which must consequently appear in equations (5.6-4) and (5.6-5), together with a modified volume integral.

All of the above formulae are brought together in Table 5, p. 439.

5.6c Divergence of the retarded vector potential beyond the source

Let the density $\bar{J}$ of a volume source have continuous first derivatives in space and time. Then at an exterior point

$$\text{div} \int_\tau \frac{[\bar{J}]}{r}\, d\tau = \sum \frac{\partial}{\partial x_o} \int_\tau \frac{[J_x]}{r}\, d\tau = \sum \int_\tau \left\{[J_x] \frac{\partial}{\partial x_o}\left(\frac{1}{r}\right) - \frac{1}{cr}\frac{\partial r}{\partial x_o}\left[\frac{\partial J_x}{\partial t}\right]\right\} d\tau$$

whence

$$\text{div} \int_\tau \frac{[\bar{J}]}{r}\, d\tau = -\int_\tau \left\{[\bar{J}].\text{grad}\,\frac{1}{r} - \frac{\partial \bar{J}}{\partial t} . \frac{\bar{r}}{cr^2}\right\} d\tau \tag{5.6-14}$$

Since

$$\operatorname{div} \frac{[\bar{J}]}{r} = \frac{1}{r} \operatorname{div} [\bar{J}] + [\bar{J}].\operatorname{grad} \frac{1}{r} \tag{5.6-15}$$

$$= \frac{1}{r} [\operatorname{div} \bar{J}] - \boxed{\frac{\partial \bar{J}}{\partial t}}.\frac{\bar{r}}{cr^2} + [\bar{J}].\operatorname{grad} \frac{1}{r} \tag{5.6-16}$$

we have also

$$\operatorname{div} \int_\tau \frac{[\bar{J}]}{r} d\tau = \int_\tau \frac{1}{r} [\operatorname{div} \bar{J}] \, d\tau - \oint_{S_{1..n}\Sigma} \frac{[\bar{J}]}{r}.d\bar{S} \tag{5.6-17}$$

and

$$\operatorname{div} \int_\tau \frac{[\bar{J}]}{r} d\tau = \int_\tau \left\{\frac{1}{r} \operatorname{div} [\bar{J}] + \boxed{\frac{\partial \bar{J}}{\partial t}}.\frac{\bar{r}}{cr^2}\right\} d\tau - \oint_{S_{1..n}\Sigma} \frac{[\bar{J}]}{r}.d\bar{S} \tag{5.6-18}$$

Similarly, for simple line and surface sources,

$$\operatorname{div} \int_\Gamma \frac{[\bar{I}]}{r} ds = - \int_\Gamma \left\{[\bar{I}].\operatorname{grad} \frac{1}{r} - \boxed{\frac{\partial \bar{I}}{\partial t}}.\frac{\bar{r}}{cr^2}\right\} ds \tag{5.6-19}$$

$$\operatorname{div} \int_S \frac{[\bar{K}]}{r} dS = - \int_S \left\{[\bar{K}].\operatorname{grad} \frac{1}{r} - \boxed{\frac{\partial \bar{K}}{\partial t}}.\frac{\bar{r}}{cr^2}\right\} dS \tag{5.6-20}$$

The discontinuity of $\operatorname{div} \int_S \frac{[\bar{K}]}{r} dS$ which accompanies movement through the surface S at an interior point is identical with that obtaining in the non-retarded case, ie

$$\Delta \operatorname{div} \int_S \frac{[\bar{K}]}{r} dS = -4\pi\hat{n}.\bar{K} \tag{5.6-21}$$

where $\hat{n}$ defines the sense of movement through the surface and $\bar{K}$ is the local surface density at the instant of evaluation.

5.6d Divergence of the retarded vector potential within a volume source

From equation (5.6-10)

$$\frac{\partial}{\partial x_o}\ (\text{partial}) \int\limits_{\tau-\tau_\delta} \frac{[J_x]}{r}\, d\tau = \int\limits_{\tau-\tau_\delta} \frac{1}{r}\left[\frac{\partial J_x}{\partial x}\right] d\tau - \oint\limits_{S_{1..n}\Sigma} \frac{[J_x]}{r}\, dS_x$$

hence

$$\text{div (partial)} \int\limits_{\tau-\tau_\delta} \frac{[\bar{J}]}{r}\, d\tau = \int\limits_{\tau-\tau_\delta} \frac{1}{r}\,[\text{div } \bar{J}]\, d\tau - \oint\limits_{S_{1..n}\Sigma} \frac{[\bar{J}]}{r}.d\bar{S} \qquad (5.6\text{-}22)$$

On substituting equations (5.6-15) and (5.6-16) in (5.6-22) we get

$$\text{div (partial)} \int\limits_{\tau-\tau_\delta} \frac{[\bar{J}]}{r}\, d\tau = -\int\limits_{\tau-\tau_\delta} \left\{[\bar{J}].\text{grad}\,\frac{1}{r} - \left[\frac{\partial \bar{J}}{\partial t}\right].\frac{\bar{r}}{cr^2}\right\} d\tau + \oint\limits_{S_\delta} \frac{[\bar{J}]}{r}.d\bar{S}$$

(5.6-23)

and

$$\text{div (partial)} \int\limits_{\tau-\tau_\delta} \frac{[\bar{J}]}{r}\, d\tau = \int\limits_{\tau-\tau_\delta} \left\{\frac{1}{r}\,\text{div } [\bar{J}] + \left[\frac{\partial \bar{J}}{\partial t}\right].\frac{\bar{r}}{cr^2}\right\} d\tau - \oint\limits_{S_{1..n}\Sigma} \frac{[\bar{J}]}{r}.d\bar{S}$$

(5.6-24)

The limiting forms of the above expressions as $\delta \to 0$ are identical with equations (5.6-17), (5.6-14) and (5.6-18) hence the latter expressions are valid for points of evaluation within and without the source.

5.6e Curl of the retarded vector potential beyond the source

At a point beyond the integration region

$$\text{curl} \int\limits_\tau \frac{[\bar{J}]}{r}\, d\tau$$

$$= \sum \bar{i} \left\{\frac{\partial}{\partial y_o} \int\limits_\tau \frac{[J_z]}{r}\, d\tau - \frac{\partial}{\partial z_o} \int\limits_\tau \frac{[J_y]}{r}\, d\tau\right\}$$

$$= \sum \bar{i} \int \left\{[J_z]\,\frac{\partial}{\partial y_o}\left(\frac{1}{r}\right) - \frac{1}{cr}\,\frac{\partial r}{\partial y_o}\left[\frac{\partial J_z}{\partial t}\right] - [J_y]\,\frac{\partial}{\partial z_o}\left(\frac{1}{r}\right) + \frac{1}{cr}\,\frac{\partial r}{\partial z_o}\left[\frac{\partial J_y}{\partial t}\right]\right\} d\tau$$

or

$$\operatorname{curl}\int_{\tau}\frac{[\bar{J}]}{r}\,d\tau = \int_{\tau}\left\{[\bar{J}]\times\operatorname{grad}\frac{1}{r} - \boxed{\frac{\partial\bar{J}}{\partial t}}\times\frac{\bar{r}}{cr^2}\right\}d\tau \qquad (5.6\text{-}25)$$

Since

$$\operatorname{curl}\frac{[\bar{J}]}{r} = \frac{1}{r}\operatorname{curl}[\bar{J}] + \operatorname{grad}\frac{1}{r}\times[\bar{J}] \qquad (5.6\text{-}26)$$

$$= \frac{1}{r}[\operatorname{curl}\bar{J}] + \boxed{\frac{\partial\bar{J}}{\partial t}}\times\frac{\bar{r}}{cr^2} + \operatorname{grad}\frac{1}{r}\times[\bar{J}] \qquad (5.6\text{-}27)$$

we have also

$$\operatorname{curl}\int_{\tau}\frac{[\bar{J}]}{r}\,d\tau = \int_{\tau}\frac{1}{r}[\operatorname{curl}\bar{J}]\,d\tau - \oint_{S_{1..n}\Sigma} d\bar{S}\times\frac{[\bar{J}]}{r} \qquad (5.6\text{-}28)$$

and

$$\operatorname{curl}\int_{\tau}\frac{[\bar{J}]}{r}\,d\tau = \int_{\tau}\left\{\frac{1}{r}\operatorname{curl}[\bar{J}] - \boxed{\frac{\partial\bar{J}}{\partial t}}\times\frac{\bar{r}}{cr^2}\right\}d\tau - \oint_{S_{1..n}\Sigma} d\bar{S}\times\frac{[\bar{J}]}{r} \qquad (5.6\text{-}29)$$

Similarly, for simple line and surface sources,

$$\operatorname{curl}\int_{\Gamma}\frac{[\bar{I}]}{r}\,ds = \int_{\Gamma}\left\{[\bar{I}]\times\operatorname{grad}\frac{1}{r} - \boxed{\frac{\partial\bar{I}}{\partial t}}\times\frac{\bar{r}}{cr^2}\right\}ds \qquad (5.6\text{-}30)$$

and

$$\operatorname{curl}\int_{S}\frac{[\bar{K}]}{r}\,dS = \int_{S}\left\{[\bar{K}]\times\operatorname{grad}\frac{1}{r} - \boxed{\frac{\partial\bar{K}}{\partial t}}\times\frac{\bar{r}}{cr^2}\right\}dS \qquad (5.6\text{-}31)$$

The discontinuity of $\operatorname{curl}\int_{S}\frac{[\bar{K}]}{r}\,dS$ which accompanies movement through S at an interior point is identical with that obtaining in the unretarded case, ie

$$\Delta \text{ curl} \int_S \frac{[\overline{K}]}{r} dS = -4\pi(\hat{\overline{n}}\times\overline{K}) \qquad (5.6\text{-}32)$$

The normal component of the curl is seen to be continuous through the surface irrespective of the orientation of $\overline{K}$.

5.6f <u>Curl of the retarded vector potential within a volume source</u>

From equation (5.6-10)

$$\text{curl (partial)} \int_{\tau-\tau_\delta} \frac{[\overline{J}]}{r} d\tau = \sum \overline{i}\left\{\frac{\partial}{\partial y_o}\text{(partial)}\int_{\tau-\tau_\delta}\frac{[J_z]}{r}d\tau - \frac{\partial}{\partial z_o}\text{(partial)}\int_{\tau-\tau_\delta}\frac{[J_y]}{r}d\tau\right\}$$

$$= \sum \overline{i}\left\{\int_{\tau-\tau_\delta}\left(\frac{1}{r}\left[\frac{\partial J_z}{\partial y}\right] - \frac{1}{r}\left[\frac{\partial J_y}{\partial z}\right]\right)d\tau - \oint_{S_{1..n}\Sigma}\left(\frac{[J_z]}{r}dS_y - \frac{[J_y]}{r}dS_z\right)\right\}$$

or

$$\text{curl (partial)} \int_{\tau-\tau_\delta} \frac{[\overline{J}]}{r} d\tau = \int_{\tau-\tau_\delta}\frac{1}{r}[\text{curl }\overline{J}]\,d\tau - \oint_{S_{1..n}\Sigma} d\overline{S}\times\frac{[\overline{J}]}{r} \qquad (5.6\text{-}33)$$

On substituting equations (5.6-26) and (5.6-27) in (5.6-33) we get

$$\text{curl (partial)} \int_{\tau-\tau_\delta} \frac{[\overline{J}]}{r} d\tau = \int_{\tau-\tau_\delta}\left\{[\overline{J}]\times\text{grad}\,\frac{1}{r} - \left[\frac{\partial\overline{J}}{\partial t}\right]\times\frac{\overline{r}}{cr^2}\right\}d\tau + \oint_{S_\delta} d\overline{S}\times\frac{[\overline{J}]}{r}$$

(5.6-34)

and

$$\text{curl (partial)} \int_{\tau-\tau_\delta} \frac{[\overline{J}]}{r} d\tau = \int_{\tau-\tau_\delta}\left\{\frac{1}{r}\text{ curl }[\overline{J}] - \left[\frac{\partial\overline{J}}{\partial t}\right]\times\frac{\overline{r}}{cr^2}\right\}d\tau - \oint_{S_{1..n}\Sigma} d\overline{S}\times\frac{[\overline{J}]}{r}$$

(5.6-35)

The limiting forms of the above expressions as $\delta\to 0$ are identical with equations (5.6-28), (5.6-25) and (5.6-29), hence the latter expressions are valid for points of evaluation within and without the source. As in the cases of grad partial pot $[\rho]$ and div partial pot $[\overline{J}]$, the same limits obtain when the δ sphere is replaced by any regular excluding region.

The formulae for divergence and curl are brought together in Table 6, p. 441.

EXERCISES

5-11. Apply equation (5.6-6) to the evaluation of grad $\int_S \frac{[\sigma]}{r}\, dS$ at a point on the axis of a circular, disc-shaped source of uniform, time-dependent density. Show by integration that for finite values of $\left[\frac{\partial\sigma}{\partial t}\right]$ the second component of the surface integral does not contribute to the discontinuity of grad $\int_S \frac{[\sigma]}{r}\, dS$ on passing through the source. Hence conclude plausibly that a discontinuity of $-4\pi\sigma\,\hat{\bar{n}}$ will obtain at interior points of any smooth surface source where the local surface density σ and its first time derivative are continuous in space and time.

Make use of a similar analysis to show that there is an increment in the retarded potential of $4\pi\mu$ on passing in a positive sense through a scalar surface doublet of local density μ. Demonstrate the equivalent result for a vector surface doublet.

5-12. For the conditions of continuity stated in the previous exercise show that

$$\Delta \frac{\partial}{\partial s} \int_S \frac{[\bar{K}]}{r}\, dS = \bar{0}$$

where $\frac{\partial}{\partial s}$ is any tangential derivative, and

$$\Delta \frac{\partial}{\partial n} \int_S \frac{[\bar{K}]}{r}\, dS = -4\pi\bar{K}$$

Show further that

$$\Delta\, \mathrm{div} \int_S \frac{[\bar{K}]}{r}\, dS = -4\pi(\hat{\bar{n}}.\bar{K})$$

and

$$\Delta\, \mathrm{curl} \int_S \frac{[\bar{K}]}{r}\, dS = -4\pi(\hat{\bar{n}}\times\bar{K})$$

Hence conclude that

$$\Delta \text{ grad} \int_S \frac{[\overline{M}]}{r} \cdot d\overline{S} = -4\pi(\overline{M}.\hat{n})\hat{n}$$

$$\Delta \text{ div} \int_S d\overline{S} \times \frac{[\overline{M}]}{r} = 0$$

$$\Delta \text{ curl} \int_S d\overline{S} \times \frac{[\overline{M}]}{r} = -4\pi\hat{n}\times(\hat{n}\times\overline{M}) = 4\pi\overline{M}_t$$

where $\overline{M}_t$ is the vector tangential component of $\overline{M}$ and $\hat{n}$ corresponds with the positive sense of $d\overline{S}$.

5-13. It has been assumed in earlier pages that the discontinuities of ϕ and $\overline{A}$ and their derivatives, which accompany movement through simple regular surface sources, may be determined by evaluation of the discontinuity obtaining with motion along the axis of a uniform disc-shaped source having the local (continuous) surface density. While this is true for the potentials and their first derivatives it may fail for higher-order derivatives. Demonstrate this by evaluating $\Delta \frac{\partial}{\partial n} \text{ grad} \oint_S \frac{\sigma}{r} dS$ for constant σ, where S is (1) a disc (2) a closed spherical surface of radius R, and noting a numerical discrepancy of $8\pi\sigma/R$ in the results.

The matter is taken up again in Ex.7-4/5, p. 633.

5-14. By making use of the relationship $\frac{\partial}{\partial x_o} \text{div } [\overline{F}] = \text{div } \frac{\partial}{\partial x_o} [\overline{F}]$ show that at exterior points of τ

$$\text{grad} \int_\tau \frac{\text{div } [\overline{F}]}{r} d\tau = \sum \overline{i} \int_\tau \frac{\partial}{\partial x_o} \frac{\text{div } [\overline{F}]}{r} d\tau$$

$$= -\int_\tau \text{div } [\overline{F}] \text{ grad } \frac{1}{r} d\tau - \int_\tau \frac{\overline{r}}{cr} \left[\frac{\partial \overline{F}}{\partial t}\right] . \text{grad } \frac{1}{r} d\tau + \oint_{S_{1..n}\Sigma} \frac{\overline{r}}{cr^2} \left[\frac{\partial \overline{F}}{\partial t}\right] . d\overline{S}$$

$$= -\int_\tau [\text{div } \overline{F}] \text{ grad } \frac{1}{r} d\tau + \oint_{S_{1..n}\Sigma} \frac{\overline{r}}{cr^2} \left[\frac{\partial \overline{F}}{\partial t}\right] . d\overline{S}$$

An alternative expansion takes the form

$$\text{grad} \int_\tau \frac{\text{div }[\bar{F}]}{r}\, d\tau$$

$$= - \int_\tau \text{div }[\bar{F}]\ \text{grad}\ \frac{1}{r}\, d\tau + \int_\tau \left[\text{div}\ \frac{\partial \bar{F}}{\partial t}\right] \frac{\bar{r}}{cr^2}\, d\tau + \int_\tau \frac{1}{r}\ \text{grad}\left(\frac{\bar{r}}{cr}\cdot\left[\frac{\partial \bar{F}}{\partial t}\right]\right) d\tau$$

Prove this

(a) by writing

$$\text{grad}\int_\tau \frac{\text{div }[\bar{F}]}{r}\, d\tau = \text{grad}\int_\tau \frac{[\text{div }\bar{F}]}{r}\, d\tau - \text{grad}\int_\tau \left[\frac{\partial \bar{F}}{\partial t}\right]\cdot\frac{\bar{r}}{cr^2}\, d\tau$$

and utilising equation (5.6-4) as the initial step in the expansion of the first term, or

(b) by writing

$$\text{grad}\int_\tau \frac{\text{div }[\bar{F}]}{r}\, d\tau = \sum \bar{i}\int_\tau \frac{\partial}{\partial x_o}\frac{\text{div }[\bar{F}]}{r}\, d\tau$$

$$= \sum \bar{i}\int_\tau \left\{\text{div }[\bar{F}]\ \frac{\partial}{\partial x_o}\left(\frac{1}{r}\right) + \left(\frac{1}{r}\right)\frac{\partial}{\partial x_o}\left([\text{div }\bar{F}] - \left[\frac{\partial \bar{F}}{\partial t}\right]\cdot\frac{\bar{r}}{cr}\right)\right\} d\tau$$

5-15. The results of the previous exercise lead to the conclusion that for an exterior origin of r

$$\int_\tau \left\{\left[\text{div}\ \frac{\partial \bar{F}}{\partial t}\right]\frac{\bar{r}}{cr^2} + \frac{1}{r}\ \text{grad}\left(\frac{\bar{r}}{cr}\cdot\left[\frac{\partial \bar{F}}{\partial t}\right]\right)\right\} d\tau = - \int_\tau \frac{\bar{r}}{cr}\left[\frac{\partial \bar{F}}{\partial t}\right]\cdot\text{grad}\ \frac{1}{r}\, d\tau + \oint_{S_{1..n}\Sigma} \frac{\bar{r}}{cr^2}\,\frac{\partial \bar{F}}{\partial t}\cdot d\bar{S}$$

Give an independent demonstration of this by proving and utilising the following relationships

$$\int_\tau \left[\text{div}\ \frac{\partial \bar{F}}{\partial t}\right]\frac{\bar{r}}{cr^2}\, d\tau = \int_\tau \left\{\text{div}\left[\frac{\partial \bar{F}}{\partial t}\right]\frac{\bar{r}}{cr^2} + \frac{\bar{r}}{cr^2}\,\frac{\bar{r}}{cr}\cdot\left[\frac{\partial^2 \bar{F}}{\partial t^2}\right]\right\} d\tau$$

$$\int_\tau \operatorname{div}\left[\frac{\partial \bar{F}}{\partial t}\right]\frac{\bar{r}}{cr^2}\,d\tau = \oint_{S_{1..n}\Sigma} \frac{\bar{r}}{cr^2}\left[\frac{\partial \bar{F}}{\partial t}\right].d\bar{S} - \sum \bar{i}\int_\tau \left[\frac{\partial \bar{F}}{\partial t}\right].\operatorname{grad}\frac{x-x_o}{cr^2}\,d\tau$$

$$= \oint_{S_{1..n}\Sigma} \frac{\bar{r}}{cr^2}\left[\frac{\partial \bar{F}}{\partial t}\right].d\bar{S} - \int_\tau \frac{1}{cr^2}\left[\frac{\partial \bar{F}}{\partial t}\right]d\tau + \int_\tau \frac{2\bar{r}}{cr^4}\,\bar{r}.\left[\frac{\partial \bar{F}}{\partial t}\right]d\tau$$

$$\int_\tau \frac{1}{r}\operatorname{grad}\left(\frac{\bar{r}}{cr}.\left[\frac{\partial \bar{F}}{\partial t}\right]\right)d\tau = \int_\tau \left\{\frac{1}{cr^2}\left[\frac{\partial \bar{F}}{\partial t}\right] - \frac{\bar{r}}{cr^4}\,\bar{r}.\left[\frac{\partial \bar{F}}{\partial t}\right] - \frac{\bar{r}}{c^2r^3}\,\bar{r}.\left[\frac{\partial^2 \bar{F}}{\partial t^2}\right]\right\}d\tau$$

5.7 The d'Alembertian of the Retarded Potentials

5.7a d'Alembertian of the retarded scalar potential beyond the source

Let the scalar source density have continuous second derivatives with respect to time. Then at an exterior point of the source

$$\operatorname{div}\operatorname{grad}\int_\tau \frac{[\rho]}{r}\,d\tau$$

$$= \operatorname{div}\int_\tau \left\{-[\rho]\operatorname{grad}\frac{1}{r} + \frac{\bar{r}}{cr^2}\left[\frac{\partial \rho}{\partial t}\right]\right\}d\tau$$

$$= \sum \frac{\partial}{\partial x_o}\int_\tau \left\{-[\rho]\frac{\partial}{\partial x}\left(\frac{1}{r}\right) + \frac{(x-x_o)}{cr^2}\left[\frac{\partial \rho}{\partial t}\right]\right\}d\tau$$

$$= \sum \int_\tau \left\{[\rho]\frac{\partial^2}{\partial x^2}\left(\frac{1}{r}\right) - \frac{(x-x_o)}{cr}\left[\frac{\partial \rho}{\partial t}\right]\frac{\partial}{\partial x}\left(\frac{1}{r}\right) - \frac{1}{cr^2}\left[\frac{\partial \rho}{\partial t}\right] + 2\,\frac{(x-x_o)^2}{cr^4}\left[\frac{\partial \rho}{\partial t}\right] + \frac{(x-x_o)^2}{c^2r^3}\left[\frac{\partial^2 \rho}{\partial t^2}\right]\right\}d\tau$$

$$= \int_\tau \left\{[\rho]\,\nabla^2\left(\frac{1}{r}\right) + \frac{1}{cr^2}\left[\frac{\partial \rho}{\partial t}\right] - \frac{3}{cr^2}\left[\frac{\partial \rho}{\partial t}\right] + \frac{2}{cr^2}\left[\frac{\partial \rho}{\partial t}\right] + \frac{1}{c^2r}\left[\frac{\partial^2 \rho}{\partial t^2}\right]\right\}d\tau$$

$$= \int_\tau \frac{1}{c^2r}\left[\frac{\partial^2 \rho}{\partial t^2}\right]d\tau$$

whence

$$\left(\nabla^2 - \frac{1}{c^2}\frac{\partial^2}{\partial t^2}\right)\int_\tau \frac{[\rho]}{r}\,d\tau = 0 \tag{5.7-1}$$

or at exterior points

$$\text{dal pot } [\rho] = 0 \tag{5.7-1(a)}$$

By similar argument

$$\text{dal}\int_S \frac{[\sigma]}{r}\,dS = \text{dal pot } [\sigma] = 0 \tag{5.7-2}$$

$$\text{dal}\int_\Gamma \frac{[\lambda]}{r}\,ds = \text{dal pot } [\lambda] = 0 \tag{5.7-3}$$

$$\text{dal}\,\frac{[a]}{r} = \text{dal pot } [a] = 0 \tag{5.7-4}$$

Since the d'Alembertian of the retarded potential of a set of individual simple sources is zero beyond those sources it is fairly obvious that dal ϕ will be zero outside any doublet source. This may be demonstrated by direct differentiation of the doublet potential. (See Ex.5-17., p. 432.)

5.7b <u>d'Alembertian of the retarded scalar potential within a volume source</u>

If the source density has continuous second derivatives in space and time, then from equations (5.6-11) and (5.6-22)

$$\text{div grad (partial)}\int_{\tau-\tau_\delta} \frac{[\rho]}{r}\,d\tau$$

$$= \text{div (partial)}\int_{\tau-\tau_\delta} \frac{[\text{grad }\rho]}{r}\,d\tau - \text{div}\oint_{S_{1..n}\Sigma} \frac{[\rho]}{r}\,d\bar{S}$$

$$= \int_{\tau-\tau_\delta} \frac{[\text{div grad }\rho]}{r}\,d\tau - \oint_{S_{1..n}\Sigma} \frac{[\text{grad }\rho]}{r}\cdot d\bar{S} - \sum \frac{\partial}{\partial x_o}\oint_{S_{1..n}\Sigma} \frac{[\rho]}{r}\,dS_x$$

$$= \int_{\tau-\tau_\delta} \frac{1}{r}\,[\text{div grad } \rho]\,d\tau - \oint_{S_{1..n}\Sigma} \frac{1}{r}\,\text{grad }[\rho].d\bar{S} - \oint_{S_{1..n}\Sigma} \frac{\bar{r}}{cr^2}\left[\frac{\partial\rho}{\partial t}\right].d\bar{S}$$

$$+ \sum \oint_{S_{1..n}\Sigma} \left\{[\rho]\,\frac{\partial}{\partial x}\left(\frac{1}{r}\right) - \frac{x-x_o}{cr^2}\left[\frac{\partial\rho}{\partial t}\right]\right\} dS_x$$

$$= \int_{\tau-\tau_\delta} \frac{1}{r}\,[\text{div grad } \rho]\,d\tau - \oint_{S_{1..n}\Sigma} \frac{1}{r}\,\text{grad }[\rho].d\bar{S} - 2\oint_{S_{1..n}\Sigma} \frac{\bar{r}}{cr^2}\left[\frac{\partial\rho}{\partial t}\right].d\bar{S} + \oint_{S_{1..n}\Sigma} [\rho]\,\text{grad }\frac{1}{r}.d\bar{S} \qquad (5.7\text{-}5)$$

On substituting for [div grad ρ] in accordance with equation (5.3-1), (5.7-5) becomes

$$\int_{\tau-\tau_\delta} \frac{1}{r}\left(\nabla^2[\rho] + \frac{2\bar{r}}{cr}\,.\,\text{grad}\left[\frac{\partial\rho}{\partial t}\right] + \frac{1}{c^2}\left[\frac{\partial^2\rho}{\partial t^2}\right] + \frac{2}{cr}\left[\frac{\partial\rho}{\partial t}\right]\right) d\tau$$

$$- \oint_{S_{1..n}\Sigma} \left(\frac{1}{r}\,\frac{\partial[\rho]}{\partial n} - [\rho]\,\frac{\partial}{\partial n}\left(\frac{1}{r}\right)\right) dS - 2\oint_{S_{1..n}\Sigma} \left[\frac{\partial\rho}{\partial t}\right]\frac{\bar{r}}{cr^2}.d\bar{S}$$

Now Green's formula when applied to [ρ] in the region $\tau-\tau_\delta$ yields

$$0 = \oint_{S_{1..n}\Sigma,\,S_\delta} \left(\frac{1}{r}\,\frac{\partial[\rho]}{\partial n} - [\rho]\,\frac{\partial}{\partial n}\left(\frac{1}{r}\right)\right) dS - \int_{\tau-\tau_\delta} \frac{1}{r}\,\nabla^2[\rho]\,d\tau$$

hence the previous expression may be written as

$$\int_{\tau-\tau_\delta} \left\{\frac{2\bar{r}}{cr^2}.\text{grad}\left[\frac{\partial\rho}{\partial t}\right] + \frac{1}{c^2 r}\left[\frac{\partial^2\rho}{\partial t^2}\right] + \frac{2}{cr^2}\left[\frac{\partial\rho}{\partial t}\right]\right\} d\tau - 2\oint_{S_{1..n}\Sigma} \left[\frac{\partial\rho}{\partial t}\right]\frac{\bar{r}}{cr^2}.d\bar{S} + \oint_{S_\delta} \left\{\frac{1}{r}\,\frac{\partial[\rho]}{\partial n} - [\rho]\,\frac{\partial}{\partial n}\left(\frac{1}{r}\right)\right\} dS$$

But

$$\text{div}\,\frac{2\bar{r}}{cr^2}\left[\frac{\partial\rho}{\partial t}\right] = \left[\frac{\partial\rho}{\partial t}\right]\frac{2}{cr^2} + \frac{2\bar{r}}{cr^2}.\text{grad}\left[\frac{\partial\rho}{\partial t}\right]$$

so that

$$\oint_{S_{1..n}\Sigma,\ S_\delta} \frac{2\bar{r}}{cr^2}\left[\frac{\partial\rho}{\partial t}\right] \cdot d\bar{S} = \int_{\tau-\tau_\delta} \left\{\left[\frac{\partial\rho}{\partial t}\right]\frac{2}{cr^2} + \frac{2\bar{r}}{cr^2}\cdot\text{grad}\left[\frac{\partial\rho}{\partial t}\right]\right\} d\tau$$

Substitution then yields

$$\text{div grad (partial)} \int_{\tau-\tau_\delta} \frac{[\rho]}{r}\, d\tau$$

$$= \oint_{S_\delta} \left\{\frac{1}{r}\frac{\partial[\rho]}{\partial n} - [\rho]\frac{\partial}{\partial n}\left(\frac{1}{r}\right) + \frac{2}{cr}\frac{\partial r}{\partial n}\left[\frac{\partial\rho}{\partial t}\right]\right\} dS + \int_{\tau-\tau_\delta} \frac{1}{c^2 r}\left[\frac{\partial^2\rho}{\partial t^2}\right] d\tau$$

$$= \oint_{S_\delta} \left\{\frac{1}{r}\left[\frac{\partial\rho}{\partial n}\right] - [\rho]\frac{\partial}{\partial n}\left(\frac{1}{r}\right) + \frac{1}{cr}\frac{\partial r}{\partial n}\left[\frac{\partial\rho}{\partial t}\right]\right\} dS + \frac{1}{c^2}\frac{\partial^2}{\partial t^2}\int_{\tau-\tau_\delta} \frac{[\rho]}{r}\, d\tau$$

or

$$\text{dal (partial)} \int_{\tau-\tau_\delta} \frac{[\rho]}{r}\, d\tau = \oint_{S_\delta} \left\{\frac{1}{r}\left[\frac{\partial\rho}{\partial n}\right] - [\rho]\frac{\partial}{\partial n}\left(\frac{1}{r}\right) + \frac{1}{cr}\frac{\partial r}{\partial n}\left[\frac{\partial\rho}{\partial t}\right]\right\} dS \qquad (5.7\text{-}6)$$

On taking limits we obtain

$$\text{dal} \int_\tau \frac{[\rho]}{r}\, d\tau = -4\pi\rho \qquad (5.7\text{-}7)$$

where ρ is the instantaneous value of the source density at the point of evaluation.

5.7c <u>d'Alembertian of the retarded vector potential beyond the source</u>

It follows from equation (5.7-1) that if the Cartesian components of the vector density $\bar{J}$ have the required degree of continuity then

$$\text{dal} \int_\tau \frac{[J_x]}{r}\, d\tau = \text{dal} \int_\tau \frac{[J_y]}{r}\, d\tau = \text{dal} \int_\tau \frac{[J_z]}{r}\, d\tau = 0$$

But

$$\nabla^2 \int_\tau \frac{[\bar{J}]}{r}\, d\tau = \sum \bar{i}\, \nabla^2 \int_\tau \frac{[J_x]}{r}\, d\tau$$

hence

$$\nabla^2 \int_\tau \frac{[\bar{J}]}{r}\, d\tau = \sum \bar{i}\, \frac{1}{c^2}\frac{\partial^2}{\partial t^2} \int_\tau \frac{[J_x]}{r}\, d\tau = \frac{1}{c^2}\frac{\partial^2}{\partial t^2} \int_\tau \frac{[\bar{J}]}{r}\, d\tau$$

whence

$$\text{dal} \int_\tau \frac{[\bar{J}]}{r}\, d\tau = \bar{0} \qquad (5.7\text{-}8)$$

at exterior points

or

$$\text{dal pot } [\bar{J}] = \bar{0} \qquad (5.7\text{-}8(a))$$

Similarly,

$$\text{dal} \int_S \frac{[\bar{K}]}{r}\, dS = \bar{0} \qquad (5.7\text{-}9)$$

$$\text{dal} \int_\Gamma \frac{[\bar{I}]}{r}\, ds = \bar{0} \qquad (5.7\text{-}10)$$

5.7d d'Alembertian of the retarded vector potential within a volume source

Proceeding as above, and for a density function having continuous second derivatives in space and time, we have

$$\nabla^2 \text{ (partial)} \int_{\tau-\tau_\delta} \frac{[\bar{J}]}{r}\, d\tau$$

$$= \sum \bar{i}\, \nabla^2 \text{ (partial)} \int_{\tau-\tau_\delta} \frac{[J_x]}{r}\, d\tau$$

$$= \oint_{S_\delta} \left\{ \frac{1}{r} \left[\frac{\partial \bar{J}}{\partial n}\right] - [\bar{J}] \frac{\partial}{\partial n}\left(\frac{1}{r}\right) + \frac{1}{cr}\frac{\partial r}{\partial n}\left[\frac{\partial \bar{J}}{\partial t}\right] \right\} dS + \frac{1}{c^2}\frac{\partial^2}{\partial t^2} \int_{\tau-\tau_\delta} \frac{[\bar{J}]}{r}\, d\tau$$

or

$$\text{dal (partial)} \int_{\tau-\tau_\delta} \frac{[\bar{J}]}{r}\, d\tau = \oint_{S_\delta} \left\{ \frac{1}{r} \left[\frac{\partial \bar{J}}{\partial n}\right] - [\bar{J}] \frac{\partial}{\partial n}\left(\frac{1}{r}\right) + \frac{1}{cr}\frac{\partial r}{\partial n}\left[\frac{\partial \bar{J}}{\partial t}\right] \right\} dS \qquad (5.7\text{-}11)$$

and

$$\text{dal} \int_{\tau} \frac{[\bar{J}]}{r}\, d\tau = -4\pi \bar{J} \qquad (5.7\text{-}12)$$

An alternative derivation of (5.7-11) and (5.7-12) involves the expansion of curl curl (partial) $\int_{\tau-\tau_\delta} \frac{[\bar{J}]}{r}\, d\tau$ and grad div (partial) $\int_{\tau-\tau_\delta} \frac{[\bar{J}]}{r}\, d\tau$. This is the subject of Ex.5-20. to 5-22., pp. 434-8.

Like their unretarded Laplacian equivalents, dal pot $[\rho]$ and dal pot $[\bar{J}]$ are convergent since equations (5.7-6) and (5.7-11) continue to hold for any regular excluding region.

EXERCISES

5-16. Make use of equation (5.3-3) to show that

$$\operatorname{dal} \oint_{S_{1..n}\Sigma} \left\{ \frac{1}{r} \left[\frac{\partial \phi}{\partial n}\right] - [\phi] \frac{\partial}{\partial n}\left(\frac{1}{r}\right) + \frac{1}{cr}\frac{\partial r}{\partial n}\left[\frac{\partial \phi}{\partial t}\right] \right\} dS = 0$$

within and without the region $\underline{R}$ bounded by $S_{1..n}\Sigma$, when dal $\phi = 0$ in $\underline{R}$. Show that this continues to hold when dal $\phi \neq 0$ in $\underline{R}$ provided that the second derivatives of $\frac{\partial V}{\partial n}$ and $\frac{\partial V}{\partial t}$ with respect to time exist everywhere upon $S_{1..n}\Sigma$, and that the relationship subsists when referred to an element of surface dS.

5-17. From equation (5.5-7(a)) the retarded potential of a scalar point doublet is given at an exterior point by

$$\phi = -[\bar{p}] \cdot \frac{\bar{r}}{r^3} - \left[\frac{d\bar{p}}{dt}\right] \cdot \frac{\bar{r}}{cr^2}$$

where $\bar{p}$ is the vector moment of the doublet.

Show that

$$\operatorname{grad} \phi = \frac{[\bar{p}]}{r^3} - \frac{3\bar{r}}{r^5}\,\bar{r}.[\bar{p}] + \frac{1}{cr^2}\left[\frac{d\bar{p}}{dt}\right] - \frac{3\bar{r}}{cr^4}\,\bar{r}.\left[\frac{d\bar{p}}{dt}\right] - \frac{\bar{r}}{c^2r^3}\,\bar{r}.\left[\frac{d^2\bar{p}}{dt^2}\right]$$

Hence prove that

$$\nabla^2\phi = -\frac{\bar{r}}{c^2r^3}.\left[\frac{d^2\bar{p}}{dt^2}\right] - \frac{1}{c^3r^2}\,\bar{r}.\left[\frac{d^3\bar{p}}{dt^3}\right] = \frac{1}{c^2}\frac{\partial^2\phi}{\partial t^2}$$

or dal $\phi = 0$

Show further that

$$\operatorname{curl}\left\{-[\bar{p}] \times \frac{\bar{r}}{r^3} - \left[\frac{d\bar{p}}{dt}\right] \times \frac{\bar{r}}{cr^2}\right\} = -\operatorname{grad}\left\{-[\bar{p}].\frac{\bar{r}}{r^3} - \left[\frac{d\bar{p}}{dt}\right].\frac{\bar{r}}{cr^2}\right\} - \frac{1}{c^2r}\left[\frac{d^2\bar{p}}{dt^2}\right]$$

5-18. Proceed in the following way to evaluate the d'Alembertian of the retarded potential at an interior point of a volume source.

Expand grad div (partial) $\int_{\tau-\tau_*} \frac{[\bar{J}]}{r}\, d\tau$ and curl curl (partial) $\int_{\tau-\tau_*} \frac{[\bar{J}]}{r}\, d\tau$

in accordance with equations (5.6-11), (5.6-22) and (5.6-33), and after further expansion and cancellation of terms derive

$$\nabla^2 \text{ (partial)} \int_{\tau-\tau_\delta} \frac{[\bar{J}]}{r}\, d\tau$$

$$= \int_{\tau-\tau_\delta} \frac{1}{r}[\nabla^2\bar{J}]\, d\tau + \oint_{S_{1..n}\Sigma} \left\{ \frac{1}{r}(d\bar{S} \times \text{curl } [\bar{J}]) - \frac{1}{r}\text{div } [\bar{J}]\, d\bar{S} - 2\left[\frac{\partial\bar{J}}{\partial t}\right] \frac{\bar{r}}{cr^2} \cdot d\bar{S} \right\}$$

$$+ \oint_{S_{1..n}\Sigma} \left\{ [\bar{J}].dS \text{ grad } \frac{1}{r} - [\bar{J}].\text{grad } \frac{1}{r}\, d\bar{S} + [\bar{J}] \text{ grad } \frac{1}{r}.d\bar{S} \right\}$$

Bring this into the form

$$\int_{\tau-\tau_\delta} \frac{1}{r}[\nabla^2\bar{J}]\, d\tau + \oint_{S_{1..n}\Sigma} \left\{ d\bar{S} \times \text{curl } \frac{[\bar{J}]}{r} - \text{div } \frac{[\bar{J}]}{r}\, d\bar{S} + 2[\bar{J}] \text{ grad } \frac{1}{r}.d\bar{S} - 2\left[\frac{\partial\bar{J}}{\partial t}\right] \frac{\bar{r}}{cr^2}.d\bar{S} \right\}$$

and apply equation (1.17-13) to obtain

$$\int_{\tau-\tau_\delta} \frac{1}{r}[\nabla^2\bar{J}]\, d\tau + \oint_{S_{1..n}\Sigma} \left\{ 2[\bar{J}] \text{ grad } \frac{1}{r}.d\bar{S} - (d\bar{S}.\nabla)\frac{[\bar{J}]}{r} - 2\left[\frac{\partial\bar{J}}{\partial t}\right] \frac{\bar{r}}{cr^2}.d\bar{S} \right\}$$

Reduce this to

$$\int_{\tau-\tau_\delta} \frac{1}{r}[\nabla^2\bar{J}]\, d\tau + \oint_{S_{1..n}\Sigma} \left\{ [\bar{J}]\, d\bar{S}.\text{grad } \frac{1}{r} - \frac{1}{r}(d\bar{S}.\nabla)[\bar{J}] - 2\left[\frac{\partial\bar{J}}{\partial t}\right] \frac{\bar{r}}{cr^2}.d\bar{S} \right\}$$

and make use of the analysis following equation (5.7-5), of which the above expression is the vector analogue, to arrive at

$$\text{dal (partial)} \int_{\tau-\tau_\delta} \frac{[\bar{J}]}{r}\, d\tau = \oint_{S_\delta} \left\{ \frac{1}{r}\left[\frac{\partial\bar{J}}{\partial n}\right] - [\bar{J}]\frac{\partial}{\partial n}\left(\frac{1}{r}\right) + \frac{1}{cr}\frac{\partial r}{\partial n}\left[\frac{\partial\bar{J}}{\partial t}\right] \right\} dS$$

and

$$\operatorname{dal}\int_\tau \frac{[\bar{J}]}{r}\,d\tau = -4\pi\bar{J}$$

5-19. By working from equations (5.6-14) and (5.6-25) show that at points beyond the source

$$\operatorname{grad}\operatorname{div}\int_\tau \frac{[\bar{J}]}{r}\,d\tau$$

$$= \int_\tau \left\{-\frac{[\bar{J}]}{r^3} + \frac{3\bar{r}}{r^5}\,\bar{r}.[\bar{J}] - \frac{1}{cr^2}\left[\frac{\partial\bar{J}}{\partial t}\right] + \frac{3\bar{r}}{cr^4}\,\bar{r}.\left[\frac{\partial\bar{J}}{\partial t}\right] + \frac{\bar{r}}{c^2r^3}\,\bar{r}.\left[\frac{\partial^2\bar{J}}{\partial t^2}\right]\right\} d\tau$$

$$= \int_\tau \left\{([\bar{J}].\nabla)\operatorname{grad}\frac{1}{r} - \left(\left[\frac{\partial\bar{J}}{\partial t}\right].\nabla\right)\frac{\bar{r}}{cr^2} + \frac{\bar{r}}{cr^4}\,\bar{r}.\left[\frac{\partial\bar{J}}{\partial t}\right] + \frac{\bar{r}}{c^2r^3}\,\bar{r}.\left[\frac{\partial^2\bar{J}}{\partial t^2}\right]\right\} d\tau$$

and

$$\operatorname{curl}\operatorname{curl}\int_\tau \frac{[\bar{J}]}{r}\,d\tau$$

$$= \int_\tau \left\{([\bar{J}].\nabla)\operatorname{grad}\frac{1}{r} - \left(\left[\frac{\partial\bar{J}}{\partial t}\right].\nabla\right)\frac{\bar{r}}{cr^2} + \frac{\bar{r}}{cr^4}\,\bar{r}.\left[\frac{\partial\bar{J}}{\partial t}\right] + \frac{\bar{r}}{c^2r^3}\,\bar{r}.\left[\frac{\partial^2\bar{J}}{\partial t^2}\right]\right\} d\tau$$

$$- \frac{1}{c^2}\frac{\partial^2}{\partial t^2}\int_\tau \frac{[\bar{J}]}{r}\,d\tau$$

5-20. Prove that

$$\operatorname{grad}\operatorname{div}\operatorname{pot}[\bar{J}]$$

$$= \int_\tau \left\{-\left(\left[\frac{\partial\bar{J}}{\partial t}\right].\nabla\right)\frac{\bar{r}}{cr^2} + \frac{\bar{r}}{cr^4}\,\bar{r}.\left[\frac{\partial\bar{J}}{\partial t}\right] + \frac{\bar{r}}{c^2r^3}\,\bar{r}.\left[\frac{\partial^2\bar{J}}{\partial t^2}\right]\right\} d\tau$$

$$+ \lim_{\tau'\to 0}\int_{\tau-\tau'} ([\bar{J}].\nabla)\operatorname{grad}\frac{1}{r}\,d\tau - \lim_{S'\to 0}\oint_{S'} [\bar{J}].d\bar{S}\,\operatorname{grad}\frac{1}{r}$$

at interior points of τ,

by proceeding in the manner outlined below.

grad div partial pot $[\bar{J}]$

$$= \int_{\tau-\tau_\delta} \left\{ \frac{1}{r}\,\mathrm{grad\ div}\,[\bar{J}] + \frac{1}{r}\,\mathrm{grad}\,\frac{\bar{r}}{cr}\cdot\boxed{\frac{\partial \bar{J}}{\partial t}} + \frac{\bar{r}}{cr^2}\,\mathrm{div}\,\boxed{\frac{\partial \bar{J}}{\partial t}} + \frac{\bar{r}}{c^2 r^3}\,\bar{r}\cdot\boxed{\frac{\partial^2 \bar{J}}{\partial t^2}}\right\} d\tau$$

$$+ \oint_{S_{1..n}\Sigma} \left\{ [\bar{J}].d\bar{S}\ \mathrm{grad}\,\frac{1}{r} - \frac{1}{r}\,\mathrm{div}\,[\bar{J}]\,d\bar{S} - \frac{\bar{r}}{cr^2}\cdot\boxed{\frac{\partial \bar{J}}{\partial t}}\,d\bar{S} - \frac{\bar{r}}{cr^2}\,\boxed{\frac{\partial \bar{J}}{\partial t}}.d\bar{S}\right\}$$

$$= \int_{\tau-\tau_\delta} \left\{ -\,\mathrm{div}\,[\bar{J}]\ \mathrm{grad}\,\frac{1}{r} + \frac{\bar{r}}{cr^4}\,\bar{r}\cdot\boxed{\frac{\partial \bar{J}}{\partial t}} + \frac{\bar{r}}{cr^2}\,\mathrm{div}\,\boxed{\frac{\partial \bar{J}}{\partial t}} + \frac{\bar{r}}{c^2 r^3}\,\bar{r}\cdot\boxed{\frac{\partial^2 \bar{J}}{\partial t^2}}\right\} d\tau$$

$$+ \oint_{S_{1..n}\Sigma} \left\{ [\bar{J}].d\bar{S}\ \mathrm{grad}\,\frac{1}{r} - \frac{\bar{r}}{cr^2}\,\boxed{\frac{\partial \bar{J}}{\partial t}}.d\bar{S}\right\} + \oint_{S_\delta} \left\{ \frac{1}{r}\,\mathrm{div}\,[\bar{J}]\,d\bar{S} + \frac{\bar{r}}{cr^2}\cdot\boxed{\frac{\partial \bar{J}}{\partial t}}\,d\bar{S}\right\}$$

$$= \int_{\tau-\tau_\delta} \left\{ ([\bar{J}].\nabla)\ \mathrm{grad}\,\frac{1}{r} - \left(\boxed{\frac{\partial \bar{J}}{\partial t}}.\nabla\right)\frac{\bar{r}}{cr^2} + \frac{\bar{r}}{cr^4}\,\bar{r}\cdot\boxed{\frac{\partial \bar{J}}{\partial t}} + \frac{\bar{r}}{c^2 r^3}\,\bar{r}\cdot\boxed{\frac{\partial^2 \bar{J}}{\partial t^2}}\right\} d\tau$$

$$+ \oint_{S_\delta} \left\{ \frac{1}{r}\,[\mathrm{div}\,\bar{J}]\,d\bar{S} - [\bar{J}].d\bar{S}\ \mathrm{grad}\,\frac{1}{r} + \frac{\bar{r}}{cr^2}\,\boxed{\frac{\partial \bar{J}}{\partial t}}.d\bar{S}\right\}$$

5-21. Prove that

curl curl pot $[\bar{J}]$

$$= \int_{\tau} \left\{ -\left(\boxed{\frac{\partial \bar{J}}{\partial t}}.\nabla\right)\frac{\bar{r}}{cr^2} + \frac{\bar{r}}{cr^4}\,\bar{r}\cdot\boxed{\frac{\partial \bar{J}}{\partial t}} + \frac{\bar{r}}{c^2 r^3}\,\bar{r}\cdot\boxed{\frac{\partial^2 \bar{J}}{\partial t^2}}\right\} d\tau$$

$$- \frac{1}{c^2}\frac{\partial^2}{\partial t^2}\int \frac{[\bar{J}]}{r}\,d\tau + \lim_{\tau'\to 0}\int ([\bar{J}].\nabla)\ \mathrm{grad}\,\frac{1}{r}\,d\tau + \lim_{S'\to 0}\oint ([\bar{J}]\times d\bar{S}) \times \frac{\bar{r}}{r^3}$$

at interior points of τ,

by establishing the following relationships:

(a)

curl curl partial pot $[\bar{J}]$

$$= \int_{\tau-\tau_\delta} \frac{1}{r}\,[\text{curl curl } \bar{J}]\, d\tau$$

$$+ \oint_{S_{1..n}\Sigma} \left\{ \text{grad}\,\frac{1}{r} \times (d\bar{S}\times[\bar{J}]) - d\bar{S} \times \frac{1}{r}\,\text{curl}\,[\bar{J}] - d\bar{S} \times \left(\frac{\bar{r}}{cr^2} \times \boxed{\frac{\partial \bar{J}}{\partial t}}\right) - \frac{\bar{r}}{cr^2} \times \left(d\bar{S} \times \boxed{\frac{\partial \bar{J}}{\partial t}}\right)\right\}$$

(b)

$$\int_{\tau-\tau_\delta} \frac{1}{r}\,[\text{curl curl } \bar{J}]\, d\tau$$

$$= \int_{\tau-\tau_\delta} \left\{ \frac{\bar{r}}{cr^2}\times\text{curl}\,\boxed{\frac{\partial \bar{J}}{\partial t}} - \text{grad}\,\frac{1}{r}\times\left(\frac{\bar{r}}{cr}\times\boxed{\frac{\partial \bar{J}}{\partial t}}\right) - \text{grad}\,\frac{1}{r}\times\text{curl}\,[\bar{J}] + \frac{\bar{r}}{cr^2}\times\left(\frac{\bar{r}}{cr}\times\boxed{\frac{\partial^2 \bar{J}}{\partial t^2}}\right)\right\} d\tau$$

$$+ \oint_{S_{1..n}\Sigma,\ S_\delta} \left\{ d\bar{S} \times \frac{1}{r}\,\text{curl}\,[\bar{J}] + d\bar{S} \times \left(\frac{\bar{r}}{cr^2} \times \boxed{\frac{\partial \bar{J}}{\partial t}}\right)\right\}$$

(c)

$$\text{grad}\,\frac{\bar{r}}{cr^2}\cdot\boxed{\frac{\partial \bar{J}}{\partial t}} = \left(\frac{\bar{r}}{cr^2}\cdot\nabla\right)\boxed{\frac{\partial \bar{J}}{\partial t}} + \left(\boxed{\frac{\partial \bar{J}}{\partial t}}\cdot\nabla\right)\frac{\bar{r}}{cr^2} + \frac{\bar{r}}{cr^2}\times\text{curl}\,\boxed{\frac{\partial \bar{J}}{\partial t}}$$

at interior points of τ,

by proceeding in the manner outlined below.

grad div partial pot $[\bar{J}]$

$$= \int_{\tau-\tau_\delta} \left\{ \frac{1}{r} \operatorname{grad} \operatorname{div} [\bar{J}] + \frac{1}{r} \operatorname{grad} \frac{\bar{r}}{cr} \cdot \left[\frac{\partial \bar{J}}{\partial t}\right] + \frac{\bar{r}}{cr^2} \operatorname{div} \left[\frac{\partial \bar{J}}{\partial t}\right] + \frac{\bar{r}}{c^2 r^3}\, \bar{r} \cdot \left[\frac{\partial^2 \bar{J}}{\partial t^2}\right] \right\} d\tau$$

$$+ \oint_{S_{1..n}\Sigma} \left\{ [\bar{J}] \cdot d\bar{S} \operatorname{grad} \frac{1}{r} - \frac{1}{r} \operatorname{div} [\bar{J}]\, d\bar{S} - \frac{\bar{r}}{cr^2} \cdot \left[\frac{\partial \bar{J}}{\partial t}\right] d\bar{S} - \frac{\bar{r}}{cr^2} \left[\frac{\partial \bar{J}}{\partial t}\right] \cdot d\bar{S} \right\}$$

$$= \int_{\tau-\tau_\delta} \left\{ - \operatorname{div} [\bar{J}] \operatorname{grad} \frac{1}{r} + \frac{\bar{r}}{cr^4}\, \bar{r} \cdot \left[\frac{\partial \bar{J}}{\partial t}\right] + \frac{\bar{r}}{cr^2} \operatorname{div} \left[\frac{\partial \bar{J}}{\partial t}\right] + \frac{\bar{r}}{c^2 r^3}\, \bar{r} \cdot \left[\frac{\partial^2 \bar{J}}{\partial t^2}\right] \right\} d\tau$$

$$+ \oint_{S_{1..n}\Sigma} \left\{ [\bar{J}] \cdot d\bar{S} \operatorname{grad} \frac{1}{r} - \frac{\bar{r}}{cr^2} \left[\frac{\partial \bar{J}}{\partial t}\right] \cdot d\bar{S} \right\} + \oint_{S_\delta} \left\{ \frac{1}{r} \operatorname{div} [\bar{J}]\, d\bar{S} + \frac{\bar{r}}{cr^2} \cdot \left[\frac{\partial \bar{J}}{\partial t}\right] d\bar{S} \right\}$$

$$= \int_{\tau-\tau_\delta} \left\{ ([\bar{J}] \cdot \nabla) \operatorname{grad} \frac{1}{r} - \left(\left[\frac{\partial \bar{J}}{\partial t}\right] \cdot \nabla \right) \frac{\bar{r}}{cr^2} + \frac{\bar{r}}{cr^4}\, \bar{r} \cdot \left[\frac{\partial \bar{J}}{\partial t}\right] + \frac{\bar{r}}{c^2 r^3}\, \bar{r} \cdot \left[\frac{\partial^2 \bar{J}}{\partial t^2}\right] \right\} d\tau$$

$$+ \oint_{S_\delta} \left\{ \frac{1}{r} [\operatorname{div} \bar{J}]\, d\bar{S} - [\bar{J}] \cdot d\bar{S} \operatorname{grad} \frac{1}{r} + \frac{\bar{r}}{cr^2} \left[\frac{\partial \bar{J}}{\partial t}\right] \cdot d\bar{S} \right\}$$

5-21. Prove that

curl curl pot $[\bar{J}]$

$$= \int_\tau \left\{ - \left(\left[\frac{\partial \bar{J}}{\partial t}\right] \cdot \nabla \right) \frac{\bar{r}}{cr^2} + \frac{\bar{r}}{cr^4}\, \bar{r} \cdot \left[\frac{\partial \bar{J}}{\partial t}\right] + \frac{\bar{r}}{c^2 r^3}\, \bar{r} \cdot \left[\frac{\partial^2 \bar{J}}{\partial t^2}\right] \right\} d\tau$$

$$- \frac{1}{c^2} \frac{\partial^2}{\partial t^2} \int \frac{[\bar{J}]}{r}\, d\tau + \lim_{\tau' \to 0} \int ([\bar{J}] \cdot \nabla) \operatorname{grad} \frac{1}{r}\, d\tau + \lim_{S' \to 0} \oint ([\bar{J}] \times d\bar{S}) \times \frac{\bar{r}}{r^3}$$

at interior points of τ,

by establishing the following relationships:

(a)

$$\text{curl curl partial pot } [\bar{J}]$$

$$= \int_{\tau-\tau_\delta} \frac{1}{r}\,[\text{curl curl } \bar{J}]\, d\tau$$

$$+ \oint_{S_{1..n}\Sigma} \left\{ \text{grad } \frac{1}{r} \times (d\bar{S}\times[\bar{J}]) - d\bar{S} \times \frac{1}{r}\,\text{curl } [\bar{J}] - d\bar{S} \times \left(\frac{\bar{r}}{cr^2} \times \left[\frac{\partial \bar{J}}{\partial t}\right]\right) - \frac{\bar{r}}{cr^2} \times \left(d\bar{S} \times \left[\frac{\partial \bar{J}}{\partial t}\right]\right)\right\}$$

(b)

$$\int_{\tau-\tau_\delta} \frac{1}{r}\,[\text{curl curl } \bar{J}]\, d\tau$$

$$= \int_{\tau-\tau_\delta} \left\{ \frac{\bar{r}}{cr^2} \times \text{curl} \left[\frac{\partial \bar{J}}{\partial t}\right] - \text{grad } \frac{1}{r} \times \left(\frac{\bar{r}}{cr} \times \left[\frac{\partial \bar{J}}{\partial t}\right]\right) - \text{grad } \frac{1}{r} \times \text{curl } [\bar{J}] + \frac{\bar{r}}{cr^2} \times \left(\frac{\bar{r}}{cr} \times \left[\frac{\partial^2 \bar{J}}{\partial t^2}\right]\right)\right\} d\tau$$

$$+ \oint_{S_{1..n}\Sigma,\ S_\delta} \left\{ d\bar{S} \times \frac{1}{r}\,\text{curl } [\bar{J}] + d\bar{S} \times \left(\frac{\bar{r}}{cr^2} \times \left[\frac{\partial \bar{J}}{\partial t}\right]\right)\right\}$$

(c)

$$\text{grad } \frac{\bar{r}}{cr^2} \cdot \left[\frac{\partial \bar{J}}{\partial t}\right] = \left(\frac{\bar{r}}{cr^2} \cdot \nabla\right) \left[\frac{\partial \bar{J}}{\partial t}\right] + \left(\left[\frac{\partial \bar{J}}{\partial t}\right] \cdot \nabla\right) \frac{\bar{r}}{cr^2} + \frac{\bar{r}}{cr^2} \times \text{curl} \left[\frac{\partial \bar{J}}{\partial t}\right]$$

whence

$$\oint_{S_{1..n}^{\Sigma},\, S_\delta} \frac{\bar{r}}{cr^2}\cdot\left[\frac{\partial \bar{J}}{\partial t}\right] d\bar{S}$$

$$= \int_{\tau-\tau_\delta} \left\{\left(\frac{\bar{r}}{cr^2}\cdot\nabla\right)\left[\frac{\partial \bar{J}}{\partial t}\right] + \left(\left[\frac{\partial \bar{J}}{\partial t}\right]\cdot\nabla\right)\frac{\bar{r}}{cr^2} + \frac{\bar{r}}{cr^2}\times \text{curl}\left[\frac{\partial \bar{J}}{\partial t}\right]\right\} d\tau$$

$$= \int_{\tau-\tau_\delta} \left\{\left(\left[\frac{\partial \bar{J}}{\partial t}\right]\cdot\nabla\right)\frac{\bar{r}}{cr^2} - \frac{1}{cr^2}\left[\frac{\partial \bar{J}}{\partial t}\right] + \frac{\bar{r}}{cr^2}\times \text{curl}\left[\frac{\partial \bar{J}}{\partial t}\right]\right\} d\tau + \oint_{S_{1..n}^{\Sigma},\, S_\delta}\left[\frac{\partial \bar{J}}{\partial t}\right]\frac{\bar{r}}{cr^2}\cdot d\bar{S}$$

(d)

$$\text{grad}\left(\nabla\left(\frac{1}{r}\right)\cdot[\bar{J}]\right) = \left(\nabla\left(\frac{1}{r}\right)\cdot\nabla\right)[\bar{J}] + ([\bar{J}]\cdot\nabla)\,\nabla\left(\frac{1}{r}\right) + \nabla\left(\frac{1}{r}\right)\times \text{curl}\,[\bar{J}]$$

whence

$$\oint_{S_{1..n}^{\Sigma},\, S_\delta} [\bar{J}]\cdot\text{grad}\,\frac{1}{r}\, d\bar{S}$$

$$= \int_{\tau-\tau_\delta} \left\{\left(\nabla\left(\frac{1}{r}\right)\cdot\nabla\right)[\bar{J}] + ([\bar{J}]\cdot\nabla)\,\text{grad}\,\frac{1}{r} + \text{grad}\,\frac{1}{r}\times \text{curl}\,[\bar{J}]\right\} d\tau$$

$$= \int_{\tau-\tau_\delta} \left\{([\bar{J}]\cdot\nabla)\,\text{grad}\,\frac{1}{r} + \text{grad}\,\frac{1}{r}\times \text{curl}\,[\bar{J}]\right\} d\tau + \oint_{S_{1..n}^{\Sigma},\, S_\delta} [\bar{J}]\,\text{grad}\,\frac{1}{r}\cdot d\bar{S}$$

(e)

curl curl partial pot $[\bar{J}]$

$$= \int_{\tau-\tau_\delta} \left\{ ([J].\nabla) \text{ grad } \frac{1}{r} - \left(\left[\frac{\partial \bar{J}}{\partial t}\right].\nabla \right) \frac{\bar{r}}{cr^2} + \frac{\bar{r}}{cr^4} \bar{r}.\left[\frac{\partial \bar{J}}{\partial t}\right] + \frac{\bar{r}}{c^2 r^3} \bar{r}.\left[\frac{\partial^2 \bar{J}}{\partial t^2}\right] \right\} d\tau$$

$$- \frac{1}{c^2} \frac{\partial^2}{\partial t^2} \int_{\tau-\tau_\delta} \frac{[\bar{J}]}{r} d\tau + \oint_{S_\delta} \left\{ d\bar{S}\times\frac{1}{r} [\text{curl } \bar{J}] + \text{grad } \frac{1}{r} \times ([\bar{J}]\times d\bar{S}) - \frac{\bar{r}}{cr^2} \times \left(\left[\frac{\partial \bar{J}}{\partial t}\right] \times d\bar{S} \right) \right\}$$

5-22. Use the expressions derived in Ex.5-20. and 5-21. above for grad div partial pot $[\bar{J}]$ and curl curl partial pot $[\bar{J}]$ to show that

$$\text{dal partial pot } [\bar{J}] = \oint_{S_\delta} \left\{ \text{div } \frac{[\bar{J}]}{r} d\bar{S} - d\bar{S}\times\text{curl } \frac{[\bar{J}]}{r} + 2\left[\frac{\partial \bar{J}}{\partial t}\right] \frac{\bar{r}}{cr^2}.d\bar{S} - 2[\bar{J}]d\bar{S}.\text{grad } \frac{1}{r} \right\}$$

and transform this by means of equation (1.17-13) into

$$\text{dal partial pot } [\bar{J}] = \oint_{S_\delta} \left\{ \frac{1}{r} \left[\frac{\partial \bar{J}}{\partial n}\right] - [\bar{J}] \frac{\partial}{\partial n}\left(\frac{1}{r}\right) + \frac{1}{cr} \frac{\partial r}{\partial n} \left[\frac{\partial \bar{J}}{\partial t}\right] \right\} dS$$

5-23. By proceeding via equations (5.6-22), (5.6-11) and (5.6-6) show that

grad div partial pot $[\bar{J}]$

$$= \int_{\tau-\tau_\delta} \left\{ [\text{div } \bar{J}] \frac{\bar{r}}{r^3} + \left[\frac{\partial}{\partial t} \text{ div } \bar{J}\right] \frac{\bar{r}}{cr^2} \right\} d\tau - \oint_{S_{1..n}\Sigma} \left\{ [\bar{J}].d\bar{S} \frac{\bar{r}}{r^3} + \left[\frac{\partial \bar{J}}{\partial t}\right].d\bar{S} \frac{\bar{r}}{cr^2} \right\}$$

$$+ \oint_{S_\delta} \frac{1}{r} [\text{div } \bar{J}] d\bar{S}$$

Show similarly that

curl curl partial pot $[\bar{J}]$

$$= - \int_{\tau-\tau_\delta} \left\{ [\text{curl } \bar{J}] \times \frac{\bar{r}}{r^3} + \left[\frac{\partial}{\partial t} \text{curl } \bar{J}\right] \times \frac{\bar{r}}{cr^2} \right\} d\tau$$

$$+ \oint_{S_{1..n}^{\Sigma}} \left\{ (d\bar{S}\times[\bar{J}]) \times \frac{\bar{r}}{r^3} + \left(d\bar{S} \times \left[\frac{\partial \bar{J}}{\partial t}\right]\right) \times \frac{\bar{r}}{cr^2} \right\} + \oint_{S_\delta} d\bar{S} \times \frac{1}{r} [\text{curl } \bar{J}]$$

TABLE 5

The Retarded Scalar Potential Function $\int \frac{[\rho]}{r} d\tau$ and its Derivatives

(1)

pot $[\rho]$ $= \int_\tau \frac{[\rho]}{r} d\tau$ (interior and exterior points of τ)

(2)

partial pot $[\rho]$ $= \int_{\tau-\tau_\delta} \frac{[\rho]}{r} d\tau$ (evaluated at centre of moving δ sphere within τ)

(3)

cavity pot $[\rho]$ $= \int_{\tau-\tau_\delta} \frac{[\rho]}{r} d\tau$ (defined throughout fixed δ sphere within τ)

TABLE 5(CONTD)

(4)

$$\left.\begin{aligned}\text{grad pot }[\rho] &= \int_{\tau} \frac{1}{r}\,[\text{grad }\rho]\,d\tau - \oint_{S_{1..n}\Sigma} \frac{[\rho]}{r}\,d\bar{S} \\ &= -\int_{\tau} \left\{[\rho]\,\text{grad}\,\frac{1}{r} - \boxed{\frac{\partial\rho}{\partial t}}\,\frac{\bar{r}}{cr^2}\right\} d\tau \\ &= \int_{\tau} \left\{\frac{1}{r}\,\text{grad }[\rho] + \boxed{\frac{\partial\rho}{\partial t}}\,\frac{\bar{r}}{cr^2}\right\} d\tau - \oint_{S_{1..n}\Sigma} \frac{[\rho]}{r}\,d\bar{S}\end{aligned}\right\}\text{ interior and exterior points of }\tau$$

(5)

$$\begin{aligned}\text{grad partial pot }[\rho] &= \int_{\tau-\tau_\delta} \frac{1}{r}\,[\text{grad }\rho]\,d\tau - \oint_{S_{1..n}\Sigma} \frac{[\rho]}{r}\,d\bar{S} \\ &= -\int_{\tau-\tau_\delta} \left\{[\rho]\,\text{grad}\,\frac{1}{r} - \boxed{\frac{\partial\rho}{\partial t}}\,\frac{\bar{r}}{cr^2}\right\} d\tau + \oint_{S_\delta} \frac{[\rho]}{r}\,d\bar{S} \\ &= \int_{\tau-\tau_\delta} \left\{\frac{1}{r}\,\text{grad }[\rho] + \boxed{\frac{\partial\rho}{\partial t}}\,\frac{\bar{r}}{cr^2}\right\} d\tau - \oint_{S_{1..n}\Sigma} \frac{[\rho]}{r}\,d\bar{S}\end{aligned}$$

(6)

$$\begin{aligned}\text{grad cavity pot }[\rho] &= \int_{\tau-\tau_\delta} \frac{1}{r}\,[\text{grad }\rho]\,d\tau - \oint_{S_{1..n}\Sigma,\ S_\delta} \frac{[\rho]}{r}\,d\bar{S} \\ &= -\int_{\tau-\tau_\delta} \left\{[\rho]\,\text{grad}\,\frac{1}{r} - \boxed{\frac{\partial\rho}{\partial t}}\,\frac{\bar{r}}{cr^2}\right\} d\tau \\ &= \int_{\tau-\tau_\delta} \left\{\frac{1}{r}\,\text{grad }[\rho] + \boxed{\frac{\partial\rho}{\partial t}}\,\frac{\bar{r}}{cr^2}\right\} d\tau - \oint_{S_{1..n}\Sigma,\ S_\delta} \frac{[\rho]}{r}\,d\bar{S}\end{aligned}$$

TABLE 5(CONTD)

(7)

dal pot $[\rho]$ = 0 at exterior points of τ

(8)

dal pot $[\rho]$ = $-4\pi\rho$ at interior points of τ

(9)

$$\nabla^2 \text{ partial pot } [\rho] = \text{partial pot } [\nabla^2\rho] - \oint_{S_{1..n}\Sigma} \left\{ \frac{1}{r}\left[\frac{\partial\rho}{\partial n}\right] - [\rho]\frac{\partial}{\partial n}\left(\frac{1}{r}\right) + \frac{1}{cr}\frac{\partial r}{\partial n}\left[\frac{\partial\rho}{\partial t}\right]\right\} dS$$

(10)

$$\text{dal partial pot } [\rho] = \oint_{S_\delta} \left\{ \frac{1}{r}\left[\frac{\partial\rho}{\partial n}\right] - [\rho]\frac{\partial}{\partial n}\left(\frac{1}{r}\right) + \frac{1}{cr}\frac{\partial r}{\partial n}\left[\frac{\partial\rho}{\partial t}\right]\right\} dS$$

(11)

dal cavity pot $[\rho]$ = 0

TABLE 6

The Retarded Vector Potential Function $\int \frac{[\bar{J}]}{r}\, d\tau$ and its Derivatives

(1)

$$\text{pot } [\bar{J}] = \int_\tau \frac{[\bar{J}]}{r}\, d\tau \qquad \text{(interior and exterior points of } \tau\text{)}$$

(2)

$$\text{partial pot } [\bar{J}] = \int_{\tau-\tau_\delta} \frac{[\bar{J}]}{r}\, d\tau$$

(evaluated at centre of moving δ sphere within τ)

TABLE 6(CONTD)

(3)

$$\text{cavity pot } [\bar{J}] = \int_{\tau-\tau_\delta} \frac{[\bar{J}]}{r}\, d\tau \qquad \text{(defined throughout fixed } \delta \text{ sphere within } \tau\text{)}$$

(4)

$$\text{div pot } [\bar{J}] = \int_{\tau} \frac{1}{r}\,[\text{div } \bar{J}]\, d\tau - \oint_{S_{1..n}\Sigma} \frac{[\bar{J}]}{r}.d\bar{S}$$

$$= -\int_{\tau} \left\{[\bar{J}].\text{grad } \frac{1}{r} - \boxed{\frac{\partial \bar{J}}{\partial t}}.\frac{\bar{r}}{cr^2}\right\} d\tau$$

$$= \int_{\tau} \left\{\frac{1}{r}\,\text{div } [\bar{J}] + \boxed{\frac{\partial \bar{J}}{\partial t}}.\frac{\bar{r}}{cr^2}\right\} d\tau - \oint_{S_{1..n}\Sigma} \frac{[\bar{J}]}{r}.d\bar{S}$$

(interior and exterior points of τ)

(5)

$$\text{div partial pot } [\bar{J}] = \int_{\tau-\tau_\delta} \frac{1}{r}\,[\text{div } \bar{J}]\, d\tau - \oint_{S_{1..n}\Sigma} \frac{[\bar{J}]}{r}.d\bar{S}$$

$$= -\int_{\tau-\tau_\delta} \left\{[\bar{J}].\text{grad } \frac{1}{r} - \boxed{\frac{\partial \bar{J}}{\partial t}}.\frac{\bar{r}}{cr^2}\right\} d\tau + \oint_{S_\delta} \frac{[\bar{J}]}{r}.d\bar{S}$$

$$= \int_{\tau-\tau_\delta} \left\{\frac{1}{r}\,\text{div } [\bar{J}] + \boxed{\frac{\partial \bar{J}}{\partial t}}.\frac{\bar{r}}{cr^2}\right\} d\tau - \oint_{S_{1..n}\Sigma} \frac{[\bar{J}]}{r}.d\bar{S}$$

TABLE 6(CONTD)

(6)

$$\operatorname{div}\ \text{cavity pot}\ [\bar{J}] = \int_{\tau-\tau_\delta} \frac{1}{r}\,[\operatorname{div}\bar{J}]\,d\tau - \oint_{S_{1..n}\Sigma,\ S_\delta} \frac{[\bar{J}]}{r}.d\bar{S}$$

$$= -\int_{\tau-\tau_\delta} \left\{[\bar{J}].\operatorname{grad}\frac{1}{r} - \left[\frac{\partial\bar{J}}{\partial t}\right].\frac{\bar{r}}{cr^2}\right\} d\tau$$

$$= \int_{\tau-\tau_\delta} \left\{\frac{1}{r}\operatorname{div}[\bar{J}] + \left[\frac{\partial\bar{J}}{\partial t}\right].\frac{\bar{r}}{cr^2}\right\} d\tau - \oint_{S_{1..n}\Sigma,\ S_\delta} \frac{[\bar{J}]}{r}.d\bar{S}$$

(7)

$$\left.\begin{aligned}
\operatorname{curl}\ \text{pot}\ [\bar{J}] &= \int_\tau \frac{1}{r}\,[\operatorname{curl}\bar{J}]\,d\tau - \oint_{S_{1..n}\Sigma} d\bar{S}\times\frac{[\bar{J}]}{r} \\
&= \int_\tau \left\{[\bar{J}]\times\operatorname{grad}\frac{1}{r} - \left[\frac{\partial\bar{J}}{\partial t}\right]\times\frac{\bar{r}}{cr^2}\right\} d\tau \\
&= \int_\tau \left\{\frac{1}{r}\operatorname{curl}[\bar{J}] - \left[\frac{\partial\bar{J}}{\partial t}\right]\times\frac{\bar{r}}{cr^2}\right\} d\tau - \oint_{S_{1..n}\Sigma} d\bar{S}\times\frac{[\bar{J}]}{r}
\end{aligned}\right\}\ \text{interior and exterior points of } \tau$$

TABLE 6(CONTD)

(8)

$$\text{curl partial pot }[\bar{J}] = \int_{\tau-\tau_\delta} \frac{1}{r}[\text{curl }\bar{J}]\,d\tau - \oint_{S_{1..n}\Sigma} d\bar{S}\times\frac{[\bar{J}]}{r}$$

$$= \int_{\tau-\tau_\delta} \left\{[\bar{J}]\times\text{grad }\frac{1}{r} - \left[\frac{\partial\bar{J}}{\partial t}\right]\times\frac{\bar{r}}{cr^2}\right\}d\tau + \oint_{S_\delta} d\bar{S}\times\frac{[\bar{J}]}{r}$$

$$= \int_{\tau-\tau_\delta} \left\{\frac{1}{r}\text{ curl }[\bar{J}] - \left[\frac{\partial\bar{J}}{\partial t}\right]\times\frac{\bar{r}}{cr^2}\right\}d\tau - \oint_{S_{1..n}\Sigma} d\bar{S}\times\frac{[\bar{J}]}{r}$$

(9)

$$\text{curl cavity pot }[\bar{J}] = \int_{\tau-\tau_\delta} \frac{1}{r}[\text{curl }\bar{J}]\,d\tau - \oint_{S_{1..n}\Sigma,\ S_\delta} d\bar{S}\times\frac{[\bar{J}]}{r}$$

$$= \int_{\tau-\tau_\delta} \left\{[\bar{J}]\times\text{grad }\frac{1}{r} - \left[\frac{\partial\bar{J}}{\partial t}\right]\times\frac{\bar{r}}{cr^2}\right\}d\tau$$

$$= \int_{\tau-\tau_\delta} \left\{\frac{1}{r}\text{ curl }[\bar{J}] - \left[\frac{\partial\bar{J}}{\partial t}\right]\times\frac{\bar{r}}{cr^2}\right\}d\tau - \oint_{S_{1..n}\Sigma,\ S_\delta} d\bar{S}\times\frac{[\bar{J}]}{r}$$

(10)

$\text{dal pot }[\bar{J}] = \bar{0}$ at exterior points of τ

(11)

$\text{dal pot }[\bar{J}] = -4\pi\bar{J}$ at interior points of τ

(12)

$\text{dal cavity pot }[\bar{J}] = \bar{0}$

TABLE 6(CONTD)

(13)

$$\text{dal partial pot } [\bar{J}] = \oint_{S_\delta} \left\{ \text{div } \frac{[\bar{J}]}{r} d\bar{S} - d\bar{S} \times \text{curl } \frac{[\bar{J}]}{r} + 2 \left[\frac{\partial \bar{J}}{\partial t}\right] \frac{\bar{r}}{cr^2} . d\bar{S} - 2[\bar{J}] d\bar{S} . \text{grad } \frac{1}{r} \right\}$$

$$= \oint_{S_\delta} \left\{ \frac{1}{r} \left[\frac{\partial \bar{J}}{\partial n}\right] - [\bar{J}] \frac{\partial}{\partial n}\left(\frac{1}{r}\right) + \frac{1}{cr} \frac{\partial r}{\partial n} \left[\frac{\partial \bar{J}}{\partial t}\right] \right\} dS$$

(14)

$$\nabla^2 \text{ partial pot } [\bar{J}] = \text{partial pot } [\nabla^2 \bar{J}] - \oint_{S_{1..n}\Sigma} \left\{ \frac{1}{r} \left[\frac{\partial \bar{J}}{\partial n}\right] - [\bar{J}] \frac{\partial}{\partial n}\left(\frac{1}{r}\right) + \frac{1}{cr} \frac{\partial r}{\partial n} \left[\frac{\partial \bar{J}}{\partial t}\right] \right\} dS$$

(15)

grad div pot $[\bar{J}]$

$$= \int_\tau \left\{ [\text{div } \bar{J}] \frac{\bar{r}}{r^3} + \left[\frac{\partial}{\partial t} \text{div } \bar{J}\right] \frac{\bar{r}}{cr^2} \right\} d\tau - \oint_{S_{1..n}\Sigma} \left\{ [\bar{J}] . d\bar{S} \frac{\bar{r}}{r^3} + \left[\frac{\partial \bar{J}}{\partial t}\right] . d\bar{S} \frac{\bar{r}}{cr^2} \right\}$$

(interior and exterior points of τ)

(16)

grad div pot $[\bar{J}]$

$$= \int_\tau \left\{ ([\bar{J}] . \nabla) \text{ grad } \frac{1}{r} - \left(\left[\frac{\partial \bar{J}}{\partial t}\right] . \nabla \right) \frac{\bar{r}}{cr^2} + \frac{\bar{r}}{cr^4} \bar{r} . \left[\frac{\partial \bar{J}}{\partial t}\right] + \frac{\bar{r}}{c^2 r^3} \bar{r} . \left[\frac{\partial^2 \bar{J}}{\partial t^2}\right] \right\} d\tau$$

(exterior points of τ)

TABLE 6 (CONTD)

(17)

grad div pot $[\bar{J}]$

$$= \int_{\tau} \left\{ -\left(\left[\frac{\partial \bar{J}}{\partial t}\right].\nabla\right) \frac{\bar{r}}{cr^2} + \frac{\bar{r}}{cr^4}\,\bar{r}.\left[\frac{\partial \bar{J}}{\partial t}\right] + \frac{\bar{r}}{c^2 r^3}\,\bar{r}.\left[\frac{\partial^2 \bar{J}}{\partial t^2}\right] \right\} d\tau$$

(interior points of τ)

$$+ \lim_{\tau' \to 0} \int_{\tau - \tau'} ([\bar{J}].\nabla)\,\text{grad}\,\frac{1}{r}\,d\tau - \lim_{S' \to 0} \oint_{S'} [\bar{J}].d\bar{S}\,\text{grad}\,\frac{1}{r}$$

(18)

grad div partial pot $[\bar{J}]$

$$= \int_{\tau - \tau_\delta} \left\{ [\text{div}\,\bar{J}]\,\frac{\bar{r}}{r^3} + \left[\frac{\partial}{\partial t}\,\text{div}\,\bar{J}\right] \frac{\bar{r}}{cr^2} \right\} d\tau - \oint_{S_{1..n}^{\Sigma}} \left\{ [\bar{J}].d\bar{S}\,\frac{\bar{r}}{r^3} + \left[\frac{\partial \bar{J}}{\partial t}\right].d\bar{S}\,\frac{\bar{r}}{cr^2} \right\}$$

$$+ \oint_{S_\delta} \frac{1}{r}\,[\text{div}\,\bar{J}]\,d\bar{S}$$

(19)

grad div partial pot $[\bar{J}]$

$$= \int_{\tau - \tau_\delta} \left\{ ([\bar{J}].\nabla)\,\text{grad}\,\frac{1}{r} - \left(\left[\frac{\partial \bar{J}}{\partial t}\right].\nabla\right) \frac{\bar{r}}{cr^2} + \frac{\bar{r}}{cr^4}\,\bar{r}.\left[\frac{\partial \bar{J}}{\partial t}\right] + \frac{\bar{r}}{c^2 r^3}\,\bar{r}.\left[\frac{\partial^2 \bar{J}}{\partial t^2}\right] \right\} d\tau$$

$$+ \oint_{S_\delta} \left\{ \frac{1}{r}\,[\text{div}\,\bar{J}]\,d\bar{S} - [\bar{J}].d\bar{S}\,\text{grad}\,\frac{1}{r} + \frac{\bar{r}}{cr^2}\left[\frac{\partial \bar{J}}{\partial t}\right].d\bar{S} \right\}$$

(20)

$$\text{curl curl pot}\,[\bar{J}] = \text{grad div pot}\,[\bar{J}] - \frac{1}{c^2}\frac{\partial^2}{\partial t^2}\int_{\tau} \frac{[\bar{J}]}{r}\,d\tau$$

(exterior points of τ)

TABLE 6 (CONTD)

(21)

$$\text{curl curl pot } [\bar{J}] = \text{grad div pot } [\bar{J}] - \frac{1}{c^2}\frac{\partial^2}{\partial t^2}\int_\tau \frac{[\bar{J}]}{r}\, d\tau + 4\pi\bar{J}$$

(interior points of τ)

(22)

curl curl pot $[\bar{J}]$

$$= -\int_\tau \left\{[\text{curl } \bar{J}] \times \frac{\bar{r}}{r^3} + \left[\frac{\partial}{\partial t}\text{ curl } \bar{J}\right] \times \frac{\bar{r}}{cr^2}\right\} d\tau$$

$$+ \oint_{S_{1..n}\Sigma} \left\{(d\bar{S}\times[\bar{J}]) \times \frac{\bar{r}}{r^3} + \left(d\bar{S} \times \left[\frac{\partial\bar{J}}{\partial t}\right]\right) \times \frac{\bar{r}}{cr^2}\right\}$$

(interior and exterior points of τ)

(23)

curl curl partial pot $[\bar{J}]$

$$= -\int_{\tau-\tau_\delta} \left\{[\text{curl } \bar{J}] \times \frac{\bar{r}}{r^3} + \left[\frac{\partial}{\partial t}\text{ curl } \bar{J}\right] \times \frac{\bar{r}}{cr^2}\right\} d\tau$$

$$+ \oint_{S_{1..n}\Sigma} \left\{(d\bar{S}\times[\bar{J}]) \times \frac{\bar{r}}{r^3} + \left(d\bar{S} \times \left[\frac{\partial\bar{J}}{\partial t}\right]\right) \times \frac{\bar{r}}{cr^2}\right\} + \oint_{S_\delta} d\bar{S} \times \frac{1}{r}\,[\text{curl } \bar{J}]$$

(24)

curl curl partial pot $[\bar{J}]$

$$= \int_{\tau-\tau_\delta} \left\{([\bar{J}].\nabla)\text{ grad }\frac{1}{r} - \left(\left[\frac{\partial\bar{J}}{\partial t}\right].\nabla\right)\frac{\bar{r}}{cr^2} + \frac{\bar{r}}{cr^4}\,\bar{r}.\left[\frac{\partial\bar{J}}{\partial t}\right] + \frac{\bar{r}}{c^2r^3}\,\bar{r}.\left[\frac{\partial^2\bar{J}}{\partial t^2}\right]\right\} d\tau$$

$$- \frac{1}{c^2}\frac{\partial^2}{\partial t^2}\int_{\tau-\tau_\delta}\frac{[\bar{J}]}{r}\, d\tau + \oint_{S_\delta}\left\{d\bar{S} \times \frac{1}{r}\,[\text{curl } \bar{J}] + ([\bar{J}]\times d\bar{S}) \times \frac{\bar{r}}{r^3} + \left(\left[\frac{\partial\bar{J}}{\partial t}\right] \times d\bar{S}\right) \times \frac{\bar{r}}{cr^2}\right\}$$

5.8 The Gradient and d'Alembertian of the Scalar Point Function

$$\int \left\{ [\overline{P}].\text{grad}\,\frac{1}{r} - \left[\frac{\partial \overline{P}}{\partial t}\right].\frac{\overline{r}}{cr^2} \right\} d\tau$$

This section, and that which follows, parallels the developments of Sec. 4.18 and 4.19, to which the reader should refer. The function under consideration is identical with the retarded potential, at exterior points, of a limiting configuration of point doublets, $\overline{P}$ being a well-behaved point function derived from the doublet moment per unit volume as discussed in Sec. 4.20a. The integral is known as the macroscopic retarded potential of the doublet system. It continues to be well-behaved at interior points of the source, whereas the true or microscopic potential, viz $\sum \left\{ [\overline{p}].\text{grad}\,\frac{1}{r} - \left[\frac{d\overline{p}}{dt}\right].\frac{\overline{r}}{cr^2} \right\}$ becomes discontinuous at the doublets themselves.

It is easily shown by expansion of div $\frac{[\overline{P}]}{r}$, and subsequent volume integration, that for an exterior point of evaluation

$$\int_\tau \left\{ [\overline{P}].\text{grad}\,\frac{1}{r} - \left[\frac{\partial \overline{P}}{\partial t}\right].\frac{\overline{r}}{cr^2} \right\} d\tau = \int_\tau \frac{[-\,\text{div}\,\overline{P}]}{r}\, d\tau + \oint_{S_{1..n}\Sigma} \frac{[\overline{P}]}{r}.d\overline{S} \qquad (5.8\text{-}1)$$

The same relationship holds for an interior point of evaluation provided that we interpret the volume integrals as the limiting values of such integrals when taken over the region $\tau-\tau'$, where τ' is a subregion of τ which includes 0 and shrinks uniformly about it. It follows that the macroscopic potential is equivalent, at interior and exterior points of τ, to the combined retarded potential of a volume source throughout τ of density $-\text{div}\,\overline{P}$ and a surface source over $S_{1..n}\Sigma$ of density $\overline{P}.\hat{n}$. It is consequently convergent and continuous everywhere, while its gradient is discontinuous through $S_{1..n}\Sigma$ by $-4\pi\overline{P}.\hat{n}\,\hat{n}$.

The partial and cavity macroscopic potentials are defined in the usual way in terms of an excluding δ sphere.

5.8a Gradient of $\int_\tau \left\{[\bar{P}].\text{grad}\,\frac{1}{r} - \left[\frac{\partial \bar{P}}{\partial t}\right].\frac{\bar{r}}{cr^2}\right\} d\tau$ at interior and exterior points of τ

Since equations (5.8-1) and (5.6-1) are valid at interior and exterior points of τ, we have in each case

$$\text{grad} \int_\tau \left\{[\bar{P}].\text{grad}\,\frac{1}{r} - \left[\frac{\partial \bar{P}}{\partial t}\right].\frac{\bar{r}}{cr^2}\right\} d\tau$$

$$= \text{grad} \int_\tau \frac{[-\,\text{div}\,\bar{P}]}{r}\, d\tau + \text{grad} \oint_{S_{1..n}\Sigma} \frac{[\bar{P}]}{r}.d\bar{S}$$

$$= \int_\tau \left\{[\text{div}\,\bar{P}]\,\text{grad}\,\frac{1}{r} - \left[\frac{\partial}{\partial t}\,\text{div}\,\bar{P}\right]\frac{\bar{r}}{cr^2}\right\} d\tau - \oint_{S_{1..n}\Sigma} \left\{[\bar{P}].d\bar{S}\,\text{grad}\,\frac{1}{r} - \frac{\bar{r}}{cr^2}\left[\frac{\partial \bar{P}}{\partial t}\right].d\bar{S}\right\} \tag{5.8-2}$$

Reference to equation (5.6-17) reveals that the right hand side of (5.8-1) is identical with $-\text{div} \int_\tau \frac{[\bar{P}]}{r}\, d\tau$ at interior and exterior points of τ, so that we have also

$$\text{grad} \int_\tau \left\{[\bar{P}].\text{grad}\,\frac{1}{r} - \left[\frac{\partial \bar{P}}{\partial t}\right].\frac{\bar{r}}{cr^2}\right\} d\tau$$

$$= -\,\text{grad div} \int_\tau \frac{[\bar{P}]}{r}\, d\tau \tag{5.8-3}$$

$$= -\,\text{curl curl} \int_\tau \frac{[\bar{P}]}{r}\, d\tau - \nabla^2 \int_\tau \frac{[\bar{P}]}{r}\, d\tau \tag{5.8-4}$$

The curl curl term may be variously transformed as follows:

$$\operatorname{curl}\operatorname{curl}\int_\tau \frac{[\bar{P}]}{r}\,d\tau = \operatorname{curl}\int_\tau \left\{[\bar{P}] \times \operatorname{grad}\frac{1}{r} - \left[\frac{\partial\bar{P}}{\partial t}\right] \times \frac{\bar{r}}{cr^2}\right\} d\tau \tag{5.8-5}$$

$$= \operatorname{curl}\int_\tau \frac{[\operatorname{curl}\bar{P}]}{r}\,d\tau - \operatorname{curl}\oint_{S_{1..n}\Sigma} d\bar{S} \times \frac{[\bar{P}]}{r} \tag{5.8-6}$$

$$= \int_\tau \left\{[\operatorname{curl}\bar{P}]\times\operatorname{grad}\frac{1}{r} - \left[\frac{\partial}{\partial t}\operatorname{curl}\bar{P}\right]\times\frac{\bar{r}}{cr^2}\right\} d\tau + \oint_{S_{1..n}\Sigma}\left\{\operatorname{grad}\frac{1}{r}\times(d\bar{S}\times[\bar{P}]) - \frac{\bar{r}}{cr^2}\times\left(d\bar{S}\times\left[\frac{\partial\bar{P}}{\partial t}\right]\right)\right\} \tag{5.8-7}$$

In particular, a combination of equations (5.8-4) and (5.8-5) with (5.7-8) or (5.7-12) leads to the relationship

$$\operatorname{grad}\int_\tau \left\{[\bar{P}].\operatorname{grad}\frac{1}{r} - \left[\frac{\partial\bar{P}}{\partial t}\right].\frac{\bar{r}}{cr^2}\right\} d\tau$$

$$= -\operatorname{curl}\int_\tau \left\{[\bar{P}]\times\operatorname{grad}\frac{1}{r} - \left[\frac{\partial\bar{P}}{\partial t}\right]\times\frac{\bar{r}}{cr^2}\right\} d\tau - \frac{1}{c^2}\frac{\partial^2}{\partial t^2}\int_\tau \frac{[\bar{P}]}{r}\,d\tau + \begin{cases}\bar{0} & \text{(exterior}\\ 4\pi\bar{P} & \text{(interior points of } \tau)\end{cases} \tag{5.8-8}$$

It will be seen that equation (5.8-8) reduces to (4.18-8) or (4.18-9) when $c=\infty$ or $\left[\frac{\partial\bar{P}}{\partial t}\right] = \bar{0}$, ie, when retardation is absent or the system is time-invariant over the relevant interval.

The corresponding expressions for the gradient of the cavity potential are derived from those of the above formulae which apply to exterior points of τ by modification of the volume integral and the addition of S_δ to the surface integral.

5.8b <u>d'Alembertian of $\int_\tau \left\{[\bar{P}].\operatorname{grad}\frac{1}{r} - \left[\frac{\partial\bar{P}}{\partial t}\right].\frac{\bar{r}}{cr^2}\right\} d\tau$ at interior and exterior points of τ</u>

From equation (5.8-1)

$$\operatorname{dal}\int_\tau \left\{[\bar{P}].\operatorname{grad}\frac{1}{r} - \left[\frac{\partial\bar{P}}{\partial t}\right].\frac{\bar{r}}{cr^2}\right\} d\tau = \operatorname{dal}\int_\tau \frac{[-\operatorname{div}\bar{P}]}{r}\,d\tau - \operatorname{dal}\oint_{S_{1..n}\Sigma} \frac{[P_n]}{r}\,dS$$

within and without the source,

hence from equations (5.7-1(a)), (5.7-2) and (5.7-7)

$$\text{dal} \int_\tau \left\{ [\overline{P}].\text{grad}\,\frac{1}{r} - \boxed{\frac{\partial \overline{P}}{\partial t}} . \frac{\overline{r}}{cr^2} \right\} d\tau = \begin{cases} 0 & \text{(exterior points of } \tau) \\ 4\pi \text{div}\,\overline{P} & \text{(interior points of } \tau) \end{cases} \tag{5.8-9}$$

It follows at once that dal cavity pot = 0

The above relationships, together with the corresponding formulae for partial potentials, are brought together in Table 7, p. 460.

5.9 The Divergence, Curl and d'Alembertian of the Vector Point Function

$$\int \left\{ [\overline{M}] \times \text{grad}\,\frac{1}{r} - \boxed{\frac{\partial \overline{M}}{\partial t}} \times \frac{\overline{r}}{cr^2} \right\} d\tau$$

The function under consideration is identical with the retarded vector potential, at exterior points, of a limiting volume configuration of whirls, $\overline{M}$ being a well-behaved point function derived from the doublet moment per unit volume as discussed in Sec. 4.20b. It is known as the macroscopic retarded potential of the system of whirls and, unlike the true or microscopic potential, viz $\sum \left\{ [\overline{m}] \times \text{grad}\,\frac{1}{r} - \boxed{\frac{d\overline{m}}{dt}} \times \frac{\overline{r}}{cr^2} \right\}$, it is well-behaved throughout the source complex.

Expansion of curl $\frac{[\overline{M}]}{r}$ and subsequent volume integration leads to the relationship

$$\int_\tau \left\{ [\overline{M}] \times \text{grad}\,\frac{1}{r} - \boxed{\frac{\partial \overline{M}}{\partial t}} \times \frac{\overline{r}}{cr^2} \right\} d\tau = \int_\tau \frac{[\text{curl}\,\overline{M}]}{r}\, d\tau - \oint_{S_{1..n}\Sigma} d\overline{S} \times \frac{[\overline{M}]}{r} \tag{5.9-1}$$

at exterior points of τ. This holds at interior points also since the surface integral taken over an excluding surface drawn about 0 vanishes in the limit as the surface shrinks uniformly about 0, and the associated volume integral is convergent. The macroscopic potential is consequently identical with the retarded potential of a volume source throughout τ of density curl $\overline{M}$ and a surface source over $S_{1..n}\Sigma$ of density $\overline{M} \times \hat{n}$ and, as such, is everywhere convergent and continuous.

5.9a Divergence and curl of $\int_\tau \left\{[\bar{M}] \times \text{grad}\,\frac{1}{r} - \left[\frac{\partial \bar{M}}{\partial t}\right] \times \frac{\bar{r}}{cr^2}\right\} d\tau$ at interior and exterior points of τ

Since equations (5.9-1) and (5.6-28) are valid within and without τ we have in each case

$$\int_\tau \left\{[\bar{M}] \times \text{grad}\,\frac{1}{r} - \left[\frac{\partial \bar{M}}{\partial t}\right] \times \frac{\bar{r}}{cr^2}\right\} d\tau = \text{curl}\int_\tau \frac{[\bar{M}]}{r}\, d\tau \tag{5.9-2}$$

whence

$$\text{div}\int_\tau \left\{[\bar{M}] \times \text{grad}\,\frac{1}{r} - \left[\frac{\partial \bar{M}}{\partial t}\right] \times \frac{\bar{r}}{cr^2}\right\} d\tau = 0 \tag{5.9-3}$$

at interior and exterior points of τ.

The divergence of the cavity potental must likewise be zero.

From equations (5.9-1) and (5.6-25)

$$\text{curl}\int_\tau \left\{[\bar{M}] \times \text{grad}\,\frac{1}{r} - \left[\frac{\partial \bar{M}}{\partial t}\right] \times \frac{\bar{r}}{cr^2}\right\} d\tau$$

$$= \text{curl}\int_\tau \frac{[\text{curl}\,\bar{M}]}{r}\, d\tau - \text{curl}\oint_{S_{1..n}\Sigma} d\bar{S} \times \frac{[\bar{M}]}{r}$$

$$= \int_\tau \left\{[\text{curl}\,\bar{M}]\times\text{grad}\,\frac{1}{r} - \left[\frac{\partial}{\partial t}\,\text{curl}\,\bar{M}\right]\times\frac{\bar{r}}{cr^2}\right\} d\tau + \oint_{S_{1..n}\Sigma} \left\{\text{grad}\,\frac{1}{r}\times(d\bar{S}\times[\bar{M}]) - \frac{\bar{r}}{cr^2}\times\left(d\bar{S}\times\left[\frac{\partial \bar{M}}{\partial t}\right]\right)\right\} \tag{5.9-4}$$

at interior and exterior points of τ.

Also, from equation (5.9-2)

$$\text{curl}\int_\tau \left\{[\bar{M}] \times \text{grad}\,\frac{1}{r} - \left[\frac{\partial \bar{M}}{\partial t}\right] \times \frac{\bar{r}}{cr^2}\right\} d\tau$$

$$= \text{curl}\;\text{curl}\int \frac{[\bar{M}]}{r}\, d\tau \tag{5.9-5}$$

$$= \text{grad div} \int_\tau \frac{[\overline{M}]}{r} d\tau - \nabla^2 \int_\tau \frac{[\overline{M}]}{r} d\tau \tag{5.9-6}$$

The grad div term may be variously transformed as follows:

$$\text{grad div} \int_\tau \frac{[\overline{M}]}{r} d\tau$$

$$= - \text{grad} \int_\tau \left\{ [\overline{M}].\text{grad} \frac{1}{r} - \left[\frac{\partial \overline{M}}{\partial t}\right] . \frac{\overline{r}}{cr^2} \right\} d\tau \tag{5.9-7}$$

$$= \text{grad} \int_\tau \frac{[\text{div } \overline{M}]}{r} d\tau - \text{grad} \oint_{S_{1..n}\Sigma} \frac{[\overline{M}]}{r} . d\overline{S} \tag{5.9-8}$$

$$= \int_\tau \left\{ [-\text{div } \overline{M}] \text{ grad} \frac{1}{r} + \left[\frac{\partial}{\partial t} \text{div } \overline{M}\right] \frac{\overline{r}}{cr^2} \right\} d\tau + \oint_{S_{1..n}\Sigma} \left\{ [\overline{M}].d\overline{S} \text{ grad} \frac{1}{r} - \frac{\overline{r}}{cr^2} \left[\frac{\partial \overline{M}}{\partial t}\right] . d\overline{S} \right\} \tag{5.9-9}$$

A combination of equations (5.9-6) and (5.9-7) leads to (5.8-8) with $\overline{M}$ replacing $\overline{P}$. Since the divergence of the macroscopic potential is zero at both interior and exterior points of the source, its discontinuity through any bounding surface is likewise zero. This result follows independently from the observation that the divergence of the first term of the right hand side of (5.9-1) is everywhere continuous and that

$$\Delta \text{ div} \int_S d\overline{S} \times \frac{[\overline{M}]}{r} = 0$$

in accordance with a result of Ex.5-12., p. 423.

Correspondingly, the discontinuity of the curl of the potential for outward movement from τ through a point of $S_{1..n}\Sigma$ is given by

$$\Delta \text{ curl} \left\{ - \int_S d\overline{S} \times \frac{[\overline{M}]}{r} \right\} = -4\pi \overline{M}_t$$

where $\overline{M}_t$ is the local vector tangential component of $\overline{M}$.

5.9b <u>d'Alembertian of</u> $\int_\tau \left\{[\bar{M}] \times \operatorname{grad}\frac{1}{r} - \boxed{\frac{\partial \bar{M}}{\partial t}} \times \frac{\bar{r}}{cr^2}\right\} d\tau$ <u>at interior and exterior points of τ</u>

From equation (5.9-1)

$$\operatorname{dal}\int_\tau \left\{[\bar{M}] \times \operatorname{grad}\frac{1}{r} - \boxed{\frac{\partial \bar{M}}{\partial t}} \times \frac{\bar{r}}{cr^2}\right\} d\tau = \operatorname{dal}\int_\tau \frac{[\operatorname{curl}\bar{M}]}{r}\, d\tau - \operatorname{dal}\oint_{S_{1..n}\Sigma} d\bar{S} \times \frac{[\bar{M}]}{r}$$

whence from equations (5.7-8(a)), (5.7-9) and (5.7-12)

$$\operatorname{dal}\int_\tau \left\{[\bar{M}] \times \operatorname{grad}\frac{1}{r} - \boxed{\frac{\partial \bar{M}}{\partial t}} \times \frac{\bar{r}}{cr^2}\right\} d\tau = \begin{cases} \bar{0} & \text{(exterior points of } \tau) \\ -4\pi \operatorname{curl}\bar{M} & \text{(interior points of } \tau) \end{cases} \tag{5.9-10}$$

The d'Alembertian of the cavity potential is consequently zero.

Exercises involving the determination of the derivatives of the partial macroscopic potential appear in the following pages. The results are listed, together with those developed above, in Table 8, p. 464.

EXERCISES

5-24. In subsequent exercises involving partial potentials it is required that the derivatives of certain surface integrals over S_δ be calculated at the centre of the moving sphere (radius δ). Fill in the missing steps in the following transformations.

(a) $\operatorname{grad}\oint_{S_\delta} \frac{[\bar{P}]}{r}.d\bar{S}$

$$= \oint_{S_\delta} \frac{1}{r}[\operatorname{div}\bar{P}]\, d\bar{S} + \sum \bar{i}\oint_{S_\delta} \frac{1}{r}\{(d\bar{S}\times[\operatorname{grad} P_z])_y - (d\bar{S}\times[\operatorname{grad} P_y])_z\}$$

$$= \oint_{S_\delta} \frac{1}{r}[\operatorname{div}\bar{P}]\, d\bar{S} + \oint_{S_\delta} [\bar{P}] \times \left(d\bar{S}\times\operatorname{grad}\frac{1}{r}\right) - \oint_{S_\delta} \boxed{\frac{\partial \bar{P}}{\partial t}} \times \left(d\bar{S} \times \frac{\bar{r}}{cr^2}\right)$$

$$= \frac{1}{\delta}\oint_{S_\delta} [\operatorname{div}\bar{P}]\, d\bar{S}$$

(b) $$\mathrm{div} \oint_{S_\delta} d\bar{S} \times \frac{[\bar{M}]}{r}$$

$$= -\oint_{S_\delta} \frac{1}{r} \, [\mathrm{curl}\ \bar{M}].d\bar{S}$$

$$= \oint_{S_\delta} [\bar{M}].d\bar{S} \times \mathrm{grad}\ \frac{1}{r} - \oint_{S_\delta} \left[\frac{\partial \bar{M}}{\partial t}\right].d\bar{S} \times \frac{\bar{r}}{cr^2}$$

$$= 0$$

(c) $$\mathrm{curl} \oint_{S_\delta} d\bar{S} \times \frac{[\bar{M}]}{r}$$

$$= \oint_{S_\delta} \frac{1}{r} \left(d\bar{S} \times [\mathrm{curl}\ \bar{M}]\right) + \sum \bar{i} \oint_{S_\delta} \frac{1}{r} \, \{(d\bar{S}\times[\mathrm{grad}\ M_y])_z - (d\bar{S}\times[\mathrm{grad}\ M_z])_y\}$$

$$= \oint_{S_\delta} \frac{1}{r}\left(d\bar{S} \times [\mathrm{curl}\ \bar{M}]\right) - \oint_{S_\delta} [\bar{M}] \times \left(d\bar{S} \times \mathrm{grad}\ \frac{1}{r}\right) + \oint_{S_\delta} \left[\frac{\partial \bar{M}}{\partial t}\right] \times \left(d\bar{S} \times \frac{\bar{r}}{cr^2}\right)$$

$$= \frac{1}{\delta} \oint_{S_\delta} d\bar{S} \times [\mathrm{curl}\ \bar{M}]$$

(d) $$\nabla^2 \oint_{S_\delta} \frac{[\bar{P}]}{r}.d\bar{S} = \mathrm{div}\ \mathrm{grad} \oint_{S_\delta} \frac{[\bar{P}]}{r}.d\bar{S}$$

$$= \mathrm{div}\ \frac{1}{\delta} \oint_{S_\delta} [\mathrm{div}\ \bar{P}]\ d\bar{S}$$

$$= \frac{1}{\delta} \oint_{S_\delta} \left[\frac{\partial}{\partial n} \mathrm{div}\ \bar{P}\right] dS$$

(e) $$\operatorname{curl}\operatorname{curl}\oint_{S_\delta} d\bar{S}\times\frac{[\bar{M}]}{r}$$

$$=\operatorname{curl}\frac{1}{\delta}\oint_{S_\delta} d\bar{S}\times[\operatorname{curl}\bar{M}]$$

$$=\frac{1}{\delta}\sum\bar{i}\oint_{S_\delta}\left(dS_x\,[\operatorname{div}\operatorname{curl}\bar{M}]-[\operatorname{grad}(\operatorname{curl}\bar{M})_x].d\bar{S}\right)$$

$$=-\frac{1}{\delta}\oint_{S_\delta}\left[\frac{\partial}{\partial n}\operatorname{curl}\bar{M}\right]dS$$

(f) $$\nabla^2\oint_{S_\delta} d\bar{S}\times\frac{[\bar{M}]}{r}=\frac{1}{\delta}\oint_{S_\delta}\left[\frac{\partial}{\partial n}\operatorname{curl}\bar{M}\right]dS$$

5-25. By writing

$$\int_{\tau-\tau_\delta}\left\{[\bar{P}].\operatorname{grad}\frac{1}{r}-\left[\frac{\partial\bar{P}}{\partial t}\right].\frac{\bar{r}}{cr^2}\right\}d\tau=\int_{\tau-\tau_\delta}\frac{1}{r}[-\operatorname{div}\bar{P}]\,d\tau+\oint_{S_{1..n}\Sigma,\,S_\delta}\frac{[\bar{P}]}{r}.d\bar{S}$$

show that

$$\operatorname{grad}\,(\text{partial})\int_{\tau-\tau_\delta}\left\{[\bar{P}].\operatorname{grad}\frac{1}{r}-\left[\frac{\partial\bar{P}}{\partial t}\right].\frac{\bar{r}}{cr^2}\right\}d\tau$$

$$=-\int_{\tau-\tau_\delta}\frac{1}{r}[\operatorname{grad}\operatorname{div}\bar{P}]\,d\tau+\oint_{S_{1..n}\Sigma}\left\{\frac{1}{r}[\operatorname{div}\bar{P}]\,d\bar{S}-[\bar{P}].d\bar{S}\operatorname{grad}\frac{1}{r}+\frac{\bar{r}}{cr^2}\left[\frac{\partial\bar{P}}{\partial t}\right].d\bar{S}\right\}$$

$$+\operatorname{grad}\oint_{S_\delta}\frac{[\bar{P}]}{r}.d\bar{S}$$

$$= \int_{\tau-\tau_\delta} \left\{[\operatorname{div} \bar{P}] \operatorname{grad} \frac{1}{r} - \frac{\bar{r}}{cr^2} \left[\frac{\partial}{\partial t} \operatorname{div} \bar{P}\right]\right\} d\tau + \oint_{S_{1..n}\Sigma} \left\{\frac{\bar{r}}{cr^2} \left[\frac{\partial \bar{P}}{\partial t}\right].d\bar{S} - [\bar{P}].d\bar{S} \operatorname{grad} \frac{1}{r}\right\}$$

5-26. By writing

$$\int_{\tau-\tau_\delta} \left\{[\bar{M}] \times \operatorname{grad} \frac{1}{r} - \left[\frac{\partial \bar{M}}{\partial t}\right] \times \frac{\bar{r}}{cr^2}\right\} d\tau = \int_{\tau-\tau_\delta} \frac{1}{r} [\operatorname{curl} \bar{M}] \, d\tau - \oint_{S_{1..n}\Sigma,\ S_\delta} d\bar{S} \times \frac{[\bar{M}]}{r}$$

show that

$$\operatorname{curl} (\text{partial}) \int_{\tau-\tau_\delta} \left\{[\bar{M}] \times \operatorname{grad} \frac{1}{r} - \left[\frac{\partial \bar{M}}{\partial t}\right] \times \frac{\bar{r}}{cr^2}\right\} d\tau$$

$$= \int_{\tau-\tau_\delta} \frac{1}{r} [\operatorname{curl} \operatorname{curl} \bar{M}] \, d\tau + \oint_{S_{1..n}\Sigma} \left\{\operatorname{grad} \frac{1}{r} \times (d\bar{S} \times [\bar{M}]) - d\bar{S} \times \frac{1}{r} [\operatorname{curl} \bar{M}] - \frac{\bar{r}}{cr^2} \times \left(d\bar{S} \times \left[\frac{\partial \bar{M}}{\partial t}\right]\right)\right\}$$

$$- \operatorname{curl} \oint_{S_\delta} d\bar{S} \times \frac{[\bar{M}]}{r}$$

$$= \int_{\tau-\tau_\delta} \left\{[\operatorname{curl} \bar{M}] \times \operatorname{grad} \frac{1}{r} + \frac{\bar{r}}{cr^2} \times \left[\frac{\partial}{\partial t} \operatorname{curl} \bar{M}\right]\right\} d\tau$$

$$+ \oint_{S_{1..n}\Sigma} \left\{\operatorname{grad} \frac{1}{r} \times (d\bar{S} \times [\bar{M}]) - \frac{\bar{r}}{cr^2} \times \left(d\bar{S} \times \left[\frac{\partial \bar{M}}{\partial t}\right]\right)\right\}$$

5-27. By writing

$$\int_{\tau-\tau_\delta} \left\{[\bar{M}].\text{grad}\,\frac{1}{r} - \left[\frac{\partial\bar{M}}{\partial t}\right].\frac{\bar{r}}{cr^2}\right\} d\tau = -\,\text{div (partial)} \int_{\tau-\tau_\delta} \frac{[\bar{M}]}{r}\, d\tau + \oint_{S_\delta} \frac{[\bar{M}]}{r}.d\bar{S}$$

and

$$\int_{\tau-\tau_\delta} \left\{[\bar{M}] \times \text{grad}\,\frac{1}{r} - \left[\frac{\partial\bar{M}}{\partial t}\right] \times \frac{\bar{r}}{cr^2}\right\} d\tau = \text{curl (partial)} \int_{\tau-\tau_\delta} \frac{[\bar{M}]}{r}\, d\tau - \oint_{S_\delta} d\bar{S} \times \frac{[\bar{M}]}{r}$$

show that

$$\text{curl (partial)} \int_{\tau-\tau_\delta} \left\{[\bar{M}] \times \text{grad}\,\frac{1}{r} - \left[\frac{\partial\bar{M}}{\partial t}\right] \times \frac{\bar{r}}{cr^2}\right\} d\tau$$

$$= -\,\text{grad (partial)} \int_{\tau-\tau_\delta} \left\{[\bar{M}].\text{grad}\,\frac{1}{r} - \left[\frac{\partial\bar{M}}{\partial t}\right].\frac{\bar{r}}{cr^2}\right\} d\tau - \nabla^2 \text{ (partial)} \int_{\tau-\tau_\delta} \frac{[\bar{M}]}{r}\, d\tau$$

$$+ \oint_{S_\delta} \frac{1}{\delta} \{[\text{div}\,\bar{M}]\, d\bar{S} - d\bar{S} \times [\text{curl}\,\bar{M}]\}$$

5-28. Use equations (1.17-13) and (5.7-11) to transform the result of the previous exercise into

$$\text{curl (partial)} \int_{\tau-\tau_\delta} \left\{ [\bar{M}] \times \text{grad}\,\frac{1}{r} - \left[\frac{\partial \bar{M}}{\partial t}\right] \times \frac{\bar{r}}{cr^2} \right\} d\tau$$

$$= -\,\text{grad (partial)} \int_{\tau-\tau_\delta} \left\{ [\bar{M}].\text{grad}\,\frac{1}{r} - \left[\frac{\partial \bar{M}}{\partial t}\right] . \frac{\bar{r}}{cr^2} \right\} d\tau - \frac{1}{c^2}\frac{\partial^2}{\partial t^2} \int_{\tau-\tau_\delta} \frac{[\bar{M}]}{r}\, d\tau$$

$$+ \oint_{S_\delta} \left\{ \frac{[\bar{M}]}{\delta^2} + \frac{1}{c\delta}\left[\frac{\partial \bar{M}}{\partial t}\right] \right\} dS$$

5-29. Prove that

$$\text{div (partial)} \int_{\tau-\tau_\delta} \left\{ [\bar{M}] \times \text{grad}\,\frac{1}{r} - \left[\frac{\partial \bar{M}}{\partial t}\right] \times \frac{\bar{r}}{cr^2} \right\} d\tau = 0$$

5-30. Use the results of Ex.5-28. and 5-29. to show that

$$\nabla^2 \text{ (partial)} \int_{\tau-\tau_\delta} \left\{ [\bar{M}] \times \text{grad}\,\frac{1}{r} - \left[\frac{\partial \bar{M}}{\partial t}\right] \times \frac{\bar{r}}{cr^2} \right\} d\tau$$

$$= -\,\text{curl} \oint_{S_\delta} \left\{ \frac{[\bar{M}]}{\delta^2} + \frac{1}{c\delta}\left[\frac{\partial \bar{M}}{\partial t}\right] \right\} dS + \frac{1}{c^2}\frac{\partial^2}{\partial t^2}\,\text{curl (partial)} \int_{\tau-\tau_\delta} \frac{[\bar{M}]}{r}\, d\tau$$

and

$$\text{dal (partial)} \int_{\tau-\tau_\delta} \left\{ [\bar{M}] \times \text{grad}\,\frac{1}{r} - \left[\frac{\partial \bar{M}}{\partial t}\right] \times \frac{\bar{r}}{cr^2} \right\} d\tau$$

$$= \oint_{S_\delta} \left\{ \frac{1}{\delta^2}\,[-\,\text{curl}\,\bar{M}] - \frac{1}{c\delta}\left[\frac{\partial}{\partial t}\,\text{curl}\,\bar{M}\right] \right\} dS + \frac{1}{c^2}\frac{\partial^2}{\partial t^2} \oint_{S_\delta} d\bar{S} \times \frac{[\bar{M}]}{r}$$

Confirm this by writing

$$\text{dal (partial)} \int_{\tau-\tau_\delta} \left\{[\bar{M}] \times \text{grad}\, \frac{1}{r} - \left[\frac{\partial \bar{M}}{\partial t}\right] \times \frac{\bar{r}}{cr^2}\right\} d\tau$$

$$= \text{dal (partial)} \int_{\tau-\tau_\delta} \frac{1}{r}\, [\text{curl}\, \bar{M}]\, d\tau - \text{dal} \oint_{S_{1..n}\Sigma,\, S_\delta} d\bar{S} \times \frac{[\bar{M}]}{r}$$

5-31. Show that

$$\text{dal (partial)} \int_{\tau-\tau_\delta} \left\{[\bar{P}].\text{grad}\, \frac{1}{r} - \left[\frac{\partial \bar{P}}{\partial t}\right].\frac{\bar{r}}{cr^2}\right\} d\tau$$

$$= \oint_{S_\delta} \left\{\frac{1}{\delta^2}\, [\text{div}\, \bar{P}] + \frac{1}{c\delta} \left[\frac{\partial}{\partial t}\, \text{div}\, \bar{P}\right]\right\} dS - \frac{1}{c^2} \frac{\partial^2}{\partial t^2} \oint_{S_\delta} \frac{[\bar{P}]}{r}.d\bar{S}$$

5-32. The analytical treatment of the above exercises has been simplified by the assumption that the excluding region is spherical and centred upon the point of evaluation 0. Observing that the factors $\left(d\bar{S} \times \text{grad}\, \frac{1}{r}\right)$ and $\left(d\bar{S} \times \frac{\bar{r}}{cr^2}\right)$, which appear in the expansions of Ex.5-24. (a), (b) and (c), remain intact when the operators grad div and curl, as appropriate, are re-applied, show that the same limiting expressions obtain in Ex.5-25. to 5-31. when the δ sphere is replaced by any regular region which shrinks uniformly about 0.

TABLE 7

The Scalar Potential Function $\int \left\{[\bar{P}].\text{grad}\, \frac{1}{r} - \left[\frac{\partial \bar{P}}{\partial t}\right].\frac{\bar{r}}{cr^2}\right\} d\tau$ and its Derivatives

(1)

$$\int_\tau \left\{[\bar{P}].\text{grad}\, \frac{1}{r} - \left[\frac{\partial \bar{P}}{\partial t}\right].\frac{\bar{r}}{cr^2}\right\} d\tau = \int_\tau \frac{[-\,\text{div}\, \bar{P}]}{r}\, d\tau + \oint_{S_{1..n}\Sigma} \frac{[\bar{P}]}{r}.d\bar{S}$$

(interior and exterior points of τ)

(2)

$$\text{partial} \int_{\tau-\tau_\delta} \left\{ [\bar{P}].\text{grad}\,\frac{1}{r} - \boxed{\frac{\partial \bar{P}}{\partial t}}.\frac{\bar{r}}{cr^2} \right\} d\tau = \int_{\tau-\tau_\delta} \frac{[-\,\text{div}\,\bar{P}]}{r}\, d\tau + \oint_{S_{1..n}\Sigma,\ S_\delta} \frac{[\bar{P}]}{r}.d\bar{S}$$

(evaluated at centre of moving δ sphere within τ)

(3)

$$\text{cavity} \int_{\tau-\tau_\delta} \left\{ [\bar{P}].\text{grad}\,\frac{1}{r} - \boxed{\frac{\partial \bar{P}}{\partial t}}.\frac{\bar{r}}{cr^2} \right\} d\tau = \int_{\tau-\tau_\delta} \frac{[-\,\text{div}\,\bar{P}]}{r}\, d\tau + \oint_{S_{1..n}\Sigma,\ S_\delta} \frac{[\bar{P}]}{r}.d\bar{S}$$

(defined throughout fixed δ sphere within τ)

(4)

$$\text{grad} \int_\tau \left\{ [\bar{P}].\text{grad}\,\frac{1}{r} - \boxed{\frac{\partial \bar{P}}{\partial t}}.\frac{\bar{r}}{cr^2} \right\} d\tau$$

$$= \int_\tau \left\{ [-\text{div}\,\bar{P}]\,\frac{\bar{r}}{r^3} - \left[\frac{\partial}{\partial t}\,\text{div}\,\bar{P}\right] \frac{\bar{r}}{cr^2} \right\} d\tau + \oint_{S_{1..n}\Sigma} \left\{ [\bar{P}].d\bar{S}\,\frac{\bar{r}}{r^3} + \frac{\bar{r}}{cr^2}\,\boxed{\frac{\partial \bar{P}}{\partial t}}.d\bar{S} \right\}$$

(interior and exterior points of τ)

$$= -\,\text{curl} \int_\tau \left\{ [\bar{P}] \times \text{grad}\,\frac{1}{r} - \boxed{\frac{\partial \bar{P}}{\partial t}} \times \frac{\bar{r}}{cr^2} \right\} d\tau - \frac{1}{c^2}\frac{\partial^2}{\partial t^2} \int_\tau \frac{[\bar{P}]}{r}\, d\tau + \begin{cases} \bar{0} & \text{(exterior} \\ 4\pi\bar{P} & \text{(interior points of } \tau) \end{cases}$$

$$= \int_\tau \left\{ [\text{curl}\,\bar{P}] \times \frac{\bar{r}}{r^3} + \left[\frac{\partial}{\partial t}\,\text{curl}\,\bar{P}\right] \times \frac{\bar{r}}{cr^2} \right\} d\tau - \oint_{S_{1..n}\Sigma} \left\{ (d\bar{S} \times [\bar{P}]) \times \frac{\bar{r}}{r^3} + \left(d\bar{S} \times \boxed{\frac{\partial \bar{P}}{\partial t}} \right) \times \frac{\bar{r}}{cr^2} \right\}$$

$$-\,\frac{1}{c^2}\frac{\partial^2}{\partial t^2} \int_\tau \frac{[\bar{P}]}{r}\, d\tau + \begin{cases} \bar{0} & \text{(exterior points of } \tau) \\ 4\pi\bar{P} & \text{(interior points of } \tau) \end{cases}$$

(5)

$$\text{grad (cavity)} \int_{\tau-\tau_\delta} \left\{ [\bar{P}].\text{grad}\,\frac{1}{r} - \left[\frac{\partial \bar{P}}{\partial t}\right].\frac{\bar{r}}{cr^2} \right\} d\tau$$

$$= \int_{\tau-\tau_\delta} \left\{ [-\,\text{div}\,\bar{P}]\,\frac{\bar{r}}{r^3} - \left[\frac{\partial}{\partial t}\,\text{div}\,\bar{P}\right]\frac{\bar{r}}{cr^2} \right\} d\tau + \oint_{S_{1..n}\Sigma,\ S_\delta} \left\{ [\bar{P}].d\bar{S}\,\frac{\bar{r}}{r^3} + \frac{\bar{r}}{cr^2}\left[\frac{\partial \bar{P}}{\partial t}\right].d\bar{S} \right\}$$

$$= -\,\text{curl (cavity)} \int_{\tau-\tau_\delta} \left\{ [\bar{P}] \times \text{grad}\,\frac{1}{r} - \left[\frac{\partial \bar{P}}{\partial t}\right] \times \frac{\bar{r}}{cr^2} \right\} d\tau - \frac{1}{c^2}\frac{\partial^2}{\partial t^2}\int_{\tau-\tau_\delta} \frac{[\bar{P}]}{r}\, d\tau$$

$$= \int_{\tau-\tau_\delta} \left\{ [\text{curl}\,\bar{P}] \times \frac{\bar{r}}{r^3} + \left[\frac{\partial}{\partial t}\,\text{curl}\,\bar{P}\right] \times \frac{\bar{r}}{cr^2} \right\} d\tau - \oint_{S_{1..n}\Sigma,\ S_\delta} \left\{ (d\bar{S}\times[\bar{P}]) \times \frac{\bar{r}}{r^3} + \left(d\bar{S}\times\left[\frac{\partial \bar{P}}{\partial t}\right]\right) \times \frac{\bar{r}}{cr^2} \right\}$$

$$-\,\frac{1}{c^2}\frac{\partial^2}{\partial t^2}\int_{\tau-\tau_\delta} \frac{[\bar{P}]}{r}\, d\tau$$

(6)

$$\lim_{\delta\to 0}\ \text{grad (cavity)} \int_{\tau-\tau_\delta} \left\{ [\bar{P}].\text{grad}\,\frac{1}{r} - \left[\frac{\partial \bar{P}}{\partial t}\right].\frac{\bar{r}}{cr^2} \right\} d\tau$$

$$= \text{grad} \int_{\tau} \left\{ [\bar{P}].\text{grad}\,\frac{1}{r} - \left[\frac{\partial \bar{P}}{\partial t}\right].\frac{\bar{r}}{cr^2} \right\} d\tau - \frac{4}{3}\,\pi\bar{P}$$

for evaluation at the centre of the δ sphere.

(7)

$$\text{grad (partial)} \int_{\tau-\tau_\delta} \left\{[\bar{P}].\text{grad}\,\frac{1}{r} - \boxed{\frac{\partial\bar{P}}{\partial t}}.\frac{\bar{r}}{cr^2}\right\} d\tau$$

$$= \int_{\tau-\tau_\delta} \left\{[-\,\text{div}\,\bar{P}]\,\frac{\bar{r}}{r^3} - \boxed{\frac{\partial}{\partial t}\,\text{div}\,\bar{P}}\,\frac{\bar{r}}{cr^2}\right\} d\tau + \oint_{S_{1..n^{\Sigma}}} \left\{[\bar{P}].d\bar{S}\,\frac{\bar{r}}{r^3} + \frac{\bar{r}}{cr^2}\,\boxed{\frac{\partial\bar{P}}{\partial t}}.d\bar{S}\right\}$$

$$- \oint_{S_\delta} \left\{[\bar{P}] \times \left(d\bar{S} \times \frac{\bar{r}}{r^3}\right) + \boxed{\frac{\partial\bar{P}}{\partial t}} \times \left(d\bar{S} \times \frac{\bar{r}}{cr^2}\right)\right\}$$

$$= -\,\text{curl (partial)} \int_{\tau-\tau_\delta} \left\{[\bar{P}] \times \text{grad}\,\frac{1}{r} - \boxed{\frac{\partial\bar{P}}{\partial t}} \times \frac{\bar{r}}{cr^2}\right\} d\tau - \frac{1}{c^2}\frac{\partial^2}{\partial t^2} \int_{\tau-\tau_\delta} \frac{[\bar{P}]}{r}\, d\tau$$

$$- \oint_{S_\delta} \left\{[\bar{P}]\; d\bar{S}.\frac{\bar{r}}{r^3} + \boxed{\frac{\partial\bar{P}}{\partial t}}\; d\bar{S}.\frac{\bar{r}}{cr^2} + [\bar{P}] \times \left(d\bar{S} \times \frac{\bar{r}}{r^3}\right) + \boxed{\frac{\partial\bar{P}}{\partial t}} \times \left(d\bar{S} \times \frac{\bar{r}}{cr^2}\right)\right\}$$

(8)

$$\text{dal} \int_{\tau} \left\{[\bar{P}].\text{grad}\,\frac{1}{r} - \boxed{\frac{\partial\bar{P}}{\partial t}}.\frac{\bar{r}}{cr^2}\right\} d\tau = \begin{cases} 0 & \text{(exterior points of } \tau) \\ 4\pi\,\text{div}\,\bar{P} & \text{(interior points of } \tau) \end{cases}$$

(9)

$$\text{dal (cavity)} \int_{\tau-\tau_\delta} \left\{[\bar{P}].\text{grad}\,\frac{1}{r} - \boxed{\frac{\partial\bar{P}}{\partial t}}.\frac{\bar{r}}{cr^2}\right\} d\tau = 0$$

(10)

$$\text{dal (partial)} \int_{\tau-\tau_\delta} \left\{ [\bar{P}].\text{grad}\,\frac{1}{r} - \left[\frac{\partial \bar{P}}{\partial t}\right].\frac{\bar{r}}{cr^2} \right\} d\tau$$

$$= -\oint_{S_\delta} \left\{ [\text{div}\,\bar{P}]\; d\bar{S}.\frac{\bar{r}}{r^3} + \left[\frac{\partial}{\partial t}\,\text{div}\,\bar{P}\right] d\bar{S}.\frac{\bar{r}}{cr^2} \right\}$$

$$-\oint_{S_\delta} \left\{ [\text{curl}\,\bar{P}].\left(d\bar{S} \times \frac{\bar{r}}{r^3}\right) + \left[\frac{\partial}{\partial t}\,\text{curl}\,\bar{P}\right].\left(d\bar{S} \times \frac{\bar{r}}{cr^2}\right) \right\} - \frac{1}{c^2}\frac{\partial^2}{\partial t^2}\oint_{S_\delta} \frac{[\bar{P}]}{r}.d\bar{S}$$

TABLE 8

<u>The Vector Potential Function</u> $\int \left\{ [\bar{M}] \times \text{grad}\,\frac{1}{r} - \left[\frac{\partial \bar{M}}{\partial t}\right] \times \frac{\bar{r}}{cr^2} \right\} d\tau$ <u>and its Derivatives</u>

(1)

$$\int_{\tau} \left\{ [\bar{M}] \times \text{grad}\,\frac{1}{r} - \left[\frac{\partial \bar{M}}{\partial t}\right] \times \frac{\bar{r}}{cr^2} \right\} d\tau = \int_{\tau} \frac{[\text{curl}\,\bar{M}]}{r}\, d\tau - \oint_{S_{1..n}\Sigma} d\bar{S} \times \frac{[\bar{M}]}{r}$$

(interior and exterior points of τ)

(2)

$$\text{partial} \int_{\tau-\tau_\delta} \left\{ [\bar{M}] \times \text{grad}\,\frac{1}{r} - \left[\frac{\partial \bar{M}}{\partial t}\right] \times \frac{\bar{r}}{cr^2} \right\} d\tau = \int_{\tau-\tau_\delta} \frac{[\text{curl}\,\bar{M}]}{r}\, d\tau - \oint_{S_{1..n}\Sigma,\, S_\delta} d\bar{S} \times \frac{[\bar{M}]}{r}$$

(evaluated at centre of moving δ sphere within τ)

(3)

$$\text{cavity} \int_{\tau-\tau_\delta} \left\{ [\bar{M}] \times \text{grad}\,\frac{1}{r} - \left[\frac{\partial \bar{M}}{\partial t}\right] \times \frac{\bar{r}}{cr^2} \right\} d\tau = \int_{\tau-\tau_\delta} \frac{[\text{curl}\,\bar{M}]}{r}\, d\tau - \oint_{S_{1..n}\Sigma,\, S_\delta} d\bar{S} \times \frac{[\bar{M}]}{r}$$

(defined throughout fixed δ sphere within τ)

(4)

$$\operatorname{div} \int_{\tau} \left\{ [\bar{M}] \times \operatorname{grad} \frac{1}{r} - \left[\frac{\partial \bar{M}}{\partial t}\right] \times \frac{\bar{r}}{cr^2} \right\} d\tau = 0$$

(interior and exterior points of τ)

(5)

$$\operatorname{div}\ (\text{cavity}) \int_{\tau-\tau_\delta} \left\{ [\bar{M}] \times \operatorname{grad} \frac{1}{r} - \left[\frac{\partial \bar{M}}{\partial t}\right] \times \frac{\bar{r}}{cr^2} \right\} d\tau = 0$$

(6)

$$\operatorname{div}\ (\text{partial}) \int_{\tau-\tau_\delta} \left\{ [\bar{M}] \times \operatorname{grad} \frac{1}{r} - \left[\frac{\partial \bar{M}}{\partial t}\right] \times \frac{\bar{r}}{cr^2} \right\} d\tau = \oint_{S_\delta} \left\{ [\bar{M}].d\bar{S} \times \frac{\bar{r}}{r^3} + \left[\frac{\partial \bar{M}}{\partial t}\right].d\bar{S} \times \frac{\bar{r}}{cr^2} \right\}$$

(7)

$$\operatorname{curl} \int_{\tau} \left\{ [\bar{M}] \times \operatorname{grad} \frac{1}{r} - \left[\frac{\partial \bar{M}}{\partial t}\right] \times \frac{\bar{r}}{cr^2} \right\} d\tau$$

$$= - \int_{\tau} \left\{ [\operatorname{curl} \bar{M}] \times \frac{\bar{r}}{r^3} + \left[\frac{\partial}{\partial t} \operatorname{curl} \bar{M}\right] \times \frac{\bar{r}}{cr^2} \right\} d\tau + \oint_{S_{1..n}\Sigma} \left\{ (d\bar{S} \times [\bar{M}]) \times \frac{\bar{r}}{r^3} + \left(d\bar{S} \times \left[\frac{\partial \bar{M}}{\partial t}\right] \right) \times \frac{\bar{r}}{cr^2} \right\}$$

(interior and exterior points of τ)

$$= - \operatorname{grad} \int_{\tau} \left\{ [\bar{M}].\operatorname{grad} \frac{1}{r} - \left[\frac{\partial \bar{M}}{\partial t}\right] . \frac{\bar{r}}{cr^2} \right\} d\tau - \frac{1}{c^2} \frac{\partial^2}{\partial t^2} \int_{\tau} \frac{[\bar{M}]}{r} d\tau + \begin{cases} \bar{0} & \text{(exterior} \\ 4\pi\bar{M} & \text{(interior} \end{cases}$$

points of τ)

$$= \int_{\tau} \left\{ [\operatorname{div} \bar{M}] \frac{\bar{r}}{r^3} + \left[\frac{\partial}{\partial t} \operatorname{div} \bar{M}\right] \frac{\bar{r}}{cr^2} \right\} d\tau$$

$$- \oint_{S_{1..n}\Sigma} \left\{ [\bar{M}].d\bar{S} \frac{\bar{r}}{r^3} + \frac{\bar{r}}{cr^2} \left[\frac{\partial \bar{M}}{\partial t}\right].d\bar{S} \right\} - \frac{1}{c^2} \frac{\partial^2}{\partial t^2} \int_{\tau} \frac{[\bar{M}]}{r} d\tau + \begin{cases} \bar{0} & \text{(exterior} \\ 4\pi\bar{M} & \text{(interior} \end{cases}$$

points of τ)

(8)

$$\text{curl (cavity)} \int_{\tau-\tau_\delta} \left\{ [\bar{M}] \times \text{grad}\,\frac{1}{r} - \left[\frac{\partial \bar{M}}{\partial t}\right] \times \frac{\bar{r}}{cr^2} \right\} d\tau$$

$$= - \int_{\tau-\tau_\delta} \left\{ [\text{curl}\,\bar{M}] \times \frac{\bar{r}}{r^3} + \left[\frac{\partial}{\partial t}\,\text{curl}\,\bar{M}\right] \times \frac{\bar{r}}{cr^2} \right\} d\tau + \oint_{S_{1..n}\Sigma,\ S_\delta} \left\{ (d\bar{S} \times [\bar{M}]) \times \frac{\bar{r}}{r^3} + \left(d\bar{S} \times \left[\frac{\partial \bar{M}}{\partial t}\right] \right) \times \frac{\bar{r}}{cr^2} \right\}$$

$$= - \text{grad (cavity)} \int_{\tau-\tau_\delta} \left\{ [\bar{M}].\text{grad}\,\frac{1}{r} - \left[\frac{\partial \bar{M}}{\partial t}\right].\frac{\bar{r}}{cr^2} \right\} d\tau - \frac{1}{c^2}\frac{\partial^2}{\partial t^2} \int_{\tau-\tau_\delta} \frac{[\bar{M}]}{r}\, d\tau$$

$$= \int_{\tau-\tau_\delta} \left\{ [\text{div}\,\bar{M}]\frac{\bar{r}}{r^3} + \left[\frac{\partial}{\partial t}\,\text{div}\,\bar{M}\right] \frac{\bar{r}}{cr^2} \right\} d\tau - \oint_{S_{1..n}\Sigma,\ S_\delta} \left\{ [\bar{M}].d\bar{S}\,\frac{\bar{r}}{r^3} + \frac{\bar{r}}{cr^2} \left[\frac{\partial \bar{M}}{\partial t}\right].d\bar{S} \right\} - \frac{1}{c^2}\frac{\partial^2}{\partial t^2} \int_{\tau-\tau_\delta} \frac{[\bar{M}]}{r}\, d\tau$$

(9)

$$\lim_{\delta \to 0} \text{curl (cavity)} \int_{\tau-\tau_\delta} \left\{ [\bar{M}] \times \text{grad}\,\frac{1}{r} - \left[\frac{\partial \bar{M}}{\partial t}\right] \times \frac{\bar{r}}{cr^2} \right\} d\tau$$

$$= \text{curl} \int_{\tau} \left\{ [\bar{M}] \times \text{grad}\,\frac{1}{r} - \left[\frac{\partial \bar{M}}{\partial t}\right] \times \frac{\bar{r}}{cr^2} \right\} d\tau - \frac{8}{3}\pi\bar{M}$$

for evaluation at the centre of the δ sphere.

(10)

$$\text{curl (partial)} \int_{\tau-\tau_\delta} \left\{ [\bar{M}] \times \text{grad}\,\frac{1}{r} - \left[\frac{\partial \bar{M}}{\partial t}\right] \times \frac{\bar{r}}{cr^2} \right\} d\tau$$

$$= - \int_{\tau-\tau_\delta} \left\{ [\text{curl } \bar{M}] \times \frac{\bar{r}}{r^3} + \left[\frac{\partial}{\partial t} \text{ curl } \bar{M}\right] \times \frac{\bar{r}}{cr^2} \right\} d\tau + \oint_{S_{1..n}\Sigma} \left\{ (d\bar{S} \times [\bar{M}]) \times \frac{\bar{r}}{r^3} + \left(d\bar{S} \times \left[\frac{\partial \bar{M}}{\partial t}\right] \right) \times \frac{\bar{r}}{cr^2} \right\}$$

$$- \oint_{S_\delta} \left\{ [\bar{M}] \times \left(d\bar{S} \times \frac{\bar{r}}{r^3} \right) + \left[\frac{\partial \bar{M}}{\partial t}\right] \times \left(d\bar{S} \times \frac{\bar{r}}{cr^2} \right) \right\}$$

$$= - \text{grad (partial)} \int_{\tau-\tau_\delta} \left\{ [\bar{M}] . \text{grad}\,\frac{1}{r} - \left[\frac{\partial \bar{M}}{\partial t}\right] . \frac{\bar{r}}{cr^2} \right\} d\tau - \frac{1}{c^2} \frac{\partial^2}{\partial t^2} \int_{\tau-\tau_\delta} \frac{[\bar{M}]}{r} \, d\tau$$

$$- \oint_{S_\delta} \left\{ [\bar{M}] d\bar{S} . \frac{\bar{r}}{r^3} + \left[\frac{\partial \bar{M}}{\partial t}\right] d\bar{S} . \frac{\bar{r}}{cr^2} + [\bar{M}] \times \left(d\bar{S} \times \frac{\bar{r}}{r^3} \right) + \left[\frac{\partial \bar{M}}{\partial t}\right] \times \left(d\bar{S} \times \frac{\bar{r}}{cr^2} \right) \right\}$$

(11)

$$\text{dal} \int_{\tau} \left\{ [\bar{M}] \times \text{grad}\,\frac{1}{r} - \left[\frac{\partial \bar{M}}{\partial t}\right] \times \frac{\bar{r}}{cr^2} \right\} d\tau = \begin{cases} \bar{0} & \text{(exterior points of } \tau) \\ -4\pi \text{ curl } \bar{M} & \text{(interior points of } \tau) \end{cases}$$

(12)

$$\text{dal (cavity)} \int_{\tau-\tau_\delta} \left\{ [\bar{M}] \times \text{grad}\,\frac{1}{r} - \left[\frac{\partial \bar{M}}{\partial t}\right] \times \frac{\bar{r}}{cr^2} \right\} d\tau = \bar{0}$$

(13)

$$\text{dal (partial)} \int_{\tau-\tau_\delta} \left\{[\bar{M}] \times \text{grad}\,\frac{1}{r} - \left[\frac{\partial \bar{M}}{\partial t}\right] \times \frac{\bar{r}}{cr^2}\right\} d\tau$$

$$= \oint_{S_\delta} \left\{[\text{curl}\,\bar{M}]\, d\bar{S}.\frac{\bar{r}}{r^3} + \left[\frac{\partial}{\partial t}\,\text{curl}\,\bar{M}\right] d\bar{S}.\frac{\bar{r}}{cr^2} - [\text{div}\,\bar{M}]\left(d\bar{S} \times \frac{\bar{r}}{r^3}\right)\right.$$

$$\left. - \left[\frac{\partial}{\partial t}\,\text{div}\,\bar{M}\right] d\bar{S} \times \frac{\bar{r}}{cr^2} + \left(d\bar{S} \times \frac{\bar{r}}{r^3}\right) \times [\text{curl}\,\bar{M}] + \left(d\bar{S} \times \frac{\bar{r}}{cr^2}\right) \times \left[\frac{\partial}{\partial t}\,\text{curl}\,\bar{M}\right]\right\}$$

$$+ 2\oint_{S_\delta} \left\{\left[\left(\left(d\bar{S} \times \frac{\bar{r}}{r^3}\right).\nabla\right)\bar{M}\right] + \left[\left(\left(d\bar{S} \times \frac{\bar{r}}{cr^2}\right).\nabla\right)\frac{\partial \bar{M}}{\partial t}\right]\right\} + \frac{1}{c^2}\frac{\partial^2}{\partial t^2}\oint_{S_\delta} d\bar{S} \times \frac{[\bar{M}]}{r}$$

5.10 The Liénard - Wiechert Potentials

5.10a Translation of a volume source

Let a rectangular system of axes maintain parallelism with a fixed reference system while its origin 0' moves in a straight line. Then if the density ρ of a volume source which is a function both of space and time can be expressed, for appropriate motion of 0', as a function of position alone in the moving system, the source is said to have motion of translation as a whole identical with that of 0'.

5.10b Retarded scalar potential of a translating volume source

Fig. 5.1 depicts the region of space occupied by a volume source at the instant of time $t - \frac{r_P}{c}$, where P is a point of the source. It is required to determine the retarded potential of the source at an exterior point 0 at the time t.

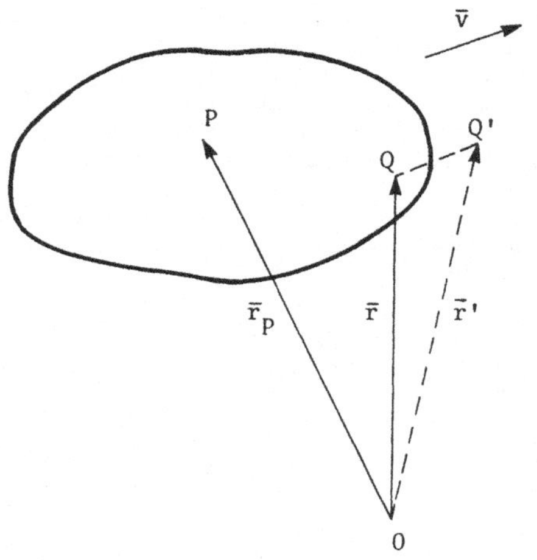

Fig. 5.1

While an element of the source at P will contribute directly to the required potential in virtue of its position, an element at Q will not so contribute unless $OQ = r_P$. Nevertheless, the latter element will contribute at the time $t - \frac{r_P}{c} + \Delta t$, when it will have moved to Q', provided that $t - \frac{r_P}{c} + \Delta t = t - \frac{r'}{c}$ where $r' = OQ'$. If the velocity of the source and its time derivatives at the instant $t - \frac{r_P}{c}$ are written as $\bar{v}, \dot{\bar{v}}, \ddot{\bar{v}}$ ... then the required condition becomes

$$\overrightarrow{QQ'} = \frac{\bar{v}}{c}(r_P - r') + \frac{1}{2!}\frac{\dot{\bar{v}}}{c^2}(r_P - r')^2 + \frac{1}{3!}\frac{\ddot{\bar{v}}}{c^3}(r_P - r')^3 + \ldots.$$

which gives rise to the three scalar equations

$$\begin{aligned} x' &= x + \frac{v_x}{c}(r_P - r') + \frac{1}{2!}\frac{\dot{v}_x}{c^2}(r_P - r')^2 + \frac{1}{3!}\frac{\ddot{v}_x}{c^3}(r_P - r')^3 + \ldots\ldots \\ y' &= y + \frac{v_y}{c}(r_P - r') + \frac{1}{2!}\frac{\dot{v}_y}{c^2}(r_P - r')^2 + \frac{1}{3!}\frac{\ddot{v}_y}{c^3}(r_P - r')^3 + \ldots\ldots \qquad (5.10\text{-}1) \\ z' &= z + \frac{v_z}{c}(r_P - r') + \frac{1}{2!}\frac{\dot{v}_z}{c^2}(r_P - r')^2 + \frac{1}{3!}\frac{\ddot{v}_z}{c^3}(r_P - r')^3 + \ldots\ldots \end{aligned}$$

where (x,y,z) and (x',y',z') are the coordinates of Q and Q' in the fixed reference system.

An appeal to the construction detailed on p. 409 suggests that, so long as $v < c$ at all times, a single solution exists for x', y', z' in terms of x, y, z, r_P, $\bar{v}$, etc, and Q' approaches Q as Q approaches P. Then if all points (x,y,z) of the distribution as shown at the instant $t - \frac{r_P}{c}$ are moved to (x',y',z'), the appropriate source configuration is obtained for the evaluation of the retarded potential at 0 at the time t. On the other hand, if no restriction is placed upon v we may expect multiple values for (x',y',z'), in which case the source will contribute to the retarded potential on more than one occasion during its movement. This possibility will be ignored in the present analysis.

A volume element centred upon Q does not retain its magnitude in passing from Q to Q'. It has been shown in Sec. 2.11 that if $\Delta\tau$ and $\Delta\tau'$ are the respective magnitudes then

$$\Delta\tau' = J\left(\frac{x',y',z'}{x,y,z}\right)\Delta\tau \tag{5.10-2}$$

$$\text{where} \quad J\left(\frac{x',y',z'}{x,y,z}\right) = \begin{vmatrix} \frac{\partial x'}{\partial x} & \frac{\partial x'}{\partial y} & \frac{\partial x'}{\partial z} \\ \frac{\partial y'}{\partial x} & \frac{\partial y'}{\partial y} & \frac{\partial y'}{\partial z} \\ \frac{\partial z'}{\partial x} & \frac{\partial z'}{\partial y} & \frac{\partial z'}{\partial z} \end{vmatrix} \tag{5.10-3}$$

Now it follows from equation (5.10-1) that

$$\frac{\partial x'}{\partial x} = 1 - \frac{v_x}{c}\frac{\partial r'}{\partial x} - \frac{\dot{v}_x}{c^2}(r_p-r')\frac{\partial r'}{\partial x} - \frac{1}{2!}\frac{\ddot{v}_x}{c^3}(r_p-r')^2\frac{\partial r'}{\partial x} \ldots$$

or

$$\frac{\partial x'}{\partial x} = 1 - l\,\frac{\partial r'}{\partial x}$$

where

$$l = \frac{v_x}{c} + \frac{\dot{v}_x}{c^2}(r_p-r') + \frac{1}{2!}\frac{\ddot{v}_x}{c^3}(r_p-r')^2 + \ldots\ldots$$

Similarly,

$$\left.\begin{aligned} &\frac{\partial x'}{\partial y} = -\,l\,\frac{\partial r'}{\partial y} \quad \text{and} \quad \frac{\partial x'}{\partial z} = -\,l\,\frac{\partial r'}{\partial z} \\ &\text{Further,} \\ &\frac{\partial y'}{\partial x} = -\,m\,\frac{\partial r'}{\partial x} \qquad \frac{\partial y'}{\partial y} = 1 - m\,\frac{\partial r'}{\partial y} \qquad \frac{\partial y'}{\partial z} = -\,m\,\frac{\partial r'}{\partial z} \\ &\frac{\partial z'}{\partial x} = -\,n\,\frac{\partial r'}{\partial x} \qquad \frac{\partial z'}{\partial y} = -\,n\,\frac{\partial r'}{\partial y} \qquad \frac{\partial z'}{\partial z} = 1 - n\,\frac{\partial r'}{\partial z} \end{aligned}\right\} \tag{5.10-4}$$

where

$$m = \frac{v_y}{c} + \frac{\dot{v}_y}{c^2}(r_p - r') + \frac{1}{2!}\frac{\ddot{v}_y}{c^3}(r_p - r')^2 + \ldots\ldots$$

and

$$n = \frac{v_z}{c} + \frac{\dot{v}_z}{c^2}(r_p - r') + \frac{1}{2!}\frac{\ddot{v}_z}{c^3}(r_p - r')^2 + \ldots\ldots$$

On multiplying out we find that

$$J\left(\frac{x',y',z'}{x,y,z}\right) = 1 - 1\frac{\partial r'}{\partial x} - m\frac{\partial r'}{\partial y} - n\frac{\partial r'}{\partial z} \qquad (5.10\text{-}5)$$

But

$$\frac{\partial r'}{\partial x} = \frac{\partial r'}{\partial x'}\frac{\partial x'}{\partial x} + \frac{\partial r'}{\partial y'}\frac{\partial y'}{\partial x} + \frac{\partial r'}{\partial z'}\frac{\partial z'}{\partial x}$$

and

$$r' = \{(x'-x_o)^2 + (y'-y_o)^2 + (z'-z_o)^2\}^{\frac{1}{2}}$$

hence

$$\frac{\partial r'}{\partial x} = \frac{(x'-x_o)}{r'}\left(1 - 1\frac{\partial r'}{\partial x}\right) - \frac{(y'-y_o)}{r'}\,m\,\frac{\partial r'}{\partial x} - \frac{(z'-z_o)}{r'}\,n\,\frac{\partial r'}{\partial x}$$

or

$$\frac{\partial r'}{\partial x}\left\{1 + 1\frac{(x'-x_o)}{r'} + m\frac{(y'-y_o)}{r'} + n\frac{(z'-z_o)}{r'}\right\} = \frac{x' - x_o}{r'} \qquad (5.10\text{-}6)$$

Similar equations hold for $\frac{\partial r'}{\partial y}$ and $\frac{\partial r'}{\partial z}$

Substitution in equation (5.10-5) then yields

$$J\left(\frac{x',y',z'}{x,y,z}\right) = \frac{1}{1 + 1\frac{(x'-x_o)}{r'} + m\frac{(y'-y_o)}{r'} + n\frac{(z'-z_o)}{r'}}$$

or

$$J\left(\frac{x',y',z'}{x,y,z}\right) = \frac{1}{1 + \frac{\bar{r}'}{cr'}\cdot\left(\bar{v} + \frac{\dot{\bar{v}}}{c}(r_p - r') + \frac{1}{2!}\frac{\ddot{\bar{v}}}{c^2}(r_p - r')^2 + \ldots\right)} \tag{5.10-7}$$

Since ρ is supposed to remain constant during the movement of an element from Q to Q', the required retarded potential may be written as

$$\phi = \int_{\tau'} \rho_\tau \frac{\partial\tau'}{r'} \tag{5.10-8}$$

where τ' is the region defined by all points Q', and this may be replaced by

$$\phi = \int_\tau \frac{\rho\, d\tau}{r'\left\{1 + \frac{\bar{r}'}{cr'}\cdot\left(\bar{v} + \frac{\dot{\bar{v}}}{c}(r_p - r') + \frac{1}{2!}\frac{\ddot{\bar{v}}}{c^2}(r_p - r')^2 + \ldots\right)\right\}} \tag{5.10-9}$$

where the integration is carried out for each element $\Delta\tau$ of the instantaneous source distribution τ, but where r' relates to the position of the corresponding element $\Delta\tau'$. (This mixed type of integration is necessary because r' has not been expressed explicitly in terms of the coordinates of $\Delta\tau$).

5.10c Retarded vector potential of a translating volume source

It has not been possible in earlier pages to associate a vector potential with a volume source of scalar density. This is effected in the case of a translating source by substituting $\frac{\rho\bar{v}}{c}$ for ρ where $\bar{v}$ is the velocity of the source. In the present instance this leads to a retarded vector potential defined by[6]

$$\bar{A} = \int_{\tau'} \frac{\rho\bar{v}'}{cr'}\, d\tau' \tag{5.10-10}$$

where $\bar{v}'$ is the velocity associated with the element $\Delta\tau'$ (ie the velocity of the source when the typical point Q has moved to its appropriately retarded position Q').

6. The presence of the retardation constant c in the denominator of the expression has no theoretical significance but permits of the development of certain relationships in the so-called Gaussian form. See also the footnote to p. 519.

Since the expression for the velocity of the source in a neighbourhood of the time $t - \frac{r_P}{c}$ is given by $\bar{v} + \dot{\bar{v}}\Delta t + \frac{1}{2!}\ddot{\bar{v}}(\Delta t)^2 + \ldots.$ we have

$$\bar{v}' = \bar{v} + \frac{\dot{\bar{v}}}{c}(r_P - r') + \frac{1}{2!}\frac{\ddot{\bar{v}}}{c^2}(r_P - r')^2 + \ldots.$$

whence

$$\bar{A} = \int_\tau \frac{\rho\left(\bar{v} + \frac{\dot{\bar{v}}}{c}(r_P - r') + \frac{1}{2!}\frac{\ddot{\bar{v}}}{c^2}(r_P - r')^2 + \ldots\right)d\tau}{cr'\left\{1 + \frac{\bar{r}'}{cr'}\cdot\left(\bar{v} + \frac{\dot{\bar{v}}}{c}(r_P - r') + \frac{1}{2!}\frac{\ddot{\bar{v}}}{c^2}(r_P - r')^2 + \ldots\right)\right\}} \qquad (5.10\text{-}11)$$

5.10d Retarded potentials of a point source

If the volume source described above is allowed to shrink uniformly about P while the source strength $\int_\tau \rho d\tau$ is maintained constant and equal to, say, a, then since $r' \to r_P$ as $\bar{r} \to \bar{r}_P$, we have in the limit

$$\phi = \frac{a}{r_P\left(1 + \frac{\bar{v}_P \cdot \bar{r}_P}{cr_P}\right)} \qquad (5.10\text{-}12)$$

$$\bar{A} = \frac{a\,\bar{v}_P}{cr_P\left(1 + \frac{\bar{v}_P \cdot \bar{r}_P}{cr_P}\right)} \qquad (5.10\text{-}13)$$

The general expressions for the potentials of a moving point source are consequently

$$\phi = \frac{a}{\left[r\left(1 + \frac{\bar{v}.\bar{r}}{cr}\right)\right]} \qquad (5.10\text{-}14)$$

$$\bar{A} = \frac{a[\bar{v}]}{c\left[r\left(1 + \frac{\bar{v}.\bar{r}}{cr}\right)\right]} \qquad (5.10\text{-}15)$$

These expressions were originally derived by Liénard and Wiechert, circa 1900, and are known as the Liénard-Wiechert potentials.

It is convenient at this stage to denote the retarded quantities associated with point sources by capital letters. This renders the square brackets redundant and makes lower-case letters available for the representation of instantaneous quantities. In addition, the positive sense of the radius vector will now be taken to be directed away from the source. The above expressions are consequently replaced by

$$\phi = \frac{a}{R\left(1 - \frac{V_R}{c}\right)} \tag{5.10-14(a)}$$

$$\bar{A} = \frac{a\bar{V}}{cR\left(1 - \frac{V_R}{c}\right)} \tag{5.10-15(a)}$$

where V_R is the resolved part of $\bar{V}$ along $\bar{R}$.

EXERCISES

5-33. Solve equation (5.10-1) for x', y', and z' in terms of x, y, z r_p and $\bar{v}$ for uniform motion with $v < c$. Hence derive a single-valued expression for $\bar{r}' - \bar{r}$ and confirm that $r' \to r_p$ as $\bar{r} \to \bar{r}_p$.

Ans: $$\bar{r}' - \bar{r} = \frac{\bar{v}}{c^2 - v^2}\left\{\bar{v}.\bar{r} + cr_p - \left((\bar{v}.\bar{r} + cr_p)^2 + (c^2 - v^2)(r^2 - r_p^2)\right)^{\frac{1}{2}}\right\}$$

5-34. By decomposing a uniformly moving source into a system of elementary prisms lying parallel to the direction of motion, and making use of the 'expanding sphere' construction detailed on p. 409, show that the contribution of each prism to the retarded potential at an exterior point is increased approximately by the factor $\frac{1}{1 - \frac{V_R}{c}}$ as a result of the motion, where V_R refers to some point of the prism. Show further that the expression becomes exact for a point source irrespective of the density distribution before condensation.

5-35. A particular source translates with constant acceleration. Show that although the time reversal of source density, as required by the 'moving sphere' construction, involves a reversal of source velocity at $t = t_o$, the acceleration is unaffected. Show also that the ratio $\Delta\tau'/\Delta\tau$ exhibits a variation across the source because of the transit time of the spherical surface and that this complication vanishes when the dimensions of the source approach zero.

5-36. If the velocity of a source having constant acceleration is unrestricted the source may contribute once, twice, three or four times, or not at all, to the retarded potential at a specified place and time, depending upon the position and velocity of the source at the instant of evaluation. Demonstrate this for the case of a source moving radially with respect to the point of evaluation.

5.11 Space and Time Derivatives of the Liénard-Wiechert Potentials

5.11a Fundamental relationships

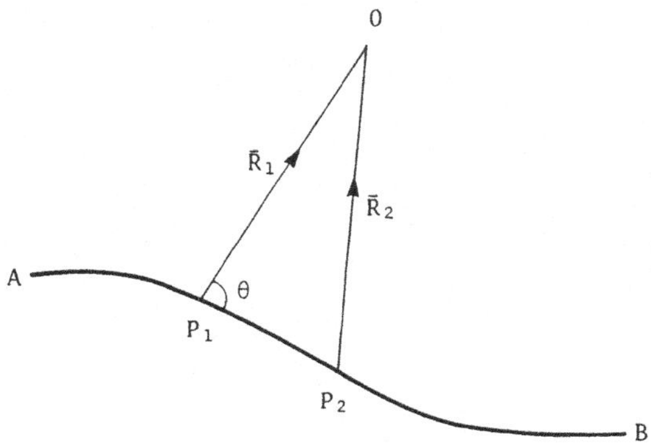

[In this and subsequent figures, arrowheads are placed in the centres of the associated vectors if required to ease congestion.]

Fig. 5.2

Fig. 5.2 depicts the path AB of a point source which occupies the position P_1 at the time t_1' and P_2 at the time t_2'. The source, when at P_1, contributes to the retarded potential at 0 at the time $t_1' + \frac{R_1}{c} = t_1$; when at P_2, at the time $t_2' + \frac{R_2}{c} = t_2$.

Hence

$$t_2 - t_1 = t_2' - t_1' + \frac{R_2 - R_1}{c}$$

If $\overline{V}_1$ is the velocity of the source when at P_1 then

$$R_2 - R_1 \approx - V_1 \, (t_2'-t_1') \cos\theta = - V_{R_1} \, (t_2'-t_1')$$

where V_{R_1} is the radial component of $\overline{V}$ at P_1, hence

$$t_2 - t_1 \approx (t_2'-t_1') \left(1 - \frac{V_{R_1}}{c}\right)$$

If we restrict our considerations to the case $V < c$ then the factor $1 - \frac{V_R}{c}$ is always positive, so that later contributions to the retarded potential at 0 are associated with later positions of the source in the path AB and a 1:1 correspondence obtains between t, t' and a point of the path.

We now take t to be the independent variable and write

$$t_2' - t_1' \approx \frac{t_2 - t_1}{1 - \frac{V_{R_1}}{c}}$$

whence in general,

$$\frac{dt'}{dt} = \frac{1}{1 - \frac{V_R}{c}} = \frac{1}{\alpha} \qquad (5.11\text{-}1)$$

where

$$\alpha = 1 - \frac{V_R}{c} = 1 - \frac{\overline{V}.\overline{R}}{cR} \qquad (5.11\text{-}2)$$

It follows that

$$\frac{dR}{dt} = \frac{dR}{dt'}\frac{1}{\alpha} = -\frac{V_R}{\alpha} = -\frac{\overline{V}.\overline{R}}{\alpha R} \qquad (5.11\text{-}3)$$

and

$$\frac{d\overline{R}}{dt} = \frac{d\overline{R}}{dt'}\frac{1}{\alpha} = -\frac{\overline{V}}{\alpha} \qquad (5.11\text{-}4)$$

Similarly

$$\frac{d\overline{V}}{dt} = \frac{d\overline{V}}{dt'}\frac{1}{\alpha} = \frac{\dot{\overline{V}}}{\alpha} \qquad (5.11\text{-}5)$$

where $\dot{\overline{V}}$ is the acceleration of the source in its retarded position.

Further

$$\frac{d\alpha}{dt} = -\frac{1}{c}\frac{d}{dt}\left(\frac{\overline{V}.\overline{R}}{R}\right)$$

whence we find from equations (5.11-3) to (5.11-5) that

$$\frac{d\alpha}{dt} = -\frac{1}{c}\left\{\frac{\dot{\overline{V}}.\overline{R}}{\alpha R} - \frac{V^2}{\alpha R} + \frac{(\overline{V}.\overline{R})^2}{\alpha R^3}\right\} \tag{5.11-6}$$

We have also

$$\frac{d^2R}{dt^2} = -\frac{d}{dt}\left(\frac{\overline{V}.\overline{R}}{\alpha R}\right) = \frac{\overline{V}.\overline{R}}{\alpha^2 R}\frac{d\alpha}{dt} - \frac{c}{\alpha}\frac{d}{dt}\left(\frac{\overline{V}.\overline{R}}{cR}\right)$$

$$= \frac{c}{\alpha^2}\frac{d\alpha}{dt}$$

or

$$\frac{d^2R}{dt^2} = -\frac{\dot{\overline{V}}.\overline{R}}{\alpha^3 R} + \frac{V^2}{\alpha^3 R} - \frac{(\overline{V}.\overline{R})^2}{\alpha^3 R^3} \tag{5.11-7}$$

Finally

$$\frac{d^2\overline{R}}{dt^2} = -\frac{d}{dt}\left(\frac{\overline{V}}{\alpha}\right) = \frac{\overline{V}}{\alpha^2}\frac{d\alpha}{dt} - \frac{1}{\alpha}\frac{d\overline{V}}{dt}$$

or

$$\frac{d^2\overline{R}}{dt^2} = -\frac{\overline{V}}{c\alpha^2}\left\{\frac{\dot{\overline{V}}.\overline{R}}{\alpha R} - \frac{V^2}{\alpha R} + \frac{(\overline{V}.\overline{R})^2}{\alpha R^3}\right\} - \frac{\dot{\overline{V}}}{\alpha^2} \tag{5.11-8}$$

We now proceed to derive expressions for the space derivatives at 0 of R, V, α etc, corresponding to a fixed time of evaluation in a neighbourhood of 0.

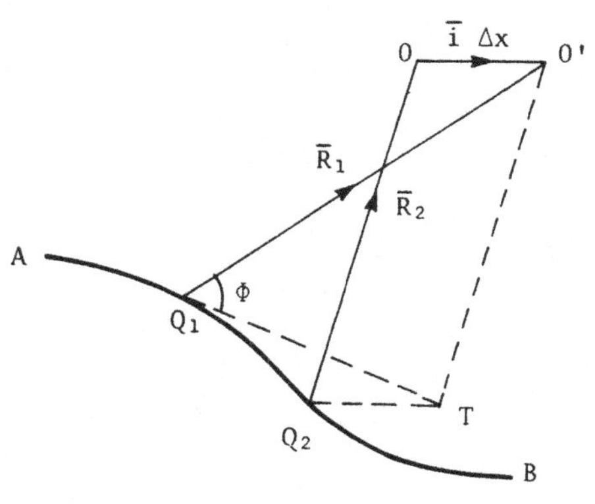

Fig. 5.3

Let Q_2 in Fig. 5.3 be the retarded position of the source corresponding to evaluation of the potential at 0 at the time t_o and let Q_1 be the retarded position for evaluation at 0' at the time t_o. Suppose that the source occupies the positions Q_1 and Q_2 at the times t_1' and t_2' respectively, and that $\overrightarrow{00'} = \bar{i}\Delta x$. Q_2T is equal and parallel to 00'.

Now

$$t_1' + \frac{R_1}{c} = t_2' + \frac{R_2}{c} = t_o$$

hence

$$t_2' - t_1' = \frac{R_1 - R_2}{c}$$

But

$$R_1 - R_2 \approx Q_1T \cos\phi \approx (\bar{V}_1(t_2'-t_1')+\bar{i}\Delta x).\frac{\bar{R}_1}{R_1}$$

where $\bar{V}_1$ is the velocity of the source at Q_1

so that

$$(R_1-R_2)\left\{1 - \frac{\bar{V}_1.\bar{R}_1}{cR_1}\right\} \approx \frac{\bar{i}.\bar{R}_1}{R_1}\Delta x$$

whence, in general

$$\frac{\partial R}{\partial x} = \frac{\bar{i}.\bar{R}}{\alpha R} \tag{5.11-9}$$

Also

$$\bar{R}_1 - \bar{R}_2 = \overrightarrow{Q_1T} \approx \bar{V}_1\frac{(R_1-R_2)}{c} + \bar{i}\Delta x$$

whence, in general,

$$\frac{\partial \bar{R}}{\partial x} = \frac{\bar{V}}{c}\frac{\partial R}{\partial x} + \bar{i}$$

or, from equation (5.11-9),

$$\frac{\partial \bar{R}}{\partial x} = \bar{i} + \frac{\bar{V}\ \bar{i}.\bar{R}}{\alpha c R} \tag{5.11-10}$$

Further,

$$\frac{\partial \bar{V}}{\partial x} = \frac{\partial \bar{V}}{\partial t'}\frac{\partial t'}{\partial x} = \dot{\bar{V}}\ \frac{\partial t'}{\partial R}\frac{\partial R}{\partial x} = -\frac{\dot{\bar{V}}}{c}\frac{\partial R}{\partial x}$$

or

$$\frac{\partial \bar{V}}{\partial x} = -\dot{\bar{V}}\frac{\bar{i}.\bar{R}}{\alpha c R} \tag{5.11-11}$$

Finally,

$$\frac{\partial \alpha}{\partial x} = -\frac{1}{c}\frac{\partial}{\partial x}\left(\frac{\bar{V}.\bar{R}}{R}\right) = -\frac{1}{c}\left(\frac{\partial \bar{V}}{\partial x}.\frac{\bar{R}}{R} + \frac{\bar{V}}{R}.\frac{\partial \bar{R}}{\partial x} - \frac{\bar{V}.\bar{R}}{R^2}\frac{\partial R}{\partial x}\right)$$

whence from equations (5.11-9) to (5.11-11)

$$\frac{\partial \alpha}{\partial x} = \frac{\dot{\bar{V}}.\bar{R}\ \bar{i}.\bar{R}}{\alpha c^2 R^2} - \frac{\bar{i}.\bar{V}}{cR} - \frac{V^2\bar{i}.\bar{R}}{\alpha c^2 R^2} + \frac{\bar{V}.\bar{R}\ \bar{i}.\bar{R}}{\alpha c R^3} \tag{5.11-12}$$

5.11b Derivatives of the potentials

It is now a straight-forward matter to differentiate the Liénard-Wiechert potentials, provided that the results are expressed in terms of retarded quantities.

Taking

$$\phi = \frac{a}{R\left(1 - \frac{V_R}{c}\right)} = \frac{a}{\alpha R}$$

we have

$$\frac{1}{a}\frac{\partial \phi}{\partial x} = -\frac{1}{\alpha^2 R}\frac{\partial \alpha}{\partial x} - \frac{1}{\alpha R^2}\frac{\partial R}{\partial x}$$

whence we obtain

$$\frac{1}{a}\frac{\partial \phi}{\partial x} = \frac{V^2\bar{i}.\bar{R}}{\alpha^3 c^2 R^3} - \frac{\bar{i}.\bar{R}}{\alpha^3 R^3} + \frac{\bar{i}.\bar{V}}{\alpha^2 c R^2} - \frac{\dot{\bar{V}}.\bar{R}\ \bar{i}.\bar{R}}{\alpha^3 c^2 R^3}$$

so that

$$\frac{1}{a}\,\mathrm{grad}\,\phi = \frac{V^2\overline{R}}{\alpha^3c^2R^3} - \frac{\overline{R}}{\alpha^3R^3} + \frac{\overline{V}}{\alpha^2cR^2} - \frac{\dot{\overline{V}}.\overline{R}\,\overline{R}}{\alpha^3c^2R^3} \qquad (5.11\text{-}13)$$

Similarly

$$\frac{\partial\phi}{\partial t} = -\frac{a}{\alpha^2R}\frac{\partial\alpha}{\partial t} - \frac{a}{\alpha R^2}\frac{\partial R}{\partial t}$$

whence

$$\frac{1}{a}\frac{\partial\phi}{\partial t} = \frac{\overline{V}.\overline{R}}{\alpha^3R^3} - \frac{V^2}{\alpha^3cR^2} + \frac{\dot{\overline{V}}.\overline{R}}{\alpha^3cR^2} \qquad (5.11\text{-}14)$$

Again, with

$$\overline{A} = \frac{a\overline{V}}{cR\left(1 - \dfrac{V_R}{c}\right)} = \frac{a\overline{V}}{c\alpha R}$$

we have

$$\mathrm{div}\,\overline{A} = \sum \frac{\partial}{\partial x}\frac{aV_x}{c\alpha R}$$

whence we find that

$$\frac{c}{a}\,\mathrm{div}\,\overline{A} = \frac{V^2}{\alpha^3cR^2} - \frac{\overline{V}.\overline{R}}{\alpha^3R^3} - \frac{\dot{\overline{V}}.\overline{R}}{\alpha^3cR^2} \qquad (5.11\text{-}15)$$

or

$$\mathrm{div}\,\overline{A} = -\frac{1}{c}\frac{\partial\phi}{\partial t} \qquad (5.11\text{-}16)$$

Also

$$\frac{c}{a}\,\mathrm{curl}\,\overline{A} = \frac{c}{a}\sum \overline{i}\left(\frac{\partial A_z}{\partial y} - \frac{\partial A_y}{\partial z}\right) = \sum \overline{i}\left\{\frac{\partial}{\partial y}\left(\frac{V_z}{\alpha R}\right) - \frac{\partial}{\partial z}\left(\frac{V_y}{\alpha R}\right)\right\}$$

which, on expansion and simplification, yields

$$\frac{c}{a}\,\mathrm{curl}\,\overline{A} = \frac{\dot{\overline{V}}\times\overline{R}}{\alpha^2cR^2} + (\overline{V}\times\overline{R})\left\{\frac{\dot{\overline{V}}.\overline{R} + c^2 - V^2}{\alpha^3c^2R^3}\right\} \qquad (5.11\text{-}17)$$

Finally

$$\frac{c}{a}\frac{\partial\bar{A}}{\partial t} = -\frac{\bar{V}}{\alpha^2 R}\frac{\partial\alpha}{\partial t} + \frac{1}{\alpha R}\frac{\partial\bar{V}}{\partial t} - \frac{\bar{V}}{\alpha R^2}\frac{\partial R}{\partial t}$$

whence

$$\frac{c}{a}\frac{\partial\bar{A}}{\partial t} = \frac{\bar{V}\,\bar{V}.\bar{R}}{\alpha^3 R^3} - \frac{V^2\bar{V}}{\alpha^3 c R^2} + \frac{\bar{V}\,\dot{\bar{V}}.\bar{R}}{\alpha^3 c R^2} + \frac{\dot{\bar{V}}}{\alpha^2 R} \qquad (5.11\text{-}18)$$

5.11c The $\bar{E}$ and $\bar{B}$ fields of a moving point source

Consider the vector point functions defined by

$$\bar{E} = -\,\text{grad}\,\phi - \frac{1}{c}\frac{\partial\bar{A}}{\partial t} \qquad (5.11\text{-}19)$$

$$\bar{B} = \text{curl}\,\bar{A} \qquad (5.11\text{-}20)$$

From equations (5.11-13) and (5.11-18) we find that

$$\frac{\bar{E}}{a} = -\frac{\dot{\bar{V}}}{\alpha^2 c^2 R} + \left(\frac{\bar{R}}{R} - \frac{\bar{V}}{c}\right)\left(\frac{\dot{\bar{V}}.\bar{R} + c^2 - V^2}{\alpha^3 c^2 R^2}\right) \qquad (5.11\text{-}21)$$

and from equation (5.11-17)

$$\frac{\bar{B}}{a} = \frac{\dot{\bar{V}} \times \bar{R}}{\alpha^2 c^2 R^2} + (\bar{V}\times\bar{R})\left(\frac{\dot{\bar{V}}.\bar{R} + c^2 - V^2}{\alpha^3 c^3 R^3}\right) \qquad (5.11\text{-}22)$$

It is evident that

$$\bar{B} = \frac{\bar{R}}{R} \times \bar{E} \qquad (5.11\text{-}23)$$

so that $\bar{B}$ is perpendicular to $\bar{E}$ and to the retarded radius vector.

Important relationships subsist between the $\bar{E}$ and $\bar{B}$ fields deriving from one or more point sources.

It follows directly from their definitions and the principle of superposition that

$$\text{curl}\,\bar{E} = -\frac{1}{c}\frac{\partial\bar{B}}{\partial t} \qquad (5.11\text{-}24)$$

$$\text{div}\,\bar{B} = 0 \qquad (5.11\text{-}25)$$

It may also be shown[7] by differentiation of equations (5.11-21) and (5.11-22) that

$$\text{div } \overline{E} = 0 \qquad (5.11\text{-}26)$$

$$\text{curl } \overline{B} = \frac{1}{c}\frac{\partial \overline{E}}{\partial t} \qquad (5.11\text{-}27)$$

By combining equations (5.11-16) and (5.11-24/27) we easily find that

$$\text{dal } \phi = \text{dal } \overline{A} = \text{dal } \overline{E} = \text{dal } \overline{B} = 0 \qquad (5.11\text{-}28)$$

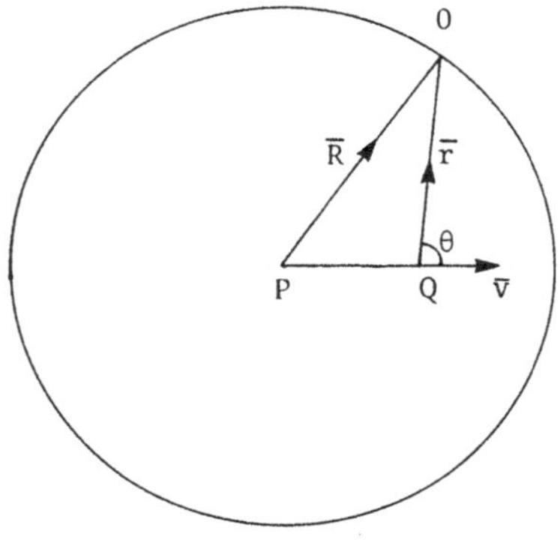

Fig. 5.4

In Fig. 5.4 P represents the retarded position of a point source corresponding to evaluation at 0 at some particular instant. Q is the actual position of the source at that instant. It is apparent that Q must lie within the spherical surface centred upon P and passing through 0 since $PQ \leq \frac{R}{c} v_{max} < R$ (the maximum source velocity during the retardation period being less than c.) If, for a particular position of P, we reduce R continuously (and alter the instant of evaluation at 0 accordingly) Q approaches P continuously. Since the various expressions derived above are clearly valid outside the spherical surface they must likewise subsist outside an arbitrarily small sphere centred upon Q ie the expressions hold everywhere beyond the instantaneous position of the source.

---

7. See Ex.5-51. and 5-52., p. 487.

At sufficiently great distance from the retarded source, ie in the so-called 'far zone', only those components of $\overline{E}$ and $\overline{B}$ which are associated with acceleration survive; these fall off as the first power of distance while the remainder fall off as the square. In this case it is easily shown from equation (5.11-21) that

$$\frac{\overline{E}}{a}\cdot\frac{\overline{R}}{R} = \frac{(1-V^2/c^2)}{a^2R^2} \qquad (5.11\text{-}29)$$

whence we see that $\frac{E_R}{E}\to 0$ as $R\to\infty$, ie $\overline{E}$ becomes transverse to the retarded radius. It then follows from equation (5.11-23) that $\overline{E}$, $\overline{B}$ and $\overline{R}$ are mutually perpendicular in the far zone and that $E = B$.

When the motion of the source is uniform over any period which exceeds the retardation time

$$\dot{\overline{V}} = \overline{0} \quad \text{and} \quad \overline{R} - \frac{\overline{V}}{c}R = \overline{r}$$

where $\overline{r}$ is the radius vector <u>directed from the instantaneous position of the source to the point of evaluation</u>.

Then

$$\frac{\overline{E}}{a} = \overline{r}\,\frac{(1-v^2/c^2)}{\alpha^3R^3} \qquad (5.11\text{-}30)$$

and, since $\overline{V}\times\overline{R} = \overline{V}\times\overline{r}$,

$$\frac{\overline{B}}{a} = \frac{(\overline{v}\times\overline{r})(1-v^2/c^2)}{\alpha^3cR^3} \qquad (5.11\text{-}31)$$

where $\overline{v}$ now replaces $\overline{V}$.

These expressions may be transformed (Ex.5-40., p. 484) into

$$\frac{\overline{E}}{a} = \frac{\overline{r}(1-v^2/c^2)}{r^3\left(1-\frac{v^2}{c^2}\sin^2\theta\right)^{3/2}} \qquad (5.11\text{-}32)$$

$$\frac{\overline{B}}{a} = \frac{(\overline{v}\times\overline{r})(1-v^2/c^2)}{cr^3\left(1-\frac{v^2}{c^2}\sin^2\theta\right)^{3/2}} \qquad (5.11\text{-}33)$$

where θ is the angle between $\overline{v}$ and $\overline{r}$.

We have also

$$\bar{B} = \frac{\bar{v}}{c} \times \bar{E} \tag{5.11-34}$$

It is seen that the $\bar{E}$ field is directed radially away from the instantaneous source position and is of an inverse-square nature. However, it is not spherically symmetrical and the asymmetry increases with increase of v.

EXERCISES

5-37. Verify equations (5.11-17) and (5.11-18).

5-38. Show that the far $\bar{E}$ field can be expressed in the form

$$\frac{\bar{E}}{a} = \frac{\bar{R}}{\alpha^3 c^2 R^2} \times \left\{\left(\frac{\bar{R}}{R} - \frac{\bar{V}}{c}\right) \times \dot{\bar{V}}\right\}$$

5-39. In terms of the current notation R.P. Feynman has given the following elegant expression for $\bar{E}$.

$$\frac{\bar{E}}{a} = \frac{\bar{R}}{R^3} + \frac{R}{c}\frac{d}{dt}\left(\frac{\bar{R}}{R^3}\right) + \frac{1}{c^2}\frac{d^2}{dt^2}\left(\frac{\bar{R}}{R}\right)$$

Show by expansion that this is identical with equation (5.11-21), but note that the last term expands into nine terms of which only three fall off inversely as retarded distance; all other terms of the expression fall off as the square.

5-40. If Fig. 5.4 is restricted to a source moving with uniform velocity $\bar{v}$, observe that

$$OP = R = Rv_R/c + r \cos \angle POQ$$

and proceed to prove equation (5.11-32) by substitution in (5.11-30).

5-41. Evaluate the surface integral of equation (5.11-32) over a spherical surface centred upon the instantaneous position of the source, and extrapolate the result via equation (5.11-26) to any simple surface which encloses the source.

Hint: Change the variable θ to x = cos θ and note that

$$\int \frac{dx}{(b^2+x^2)^{3/2}} = \frac{x}{b^2(b^2+x^2)^{\frac{1}{2}}}$$

Ans: $4\pi a$

5-42. In Fig. 5.4 P represents the common retarded position of the source for evaluation at each element of the spherical surface shown, so that $\oint \overline{E}.d\overline{S}$ may be determined for that surface by integration of (5.11-21) with R, $\overline{V}$ and $\dot{\overline{V}}$ constant. Show first that for $\dot{\overline{V}} = \overline{0}$, $\oint \overline{E}.d\overline{S} = 4\pi a$ and hence confirm the result of the previous exercise.

Then show that for $\dot{\overline{V}} \neq \overline{0}$ the additional contribution to the surface integral is zero - it is not necessary to evaluate the integral for this purpose - and complete the generalisation by means of equation (5.11-26).

5-43. Make use of equations (2.6-6), (2.6-7), (5.11-32) and (5.11-33) to confirm that in the case of a uniformly moving source

$$\text{div } \overline{E} = 0$$

and

$$(\text{curl } \overline{B})_r = \frac{1}{c}\left(\frac{\partial \overline{E}}{\partial t}\right)_r = \frac{av \cos\theta \left(1 - \frac{v^2}{c^2}\right)\left(2 + \frac{v^2}{c^2}\sin^2\theta\right)}{cr^3\left(1 - \frac{v^2}{c^2}\sin^2\theta\right)^{5/2}}$$

$$(\text{curl } \overline{B})_\theta = \frac{1}{c}\left(\frac{\partial \overline{E}}{\partial t}\right)_\theta = \frac{av \sin\theta \left(1 - \frac{v^2}{c^2}\right)}{cr^3\left(1 - \frac{v^2}{c^2}\sin^2\theta\right)^{3/2}}$$

5-44. A regular surface S encloses a point source which moves in any manner ($v<c$).

Show that

$$\oint_S \overline{B}.d\overline{S} = 0$$

5-45. A point source has motion of translation with a velocity $\overline{V}$ and acceleration $\dot{\overline{V}}$ at the time t'. The far $\overline{E}$ field is evaluated at 0 at the time $t' + \frac{R}{c}$ where $\overline{R}$ is directed from the source to 0 and makes an angle θ with $\overline{V}$. Show that for a given acceleration and for $\frac{V}{c} \ll 1$ the magnitude of $\overline{E}$ is proportional to $\sin\theta$ and independent of V, but that in the limit as $\frac{V}{c} \to 1$ the magnitude is a maximum when $\theta = \sqrt{\frac{2}{5}}\left(1 - \frac{V}{c}\right)^{\frac{1}{2}}$ radians and is then proportional to $\left(1 - \frac{V}{c}\right)^{-5/2}$.

5-46. A point source of strength a moves in a circle of radius r with angular velocity ω. At the time t_o the radius vector drawn from the centre of the circle to the source is normal to the vector $\overline{R}$ drawn from the source to a distant point 0 in the plane of the circle and defines a direction $\hat{s}$ from which the swept angle θ is measured, the positive sense of $\overline{\omega}$ being defined by $\overline{r} \times \overline{R}$.

Show that when 0 recedes to infinity

$$\overline{E}_0(t) = -\hat{s}\frac{aV^2}{c^2rd}\frac{\left(\frac{V}{c} - \cos\theta\right)}{\left(1 - \frac{V}{c}\cos\theta\right)^3}$$

where $t = t_o + \frac{\theta}{\omega} + \frac{d}{c} - \frac{r\sin\theta}{c}$, d being the distance of 0 from the centre of the circle.

5-47. For the conditions stated in Ex.5-46. plot $E_0(t)$ against θ over the range $\theta = 0$ to $\theta = 2\pi$ for $\frac{V}{c} = 0.1$, 0.5 and 0.9. In each case superimpose the plot of $E_0(t)$ against $\theta\left(1 - \frac{V}{c}\frac{\sin\theta}{\theta}\right)$ without change of scale. The latter curves represent $E_0(t)$ as a function of t with $\theta = 0$ and $\theta = 2\pi$ corresponding respectively to $t = t_o + \frac{d}{c}$ and $t_o + \frac{d}{c} + \frac{2\pi}{\omega}$. $E_0(t)$ is clearly periodic in time with a period equal to that of the source motion.

Observe that the functional forms of both $E_0(t)$ and t conspire to produce a 'pulse'-type waveform for the larger values of $\frac{V}{c}$, comprising a short-duration crest and long-duration shallow trough having vastly accentuated characteristics as V approaches c.

5-48. We may define the pulse length of the waveform discussed in the previous exercise as the time interval at 0 between zero values of $\overline{E}$ taken symmetrically about the crest. Show that this time interval approaches $\frac{8\sqrt{2}}{3\omega}\left(1 - \frac{V}{c}\right)^{3/2}$ as $\frac{V}{c} \to 1$.

Show further that the time average of $\overline{E}_0$ for all $\frac{V}{c} < 1$ is given by

$$-\hat{s}\frac{1}{2\pi}\frac{aV^2}{c^2rd}\int_0^{2\pi}\frac{\left(\frac{V}{c} - \cos\theta\right)}{\left(1 - \frac{V}{c}\cos\theta\right)^2}d\theta = \hat{s}\frac{1}{2\pi}\frac{aV^2}{c^2rd}\left[\frac{\sin\theta}{\left(1 - \frac{V}{c}\cos\theta\right)}\right]_0^{2\pi} = 0$$

5-49. In Ex.5-46. to 5-48. attention has been confined to the inverse distance component of $\overline{E}$. Now derive limiting expressions for the inverse square components of $\overline{E}$ for a single source as $d\to\infty$, and show that the time average of the transverse field is zero while that of the radial field is equal to a/d^2.

5-50. When the single source of the above exercises is replaced by a system of equal sources uniformly spaced around the circle and having a common angular velocity, the inverse distance and inverse square transverse components of $\overline{E}_0$ at great distance reduce to zero in the limit as the number of sources approaches infinity while the total source strength remains fixed. Prove this by demonstrating that the sum of the contributions from all sources at a given instant is proportional to the time integral of $\overline{E}_0$, as derived above for a single source, over the period of a cycle.

5-51. Prove equation (5.11-26) by differentiation of (5.11-21), in accordance with the identities developed in Sec. 5.11a.

(NB This procedure involves a considerable amount of 'algebraic crank-turning'.)

5-52. Prove equation (5.11-27) in the following way:

Show first that

$$\text{dal}\ \overline{A} = \text{dal}\ \phi\ \overline{V} = \overline{V}\ \text{dal}\ \phi + \phi\ \text{dal}\ \overline{V} + 2\sum \frac{\partial\phi}{\partial x}\frac{\partial\overline{V}}{\partial x} - \frac{2}{c^2}\frac{\partial\phi}{\partial t}\frac{\partial\overline{V}}{\partial t}$$

and combine equations (5.11-16), (5.11-19) and (5.11-26) to show that dal $\phi = 0$.

Expand the remaining terms of dal $\overline{A}$ in accordance with Sec. 5.11a and show that their sum is zero.

Then make use of equations (5.11-16), (5.11-19) and (5.11-20) to derive (5.11-27).

5.12 Approximations for the Liénard-Wiechert Potentials and Their Derivatives in Terms of Unretarded Quantities[8]

5.12a Transformation of the potentials

The point P_1 of Fig. 5.5 represents the retarded position of a source corresponding to some particular time of evaluation at 0, and P_2 is its instantaneous or unretarded position ie the position actually occupied by the source at the moment of evaluation.

8. Based on O'Rahilly's exposition of Ritz.

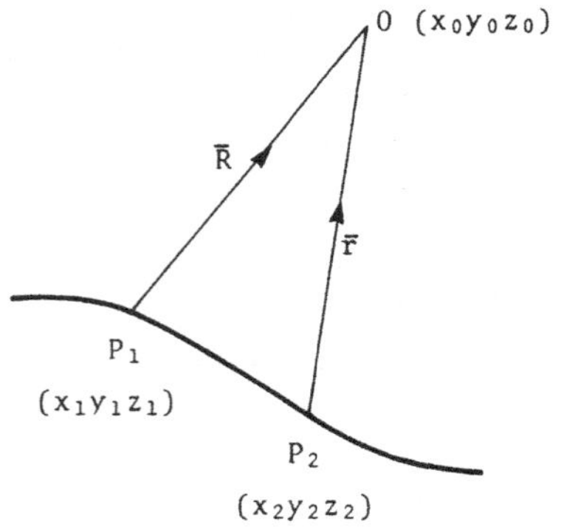

Fig. 5.5

The coordinates of O are taken to be (x_o, y_o, z_o) while those of P_1 and P_2 are (x_1, y_1, z_1) and (x_2, y_2, z_2) respectively.

Then

$$R^2 = \sum (x_o - x_1)^2 \quad \text{and} \quad r^2 = \sum (x_o - x_2)^2$$

where R and r are the retarded and instantaneous distances of the source from O.

Now

$$x_1 = x_2 + v_x\left(-\frac{R}{c}\right) + \frac{1}{2} f_x\left(\frac{R^2}{c^2}\right) \dots$$

where v_x and f_x are the x components of velocity and acceleration at P_2, since $\frac{R}{c}$ is the delay time between P_1 and P_2,

hence

$$x_o - x_1 = x_o - x_2 + \frac{R}{c} v_x - \frac{1}{2} \frac{R^2}{c^2} f_x \dots\dots$$

and

$$R^2 = \sum \left(x_o - x_2 + \frac{R}{c} v_x - \frac{1}{2} \frac{R^2}{c^2} f_x \;......\right)^2$$

$$= \sum \left\{(x_o-x_2)^2 + \frac{R^2}{c^2} v_x^2 + \frac{1}{4} \frac{R^4}{c^4} f_x{}^2 \;...+2(x_o-x_2) \frac{R}{c} v_x - (x_o-x_2) \frac{R^2}{c^2} f_x \;...\right\}$$

or

$$R^2 = r^2 + \frac{R^2}{c^2} v^2 \;..... + 2r \frac{R}{c} v_r - r \frac{R^2}{c^2} f_r \;...... \qquad (5.12\text{-}1)$$

where v_r and f_r are the radial components of instantaneous velocity and acceleration directed towards 0.[9]

We now suppose that the motion of the source is such that all terms in $\frac{1}{c^3}$, $\frac{1}{c^4}$ ... may be ignored. Then

$$R^2 \left(1 - \frac{v^2}{c^2} + \frac{r}{c^2} f_r\right) - 2 \frac{rv_r}{c} R - r^2 = 0$$

On solving the quadratic equation for R and expanding in binomial form we obtain

$$R = r \left(1 + \frac{v_r}{c} + \frac{v^2}{2c^2} + \frac{v_r^2}{2c^2} - \frac{rf_r}{2c^2} \;....\right) \qquad (5.12\text{-}2)$$

But if

$$\frac{R}{r} = 1 + \gamma \quad \text{then} \quad \frac{r}{R} = (1+\gamma)^{-1} = 1 - \gamma + \gamma^2 \;....$$

hence

$$r = R \left(1 - \frac{v_r}{c} - \frac{v^2}{2c^2} + \frac{v_r^2}{2c^2} + \frac{rf_r}{2c^2} \;....\right) \qquad (5.12\text{-}3)$$

We now proceed to express V_R in terms of instantaneous quantities.

Since

$$V_R = \sum \frac{(x_o - x_1)}{R} V_x$$

9. In this Section, as in the latter part of Sec. 5.11, we take $\overline{r}$ to be directed towards 0.

and

$$V_x = v_x + f_x\left(-\frac{R}{c}\right)\ldots$$

$$V_R = \frac{1}{R}\sum\left(x_o - x_2 + \frac{R}{c}v_x - \frac{1}{2}\frac{R^2}{c^2}f_x \ldots\ldots\right)\left(v_x - \frac{R}{c}f_x \ldots\right)$$

whence

$$\frac{V_R}{c} = \frac{r}{R}\frac{v_r}{c} + \frac{v^2}{c^2} - \frac{r}{R}\frac{Rf_r}{c^2}\ldots\ldots$$

and on substituting for $\frac{r}{R}$ in accordance with equation (5.12-3)

$$\frac{V_R}{c} = \frac{v_r}{c} + \frac{v^2}{c^2} - \frac{{v_r}^2}{c^2} - \frac{rf_r}{c^2}\ldots\ldots$$

Then

$$\left(1 - \frac{V_R}{c}\right)^{-1} = 1 + \frac{V_R}{c} + \frac{{V_R}^2}{c^2}\ldots\ldots$$

or

$$\left(1 - \frac{V_R}{c}\right)^{-1} = 1 + \frac{v_r}{c} + \frac{v^2}{c^2} - \frac{rf_r}{c^2}\ldots\ldots \qquad (5.12\text{-}4)$$

On combining equations (5.12-3) and (5.12-4) we obtain

$$\frac{a}{R\left(1 - \frac{V_R}{c}\right)} = \frac{a}{r}\left(1 - \frac{v_R}{c} - \frac{v^2}{2c^2} + \frac{{v_r}^2}{2c^2} + \frac{rf_r}{2c^2}\ldots\right)\left(1 + \frac{v_r}{c} + \frac{v^2}{c^2} - \frac{rf_r}{c^2}\ldots\ldots\right)$$

or

$$\phi = \frac{a}{r}\left(1 + \frac{v^2}{2c^2} - \frac{{v_r}^2}{2c^2} - \frac{rf_r}{2c^2}\ldots\ldots\right) \qquad (5.12\text{-}5)$$

where ϕ is the Liénard-Wiechert scalar potential.

The x component of the vector potential is

$$\frac{a\,V_x}{cR\left(1 - \frac{V_R}{c}\right)}$$

Now

$$V_x = v_x + f_x\left(-\frac{R}{c}\right) + \frac{1}{2}\, g_x\, \frac{R^2}{c^2} \ \ldots .$$

where

$$\bar{g} = \frac{d\bar{f}}{dt}$$

hence

$$A_x = \frac{a}{cr}\left(1 + \frac{v^2}{2c^2} - \frac{v_r^{\,2}}{2c^2} - \frac{rf_r}{2c^2} \ \ldots .\right)\left(v_x - \frac{R}{c}\, f_x + \frac{1}{2}\frac{R^2}{c^2}\, g_x \ \ldots .\right)$$

whence from equation (5.12-2)

$$A_x = \frac{a}{r}\left(\frac{v_x}{c} - \frac{rf_x}{c^2} + \frac{v_x v^2}{2c^3} - \frac{v_x v_r^{\,2}}{2c^3} - \frac{rv_x f_r}{2c^3} - \frac{rv_r f_x}{c^3} + \frac{r^2 g_x}{2c^3} \ \ldots .\right) \qquad (5.12\text{-}6)$$

or

$$\bar{A} = \frac{a}{r}\left(\frac{\bar{v}}{c} - \frac{r\bar{f}}{c^2} + \frac{\bar{v}v^2}{2c^3} - \frac{\bar{v}v_r^{\,2}}{2c^3} - \frac{r\bar{v}f_r}{2c^3} - \frac{rv_r\bar{f}}{c^3} + \frac{r^2\bar{g}}{2c^3} \ \ldots .\right) \qquad (5.12\text{-}7)$$

5.12b <u>Derivatives of the transformed potentials</u>

From equation (5.12-5) we have

$$\frac{1}{a}\frac{\partial\phi}{\partial x_o} = \frac{1}{r}\frac{\partial}{\partial x_o}\left(1 + \frac{v^2}{2c^2} - \frac{v_r^{\,2}}{2c^2} - \frac{rf_r}{2c^2} \ \ldots\right) + \left(1 + \frac{v^2}{2c^2} - \frac{v_r^{\,2}}{2c^2} - \frac{rf_r}{2c^2} \ \ldots\right)\frac{\partial}{\partial x_o}\left(\frac{1}{r}\right)$$

Now $\dfrac{\partial v}{\partial x_o} = 0$ since v is independent of the position of 0.

Also

$$\frac{\partial v_r}{\partial x_o} = \frac{\partial}{\partial x_o}\left(\frac{\bar{v}.\bar{r}}{r}\right) = \frac{\bar{v}}{r}.\frac{\partial\bar{r}}{\partial x_o} - \frac{\bar{v}.\bar{r}}{r^2}\frac{\partial\bar{r}}{\partial x_o}$$

or

$$\frac{\partial v_r}{\partial x_o} = \frac{v_x}{r} - \frac{v_r}{r}\cos\,(rx)$$

Similarly

$$\frac{\partial f_r}{\partial x_o} = \frac{f_x}{r} - \frac{f_r}{r} \cos(rx)$$

Substitution then yields

$$\frac{1}{a}\frac{\partial \phi}{\partial x_o} = \frac{-\cos(rx)}{r^2}\left(1 + \frac{v^2}{2c^2} - \frac{3}{2}\frac{v_r^{\,2}}{c^2}\right) - \frac{v_x v_r}{c^2 r^2} - \frac{1}{2c^2 r}(f_x - f_r \cos(rx)) \ldots$$

whence

$$\frac{1}{a}\,\text{grad}\,\phi = -\frac{\bar{r}}{r^3}\left(1 + \frac{v^2}{2c^2} - \frac{3}{2}\frac{v_r^{\,2}}{c^2}\right) - \frac{\bar{v}v_r}{c^2 r^2} - \frac{1}{2c^2 r}\left(\bar{f} - f_r\frac{\bar{r}}{r}\right) \ldots \qquad (5.12\text{-}8)$$

In like fashion we obtain

$$\frac{1}{ca}\frac{\partial \bar{A}}{\partial t} = \frac{v_r \bar{v}}{c^2 r^2} + \frac{\bar{f}}{c^2 r} \quad \text{to the order } \frac{1}{c^2} \qquad (5.12\text{-}9)$$

hence from equation (5.11-19)

$$\frac{\bar{E}}{a} = \frac{\bar{r}}{r^3}\left(1 + \frac{v^2}{2c^2} - \frac{3v_r^{\,2}}{2c^2} - \frac{rf_r}{2c^2}\right) - \frac{\bar{f}}{2c^2 r} \ldots\ldots \qquad (5.12\text{-}10)$$

On differentiating equation (5.12-7) we find that to the order $\frac{1}{c^3}$

$$\frac{1}{a}\frac{\partial A_z}{\partial y} = \frac{-\cos(ry)}{r^2}\left\{\frac{v_z}{c} + \frac{v^2 v_z}{2c^3} - \frac{3v_r^{\,2}v_z}{2c^3} - \frac{r^2 g_z}{2c^3} - \frac{rv_z f_r}{2c^3} - \frac{rv_r f_z}{c^3}\right\}$$
$$- \frac{1}{r}\left\{\frac{v_r v_y v_z}{c^3 r} + \frac{v_z f_y}{2c^3} + \frac{v_y f_z}{c^3}\right\}$$

whence we may proceed to obtain

$$\frac{\bar{B}}{a} = \frac{1}{cr^3}(\bar{v}\times\bar{r})\left(1 + \frac{v^2}{2c^2} - \frac{3v_r^{\,2}}{2c^2} - \frac{rf_r}{2c^2}\right) + \frac{1}{c^3 r^2}v_r(\bar{r}\times\bar{f}) + \frac{1}{2c^3 r}(\bar{f}\times\bar{v} + \bar{r}\times\bar{g})$$
$$(5.12\text{-}11)$$

It should be noted that the approximation

$$\frac{\bar{B}}{a} = \frac{1}{cr^3}(\bar{v}\times\bar{r}) \qquad (5.12\text{-}11(a))$$

is correct to the order $\frac{1}{c^2}$

EXERCISES

5-53. Show that

$$\frac{\partial r}{\partial t} = -v_r$$

$$\frac{\partial}{\partial t}(v_r) = f_r - \frac{v^2}{r} + \frac{v_r^2}{r} \qquad \text{(cf Ex.1-18., p. 22)}$$

$$\frac{\partial}{\partial t}(f_r) = g_r - \frac{\bar{v}.\bar{f}}{r} + \frac{v_r f_r}{r}$$

and confirm the expansion

$$\frac{1}{a}\frac{\partial \phi}{\partial t} = \frac{v_r}{r^2}\left\{1 + \frac{3v^2}{2c^2} - \frac{3v_r^2}{2c^2} - \frac{3rf_r}{2c^2}\right\} + \frac{3\bar{v}.\bar{f}}{2c^2 r} - \frac{g_r}{2c^2} \cdots$$

Then show by differentiating equation (5.12-6) that, to the degree of approximation involved,

$$\text{div}\,\bar{A} = -\frac{1}{c}\frac{\partial \phi}{\partial t}$$

5-54. Derive equation (5.12-11) from (5.12-10) by expanding $\frac{\bar{R}}{R}$ in terms of $\frac{\bar{r}}{r}$ and employing the general relationship $\bar{B} = \frac{\bar{R}}{R} \times \bar{E}$.

[The discrepancy in the term in g is due to the omission of the corresponding term in equation (5.12-10)]

Ans: $$\frac{\bar{R}}{R} = \frac{\bar{r}}{r}\left\{1 - \frac{v_r}{c} - \frac{v^2}{2c^2} + \frac{v_r^2}{2c^2} + \frac{rf_r}{2c^2}\right\} + \frac{\bar{v}}{c} - \frac{r\bar{f}}{2c^2}\left\{1 + \frac{v_r}{c}\right\} + \frac{1}{6}\frac{r^2\bar{g}}{c^3} \cdots\cdot$$

5.13 The Retarded Potentials of an Oscillating Point Doublet with Time-Dependent Orientation

Consider point sources of strength ±a and spacing 2d sin ωt centred upon a fixed point S (Fig. 5.6). The sources coincide with P' and Q' at the time $t_o - \frac{R}{c}$, where R is the distance of S from the point of evaluation, O; the line of the sources, at this instant, is rotating about S with angular velocity $\bar{\Omega}$.

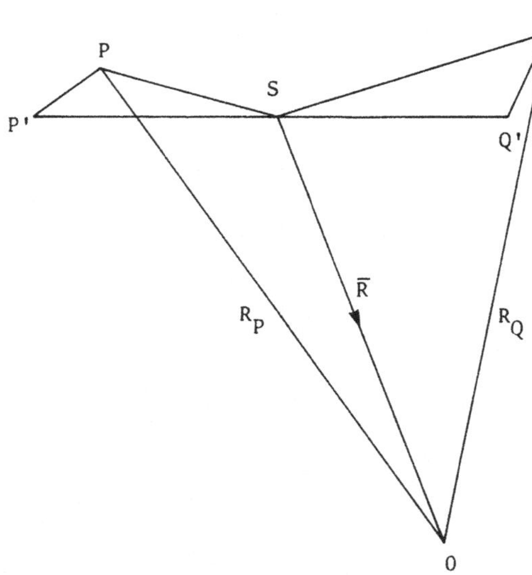

Fig. 5.6

P and Q will represent the retarded positions of -a and +a, respectively, for evaluation at O at the time t_o if the sources occupy these positions at the times $t_o - \frac{R_P}{c}$ and $t_o - \frac{R_Q}{c}$, where $R_P = OP$ and $R_Q = OQ$.

Now

$$\overrightarrow{SQ} = \overrightarrow{SQ}' + \overrightarrow{Q'Q}$$

$$= \hat{s}\, d \sin\omega\left(t_o - \frac{R}{c}\right) + \hat{s}\,\Delta t\, \omega d \cos\omega\left(t_o - \frac{R}{c}\right) + (\bar{\Omega}\times\hat{s})\Delta t\, d \sin\omega\left(t_o - \frac{R}{c}\right) + \ldots \tag{5.13-1}$$

where $\hat{s}$ is a unit vector directed from P' to Q' and

$$\Delta t = t_o - \frac{R_Q}{c} - \left(t_o - \frac{R}{c}\right) = \frac{R - R_Q}{c}$$

The remaining terms of equation (5.13-1) involve higher powers of $(R-R_Q)/c$. Since $(R-R_Q)$ is of the same order of magnitude as d, a restriction of the working to the first power of d reduces (5.13-1) to

$$\overrightarrow{SQ} = \hat{s}\, d \sin\omega\left(t_o - \frac{R}{c}\right) \tag{5.13-1(a)}$$

and this, in conjunction with

$$R_Q{}^2 = R^2 + SQ^2 - 2\bar{R}.\vec{SQ}$$

yields

$$\frac{1}{R_Q} = \frac{1}{R}\left\{1 + \frac{\bar{R}.\hat{\bar{s}}}{R^2} d \sin \omega\left(t_o - \frac{R}{c}\right)\right\} \tag{5.13-2}$$

The radial component of the velocity of +a at Q is given to the same order by

$$\frac{\bar{V}_Q.\bar{R}_Q}{R_Q} = \frac{\bar{R}}{R}.\hat{\bar{s}}\omega d \cos \omega\left(t_o - \frac{R}{c}\right) + \frac{\bar{R}}{R}.(\bar{\Omega}\times\hat{\bar{s}}) d \sin \omega\left(t_o - \frac{R}{c}\right) \tag{5.13-3}$$

which, in conjunction with equation (5.13-2), leads to

$$\left\{R_Q\left(1 - \frac{\bar{V}_Q.\bar{R}_Q}{cR_Q}\right)\right\}^{-1}$$

$$= \frac{1}{R}\left\{1 + \frac{\bar{R}.\hat{\bar{s}}}{R^2} d \sin \omega\left(t_o - \frac{R}{c}\right) + \frac{\bar{R}.\hat{\bar{s}}}{cR} \omega d \cos \omega\left(t_o - \frac{R}{c}\right) + \frac{\bar{R}.(\bar{\Omega}\times\hat{\bar{s}})}{cR} d \sin \omega\left(t_o - \frac{R}{c}\right)\right\} \tag{5.13-4}$$

Similarly

$$\left\{R_P\left(1 - \frac{\bar{V}_P.\bar{R}_P}{cR_P}\right)\right\}^{-1}$$

$$= \frac{1}{R}\left\{1 - \frac{\bar{R}.\hat{\bar{s}}}{R^2} d \sin \omega\left(t_o - \frac{R}{c}\right) - \frac{\bar{R}.\hat{\bar{s}}}{cR} \omega d \cos \omega\left(t_o - \frac{R}{c}\right) - \frac{\bar{R}.(\bar{\Omega}\times\hat{\bar{s}})}{cR} d \sin \omega\left(t_o - \frac{R}{c}\right)\right\} \tag{5.13-5}$$

Then

$$\phi_O(t_o) = -a\left\{R_P\left(1 - \frac{\bar{V}_P.\bar{R}_P}{cR_P}\right)\right\}^{-1} + a\left\{R_Q\left(1 - \frac{\bar{V}_Q.\bar{R}_Q}{cR_Q}\right)\right\}^{-1}$$

$$= \frac{\bar{R}}{R^3}.\hat{\bar{s}}\ 2ad \sin \omega\left(t_o - \frac{R}{c}\right) + \frac{\bar{R}}{cR^2}.\hat{\bar{s}}\ 2ad\omega \cos \omega\left(t_o - \frac{R}{c}\right) + \frac{\bar{R}}{cR^2}.(\bar{\Omega}\times\hat{\bar{s}})\ 2ad \sin \omega\left(t_o - \frac{R}{c}\right) \tag{5.13-6}$$

In the limit as $d\to 0$ and $a\to\infty$ in such a way as to maintain ad constant, the suppressed higher-order terms vanish and equation (5.13-6) becomes an accurate representation of the doublet scalar potential.

Since

$$(\bar{p})_{t_o-\frac{R}{c}} = \hat{s}\ 2\ ad\ \sin\omega\left(t_o - \frac{R}{c}\right) \tag{5.13-7}$$

and

$$\left(\frac{d\bar{p}}{dt}\right)_{t_o-\frac{R}{c}} = \hat{s}\ 2ad\omega\ \cos\omega\left(t_o - \frac{R}{c}\right) + (\bar{\Omega}\times\hat{s})\ 2\ ad\ \sin\omega\left(t_o - \frac{R}{c}\right) \tag{5.13-8}$$

we have

$$\phi_0(t_o) = \frac{\bar{R}}{R^3}\cdot(\bar{p})_{t_o-\frac{R}{c}} + \frac{\bar{R}}{cR^2}\cdot\left(\frac{d\bar{p}}{dt}\right)_{t_o-\frac{R}{c}}$$

or, in general,

$$\phi = \frac{\bar{R}}{R^3}\cdot[\bar{p}] + \frac{\bar{R}}{cR^2}\cdot\left[\frac{d\bar{p}}{dt}\right] \tag{5.13-9}$$

$$= [\bar{p}].\text{grad}\ \frac{1}{r} - \frac{\bar{r}}{cr^2}\cdot\left[\frac{d\bar{p}}{dt}\right] \tag{5.13-9(a)}$$

where $\bar{r}$ is directed away from the point of evaluation.

It will be observed that equation (5.13-9(a)) is identical with (5.5-7(a)) which obtains for a point doublet comprising stationary sources whose magnitudes are supposed to vary with time. In the present instance the second term in the expression for ϕ is a direct consequence of the presence of the factor $1 - \frac{V_R}{c}$ in the Liénard formula.

In the same way we find that the vector potential of the doublet is given by

$$\bar{A}_0(t_o) = \frac{2a}{cR}\left\{\hat{s}\ \omega\ d\ \cos\omega\left(t_o - \frac{R}{c}\right) + (\bar{\Omega}\times\hat{s})\ d\ \sin\omega\left(t_o - \frac{R}{c}\right)\right\}$$

$$= \frac{1}{cR}\left(\frac{d\bar{p}}{dt}\right)_{t_o-\frac{R}{c}}$$

whence, in general,

$$\bar{A} = \frac{1}{cR}\left[\frac{d\bar{p}}{dt}\right] = \frac{1}{cr}\left[\frac{d\bar{p}}{dt}\right] \tag{5.13-10}$$

Although equations (5.13-9) and (5.13-10) have been derived on the supposition that the scalar doublet moment p varies sinusoidally with time, it is clear that the expressions will remain valid for any form of smooth periodic variation, and for the case in which p is time-invariant.

5.14 The Retarded Potentials of a Point Whirl of Constant Moment

In Fig. 5.7 a point source of strength a revolves in a circle of radius ρ about a fixed point S with constant angular velocity $\bar{\omega}$.

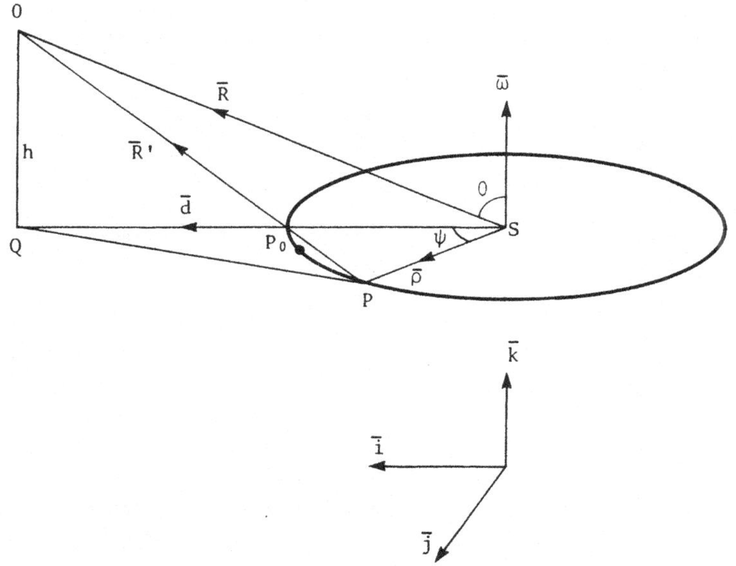

Fig. 5.7

If P is the retarded position of the source corresponding to evaluation at O at the time t_o, the source must occupy this position at the time $t_o - \frac{R'}{c}$. Suppose that the source is located at P_o at the time $t_o - \frac{R}{c}$ and that $\angle P_oSQ = \psi_o$ where Q is the end point of the normal drawn from O to the plane of the circle. Then

$$\psi = \psi_o + \omega\left(\frac{R-R'}{c}\right) \tag{5.14-1}$$

We have also

$$QP^2 = d^2 + \rho^2 - 2\rho d \cos \psi$$

and

$$R'^2 = h^2 + d^2 + \rho^2 - 2\rho d \cos\psi = R^2 + \rho^2 - 2\rho d \cos\psi$$

hence

$$R' = R\left(1 - \frac{2\rho}{R}\sin\theta\cos\psi + \frac{\rho^2}{R^2}\right)^{\frac{1}{2}}$$

A binomial expansion then yields

$$R - R' = \rho\sin\theta\cos\psi - \frac{\rho^2}{2R} + \frac{\rho^2}{4R}\sin^2\theta + \frac{\rho^2}{4R}\sin^2\theta\cos 2\psi + \ldots.$$

On expanding $\sin\psi$ and $\cos\psi$ as expressed in equation (5.14-1) and treating $(R-R')$ as a small quantity, for reasons which will become evident later, we obtain

$$R-R' = \rho\sin\theta\cos\psi_o - \frac{\rho^2\omega}{2c}\sin^2\theta\sin 2\psi_o - \frac{\rho^2}{2R} + \frac{\rho^2}{4R}\sin^2\theta + \frac{\rho^2}{4R}\sin^2\theta\cos 2\psi_o + \ldots \tag{5.14-2}$$

where the remaining terms involve higher powers of ρ, whence

$$\frac{1}{R'} = \frac{1}{R}\left\{1 + \frac{\rho}{R}\sin\theta\cos\psi_o - \frac{\rho^2\omega}{2cR}\sin^2\theta\sin 2\psi_o - \frac{\rho^2}{2R^2} + \frac{3\rho^2}{4R^2}\sin^2\theta + \frac{3\rho^2}{4R^2}\sin^2\theta\cos 2\psi_o + \ldots\right\} \tag{5.14-3}$$

to the same order.

Let the unit vectors $\bar{i}$ and $\bar{k}$ of a right-handed set be aligned with $\overrightarrow{SQ}$ and $\bar{\omega}$ respectively. Then the source velocity at P is given by

$$\bar{V} = -\bar{i}\rho\omega\sin\psi + \bar{j}\,\rho\omega\cos\psi$$

Since

$$\bar{R}' = -\bar{i}\rho\cos\psi - \bar{j}\,\rho\sin\psi + \bar{i}\,d + \bar{k}\,h$$

$$\bar{V}.\bar{R}' = -\rho\omega d\sin\psi$$

Expansion in accordance with equations (5.14-1) to (5.14-3) yields

$$\frac{\bar{V}.\bar{R}'}{cR'} = -\frac{\rho\omega}{c}\sin\theta\left\{\sin\psi_o + \frac{\rho\omega}{2c}\sin\theta + \frac{\rho\omega}{2c}\sin\theta\cos 2\psi_o + \frac{\rho}{2R}\sin\theta\sin 2\psi_o + \ldots\right\}$$

whence

$$\left\{1 - \frac{V_{R'}}{c}\right\}^{-1} = 1 - \frac{\rho\omega}{c} \sin\theta \sin \psi_o - \frac{\rho^2\omega^2}{c^2} \sin^2\theta \cos 2\psi_o - \frac{\rho^2\omega}{2cR} \sin^2\theta \sin 2\psi_o + \ldots \tag{5.14-4}$$

and

$$\left\{R'\left(1 - \frac{V_{R'}}{c}\right)\right\}^{-1} = \frac{1}{R}\left\{1 + \frac{\rho}{R} \sin \theta \cos \psi_o - \frac{\rho\omega}{c} \sin \theta \sin \psi_o - \frac{\rho^2}{2R^2} + \frac{3\rho^2}{4R^2} \sin^2 \theta \right. \tag{5.14-5}$$

$$\left. + \frac{3\rho^2}{4R^2} \sin^2\theta \cos 2\psi_o - \frac{3\rho^2\omega}{2cR} \sin^2\theta \sin 2\psi_o - \frac{\rho^2\omega^2}{c^2} \sin^2\theta \cos 2\psi_o\right\}$$

correct to the second power of ρ.

Now

$$\begin{aligned} V_x &= - \rho\omega \sin \psi \\ &= - \rho\omega \left\{\sin \psi_o + \frac{\rho\omega}{2c} \sin \theta + \frac{\rho\omega}{2c} \sin \theta \cos 2\psi_o + \ldots\right\} \end{aligned}$$

whence

$$(A_x)_{t_o} = \frac{aV_x}{cR'\left(1 - \frac{V_{R'}}{c}\right)} = \frac{-a\rho\omega}{cR}\left\{\sin \psi_o + \frac{\rho}{2R} \sin \theta \sin 2\psi_o + \frac{\rho\omega}{c} \sin \theta \cos 2\psi_o\right\} \tag{5.14-6}$$

to the order ρ^2.

Similarly,

$$V_y = \rho\omega\left\{\cos \psi_o - \frac{\rho\omega}{2c} \sin \theta \sin 2\psi_o \;\ldots\ldots\right\}$$

and

$$(A_y)_{t_o} = \frac{aV_y}{cR'\left(1 - \frac{V_{R'}}{c}\right)} = \frac{a\rho\omega}{cR}\left\{\cos \psi_o + \frac{\rho}{2R} \sin \theta + \frac{\rho}{2R} \sin \theta \cos 2\psi_o - \frac{\rho\omega}{c} \sin \theta \sin 2\psi_o\right\} \tag{5.14-7}$$

We now generate a point whirl by requiring that $\rho \to 0$ and $a \to \infty$ in such a way as to maintain $a\rho^2$ constant, and we define the vector moment of the whirl by

$$\overline{m} = \frac{a\rho^2\overline{\omega}}{2c} \tag{5.14-8}$$

Since terms of order higher than ρ^2 vanish under these conditions, the components of the vector potential at 0 are given accurately by (5.14-6) and (5.14-7)[10]. It is easily seen that to obtain the values of A_x and A_y corresponding to the time t at 0 we must substitute $\omega(t-t_o) + \psi_o$ for ψ_o in these expressions, whence it follows that the average values of A_x and A_y, when taken over any interval of duration $\frac{2\pi}{\omega}$, are respectively zero and $\frac{a\rho^2\omega}{2cR^2}\sin\theta$, ie

$$\overline{A}(\text{average}) = \overline{i}\,0 + \overline{j}\,\frac{m\sin\theta}{R^2} + \overline{k}\,0$$

$$= \overline{m} \times \frac{\overline{R}}{R^3}$$

or

$$\overline{A}(\text{average}) = \overline{m} \times \operatorname{grad}\frac{1}{r} \tag{5.14-9}$$

where r is distance measured from 0.

Reference to Sec. 4.12a reveals that the vector potential of a time-invariant continuous circular whirl is c times the average vector potential of the point whirl (provided that the sense of circulation is the same in each case) when

$$\pi\rho^2 I = \frac{a\rho^2\omega}{2} \quad \text{or} \quad I = \frac{a\omega}{2\pi} = \frac{a}{T}$$

where T is the period of revolution.

Since the average value of $\overline{A}$ is independent of the time of initiation of the associated integration cycle, it is clear that the equivalence continues to hold for a set of point sources distributed around the circle if we put

---

10. The first terms of these expressions (which become infinite in the limiting process) will be eliminated subsequently.

$$I = \frac{a_1}{T_1} + \frac{a_2}{T_2} \;\ldots. \qquad (5.14\text{-}10)$$

ie if I is made equal to the total source strength passing a given point of the circle in unit time, averaged over the composite period.

In view of the limiting requirement, $a \to \infty$, the scalar potential, viz

$$\phi = \frac{a}{R'\left(1 - \frac{V_{R'}}{c}\right)}$$

becomes infinite at 0. To render this finite we may place at the centre of the circle a point source of equal magnitude and opposite sign[11]. It then follows from equation (5.14-5) that in the limit

$$\phi(\text{average}) = \frac{a}{R}\left\{\frac{-\rho^2}{2R^2} + \frac{3\rho^2}{4R^2}\sin^2\theta\right\} = \frac{a\rho^2}{4R^3}\,(1-3\cos^2\theta) \qquad (5.14\text{-}11)$$

It will be seen from the considerations of Sec. 4.1 that (5.14-11) is identical with the potential of a static axial quadrupole of moment $\frac{a\rho^2}{2}$, placed at the centre of the circle and aligned with its axis.

A considerable simplification is introduced by supposing that three or more point sources are disposed symmetrically around the circle with common angular velocity. Of the additional terms introduced by this means into equations (5.14-6) and (5.14-7) those which involve phase angle cancel out in the sum, so that, if a now refers to total circulatory source strength, equations (5.14-9) and (5.14-11) describe the instantaneous values of $\bar{A}$ and ϕ. The potentials are then seen to be time-invariant.

5.15 The Retarded Vector Potential of a Point Whirl of Time-Dependent Orientation

Fig. 5.8 depicts the orbital plane of a point source at the time $t_o - \frac{R}{c}$. OQ is normal to this plane, as in Fig. 5.7, and the unit axes are defined in the same way. The source occupies the position P_o, where $\angle P_oSQ = \psi_o$.

---

11. However, see p. 512.

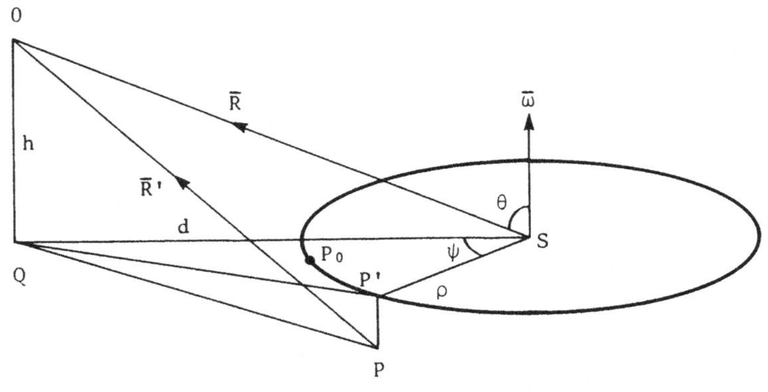

Fig. 5.8

It is supposed that the orbital plane is tilting about the y axis with angular velocity Ω, and that P is the retarded position of the source corresponding to evaluation of the potential at 0 at the time t_o. Hence

$$\psi = \psi_o + \omega \frac{(R-R')}{c} \tag{5.15-1}$$

and

$$P'P = \rho\Omega \frac{(R-R')}{c} \cos\psi + \ldots \tag{5.15-2}$$

From geometrical considerations

$$R'^2 = R^2 + \rho^2 - 2\rho d \cos\psi + 2\,\rho h\Omega \frac{(R-R')}{c} \cos\psi + \rho^2\Omega^2 \frac{(R-R')^2}{c^2} \cos^2\psi + \ldots.$$

whence

$$R - R' = \rho \sin\theta \cos\psi_o + \ldots. \tag{5.15-3}$$

and

$$\frac{1}{R'} = \frac{1}{R}\left\{1 + \frac{\rho}{R}\sin\theta\cos\psi_o + \ldots\right\} \tag{5.15-4}$$

The components of the source velocity at P are given to the second power of ρ by

$$V_x = -\rho\omega \sin\psi - \rho\Omega^2 \frac{(R-R')}{c}\cos\psi + \ldots.$$

$$= -\rho\omega\left\{\sin\psi_o + \frac{\rho\omega}{2c}\sin\theta + \frac{\rho\omega}{2c}\sin\theta\cos 2\psi_o\right\} \quad (5.15\text{-}5)$$

$$-\frac{\rho^2\Omega^2}{2c}\sin\theta - \frac{\rho^2\Omega^2}{2c}\sin\theta\cos 2\psi_o$$

$$V_y = \rho\omega\cos\psi + \ldots.$$

$$= \rho\omega\left\{\cos\psi_o - \frac{\rho\omega}{2c}\sin\theta\sin 2\psi_o\right\} \quad (5.15\text{-}6)$$

$$V_z = -\rho\Omega\cos\psi + \rho\omega\,\Omega\frac{(R-R')}{c}\sin\psi + \ldots.$$

$$= -\rho\Omega\cos\psi_o + \frac{\rho^2\omega\Omega}{c}\sin\theta\sin 2\psi_o \quad (5.15\text{-}7)$$

On combining equations (5.15-5) to (5.15-7) with

$$\bar{R}' = -\bar{i}\rho\cos\psi - \bar{j}\rho\sin\psi + \bar{i}d + \bar{k}h + \bar{k}\rho\Omega\frac{(R-R')}{c}\cos\psi + \ldots$$

we find that to the first power of ρ

$$\bar{V}.\bar{R}' = -\left(d\rho\omega\sin\psi_o + h\rho\Omega\cos\psi_o\right)$$

whence, to the same order,

$$\left\{1 - \frac{V_{R'}}{c}\right\}^{-1} = 1 - \frac{\rho\omega}{c}\sin\theta\sin\psi_o - \frac{\rho\Omega}{c}\cos\theta\cos\psi_o \quad (5.15\text{-}8)$$

and

$$\left\{cR'\left(1 - \frac{V_{R'}}{c}\right)\right\}^{-1} = \frac{1}{cR}\left\{1 + \frac{\rho}{R}\sin\theta\cos\psi_o - \frac{\rho\omega}{c}\sin\theta\sin\psi_o - \frac{\rho\Omega}{c}\cos\theta\cos\psi_o\right\} \quad (5.15\text{-}9)$$

If, now, we postulate a symmetrical source system of total circulatory strength a, the combination of equation (5.15-9) with the velocity components yields in the limit

$$(A_x)_{t_o} = \frac{-a\rho^2\Omega^2}{2c^2R}\sin\theta \quad (5.15\text{-}10)$$

$$(A_y)_{t_o} = \frac{a\rho^2\omega}{2cR^2}\sin\theta - \frac{a\rho^2\omega\Omega}{2c^2R}\cos\theta \tag{5.15-11}$$

$$(A_z)_{t_o} = \frac{-a\rho^2\Omega}{2cR^2}\sin\theta + \frac{a\rho^2\Omega^2}{2c^2R}\cos\theta \tag{5.15-12}$$

It will be observed that the magnitudes of the various components of $(\bar{A})_{t_o}$ are such that as $\Omega/\omega\to 0$, $(\bar{A})_{t_o}\to\bar{j}(A_y)_{t_o}$. But

$$[\bar{m}] = \frac{a\rho^2\bar{\omega}}{2c} \quad\text{and}\quad \left[\frac{d\bar{m}}{dt}\right] = \frac{a\rho^2}{2c}\left[\frac{d\bar{\omega}}{dt}\right] = \bar{i}\,\frac{a\rho^2\omega\Omega}{2c}$$

so that we then have

$$(\bar{A})_{t_o}\to[\bar{m}]\times\frac{\bar{R}}{R^3} + \left[\frac{d\bar{m}}{dt}\right]\times\frac{\bar{R}}{cR^2} \tag{5.15-13}$$

On carrying out a similar analysis for tilt about the x axis we obtain

$$(A_x)_{t_o} = \frac{-a\rho^2\omega\Omega}{2c^2R}\cos\theta \tag{5.15-14}$$

$$(A_y)_{t_o} = \frac{a\rho^2\omega}{2cR^2}\sin\theta \tag{5.15-15}$$

$$(A_z)_{t_o} \to \frac{a\rho^2\omega\Omega}{2c^2R}\sin\theta \tag{5.15-16}$$

In this case

$$\left[\frac{d\bar{m}}{dt}\right] = -\,\bar{j}\,\frac{a\rho^2\omega\Omega}{2c}$$

hence

$$\left[\frac{d\bar{m}}{dt}\right]\times\frac{\bar{R}}{cR^2} = -\,\bar{i}\,\frac{a\rho^2\omega\Omega}{2c^2R}\cos\theta + \bar{k}\,\frac{a\rho^2\omega\Omega}{2c^2R}\sin\theta$$

and

$$(\bar{A})_{t_o} \to [\bar{m}]\times\frac{\bar{R}}{R^3} + \left[\frac{d\bar{m}}{dt}\right]\times\frac{\bar{R}}{cR^2} \tag{5.15-17}$$

It follows that for tilt about any axis perpendicular to $\bar{\omega}$, and with $\Omega/\omega \to 0$, we have the general result

$$\bar{A} = [\bar{m}] \times \text{grad}\,\frac{1}{r} - \left[\frac{d\bar{m}}{dt}\right] \times \frac{\bar{r}}{cr^2} \tag{5.15-18}$$

This result is applicable to a variety of source systems. It holds for a symmetrical point whirl of fixed orientation and variable moment (Ex.5-57., p. 506), for a continuous whirl of the same type (See Sec. 5.5), and for a continuous whirl of time-dependent orientation (Ex.5-62., p. 508).

The derivation of the scalar potential of a symmetrical point whirl of time-dependent orientation is the subject of Ex.5-61., p. 507. It is there required to show that in the presence of an equal and opposite central source

$$\phi = \left[\left\{1 + \frac{R}{c}\frac{d}{dt}\right\} \frac{a\rho^2}{4R^3}\,(1-3\cos^2\theta)\right] \tag{5.15-19}$$

This is a generalisation of equation (5.14-11), and is closely related to the potentials of both a uniform scalar line source which shrinks about a central compensating source and of an axial quadrupole of appropriate orientation (Ex.5-63. and 5-64., p. 508).

It will be observed that, on occasion, the magnitude of a stationary point source has been postulated to be a function of time. This time-dependence has not been extended to moving point sources because the derivation of the Liénard-Wiechert potentials presupposes the invariance in time of the parent volume distributions.

It will not have escaped the notice of the reader that the retarded scalar potential of a point doublet, as expressed by equation (5.13-9), may be written in the form

$$\phi = \left[\bar{p} + \frac{R}{c}\frac{d\bar{p}}{dt}\right] \cdot \frac{\bar{R}}{R^3}$$

Hence, provided that the rate of change of $\bar{p}$ is constant over the retardation interval R/c, we have

$$\phi = \frac{\bar{p}.\bar{R}}{R^3}$$

from which we see that the retarded and unretarded potentials are identical. In like circumstance equation (5.15-17) reduces to

$$\bar{A} = \bar{m} \times \frac{\bar{R}}{R^3}$$

and it is easily shown, by reworking Sec. 5.15 without retardation and Liénard modification, that this expression represents the unretarded vector potential of the point whirl. The same behaviour is exhibited by the scalar potential[12].

When the rate of change of the relevant variable is not constant in time the unretarded potential will continue to approximate the retarded potential provided that the second time derivative is continuous over the retardation interval and R is sufficiently small. For sinusoidal variation of $\bar{p}$ or $\bar{m}$ the required correction is of the order $\frac{\omega^2 R^2}{c^2}$, where ω now denotes the angular velocity of the source moment. However, for non-zero values of ω, the approximation must inevitably fail at sufficient distance from the source.

EXERCISES

5-55. Show from first principles that equations (5.13-9(a)) and (5.13-10) continue to hold when the doublet moment is an arbitrary function of time with continuous derivatives.

5-56. Confirm that the presence of three or more symmetrically disposed sources eliminates those terms involving phase angle from the various limiting expressions for the components of potential in Secs. 5.14 and 5.15.

5-57. Show that equation (5.15-18) holds not only for a symmetrical point whirl of variable orientation but also for a whirl of fixed orientation and variable moment, where the variation of moment is due to

(1) time-dependence of angular velocity of the sources

(2) time-dependence of the orbital radius (provided that the fluctuation of the orbital radius is sufficiently small and its associated period sufficiently large).

5-58. Confirm equations (5.15-14) to (5.15-16).

5-59. It has been shown in Sec. 5.11a that the time interval Δt at the point of evaluation of the retarded potential corresponding to the time interval $\Delta t'$ at the source is given by $\Delta t = \Delta t' \left(1 - \frac{V_R}{c}\right)$. It follows that if ϕ and $\bar{A}$ are the scalar and vector potentials of the source system of Fig. 5.7, as evaluated at 0,

$$\phi dt = \frac{a}{R'\left(1 - \frac{V_{R'}}{c}\right)} dt = \frac{a}{R'} dt'$$

12. This behaviour does not extend to the vector potential of a point doublet.

$$\bar{A}dt = \frac{a\bar{V}}{cR'\left(1 - \frac{V_{R'}}{c}\right)} dt = \frac{a\bar{V}}{cR'} dt'$$

Duplicate the results of Sec. 5.14 for a single-point whirl with a central compensating source by evaluating

$$\int_{t_1}^{t_1 + \frac{2\pi}{\omega}} \frac{a}{R'} dt' \quad \text{and} \quad \int_{t_1}^{t_1 + \frac{2\pi}{\omega}} \frac{a\bar{V}}{cR'} dt'$$

at the source, where t_1 is an arbitrarily chosen point of time, and consequently showing that

$$\frac{\omega}{2\pi} \int_{t_1 + \frac{R_1}{c}}^{t_1 + \frac{R_1}{c} + \frac{2\pi}{\omega}} \phi dt = \frac{a\rho^2}{4R^3} (1-3 \cos^2\theta)$$

and

$$\frac{\omega}{2\pi} \int_{t_1 + \frac{R_1}{c}}^{t_1 + \frac{R_1}{c} + \frac{2\pi}{\omega}} A_x dt = 0 \quad ; \quad \frac{\omega}{2\pi} \int_{t_1 + \frac{R_1}{c}}^{t_1 + \frac{R_1}{c} + \frac{2\pi}{\omega}} A_y dt = \frac{a\rho^2\omega}{2cR^2} \sin\theta$$

where R_1 is the distance of the source from 0 at the time t_1.

5-60. In equation (5.12-7) the retarded vector potential of a moving point source is expressed in terms of the instantaneous distance of the source from the point of evaluation, together with time derivatives of that distance. Obtain (5.14-9) by applying (5.12-7) to a symmetrical point whirl, and observe that the term involving rate of change of acceleration must be included in the working.

5-61. A symmetrical point whirl of time-dependent orientation is supplied with a central compensating source. Show, in the notation of Fig. 5.8, that the scalar potential at 0 at the time t_o is given by the limiting value of

$$\left[\left\{1 + \frac{R}{c}\frac{d}{dt}\right\} \frac{a\rho^2}{4R^3} (1-3 \cos^2\theta) \right]$$

provided that $\Omega << \frac{c}{R}$.

5-62. Let the instantaneous orbit of the revolving point source of Fig. 5.8 be identified with a uniform tangential line source of time-invariant density $\bar{I}$. If the vector potential of such a source is defined, for present purposes, by the general retarded expression

$$\bar{A} = \oint \frac{\bar{I}\, ds}{cR\left(1 - \frac{V_R}{c}\right)}$$

where $\bar{V}$ is the velocity of the element ds, show that the vector potential at 0 at the time t_o is given by

$$[\bar{m}] \times \frac{\bar{R}}{R^3} + \left[\frac{d\bar{m}}{dt}\right] \times \frac{\bar{R}}{cR^2}$$

where

$$\bar{m} = \hat{n} \lim_{\substack{\rho \to 0 \\ I \to \infty}} \frac{\pi \rho^2 I}{c} = \lim_{\substack{S \to 0 \\ I \to \infty}} \frac{I\bar{S}}{c}$$

5-63. A uniform scalar line source of strength a coincides with the circular orbit of Fig. 5.8 and shares its motion. A point source of equal and opposite strength is located centrally. If $a\rho^2$ is maintained finite during the limiting process and is time-invariant, show that the potential at 0 at the time t_o is given by

$$\left[\left\{1 + \frac{R}{c}\frac{d}{dt}\right\} \frac{a\rho^2}{4R^3}\,(1-3\cos^2\theta)\right]$$

provided that $\Omega << \frac{c}{R}$.

Prove that the same expression holds when the contour is fixed in position and the source strength varies sinusoidally in time with frequency f, proved that $f^2 << \frac{c^2}{4\pi^2 R^2}$.

5-64. An axial quadrupole comprising the limiting configuration of stationary sources of time-dependent magnitude is located at the origin of spherical coordinates and aligned with the z axis, the inner sources being of positive sign. Show from equation (5.5-7(a)) that the retarded scalar potential at (R,θ,Φ) is given by

$$\left\{1 + \frac{R}{c}\frac{d}{dt}\right\} \frac{[p^{(2)}]}{2R^3}\,(1-3\cos^2\theta) - \frac{1}{2}\left[\frac{d^2 p^{(2)}}{dt^2}\right] \frac{\cos^2\theta}{c^2 R}$$

5.16 The $\bar{E}$ and $\bar{B}$ Fields of Time-Dependent Doublets and Whirls

5.16a The time-dependent point doublet

The retarded scalar and vector potentials of a point doublet of time-dependent moment, as derived in Sec. 5.13, are

$$\phi = [\bar{p}].\text{grad}\,\frac{1}{r} - \frac{\bar{r}}{cr^2}.\left[\frac{d\bar{p}}{dt}\right] \quad ; \quad \bar{A} = \frac{1}{cr}\left[\frac{d\bar{p}}{dt}\right]$$

The gradient of ϕ at an exterior point 0 (x_o,y_o,z_o) is consequently given by

$$\text{grad}\,\phi = \sum \bar{i}\,\frac{\partial}{\partial x_o}\left(-[\bar{p}].\frac{\bar{r}}{r^3} - \frac{\bar{r}}{cr^2}.\left[\frac{d\bar{p}}{dt}\right]\right) = \sum \bar{i}\,\frac{\partial}{\partial x}\left([\bar{p}].\frac{\bar{r}}{r^3} + \frac{\bar{r}}{cr^2}.\left[\frac{d\bar{p}}{dt}\right]\right)$$

Routine expansion yields

$$\text{grad}\,\phi = \sum \bar{i}\left\{\frac{\bar{i}.[\bar{p}]}{r^3} - \frac{3(x-x_o)}{r}\,\frac{\bar{r}}{r^4}.[\bar{p}] + \frac{\bar{i}}{cr^2}.\left[\frac{d\bar{p}}{dt}\right] - \frac{3(x-x_o)}{r}\,\frac{\bar{r}}{cr^3}.\left[\frac{d\bar{p}}{dt}\right]\right.$$
$$\left. - \frac{(x-x_o)}{cr}\,\frac{\bar{r}}{cr^2}.\left[\frac{d^2\bar{p}}{dt^2}\right]\right\}$$

whence

$$\bar{E} = \frac{-[\bar{p}]}{r^3} + 3\,\frac{\bar{r}.[\bar{p}]}{r^5}\,\bar{r} - \frac{1}{cr^2}\left[\frac{d\bar{p}}{dt}\right] + 3\,\frac{\bar{r}}{cr^4}.\left[\frac{d\bar{p}}{dt}\right]\bar{r} + \frac{\bar{r}}{c^2r^3}.\left[\frac{d^2\bar{p}}{dt^2}\right]\bar{r} - \frac{1}{c^2r}\left[\frac{d^2\bar{p}}{dt^2}\right] \tag{5.16-1}$$

It will be observed that the time-dependence of $\bar{p}$ gives rise to terms which

(1) fall off as the square of distance and are proportional to $\frac{d\bar{p}}{dt}$

(2) fall off as the first power of distance and are proportional to $\frac{d^2\bar{p}}{dt^2}$

The latter terms may be combined in the form

$$\frac{1}{c^2r}\left\{\hat{\bar{r}} \times \left(\hat{\bar{r}} \times \left[\frac{d^2\bar{p}}{dt^2}\right]\right)\right\} \tag{5.16-2}$$

This component is normal to the radius vector and lies in the plane defined by $\left[\frac{d^2\bar{p}}{dt^2}\right]$ and $\hat{\bar{r}}$. Thus for non-zero values of $\frac{d^2\bar{p}}{dt^2}$ the $\bar{E}$ field becomes transverse at great distance from the source, as would be expected from the considerations of Sec. 5.11b.

We have also

$$\bar{B} = \text{curl}\,\bar{A} = \sum \bar{i}\left\{\frac{\partial}{\partial y_o}\left(\frac{1}{cr}\left[\frac{dp_z}{dt}\right]\right) - \frac{\partial}{\partial z_o}\left(\frac{1}{cr}\left[\frac{dp_y}{dt}\right]\right)\right\}$$

$$= -\frac{1}{c}\sum \bar{i}\left\{\frac{-(y-y_o)}{r^3}\left[\frac{dp_z}{dt}\right] - \frac{(y-y_o)}{cr^2}\left[\frac{d^2p_z}{dt^2}\right] + \frac{(z-z_o)}{r^3}\left[\frac{dp_y}{dt}\right] + \frac{(z-z_o)}{cr^2}\left[\frac{d^2\bar{p}_y}{dt^2}\right]\right\}$$

whence

$$\bar{B} = \frac{\bar{r}}{cr^3} \times \left[\frac{d\bar{p}}{dt}\right] + \frac{\bar{r}}{c^2r^2} \times \left[\frac{d^2\bar{p}}{dt^2}\right] \qquad (5.16\text{-}3)$$

From equation (5.16-1)

$$\hat{\bar{R}} \times \bar{E} = -(\hat{\bar{r}}\times\bar{E}) = \frac{\bar{r}}{r^4} \times [\bar{p}] + \frac{\bar{r}}{cr^3} \times \left[\frac{d\bar{p}}{dt}\right] + \frac{\bar{r}}{c^2r^2} \times \left[\frac{d^2\bar{p}}{dt^2}\right]$$

hence

$$\bar{B} = (\hat{\bar{R}}\times\bar{E}) - \frac{\bar{r}}{r^4} \times [\bar{p}] \qquad (5.16\text{-}4)$$

It is evident that the general relationship derived for a moving (or stationary) point source, viz $\bar{B} = \hat{\bar{R}} \times \bar{E}$ may fail at close range for multiple sources. However, the relationship is re-established when the distance from the doublet is such as to render the inverse-cube component of $\bar{E}$ negligible.

Since the potentials of the component singlet sources add linearly we conclude from equation (5.11-28) that the d'Alembertians of ϕ, $\bar{A}$, $\bar{E}$ and $\bar{B}$ are zero beyond the doublet, and that equations (5.11-24/27) continue to hold.

5.16b The time-dependent symmetrical point whirl

It was shown in Sec. 5.15 that the vector potential of a time-dependent symmetrical point whirl is given by

$$\bar{A} = [\bar{m}] \times \text{grad}\,\frac{1}{r} - \left[\frac{d\bar{m}}{dt}\right] \times \frac{\bar{r}}{cr^2}$$

We may derive $\bar{B} = \text{curl}\,\bar{A}$ by direct differentiation of this expression or by proceeding in the following way.

It is easily shown that

$$[\bar{m}].\text{grad}\,\frac{1}{r} - \frac{\bar{r}}{cr^2}.\left[\frac{d\bar{m}}{dt}\right] = -\,\text{div}\left(\frac{[\bar{m}]}{r}\right) \tag{5.16-5}$$

and

$$[\bar{m}] \times \text{grad}\,\frac{1}{r} - \left[\frac{d\bar{m}}{dt}\right] \times \frac{\bar{r}}{cr^2} = \text{curl}\left(\frac{[\bar{m}]}{r}\right) \tag{5.16-6}$$

Hence

$$\begin{aligned}\bar{B} &= \text{curl curl}\left(\frac{[\bar{m}]}{r}\right) = \text{grad div}\left(\frac{[\bar{m}]}{r}\right) - \nabla^2\left(\frac{[\bar{m}]}{r}\right) \\ &= -\,\text{grad}\left\{[\bar{m}].\text{grad}\,\frac{1}{r} - \frac{\bar{r}}{cr^2}.\left[\frac{d\bar{m}}{dt}\right]\right\} - \nabla^2\left(\frac{[\bar{m}]}{r}\right)\end{aligned} \tag{5.16-7}$$

The first term of the above expression has been evaluated in the previous sub-section with $\bar{m}$ replaced by $\bar{p}$; it therefore remains to evaluate $\nabla^2\left(\frac{[\bar{m}]}{r}\right)$.

Now

$$\nabla^2\left(\frac{[\bar{m}]}{r}\right) = \sum \bar{i}\,\nabla^2\left(\frac{[m_x]}{r}\right)$$

but it follows from equation (5.7-4) that $\text{dal}\left(\frac{[m_x]}{r}\right) = 0$

hence

$$\nabla^2\left(\frac{[\bar{m}]}{r}\right) = \frac{1}{c^2}\frac{d^2}{dt^2}\left(\frac{[\bar{m}]}{r}\right) = \frac{1}{c^2 r}\left[\frac{d^2\bar{m}}{dt^2}\right] \tag{5.16-8}$$

We then find that

$$\bar{B} = \frac{-[\bar{m}]}{r^3} + 3\,\frac{\bar{r}.[\bar{m}]}{r^5}\,\bar{r} - \frac{1}{cr^2}\left[\frac{d\bar{m}}{dt}\right] + 3\,\frac{\bar{r}}{cr^4}.\left[\frac{d\bar{m}}{dt}\right]\bar{r} + \frac{\bar{r}}{c^2r^3}.\left[\frac{d^2\bar{m}}{dt^2}\right]\bar{r} - \frac{1}{c^2r}\left[\frac{d^2\bar{m}}{dt^2}\right] \tag{5.16-9}$$

This expression is formally identical with that for the $\bar{E}$ field of the point doublet ($\bar{m}$ replacing $\bar{p}$) and consequently exhibits the same asymptotic behaviour.

The component of $\overline{E}$ deriving from the vector potential is given by

$$-\frac{1}{c}\frac{\partial\overline{A}}{\partial t} = -\frac{1}{c}\left\{\left[\frac{d\overline{m}}{dt}\right] \times \text{grad}\,\frac{1}{r} - \left[\frac{d^2\overline{m}}{dt^2}\right] \times \frac{\overline{r}}{cr^2}\right\} \qquad (5.16\text{-}10)$$

It is seen from equation (5.16-3) that this is formally identical with the negative $\overline{B}$ field of the doublet.

The remaining component of $\overline{E}$ is represented by -grad ϕ. This may be shown to comprise terms which fall off as the second, third and fourth power of distance. Such additions to $\overline{E}$ upset the symmetry displayed by the $\overline{E}/\overline{B}$ relationship for doublets and whirls as they stand above, and complicate subsequent analyses. For this reason we now postulate a system of compensating sources for each whirl which, unlike the central source, reduces the scalar potential to zero at all exterior points. Reference to equation (5.14-5) reveals that this may be effected by the addition of a symmetrically-disposed set of identical orbital sources of such magnitude and sign as will reduce the aggregate source strength to zero. The vector potential of the primary sources is unaffected if the compensatory sources are stationary. Accordingly, equation (5.16-10) is now replaced by

$$\overline{E} = -\frac{1}{c}\left\{\left[\frac{d\overline{m}}{dt}\right] \times \text{grad}\,\frac{1}{r} - \left[\frac{d^2\overline{m}}{dt^2}\right] \times \frac{\overline{r}}{cr^2}\right\} \qquad (5.16\text{-}10(a))$$

As in the case of the point doublet, equations (5.11-24) to (5.11-28) continue to apply beyond the source.

EXERCISES

5-65. Prove equations (5.16-5) and (5.16-6).

Show that the $\overline{E}$ field of a point doublet may be expressed as

$$\overline{E} = \text{curl curl}\left(\frac{[\overline{p}]}{r}\right)$$

5-66. Confirm equation (5.16-9) by direct differentiation of (5.15-18).

5-67. A point doublet of time-dependent magnitude and moment $\overline{p} = \overline{k}p$ is located at the origin of spherical coordinates. Show that at the point (R,θ,Φ)

$$E_R = 2\cos\theta\left\{\frac{[p]}{R^3} + \frac{1}{cR^2}\left[\frac{dp}{dt}\right]\right\}$$

$$E_\theta = \sin\theta\left\{\frac{[p]}{R^3} + \frac{1}{cR^2}\left[\frac{dp}{dt}\right] + \frac{1}{c^2R}\left[\frac{d^2p}{dt^2}\right]\right\}$$

$$E_\Phi = B_R = B_\theta = 0$$

$$B_\Phi = \sin\theta \left\{ \frac{1}{cR^2}\left[\frac{dp}{dt}\right] + \frac{1}{c^2R}\left[\frac{d^2p}{dt^2}\right] \right\}$$

5-68. Use equation (2.6-7) to confirm that $\operatorname{div}\bar{E} = \operatorname{div}\bar{B} = 0$ in the previous exercise.

5-69. A fully-compensated point whirl of time-dependent magnitude and moment $\bar{m} = \bar{k}m$ is located at the origin of spherical coordinates. Show that at the point (R,θ,Φ)

$$B_R = 2\cos\theta \left\{ \frac{[m]}{R^3} + \frac{1}{cR^2}\left[\frac{dm}{dt}\right] \right\}$$

$$B_\theta = \sin\theta \left\{ \frac{[m]}{R^3} + \frac{1}{cR^2}\left[\frac{dm}{dt}\right] + \frac{1}{c^2R}\left[\frac{d^2m}{dt^2}\right] \right\}$$

$$B_\Phi = E_R = E_\theta = 0$$

$$E_\Phi = -\sin\theta \left\{ \frac{1}{cR^2}\left[\frac{dm}{dt}\right] + \frac{1}{c^2R}\left[\frac{d^2m}{dt^2}\right] \right\}$$

5-70. A point doublet of constant magnitude p is located at the origin of rectangular coordinates and has a constant angular velocity $\bar{i}\Omega$. At the time $t = 0$ the doublet moment is $\bar{j}p$. Determine $\bar{E}(t)$ and $\bar{B}(t)$ at the point $(0,d,0)$.

Ans: $E_x = B_y = B_z = 0$

$$E_y = 2p\left\{ \frac{\cos\theta}{d^3} - \frac{\Omega\sin\theta}{cd^2} \right\}$$

$$E_z = p\left\{ \frac{-\sin\theta}{d^3} - \frac{\Omega\cos\theta}{cd^2} + \frac{\Omega^2\sin\theta}{c^2d} \right\}$$

$$B_x = p\left\{ \frac{-\Omega\cos\theta}{cd^2} + \frac{\Omega^2\sin\theta}{c^2d} \right\}$$

where

$$\theta = \Omega\left(t - \frac{d}{c}\right)$$

5-71. Determine the values of $\bar{E}(t)$ and $\bar{B}(t)$ for a point whirl in an analysis parallel to that of the previous exercise.

Ans: $B_x = E_y = E_z = 0$

$$B_y = 2m\left\{\frac{\cos\theta}{d^3} - \frac{\Omega\sin\theta}{cd^2}\right\}$$

$$B_z = m\left\{\frac{-\sin\theta}{d^3} - \frac{\Omega\cos\theta}{cd^2} + \frac{\Omega^2\sin\theta}{c^2d}\right\}$$

$$E_x = m\left\{\frac{\Omega\cos\theta}{cd^2} - \frac{\Omega^2\sin\theta}{c^2d}\right\}$$

5.17 The Retarded Densities and Potentials of a Statistically-Regular Configuration of Point Singlets in Motion

The Equation of Continuity

The scalar density ρ, as employed in earlier sections, has for the most part been treated as a piecewise continuous function which appears in its own right in the integrand of the scalar potential function. However, in Sec. 4.20, where the macroscopic potential of a statistically-regular configuration of stationary point singlets was under consideration, it was found convenient to derive a continuous density function from the discrete values of source strength per unit volume for each of a set of subregions into which the source complex was divided. We now extend such considerations to an assemblage of moving point singlets, and endeavour to define both scalar and vector density functions such that their substitution in the standard integral expressions for retarded scalar and vector potential yield values which are in sensible agreement with the results of direct summation of individual contributions, as expressed by equations (5.10-14(a)) and (5.10-14(b)), at points outside the distribution.

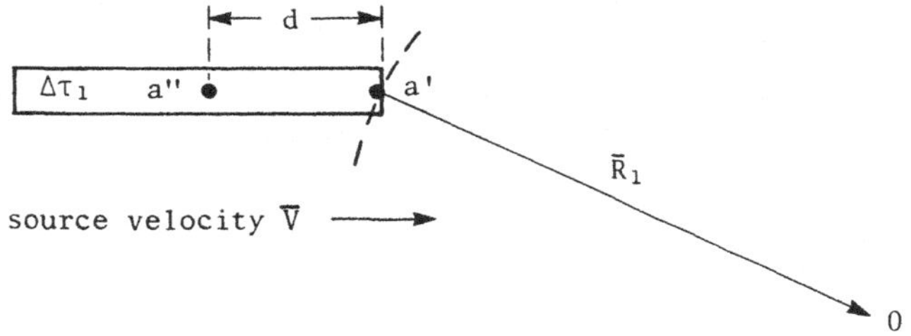

Fig. 5.9a Configuration of sources at time $t_o - \frac{R_1}{c}$

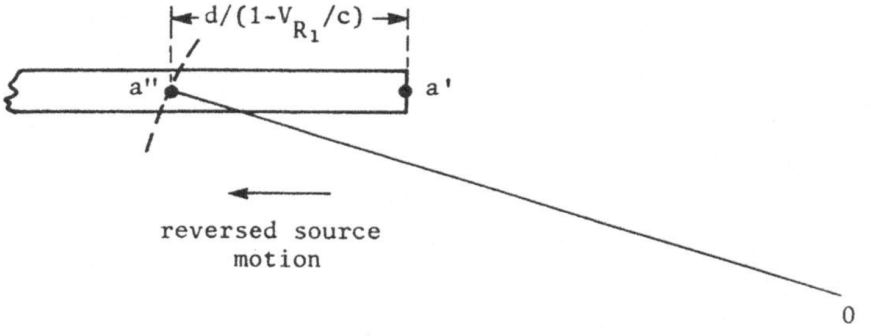

Fig. 5.9b Configuration of individually retarded sources

Consider first a set of sources of equal magnitude a and fixed velocity $\overline{V}$. To determine the retarded positions of the sources corresponding to evaluation of the potentials at the point 0 at the time t_o we reverse the motion of the sources at t_o, as discussed elsewhere, and allow a spherical surface to expand about 0 with speed c. Suppose that this surface cuts the end surface of a fixed subregion, or cell, $\Delta\tau_1$, having the form of a narrow prism whose axis is parallel to $\overline{V}$ (Fig. 5.9a), at a time when the particular source a' lies in the end surface and the source a" is a distance d in advance. Then from previous considerations it will be evident that the spherical surface will overtake a" at a distance $d/(1-V_{R_1}/c)$ from the end of the prism, where V_{R_1} is the radial component of the velocity of the reversed source motion away from 0 (and therefore of the actual motion in the direction of 0 at the retarded time). This fractional increase of spacing between sources in their individually retarded positions relative to the positions occupied at the instant t_o-R_1/c, where R_1 is the distance of the end face of the prism from 0, is seen to apply to all sources initially within $\Delta\tau_1$, irrespective of their original spacing, so that the total source strength of the individually retarded configuration within $\Delta\tau_1$ will be $(1-V_{R_1}/c)$ times that of the instantaneous configuration at the time t_o-R_1/c. If the latter source strength has the value N_1a, then the contribution of $\Delta\tau_1$ to the retarded scalar potential at 0 at the time t_o will, in accordance with equation (5.10-14(a)), be given approximately by

$$\Delta\phi = \frac{N_1a(1-V_{R_1}/c)}{R_1(1-V_{R_1}/c)} = \frac{N_1a}{R_1}$$

It is seen that the Liénard modification in the denominator of the expression for potential is just cancelled by the reduction of source strength in $\Delta\tau_1$ due to movement.

<table>
<tr><th></th><th>$\Delta\tau_5$</th><th>$\Delta\tau_4$</th><th>$\Delta\tau_3$</th><th>$\Delta\tau_2$</th><th>$\Delta\tau_1$
A</th></tr>
<tr><td>Fig. 5.10a</td><td></td><td></td><td>27 26 25 24 23 22 21 20 19</td><td>18 17 16 15 14 13 12 11 10</td><td>9 8 7 6 5 4 3 2 1</td></tr>
<tr><td>Fig. 5.10b</td><td colspan="5">27 26 25 24 23 22 21 20 19 18 17 16 15 14 13 12 11 10 9 8 7 6 5 4 3 2</td></tr>
<tr><td>Fig. 5.10c</td><td>27 26 25</td><td>27 26 25 24 23 22 21 20 19</td><td>21 20 19 18 17 16 15 14 13</td><td>15 14 13 12 11 10 9 8 7</td><td>9 8 7 6 5 4 3 2 1</td></tr>
</table>

Fig. 5.10a Instantaneous configuration of sources when expanding sphere reaches point A from right

Fig. 5.10b Retarded configuration corresponding to Fig. 5.10a for speed c/3

Fig. 5.10c Instantaneous configuration of sources in each element at the time at which the expanding sphere reaches the centre of that element

Fig. 5.10

The principle is illustrated in greater detail in Fig. 5.10 which depicts the conditions obtaining in adjacent collinear cells when equally spaced sources numbered 1 to 27 move radially with respect to the point of evaluation with speed c/3, the reversed motion being directed towards the left and away from 0.

In this case the instantaneous configuration has been shown for each cell at the time at which the spherical surface reaches the cell centre. It will be observed that for this progressive type of motion the ratio of retarded to instantaneous source population within incompletely filled boundary cells may differ from $(1-V_R/c)$. Since only boundary cells are affected and since in practice these comprise only a very small fraction of the total, the attendant error may be neglected. This difficulty does not arise when the source system fully occupies a fixed region of space at all times.

The potential at 0 of the complete source system may be approximated by dividing all space into volume cells parallel to $\overline{V}$, evaluating $Na/\Delta\tau$ in turn at the appropriately retarded time, and generating a smooth point function $\rho_{t_o - R/c}$ by interpolation between the values of $Na/\Delta\tau$ referred to the cell centres (spot densities). Then

$$\phi_0(t_o) = \int_\tau \frac{\rho_{t_o - R/c}}{R} \, d\tau$$

or, in general,

$$\phi = \int_\tau \frac{[\rho]}{r} \, d\tau \qquad (5.17\text{-}1)$$

where R has been replaced by r because all volume elements are fixed in space. The region τ must include all subregions where $[\rho]$ is non-zero.

The accuracy with which the integral formulation approaches that derived from the summation of individual contributions depends, of course, upon the degree of statistical regularity which prevails between adjacent volume cells when their dimensions are reduced to values which are small in relation to their distances from the point of evaluation.

The vector potential is given correspondingly by

$$\overline{A} = \int_\tau \frac{[\rho]\overline{v}}{cr} \, d\tau \qquad (5.17\text{-}2)$$

where the velocity, being time-invariant, is represented in its unretarded, lower-case form.

We may interpret (5.17-1) and (5.17-2) more generally as approximations at exterior points of the potentials deriving from those particular sources of a common-strength/mixed-velocity ensemble whose velocities have the given value $\overline{v}$ when overtaken by the expanding sphere, provided that the velocity of any individual source varies negligibly during the transit interval of the surface across the associated volume element and statistical regularity continues to prevail. It follows that if we specify a complete set of velocities $\bar{v}_1$, $\bar{v}_2$.... and determine the corresponding retarded density fields $[\rho_1]$, $[\rho_2]$......, the potentials of the ensemble become

$$\phi = \int_\tau \frac{1}{r}\{[\rho_1]+[\rho_2].....\}\,d\tau \qquad (5.17\text{-}3)$$

and

$$\bar{A} = \int_\tau \frac{1}{cr}\{[\rho_1]\bar{v}_1+[\rho_2]\bar{v}_2....\}\,d\tau \qquad (5.17\text{-}4)$$

While this approach requires that statistical regularity obtain for each discrete velocity population, it is plausible to assume that the expressions for ϕ and $\bar{A}$ will continue to represent valid approximations provided only that statistical regularity prevail within each of a set of velocity 'slots' into which the velocity distribution may be divided, and that $\bar{v}_1$, $\bar{v}_2$.... are indentified with the mean values of source velocity within these 'slots'. When the ensemble comprises sources of strengths a_1, a_2... the component distributions of common strength are treated separately and the results summed.

Thus, in general,

$$\phi = \int_\tau \frac{1}{r}\left\{\left(\sum [\rho_i]\right)_{a_1} + \left(\sum [\rho_i]\right)_{a_2} \cdots\right\} d\tau = \int_\tau \frac{[\rho]}{r}\,d\tau \qquad (5.17\text{-}5)$$

$$\bar{A} = \int_\tau \frac{1}{cr}\left\{\left(\sum [\rho_i]\bar{v}_i\right)_{a_1} + \left(\sum [\rho_i]\bar{v}_i\right)_{a_2} \cdots\cdots\right\} d\tau \qquad (5.17\text{-}6)$$

If we write

$$\bar{J} \equiv \left(\sum \rho_i\bar{v}_i\right)_{a_1} + \left(\sum \rho_i\bar{v}_i\right)_{a_2} \cdots\cdots \qquad (5.17\text{-}7)$$

the vector potential assumes the familiar form of earlier treatments (apart from the factor c)[13], viz,

$$\bar{A} = \frac{1}{c}\int_{\tau} \frac{[\bar{J}]}{r}\, d\tau \qquad (5.17\text{-}8)$$

The moving sources are said to constitute a volume current and $\bar{J}$ is known as the current density. The current is said to be neutral when $\rho = 0$, in which case the scalar potential is zero. This is possible only when suorces of positive and negative magnitude are present in appropriate numbers in each volume element. When all sources are of common sign ρ can be nowhere zero but $\bar{J}$ may vanish everywhere, as in the case of a configuration of common-magnitude sources with a random velocity distribution in each element (see Ex.5-72., p. 523).

The scalar and vector potentials developed above not only approximate the true or microscopic potentials (based upon (5.10-14/15) at exterior points of an assemblage of singlets, but also serve to define the macroscopic potentials at all points. Unlike the microscopic potentials with their attendant singularities within the distribution, the macroscopic potentials and their first space derivatives are continuous everywhere, and higher-order derivatives exist where ρ and $\bar{J}$ are continuous.

Similar analyses apply to surface and line distributions in motion upon fixed surfaces and along fixed contours. Thus we may define the scalar macroscopic surface and line densities by smooth interpolation between spot values of areal and linear density, and so obtain

$$\phi = \int_S \frac{1}{r}\left\{\left(\sum [\sigma_i]\right)_{a_1} + \left(\sum [\sigma_i]\right)_{a_2} \cdots\right\} dS \equiv \int_S \frac{[\sigma]}{r}\, dS \qquad (5.17\text{-}9)$$

13.
In earlier Sections the symbols $\bar{J}$, $\bar{K}$ and $\bar{I}$ were employed to denote vectorial source densities when treated as underived quantities, and the associated vector potentials were expressed in the basic form typified by $\bar{A} = \int \frac{\bar{J}}{r}\, d\tau$. This convention will be adhered to in later pages. However, when $\bar{J}$, $\bar{K}$ and $\bar{I}$ are employed to denote macroscopic current densities it is necessary to import the constant c into the denominator of the vector potential to permit of the subsequent expression of the macroscopic $\bar{E}$ and $\bar{B}$ fields, and of Maxwell's equations, in Gaussian form.

It should be noted that when $\bar{K}$ and $\bar{I}$ are treated as primary quantities they are not constrained to be orientated parallel to the associated surface or contour; when denoting current density, they are so constrained.

and

$$\phi = \int_{\Gamma} \frac{1}{r}\left\{\left(\sum [\lambda_i]\right)_{a_1} + \left(\sum [\lambda_i]\right)_{a_2} \cdots\cdots\right\} ds \equiv \int_{\Gamma} \frac{[\lambda]}{r}\, ds \quad (5.17\text{-}10)$$

The corresponding vector potentials are

$$\bar{A} = \int_{S} \frac{1}{cr}\left\{\left(\sum [\sigma_i]\bar{v}_i\right)_{a_1} + \left(\sum [\sigma_i]\bar{v}_i\right)_{a_2} \cdots\cdots\right\} dS \equiv \frac{1}{c}\int_{S} \frac{[\bar{K}]}{r}\, dS \quad (5.17\text{-}11)$$

and

$$\bar{A} = \int_{\Gamma} \frac{1}{cr}\left\{\left(\sum [\lambda_i]\bar{v}_i\right)_{a_1} + \left(\sum [\lambda_i]\bar{v}_i\right)_{a_2} \cdots\cdots\right\} ds \equiv \frac{1}{c}\int_{\Gamma} \frac{[\bar{I}]}{r}\, ds \quad (5.17\text{-}12)$$

where $\bar{K}$ and $\bar{I}$ are known respectively as the densities of surface and line currents.

An important relationship, known as the Equation of Continuity, subsists between the macroscopic vector and scalar density functions. We may proceed to develop this relationship by way of the following considerations.

When a source system comprising unaccelerated singlets is divided into equal-velocity sets, the unretarded macroscopic source density at any point is that resulting from the superposition of an ensemble of time-invariant scalar density fields which move as a whole with the corresponding set velocities. However, when individual sources are subject to acceleration there will be a transference of sources between sets and the moving density fields will exhibit spatial variations in time. We will proceed on the assumption that such variations are smooth.

Consider first a common-source-strength set having the velocity $\bar{v}_1$ and spatial density distribution ρ_1. It has been shown in Sec. 1.20 that for any well-behaved scalar field having both spatial and temporal dependence the rate of change of field strength at a point which moves with velocity $\bar{v}$ is related to that at a fixed, coincident point by

$$\frac{d}{dt} = \frac{\partial}{\partial t} + \bar{v}.\nabla$$

On applying this to the density field ρ_1 and putting $\bar{v} = \bar{v}_1$ we get

$$\frac{d\rho_1}{dt} = \frac{\partial\rho_1}{\partial t} + \bar{v}_1.\nabla\rho_1$$

But

$$\text{div}\ \rho_1\bar{v}_1 = \rho_1\ \text{div}\ \bar{v}_1 + \bar{v}_1.\nabla\rho_1 = \bar{v}_1.\nabla\rho_1$$

since $\bar{v}_1$ is a constant vector field,

hence

$$\text{div}\ \rho_1\bar{v}_1 = -\frac{\partial\rho_1}{\partial t} + \frac{d\rho_1}{dt}$$

The remaining velocity sets of equal source strength yield similar equations so that

$$\text{div}\ (\rho_1\bar{v}_1+\rho_2\bar{v}_2...) = -\frac{\partial}{\partial t}(\rho_1+\rho_2....) + \frac{d\rho_1}{dt} + \frac{d\rho_2}{dt}.... \qquad (5.17\text{-}13)$$

We now consider the source population enclosed by small spherical surfaces of equal radii which are carried along by the moving fields and are momentarily coincident. The strength of this population will approximate $(\rho_1+\rho_2..)\ \Delta\tau$, provided that the volume $\Delta\tau$ of the spheres is sufficiently large to accomodate a statistically-regular configuration for each velocity. Since the same sources continue to be contained by the spheres as they go their respective ways at the instant under consideration, it is evident that the individual density fields must be related at least approximately by

$$\frac{d\rho_1}{dt} + \frac{d\rho_2}{dt}...... = 0 \qquad (5.17\text{-}14)$$

This appeal to the conservation of population, and therefore of source strength, cannot establish equation (5.17-14) precisely, although it is intuitively obvious that the finer the source structure, and the greater the restraint thereby imposed in the smooth interpolation of the ρ fields, the more nearly will (5.17-14) hold at interior points of the source complex. We now postulate that the time and space smoothing undertaken in the generation of the macroscopic scalar and vector density functions is to be effected in such a way as to ensure that (5.17-14) holds <u>exactly</u>. Then equation (5.17-13) reduces to

$$\text{div}\ (\rho_1\bar{v}_1+\rho_2\bar{v}_2....) = -\frac{\partial}{\partial t}(\rho_1+\rho_2...)$$

Similar equations hold for each component of a mixed source strength ensemble hence in general

$$\text{div}\left\{\left(\sum \rho_i\bar{v}_i\right)_{a_1} + \left(\sum \rho_i\bar{v}_i\right)_{a_2} \cdots\right\} = -\frac{\partial}{\partial t}\left\{\left(\sum \rho_i\right)_{a_1} + \left(\sum \rho_i\right)_{a_2} \cdots\cdots\right\}$$

or

$$\text{div } \bar{J} = -\frac{\partial\rho}{\partial t} \tag{5.17-15}$$

This is the Equation of Continuity for a volume distribution of singlets. The corresponding equations for surface and line distributions are

$$\text{divs } \bar{K} = -\frac{\partial\sigma}{\partial t} \tag{5.17-16}$$

and

$$\frac{\partial I}{\partial s} = -\frac{\partial\lambda}{\partial t} \tag{5.17-17}$$

where

$$\bar{I} = I\,\hat{s}$$

It follows from the definition of $\bar{J}$ that the total source strength which passes in unit time through an element of surface ΔS orientated normally with respect to $\bar{J}$ in a region where $\bar{J}$ is continuous is approximately equal to $J\Delta S$, provided that the dimensions of ΔS are comparable with or greater than those of the volume cells employed in the determination of the parent scalar density functions. More generally, the rate of transfer of source strength in the positive sense through $\Delta\bar{S}$ is given by $\bar{J}.\hat{n}\ \Delta S$.

When the normal component of $\bar{J}$ is discontinuous through the surface element, the rate of increase of source strength upon it is approximately equal to $-\Delta\bar{J}.\hat{n}\ \Delta S$, where $\Delta\bar{J}$ is the increment of $\bar{J}$ in the direction of the arbitrarily-assigned common normal $\hat{n}$. Hence

$$-\Delta\bar{J}.\hat{n}\ \Delta S \approx \frac{\partial\sigma}{\partial t}\Delta S$$

where σ is the macroscopic surface density deriving from the flow of volume current. We now postulate that the time and space averaging of σ is to be such that at all interior points of any surface of discontinuity of $\bar{J}$ there holds the exact relationship

$$\Delta\bar{J}.\hat{n} = -\frac{\partial\sigma}{\partial t}$$

Then in the presence of a continuous current $\bar{K}$ upon the surface of discontinuity we have the general result

$$\text{divs }\bar{K} + \Delta\bar{J}.\hat{n} = -\frac{\partial\sigma}{\partial t} \qquad (5.17\text{-}18)$$

where σ now includes both volume and surface current contributions.

The corresponding relationship for a line of discontinuity of a surface current $\bar{K}$ which also carries a continuous current of density $\bar{I}$ is

$$\frac{\partial I}{\partial s} + \Delta\bar{K}.\hat{n}' = -\frac{\partial\lambda}{\partial t} \qquad (5.17\text{-}19)$$

where $\hat{n}'$ is tangential to the surface and normal to the line of discontinuity.

Before proceeding further, attention should be directed towards the anomalous nomenclature which surrounds the description of current flow. By associating the terms 'density of a volume source' (or volume density), 'density of a surface source' (or surface density) and 'density of a line source' (or line density) with the symbols $\bar{J}$, $\bar{K}$ and $\bar{I}$ it has been implied that current may be expressed in the forms $\int\bar{J}d\tau$, $\int\bar{K}dS$ and $\int\bar{I}ds$. However, the term 'current', when standing alone, is conventionally defined to be the rate of transfer of source strength through a surface, or across a line, or past a point, and is consequently a scalar quantity represented by $\int\bar{J}.d\bar{S}$, $\int\bar{K}.\hat{n}'\,ds$ and I.

Provided that the anomaly is borne in mind it should not cause any confusion.

EXERCISES

5-72. Show that $\bar{J}$ may be defined by smooth interpolation between spot values of

$$(\rho'\bar{v}_m)_{a_1} + (\rho'\bar{v}_m)_{a_2} + \ldots$$

where $(\rho')_{a_1}\,\Delta\tau$ is the total source strength within $\Delta\tau$ of sources of individual strength a_1 and $(\bar{v}_m)_{a_1}$ is their mean velocity.

5-73. Prove equations (5.17-16) and (5.17-17).

5.18 Construction of the Macroscopic Density and Potential Functions for Singlet, Doublet and Whirl Distributions

It is clear from the considerations of Secs. 4.20 and 5.17 that the manner in which the macroscopic density functions are constructed will largely determine the accuracy with which the macroscopic and microscopic potentials of statistically-regular distributions of discrete sources are matched at exterior points. We now examine this aspect in greater detail and extend considerations to configurations of stationary doublets and whirls.

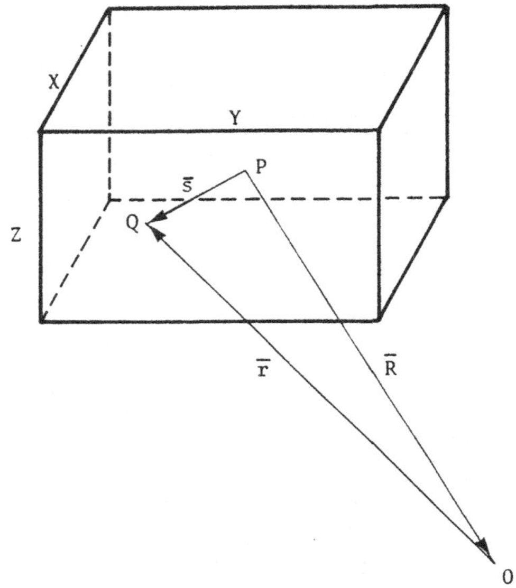

Fig. 5.11

Consider first a distribution of singlets having the common velocity $\bar{V} = \bar{i}V$, and within this distribution an elementary cell $\Delta\tau$ taking the form of a parallelepiped of edges X, Y, Z, aligned respectively with the x, y, z axes of coordinates, and of such dimensions that statistical continuity prevails between adjacent cells of comparable size. The cell centre is P (Fig. 5.11) and the point of evaluation of potential is O.

To maintain uniformity with previous notation we write

$$\vec{PO} = \bar{R} \qquad \vec{OQ} = \bar{r} \qquad \vec{PQ} = \bar{s}$$

where Q is some point of $\Delta\tau$.

In Sec. 5.17 the contribution of sources within $\Delta\tau$ to the retarded scalar potential at 0 was written as $\frac{Na}{R}$ ie $\frac{1}{R}\sum a_i$; we now sharpen the approximation by replacing $\frac{1}{R}\sum a_i$ with $\sum\left(\frac{a_i}{r_i}\right)$ ie we identify $\sum' a_i\left\{r_i\left(1+\frac{\bar{V}.\bar{r}_i}{cr_i}\right)\right\}^{-1}$ with $\sum\left(\frac{a_i}{r_i}\right)$ where $\sum'$ sums the contributions of the singlets in their individually retarded positions complete with Liénard modification, while $\sum$ sums the contributions at the instant $t_o - \frac{R}{c}$ without modification. This substitution presupposes that the variation of population due to movement in a subregion of $\Delta\tau$ just balances the Liénard factor as it applies to that subregion, and that the aggregate effect of the positional discrepancy at the time $t_o - \frac{R}{c}$ is negligible when taken over the complete cell. Once this approximation is accepted - and we will restrict V/c to values very small compared with unity in order to assist it - our concern lies with the degree of precision with which $\sum\left(\frac{a_i}{r_i}\right)$ may be replaced by $\int_{\Delta\tau}\frac{[\rho]}{r}\,d\tau$, where $[\rho]$ is a continuous density function having a statistical basis for evaluation.

Now from equation (1.2-9)

$$\left(\frac{[\rho]}{r}\right)_Q = \left(\frac{[\rho]}{r}\right)_P + \left(\bar{s}.\nabla\,\frac{[\rho]}{r}\right)_P + \frac{1}{2!}\left((\bar{s}.\nabla)^2\,\frac{[\rho]}{r}\right)_P + \ldots \qquad (5.18\text{-}1)$$

hence

$$\int_{\Delta\tau}\frac{[\rho]}{r}\,d\tau = \frac{[\rho]_P}{R}\,\Delta\tau + \left(\nabla\,\frac{[\rho]}{r}\right)_P\cdot\int_{\Delta\tau}\bar{s}\,d\tau$$

$$+\frac{1}{2}\int_{\Delta\tau}\left\{\left(s_x^2\frac{\partial^2}{\partial x^2}+s_y^2\frac{\partial^2}{\partial y^2}+s_z^2\frac{\partial^2}{\partial z^2}+2s_xs_y\frac{\partial^2}{\partial x\partial y}+2s_xs_z\frac{\partial^2}{\partial x\partial z}+2s_ys_z\frac{\partial^2}{\partial y\partial z}\right)\frac{[\rho]}{r}\right\}_P d\tau + \ldots$$

From symmetry,

$$\int_{\Delta\tau}\bar{s}\,d\tau = \bar{0}$$

The cross-terms of the second integral vanish likewise, leaving

$$\int_{\Delta\tau} \frac{[\rho]}{r}\, d\tau = \frac{[\rho]_P}{R}\,\Delta\tau + \frac{1}{2}\int_{\Delta\tau}\left\{s_x^2\left(\frac{\partial^2}{\partial x^2}\frac{[\rho]}{r}\right)_P + s_y^2\left(\frac{\partial^2}{\partial y^2}\frac{[\rho]}{r}\right)_P + s_z^2\left(\frac{\partial^2}{\partial z^2}\frac{[\rho]}{r}\right)_P\right\} d\tau + \ldots$$

$$= \frac{[\rho]_P}{R}\,\Delta\tau + \frac{\Delta\tau}{24}\left\{X^2\left(\frac{\partial^2}{\partial x^2}\frac{[\rho]}{r}\right)_P + Y^2\left(\frac{\partial^2}{\partial y^2}\frac{[\rho]}{r}\right)_P + Z^2\left(\frac{\partial^2}{\partial z^2}\frac{[\rho]}{r}\right)_P\right\} + \ldots$$

which, on expansion, becomes

$$\int_{\Delta\tau} \frac{[\rho]}{r}\, d\tau = \frac{[\rho]_P}{R}\,\Delta\tau\left\{1 + \frac{1}{24}\left(\frac{X^2}{R^2}\left(3\,\frac{R_x^2}{R^2} - 1\right) + \frac{Y^2}{R^2}\left(3\,\frac{R_y^2}{R^2} - 1\right) + \frac{Z^2}{R^2}\left(3\,\frac{R_z^2}{R^2} - 1\right)\right)\right\}$$

$$+ \frac{\Delta\tau}{12R^3}\left\{X^2 R_x\left(\frac{\partial[\rho]}{\partial x}\right)_P + Y^2 R_y\left(\frac{\partial[\rho]}{\partial y}\right)_P + Z^2 R_z\left(\frac{\partial[\rho]}{\partial z}\right)_P\right\} \qquad (5.18\text{-}2)$$

$$+ \frac{\Delta\tau}{24R}\left\{X^2\left(\frac{\partial^2[\rho]}{\partial x^2}\right)_P + Y^2\left(\frac{\partial^2[\rho]}{\partial y^2}\right)_P + Z^2\left(\frac{\partial^2[\rho]}{\partial z^2}\right)_P\right\} + \ldots$$

We now attempt to match (5.18-2) with the discrete summation $\sum \frac{a_i}{r_i}$ which may be expanded in accordance with equation (4.1-17) as

$$\sum\left(\frac{a_i}{r_i}\right) = \frac{1}{R}\sum a_i + \left(\sum a_i\bar{s}_i\right)\cdot\frac{\bar{R}}{R^3} + \ldots. \qquad (5.18\text{-}3)$$

If we retain only zeroth and first order terms in s/R, the terms in $\frac{X^2}{R^2}$ etc in the top line of (5.18-2) are deleted and (5.18-3) is truncated as shown. A match may then be effected by putting

$$[\rho]_P + \frac{1}{24}\left\{X^2\left(\frac{\partial^2[\rho]}{\partial x^2}\right)_P + Y^2\left(\frac{\partial^2[\rho]}{\partial y^2}\right)_P + Z^2\left(\frac{\partial^2[\rho]}{\partial z^2}\right)_P\right\} = \frac{1}{\Delta\tau}\sum a_i \qquad (5.18\text{-}4)$$

and

$$\frac{1}{12}\left\{\bar{i}\,X^2\left(\frac{\partial[\rho]}{\partial x}\right)_P + \bar{j}\,Y^2\left(\frac{\partial[\rho]}{\partial y}\right)_P + \bar{k}\,Z^2\left(\frac{\partial[\rho]}{\partial z}\right)_P\right\} = \frac{1}{\Delta\tau}\sum a_i\bar{s}_i \qquad (5.18\text{-}5)$$

While equation (5.18-4) can be satisfied, at least in principle, by an initial choice of $[\rho]_P = \frac{1}{\Delta\tau}\sum a_i$, as previously suggested, with some subsequent process of successive approximation, the derivatives of $[\rho]$, as they appear in equation (5.18-5), are thereby preempted, and the extent to which the two sides of (5.18-5) then balance is a measure of the degree of precision with which the discrete formulation may be replaced by the integral. It should be noted in this connection that the order of magnitude of the various terms in (5.18-2) is dependent upon both the order of the s/R factor involved and the rate of change of [ρ] and its derivatives. We will suppose, for present purposes, that the variation of [ρ] between adjacent cell centres (ie Δ[ρ]) is not necessarily small compared with [ρ] but that the variation of Δ[ρ] is an order smaller than Δ[ρ] itself, and so on for higher derivatives. In this case the first and second derivative terms of (5.18-2) are each of the first order of smallness. It is only when [ρ] varies rapidly from cell to cell that first order corrections appear at all. For slow variations of [ρ] all terms of (5.18-2), other than the leading term, are of second order.

It will be observed that equation (5.18-4) has been developed as a condition for the matching of microscopic and macroscopic potentials at exterior points, rather than the matching of discrete and continuous source strengths within an elementary cell. It may be shown, however, that to first order accuracy equation (5.18-4) represents a common criterion. (See Ex.5-74. and 5-77., pp. 530-1).

So far, considerations have been restricted to sources of common velocity, or those filling a narrow velocity slot. If a similar treatment is afforded each velocity slot the overall singlet density may be computed by scalar addition. Since integration is extended over the complete source system there is no requirement to employ the same elementary cells for each evaluation. The determination of an appropriate scalar density function for the individual slots leads immediately to corresponding values for the volume density of current via equation (5.17-7).

We turn now to the scalar potential of a volume distribution of time-dependent doublets whose source centres are fixed in space[14]. Since the potential of an individual doublet takes the form

$$[\bar{p}].\text{grad}\,\frac{1}{r} - \frac{\bar{r}}{cr^2}.\left[\frac{d\bar{p}}{dt}\right]$$

the macroscopic potential of a distribution may be assumed to be

$$-\int \left\{[\bar{P}].\frac{\bar{r}}{r^3} + \frac{\bar{r}}{cr^2}.\left[\frac{\partial\bar{P}}{\partial t}\right]\right\} d\tau$$

14. The potential functions developed in Secs. 5.13 to 5.15 assume that centroids of the elementary source systems remain at rest.

Expansion about the point P and integration over $\Delta\tau$ yields

$$[\bar{P}]_P \cdot \frac{\bar{R}}{R^3}\,\Delta\tau + \frac{\Delta\tau}{24R^3}\left\{X^2\bar{R}.\left(\frac{\partial^2[\bar{P}]}{\partial x^2}\right)_P + Y^2\bar{R}.\left(\frac{\partial^2[\bar{P}]}{\partial y^2}\right)_P + Z^2\bar{R}.\left(\frac{\partial^2[\bar{P}]}{\partial z^2}\right)_P\right.$$

$$+ \frac{\Delta\tau}{4R^3}\left\{\frac{X^2}{R^2}\,R_x\bar{R}.\left(\frac{\partial[\bar{P}]}{\partial x}\right)_P + \frac{Y^2}{R^2}\,R_y\bar{R}.\left(\frac{\partial[\bar{P}]}{\partial y}\right)_P + \frac{Z^2}{R^2}\,R_z\bar{R}.\left(\frac{\partial[\bar{P}]}{\partial z}\right)_P\right\}$$

$$- \frac{\Delta\tau}{12R^3}\left\{X^2\left(\frac{\partial}{\partial x}[P_x]\right)_P + Y^2\left(\frac{\partial}{\partial y}[P_y]\right)_P + Z^2\left(\frac{\partial}{\partial z}[P_z]\right)_P\right\} + \ldots \qquad (5.18\text{-}6)$$

$$+ \frac{\bar{R}}{cR^2}\cdot\left[\frac{\partial\bar{P}}{\partial t}\right]_P\Delta\tau + \frac{\Delta\tau}{24cR^2}\left\{X^2\bar{R}.\left(\frac{\partial^2}{\partial x^2}\left[\frac{\partial\bar{P}}{\partial t}\right]\right)_P + Y^2\bar{R}.\left(\frac{\partial^2}{\partial y^2}\left[\frac{\partial\bar{P}}{\partial t}\right]\right)_P + Z^2\bar{R}.\left(\frac{\partial^2}{\partial z^2}\left[\frac{\partial\bar{P}}{\partial t}\right]\right)_P\right\}$$

$$+ \frac{\Delta\tau}{6cR^2}\left\{\frac{X^2}{R^2}\,R_x\bar{R}.\left(\frac{\partial}{\partial x}\left[\frac{\partial\bar{P}}{\partial t}\right]\right)_P + \frac{Y^2}{R^2}\,R_y\bar{R}.\left(\frac{\partial}{\partial y}\left[\frac{\partial\bar{P}}{\partial t}\right]\right)_P + \frac{Z^2}{R^2}\,R_z\bar{R}.\left(\frac{\partial}{\partial z}\left[\frac{\partial\bar{P}}{\partial t}\right]\right)_P\right\}$$

$$- \frac{\Delta\tau}{12cR^2}\left\{X^2\left(\frac{\partial}{\partial x}\left[\frac{\partial P_x}{\partial t}\right]\right)_P + Y^2\left(\frac{\partial}{\partial y}\left[\frac{\partial P_y}{\partial t}\right]\right)_P + Z^2\left(\frac{\partial}{\partial z}\left[\frac{\partial P_z}{\partial t}\right]\right)_P\right\} + \ldots$$

The potential at O of the discrete system, to the first order in s/R, is easily shown to be

$$\sum[\bar{p}_i].\frac{\bar{R}}{R^3} + \sum 3[\bar{p}_i].\frac{\bar{R}}{R^3}\,\frac{\bar{s}_i.\bar{R}}{R^2} - \sum\frac{[\bar{p}_i].\bar{s}_i}{R^3}$$
$$+ \sum\left[\frac{d\bar{p}_i}{dt}\right].\frac{\bar{R}}{cR^2} + \sum 2\left[\frac{d\bar{p}_i}{dt}\right].\frac{\bar{R}}{cR^2}\,\frac{\bar{s}_i.\bar{R}}{R^2} - \sum\left[\frac{d\bar{p}_i}{dt}\right].\frac{\bar{s}_i}{cR^2}\,\ldots \qquad (5.18\text{-}7)$$

where $[\;] \equiv (\;)\; t_o - \frac{r}{c}$

Expressions (5.18-6) and (5.18-7) are matched precisely if

$$[\bar{P}]_P + \frac{1}{24}\left\{X^2\left(\frac{\partial^2[\bar{P}]}{\partial x^2}\right)_P + Y^2\left(\frac{\partial^2[\bar{P}]}{\partial y^2}\right)_P + Z^2\left(\frac{\partial^2[\bar{P}]}{\partial z^2}\right)_P\right\} = \frac{1}{\Delta\tau}\sum[\bar{p}_i] \qquad (5.18\text{-}8)$$

and

$$\frac{X^2}{12}\left(\frac{\partial[\bar{P}]}{\partial x}\right)_P = \frac{1}{\Delta\tau}\sum[\bar{p}_i]s_{ix}\;;\quad \frac{Y^2}{12}\left(\frac{\partial[\bar{P}]}{\partial y}\right)_P = \frac{1}{\Delta\tau}\sum[\bar{p}_i]s_{iy}\;; \qquad (5.18\text{-}9)$$

$$\frac{Z^2}{12}\left(\frac{\partial[\bar{P}]}{\partial z}\right)_P = \frac{1}{\Delta\tau}\sum[\bar{p}_i]s_{iz}$$

Similar relationships are required in which $\left[\frac{\partial \bar{P}}{\partial t}\right]$ and $\left[\frac{d\bar{p}_i}{dt}\right]$ replace $[\bar{P}]$ and $[\bar{p}_i]$, but these are automatically satisfied if the above equations hold.

Thus

$$\frac{\partial}{\partial x}\left[\frac{\partial \bar{P}}{\partial t}\right] = \frac{\partial}{\partial x}\frac{\partial}{\partial t}[\bar{P}] = \frac{\partial}{\partial t}\frac{\partial}{\partial x}[\bar{P}]$$

hence if

$$\frac{X^2}{12}\left(\frac{\partial}{\partial x}[\bar{P}]\right)_P = \sum [\bar{p}_i]s_{ix}\Big/\Delta\tau$$

then

$$\frac{X^2}{12}\left(\frac{\partial}{\partial x}\left[\frac{\partial \bar{P}}{\partial t}\right]\right)_P = \sum\left(\left[\frac{d\bar{p}_i}{dt}\right] s_{ix}\right)\Big/\Delta\tau$$

since s_i is independent of t.

An entirely similar result holds for the vector potential of a volume distribution of symmetrical point whirls (or continuous whirls). The integral formulation, to the order of accuracy discussed above, is

$$\int\left\{[\bar{M}] \times \text{grad}\,\frac{1}{r} - \left[\frac{\partial \bar{M}}{\partial t}\right] \times \frac{\bar{r}}{cr^2}\right\} d\tau$$

provided that equations (5.18-8) and (5.18-9) are satisfied when $[\bar{P}]$ and $[\bar{p}_i]$ are replaced by $[\bar{M}]$ and $[\bar{m}_i]$.

There remains to be considered the vector potential of a volume distribution of time-dependent doublets.

It follows from equation (5.13-10) that the integral form is $\frac{1}{c}\int\frac{1}{r}\left[\frac{\partial \bar{P}}{\partial t}\right]d\tau$, whence the required first-order matching conditions are found by substituting the rectangular components of $\left[\frac{\partial \bar{P}}{\partial t}\right]$ and $\left[\frac{d\bar{p}_i}{dt}\right]$ for $[\rho]$ and a_i in equations (5.18-4) and (5.18-5).

The vector relationships are then found to be identical with (5.18-8) and (5.18-9) when $\left[\frac{\partial \bar{P}}{\partial t}\right]$ and $\left[\frac{d\bar{p}_i}{dt}\right]$ are substituted for $[\bar{P}]$ and $[\bar{p}_i]$, and are consequently satisfied when equations (5.18-8) and (5.18-9) are satisified.

The above analyses are readily modified to cover surface and line distributions.

While we are not concerned here with physical applications of the foregoing treatment, it may be remarked in passing that in the field of electromagnetics the nature of the distributions encountered is such that cell sizes may usually be chosen sufficiently small as to render density variations across them negligible, and that, in consequence, zeroth order approximations suffice, both in the calculation of potential at exterior points and in the computation of ponderomotive interaction between non-overlapping configurations. Nevertheless, it should be borne in mind that the replacement of a discrete summation by an integral is a convenience which necessarily involves approximation, both in respect of the number of terms adopted and the construction of the corresponding density functions.

At interior points of discrete distributions, where the macroscopic and microscopic potentials cease to correspond, the relationship between their derivatives (as touched upon in Sec. 4.21 for time-invariant doublets and whirls) is clearly an additional subject for investigation. This will be considered briefly in Sec. 5.20.

EXERCISES

In the following exercises it will be supposed that the variation of density from cell to cell, $\Delta\rho$, is not necessarily small compared with ρ, but that second differences are small compared with first, and so on.

5-74. Show that the unretarded form of equation (5.18-4) is consistent with the first order matching of $\sum a_i$ and $\int_{\Delta\tau} \rho d\tau$. Hence conclude that the argument leading to the equation of continuity is unaffected by the modified construction of the density field.

5-75. Any discrepancy between the left and right hand sides of equation (5.18-4), when applied to a distribution of stationary singlets, may be corrected in an ad hoc manner by assigning to the distribution a polarisation density $\overline{P}$, where

$$(\overline{P})_P = \frac{1}{\Delta\tau}\sum a_i\overline{s}_i - \frac{1}{12}\left\{\overline{i}X^2\left(\frac{\partial\rho}{\partial x}\right)_P + \overline{j}Y^2\left(\frac{\partial\rho}{\partial y}\right)_P + \overline{k}Z^2\left(\frac{\partial\rho}{\partial z}\right)_P\right\}$$

Prove this, and argue plausibly that the more finely-grained the structure of the system the smaller the value of $\overline{P}$ that needs to be invoked for a given value of polarisation $\sum a_i\overline{s}_i$ within the cell.

Show that it is not possible to supplement $\overline{J}$ with $\overline{M}$ to force a fit in the equivalent expression for steady current flow with rapid transverse variation.

5-76. A point function ϕ varies slowly from cell to cell within a distribution of stationary singlets. Confirm that $\sum a_i\phi_i$ and $\int_{\Delta\tau} \rho\phi d\tau$ are precisely matched to the first order of smallness if ρ can be chosen to satisfy the unretarded forms of equations (5.18-4) and (5.18-5).

Investigate the corresponding requirements for the matching of $\sum (\bar{r}_i \times a_i \bar{E}_i)$ and $\int_{\Delta\tau} (\bar{r}\times\rho\bar{E})\, d\tau$ where $\bar{E}$ is a slowly-varying point function.

5-77. For the mode of construction adopted above, the unretarded density function ρ is required to satisfy the following relationship at the point P and time $t_o - R/c$

$$(\rho)_{P\atop t_o-\frac{R}{c}} + \frac{1}{24}\left\{X^2\left(\frac{\partial^2\rho}{\partial x^2}\right)_{P\atop t_o-\frac{R}{c}} + Y^2\left(\frac{\partial^2\rho}{\partial y^2}\right)_{P\atop t_o-\frac{R}{c}} + Z^2\left(\frac{\partial^2\rho}{\partial z^2}\right)_{P\atop t_o-\frac{R}{c}}\right\} = \frac{1}{\Delta\tau}\left(\sum a_i\right)_{t_o-R/c}$$

ie

$$[\rho]_P + \frac{1}{24}\left\{X^2\left[\frac{\partial^2\rho}{\partial x^2}\right]_P + Y^2\left[\frac{\partial^2\rho}{\partial y^2}\right]_P + Z^2\left[\frac{\partial^2\rho}{\partial z^2}\right]_P\right\} = \frac{1}{\Delta\tau}\left(\sum a_i\right)_{t_o-R/c}$$

while the retarded density function is required to satisfy

$$[\rho]_P + \frac{1}{24}\left\{X^2\left(\frac{\partial^2}{\partial x^2}[\rho]\right)_P + Y^2\left(\frac{\partial^2}{\partial y^2}[\rho]\right)_P + Z^2\left(\frac{\partial^2}{\partial z^2}[\rho]\right)_P\right\} = \frac{1}{\Delta\tau}\left(\sum a_i\right)_{t_o-R/c}$$

Show by expansion of $\frac{\partial^2}{\partial x^2}[\rho]$ etc into its five constituents that these relationships are consistent to within the first order of smallness, provided that there is negligible variation of ρ during the transit time of the expanding sphere across $\Delta\tau$.

5.19 The Macroscopic Potentials of a Composite Source System

The Polarisation Potentials

The Lorentz Gauge

5.19a The macroscopic potentials of a composite source system

It follows from Secs. 5.17 and 5.18 that the macroscopic retarded scalar potential of a system comprising volume, surface and line distributions of singlets and volume distributions of doublets is given by

$$\phi = \int_{\infty} \frac{[\rho]}{r}\,d\tau + \int_{S} \frac{[\sigma]}{r}\,dS + \int_{\Gamma} \frac{[\lambda]}{r}\,ds + \int_{\infty} \left\{ [\bar{P}].\text{grad}\,\frac{1}{r} - \left[\frac{\partial \bar{P}}{\partial t}\right].\frac{\bar{r}}{cr^2} \right\} d\tau \qquad (5.19\text{-}1)$$

provided that the surface S, the contours Γ and the centres of the doublets are fixed in space.

Surfaces and contours may be open or closed and are currently represented by a single subscript in the associated integral. Since, for present purposes, the volume distributions of singlets and doublets may or may not overlap and may involve several distinct regions, the volume integral is represented as taken over all space; it is supposed, however, that volume densities are zero outside a spherical surface of finite radius centred upon a local origin. The surface density σ may include a time-invariant component, a time-dependent component associated with a discontinuity of volume current $\bar{J}$, and a time-dependent component deriving from a current $\bar{K}$ within the surface where $\text{divs}\,\bar{K} \neq 0$. The line density λ, in turn, may comprise a time-invariant component, a time-dependent component associated with discontinuity of surface current $\bar{K}$, and a time-dependent component deriving from a variation of line current density I along Γ.

When I is discontinuous, or Γ is open and I is non-zero at the end points, it becomes necessary to admit one or more 'macroscopic' point sources in order to keep faith with the model.

Thus if a denotes the strength of the point source, it is clear that at an interior discontinuity of I

$$-\Delta I = \frac{da}{dt} \qquad (5.19\text{-}2(a))$$

where ΔI is the scalar line current increment in passing in a positive sense through the discontinuity.

We see also that if positive movement between open ends involves movement from P to Q then

$$I_P = -\left(\frac{da}{dt}\right)_P \quad \text{and} \quad I_Q = +\left(\frac{da}{dt}\right)_Q \qquad (5.19\text{-}2(b))$$

It should be noted that this type of point source is wedded to the concept of continuity so that time dependence of source strength can derive only from singlet motion. The 'non-macroscopic' point source considered in Sec. 5.5 had a different connotation.

Equation (5.19-1) can be further generalised by the addition of terms representing the potentials of surface and line doublets. The latter are of little significance in applications of the theory and will consequently be ignored. The surface doublet potential takes the form

$$\int_S \left\{ [\overline{P}'].\text{grad}\,\frac{1}{r} - \left[\frac{\partial \overline{P}'}{\partial t}\right].\frac{\overline{r}}{cr^2} \right\} dS$$

where $\overline{P}'$ is the macroscopic doublet moment per unit area.

The doublets may or may not be orientated normally to the surfaces, but their centres are supposed to be fixed in space.

More generally, then,

$$\phi = \int_\infty \frac{[\rho]}{r}\,d\tau + \int_S \frac{[\sigma]}{r}\,dS + \int_\Gamma \frac{[\lambda]}{r}\,ds + \sum \frac{[a]}{r}$$

$$+ \int_\infty \left\{ [\overline{P}].\text{grad}\,\frac{1}{r} - \left[\frac{\partial \overline{P}}{\partial t}\right].\frac{\overline{r}}{cr^2} \right\} d\tau + \int_S \left\{ [\overline{P}'].\text{grad}\,\frac{1}{r} - \left[\frac{\partial \overline{P}'}{\partial t}\right].\frac{\overline{r}}{cr^2} \right\} dS \tag{5.19-1(a)}$$

The corresponding expression for the vector potential, with the addition of terms deriving from volume and surface distributions of whirls, is

$$\overline{A} = \frac{1}{c}\int_\infty \frac{[\overline{J}]}{r}\,d\tau + \frac{1}{c}\int_S \frac{[\overline{K}]}{r}\,dS + \frac{1}{c}\int_\Gamma \frac{[\overline{I}]}{r}\,ds + \frac{1}{c}\int_\infty \frac{1}{r}\left[\frac{\partial \overline{P}}{\partial t}\right] d\tau + \frac{1}{c}\int_S \frac{1}{r}\left[\frac{\partial \overline{P}'}{\partial t}\right] dS$$

$$+ \int_\infty \left\{ [\overline{M}]\times\text{grad}\,\frac{1}{r} - \left[\frac{\partial \overline{M}}{\partial t}\right] \times \frac{\overline{r}}{cr^2} \right\} d\tau + \int_S \left\{ [\overline{M}']\times\text{grad}\,\frac{1}{r} - \left[\frac{\partial \overline{M}'}{\partial t}\right] \times \frac{\overline{r}}{cr^2} \right\} dS \tag{5.19-3}$$

where $\overline{M}'$ is the macroscopic whirl moment per unit area.

We are sometimes concerned with only a portion of the total source complex, such as that bounded externally by the closed surface Σ. This will be referred to as a subsource. A subsource is said to be complete if no volume, surface or line current crosses Σ. In terms of the microscopic model such completeness implies the existence of scalar, time-dependent surface and line densities and point source strengths upon Σ, wherever there is a normal component of volume or surface current or an intersection with line current, in accordance with the equations of continuity:

$$\frac{\partial \sigma}{\partial t} = \overline{J}.\hat{\overline{n}} \quad ; \quad \frac{\partial \lambda}{\partial t} = \overline{K}.\hat{\overline{n}}' \quad ; \quad \frac{da}{dt} = I \tag{5.19-4}$$

where $\hat{\overline{n}}$ is the unit outward normal to Σ, $\hat{\overline{n}}'$ is that outwardly-directed normal to the line of intersection of S and Σ which is tangential to S, and I is positive when directed outwards.

When these equations are not satisfied (as when current is continuous through Σ) we may render the subsource complete by allocating source densities in accordance with equation (5.19-4) to the inner side of Σ and equal and opposite densities to the outer side. Since the paired densities are actually superimposed the construction does not affect the potentials of the overall source complex, but it enables the completeness criterion to be met for the subsource bounded externally by Σ and also for that particular bounding surface of the exterior complex. The concept of completeness is not relevant to doublet or whirl distributions; their macroscopic density fields may be subdivided without restriction.

The application of the grad, div and curl operators to multiple volume sources does not present any difficulty because the formulae previously developed apply at all points where the derivatives are defined. It should be borne in mind, however, that those formulae involving integrals over bounding surfaces have been based on the assumption that the density functions have continuous derivatives throughout the bounded region. Hence surfaces of discontinuity must be excluded from the region of volume integration in such cases. As shown elsewhere, the effect of such discontinuities is represented by the resultant of paired surface integrals taken over the excluding surfaces. For an unbounded system of volume sources such surface integrals alone survive. Since dal operates upon a bounded scalar or vector volume source to yield a value at an interior point which depends upon the local source density (or its derivatives) and which is zero beyond the source, and since it is unaffected by the presence of discontinuities which lie beyond a neighbourhood of the point of evaluation, it is possible to employ a single formulae for dal ϕ and dal $\bar{A}$ at all points where it is defined, eg dal pot $[\bar{J}] = -4\pi\bar{J}$. This is accomplished by supposing that the volume source extends to infinity but has zero density outside the source proper. All points then become interior points of the source and dal is additive for multiple sources without regard for position.

The d'Alembertian of equation (5.19-1(a)) assumes a simple form. It may be shown that the component deriving from the surface doublet potential is zero at exterior points of the surfaces; reference to equations (5.7-2/3/4), (5.7-7) and (5.8-9) then allows us to write

$$\text{dal } \phi = -4\pi(\rho - \text{div } \bar{P}) \tag{5.19-5}$$

Similarly, surface whirl distributions contribute nothing to dal $\bar{A}$ at exterior points. Hence from equations (5.7-9/10), (5.7-12) and (5.9-10)

$$\text{dal } \bar{A} = \frac{-4\pi}{c}\left(\bar{J} + \frac{\partial \bar{P}}{\partial t} + c \text{ curl } \bar{M}\right) \tag{5.19-6}$$

Boundary conditions for the macroscopic ϕ and $\bar{A}$ may be summarised as follows:

(1) ϕ is continuous through and upon

(a) the bounding surface of a volume distribution of singlets.

(b) a surface distribution of singlets.

(c) the bounding surface of a volume distribution of doublets.

(d) a tangentially-orientated surface distribution of doublets at interior points where P' is continuous.

(1a) and (1b) follow from the arguments developed in Secs. 4.3 and 4.4 for the non-retarded case.

(1c) follows from the transformation (5.8-1) coupled with the arguments leading to (1a) and (1b).

(1d) is the subject of Ex.5-82., p. 542.

ϕ is discontinuous by $4\pi P'$ for movement in a positive sense through a normally-orientated surface distribution of doublets at interior points where P' is continuous.

This result follows from integration of $\int_S \left\{[\bar{P}'].\text{grad}\,\frac{1}{r} - \left[\frac{\partial \bar{P}'}{\partial t}\right].\frac{\bar{r}}{cr^2}\right\} dS$

along the lines suggested in Ex.4-14., p. 242.

(2) $\bar{A}$ is continuous through and upon

(a) the bounding surface of a volume distribution of current.

(b) a surface distribution of current.

(c) the bounding surface of a volume distribution of whirls.

(d) a normally-orientated surface distribution of whirls at interior points where M' is continuous.

(2a) and (2b) are the vector analogues of (1a) and (1b).

(2c) follows from the transformation (5.9-1) coupled with the arguments leading to (2a) and (2b).

(2d) is the subject of Ex.5-83., p. 542.

5.19b The polarisation potentials[15]

When doublet and whirl distributions alone are present the scalar and vector potentials are given by

$$\phi = \int_\infty \left\{[\bar{P}].\text{grad}\,\frac{1}{r} - \left[\frac{\partial \bar{P}}{\partial t}\right].\frac{\bar{r}}{cr^2}\right\} d\tau + \int_S \left\{[\bar{P}'].\text{grad}\,\frac{1}{r} - \left[\frac{\partial \bar{P}'}{\partial t}\right].\frac{\bar{r}}{cr^2}\right\} dS \tag{5.19-7}$$

15. In electromagnetic theory the word 'polarisation' is associated, for historical reasons, both with doublet and whirl distributions. But according to the definition given in Sec. 4.1 a compensated point whirl exhibits zero polarisation.

$$\bar{A} = \int_\infty \left\{[\bar{M}]\times\text{grad}\,\frac{1}{r} - \left[\frac{\partial\bar{M}}{\partial t}\right] \times \frac{\bar{r}}{cr^2}\right\} d\tau + \int_S \left\{[\bar{M}']\times\text{grad}\,\frac{1}{r} - \left[\frac{\partial\bar{M}'}{\partial t}\right] \times \frac{\bar{r}}{cr^2}\right\} dS \tag{5.19-8}$$

$$+ \frac{1}{c}\int_\infty \frac{1}{r}\left[\frac{\partial\bar{P}}{\partial t}\right] d\tau + \frac{1}{c}\int_S \frac{1}{r}\left[\frac{\partial\bar{P}'}{\partial t}\right] dS$$

It follows from the considerations of Secs. 5.6d and 5.6f (with appropriate change of symbols) that we may replace equations (5.19-7) and (5.19-8) by

$$\phi = -\,\text{div}\int_\infty \frac{[\bar{P}]}{r}\,d\tau - \text{div}\int_S \frac{[\bar{P}']}{r}\,dS \tag{5.19-9}$$

$$\bar{A} = \text{curl}\int_\infty \frac{[\bar{M}]}{r}\,d\tau + \text{curl}\int_S \frac{[\bar{M}']}{r}\,dS + \frac{1}{c}\frac{\partial}{\partial t}\int_\infty \frac{[\bar{P}]}{r}\,d\tau + \frac{1}{c}\frac{\partial}{\partial t}\int_S \frac{[\bar{P}']}{r}\,dS \tag{5.19-10}$$

On writing

$$\int_\infty \frac{[\bar{P}]}{r}\,d\tau + \int_S \frac{[\bar{P}']}{r}\,dS = \bar{\Pi}_e \tag{5.19-11}$$

$$\int_\infty \frac{[\bar{M}]}{r}\,d\tau + \int_S \frac{[\bar{M}']}{r}\,dS = \bar{\Pi}_m \tag{5.19-12}$$

we have

$$\phi = -\,\text{div}\,\bar{\Pi}_e \tag{5.19-13}$$

$$\bar{A} = \text{curl}\,\bar{\Pi}_m + \frac{1}{c}\frac{\partial}{\partial t}\,\bar{\Pi}_e \tag{5.19-14}$$

$\bar{\Pi}_e$ and $\bar{\Pi}_m$ are known as the polarisation potentials[16].

16. They are also known as Hertzian vectors.

It follows from equations (5.7-9) and (5.7-12) (with appropriate substitution) that

$$\text{dal}\ \overline{\Pi}_e = -4\pi\overline{P} \tag{5.19-15}$$

$$\text{dal}\ \overline{\Pi}_m = -4\pi\overline{M} \tag{5.19-16}$$

5.19c The Lorentz gauge[17]

An important relationship subsists between certain derivatives of the macroscopic scalar and vector potentials of any complete source, viz

$$\text{div}\ \overline{A} = -\frac{1}{c}\frac{\partial\phi}{\partial t} \tag{5.19-17}$$

This relationship has already been shown to hold for the Liénard-Wiechert potentials of a point singlet moving in any manner with $V < c$, and must therefore hold for the microscopic potentials of any combination of point sources. We now proceed to demonstrate its validity for the macroscopic potentials. This is most easily accomplished by splitting the total source complex into associated scalar and vector densities and treating each separately.

We can dispose of doublet and whirl components immediately by reference to equations (5.19-13) and (5.19-14), whence the required relationship follows.

For those components of potential deriving from the volume distribution and associated surface distribution of singlets we have

$$\phi = \int_\tau \frac{[\rho]}{r}\,d\tau + \int_{S\Sigma} \frac{[\sigma]}{r}\,dS$$

$$\overline{A} = \frac{1}{c}\int_\tau \frac{[\overline{J}]}{r}\,d\tau$$

Here τ represents the region of subsource, S refers to interior surfaces of discontinuity of $\overline{J}$, and Σ denotes the bounding surface or surfaces.

17. In the present work the relationship expressed by the Lorentz gauge follows from the definitions of the macroscopic potentials. This is not always the case (see Sec. 7.9).

Then at interior and exterior points of τ

$$\operatorname{div} \bar{A} = \frac{1}{c}\int_\tau \frac{[\operatorname{div} \bar{J}]}{r}\, d\tau - \frac{1}{c}\int_S \frac{[-\Delta\bar{J}]}{r}.\hat{n}\, dS - \frac{1}{c}\oint_\Sigma \frac{[\bar{J}]}{r}.\hat{n}\, dS$$

$$= \frac{1}{c}\int_\tau \frac{1}{r}\left[-\frac{\partial\rho}{\partial t}\right] d\tau - \frac{1}{c}\int_S \frac{1}{r}\left[\frac{\partial\sigma}{\partial t}\right] dS - \frac{1}{c}\oint_\Sigma \frac{1}{r}\left[\frac{\partial\sigma}{\partial t}\right] dS$$

$$= -\frac{1}{c}\frac{\partial\phi}{\partial t}$$

We next consider the potentials deriving from the surface currents (including possible surface currents upon Σ), the scalar surface densities arising from the non-zero surface divergence of these currents and the line densities arising from their discontinuities. In this case

$$\phi = \int_{S\Sigma} \frac{[\sigma]}{r}\, dS + \int_\Gamma \frac{[\lambda]}{r}\, dS$$

$$\bar{A} = \frac{1}{c}\int_{S\Sigma} \frac{[\bar{K}]}{r}\, dS$$

where the contours may be closed or unclosed and include those defined by the intersection of S with Σ.

Then from equation (5.6-20)

$$\operatorname{div} \bar{A} = -\frac{1}{c}\int_{S\Sigma} \left\{[\bar{K}].\operatorname{grad}\frac{1}{r} - \left[\frac{\partial\bar{K}}{\partial t}\right].\frac{\bar{r}}{cr^2}\right\} dS$$

But from equation (2.12-12)

$$\operatorname{grad}\frac{1}{r} = \operatorname{grads}\frac{1}{r} + \hat{n}\frac{\partial}{\partial n}\left(\frac{1}{r}\right)$$

hence

$$[\bar{K}].\operatorname{grad}\frac{1}{r} = [\bar{K}].\operatorname{grads}\frac{1}{r} \quad \text{since} \quad \bar{K}.\hat{n} = 0$$

Also, from equation (2.12-7),

$$\operatorname{divs}\frac{[\bar{K}]}{r} = \frac{1}{r}\operatorname{divs}[\bar{K}] + [\bar{K}].\operatorname{grads}\frac{1}{r}$$

whence, from a result of Ex.5-6., p. 404.

$$\operatorname{divs}\frac{[\bar{K}]}{r} = \frac{1}{r}[\operatorname{divs}\bar{K}] - \frac{\bar{r}}{cr^2}\cdot\left[\frac{\partial\bar{K}}{\partial t}\right] + \frac{\hat{\bar{n}}}{cr}\cdot\left[\frac{\partial\bar{K}}{\partial t}\right]\frac{\partial r}{\partial n} + [\bar{K}].\operatorname{grads}\frac{1}{r}$$

so that

$$[\bar{K}].\operatorname{grad}\frac{1}{r} - \frac{\bar{r}}{cr^2}\cdot\left[\frac{\partial\bar{K}}{\partial t}\right] = \operatorname{divs}\frac{[\bar{K}]}{r} - \frac{1}{r}[\operatorname{divs}\bar{K}]$$

and

$$\operatorname{div}\bar{A} = -\frac{1}{c}\int_{S\Sigma}\operatorname{divs}\frac{[\bar{K}]}{r}\,dS + \frac{1}{c}\int_{S\Sigma}\frac{1}{r}[\operatorname{divs}\bar{K}]\,dS$$

$$= -\frac{1}{c}\oint_{\Gamma}\frac{[\bar{K}]}{r}.\hat{\bar{n}}'\,ds + \frac{1}{c}\int_{S\Sigma}\frac{1}{r}[\operatorname{divs}\bar{K}]\,dS \qquad \text{from (2.12-27(a))}$$

where Γ denotes closed contours which embrace regions of the surfaces within which $\bar{K}$ is well-behaved. These contours comprise paired sections immediately adjacent to lines of discontinuity of $\bar{K}$ at interior points, and unpaired sections comprising the boundaries of open surfaces within τ or lines of intersection of S with Σ. In each case $\bar{K}.\hat{\bar{n}}'$ may be replaced by $\frac{\partial\lambda}{\partial t}$ along a closed or open contour, and since, in addition, $\operatorname{divs}\bar{K} = -\frac{\partial\sigma}{\partial t}$, we have

$$\operatorname{div}\bar{A} = -\frac{1}{c}\int_{\Gamma}\frac{1}{r}\left[\frac{\partial\lambda}{\partial t}\right]ds + \frac{1}{c}\int_{S\Sigma}\frac{1}{r}\left[-\frac{\partial\sigma}{\partial t}\right]dS = -\frac{1}{c}\frac{\partial\phi}{\partial t}$$

Finally, we consider the potentials deriving from line currents, their associated line densities, and the point source strengths resulting from their discontinuities.

$$\phi = \int_{\Gamma}\frac{[\lambda]}{r}\,ds + \sum\frac{[a]}{r}$$

$$\bar{A} = \frac{1}{c}\int_{\Gamma}\frac{[\bar{I}]}{r}\,ds$$

At points outside Γ equation (5.6-19) yields

$$\mathrm{div}\,\bar{A} = -\frac{1}{c}\int_\Gamma \left\{[\bar{I}].\mathrm{grad}\,\frac{1}{r} - \left[\frac{\partial\bar{I}}{\partial t}\right].\frac{\bar{r}}{cr^2}\right\} ds$$

Now

$$\frac{\partial}{\partial s}\frac{[I]}{r} = \frac{1}{r}\frac{\partial}{\partial s}[I] + [I]\frac{\partial}{\partial s}\left(\frac{1}{r}\right)$$

$$= \frac{1}{r}\left\{\left[\frac{\partial I}{\partial s}\right] - \frac{1}{c}(\hat{s}.\nabla r)\left[\frac{\partial I}{\partial t}\right]\right\} + [I]\left(\hat{s}.\nabla\frac{1}{r}\right)$$

whence

$$[\bar{I}].\mathrm{grad}\,\frac{1}{r} - \left[\frac{\partial\bar{I}}{\partial t}\right].\frac{\bar{r}}{cr^2} = \frac{\partial}{\partial s}\frac{[I]}{r} - \frac{1}{r}\left[\frac{\partial I}{\partial s}\right]$$

and

$$\mathrm{div}\,\bar{A} = -\frac{1}{c}\int_\Gamma \frac{\partial}{\partial s}\frac{[I]}{r}\,ds + \frac{1}{c}\int_\Gamma \frac{1}{r}\left[-\frac{\partial\lambda}{\partial t}\right] ds$$

where Γ comprises a set of line segments having continuous values of I at interior points.

On applying equations (5.19-2(a)/(b)) (and (5.19-4) as required) to an open contour composed of two such segments, AB and BC, we get

$$\int_A^{C} \frac{\partial}{\partial s}[I]\,ds = \left(\frac{[I]}{r}\right)_C + \left(\frac{[-\Delta I]}{r}\right)_B - \left(\frac{[I]}{r}\right)_A$$

$$= \left(\frac{1}{r}\left[\frac{da}{dt}\right]\right)_C + \left(\frac{1}{r}\left[\frac{da}{dt}\right]\right)_B + \left(\frac{1}{r}\left[\frac{da}{dt}\right]\right)_A$$

so that in general

$$\mathrm{div}\,\bar{A} = -\frac{1}{c}\sum\frac{1}{r}\left[\frac{da}{dt}\right] - \frac{1}{c}\int_\Gamma \frac{1}{r}\left[\frac{\partial\lambda}{\partial t}\right] ds = -\frac{1}{c}\frac{\partial\phi}{\partial t}$$

It now follows from superposition that $\mathrm{div}\,\bar{A} = -\frac{1}{c}\frac{\partial\phi}{\partial t}$ for the overall configuration. It will be observed that densities associated with static assemblages of point singlets have been ignored; these contribute neither to $\frac{\partial\phi}{\partial t}$ nor to $\bar{A}$. In the case of an incomplete subsource the relationship fails because the surface integrals over Σ which appear in the expressions for divergence are not matched by corresponding integrals in the scalar potential. For unbounded sources the problem does not arise because such sources are necessarily complete.

EXERCISES

5-78. Let the macroscopic point functions $\overline{J}$, ρ and σ define a complete subsource. If considerations are restricted to time-dependent components alone, show that the retarded potentials deriving from the subsource are identical with those which derive from a doublet distribution occupying the same region, provided that

$$(\overline{P})_t = \int_{t_o}^{t} \overline{J}\, dt$$

where t_o is an arbitrary datum.

Hence or otherwise show that a complete linear subsource characterised by I, λ and a may be replaced by an axial doublet distribution of macroscopic moment per unit length $\int_{t_o}^{t} I\, dt$.

Demonstrate that the equivalence fails when the subsource is incomplete because of the emergence of fictitious point sources at the extremities of the surrogate system.

5-79. Establish the value of the retarded scalar potential upon the axis of a disc-shaped, normally-polarised, surface doublet distribution of uniform scalar density P', and so plausibly demonstrate that the potential changes by $4\pi P'$ for positive movement through a regular, normally-polarised surface distribution at an interior point where P' is the local value of (continuous) surface density.

5-80. Derive the surface equivalent of equation (5.8-8), viz

$$\text{grad} \int_S \left\{[\overline{P}'].\text{grad}\,\frac{1}{r} - \left[\frac{\partial \overline{P}'}{\partial t}\right].\frac{\overline{r}}{cr^2}\right\} dS = -\,\text{curl} \int_S \left\{[\overline{P}']\times\text{grad}\,\frac{1}{r} - \left[\frac{\partial \overline{P}'}{\partial t}\right]\times\frac{\overline{r}}{cr^2}\right\} dS$$

$$- \frac{1}{c^2}\frac{\partial^2}{\partial t^2} \int_S \frac{[\overline{P}']}{r}\, dS$$

at points off the surface.

Hence demonstrate that the macroscopic $\overline{E}$ field of a doublet surface distribution (not necessarily uniform or normally polarised) is equal, at exterior points, to the $\overline{B}$ field of a surface distribution of whirls of equal vector density.

5-81. Utilise the relationship of Ex.5-80. to prove that

$$\left.\begin{aligned} \text{dal}\int_S \left\{[\overline{P}'].\text{grad}\,\frac{1}{r} - \left[\frac{\partial \overline{P}'}{\partial t}\right].\frac{\overline{r}}{cr^2}\right\} dS &= 0 \\ \text{dal}\int_S \left\{[M']\times\text{grad}\,\frac{1}{r} - \left[\frac{\partial \overline{M}'}{\partial t}\right]\times\frac{\overline{r}}{cr^2}\right\} dS &= 0 \end{aligned}\right\}\ \text{at exterior points}$$

Employ the identities dal div ≡ div dal and dal curl ≡ curl dal, together with equation (5.7-9), to arrive at the same results.

5-82. Show that the retarded potential of a tangentially-orientated, non-uniform, open surface distribution of doublets is equal, at exterior points, to

$$\oint_\Gamma \frac{[\overline{P}']}{r}.\hat{\overline{n}}'\,ds - \int_S \frac{1}{r}\,[\text{divs}\,\overline{P}']\,dS$$

where $\hat{\overline{n}}' = \hat{\overline{s}} \times \hat{\overline{n}}$.

Hence conclude that the potential of the doublet source is duplicated by that of a peripheral singlet line source of density $\overline{P}'.\hat{\overline{n}}'$ together with a singlet surface source of density $-\text{divs}\,\overline{P}'$, and deduce the associated continuity condition for ϕ.

5-83. Make use of equations (2.12-12), (2.12-6), (2.12-28) and a result of Ex.5-6., p. 404 in turn to show that

$$\int_S \left\{[\overline{M}']\times\text{grad}\,\frac{1}{r} - \left[\frac{\partial \overline{M}'}{\partial t}\right]\times\frac{\overline{r}}{cr^2}\right\} dS = \oint_\Gamma \frac{[M']}{r}\,d\overline{r} - \int_S \frac{d\overline{S}\times[\text{grads}\,M']}{r}$$

if $\overline{M}'$ is everywhere normal to S.

Hence conclude that an open, non-uniform surface distribution of normally-orientated whirls gives rise to a vector potential at exterior points which is equal to that associated with a line current of scalar density M' around the periphery together with a surface current of density (grads M') × $\hat{\overline{n}}$. (This is a generalisation of the uniform, time-invariant case discussed in Sec. 4.20.) Deduce that the vector potential of a piecewise-continuous surface distribution of normally-orientated whirls is continuous for movement through the surface at interior points where M' is continuous.

5-84. Show that the macroscopic vector potential of a tangentially-orientated, plane-surface whirl distribution is given, for an arbitrary positive sense of $\hat{\bar{n}}$, by

$$\bar{A} = \int_S \left\{ ([\bar{M}']\times\hat{\bar{n}}) \frac{\partial}{\partial n}\left(\frac{1}{r}\right) - \frac{1}{cr}\frac{\partial r}{\partial n}\frac{\partial}{\partial t}([\bar{M}']\times\hat{\bar{n}}) \right\} dS$$

$$+ \int_S \frac{1}{r}[\text{curls } \bar{M}'] \, dS + \oint_\Gamma \frac{[\bar{M}']\times\hat{\bar{n}}'}{r} \, ds$$

Hence demonstrate that this surface source is equivalent to a tangential double surface source of density $\bar{M}' \times \hat{\bar{n}}$ (see equation 5.5-11), a normal vector surface source of density curls $\bar{M}'$, and a normal vector line source of density $\bar{M}' \times \hat{\bar{n}}'$.

Show that $\bar{A}$ is discontinuous by $4\pi(\bar{M}'\times\hat{\bar{n}})$ for movement through the surface at an interior point where $\bar{M}'$ is continuous.

[While the double surface source is readily identifiable with a double surface current, the normal orientation of curls $\bar{M}'$ and $\bar{M}' \times \hat{\bar{n}}'$ to surface and contour respectively renders their interpretation as current flow less satisfactory.]

5-85. In the absence of surfaces of discontinuity it is possible to express the retarded vector potential $\bar{A}$ of a volume distribution of singlets, doublets and whirls as a potential function with an unretarded integrand. In addition to the unretarded source densities the integrand includes the point functions $\bar{E} = -\text{grad}\,\phi - \frac{1}{c}\frac{\partial \bar{A}}{\partial t}$ and ϕ, where ϕ is the retarded scalar potential of the system.

If the source densities are zero outside a spherical surface of finite radius, show that

$$\bar{A} = \frac{1}{c}\int_\infty \frac{\bar{J}}{r}\,d\tau + \frac{1}{c}\int_\infty \frac{1}{r}\frac{\partial \bar{P}}{\partial t}\,d\tau + \int_\infty \frac{\text{curl } \bar{M}}{r}\,d\tau + \frac{1}{c}\int_\infty \frac{1}{r}\frac{1}{4\pi}\frac{\partial \bar{E}}{\partial t}\,d\tau$$

$$+ \frac{1}{c}\int_\infty \frac{1}{r}\frac{1}{4\pi}\,\text{grad}\,\frac{\partial \phi}{\partial t}\,d\tau$$

Note that the unretarded vector potential of a volume distribution of continuous or symmetrical-point whirls is given by

$$\int_\infty \left\{ \bar{M} \times \operatorname{grad} \frac{1}{r} \right\} d\tau = \int_\infty \frac{\operatorname{curl} \bar{M}}{r} d\tau$$

and not by $\displaystyle\int_\infty \left\{ \bar{M} \times \operatorname{grad} \frac{1}{r} - \frac{\partial \bar{M}}{\partial t} \times \frac{\bar{r}}{cr^2} \right\} d\tau$

5-86. By making use of equations (2.12-12), (2.12-27) and a result of Ex.5-6., p. 404 show that the retarded potential of an oblique surface doublet may be brought, at exterior points, into the form

$$\phi = \int_S [P'_n] \operatorname{grad} \frac{1}{r} . d\bar{S} - \int_S \left[\frac{\partial P'_n}{\partial t}\right] \frac{\bar{r}}{cr^2} . d\bar{S} + \int_S (\operatorname{divs} \hat{n}) \frac{[\bar{P}']}{r} . d\bar{S}$$

$$- \int_S \frac{1}{r} [\operatorname{divs} \bar{P}'] \, dS + \oint_\Gamma \frac{[\bar{P}']}{r} . \hat{n}' \, ds$$

Observe that this reduces to the result of Ex.5-82. when $\bar{P}'$ is tangential to the surface, and to the standard form when $\bar{P}'$ is normal.

Suppose now that the surface is plane and that $\bar{P}'$ is oblique and everywhere the same. Show that the third and fourth terms of the above expression vanish, and justify the remaining terms by treating the source as the limiting configuration of two plane singlet layers of equal and opposite density subject to a tangential slip.

5.20 Microscopic/Macroscopic Relationships for $\bar{E}$ and $\bar{B}$ Fields within Volume Distributions of Doublets and Whirls

Let a spherical surface of radius δ be centred upon a point O within a volume distribution of doublets. It will be supposed that δ is sufficiently large to ensure that the microscopic and macroscopic potentials deriving from doublets beyond the surface are sensibly equal throughout a neighbourhood of O. Then it follows from equation (5.11-19) that the microscopic $\bar{E}$ field at O due to the complete distribution has the value

$$\bar{E}_{mic} = - \operatorname{grad}(\text{cavity}) \int_{\tau-\tau_\delta} \left\{ [\bar{P}] . \operatorname{grad} \frac{1}{r} - \left[\frac{\partial \bar{P}}{\partial t}\right] . \frac{\bar{r}}{cr^2} \right\} d\tau - \frac{1}{c} \frac{\partial}{\partial t} \frac{1}{c} \int_{\tau-\tau_\delta} \frac{1}{r} \left[\frac{\partial \bar{P}}{\partial t}\right] d\tau + \bar{E}_{int} \qquad (5.20\text{-}1)$$

where τ is the region occupied by all sources, τ_δ is the region of the sphere, and $\overline{E}_{int}$ is the contribution from sources within the sphere.

Equations (5.11-19) and (5.11-20) serve to define the macroscopic $\overline{E}$ and $\overline{B}$ fields when the microscopic potentials are replaced by their macroscopic counterparts. Hence, corresponding to equation (5.20-1), we have

$$\overline{E}_{mac} = -\,\text{grad}\int_\tau \left\{[\overline{P}].\text{grad}\,\frac{1}{r} - \left[\frac{\partial\overline{P}}{\partial t}\right].\frac{\overline{r}}{cr^2}\right\} d\tau - \frac{1}{c}\frac{\partial}{\partial t}\frac{1}{c}\int_\tau \frac{1}{r}\left[\frac{\partial\overline{P}}{\partial t}\right] d\tau \qquad (5.20\text{-}2)$$

By combining the above expressions with those given for grad cavity pot and grad pot in Table 7, pp. 461-2 we obtain

$$\overline{E}_{mic} = \overline{E}_{mac} + \int_{\tau_\delta}\left\{[-\text{div}\,\overline{P}]\,\frac{\overline{r}}{r^3} - \left[\frac{\partial}{\partial t}\,\text{div}\,\overline{P}\right]\frac{\overline{r}}{cr^2}\right\} d\tau - \oint_{S_\delta}\left\{[\overline{P}].d\overline{S}\,\frac{\overline{r}}{r^3} + \frac{\overline{r}}{cr^2}\left[\frac{\partial\overline{P}}{\partial t}\right].d\overline{S}\right\} \qquad (5.20\text{-}3)$$

$$+\frac{1}{c^2}\frac{\partial^2}{\partial t^2}\int_{\tau_\delta}\frac{[\overline{P}]}{r}\,d\tau + \overline{E}_{int}$$

The restriction imposed upon the minimum value of δ implies that the region τ_δ may be subdivided into a large number of statistically-regular volume cells. Correspondingly, we admit the possibility that $\overline{P}$ may vary from point to point of the region at any instant, but stipulate that $\overline{P}$ can be sufficiently represented by

$$\begin{aligned} P_x &= f_x(t)\ (1 + \alpha_x x + \beta_x y + \gamma_x z) \\ P_y &= f_y(t)\ (1 + \alpha_y x + \beta_y y + \gamma_y z) \\ P_z &= f_z(t)\ (1 + \alpha_z x + \beta_z y + \gamma_z z) \end{aligned} \qquad (5.20\text{-}4)$$

where 0 is identified with the origin of rectangular coordinates and α, β, γ are constants.

Thus it is supposed that the variation of the gradient of each scalar component of $\overline{P}$ across τ_δ is negligible compared with the gradient itself so that second space derivatives may be ignored. Then at any instant, div $\overline{P}$ and $\frac{\partial}{\partial t}$ div $\overline{P}$ are constant (or zero) throughout τ_δ. In these circumstances the associated volume integral in equation (5.20-3) vanishes from symmetry.

The values of $[\overline{P}]$ and $\left[\frac{\partial\overline{P}}{\partial t}\right]$ upon the spherical surface may be expressed in terms of $\overline{P}$, $\frac{\partial\overline{P}}{\partial t}$ and further derivatives at 0 for any particular time as follows:

$$[\bar{P}]_{S_\delta} = \bar{P} - \frac{\delta}{c}\frac{\partial \bar{P}}{\partial t} + \frac{\delta^2}{2c^2}\frac{\partial^2 \bar{P}}{\partial t^2} + (\bar{r}.\nabla)\,\bar{P} - \frac{\delta}{c}\frac{\partial}{\partial t}(\bar{r}.\nabla)\,\bar{P} + \frac{\delta^2}{2c^2}\frac{\partial^2}{\partial t^2}(\bar{r}.\nabla)\,\bar{P} \tag{5.20-5}$$

$$\left[\frac{\partial \bar{P}}{\partial t}\right]_{S_\delta} = \frac{\partial \bar{P}}{\partial t} - \frac{\delta}{c}\frac{\partial^2 \bar{P}}{\partial t^2} + \frac{\partial}{\partial t}(\bar{r}.\nabla)\,\bar{P} - \frac{\delta}{c}\frac{\partial^2}{\partial t^2}(\bar{r}.\nabla)\,\bar{P} \tag{5.20-6}$$

Time derivatives of order higher than the second have been omitted from these expansions.

In the evaluation of $\oint_{S_\delta} [\bar{P}].d\bar{S}\,\frac{\bar{r}}{r^3}$ it will be observed that radially opposite surface elements are associated with reversed directions of $\bar{r}$ and $d\bar{S}$.

Hence each term in the expansion of $[\bar{P}]_{S_\delta}$ which suffers a like reversal of sign will disappear upon integration. This eliminates the fourth, fifth and sixth terms of equation (5.20-5). For the same reason the third and fourth terms of equation (5.20-6) are eliminated in the integral $\oint_{S_\delta} \frac{\bar{r}}{cr^2}\left[\frac{\partial \bar{P}}{\partial t}\right].d\bar{S}$

We find that

$$-\oint_{S_\delta} [\bar{P}].d\bar{S}\,\frac{\bar{r}}{r^3} + \frac{\bar{r}}{cr^2}\left[\frac{\partial \bar{P}}{\partial t}\right].d\bar{S}$$

$$= \frac{4}{3}\pi\bar{P} - \frac{4\pi\delta}{3c}\frac{\partial \bar{P}}{\partial t} + \frac{2\pi\delta^2}{3c^2}\frac{\partial^2 \bar{P}}{\partial t^2} + \frac{4\pi\delta}{3c}\frac{\partial \bar{P}}{\partial t} - \frac{4\pi\delta^2}{3c^2}\frac{\partial^2 \bar{P}}{\partial t^2}$$

$$= \frac{4}{3}\pi\bar{P} - \frac{2\pi\delta^2}{3c^2}\frac{\partial^2 \bar{P}}{\partial t^2}$$

To the same order of approximation

$$\frac{1}{c^2}\frac{\partial^2}{\partial t^2}\int_{\tau_\delta} \frac{[\bar{P}]}{r}\,d\tau = \frac{1}{c^2}\int_{\tau_\delta} \frac{1}{r}\frac{\partial^2 \bar{P}}{\partial t^2}\,d\tau = \frac{2\pi\delta^2}{c^2}\frac{\partial^2 \bar{P}}{\partial t^2}$$

so that

$$\bar{E}_{mic} = \bar{E}_{mac} + \frac{4}{3}\pi\bar{P} + \frac{4\pi\delta^2}{3c^2}\frac{\partial^2\bar{P}}{\partial t^2} + \bar{E}_{int} \qquad (5.20\text{-}7)$$

$\bar{E}_{int}$ must be evaluated by direct summation of the contributions of individual doublets as expressed by equation (5.16-1), viz

$$\bar{E}_{int} = \sum\left\{\frac{-[\bar{p}]}{r^3} + \frac{3\bar{r}.[\bar{p}]\bar{r}}{r^5} - \frac{1}{cr^2}\left[\frac{d\bar{p}}{dt}\right] + \frac{3\bar{r}}{cr^4}.\left[\frac{d\bar{p}}{dt}\right]\bar{r} + \frac{\bar{r}}{c^2r^3}.\left[\frac{d^2\bar{p}}{dt^2}\right]\bar{r} - \frac{1}{c^2r}\left[\frac{d^2\bar{p}}{dt^2}\right]\right\} \qquad (5.20\text{-}8)$$

It is clear that this component of $\bar{E}_{mic}$ will vary wildly from one doublet to the next and that it is not possible to proceed analytically until some choice of configuration is made.

We will consider only the simplest case - that in which the doublets are located at the vertices of a cubic lattice and are of equal instantaneous vector moment. The point 0 will be taken to coincide with one of the doublets and $\bar{E}_{int}$ will be determined for all doublets other than that at 0. On expanding equation (5.20-8) in rectangular coordinates and taking the terms in consecutive pairs we find that the x component of $\bar{E}_{int}$ may be expressed as

$$\begin{aligned}&\sum\left\{\frac{(2x^2-y^2-z^2)}{r^5}[p_x] + \frac{3xy}{r^5}[p_y] + \frac{3xz}{r^5}[p_z]\right\}\\ &+\sum\left\{\frac{(2x^2-y^2-z^2)}{cr^4}\left[\frac{dp_x}{dt}\right] + \frac{3xy}{cr^4}\left[\frac{dp_y}{dt}\right] + \frac{3xz}{cr^4}\left[\frac{dp_z}{dt}\right]\right\} \qquad (5.20\text{-}9)\\ &+\sum\left\{\frac{(-y^2-z^2)}{c^2r^3}\left[\frac{d^2p_x}{dt^2}\right] + \frac{xy}{c^2r^3}\left[\frac{d^2p_y}{dt^2}\right] + \frac{xz}{c^2r^3}\left[\frac{d^2p_z}{dt^2}\right]\right\}\end{aligned}$$

The x and y components follows from cyclic permutation.

By pairing doublets having equal values of x and r and equal and opposite values of y and z it is found that the cross products in each line cancel in the sum. Further, since the rotation of one coordinate plane into another about 0 leaves the bounded lattice occupying the same points of space, we have

$$\sum\frac{(2x^2-y^2-z^2)}{r^5} = \frac{1}{3}\sum\left\{\frac{(2x^2-y^2-z^2)}{r^5} + \frac{(2y^2-z^2-x^2)}{r^5} + \frac{(2z^2-x^2-y^2)}{r^5}\right\} = 0$$

and

$$\sum\frac{(-y^2-z^2)}{c^2r^3} = \frac{1}{3}\sum\left\{\frac{(-y^2-z^2)}{c^2r^3} + \frac{(-z^2-x^2)}{c^2r^3} + \frac{(-x^2-y^2)}{c^2r^3}\right\} = -\frac{2}{3c^2}\sum\frac{1}{r}$$

Hence

$$\bar{E}_{int} = \frac{-2}{3c^2}\frac{d^2\bar{p}}{dt^2}\sum\frac{1}{r} \qquad (5.20\text{-}10)$$

if we restrict $\left[\frac{d^2\bar{p}}{dt^2}\right]$ to the leading term.

Then

$$\bar{E}_{mic} = \bar{E}_{mac} + \frac{4}{3}\pi\bar{P} + \frac{4\pi\delta^2}{3c^2}\frac{\partial^2\bar{P}}{\partial t^2} - \frac{2}{3c^2}\frac{d^2\bar{p}}{dt^2}\sum\frac{1}{r} \qquad (5.20\text{-}11)$$

where all terms are evaluated at 0.

Since equation (5.20-11) holds independently of the value assigned to δ, provided that τ_δ contains a large number of elementary cells, the sum of the last two terms must be constant. Thus if $\delta + \Delta\delta$ replaces δ, the increment in the final term may be written as the integral

$$-\frac{2}{3c^2}\frac{\partial^2\bar{P}}{\partial t^2}\int_\delta^{\delta+\Delta\delta}\frac{4\pi r^2}{r}\,dr = \frac{-8\pi\delta\Delta\delta}{3c^2}\frac{\partial^2\bar{P}}{\partial t^2}$$

This is seen to be equal and opposite to the increment in the penultimate term, as required. This procedure does not, of course, legitimise the extrapolation of the integral form of the final term beyond some minimum value of δ; it is ultimately necessary to evaluate the series $\sum\frac{1}{r}$. When this is carried out for a relatively small number of doublets centred upon 0 it is found that the last two terms of equation (5.20-11) very nearly cancel. (Ex.5-91. and 5-92., p. 552.) Hence for this particular configuration of doublets within τ_δ the microscopic field strength at 0 is given by

$$\bar{E}_{mic} = \bar{E}_{mac} + \frac{4}{3}\pi\bar{P} \qquad (5.20\text{-}12)$$

The microscopic $\bar{B}$ field within a volume distribution of whirls may be determined in a similar manner.

Since

$$\bar{B}_{mic} = \text{curl (cavity)} \int_{\tau-\tau_\delta}\left\{[\bar{M}]\times\text{grad}\,\frac{1}{r} - \left[\frac{\partial\bar{M}}{\partial t}\right]\times\frac{\bar{r}}{cr^2}\right\}d\tau + \bar{B}_{int} \qquad (5.20\text{-}13)$$

and

$$\overline{B}_{mac} = \text{curl} \int_{\tau} \left\{ [\overline{M}]\times\text{grad}\,\frac{1}{r} - \left[\frac{\partial \overline{M}}{\partial t}\right] \times \frac{\overline{r}}{cr^2} \right\} d\tau \qquad (5.20\text{-}14)$$

it follows from Table 8, pp. 465-6 that

$$\overline{B}_{mic} = \overline{B}_{mac} + \int_{\tau_\delta} \left\{ [\text{curl}\,\overline{M}] \times \frac{\overline{r}}{r^3} + \left[\frac{\partial}{\partial t}\,\text{curl}\,\overline{M}\right] \times \frac{\overline{r}}{cr^2} \right\} d\tau$$

$$+ \oint_{S_\delta} \left\{ (d\overline{S}\times[\overline{M}]) \times \frac{\overline{r}}{r^3} + \left(d\overline{S} \times \left[\frac{\partial \overline{M}}{\partial t}\right]\right) \times \frac{\overline{r}}{cr^2} \right\} + \overline{B}_{int} \qquad (5.20\text{-}15)$$

When the scalar components of $\overline{M}$ can be adequately represented within τ_δ by expressions of the type (5.20-4), curl $\overline{M}$ and $\frac{\partial}{\partial t}$ curl $\overline{M}$ have constant values at each instant and the volume integral of (5.20-15) vanishes from symmetry. By expanding $[\overline{M}]$ and $\left[\frac{\partial \overline{M}}{\partial t}\right]$ about 0 as shown for $[\overline{P}]$ and $\left[\frac{\partial \overline{P}}{\partial t}\right]$ in equations (5.20-5) and (5.20-6) and evaluating the surface integral, we obtain

$$-\left\{ \frac{8\pi}{3}\left(\overline{M} - \frac{\delta}{c}\frac{\partial \overline{M}}{\partial t} + \frac{\delta^2}{2c^2}\frac{\partial^2 \overline{M}}{\partial t^2}\right) + \frac{8\pi}{3}\,\frac{\delta}{c}\left(\frac{\partial \overline{M}}{\partial t} - \frac{\delta}{c}\frac{\partial^2 \overline{M}}{\partial t^2}\right)\right\}$$

whence

$$\overline{B}_{mic} = \overline{B}_{mac} - \frac{8}{3}\,\pi\overline{M} + \frac{4\pi\delta^2}{3c^2}\,\frac{\partial^2 \overline{M}}{\partial t^2} + \overline{B}_{int} \qquad (5.20\text{-}16)$$

On comparing this result with equation (5.20-7) and noting that the value of $\overline{B}_{int}$ deriving from a single whirl is identical with the value of $\overline{E}_{int}$ deriving from a single doublet (with $\overline{m}$ replacing $\overline{p}$), we see that for a cubic lattice distribution of whirls within τ_δ the microscopic $\overline{B}$ at the central whirl due to all others is given by

$$\overline{B}_{mic} = \overline{B}_{mac} - \frac{8}{3}\,\pi\overline{M} \qquad (5.20\text{-}17)$$

We turn now to a consideration of the microscopic $\overline{B}$ field of a volume distribution of doublets.

Since

$$\overline{B}_{mic} = \text{curl (cavity)} \ \frac{1}{c} \int_{\tau-\tau_\delta} \frac{1}{r} \left[\frac{\partial \overline{P}}{\partial t}\right] d\tau + \overline{B}_{int} \qquad (5.20\text{-}18)$$

and

$$\overline{B}_{mac} = \text{curl} \ \frac{1}{c} \int_{\tau} \frac{1}{r} \left[\frac{\partial \overline{P}}{\partial t}\right] d\tau \qquad (5.20\text{-}19)$$

it follows from Table 6, pp. 443-4, with $\frac{\partial \overline{P}}{\partial t}$ replacing $\overline{J}$, that

$$\overline{B}_{mic} = \overline{B}_{mac} - \frac{1}{c} \int_{\tau_\delta} \left\{ \left[\frac{\partial \overline{P}}{\partial t}\right] \times \text{grad} \ \frac{1}{r} - \left[\frac{\partial^2 \overline{P}}{\partial t^2}\right] \times \frac{\overline{r}}{cr^2} \right\} d\tau + \overline{B}_{int} \qquad (5.20\text{-}20)$$

On expanding $\left[\frac{\partial \overline{P}}{\partial t}\right]$ in a Taylor series about 0 as far as the second space and time derivatives, we find that the volume integral becomes

$$- \frac{1}{c} \int_{\tau_\delta} \left\{ \frac{\overline{r}}{r^3} \times (\overline{r}.\nabla) \frac{\partial \overline{P}}{\partial t} - \frac{\overline{r}}{cr^2} \times (\overline{r}.\nabla) \frac{\partial^2 \overline{P}}{\partial t^2} + \frac{\overline{r}}{cr^2} \times (\overline{r}.\nabla) \frac{\partial^2 \overline{P}}{\partial t^2} \right\} d\tau$$

$$= - \frac{1}{c} \int_{\tau_\delta} \left(\frac{\overline{r}}{r^3} \times (\overline{r}.\nabla) \frac{\partial \overline{P}}{\partial t} \right) d\tau$$

$$= \frac{-2\pi\delta^2}{3c} \ \text{curl} \ \frac{\partial \overline{P}}{\partial t}$$

whence

$$\overline{B}_{mic} = \overline{B}_{mac} - \frac{2\pi\delta^2}{3c} \ \text{curl} \ \frac{\partial \overline{P}}{\partial t} + \overline{B}_{int} \qquad (5.20\text{-}21)$$

From equation (5.16-3)

$$\overline{B}_{int} = \sum \left\{ \frac{\overline{r}}{cr^3} \times \left[\frac{d\overline{p}}{dt}\right] + \frac{\overline{r}}{c^2 r^2} \times \left[\frac{d^2 \overline{p}}{dt^2}\right] \right\} \qquad (5.20\text{-}22)$$

As required, the integral form of (5.20-22) is of equal value and opposite sign to the middle term of (5.20-20).

Since the possibility of non-zero second space derivatives of $\frac{\partial \overline{P}}{\partial t}$ has been entertained in the derivation of equation (5.20-21) the expression remains valid if, at each instant, the scalar components of curl $\frac{\partial \overline{P}}{\partial t}$ maintain a constant slope within τ_δ

For the lattice type of doublet distribution previously considered, $\overline{B}_{int}$ vanishes at 0 from symmetry and curl $\overline{P} = \overline{0}$. Hence equation (5.20-21) reduces to

$$\overline{B}_{mic} = \overline{B}_{mac} \qquad (5.20\text{-}23)$$

Finally, we investigate the microscopic $\overline{E}$ field of a volume distribution of whirls. In this case $\overline{E} = -\frac{1}{c}\frac{\partial \overline{A}}{\partial t}$, since a fully-compensated whirl does not give rise to a scalar potential, hence

$$\overline{E}_{mic} = -\frac{1}{c}\frac{\partial}{\partial t}\int_{\tau-\tau_\delta}\left\{[\overline{M}]\times \text{grad}\,\frac{1}{r} - \left[\frac{\partial \overline{M}}{\partial t}\right]\times\frac{\overline{r}}{cr^2}\right\}d\tau + \overline{E}_{int}$$

or

$$\overline{E}_{mic} = \overline{E}_{mac} + \frac{1}{c}\int_{\tau_\delta}\left\{\left[\frac{\partial \overline{M}}{\partial t}\right]\times \text{grad}\,\frac{1}{r} - \left[\frac{\partial^2 \overline{M}}{\partial t^2}\right]\times\frac{\overline{r}}{cr^2}\right\}d\tau + \overline{E}_{int} \qquad (5.20\text{-}24)$$

From equation (5.16-10)

$$\overline{E}_{int} = -\frac{1}{c}\sum\left\{\left[\frac{d\overline{m}}{dt}\right]\times \text{grad}\,\frac{1}{r} - \left[\frac{d^2\overline{m}}{dt^2}\right]\times\frac{\overline{r}}{cr^2}\right\} \qquad (5.20\text{-}25)$$

It is seen that equations (5.20-24) and (5.20-25) are formally identical with (5.20-20) and (5.20-22) where $\overline{E}$ replaces $\overline{B}$, and $(-\overline{M})$ and $(-\overline{m})$ replace $\overline{P}$ and $\overline{p}$, consequently equations (5.20-21) and (5.20-23) transform to

$$\overline{E}_{mic} = \overline{E}_{mac} + \frac{2\pi\delta^2}{3c}\,\text{curl}\,\frac{\partial \overline{M}}{\partial t} + \overline{E}_{int} \qquad (5.20\text{-}26)$$

and

$$\overline{E}_{mic} = \overline{E}_{mac} \qquad (5.20\text{-}27)$$

EXERCISES

5-87. Derive the following alternative to equation (5.20-20):

$$\overline{B}_{mic} = \overline{B}_{mac} - \frac{1}{c}\int_{\tau_\delta} \frac{1}{r}\left[\text{curl}\,\frac{\partial \overline{P}}{\partial t}\right] d\tau - \frac{1}{c}\oint_{S_\delta}\left\{d\overline{S} \times \frac{1}{r}\left[\frac{\partial \overline{P}}{\partial t}\right]\right\} + \overline{B}_{int}$$

Expand $\left[\text{curl}\,\frac{\partial \overline{P}}{\partial t}\right]$ and $\left[\frac{\partial \overline{P}}{\partial t}\right]$ about 0, assuming that the second and higher space derivatives of curl $\frac{\partial \overline{P}}{\partial t}$ are negligible, and show that the volume and surface integrals reduce to

$$\frac{-2\pi\delta^2}{c}\,\text{curl}\,\frac{\partial \overline{P}}{\partial t} + \frac{4\pi\delta^3}{3c^2}\,\text{curl}\,\frac{\partial^2 \overline{P}}{\partial t^2}$$

and

$$\frac{4\pi\delta^2}{3c}\,\text{curl}\,\frac{\partial \overline{P}}{\partial t} - \frac{4\pi\delta^3}{3c^2}\,\text{curl}\,\frac{\partial^2 \overline{P}}{\partial t^2}$$

Hence confirm equation (5.20-21).

5-88. Show that equations (4.21-11) and (4.21-27) are time-invariant forms of equations (5.20-12) and (5.20-17).

5-89. Extend equation (5.20-11) and the equivalent form of equation (5.20-16) to include third order time derivatives, and show that in each case the additional terms cancel.

5-90. Show that the neglect of third and higher-order time derivatives in equations (5.20-21) and (5.20-26) imposes an upper limit on the permissible values of $\frac{\partial \overline{P}}{\partial t}$ and $\frac{\partial \overline{M}}{\partial t}$ which, in terms of the period T of sinusoidal oscillation, may be expressed as $T^2 \gg (\delta/c)^2$.

5-91. If the edge of a simple cubic lattice element is of length d and there are N doublets within the region τ_δ, show that equation (5.20-11) may be approximated by

$$\overline{E}_{mic} = \overline{E}_{mac} + \frac{4}{3}\pi\overline{P} + \frac{1}{c^2}\frac{d^2\overline{p}}{dt^2}\left\{\frac{N}{d}\left(\frac{4\pi}{3N}\right)^{1/3} - \frac{2}{3}\sum \frac{1}{r}\right\}$$

5-92. A cubic lattice structure of overall dimensions 4d × 4d × 4d has half of its constituent cubes removed to give it a roughly spherical outline as represented in plan, front and side elevation by the accompanying figure. Doublets of equal instantaneous moment are located at each of the vertices.

Evaluate the third term of the expression in the previous exercise and so demonstrate that its two components cancel to within a few percent of their individual values.

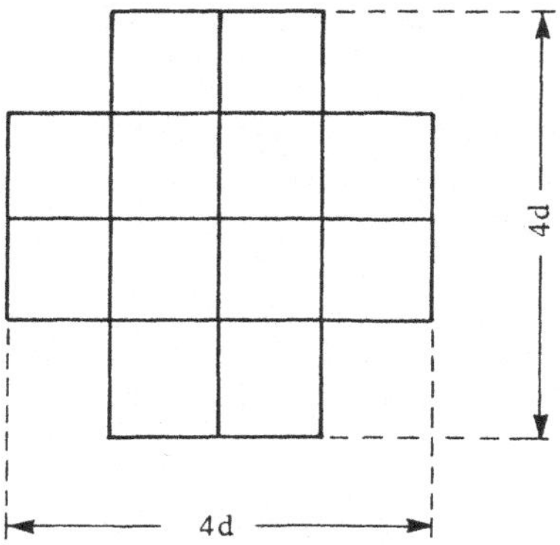

Figure to Ex.5-92.

5.21 Maxwell's Equations

It is evident from earlier considerations that macroscopic source densities are so defined as to render the microscopic and macroscopic potentials sensibly equal at points sufficiently removed from the associated distributions. The approximations fail at interior, boundary, and immediately adjacent exterior points. Nevertheless, the macroscopic potentials exist in their own right once the corresponding source densities have been defined, and the results of Sec. 5.19 continue to apply whether or not the density functions derive from any 'real' point distributions, provided that the appropriate equations of continuity are satisfied. In particular, at points exterior to surface, line and point discontinuities we have the following macroscopic relationships[18].

$$\text{dal}\ \phi = -4\pi\,(\rho - \text{div}\ \overline{P}) \tag{5.21-1}$$

$$\text{dal}\ \overline{A} = \frac{-4\pi}{c}\left(\overline{J} + \frac{\partial \overline{P}}{\partial t} + c\ \text{curl}\ \overline{M}\right) \tag{5.21-2}$$

$$\text{div}\ \overline{A} = -\frac{1}{c}\frac{\partial \phi}{\partial t} \tag{5.21-3}$$

Since the macroscopic point functions $\overline{E}$ and $\overline{B}$ continue to be defined by equations (5.11-19) and (5.11-20), where ϕ and $\overline{A}$ are macroscopic potentials, it follows immediately that

---

18. Unless stated otherwise it will be supposed that we are dealing with an undivided source complex or a complete subsource.

$$\operatorname{div} \bar{B} = 0 \tag{5.21-4}$$

$$\operatorname{curl} \bar{E} = -\frac{1}{c}\frac{\partial \bar{B}}{\partial t} \tag{5.21-5}$$

These equations duplicate those derived previously for the microscopic $\bar{B}$ and $\bar{E}$ fields. (Since scalar densities are not involved they continue to hold for an incomplete source.)

In addition,

$$\operatorname{div} \bar{E} = \operatorname{div}\left(-\operatorname{grad}\phi - \frac{1}{c}\frac{\partial \bar{A}}{\partial t}\right)$$

$$= -\nabla^2\phi + \frac{1}{c^2}\frac{\partial^2\phi}{\partial t^2} = -\operatorname{dal}\phi$$

or

$$\operatorname{div} \bar{E} = 4\pi(\rho - \operatorname{div} \bar{P}) \tag{5.21-6}$$

and

$$\operatorname{curl} \bar{B} = \operatorname{curl}\operatorname{curl} \bar{A} = \operatorname{grad}\operatorname{div} \bar{A} - \nabla^2\bar{A}$$

$$= \frac{1}{c}\frac{\partial}{\partial t}(-\operatorname{grad}\phi) - \nabla^2\bar{A}$$

$$= -\operatorname{dal}\bar{A} + \frac{1}{c}\frac{\partial \bar{E}}{\partial t}$$

or

$$\operatorname{curl} \bar{B} = \frac{4\pi}{c}\left(\bar{J} + \frac{\partial \bar{P}}{\partial t} + c \operatorname{curl} \bar{M}\right) + \frac{1}{c}\frac{\partial \bar{E}}{\partial t} \tag{5.21-7}$$

Equations (5.21-4) to (5.21-7) are known as Maxwell's equations. If we define new vector point functions $\bar{D}$ and $\bar{H}$ by

$$\bar{D} = \bar{E} + 4\pi\bar{P} \tag{5.21-8}$$

$$\bar{H} = \bar{B} - 4\pi\bar{M} \tag{5.21-9}$$

then Maxwell's equations take the form

$$\operatorname{div} \overline{B} = 0 \tag{5.21-4}$$

$$\operatorname{curl} \overline{E} = -\frac{1}{c}\frac{\partial \overline{B}}{\partial t} \tag{5.21-5}$$

$$\operatorname{div} \overline{D} = 4\pi\rho \tag{5.21-10}$$

$$\operatorname{curl} \overline{H} = \frac{4\pi}{c}\overline{J} + \frac{1}{c}\frac{\partial \overline{D}}{\partial t} \tag{5.21-11}$$

It will be seen that $\overline{D}$ and $\overline{H}$ are mixed functions in the sense that they combine density fields with potential derivatives. They are essentially macroscopic functions and reduce to $\overline{E}$ and $\overline{B}$ at points beyond the source (or subsource).

It is common practice to associate with Maxwell's equations a set of boundary conditions. These describe the behaviour of $\overline{E}$, $\overline{D}$, $\overline{B}$ and $\overline{H}$ at surfaces of discontinuity of the volume densities ρ, $\overline{J}$, $\overline{P}$ and $\overline{M}$, and at surfaces of non-zero σ and $\overline{K}$; line sources and surface distributions of doublets and whirls are not taken into account. The relevant relationships have been derived in earlier Sections - in particular, Secs. 5.5, 5.6 and 5.8, to which the reader should refer.

If the regions bounded by the two sides of the surface of discontinuity are denoted by the subscripts 1 and 2, and the corresponding outward normals likewise, then

$$\hat{\overline{n}}_1.\overline{E}_1 + \hat{\overline{n}}_2.\overline{E}_2 = -4\pi\,(\sigma + \hat{\overline{n}}_1.\overline{P}_1 + \hat{\overline{n}}_2.\overline{P}_2) \tag{5.21-12}$$

or

$$\hat{\overline{n}}_1.\overline{D}_1 + \hat{\overline{n}}_2.\overline{D}_2 = -4\pi\sigma \tag{5.21-12(a)}$$

$$\hat{\overline{n}}_1 \times \overline{E}_1 + \hat{\overline{n}}_2 \times \overline{E}_2 = \overline{0} \tag{5.21-13}$$

$$\hat{\overline{n}}_1.\overline{B}_1 + \hat{\overline{n}}_2.\overline{B}_2 = 0 \tag{5.21-14}$$

$$\hat{\overline{n}}_1 \times \overline{B}_1 + \hat{\overline{n}}_2 \times \overline{B}_2 = \frac{-4\pi}{c}\,\{\overline{K}+c(\overline{M}_1\times\hat{\overline{n}}_1)+c(\overline{M}_2\times\hat{\overline{n}}_2)\} \tag{5.21-15}$$

or

$$\hat{n}_1 \times \overline{H}_1 + \hat{n}_2 \times \overline{H}_2 = \frac{-4\pi}{c}\overline{K} \qquad (5.21\text{-}15(a))$$

It will be appreciated that since no restriction has been imposed, in the present work, upon the relative configuration of the various components of a composite source, a surface of non-zero σ need not be a surface of discontinuity of $\overline{P}$ nor need a surface of discontinuity of $\overline{M}$ be associated with a surface current $\overline{K}$. Appropriate substitution in (5.21-12) and (5.21-15) modifies these equations as required.

The boundary relationships may also be expressed in terms of the previously adopted Δ notation which assigns a common arbitrary sense of the normal to each side of a given surface. We then have

$$\Delta\,(\hat{n}.\overline{E}) = 4\pi\,(\sigma-\Delta(\hat{n}.\overline{P})) \qquad (5.21\text{-}16)$$

or

$$\Delta(\hat{n}.\overline{D}) = 4\pi\sigma \qquad (5.21\text{-}16(a))$$

$$\Delta(\hat{n}\times\overline{E}) = \overline{0} \qquad (5.21\text{-}17)$$

$$\Delta(\hat{n}.\overline{B}) = 0 \qquad (5.21\text{-}18)$$

$$\Delta(\hat{n}\times\overline{B}) = \frac{4\pi}{c}(\overline{K}-\Delta c(\overline{M}\times\hat{n})) \qquad (5.21\text{-}19)$$

or

$$\Delta(\hat{n}\times\overline{H}) = \frac{4\pi}{c}\overline{K} \qquad (5.21\text{-}19(a))$$

The above equations continue to hold at bounding surfaces of incomplete sources.

Integral forms of Maxwell's equations may be derived by appropriate application of the divergence theorem or Stokes's theorem. In general it becomes necessary to divide the region of integration into subregions in which the first field derivatives are continuous, and so to express the required integral as the sum of a set of integrals. These will include paired integrals over interior lines or surfaces of discontinuity which must be evaluated in accordance with the relevant boundary conditions.

On proceeding in this way we find that

$$\oint \bar{B}.d\bar{S} = 0 \tag{5.21-20}$$

ie $\bar{B}$ is solenoidal,

and

$$\oint_S \bar{D}.d\bar{S} = 4\pi \int_\tau \rho d\tau + 4\pi \int_{S_1,S_2..} \sigma \, dS \tag{5.21-21}$$

where τ comprises the subregions τ_1, τ_2.. enclosed by S and bounded internally by S_1, S_2... .

We have also

$$\oint_\Gamma \bar{E}.d\bar{r} = -\frac{1}{c}\frac{\partial}{\partial t}\int_S \bar{B}.d\bar{S} \tag{5.21-22}$$

where S spans Γ and comprises the subregions S_1, S_2.. bounded internally by the contours Γ_1, Γ_2.. which are defined by the intersection of the surfaces of discontinuity with S. Since the vector tangential component of $\bar{E}$ is continuous through a surface of discontinuity in accordance with equation (5.21-13) the resolved parts of $\bar{E}$ along Γ_1, Γ_2.. are equal on either side of each contour, but the line integrals cancel in the sum because of reversed currencies.

Finally,

$$\oint_\Gamma \bar{H}.d\bar{r} = \int_S \left(\frac{4\pi}{c}\bar{J} + \frac{1}{c}\frac{\partial \bar{D}}{\partial t}\right).d\bar{S} + \int_{\Gamma_1,\Gamma_2..} \frac{4\pi}{c} K_n \, ds \tag{5.21-23}$$

where K_n is the component of $\bar{K}$ normal to Γ_1, Γ_2.. in the direction of the positive sense of S as defined by circulation around Γ.

The equality may be expressed in the form

$$\oint_\Gamma \bar{H}.d\bar{r} = \frac{4\pi}{c}\left\{C + \frac{1}{4\pi}\int_S \frac{\partial \bar{D}}{\partial t}.d\bar{S}\right\} \tag{5.21-24}$$

where C is the total rate of transfer of source strength through S ie the macroscopic current through S (p. 522).

It is clear that equation (5.21-24) will continue to hold in the presence of volume currents having such density and form as to approximate line currents which cut S. Line currents, per se, are awkward to handle because $\overline{E}$ is infinite along the line.

A relationship of historical interest[19] takes the form

$$\operatorname{div}\left(\overline{J} + \frac{1}{4\pi}\frac{\partial\overline{D}}{\partial t}\right) = 0 \qquad (5.21\text{-}25)$$

This follows immediately from equation (5.21-11) and indirectly from equations (5.17-15) and (5.21-10).

In the absence of interior line and surface currents

$$\oint_S \left(\overline{J} + \frac{1}{4\pi}\frac{\partial\overline{D}}{\partial t}\right).d\overline{S} = 0 \qquad (5.21\text{-}26)$$

since a combination of the time derivative of equation (5.21-12(a)) with

$$\hat{n}_1.\overline{J}_1 + \hat{n}_2.\overline{J}_2 = \frac{\partial\sigma}{\partial t}$$

yields the required boundary condition for the cancellation of contributions from surfaces of discontinuity. In the presence of line current approximations or of surface currents which cut S, equation (5.21-26) is replaced by

$$\Delta C + \oint_S \frac{1}{4\pi}\frac{\partial\overline{D}}{\partial t}.d\overline{S} = 0 \qquad (5.21\text{-}27)$$

where ΔC is the net macroscopic current entering the enclosure. It then follows that two simple open surfaces bounded by the same contour exhibit equal values of $C + \frac{1}{4\pi}\int\frac{\partial\overline{D}}{\partial t}.d\overline{S}$ when assigned a common sense of the normal. This will be seen to be a necessary condition for the validity of equation (5.21-24), since no restriction is imposed on the shape of the surface S which spans Γ.

---

19. Possibly the most famous of all in an electromagnetic context: the component $\frac{1}{4\pi}\frac{\partial\overline{E}}{\partial t}$ is Maxwell's so-called 'displacement current in free space'.

EXERCISES

5-93. Let ϕ be defined by equation (5.19-1(a)) and $\bar{A}$ by (5.19-3). In addition, let a scalar point function ϕ_m be defined by

$$\phi_m = \int \frac{[\rho_m]}{r} d\tau + \int \frac{[\sigma_m]}{r} dS + \int \frac{[\lambda_m]}{r} ds + \sum \frac{[a_m]}{r}$$

$$+ \int \left\{ [\bar{P}_m].\text{grad}\, \frac{1}{r} - \left[\frac{\partial \bar{P}_m}{\partial t}\right].\frac{\bar{r}}{cr^2} \right\} d\tau + \int \left\{ [\bar{P}'_m].\text{grad}\, \frac{1}{r} - \left[\frac{\partial \bar{P}'_m}{\partial t}\right].\frac{\bar{r}}{cr^2} \right\} dS$$

where $\rho_m \ldots \bar{P}_m$ are the macroscopic densities of certain singlet and doublet distributions which have no connection with $\rho \ldots \bar{P}$.

Further, let

$$\bar{A}_m = \frac{1}{c} \int \frac{[\bar{J}_m]}{r} d\tau + \frac{1}{c} \int \frac{[\bar{K}_m]}{r} dS + \frac{1}{c} \int \frac{[\bar{I}_m]}{r} ds + \frac{1}{c} \int \frac{1}{r} \left[\frac{\partial \bar{P}_m}{\partial t}\right] d\tau + \frac{1}{c} \int \frac{1}{r} \left[\frac{\partial \bar{P}'_m}{\partial t}\right] dS$$

where $\bar{J}_m = \rho_m \bar{v}$ etc, so that identical equations of continuity hold for the m-type distributions as for the conventional.

Show that if

$$\bar{E} = -\text{grad}\, \phi - \frac{1}{c}\frac{\partial \bar{A}}{\partial t} - \text{curl}\, \bar{A}_m$$

$$\bar{D} = \bar{E} + 4\pi \bar{P}$$

$$\bar{B} = \text{curl}\, \bar{A} - \text{grad}\, \phi_m - \frac{1}{c}\frac{\partial \bar{A}_m}{\partial t} + 4\pi \bar{P}_m$$

$$\bar{H} = \bar{B} - 4\pi \bar{P}_m - 4\pi \bar{M}$$

then at points displaced from point, line and surface sources

$$\text{div}\, \bar{B} = 4\pi \rho_m$$

$$\text{curl}\, \bar{E} = -\frac{1}{c}\frac{\partial \bar{B}}{\partial t} - \frac{4\pi}{c} \bar{J}_m$$

$$\text{div}\, \bar{D} = 4\pi \rho$$

$$\text{curl } \bar{H} = \frac{4\pi}{c} \bar{J} + \frac{1}{c} \frac{\partial \bar{D}}{\partial t}$$

These are known as the Heaviside-Maxwell equations.

5-94. In a composite source system characterised by the density functions ρ, σ, $\bar{J}$, $\bar{K}$, $\bar{P}$ together with their m-type equivalents, show that when $\bar{E}$, $\bar{D}$, $\bar{B}$, $\bar{H}$ are defined as in the previous exercise, the boundary conditions become

$$\hat{\bar{n}}_1 . \bar{D}_1 + \hat{\bar{n}}_2 . \bar{D}_2 = -4\pi\sigma$$

$$\hat{\bar{n}}_1 \times \bar{E}_1 + \hat{\bar{n}}_2 \times \bar{E}_2 = \frac{4\pi}{c} \bar{K}_m$$

$$\hat{\bar{n}}_1 . \bar{B}_1 + \hat{\bar{n}}_2 . \bar{B}_2 = -4\pi\sigma_m$$

$$\hat{\bar{n}}_1 \times \bar{H}_1 + \hat{\bar{n}}_2 \times \bar{H}_2 = \frac{-4\pi}{c} \bar{K}$$

5-95. Deduce from the results of Ex.5-93. that if, in a conventional composite source, whirls are replaced by m-type doublets, Maxwell's equations continue to hold provided that

$$\bar{E} = -\text{ grad } \phi - \frac{1}{c} \frac{\partial \bar{A}}{\partial t} - \text{curl } \bar{A}_m$$

$$\bar{D} = \bar{E} + 4\pi\bar{P}$$

$$\bar{H} = \text{curl } \bar{A} - \text{grad } \phi_m - \frac{1}{c} \frac{\partial \bar{A}_m}{\partial t}$$

$$\bar{B} = \bar{H} + 4\pi\bar{P}_m$$

where

$$\bar{A}_m = \frac{1}{c} \int \frac{1}{r} \left[\frac{\partial \bar{P}_m}{\partial t}\right] d\tau + \frac{1}{c} \int \frac{1}{r} \left[\frac{\partial \bar{P}'_m}{\partial t}\right] dS$$

5-96. A short cylindrical source, which is characterised by the density functions ρ, σ, $\overline{J}$, is cut once by a simple surface S_1 bounded by the contour Γ. A second surface S_2 bounded by Γ does not cut the cylinder at all. Show that

$$\oint_\Gamma \overline{H}.d\overline{r} = \frac{4\pi}{c}\int_{S_1} \overline{J}.d\overline{S} + \frac{1}{c}\int_{S_1} \frac{\partial \overline{E}}{\partial t}.d\overline{S} = \frac{1}{c}\int_{S_2} \frac{\partial \overline{E}}{\partial t}.d\overline{S}$$

Now replace ρ, σ, $\overline{J}$ by $\overline{P}$, as discussed in Ex.5-78., p. 541, and arrive at

$$\oint_\Gamma \overline{H}.d\overline{r} = \frac{1}{c}\int_{S_1} \frac{\partial \overline{D}}{\partial t}.d\overline{S} = \frac{1}{c}\int_{S_2} \frac{\partial \overline{D}}{\partial t}.d\overline{S}$$

Demonstrate that the two equations are consistent.

[Note that the two source complexes are equivalent with respect to $\frac{\partial \overline{E}}{\partial t}$ at interior points but not $\frac{\partial \overline{D}}{\partial t}$].

5-97. A source-free region τ is bounded by the surfaces $S_{1..n}\Sigma$. Make use of the volume integral of the expansion of $\mathrm{div}(\overline{E}\times\overline{B})$ and Maxwell's equations to show that $\overline{E}$ and $\overline{B}$ are uniquely defined throughout τ for $t \geq 0$ if $\overline{E}$ and $\overline{B}$ are specified throughout τ at $t = 0$, and either $\hat{n} \times \overline{E}$ or $\hat{n} \times \overline{B}$ is specified upon $S_{1..n}\Sigma$ for $t \geq 0$.

5.22 The Macroscopic Vector Fields $\overline{E}$, $\overline{D}$, $\overline{B}$, $\overline{H}$

5.22a Fields of macroscopic singlet distributions

The $\overline{E}$ and $\overline{D}$ fields of a complete source comprising volume and surface singlets is given by

$$\overline{E} = \overline{D} = -\,\mathrm{grad}\,\phi - \frac{1}{c}\frac{\partial \overline{A}}{\partial t}$$

$$= -\,\mathrm{grad}\int_\tau \frac{[\rho]}{r}\,d\tau - \mathrm{grad}\int_S \frac{[\sigma]}{r}\,dS - \frac{1}{c^2}\frac{\partial}{\partial t}\int_\tau \frac{[\overline{J}]}{r}\,d\tau - \frac{1}{c^2}\frac{\partial}{\partial t}\int_S \frac{[\overline{K}]}{r}\,dS$$

whence, at points exterior to the surfaces,

$$\bar{E} = \bar{D} = -\int_\tau [\rho]\frac{\bar{r}}{r^3}\,d\tau - \int_S [\sigma]\frac{\bar{r}}{r^3}\,dS - \int_\tau \frac{\bar{r}}{cr^2}\left[\frac{\partial\rho}{\partial t}\right] d\tau - \int_S \frac{\bar{r}}{cr^2}\left[\frac{\partial\sigma}{\partial t}\right] dS$$

$$- \int_\tau \frac{1}{c^2 r}\left[\frac{\partial\bar{J}}{\partial t}\right] d\tau - \int_S \frac{1}{c^2 r}\left[\frac{\partial\bar{K}}{\partial t}\right] dS \qquad (5.22\text{-}1)$$

The subscript τ takes account of all volume distributions involving ρ and/or $\bar{J}$, and the subscript S, all surfaces involving σ and/or $\bar{K}$, whether open or closed. It is supposed that $\bar{K}$ is continuous at interior points of S and that the component of $\bar{K}$ normal to the bounding contour of an open surface is zero. This eliminates the need to consider line sources in the present context.

It is seen that equation (5.22-1) expresses $\bar{E}$ as the sum of inverse-square-distance terms involving the scalar source densities, and inverse-distance terms involving the first time derivatives of the scalar source densities and current densities. When the source is complete, (5.22-1) may be transformed into an expression in which the terms in $\frac{\partial\rho}{\partial t}$ and $\frac{\partial\sigma}{\partial t}$ are replaced by terms in $\bar{J}$ and $\bar{K}$.

This is accomplished as follows:

$$- \int_\tau \frac{\bar{r}}{cr^2}\left[\frac{\partial\rho}{\partial t}\right] d\tau - \int_S \frac{\bar{r}}{cr^2}\left[\frac{\partial\sigma}{\partial t}\right] dS$$

$$= \int_\tau \frac{\bar{r}}{cr^2}[\operatorname{div}\bar{J}]\,d\tau + \int_S \frac{\bar{r}}{cr^2}\left([\operatorname{divs}\bar{K}]-[\bar{J}].\hat{\bar{n}}\right) dS$$

$$= \int_\tau \frac{\bar{r}}{cr^2}\operatorname{div}[\bar{J}]\,d\tau + \int_\tau \frac{\bar{r}}{cr^2}\frac{\bar{r}}{cr}.\left[\frac{\partial\bar{J}}{\partial t}\right] d\tau + \int_S \frac{\bar{r}}{cr^2}\operatorname{divs}[\bar{K}]\,dS$$

$$+ \int_S \frac{\bar{r}}{cr^2}\frac{\bar{r}}{cr}.\left[\frac{\partial\bar{K}}{\partial t}\right] dS - \int_S \frac{\bar{r}}{cr^2}[\bar{J}].d\bar{S} \qquad (5.22\text{-}2)$$

The final surface integral is double-sided when associated with an interior surface of discontinuity of $\bar{J}$ and single-sided when associated with a bounding surface of the source or subsource.

By expansion of div $\frac{(x-x_o)}{cr^2}[\bar{J}]$, subsequent volume integration, and conversion to vectorial form, we find that

$$\oint_{S'} \frac{\bar{r}}{cr^2} [\bar{J}].d\bar{S} = \int_{\tau'} \frac{\bar{r}}{cr^2} \operatorname{div} [\bar{J}]\, d\tau + \int_{\tau'} \frac{1}{cr^2} [\bar{J}]\, d\tau - 2\int_{\tau'} \frac{\bar{r}}{cr^4} \bar{r}.[\bar{J}]\, d\tau \qquad (5.22\text{-}3)$$

where S' is the surface or surfaces bounding some particular region τ'.

This result holds both within and without τ' since it is easily shown that at interior points of τ' the individual volume integrals are convergent and the surface integral, when taken over a δ sphere about the origin of r, approaches zero as δ→0.

By a similar transformation

$$\oint_{\Gamma''} \frac{\bar{r}}{cr^2} [\bar{K}].\hat{n}'\, ds = \int_{S''} \frac{\bar{r}}{cr^2} \operatorname{divs} [\bar{K}]\, dS + \int_{S''} \frac{1}{cr^2} [\bar{K}]\, dS - 2\int_{S''} \frac{\bar{r}}{cr^4} \bar{r}.[\bar{K}]\, dS \qquad (5.22\text{-}4)$$

Here, $\hat{n}'$ is tangential to the surface S" and normal to its bounding contour or contours Γ". Since $\bar{K}.\hat{n}'$ has been supposed to be zero when S" is open, and $\bar{K}$ is continuous upon a closed surface, the left hand side of equation (5.22-4) vanishes in all cases.

Upon substituting equations (5.22-3) and (5.22-4) in (5.22-2) we obtain

$$-\int_{\tau} \frac{1}{cr^2} [\bar{J}]\, d\tau + 2\int_{\tau} \frac{\bar{r}}{cr^4} \bar{r}.[\bar{J}]\, d\tau + \int_{\tau} \frac{\bar{r}}{c^2r^3} \bar{r}.\left[\frac{\partial \bar{J}}{\partial t}\right] d\tau - \int_{S} \frac{1}{cr^2} [\bar{K}]\, dS$$

$$+ 2\int_{S} \frac{\bar{r}}{cr^4} \bar{r}.[\bar{K}]\, dS + \int_{S} \frac{\bar{r}}{c^2r^3} \bar{r}.\left[\frac{\partial \bar{K}}{\partial t}\right] dS$$

whence, at points exterior to the surfaces,

$$\bar{E} = \bar{D} = -\int_{\tau} [\rho] \frac{\bar{r}}{r^3} d\tau - \int_{S} [\sigma] \frac{\bar{r}}{r^3} dS - \int_{\tau} \frac{1}{cr^2} [\bar{J}_t]\, d\tau + \int_{\tau} \frac{1}{cr^2} [\bar{J}_r]\, d\tau \qquad (5.22\text{-}5)$$

$$- \int_{S} \frac{1}{cr^2} [\bar{K}_t]\, dS + \int_{S} \frac{1}{cr^2} [\bar{K}_r]\, dS - \int_{\tau} \frac{1}{c^2r} \left[\frac{\partial \bar{J}_t}{\partial t}\right] d\tau - \int_{S} \frac{1}{c^2r} \left[\frac{\partial \bar{K}_t}{\partial t}\right] dS$$

where $\bar{J}_r = \bar{J}.\hat{r}\,\hat{r}$ and $\bar{J}_t = \bar{J} - \bar{J}_r$

$\overline{J}_r$ is seen to be the vector projection of $\overline{J}$ upon the line joining $d\tau$ to the point of evaluation of $\overline{E}$, and $\overline{J}_t$ is the corresponding transverse component[20].

It is noteworthy that the transformation of equation (5.22-1) into (5.22-5) has allowed us to express the inverse-distance component of $\overline{E}$ in terms of the first time derivatives of the transverse current densities alone.

The contribution to $\overline{E}$ of line currents (and associated point sources, if any) may be shown to be

$$\overline{E} = \overline{D} = -\int_\Gamma [\lambda]\frac{\overline{r}}{r^3}\,ds - \sum [a]\frac{\overline{r}}{r^3} - \int_\Gamma \frac{1}{cr^2}[\overline{I}_t]\,ds + \int_\Gamma \frac{1}{cr^2}[\overline{I}_r]\,ds \tag{5.22-6}$$

$$- \int_\Gamma \frac{1}{c^2 r}\left[\frac{\partial \overline{I}_t}{\partial t}\right] ds$$

It is also possible to express the time-dependent component of $\overline{E}$ entirely in terms of the current densities, their time derivatives and their time integrals from an arbitrary time datum. This is the subject of Ex.5-101., p. 571.

The evaluation of $\overline{B}$ is straightforward. For the general case

$$\overline{B} = \overline{H} = \text{curl}\,\frac{1}{c}\int_\tau \frac{[\overline{J}]}{r}\,d\tau + \text{curl}\,\frac{1}{c}\int_S \frac{[\overline{K}]}{r}\,dS + \text{curl}\,\frac{1}{c}\int_\Gamma \frac{[\overline{I}]}{r}\,ds$$

---

20. It should be noted that the term 'transverse current density' is sometimes employed by other writers with the following connotation:

$$\overline{J}_t = \overline{J} - \frac{1}{4\pi}\,\text{grad}\,\frac{\partial\phi}{\partial t} \quad \text{where} \quad \phi = \int \frac{\rho}{r}\,d\tau$$

whence, at points exterior to the line and surface sources,

$$\overline{B} = \overline{H} = -\frac{1}{c}\int_{\tau} [\overline{J}] \times \frac{\overline{r}}{r^3}\, d\tau - \frac{1}{c}\int_{S} [\overline{K}] \times \frac{\overline{r}}{r^3}\, dS - \frac{1}{c}\int_{\Gamma} [\overline{I}] \times \frac{\overline{r}}{r^3}\, ds$$

$$- \frac{1}{c}\int_{\tau} \left[\frac{\partial \overline{J}}{\partial t}\right] \times \frac{\overline{r}}{cr^2}\, d\tau - \frac{1}{c}\int_{S} \left[\frac{\partial \overline{K}}{\partial t}\right] \times \frac{\overline{r}}{cr^2}\, dS - \frac{1}{c}\int_{\Gamma} \left[\frac{\partial \overline{I}}{\partial t}\right] \times \frac{\overline{r}}{cr^2}\, ds \qquad (5.22\text{-}7)$$

It is clear that $\overline{J}$ and $\frac{\partial \overline{J}}{\partial t}$ may be replaced by $\overline{J}_t$ and $\frac{\partial \overline{J}_t}{\partial t}$ (and likewise for $\overline{K}$ and $\overline{I}$) in the above formula.

Since scalar source densities do not enter into these expressions they hold equally for complete and incomplete subsources.

Under certain conditions of symmetry the evaluation of $\overline{B}$ or $\overline{H}$ in accordance with (5.22-7), and the subsequent evaluation of curl $\overline{H}$, permits of a determination of the time-dependent component of $\overline{E}$ via

$$\text{curl}\ \overline{H} = \frac{4\pi}{c}\overline{J} + \frac{1}{c}\frac{\partial \overline{E}}{\partial t}$$

However, the latter relationship was derived on the assumption of subsource completeness, and the value of $\overline{E}$ obtained in this way is consequently that appropriate to an undivided source or complete subsource although the parent source may have been incomplete.

5.22b Fields of macroscopic doublet distributions

In the presence of volume and surface distributions of doublets the value of $\overline{E}$ at exterior points of the surfaces is given by

$$\overline{E} = -\text{grad}\int_{\tau} \left\{[\overline{P}].\text{grad}\,\frac{1}{r} - \left[\frac{\partial \overline{P}}{\partial t}\right].\frac{\overline{r}}{cr^2}\right\} d\tau - \text{grad}\int_{S} \left\{[\overline{P}'].\text{grad}\,\frac{1}{r} - \left[\frac{\partial \overline{P}'}{\partial t}\right].\frac{\overline{r}}{cr^2}\right\} dS$$

$$- \frac{1}{c^2}\frac{\partial}{\partial t}\int_{\tau} \frac{1}{r}\left[\frac{\partial \overline{P}}{\partial t}\right] d\tau - \frac{1}{c^2}\frac{\partial}{\partial t}\int_{S} \frac{1}{r}\left[\frac{\partial \overline{P}'}{\partial t}\right] dS$$

$$= \text{grad div}\int_{\tau} \frac{[\overline{P}]}{r}\, d\tau + \text{grad div}\int_{S} \frac{[\overline{P}']}{r}\, dS - \frac{1}{c^2}\frac{\partial^2}{\partial t^2}\int_{\tau} \frac{[\overline{P}]}{r}\, d\tau$$

$$- \frac{1}{c^2}\frac{\partial^2}{\partial t^2}\int_{S} \frac{[\overline{P}']}{r}\, dS \qquad (5.22\text{-}8)$$

or

$$\overline{E} = \text{grad div } \overline{\Pi}_e - \frac{1}{c^2}\frac{\partial^2}{\partial t^2}\overline{\Pi}_e \qquad (5.22\text{-}8(a))$$

Also

$$\overline{D} = \overline{E} + 4\pi\overline{P} = \text{grad div } \overline{\Pi}_e - \frac{1}{c^2}\frac{\partial^2}{\partial t^2}\overline{\Pi}_e - \text{dal } \overline{\Pi}_e$$

$$= \text{grad div } \overline{\Pi}_e - \nabla^2\overline{\Pi}_e$$

or

$$\overline{D} = \text{curl curl } \overline{\Pi}_e \qquad (5.22\text{-}9)$$

It follows from equation (5.22-8) and equation (17) of Table 6, p. 446, with $\overline{P}$ replacing $\overline{J}$, that the contribution of the volume distribution to $\overline{E}$ at interior points may be expressed in the non-convergent form

$$\overline{E} = \int_\tau \left\{\left(-\left[\frac{\partial\overline{P}}{\partial t}\right].\nabla\right)\frac{\overline{r}}{cr^2} + \frac{\overline{r}}{cr^4}\,\overline{r}.\left[\frac{\partial\overline{P}}{\partial t}\right] + \frac{\overline{r}}{c^2r^3}\,\overline{r}.\left[\frac{\partial^2\overline{P}}{\partial t^2}\right] - \frac{1}{c^2r}\left[\frac{\partial^2\overline{P}}{\partial t^2}\right]\right\} d\tau$$

$$+ \lim_{\tau'\to 0}\int_{\tau-\tau'} ([\overline{P}].\nabla)\text{ grad }\frac{1}{r}\,d\tau - \lim_{S'\to 0}\oint_{S'}[\overline{P}].d\overline{S}\text{ grad }\frac{1}{r} \qquad (5.22\text{-}10)$$

At exterior points this reduces to

$$\overline{E} = \int_\tau \left\{-\frac{[\overline{P}]}{r^3} + \frac{3\overline{r}}{r^5}\,\overline{r}.[\overline{P}] - \frac{1}{cr^2}\left[\frac{\partial\overline{P}}{\partial t}\right] + \frac{3\overline{r}}{cr^4}\,\overline{r}.\left[\frac{\partial\overline{P}}{\partial t}\right] + \frac{\overline{r}}{c^2r^3}\,\overline{r}.\left[\frac{\partial^2\overline{P}}{\partial t^2}\right] - \frac{1}{c^2r}\left[\frac{\partial^2\overline{P}}{\partial t^2}\right]\right\}d\tau \qquad (5.22\text{-}11)$$

It will be observed that this expression is the integral form of the field of a microscopic doublet (5.16-1), as would be expected.

The contribution to $\overline{E}$ of surface sources, at exterior points of the surfaces, takes the form of equation (5.22-11) with $\overline{P}$ replaced by $\overline{P}'$ and $d\tau$ replaced by dS. Equation (15), p. 445, allows us to express the $\overline{E}$ field of a volume distribution of doublets, at interior and exterior points, as a combination of inverse-square-distance and inverse-distance terms. We obtain

$$\bar{E} = \int_\tau \left\{[\mathrm{div}\,\bar{P}]\frac{\bar{r}}{r^3} + \left[\frac{\partial}{\partial t}\mathrm{div}\,\bar{P}\right]\frac{\bar{r}}{cr^2}\right\} d\tau - \frac{1}{c^2}\int_\tau \frac{1}{r}\left[\frac{\partial^2\bar{P}}{\partial t^2}\right] d\tau$$

$$- \oint_S \left\{\frac{\bar{r}}{r^3}[\bar{P}].d\bar{S} + \frac{\bar{r}}{cr^2}\left[\frac{\partial\bar{P}}{\partial t}\right].d\bar{S}\right\} \qquad (5.22\text{-}12)$$

where S includes interior surfaces of discontinuity in addition to bounding surfaces.

The $\bar{B}$ and $\bar{H}$ fields of volume and surface distributions of doublets are given at exterior points of the surfaces by

$$\bar{B} = \bar{H} = \mathrm{curl}\,\frac{1}{c}\int_\tau \frac{1}{r}\left[\frac{\partial\bar{P}}{\partial t}\right] d\tau + \mathrm{curl}\,\frac{1}{c}\int_S \frac{1}{r}\left[\frac{\partial\bar{P}'}{\partial t}\right] dS$$

$$= \frac{1}{c}\,\mathrm{curl}\,\frac{\partial}{\partial t}\bar{\Pi}_e \qquad (5.22\text{-}13)$$

$$= -\frac{1}{c}\int_\tau \left\{\left[\frac{\partial\bar{P}}{\partial t}\right]\times\frac{\bar{r}}{r^3} + \left[\frac{\partial^2\bar{P}}{\partial t^2}\right]\times\frac{\bar{r}}{cr^2}\right\} d\tau - \frac{1}{c}\int_S \left\{\left[\frac{\partial\bar{P}'}{\partial t}\right]\times\frac{\bar{r}}{r^3} + \left[\frac{\partial^2\bar{P}'}{\partial t^2}\right]\times\frac{\bar{r}}{cr^2}\right\} dS$$

$$(5.22\text{-}14)$$

The contribution of a volume distribution alone may also be written for all points as

$$\bar{B} = \bar{H} = \frac{1}{c}\int_\tau \frac{1}{r}\left[\mathrm{curl}\,\frac{\partial\bar{P}}{\partial t}\right] d\tau - \frac{1}{c}\oint_S d\bar{S}\times\frac{1}{r}\left[\frac{\partial\bar{P}}{\partial t}\right] \qquad (5.22\text{-}15)$$

where S comprises interior surfaces of discontinuity together with bounding surfaces.

5.22c Fields of macroscopic whirl distributions

In the presence of volume and surface distributions of whirls the value of $\overline{B}$ at exterior points of the surfaces is given by

$$\overline{B} = \text{curl} \int_\tau \left\{[\overline{M}]\times\text{grad}\,\frac{1}{r} - \left[\frac{\partial \overline{M}}{\partial t}\right] \times \frac{\overline{r}}{cr^2}\right\} d\tau$$

$$+ \text{curl} \int_S \left\{[\overline{M}']\times\text{grad}\,\frac{1}{r} - \left[\frac{\partial \overline{M}'}{\partial t}\right] \times \frac{\overline{r}}{cr^2}\right\} dS$$

$$= \text{curl curl} \int_\tau \frac{[\overline{M}]}{r}\, d\tau + \text{curl curl} \int_S \frac{[\overline{M}']}{r}\, dS \qquad (5.22\text{-}16)$$

or

$$\overline{B} = \text{curl curl}\, \overline{\Pi}_m \qquad (5.22\text{-}17)$$

$$= \text{grad div}\, \overline{\Pi}_m - \frac{1}{c^2}\frac{\partial^2}{\partial t^2}\overline{\Pi}_m - \text{dal}\, \overline{\Pi}_m$$

whence, at interior points of the volume distribution,

$$\overline{B} = \text{grad div}\, \overline{\Pi}_m - \frac{1}{c^2}\frac{\partial^2}{\partial t^2}\overline{\Pi}_m + 4\pi\overline{M} \qquad (5.22\text{-}18)$$

and at both interior and exterior points

$$\overline{H} = \text{grad div}\, \overline{\Pi}_m - \frac{1}{c^2}\frac{\partial^2}{\partial t^2}\overline{\Pi}_m \qquad (5.22\text{-}19)$$

Alternatively, from equation (5.22-16) and equations (17) and (21) of Table 6, pp. 446-7, with $\overline{M}$ replacing $\overline{J}$, the contribution of the volume distribution to $\overline{B}$ at interior points becomes

$$\overline{B} = \int_\tau \left\{\left(-\left[\frac{\partial \overline{M}}{\partial t}\right].\nabla\right)\frac{\overline{r}}{cr^2} + \frac{\overline{r}}{cr^4}\,\overline{r}.\left[\frac{\partial \overline{M}}{\partial t}\right] + \frac{\overline{r}}{c^2r^3}\,\overline{r}.\left[\frac{\partial^2 \overline{M}}{\partial t^2}\right] - \frac{1}{c^2r}\left[\frac{\partial^2 \overline{M}}{\partial t^2}\right]\right\} d\tau$$

$$(5.22\text{-}20)$$

$$+ 4\pi\overline{M} + \lim_{\tau'\to 0}\int_{\tau-\tau'} ([\overline{M}].\nabla)\,\text{grad}\,\frac{1}{r}\, d\tau - \lim_{S'\to 0}\oint_{S'} [\overline{M}].d\overline{S}\,\text{grad}\,\frac{1}{r}$$

Beyond the source

$$\bar{B} = \int_\tau \left\{([\bar{M}].\nabla)\ \mathrm{grad}\ \frac{1}{r} - \left(\left[\frac{\partial \bar{M}}{\partial t}\right].\nabla\right) \frac{\bar{r}}{cr^2} + \frac{\bar{r}}{cr^4}\ \bar{r}.\left[\frac{\partial \bar{M}}{\partial t}\right]\right.$$

$$\left. + \frac{\bar{r}}{c^2r^3}\ \bar{r}.\left[\frac{\partial^2 \bar{M}}{\partial t^2}\right] - \frac{1}{c^2 r}\left[\frac{\partial^2 \bar{M}}{\partial t^2}\right]\right\} d\tau$$

$$= \int_\tau \left\{-\frac{[\bar{M}]}{r^3} + \frac{3\bar{r}}{r^5}\ \bar{r}.[\bar{M}] - \frac{1}{cr^2}\left[\frac{\partial \bar{M}}{\partial t}\right] + \frac{3\bar{r}}{cr^4}\ \bar{r}.\left[\frac{\partial \bar{M}}{\partial t}\right] + \frac{\bar{r}}{c^2r^3}\ \bar{r}.\left[\frac{\partial^2 \bar{M}}{\partial t^2}\right] - \frac{1}{c^2 r}\left[\frac{\partial^2 \bar{M}}{\partial t^2}\right]\right\} d\tau$$

(5.22-21)

The contribution of surface sources to $\bar{B}$ at points exterior to the surfaces takes the form of equation (5.22-21) with $\bar{M}'$ replacing $\bar{M}$ and dS replacing $d\tau$.

Further, it is seen from (5.22-16) and equation (22), p. 447, that at both interior and exterior points of a volume distribution of whirls

$$\bar{B} = -\int_\tau \left\{[\mathrm{curl}\ \bar{M}] \times \frac{\bar{r}}{r^3} + \left[\frac{\partial}{\partial t}\ \mathrm{curl}\ \bar{M}\right] \times \frac{\bar{r}}{cr^2}\right\} d\tau$$

(5.22-22)

$$+ \oint_S \left\{\left(d\bar{S}\times[\bar{M}]\right) \times \frac{\bar{r}}{r^3} + \left(d\bar{S} \times \left[\frac{\partial \bar{M}}{\partial t}\right]\right) \times \frac{\bar{r}}{cr^2}\right\}$$

where S includes both interior surfaces of discontinuity and bounding surfaces.

The $\bar{E}$ and $\bar{D}$ fields of volume and surface distributions of whirls are given at exterior points of the surfaces by

$$\bar{E} = \bar{D} = -\frac{1}{c}\frac{\partial}{\partial t}\int_\tau \left\{[\bar{M}]\times\mathrm{grad}\ \frac{1}{r} - \left[\frac{\partial \bar{M}}{\partial t}\right] \times \frac{\bar{r}}{cr^2}\right\} d\tau$$

$$-\frac{1}{c}\frac{\partial}{\partial t}\int_S \left\{[\bar{M}']\times\mathrm{grad}\ \frac{1}{r} - \left[\frac{\partial \bar{M}'}{\partial t}\right] \times \frac{\bar{r}}{cr^2}\right\} dS$$

$$= -\frac{1}{c}\frac{\partial}{\partial t}\ \mathrm{curl}\int_\tau \frac{[\bar{M}]}{r}\ d\tau - \frac{1}{c}\frac{\partial}{\partial t}\ \mathrm{curl}\int_S \frac{[\bar{M}']}{r}\ dS$$

or

$$\bar{E} = \bar{D} = -\frac{1}{c}\,\mathrm{curl}\,\frac{\partial}{\partial t}\,\bar{\Pi}_m \qquad (5.22\text{-}23)$$

The contribution of a volume source alone may be expressed everywhere as

$$\bar{E} = \bar{D} = -\frac{1}{c}\int_\tau \frac{1}{r}\left[\frac{\partial}{\partial t}\,\mathrm{curl}\,\bar{M}\right] d\tau + \frac{1}{c}\oint_S d\bar{S} \times \frac{1}{r}\left[\frac{\partial \bar{M}}{\partial t}\right]$$

where S includes both interior surfaces of discontinuity and bounding surfaces.

5.22d Summary of formulae involving polarisation potentials

It is of interest to compare the field formulae for doublet and whirl distributions when expressed in terms of the associated polarisation potentials. We have

doublet distribution	whirl distribution
$\bar{E} = \mathrm{grad\,div}\,\bar{\Pi}_e - \frac{1}{c^2}\frac{\partial^2}{\partial t^2}\bar{\Pi}_e$	$\bar{H} = \mathrm{grad\,div}\,\bar{\Pi}_m - \frac{1}{c^2}\frac{\partial^2}{\partial t^2}\bar{\Pi}_m$
$\bar{D} = \mathrm{curl\,curl}\,\bar{\Pi}_e$	$\bar{B} = \mathrm{curl\,curl}\,\bar{\Pi}_m$
$\bar{B} = \bar{H} = \frac{1}{c}\,\mathrm{curl}\,\frac{\partial}{\partial t}\,\bar{\Pi}_e$	$\bar{D} = \bar{E} = -\frac{1}{c}\,\mathrm{curl}\,\frac{\partial}{\partial t}\,\bar{\Pi}_m$

For a mixed distribution

$$\bar{E} = \mathrm{grad\,div}\,\bar{\Pi}_e - \frac{1}{c^2}\frac{\partial^2}{\partial t^2}\bar{\Pi}_e - \frac{1}{c}\,\mathrm{curl}\,\frac{\partial}{\partial t}\,\bar{\Pi}_m \qquad (5.22\text{-}24)$$

$$\bar{H} = \mathrm{grad\,div}\,\bar{\Pi}_m - \frac{1}{c^2}\frac{\partial^2}{\partial t^2}\bar{\Pi}_m + \frac{1}{c}\,\mathrm{curl}\,\frac{\partial}{\partial t}\,\bar{\Pi}_e \qquad (5.22\text{-}25)$$

$$\bar{D} = \mathrm{curl\,curl}\,\bar{\Pi}_e - \frac{1}{c}\,\mathrm{curl}\,\frac{\partial}{\partial t}\,\bar{\Pi}_m \qquad (5.22\text{-}26)$$

$$\bar{B} = \mathrm{curl\,curl}\,\bar{\Pi}_m + \frac{1}{c}\,\mathrm{curl}\,\frac{\partial}{\partial t}\,\bar{\Pi}_e \qquad (5.22\text{-}27)$$

EXERCISES

5-98. Prove equation (5.22-4).

5-99. Consider in some detail the possible geometrical disposition of the surfaces which appear in equations (5.22-2/4). Why may the final surface integral in equation (5.22-2) include an open surface when the corresponding surface in equation (5.22-3) is necessarily closed?

5-100. Consider a volume distribution of neutral current which closes upon itself. According to equation (5.22-1) the associated $\bar{E}$ field is given by

$$\bar{E} = -\int \frac{1}{c^2 r} \boxed{\frac{\partial \bar{J}}{\partial t}}\, d\tau$$

and according to (5.22-5) it is given by

$$\bar{E} = -\int \frac{1}{cr^2} [\bar{J}_t]\, d\tau + \int \frac{1}{cr^2} [\bar{J}_r]\, d\tau - \int \frac{1}{c^2 r} \left[\frac{\partial \bar{J}_t}{\partial t}\right] d\tau$$

Demonstrate that the two expressions are equal by splitting the current flow into a system of closed stream tubes and showing that the difference of the expressions may be represented as the sum of a set of closed line integrals of exact differentials.

5-101. By combining the result of Ex.5-78., p. 541 with equation (5.22-11) show that the time-dependent component of the $\bar{E}$ field of a complete source characterised by $\bar{J}$, ρ and σ may be expressed at exterior points by

$$\bar{E} = \int_\tau \left\{ -\frac{1}{r^3} \int_{t_o}^{t} [\bar{J}]\, dt + \frac{3\bar{r}}{r^5}\, \bar{r}.\int_{t_o}^{t} [\bar{J}]\, dt - \frac{1}{cr^2} [\bar{J}] + \frac{3\bar{r}}{cr^4}\, \bar{r}.[\bar{J}] + \frac{\bar{r}}{c^2 r^3}\, \bar{r}.\boxed{\frac{\partial \bar{J}}{\partial t}} \right.$$

$$\left. - \frac{1}{c^2 r} \boxed{\frac{\partial \bar{J}}{\partial t}} \right\} d\tau$$

$$= \int_\tau \left\{ -\frac{1}{r^3} \int_{t_o}^{t} [\bar{J}_t]\, dt + \frac{2}{r^3} \int_{t_o}^{t} [\bar{J}_r]\, dt - \frac{1}{cr^2} [\bar{J}_t] + \frac{2}{cr^2} [\bar{J}_r] - \frac{1}{c^2 r} \left[\frac{\partial \bar{J}_t}{\partial t}\right] \right\} d\tau$$

Show further that at interior points

$$\overline{E} = \int_\tau \left\{ \left(\int_{t_o}^{t} [\overline{J}]\, dt.\nabla \right) \operatorname{grad} \frac{1}{r} - \frac{1}{cr^2} [\overline{J}_t] + \frac{2}{cr^2} [\overline{J}_r] - \frac{1}{c^2 r} \left[\frac{\partial \overline{J}_t}{\partial t} \right] \right\} d\tau$$

$$- \frac{4}{3} \pi \int_{t_o}^{t} \overline{J}\, dt$$

where the first component of the volume integral signifies the limit which obtains when the excluding surface is spherical and centred upon the point of evaluation.

5-102. Derive the following relationship for a complete line/point singlet source system.

$$\overline{E} = - \int_\Gamma [\lambda] \frac{\overline{r}}{r^3}\, ds - \sum [a] \frac{\overline{r}}{r^3} - \int_\Gamma \frac{1}{cr^2} [\overline{I}_t]\, ds + \int_\Gamma \frac{1}{cr^2} [\overline{I}_r]\, ds$$

$$- \int_\Gamma \frac{1}{c^2 r} \left[\frac{\partial \overline{I}_t}{\partial t} \right] ds$$

5-103. A complete straight-line source is characterised by a uniform macroscopic current density $\overline{I}$ which is a sinusoidal function of time. If the length of the source is made to approach zero while the current density is increased to maintain the product of peak current density and length constant, the limiting configuration is known as a Hertzian dipole.

By analogy with the point doublet, the scalar moment of the dipole is defined to be the product of its length and the scalar point source strength developed at the ends. Use the result of Ex.5-102., together with equation (5.6-30), to show that the time-dependent $\overline{E}$ and $\overline{B}$ fields of the Hertzian dipole are identical, at exterior points, with the fields of a microscopic doublet of equal instantaneous moment, as represented by equations (5.16-1) and (5.16-3).

5-104. Use the results of Sec. 5.22 to show that $\bar{B} = \hat{\bar{R}} \times \bar{E}$ at infinite distance from a mixed source of finite dimensions, where $\hat{\bar{R}}$ is the unit vector directed from some point of the source to the point of evaluation. Hence show that $\oint \bar{E} \times \bar{B}.d\bar{S}$, when evaluated at infinity, can never be negative, and use this result to extend the uniqueness theorem of Ex.5-97., p. 561 to the case where Σ recedes to infinity and $\hat{\bar{n}} \times \bar{E}$ or $\hat{\bar{n}} \times \bar{B}$ is specified over $S_{1..n}$ alone.

CHAPTER 6

HELMHOLTZ'S FORMULA AND ALLIED TOPICS

6.1 Helmholtz's Equation

Helmholtz's Formula

Conditions for Uniqueness

6.1a The bounded scalar field

It is easily shown that the asymmetrical and symmetrical forms of Green's theorem (1.17-10/11) continue to hold when V and U are complex scalar point functions having continuous second derivatives within the integration space. Thus, in the notation of Sec. 1.22,

$$\oint_{S_{1..n}\Sigma} \tilde{U}\frac{\partial\tilde{V}}{\partial n}\,dS = \int_{\tau} \tilde{U}\,\nabla^2\tilde{V}\,d\tau + \int_{\tau} \text{grad }\tilde{U}.\text{grad }\tilde{V}\,d\tau \tag{6.1-1}$$

and

$$\oint_{S_{1..n}\Sigma} \left\{\tilde{U}\frac{\partial\tilde{V}}{\partial n} - \tilde{V}\frac{\partial\tilde{U}}{\partial n}\right\} dS = \int_{\tau} (\tilde{U}\,\nabla^2\tilde{V} - \tilde{V}\,\nabla^2\tilde{U})\,d\tau \tag{6.1-2}$$

If $\tilde{V} = a + jb$ where a and b are real, the complex conjugate of $\tilde{V}$, viz $\tilde{V}^*$, is defined by $\tilde{V}^* = a - jb$.

Let $\tilde{U} = \tilde{V}^*$. Then

$$\oint_{S_{1..n}\Sigma} \tilde{V}^*\frac{\partial\tilde{V}}{\partial n}\,dS = \int_{\tau} \tilde{V}^*\,(\nabla^2+\tilde{k}^2)\tilde{V}\,d\tau - \tilde{k}^2\int_{\tau} \tilde{V}^*\tilde{V}\,d\tau + \int_{\tau} \text{grad }\tilde{V}^*.\text{grad }\tilde{V}\,d\tau \tag{6.1-3}$$

where $\tilde{k}^2$ is a real, imaginary or complex constant.

Suppose that within the integration region R

$$(\nabla^2+\tilde{k}^2)\tilde{V} = 0 \tag{6.1-4}$$

and that, in addition, $\tilde{V}$ (and therefore $\tilde{V}^*$) or $\frac{\partial \tilde{V}}{\partial n}$ is zero upon $S_{1..n}\Sigma$. Then

$$0 = -\tilde{k}^2 \int_\tau \tilde{V}^*\tilde{V}\, d\tau + \int_\tau \text{grad}\,\tilde{V}^*.\text{grad}\,\tilde{V}\, d\tau \qquad (6.1\text{-}5)$$

Now $\tilde{V}^*\tilde{V}$ and grad $\tilde{V}^*$.grad $\tilde{V}$ are everywhere real and positive or zero since they are respectively equal to (a^2+b^2) and $\{|\text{grad a}|^2+|\text{grad b}|^2\}$. Hence if $\tilde{k}$ is imaginary or complex, equation (6.1-5) can hold only if $\tilde{V}$ is zero throughout $\underline{R}$. This result parallels that obtained for a harmonic function in Sec. 3.2. However, the argument fails when $\tilde{k}$ is real since non-zero integrals may cancel. Correspondingly, point functions are known to exist which satisfy equation (6.1-4) within $\underline{R}$ and are zero or have zero normal derivatives upon the bounding surfaces. These are called eigenfunctions and the associated values of $\tilde{k}$ are known as eigenvalues.

Equation (6.1-4) is the homogeneous scalar Helmholtz equation. It is of considerable importance because many of the differential equations of mathematical physics can be expressed in this form. When $\tilde{k}^2 = 0$ it reduces to Laplace's equation.

If equation (6.1-4), with $\tilde{k}$ imaginary or complex, is satisfied everywhere outside $S_{1..n}\Sigma$ and the surface integral over Σ in (6.1-3) vanishes as Σ recedes to infinity, it follows that $\tilde{V}$ will be zero everywhere outside $S_{1..n}$ if $\tilde{V}$ or $\frac{\partial \tilde{V}}{\partial n}$ is zero upon $S_{1..n}$. For this exterior case, however, it may be shown that the result continues to hold when $\tilde{k}$ is real, provided that certain boundary conditions are satisfied at infinity. The matter is discussed in detail in Sec. 6.1c. Meanwhile, the above considerations lead directly to the following uniqueness theorem:

If $\tilde{k}$ is an imaginary or complex constant and if $(\nabla^2+\tilde{k}^2)\tilde{V}$ is a specified function of position in the region $\underline{R}$ bounded by the surfaces $S_{1..n}\Sigma$, then $\tilde{V}$ is uniquely defined within $\underline{R}$ provided that any one of the following conditions is satisfied.

(1) $\tilde{V}$ or $\frac{\partial \tilde{V}}{\partial n}$ is a specified function of position upon $S_{1..n}\Sigma$.

(2) $\tilde{V}$ is constant over each surface in turn and $\oint \frac{\partial \tilde{V}}{\partial n}\, dS$ is specified for each surface.

(3) The vector tangential component of grad $\tilde{V}$ is specified at each point of $S_{1..n}\Sigma$ and $\oint \frac{\partial \tilde{V}}{\partial n}\, dS$ is specified for each surface.

(4) $\frac{\partial \tilde{V}}{\partial n} + p\,\tilde{V} = \tilde{q}$ where p and $\tilde{q}$ are specified functions of position upon $S_{1..n}\Sigma$ and p is everywhere real and nowhere negative.

(5) One or other of conditions (1) to (4) applies to each surface.

6.1b The bounded vector field

The results for the scalar field are applicable to each rectangular component of a well-behaved vector field $\tilde{\bar{F}}$ so that the specification of $(\nabla^2+\tilde{k}^2)\tilde{\bar{F}}$ throughout $\underline{R}$, together with boundary conditions corresponding to those set out above, render $\tilde{\bar{F}}$ unique within $\underline{R}$ provided that $\tilde{k}$ is imaginary or complex. However, we may also proceed in the manner suggested by Sec. 3.7 by writing

$$\operatorname{div}(\tilde{\bar{F}}^* \times \operatorname{curl}\tilde{\bar{F}}) = \operatorname{curl}\tilde{\bar{F}}.\operatorname{curl}\tilde{\bar{F}}^* - \tilde{\bar{F}}^*.\operatorname{curl}\operatorname{curl}\tilde{\bar{F}}$$

$$= \operatorname{curl}\tilde{\bar{F}}.\operatorname{curl}\tilde{\bar{F}}^* - \tilde{\bar{F}}^*.\operatorname{grad}\operatorname{div}\tilde{\bar{F}} + \tilde{\bar{F}}^*.\nabla^2\tilde{\bar{F}}$$

where $\tilde{\bar{F}} = \bar{a} + j\bar{b}$ and $\tilde{\bar{F}}^* = \bar{a} - j\bar{b}$

whence

$$\oint_{S_{1..n}\Sigma} (\tilde{\bar{F}}^* \times \operatorname{curl}\tilde{\bar{F}}).d\bar{S} = \int_\tau \operatorname{curl}\tilde{\bar{F}}.\operatorname{curl}\tilde{\bar{F}}^*\,d\tau - \int_\tau \tilde{\bar{F}}^*.\operatorname{grad}\operatorname{div}\tilde{\bar{F}}\,d\tau$$
$$+ \int_\tau \tilde{\bar{F}}^*.(\nabla^2+\tilde{k}^2)\tilde{\bar{F}}\,d\tau - \tilde{k}^2\int_\tau \tilde{\bar{F}}^*.\tilde{\bar{F}}\,d\tau \tag{6.1-6}$$

By proceeding as in Sec. 6.1a we may show that, for imaginary or complex values of $\tilde{k}$, $\tilde{\bar{F}}$ vanishes throughout $\underline{R}$ if

(1) $\tilde{\bar{F}}$ satisfies the homogeneous vector Helmholtz equation, $(\nabla^2+\tilde{k}^2)\tilde{\bar{F}} = 0$, throughout $\underline{R}$.

(2) $\operatorname{div}\tilde{\bar{F}}$ is constant or zero throughout $\underline{R}$.

(3) $\hat{n} \times \tilde{\bar{F}}$ or $\hat{n} \times \operatorname{curl}\tilde{\bar{F}}$ is zero upon $S_{1..n}\Sigma$.

Further, $\tilde{\bar{F}}$ is uniquely defined throughout $\underline{R}$ if $(\nabla^2+\tilde{k}^2)\tilde{\bar{F}}$ and $\operatorname{div}\tilde{\bar{F}}$ are specified functions of position within $\underline{R}$ and $\hat{n} \times \tilde{\bar{F}}$ or $\hat{n} \times \operatorname{curl}\tilde{\bar{F}}$ is a specified function of position upon $S_{1..n}\Sigma$.

The individual specifications of $(\nabla^2+\tilde{k}^2)\tilde{\bar{F}}$ and $\operatorname{div}\tilde{\bar{F}}$ may be replaced by the specification of $\operatorname{curl}\operatorname{curl}\tilde{\bar{F}} - \tilde{k}^2\tilde{\bar{F}}$.

6.1c The externally-unbounded field

Sommerfeld's conditions

By substituting $\frac{1}{r}\, e^{j\tilde{k}r}$ for $\tilde{U}$ in equation (6.1-2), where r is distance measured from 0, and proceeding in the manner of Sec. 3.3, we find that

$$\left.\begin{matrix} 4\pi\tilde{V}_o \\ 0 \end{matrix}\right\} = \oint_{S_{1..n}\Sigma} \left\{ \frac{1}{r}\, e^{j\tilde{k}r}\, \frac{\partial \tilde{V}}{\partial n} - \tilde{V}\, \frac{\partial}{\partial n}\left(\frac{1}{r}\, e^{j\tilde{k}r}\right)\right\} dS - \int_{\tau} \frac{1}{r}\, e^{j\tilde{k}r}\, (\nabla^2+\tilde{k}^2)\tilde{V}\, d\tau \tag{6.1-7}$$

according as 0 lies within or without τ.

No restriction is placed upon the (constant) value of $\tilde{k}$.

If $(\nabla^2+\tilde{k}^2)\tilde{V} = 0$ throughout τ, the above relationship reduces to Helmholtz's formula, viz

$$\left.\begin{matrix} 4\pi\tilde{V}_o \\ 0 \end{matrix}\right\} = \oint_{S_{1..n}\Sigma} \left\{ \frac{1}{r}\, e^{j\tilde{k}r}\, \frac{\partial \tilde{V}}{\partial n} - \tilde{V}\frac{\partial}{\partial n}\left(\frac{1}{r}\, e^{j\tilde{k}r}\right)\right\} dS \tag{6.1-8}$$

When $\tilde{V}$ satisfies the homogeneous Helmholtz equation everywhere outside the closed surfaces $S_{1..n}$, the surface Σ may be removed to infinity and made spherically symmetrical, with radius R, about some fixed point P in the vicinity of $S_{1..n}$. Then provided that OP is finite, the associated surface integral approaches

$$\oint_{\Sigma} e^{j\tilde{k}r} \left\{ \frac{1}{R}\, \frac{\partial \tilde{V}}{\partial R} - \frac{j\tilde{k}\tilde{V}}{R} + \frac{\tilde{V}}{R^2} \right\} R^2\, d\Omega$$

This will vanish if

$$R e^{j\tilde{k}r} \left\{ \frac{\partial \tilde{V}}{\partial R} - j\tilde{k}\tilde{V} + \frac{\tilde{V}}{R} \right\} \to 0 \tag{6.1-9}$$

uniformly in all directions as $R\to\infty$.

When $\tilde{k}$ is real, (6.1-9) is satisfied, inter alia, by

$$R\left(\frac{\partial \tilde{V}}{\partial R} - \widetilde{jkV}\right) \to 0 \quad ; \quad R\,\tilde{V} \text{ bounded} \quad \text{as} \quad R\to\infty \tag{6.1-10}$$

These are known as Sommerfeld's conditions; the first is the 'radiation condition'.

We now proceed to show that if $(\nabla^2+\tilde{k}^2)\tilde{V} = 0$ everywhere outside $S_{1..n}$, and if $\tilde{V}$ or $\frac{\partial\tilde{V}}{\partial n}$ is zero upon $S_{1..n}$, then $\tilde{V}$ vanishes everywhere outside (and upon) $S_{1..n}$ provided that appropriate Sommerfeld conditions obtain.

In the present case equation (6.1-3) reduces to

$$\oint_\Sigma \tilde{V}^* \frac{\partial\tilde{V}}{\partial n}\, dS = -\tilde{k}^2 \int_\tau \tilde{V}^*\tilde{V}\, d\tau + \int_\tau \text{grad}\, \tilde{V}^*.\text{grad}\, \tilde{V}\, d\tau \qquad (6.1\text{-}11)$$

As Σ recedes to infinity, the left hand side of equation (6.1-11) may be replaced by

$$\oint_\Sigma \tilde{V}^* (j\tilde{k}\tilde{V}+\tilde{\varepsilon})\, dS = j\tilde{k} \oint_\Sigma \tilde{V}^*\tilde{V}\, dS + \oint_\Sigma R\tilde{V}^* R\tilde{\varepsilon}\, d\Omega$$

where $\frac{\partial\tilde{V}}{\partial R} - j\tilde{k}\tilde{V} \equiv \tilde{\varepsilon}$.

Suppose first that $\tilde{k}^2$ is complex so that $\tilde{k} = \pm(p+jq)$ where p and q are real. We choose the sign in such a way that $\tilde{k}$ may be written as $\alpha + j\beta$ where β is positive.

If the Sommerfeld conditions hold for this value of $\tilde{k}$ it is evident that $R\tilde{\varepsilon}\to 0$ as Σ recedes to infinity, and, since $R\tilde{V}$ (and therefore $R\tilde{V}^*$) is bounded, the limiting form of equation (6.1-11) becomes

$$j(\alpha+j\beta) \oint_\infty \tilde{V}^*\tilde{V}\, dS + (\alpha^2-\beta^2+2j\alpha\beta) \int_\tau \tilde{V}^*\tilde{V}\, d\tau = \int_\tau \text{grad}\, \tilde{V}^*.\text{grad}\, \tilde{V}\, d\tau$$

where τ now represents all space outside $S_{1..n}$.

The imaginary component of the equation is

$$\alpha\left\{ \oint_\infty \tilde{V}^*\tilde{V}\, dS + 2\beta \int_\tau \tilde{V}^*\tilde{V}\, d\tau \right\} = 0$$

Since β is positive and $\tilde{V}^*\tilde{V}$ is always positive or zero, it follows that $\tilde{V}$ is zero everywhere outside (and upon) $S_{1..n}$.

Now suppose that $\tilde{k}^2$ is real and negative. Let $\tilde{k} = +j\beta$ where β is positive. If Sommerfeld's conditions hold for this value of $\tilde{k}$, substitution in (6.1-11) yields

$$-\beta \oint_\infty \tilde{V}^*\tilde{V}\, dS - \beta^2 \int_\tau \tilde{V}^*\tilde{V}\, d\tau = \int_\tau \text{grad}\, \tilde{V}^*.\text{grad}\, \tilde{V}\, d\tau$$

Here the signs are such that $\tilde{V}$ must again be zero throughout τ.

Finally, suppose that $\tilde{k}^2$ is real and positive.

In this case a new approach is required. When (6.1-9) is satisfied, it follows from (6.1-8) that in any finite region surrounding $S_{1..n}$

$$4\pi\tilde{V}_o = \oint_{S_{1..n}} \frac{1}{r} e^{j\tilde{k}r} \left\{\frac{\partial\tilde{V}}{\partial n} - \left(j\tilde{k} - \frac{1}{r}\right) \frac{\partial r}{\partial n} \tilde{V}\right\} dS \qquad (6.1\text{-}12)$$

It can be shown[1] that when O is sufficiently removed from P, (6.1-12) may be expressed as an absolutely and uniformly convergent series in powers of $OP(R_o)$, viz

$$4\pi\tilde{V}_o = e^{j\tilde{k}R_o} \sum_{m=1}^{\infty} \tilde{a}_m R_o^{-m} \qquad (6.1\text{-}13(a))$$

and, correspondingly,

$$4\pi\tilde{V}_o^* = e^{-j\tilde{k}^*R_o} \sum_{m=1}^{\infty} \tilde{a}_m^* R_o^{-m} \qquad (6.1\text{-}13(b))$$

Here $\tilde{a}_m$ is a function of both $\tilde{k}$ and the orientation of PO but is independent of R_o. The expressions are valid when $R_o \geq 3R'$, where R' is the greatest distance of P from any point of $S_{1..n}$.

Let Σ' be a spherical surface centred upon P and of radius $R_1 > 3R'$.

On substituting $\tilde{V}^*$ for $\tilde{U}$ in equation (6.1-2) we obtain

$$\oint_{\Sigma'} \left(\tilde{V}^* \frac{\partial\tilde{V}}{\partial n} - \tilde{V} \frac{\partial\tilde{V}^*}{\partial n}\right) dS = \int_{\tau'} (\tilde{V}^*(\nabla^2+\tilde{k}^2)\tilde{V} - \tilde{V}\,(\nabla^2+\tilde{k}^{*2})\tilde{V}^*\}\, d\tau + (\tilde{k}^{*2}-\tilde{k}^2) \int_{\tau'} \tilde{V}^*\tilde{V}\, d\tau \qquad (6.1\text{-}14)$$

where τ' is the region bounded by $S_{1..n}$ and Σ'.

But $\tilde{k}^2 = \alpha^2$ hence $\tilde{k} = \tilde{k}^* = \pm\alpha$ and $\tilde{k}^2 = \tilde{k}^{*2}$ so that equation (6.1-14) reduces to

$$\oint_{\Sigma'} \left(\tilde{V}^* \frac{\partial\tilde{V}}{\partial n} - \tilde{V} \frac{\partial\tilde{V}^*}{\partial n}\right) dS = 0 \qquad (6.1\text{-}15)$$

since $(\nabla^2+\tilde{k}^{*2})\tilde{V}^* = 0$ when $(\nabla^2+\tilde{k}^2)\tilde{V} = 0$.

---

1. F.V. Atkinson, Phil. Mag. 40, pp 646-7 (1949). See also Ex.6-3., p 584.

On substituting (6.1-13(a)), (6.1-13(b) and their derivatives in (6.1-15) we find that

$$0 = \oint_{\Sigma'} \left\{ 2j(\alpha^2)^{\frac{1}{2}} \left(\tilde{a}_1\tilde{a}_1^* + \frac{\tilde{a}_1\tilde{a}_2^* + \tilde{a}_2\tilde{a}_1^*}{R_1} + \frac{\tilde{a}_1\tilde{a}_3^* + \tilde{a}_3\tilde{a}_1^*}{R_1^2} + \frac{\tilde{a}_2\tilde{a}_2^*}{R_1^2} + --- \right) + \frac{\tilde{a}_1\tilde{a}_2^* - \tilde{a}_2\tilde{a}_1^*}{R_1^2} + \frac{2(\tilde{a}_1\tilde{a}_3^* - \tilde{a}_3\tilde{a}_1^*)}{R_1^3} + \frac{\tilde{a}_2\tilde{a}_3^* - \tilde{a}_3\tilde{a}_2^*}{R_1^4} ----- \right\} d\Omega \qquad (6.1\text{-}16)$$

The terms within the round brackets are real while those outside them are imaginary.

Since the equation holds for all finite values of $R_1 > 3R'$ and since $\tilde{a}_1$, $\tilde{a}_2$ --- are independent of R_1, the surface integrals associated with the different powers of R_1 must be individually zero. Thus $\oint_{\Sigma'} \tilde{a}_1\tilde{a}_1^* \, dS = 0$.

But since $\tilde{a}_1\tilde{a}_1^*$ is always real and positive, or zero, it follows that $\tilde{a}_1$ is zero for each surface element of Σ', ie for all orientations of the radius vector from P. This leads to the requirement $\oint_{\Sigma'} \tilde{a}_2\tilde{a}_2^* = 0$, whence $\tilde{a}_2$ is zero for all orientations, and so on. Hence $\tilde{V}$ and $\frac{\partial\tilde{V}}{\partial n}$ are zero upon Σ' for all finite values of $R_1 > 3R'$, provided that Sommerfeld's conditions hold for $\tilde{k} = +\alpha$ or $\tilde{k} = -\alpha$.

However, it can be shown[2] that if both $\tilde{V}$ and $\frac{\partial\tilde{V}}{\partial n}$ are zero upon any small piece of a regular surface which bounds in part, or lies within, a region in which $(\nabla^2+\alpha^2)\tilde{V} = 0$, as in the present case, then $\tilde{V}$ is zero throughout that region. Hence $\tilde{V}$ is zero everywhere outside $S_{1..n}$.

The demonstration is therefore complete.

The corresponding uniqueness theorem follows immediately:

If $(\nabla^2+\tilde{k}^2)\tilde{V}$ is specified at all points of the open region bounded internally by the closed surfaces $S_{1..n}$, and if $\tilde{V}$ or $\frac{\partial\tilde{V}}{\partial n}$ is a specified function of position upon these surfaces, then $\tilde{V}$ is uniquely defined throughout the region, provided that it satisfies the relationship (6.1-10) when the real part of $j\tilde{k}$ is negative or zero.

---

2. See Ex.6-2., p. 584, for an elementary approach. The rigorous argument revolves around the analytic nature of the continuous solution of an elliptic differential equation.

The results of this section may be extended directly to the complex vector field $\tilde{\bar{F}}$ since $\tilde{V}$ may be identified with the rectangular components of $\tilde{\bar{F}}$ in turn.

6.2 Scalar Green's Functions for Helmholtz's Equation

On combining equation (6.1-2) with (6.1-7) we obtain

$$\left.\begin{matrix} 4\pi\tilde{V}_o \\ 0 \end{matrix}\right\} = \oint_{S_{1..n}\Sigma} \left\{\left(\tilde{U} + \frac{1}{r}\, e^{j\tilde{k}r}\right) \frac{\partial\tilde{V}}{\partial n} - \tilde{V}\frac{\partial}{\partial n}\left(\tilde{U} + \frac{1}{r}\, e^{j\tilde{k}r}\right)\right\} dS - \int_\tau \left\{\left(\tilde{U} + \frac{1}{r}\, e^{j\tilde{k}r}\right)(\nabla^2+\tilde{k}^2)\tilde{V} - \tilde{V}\,(\nabla^2+\tilde{k}^2)\tilde{U}\right\} d\tau \qquad (6.2\text{-}1)$$

Then if $(\nabla^2+\tilde{k}^2)\tilde{U} = 0$ throughout τ, the value of $\tilde{V}$ at interior points of τ is given by

$$4\pi\tilde{V}_o = \oint_{S_{1..n}\Sigma} \left\{\tilde{G}\frac{\partial\tilde{V}}{\partial n} - \tilde{V}\frac{\partial\tilde{G}}{\partial n}\right\} dS - \int_\tau \tilde{G}\,(\nabla^2+\tilde{k}^2)\tilde{V}\, d\tau \qquad (6.2\text{-}2)$$

where $\tilde{G} \equiv \tilde{U} + \frac{1}{r}\, e^{j\tilde{k}r}$

This relationship reduces to equation (3.8-2) when $\tilde{k} = 0$.

If $\tilde{U}_1$ can be found such that $\tilde{U}_1 + \frac{1}{r}\, e^{j\tilde{k}r}$ $(\equiv \tilde{G}_1)$ is zero upon $S_{1..n}\Sigma$, ie if $\tilde{G}_1$ satisfies the homogeneous Dirichlet condition,

$$4\pi\tilde{V}_o = -\oint_{S_{1..n}\Sigma} \tilde{V}\frac{\partial\tilde{G}_1}{\partial n}\, dS - \int_\tau \tilde{G}_1\,(\nabla^2+\tilde{k}^2)\tilde{V}\, d\tau \qquad (6.2\text{-}3)$$

In particular, if $\tilde{G}_1$ exists and $\tilde{V}$ satisfies the homogeneous Helmholtz equation, then $\tilde{V}$ can be expressed at interior points of τ in terms of its value on the bounding surfaces. This result is consistent with the demonstration in Sec. 6.1a that the specification of $\tilde{V}$ over $S_{1..n}\Sigma$ renders $\tilde{V}$ unique within τ when $(\nabla^2+\tilde{k}^2)\tilde{V} = 0$ throughout τ, provided that the value of $\tilde{k}$ does not admit of the existence of an eigenfunction. It is evident that $\tilde{G}_1$ cannot exist in the presence of an eigenfunction.

$\tilde{G}_1$ is known as Green's function of the first kind for Helmholtz's equation; it is zero upon $S_{1..n}\Sigma$ and satisfies $(\nabla^2+\tilde{k}^2)\tilde{G}_1 = 0$ at all points of τ except the pole 0 where it becomes infinite like $\frac{1}{r}$. $\tilde{U}$ (and consequently $\tilde{G}_1$) is unique for any given position of 0.

Similarly, if $\widetilde{U}_2$ can be found such that $\widetilde{U}_2 + \frac{1}{r} e^{j\widetilde{k}r}$ ($\equiv \widetilde{G}_2$) satisfies the homogeneous Neumann condition on $S_{1..n}\Sigma$ then

$$4\pi\widetilde{V}_o = \oint_{S_{1..n}\Sigma} \widetilde{G}_2 \frac{\partial\widetilde{V}}{\partial n} dS - \int_\tau \widetilde{G}_2 (\nabla^2+\widetilde{k}^2)\widetilde{V} d\tau \qquad (6.2\text{-}4)$$

so that $\widetilde{V}_o$ can be expressed in terms of its normal derivative upon the bounding surfaces when $(\nabla^2+\widetilde{k}^2)\widetilde{V} = 0$ in τ. $\widetilde{G}_2$ does not exist when the corresponding eigenfunction exists.

When $\widetilde{V}$ satisfies the same boundary conditions as $\widetilde{G}$ and $(\nabla^2+\widetilde{k}^2)\widetilde{V} \neq 0$, the value of $\widetilde{V}$ at interior points of τ may be expressed as a volume integral alone. Thus equations (6.2-3) and (6.2-4) lead to

$$4\pi\widetilde{V}_o = -\int_\tau \widetilde{G}_1 (\nabla^2+\widetilde{k}^2)\widetilde{V} d\tau \qquad (6.2\text{-}5)$$

and

$$4\pi\widetilde{V}_o = -\int_\tau \widetilde{G}_2 (\nabla^2+\widetilde{k}^2)\widetilde{V} d\tau \qquad (6.2\text{-}6)$$

It is easily confirmed that if $\widetilde{G}_3$ can be found such as to satisfy homogeneous mixed conditions on the bounding surfaces ie $\frac{\partial\widetilde{G}_3}{\partial n} + p\,\widetilde{G}_3 = 0$, then

$$4\pi\widetilde{V}_o = \oint_{S_{1..n}\Sigma} \widetilde{G}_3 \left(\frac{\partial\widetilde{V}}{\partial n} + p\widetilde{V}\right) dS - \int_\tau \widetilde{G}_3 (\nabla^2+\widetilde{k}^2)\widetilde{V} d\tau \qquad (6.2\text{-}7)$$

Hence if $\widetilde{V}$ satisfies the same mixed conditions on $S_{1..n}\Sigma$, $\widetilde{V}_o$ may again be expressed solely as a volume integral.

When all surfaces of discontinuity are absent and $(\nabla^2+\widetilde{k}^2)\widetilde{V}$ is zero outside a finite region of space, $S_{1..n}$ may be eliminated from (6.1-7) and Σ removed to infinity. Then at all points of space

$$4\pi\widetilde{V}_o = \oint_\Sigma \left\{\frac{1}{r} e^{j\widetilde{k}r} \frac{\partial\widetilde{V}}{\partial n} - \widetilde{V}\frac{\partial}{\partial n}\left(\frac{1}{r} e^{j\widetilde{k}r}\right)\right\} dS - \int_\tau \frac{1}{r} e^{j\widetilde{k}r} (\nabla^2+\widetilde{k}^2)\widetilde{V} d\tau$$

$$(6.2\text{-}8)$$

If, in addition, $\tilde{V}$ satisfies the Sommerfeld conditions at infinity it may be shown[3] that equation (6.2-8) reduces to

$$4\pi\tilde{V}_o = -\int_\tau \frac{1}{r} e^{j\tilde{k}r} (\nabla^2+\tilde{k}^2)\tilde{V}\, d\tau \qquad (6.2\text{-}9)$$

for all locations of 0.

Since the function $\tilde{U}$ is not invoked, the factor $\tilde{G}$ in equations (6.2-5) and (6.2-6) is here reduced to the 'characteristic function' $\frac{1}{r} e^{j\tilde{k}r}$. It is then known as a 'free-space Green's function'.

The proof of the existence of Green's functions for Helmholtz's equation and methods for their calculation are beyond the scope of this document.

EXERCISES

6-1. Confirm equation (6.1-1)

6-2. Extend the analysis of Ex.3-18., p. 184 to show that if $\tilde{V}$ and $\frac{\partial\tilde{V}}{\partial n}$ are zero upon a regular surface element which bounds in part or lies within a region $\underline{R}$, and $(\nabla^2+\alpha^2)\tilde{V} = 0$ throughout $\underline{R}$ (α real), then $\tilde{V}$ is zero throughout $\underline{R}$.

6-3. P is a fixed point in the vicinity of the closed surface S but not coincident with it, and Q is a point of S. O lies outside S at a distance R_o from P. If PQ = ρ, OQ = r and $\angle$OPQ = ψ show that

$$\frac{e^{j\tilde{k}r}}{r} \Big/ \frac{e^{j\tilde{k}R_o}}{R_o} = (1-2\rho g \cos\psi+\rho^2 g^2)^{-\frac{1}{2}} \exp \frac{j\tilde{k}}{g} \{(1-2\rho g \cos\psi+\rho^2 g^2)^{\frac{1}{2}}-1\}$$

where $g = \frac{1}{R_o}$.

This expression may be expanded as the product of power series if $R_o > (1+\surd 2)\rho$. Prove this and deduce that

$$\frac{e^{j\tilde{k}r}}{r} \Big/ \frac{e^{j\tilde{k}R_o}}{R_o} = (1 + b_1 g + b_2 g^2 --)(\tilde{d}_o + \tilde{d}_1 g + \tilde{d}_2 g^2 ---)$$

where b_1, b_2--- are real functions of ρ and $\cos\psi$ and $\tilde{d}_o$, $\tilde{d}_1$-- are complex functions of $\tilde{k}$, ρ and $\cos\psi$.

---

3. See Ex.7-11., p. 636.

Hence show that

$$\frac{e^{j\tilde{k}r}}{r} = e^{j\tilde{k}R_o} \sum_{m=1}^{\infty} \tilde{q}_m R_o^{-m}$$

where $\tilde{q}_m$ is dependent upon the orientation of PO and the position of Q but is independent of R_o.

Make use of this result to show that equation (6.1-12) may be transformed into (6.1-13).

6-4. Let $\tilde{V}$ have continuous second derivatives throughout τ except in a neighbourhood of the point P. Here it takes the form $\tilde{V}' + \frac{a}{r'} e^{j\tilde{k}r'}$, where $\tilde{V}'$ is a well-behaved point function, a is a constant, and r' is distance measured from P. Show that when O is not coincident with P, equation (6.2-2) is replaced by

$$4\pi\tilde{V}_o = \oint_{S_{1..n}\Sigma} \left(\tilde{G}\frac{\partial\tilde{V}}{\partial n} - \tilde{V}\frac{\partial\tilde{G}}{\partial n}\right) dS - \int_{\tau} \tilde{G}\,(\nabla^2+\tilde{k}^2)\tilde{V}\,d\tau + 4\pi a\,\tilde{G}_P$$

where the volume integral is convergent.

6-5. Suppose that $\tilde{G}$ represents $\tilde{G}_1$ or $\tilde{G}_2$ or $\tilde{G}_3$ as defined in Sec. 6.2. Then if $\tilde{G}$ (O'O) is the value of the Green's function at the point O' for a pole at O and $\tilde{G}$ (OO') signifies the reverse, show that $\tilde{G}$ (O'O) = $\tilde{G}$ (OO').

6.3 Vector Green's Functions for the Equation: curl curl $\tilde{\bar{F}} - \tilde{k}^2\tilde{\bar{F}} = \bar{0}$

When equation (3.7-2) is adjusted to exclude from the region of volume integration a δ sphere centred upon the point O, and when complex vectors replace real vectors, we obtain

$$\oint_{S_{1..n}\Sigma,\ S_\delta} (\tilde{\bar{F}} \times \text{curl}\,\tilde{\bar{C}} - \tilde{\bar{C}} \times \text{curl}\,\tilde{\bar{F}}).d\bar{S} = \int_{\tau-\tau_\delta} (\tilde{\bar{C}}.\text{curl curl}\,\tilde{\bar{F}} - \tilde{\bar{F}}.\text{curl curl}\,\tilde{\bar{C}})\,d\tau \qquad (6.3\text{-}1)$$

Suppose that $\tilde{\bar{F}}$ is some vector point function which is well-behaved in τ and satisfies the relationship

$$\text{curl curl}\,\tilde{\bar{F}} - \tilde{k}^2\tilde{\bar{F}} = \bar{0} \qquad (6.3\text{-}2)$$

This equation holds if div $\tilde{\bar{F}} = 0$ and $\tilde{\bar{F}}$ satisfies the homogeneous vector Helmholtz equation:

$$(\nabla^2+\tilde{k}^2)\tilde{\bar{F}} = \bar{0} \qquad (6.3\text{-}3)$$

Suppose further that

$$\tilde{\bar{C}} = \text{curl curl } \bar{a}\tilde{\gamma} \tag{6.3-4}$$

where $\bar{a}$ is a constant vector field and

$$\tilde{\gamma} \equiv \frac{1}{r} e^{j\tilde{k}r} \tag{6.3-5}$$

Then at any point of $\tau-\tau_\delta$

$$\tilde{\bar{C}} = \text{grad div } \bar{a}\tilde{\gamma} - \nabla^2\bar{a}\tilde{\gamma} = \nabla(\bar{a}.\nabla\tilde{\gamma}) + \tilde{k}^2\bar{a}\tilde{\gamma} \tag{6.3-6}$$

Hence

$$\text{curl } \tilde{\bar{C}} = \tilde{k}^2(\nabla\tilde{\gamma} \times \bar{a})$$

and, from equations (1.16-5) and (1.16-6),

$$\begin{aligned}\text{curl curl } \tilde{\bar{C}} &= \tilde{k}^2 \{(\bar{a}.\nabla)\nabla\tilde{\gamma} - \bar{a}\,\nabla^2\tilde{\gamma}\} \\ &= \tilde{k}^2 \{\nabla(\bar{a}.\nabla\tilde{\gamma}) + \tilde{k}^2\bar{a}\,\tilde{\gamma}\} \\ &= \tilde{k}^2\,\tilde{\bar{C}}\end{aligned}$$

so that curl curl $\tilde{\bar{C}} - \tilde{k}^2\tilde{\bar{C}} = \bar{0}$ in $\tau-\tau_\delta$.

Substitution in equation (6.3-1) then yields

$$\begin{aligned}&\oint_{S_\delta} \{\tilde{\bar{F}}\times\tilde{k}^2(\nabla\tilde{\gamma}\times\bar{a}).d\bar{S} - \nabla(\bar{a}.\nabla\tilde{\gamma})\times\text{curl } \tilde{\bar{F}}.d\bar{S} - \tilde{k}^2\bar{a}\tilde{\gamma}\times\text{curl } \tilde{\bar{F}}.d\bar{S}\} \\ &= \oint_{S_{1..n}\Sigma} (\tilde{\bar{C}} \times \text{curl } \tilde{\bar{F}} - \tilde{\bar{F}} \times \text{curl } \tilde{\bar{C}}).d\bar{S}\end{aligned} \tag{6.3-7}$$

By expansion of curl $(\bar{a}.\nabla\tilde{\gamma}\,\text{curl } \tilde{\bar{F}})$ and scalar integration over S_δ we obtain

$$-\oint_{S_\delta} \nabla(\bar{a}.\nabla\tilde{\gamma})\times\text{curl } \tilde{\bar{F}}.d\bar{S} = \oint_{S_\delta} \bar{a}.\nabla\tilde{\gamma} \text{ curl curl } \tilde{\bar{F}}.d\bar{S} = \oint_{S_\delta} \tilde{k}^2\bar{a}.\nabla\tilde{\gamma}\,\tilde{\bar{F}}.d\bar{S}$$

whence the surface integrals over S_δ in equation (6.3-7) may be brought into the form

$$\tilde{k}^2\bar{a}.\oint_{S_\delta}\{\tilde{\bar{F}}\;\nabla\tilde{\gamma}.d\bar{S} - d\bar{S}\;\tilde{\bar{F}}.\nabla\tilde{\gamma} + \nabla\tilde{\gamma}\;\tilde{\bar{F}}.d\bar{S} - \tilde{\gamma}(\mathrm{curl}\;\tilde{\bar{F}})\times d\bar{S}\}$$

$$= \tilde{k}^2\bar{a}.\oint_{S_\delta}\{\tilde{\bar{F}}\;\nabla\tilde{\gamma}.d\bar{S} - \tilde{\bar{F}}\times(d\bar{S}\times\nabla\tilde{\gamma}) - \tilde{\gamma}(\mathrm{curl}\;\tilde{\bar{F}})\times d\bar{S}\}$$

As $\delta\to 0$ the final term vanishes for dimensional reasons and the remainder reduces to

$$\tilde{k}^2\bar{a}.\left\{\tilde{\bar{F}}_o\oint_{S_\delta}\nabla\tilde{\gamma}.d\bar{S} - \tilde{\bar{F}}_o\times\oint_{S_\delta}d\bar{S}\times\nabla\tilde{\gamma}\right\}$$

The second term within brackets is zero from (1.17-2) and the first term may be shown to be $4\pi\tilde{\bar{F}}_o$ - a result independent of the shape of S_δ. Hence from equation (6.3-7)

$$\begin{aligned} 4\pi\tilde{k}^2\tilde{\bar{F}}_o.\bar{a} &= \oint_{S_{1..n}^{\Sigma}}(\tilde{\bar{C}}\times\mathrm{curl}\;\tilde{\bar{F}} - \tilde{\bar{F}}\times\mathrm{curl}\;\tilde{\bar{C}}).d\bar{S} \\ &= -\oint_{S_{1..n}^{\Sigma}}(\tilde{\bar{C}}.d\bar{S}\times\mathrm{curl}\;\tilde{\bar{F}} + d\bar{S}\times\tilde{\bar{F}}.\mathrm{curl}\;\tilde{\bar{C}}) \end{aligned} \qquad (6.3\text{-}8)$$

Here $\tilde{\bar{F}}_o$ is expressed in terms of $\hat{\bar{n}}\times\tilde{\bar{F}}$ and $\hat{\bar{n}}\times\mathrm{curl}\;\tilde{\bar{F}}$ on the bounding surfaces. From the considerations of Sec. 6.1b it would be expected that an expression for $\tilde{\bar{F}}_o$ could be found involving only $\hat{\bar{n}}\times\tilde{\bar{F}}$ or $\hat{\bar{n}}\times\mathrm{curl}\;\tilde{\bar{F}}$ since, in general, the specification of either upon $S_{1..n}^{\Sigma}$ renders $\tilde{\bar{F}}$ unique in τ.

Let $\tilde{\bar{C}}'$ be a vector point function which satisfies the relationship $\mathrm{curl}\;\mathrm{curl}\;\tilde{\bar{C}}' - \tilde{k}^2\tilde{\bar{C}}' = \bar{0}$ at all points of τ. Then it follows directly from equation (3.7-2) that

$$0 = \oint_{S_{1..n}^{\Sigma}}(\tilde{\bar{C}}'\times\mathrm{curl}\;\tilde{\bar{F}} - \tilde{\bar{F}}\times\mathrm{curl}\;\tilde{\bar{C}}').d\bar{S}$$

By addition with (6.3-8) we obtain

$$4\pi\tilde{k}^2\tilde{\bar{F}}_o.\bar{a} = \oint_{S_{1..n}\Sigma} (\tilde{\bar{G}} \times \text{curl } \tilde{\bar{F}} - \tilde{\bar{F}} \times \text{curl } \tilde{\bar{G}}).d\bar{S} \qquad (6.3\text{-}9)$$

where $\tilde{\bar{G}} \equiv \tilde{\bar{C}} + \tilde{\bar{C}}'$

If $\tilde{\bar{C}}_1'$ can be found such that $\hat{n} \times \tilde{\bar{G}}_1 = \bar{0}$ on $S_{1..n}\Sigma$, then

$$4\pi\tilde{k}^2\tilde{\bar{F}}_o.\bar{a} = -\oint_{S_{1..n}\Sigma} d\bar{S} \times \tilde{\bar{F}}.\text{curl } \tilde{\bar{G}}_1 \qquad (6.3\text{-}10)$$

If, on the other hand, $\tilde{\bar{C}}_2'$ can be found such that $\hat{n} \times \text{curl } \tilde{\bar{G}}_2 = \bar{0}$ on $S_{1..n}\Sigma$, then

$$4\pi\tilde{k}^2\tilde{\bar{F}}_o.\bar{a} = -\oint_{S_{1..n}\Sigma} \tilde{\bar{G}}_2.d\bar{S} \times \text{curl } \tilde{\bar{F}} \qquad (6.3\text{-}11)$$

A determination of $\tilde{\bar{G}}_1$ or $\tilde{\bar{G}}_2$ therefore permits of the expression of $\tilde{\bar{F}}_o.\bar{a}$ in terms of the tangential component of either $\tilde{\bar{F}}$ or curl $\tilde{\bar{F}}$ over $S_{1..n}\Sigma$. $\tilde{\bar{G}}_1$ and $\tilde{\bar{G}}_2$ are known as vector Green's functions for the equation curl curl $\tilde{\bar{F}} - \tilde{k}^2\tilde{\bar{F}} = \bar{0}$ (or for Helmholtz's equation). Certain physical considerations suggest that these functions exist - at least for all real and some complex values of $\tilde{k}$ - in the absence of eigenfunctions, but their evaluation in all but the simplest cases is of surpassing difficulty. Other Green's functions exist corresponding to different choices of $\tilde{\bar{C}}$ (eg $\bar{a}\tilde{\gamma}$ or curl $\bar{a}\tilde{\gamma}$).

6.4 Surface/Volume Integral Formulations for Complex Vector Fields

When it is permissible to express $\tilde{\bar{F}}_o$ in terms of more than a single boundary condition, the task of discovering a Green's function appropriate to the surfaces under consideration does not arise and the treatment becomes straightforward. Thus, we may express $\tilde{\bar{F}}_o$ directly in terms of surface integrals involving $\hat{n} \times \tilde{\bar{F}}$ and $\hat{n} \times$ curl $\tilde{\bar{F}}$ by substituting curl curl $\bar{a}\tilde{\gamma}$ for $\tilde{\bar{C}}$ in equation (6.2-7) and subsequently eliminating $\bar{a}$. However, for reasons which will become apparent later, it is easier, in the first instance, to identify $\tilde{\bar{C}}$ with $\bar{a}\tilde{\gamma}$ and return to the basic equation (6.3-1). When this is done we arrive without difficulty at the relationship

$$\oint_{S_{1..n}\Sigma,\ S_\delta} \{\tilde{\bar{\bar{F}}}\times(\nabla\tilde{\gamma}\times\bar{a}).d\bar{S} - \bar{a}\tilde{\gamma}\times\text{curl}\ \tilde{\bar{\bar{F}}}.d\bar{S}\}$$

$$= \int_{\tau-\tau_\delta} \bar{a}.\tilde{\gamma}\ (\text{curl curl}\ \tilde{\bar{\bar{F}}} - \tilde{k}^2\tilde{\bar{\bar{F}}})\ d\tau - \int_{\tau-\tau_\delta} \tilde{\bar{\bar{F}}}.\text{grad div}\ \bar{a}\tilde{\gamma}\ d\tau$$

Upon expanding the triple vector product, interchanging dot and cross, expanding div ($\tilde{\bar{\bar{F}}}$ div $\bar{a}\tilde{\gamma}$) and applying the divergence theorem we obtain

$$\bar{a}.\oint_{S_{1..n}\Sigma,\ S_\delta} \{(\tilde{\bar{\bar{F}}}\ \nabla\tilde{\gamma}.d\bar{S} - d\bar{S}\ \tilde{\bar{\bar{F}}}.\nabla\tilde{\gamma} + d\bar{S}\times\tilde{\gamma}\ \text{curl}\ \tilde{\bar{\bar{F}}} + \nabla\tilde{\gamma}\ \tilde{\bar{\bar{F}}}.d\bar{S}\}$$

$$= \bar{a}.\int_{\tau-\tau_\delta} \{\tilde{\gamma}(\text{curl curl}\ \tilde{\bar{\bar{F}}} - \tilde{k}^2\tilde{\bar{\bar{F}}}) + \nabla\tilde{\gamma}\ \text{div}\ \tilde{\bar{\bar{F}}}\}\ d\tau \qquad (6.4\text{-}1)$$

Since $\bar{a}$ is arbitrary, this leads to

$$\oint_{S_{1..n}\Sigma,\ S_\delta} \{-\tilde{\bar{\bar{F}}}\times(d\bar{S}\times\nabla\tilde{\gamma}) + \tilde{\bar{\bar{F}}}\ \nabla\tilde{\gamma}.d\bar{S} + d\bar{S}\times\tilde{\gamma}\ \text{curl}\ \tilde{\bar{\bar{F}}}\}$$

$$= \int_{\tau-\tau_\delta} \{\tilde{\gamma}(\text{curl curl}\ \tilde{\bar{\bar{F}}} - \tilde{k}^2\tilde{\bar{\bar{F}}}) + \nabla\tilde{\gamma}\ \text{div}\ \tilde{\bar{\bar{F}}}\}\ d\tau$$

which, on taking limits as S_δ shrinks uniformly about 0, yields

$$4\pi\tilde{\bar{\bar{F}}}_o = \oint_{S_{1..n}\Sigma} \{\nabla\tilde{\gamma}\times(d\bar{S}\times\tilde{\bar{\bar{F}}}) - d\bar{S}\times\tilde{\gamma}\ \text{curl}\ \tilde{\bar{\bar{F}}} - \nabla\tilde{\gamma}\ d\bar{S}.\tilde{\bar{\bar{F}}}\}$$

$$+ \int_{\tau} \{\gamma(\text{curl curl}\ \tilde{\bar{\bar{F}}} - \tilde{k}^2\tilde{\bar{\bar{F}}}) + \nabla\tilde{\gamma}\ \text{div}\ \tilde{\bar{\bar{F}}}\}\ d\tau \qquad (6.4\text{-}2)$$

Transformation of the right hand side of equation (6.4-1) into

$$\bar{a}.\int_{\tau-\tau_\delta} \{-\tilde{\gamma}\ (\nabla^2+\tilde{k}^2)\tilde{\bar{\bar{F}}} + \text{grad}\ (\tilde{\gamma}\ \text{div}\ \tilde{\bar{\bar{F}}})\}\ d\tau$$

and application of equation (1.17-5) leads in turn to

$$4\pi\tilde{\bar{F}}_o = \oint_{S_{1..n}\Sigma} \{\nabla\tilde{\gamma}\times(d\bar{S}\times\tilde{\bar{F}}) - d\bar{S}\times\tilde{\gamma}\ \text{curl}\ \tilde{\bar{F}} - \nabla\tilde{\gamma}\ d\bar{S}.\tilde{\bar{F}} + \tilde{\gamma}\ \text{div}\ \tilde{\bar{F}}\ d\bar{S}\} - \int_\tau \tilde{\gamma}\ (\nabla^2+\tilde{k}^2)\tilde{\bar{F}}\ d\tau \tag{6.4-3}$$

Alternatively, the volume integral of equation (6.4-1) may be transformed by expanding curl ($\tilde{\gamma}$ curl $\tilde{\bar{F}}$) and applying (1.17-3). We then obtain a generalised form of the grad-curl theorem, viz

$$4\pi\tilde{\bar{F}}_o = \int_\tau \text{div}\ \tilde{\bar{F}}\ \nabla\tilde{\gamma}\ d\tau - \oint_{S_{1..n}\Sigma} \tilde{\bar{F}}.d\bar{S}\ \nabla\tilde{\gamma} + \int_\tau (\text{curl}\ \tilde{\bar{F}})\times\nabla\tilde{\gamma}\ d\tau - \oint_{S_{1..n}\Sigma} (d\bar{S}\times\tilde{\bar{F}})\times\nabla\tilde{\gamma} - \tilde{k}^2\int_\tau \tilde{\gamma}\ \tilde{\bar{F}}\ d\tau \tag{6.4-4}$$

When $\tilde{k} = 0$ this reduces to equation (4.17-4).

We now proceed to obtain a set of integral expressions for (curl $\tilde{\bar{F}})_o$ by identifying $\tilde{\bar{C}}$ in equation (6.3-1) with curl $\bar{a}\tilde{\gamma}$.

In this case

$$\text{curl}\ \tilde{\bar{C}} = \text{grad div}\ \bar{a}\tilde{\gamma} + \tilde{k}^2\bar{a}\tilde{\gamma}$$

and

$$\text{curl curl}\ \tilde{\bar{C}} = \tilde{k}^2\tilde{\bar{C}}$$

Substitution in equation (6.3-1) together with surface integration of the expansion of curl ($\tilde{\bar{F}}$ div $\bar{a}\tilde{\gamma}$) leads to the relationship

$$4\pi(\text{curl}\ \tilde{\bar{F}})_o = \oint_{S_{1..n}\Sigma} \{\nabla\tilde{\gamma}\times(d\bar{S}\times\text{curl}\ \tilde{\bar{F}}) - \nabla\tilde{\gamma}\ d\bar{S}.\text{curl}\ \tilde{\bar{F}} - \tilde{k}^2\tilde{\gamma}(d\bar{S}\times\tilde{\bar{F}})\} - \int_\tau \nabla\tilde{\gamma}\times(\text{curl curl}\ \tilde{\bar{F}} - \tilde{k}^2\tilde{\bar{F}})\ d\tau \tag{6.4-5}$$

Volume integration of the expansion of div $(\tilde{\bar{C}}$ div $\tilde{\bar{\bar{F}}})$ allows (6.4-5) to be brought into the form

$$4\pi(\mathrm{curl}\ \tilde{\bar{\bar{F}}})_o = \oint_{S_{1..n}\Sigma} \{\nabla\tilde{\gamma}\times(d\bar{S}\times \mathrm{curl}\ \tilde{\bar{\bar{F}}}) - \nabla\tilde{\gamma}\ d\bar{S}.\mathrm{curl}\ \tilde{\bar{\bar{F}}} - \tilde{k}^2\tilde{\gamma}(d\bar{S}\times\tilde{\bar{\bar{F}}}) + \mathrm{div}\ \tilde{\bar{\bar{F}}}\ (d\bar{S}\times\nabla\tilde{\gamma})\}$$

$$+ \int_\tau \nabla\tilde{\gamma}\times(\nabla^2+\tilde{k}^2)\tilde{\bar{\bar{F}}}\ d\tau \qquad (6.4\text{-}6)$$

Expansion of curl $\frac{\partial\tilde{\gamma}}{\partial x}\tilde{\bar{\bar{F}}}$ and scalar surface integration over $S_{1..n}\Sigma$ yields

$$\oint_{S_{1..n}\Sigma} \frac{\partial\tilde{\gamma}}{\partial x}(\mathrm{curl}\ \tilde{\bar{\bar{F}}}).d\bar{S} = \oint_{S_{1..n}\Sigma} d\bar{S}\times\tilde{\bar{\bar{F}}}.\nabla\left(\frac{\partial\tilde{\gamma}}{\partial x}\right)$$

hence

$$\oint_{S_{1..n}\Sigma} \nabla\tilde{\gamma}\ (\mathrm{curl}\ \tilde{\bar{\bar{F}}}).d\bar{S} = \oint_{S_{1..n}\Sigma} ((d\bar{S}\times\tilde{\bar{\bar{F}}}).\nabla)\nabla\tilde{\gamma}$$

so that equation (6.4-6) may be replaced by

$$4\pi(\mathrm{curl}\ \tilde{\bar{\bar{F}}})_o = \oint_{S_{1..n}\Sigma} \{\nabla\tilde{\gamma}\times(d\bar{S}\times \mathrm{curl}\ \tilde{\bar{\bar{F}}}) - ((d\bar{S}\times\tilde{\bar{\bar{F}}}).\nabla)\nabla\tilde{\gamma} - \tilde{k}^2\tilde{\gamma}(d\bar{S}\times\tilde{\bar{\bar{F}}}) + \mathrm{div}\ \tilde{\bar{\bar{F}}}\ (d\bar{S}\times\nabla\tilde{\gamma})\}$$

$$+ \int_\tau \nabla\tilde{\gamma}\times(\nabla^2+\tilde{k}^2)\tilde{\bar{\bar{F}}}\ d\tau \qquad (6.4\text{-}7)$$

and equation (6.4-5) may be replaced by

$$4\pi(\mathrm{curl}\ \tilde{\bar{\bar{F}}})_o = \oint_{S_{1..n}\Sigma} \{\nabla\tilde{\gamma}\times(d\bar{S}\times \mathrm{curl}\ \tilde{\bar{\bar{F}}}) - ((d\bar{S}\times\tilde{\bar{\bar{F}}}).\nabla)\nabla\tilde{\gamma} - \tilde{k}^2\tilde{\gamma}(d\bar{S}\times\tilde{\bar{\bar{F}}})\}$$

$$- \int_\tau \nabla\tilde{\gamma}\times(\mathrm{curl\ curl}\ \tilde{\bar{\bar{F}}} - \tilde{k}^2\tilde{\bar{\bar{F}}})\ d\tau \qquad (6.4\text{-}8)$$

Students of the theory of dyadics will recognise that the terms

$$((d\bar{S}\times\tilde{\bar{\bar{F}}}).\nabla)\nabla\tilde{\gamma} + \tilde{k}^2\tilde{\gamma}(d\bar{S}\times\tilde{\bar{\bar{F}}})$$

may be written in the form

$$\nabla\nabla\tilde{\gamma}.(d\bar{S}\times\tilde{\bar{F}}) + \tilde{k}^2\ \bar{\bar{I}}\tilde{\gamma}.(d\bar{S}\times\tilde{\bar{F}})$$

$$= \tilde{k}^2 \left\{\left(\bar{\bar{I}} + \frac{1}{\tilde{k}^2}\ \nabla\nabla\right)\tilde{\gamma}\right\}.(d\bar{S}\times\tilde{\bar{F}})$$

$$\equiv \tilde{k}^2\ \bar{\bar{\Gamma}}.(d\bar{S}\times\tilde{\bar{F}})$$

where $\bar{\bar{I}}$ is the idemfactor $\bar{i}\bar{i} + \bar{j}\bar{j} + \bar{k}\bar{k}$.

$\bar{\bar{\Gamma}}$ is known as Green's dyadic.

Since it is not proposed to make use of dyadic algebra (as distinct from dyadic notation) in the present work the reader need only remember that for a well-behaved point function $\tilde{\bar{V}}$

$$(\tilde{\bar{V}}.\nabla)\nabla\tilde{\gamma} + \tilde{k}^2\tilde{\gamma}\tilde{\bar{V}} \text{ may be replaced by } \tilde{k}^2\bar{\bar{\Gamma}}.\tilde{\bar{V}} \text{ or } \tilde{k}^2\ \tilde{\bar{V}}.\bar{\bar{\Gamma}}$$

and (6.4-9)

$$\nabla\tilde{\gamma}\times\tilde{\bar{V}} \text{ may be replaced by } (\nabla\times\bar{\bar{\Gamma}}).\tilde{\bar{V}} \text{ or } -\tilde{\bar{V}}.(\nabla\times\bar{\bar{\Gamma}})$$

In these circumstances equation (6.4-8) becomes

$$4\pi(\text{curl }\tilde{\bar{F}})_o = \oint_{S_{1..n}\Sigma} \{(\nabla\times\bar{\bar{\Gamma}}).(d\bar{S}\times\text{curl }\tilde{\bar{F}}) - \tilde{k}^2\bar{\bar{\Gamma}}.(d\bar{S}\times\tilde{\bar{F}})\}$$

$$- \int_\tau (\nabla\times\bar{\bar{\Gamma}}).(\text{curl curl }\tilde{\bar{F}} - \tilde{k}^2\tilde{\bar{F}})\ d\tau \qquad (6.4\text{-}10)$$

We now return to the case considered in Sec. 6.3 where $\tilde{\bar{C}}$ was identified with curl curl $\bar{a}\tilde{\gamma}$, but we no longer demand that curl curl $\tilde{\bar{F}} - \tilde{k}^2\tilde{\bar{F}} = \bar{0}$ throughout τ.

After substitution in equation (6.3-1) and some manipulation, we find that

$$\oint_{S_{1..n}\Sigma,\ S_\delta} \{\bar{a}.\breve{k}^2\nabla\breve{\gamma}\times(\tilde{\bar{\bar{F}}}\times d\bar{S}) - \nabla(\bar{a}.\nabla\breve{\gamma}) \times \text{curl}\ \tilde{\bar{\bar{F}}}.d\bar{S} - \bar{a}.\breve{k}^2\breve{\gamma}(\text{curl}\ \tilde{\bar{\bar{F}}})\times d\bar{S}\}$$

$$= \int_{\tau-\tau_\delta} \{\nabla(\bar{a}.\nabla\breve{\gamma})+\breve{k}^2\bar{a}\breve{\gamma}\}.(\text{curl curl}\ \tilde{\bar{\bar{F}}} - \breve{k}^2\tilde{\bar{\bar{F}}})\ d\tau \qquad (6.4\text{-}11)$$

Now

$$\oint_{S_{1..n}\Sigma} \{-\nabla(\bar{a}.\nabla\breve{\gamma}) \times \text{curl}\ \tilde{\bar{\bar{F}}}.d\bar{S} - \bar{a}.\breve{k}^2\breve{\gamma}\ (\text{curl}\ \tilde{\bar{\bar{F}}})\times d\bar{S}\}$$

$$= \oint_{S_{1..n}\Sigma} \{(\bar{a}.\nabla)\nabla\breve{\gamma}.d\bar{S} \times \text{curl}\ \tilde{\bar{\bar{F}}} + \bar{a}.\breve{k}^2\breve{\gamma}\ (d\bar{S} \times \text{curl}\ \tilde{\bar{\bar{F}}})\}$$

$$= \bar{a}.\oint_{S_{1..n}\Sigma} \{((d\bar{S} \times \text{curl}\ \tilde{\bar{\bar{F}}}).\nabla)\nabla\breve{\gamma} + \breve{k}^2\breve{\gamma}(d\bar{S} \times \text{curl}\ \tilde{\bar{\bar{F}}})\}$$

since it may be shown by expansion in rectangular coordinates that

$$((\bar{a}.\nabla)\nabla\breve{\gamma}).\tilde{\bar{\bar{V}}} = \bar{a}.((\tilde{\bar{\bar{V}}}.\nabla)\nabla\breve{\gamma})$$

In dyadic notation this becomes $\bar{a}.\oint_{S_{1..n}\Sigma} \breve{k}^2\ \bar{\bar{\Gamma}}.(d\bar{S} \times \text{curl}\ \tilde{\bar{\bar{F}}})$

It was shown in Sec. 6.3 that

$$-\oint_{S_\delta} \nabla(\bar{a}.\nabla\breve{\gamma}) \times \text{curl}\ \tilde{\bar{\bar{F}}}.d\bar{S} = \oint_{S_\delta} \bar{a}.\nabla\breve{\gamma}\ \text{curl curl}\ \tilde{\bar{\bar{F}}}.d\bar{S}$$

hence equation (6.4-11) transforms into

$$\oint_{S_\delta} \{\tilde{k}^2\nabla\tilde{\gamma}\times(\tilde{\overline{F}}\times d\overline{S}) + \nabla\tilde{\gamma}\ \mathrm{curl}\ \mathrm{curl}\ \tilde{\overline{F}}.d\overline{S} - \tilde{k}^2\tilde{\gamma}\ (\mathrm{curl}\tilde{\overline{F}})\times d\overline{S}\}$$

$$= \oint_{S_{1..n}\Sigma} \{\tilde{k}^2\nabla\tilde{\gamma}\times(d\overline{S}\times\tilde{\overline{F}}) - \tilde{k}^2\overline{\overline{\Gamma}}.(d\overline{S} \times \mathrm{curl}\ \tilde{\overline{F}})\}$$

$$+ \int_{\tau-\tau_\delta} \tilde{k}^2\overline{\overline{\Gamma}}.(\mathrm{curl}\ \mathrm{curl}\ \tilde{\overline{F}} - \tilde{k}^2\tilde{\overline{F}})\ d\tau$$

The surface integral over S_δ may be written as

$$\oint_{S_\delta} \{\tilde{k}^2(\tilde{\overline{F}}\ d\overline{S}.\nabla\tilde{\gamma} - \nabla\tilde{\gamma}\ \tilde{\overline{F}}.d\overline{S} - \tilde{\overline{F}}\times(d\overline{S}\times\nabla\tilde{\gamma})) + \nabla\tilde{\gamma}\ \mathrm{curl}\ \mathrm{curl}\ \tilde{\overline{F}}.d\overline{S} - \tilde{k}^2\tilde{\gamma}(\mathrm{curl}\ \tilde{\overline{F}})\times d\overline{S}\}$$

As S_δ shrinks uniformly about 0 the final term vanishes for dimensional reasons and $\oint_{S_\delta} \tilde{\overline{F}}\times(d\overline{S}\times\nabla\tilde{\gamma})\to 0$ in accordance with equation (1.17-2). Of the remaining terms, the limiting value of $\oint_{S_\delta} \tilde{\overline{F}}\ d\overline{S}.\nabla\tilde{\gamma}$ is independent of the shape of S_δ, but that of $\oint_{S_\delta} - \nabla\tilde{\gamma}\ \tilde{\overline{F}}.d\overline{S}$ is not. The volume integral is consequently non-convergent and the equation takes the form

$$4\pi\tilde{\overline{F}}_o + \lim_{S'\to 0} \frac{1}{\tilde{k}^2} \oint_{S'} \nabla\tilde{\gamma}\ (\mathrm{curl}\ \mathrm{curl}\ \tilde{\overline{F}} - \tilde{k}^2\tilde{\overline{F}}).d\overline{S}$$

$$= \oint_{S_{1..n}\Sigma} \{(\nabla\times\overline{\overline{\Gamma}}).(d\overline{S}\times\tilde{\overline{F}}) - \overline{\overline{\Gamma}}.(d\overline{S} \times \mathrm{curl}\ \tilde{\overline{F}})\} \qquad (6.4\text{-}12)$$

$$+ \lim_{\tau'\to 0} \int_{\tau-\tau'} \overline{\overline{\Gamma}}.(\mathrm{curl}\ \mathrm{curl}\ \tilde{\overline{F}} - \tilde{k}^2\tilde{\overline{F}})\ d\tau$$

For the particular case where S' is spherical and centred upon 0 this becomes

$$4\pi\tilde{\bar{\bar{F}}}_o = \oint_{S_{1..n}\Sigma} \{(\nabla\times\bar{\bar{\Gamma}}).(d\bar{S}\times\tilde{\bar{\bar{F}}}) - \bar{\bar{\Gamma}}.(d\bar{S}\times \text{curl } \tilde{\bar{\bar{F}}})\}$$

$$+ \lim_{\tau'\to 0}\int_{\tau-\tau'} \bar{\bar{\Gamma}}.(\text{curl curl } \tilde{\bar{\bar{F}}} - \tilde{k}^2\tilde{\bar{\bar{F}}})\, d\tau - \frac{4\pi}{3\tilde{k}^2}(\text{curl curl } \tilde{\bar{\bar{F}}} - \tilde{k}^2\tilde{\bar{\bar{F}}})_o \tag{6.4-12(a)}$$

The non-convergence of the volume integral in equation (6.4-12) derives from the presence in the integrand of $\bar{\bar{\Gamma}}$ as a multiplying factor. This behaviour does not occur in equation (6.4-10) where the dyadic takes the form $\nabla\times\bar{\bar{\Gamma}}$, nor indeed in any of the non-dyadic equations (6.4-2) to (6.4-8), as evidenced by the fact that in each case the limiting value of the surface integral over S_δ is independent of the shape S_δ.

Finally, we note that equation (6.1-7) leads to a particularly simple form of expression for $\tilde{\bar{\bar{F}}}$ within τ, viz

$$4\pi\tilde{\bar{\bar{F}}}_o = \oint_{S_{1..n}\Sigma} \left(\tilde{\gamma}\frac{\partial\tilde{\bar{\bar{F}}}}{\partial n} - \tilde{\bar{\bar{F}}}\frac{\partial\tilde{\gamma}}{\partial n}\right) dS - \int_\tau \tilde{\gamma}\,(\nabla^2+\tilde{k}^2)\tilde{\bar{\bar{F}}}\, d\tau \tag{6.4-13}$$

EXERCISES

6-6. Let $\tilde{\bar{\bar{C}}} = \bar{a}\tilde{\gamma}$. By proceeding in the manner of Sec. 6.3 show that

$$4\pi\tilde{\bar{\bar{F}}}_o.\bar{a} = \oint_{S_{1..n}\Sigma} (\tilde{\bar{\bar{C}}}\times \text{curl } \tilde{\bar{\bar{F}}} - \tilde{\bar{\bar{F}}}\times\text{curl } \tilde{\bar{\bar{C}}} - \tilde{\bar{\bar{F}}}\text{ div } \tilde{\bar{\bar{C}}}).d\bar{S}$$

$$+\int_\tau \{(\tilde{\bar{\bar{C}}}.(\text{curl curl } \tilde{\bar{\bar{F}}} - \tilde{k}^2\tilde{\bar{\bar{F}}}) + \text{div } \tilde{\bar{\bar{F}}}\text{ div } \tilde{\bar{\bar{C}}}\}\, d\tau$$

Hence conclude that if $\tilde{\bar{\bar{C}}}'$ satisfies $(\nabla^2+\tilde{k}^2)\tilde{\bar{\bar{C}}}' = \bar{0}$ in τ, and $\hat{n}\times\text{curl}(\tilde{\bar{\bar{C}}}+\tilde{\bar{\bar{C}}}') = \bar{0}$ on $S_{1..n}\Sigma$, then $\tilde{\bar{\bar{F}}}$ may be expressed in terms of $\hat{n}\times\text{curl } \tilde{\bar{\bar{F}}}$ and $\hat{n}.\tilde{\bar{\bar{F}}}$ on $S_{1..n}\Sigma$ and curl curl $\tilde{\bar{\bar{F}}} - \tilde{k}^2\tilde{\bar{\bar{F}}}$ and div $\tilde{\bar{\bar{F}}}$ in τ.

6-7. Let $\tilde{\bar{C}} = \text{curl}\ \bar{a}\tilde{\gamma}$. Show that

$$4\pi\ (\text{curl}\ \tilde{\bar{F}})_o.\bar{a} = \oint_{S_{1..n}\Sigma} (\tilde{\bar{C}} \times \text{curl}\ \tilde{\bar{F}} - \tilde{\bar{F}} \times \text{curl}\ \tilde{\bar{C}}).d\bar{S} + \int_\tau \tilde{\bar{C}}.(\text{curl curl}\ \tilde{\bar{F}} - \tilde{k}^2\tilde{\bar{F}})\ d\tau$$

Hence conclude that if $\tilde{\bar{C}}'$ satisfies curl curl $\tilde{\bar{C}}' - \tilde{k}^2\tilde{\bar{C}}' = \bar{0}$ throughout τ and $\hat{n} \times (\tilde{\bar{C}}+\tilde{\bar{C}}')$ or $\hat{n} \times \text{curl}\ (\tilde{\bar{C}}+\tilde{\bar{C}}')$ is zero over $S_{1..n}\Sigma$ then curl $\tilde{\bar{F}}$ may be expressed in terms of $\hat{n} \times \tilde{\bar{F}}$ or $\hat{n} \times \text{curl}\ \tilde{\bar{F}}$ over $S_{1..n}\Sigma$ and curl curl $\tilde{\bar{F}} - \tilde{k}^2\tilde{\bar{F}}$ in τ.

6-8. A region τ' is bounded by a regular closed surface S'. A point 0 is the centre of a spherical surface S'_δ lying within τ'. If r is distance measured from 0 show that

$$\oint_{S'} \frac{\partial\tilde{\gamma}}{\partial n}\ dS = 4\pi + \tilde{k}^2 \lim_{\delta'\to 0} \int_{\tau'-\tau'_\delta} \tilde{\gamma}\ d\tau$$

where $\tilde{\gamma} = \frac{1}{r}\ e^{j\tilde{k}r}$ and the positive sense of the normal is directed into τ'.

Hence show that if S' shrinks uniformly about 0

$$\lim \oint_{S'} \nabla\tilde{\gamma}.d\bar{S} = 4\pi$$

6-9. Confirm equations (6.4-3) to (6.4-6) and (6.4-11).

6-10. Show that equations (6.4-3) and (6.4-13) are identical by making use of (1.17-13) with $\bar{F}$ replaced by $\tilde{\gamma}\tilde{\bar{F}}$.

6-11. Show that equation (6.4-11) may be transformed into (6.4-2) by applying the divergence theorem to the expansion of div $\{\bar{a}.\nabla\tilde{\gamma}(\text{curl curl}\ \tilde{\bar{F}} - \tilde{k}^2\tilde{\bar{F}})\}$

6.5 Time - Harmonic Fields and their Representation by Complex Quantities

6.5a The time-harmonic scalar field

Let V be a scalar point function which takes the form

$$V = V' \cos(\omega t+\theta) \tag{6.5-1}$$

where V' and θ are functions of position and ω is a constant.

V is said to be time-harmonic.

Then

$$V = \text{Re}\,\{V' e^{j(\omega t+\theta)}\} = \text{Re}\,\{\tilde{V} e^{j\omega t}\} \tag{6.5-2}$$

where

$$\tilde{V} = V'\,e^{j\theta} \tag{6.5-3}$$

$\tilde{V}$ is a complex scalar function of position alone and is known as a space phasor.

If $U = \text{Re}\,\{\tilde{g}\tilde{V}e^{j\omega t}\}$ where $\tilde{g} = ge^{j\alpha}$, then U represents a scalar field whose maximum value at any point is g times that of V and which leads V by the angle α. The factors g and α may or may not be functions of position.

Space and time differentiation are straightforward. If V is well-behaved in space and time, we have

$$\frac{\partial V}{\partial x} = \frac{\partial}{\partial x}\text{Re}\,\{\tilde{V}e^{j\omega t}\} = \frac{\partial}{\partial x}(V_1 \cos \omega t - V_2 \sin \omega t)$$

where

$$\tilde{V} = V_1 + jV_2$$

hence

$$\frac{\partial V}{\partial x} = \frac{\partial V_1}{\partial x}\cos \omega t - \frac{\partial V_2}{\partial x}\sin \omega t$$

$$= \text{Re}\left\{\frac{\partial \tilde{V}}{\partial x}e^{j\omega t}\right\}$$

whence

$$\text{grad}\, V = \text{grad}\, V_1 \cos \omega t - \text{grad}\, V_2 \sin \omega t \tag{6.5-4}$$

or

$$\text{grad } V = \text{Re } \{(\text{grad } \tilde{V})e^{j\omega t}\} \tag{6.5-5}$$

and

$$\nabla^2 V = \text{Re } \{(\nabla^2 \tilde{V})e^{j\omega t}\}$$

Similarly

$$\frac{\partial V}{\partial t} = \frac{\partial}{\partial t} \text{Re } \{\tilde{V}e^{j\omega t}\} = -(V_1\omega \sin \omega t + V_2\omega \cos \omega t)$$

or

$$\frac{\partial V}{\partial t} = \text{Re } \{j\omega\tilde{V}e^{j\omega t}\} \tag{6.5-6}$$

and

$$\frac{\partial^2 V}{\partial t^2} = \text{Re } \{-\omega^2\tilde{V}e^{j\omega t}\}$$

Suppose that

$$\nabla^2 V - \frac{1}{c^2}\frac{\partial^2 V}{\partial t^2} = 0$$

Then

$$\text{Re}\left\{\left(\nabla^2\tilde{V} + \frac{\omega^2}{c^2}\tilde{V}\right) e^{j\omega t}\right\} = 0$$

But since this equation holds for all values of t it follows that

$$\nabla^2\tilde{V} + \frac{\omega^2}{c^2}\tilde{V} = 0$$

Thus for each real field relationship there is a corresponding space phasor relationship. The transformation of the real into the complex form is effected simply by the substitution of the complex scalar for the real scalar, and multiplication or division by $j\omega$ when time differentiation or integration is indicated.

We will have occasion in Ch. 7 to consider the complex form of the retarded scalar function.

Since

$$V = V_1 \cos \omega t - V_2 \sin \omega t$$

$$[V] = V_1 \cos \omega\left(t - \frac{r}{c}\right) - V_2 \sin \omega\left(t - \frac{r}{c}\right)$$

or

$$[V] = \text{Re}\,\{\tilde{V}e^{j\omega(t-r/c)}\} = \text{Re}\,\{\tilde{V}e^{-j\omega r/c}\,e^{j\omega t}\} \tag{6.5-7}$$

From equation (6.5-4)

$$[\text{grad } V] = \text{grad } V_1 \cos \omega\left(t - \frac{r}{c}\right) - \text{grad } V_2 \sin \omega\left(t - \frac{r}{c}\right)$$

or

$$[\text{grad } V] = \text{Re}\,\{(\text{grad } \tilde{V})\, e^{-j\omega r/c}\, e^{j\omega t}\} \tag{6.5-8}$$

and from (6.5-7)

$$\text{grad } [V] = \text{grad Re}\,\{\tilde{V}e^{-j\omega r/c}\, e^{j\omega t}\}$$

or

$$\text{grad } [V] = \text{Re}\,\{(\text{grad } \tilde{V}e^{-j\omega r/c})e^{j\omega t}\} \tag{6.5-9}$$

Finally,

$$\left[\frac{\partial V}{\partial t}\right] = \frac{\partial [V]}{\partial t} = \text{Re}\,\{j\omega\tilde{V}e^{-j\omega r/c}\, e^{j\omega t}\} \tag{6.5-10}$$

6.5b The time-harmonic vector field

Let the vector point function $\overline{F}$ be defined by

$$\overline{F} = \overline{i}F'_x \cos(\omega t+\theta_x) + \overline{j}F'_y \cos(\omega t+\theta_y) + \overline{k}F'_z \cos(\omega t+\theta_z) \tag{6.5-11}$$

where F'_x, F'_y, F'_z, θ_x, θ_y, θ_z are functions of position.

This is the most general form of expression for a time-harmonic vector field, since each scalar component has its individual amplitude and phase.

We may replace (6.5-11) by[4]

$$\bar{F} = \text{Re}\left\{\bar{i}F'_x e^{j(\omega t+\theta_x)} + \bar{j}F'_y e^{j(\omega t+\theta_y)} + \bar{k}F'_z e^{j(\omega t+\theta_z)}\right\}$$

or

$$\bar{F} = \text{Re}\,\{\tilde{\bar{F}}e^{j\omega t}\} \tag{6.5-12}$$

where

$$\tilde{\bar{F}} = \bar{i}F'_x\, e^{j\theta_x} + \bar{j}F'_y\, e^{j\theta_y} + \bar{k}F'_z\, e^{j\theta_z} \tag{6.5-13}$$

$\tilde{\bar{F}}$ is a vector space phasor. We may write

$$\tilde{\bar{F}} = \bar{F}_1 + j\bar{F}_2 \tag{6.5-14}$$

where $\bar{F}_1$ and $\bar{F}_2$ are of fixed orientation at any one point and are given by

$$\begin{aligned}\bar{F}_1 &= \bar{i}F'_x \cos\theta_x + \bar{j}F'_y \cos\theta_y + \bar{k}F'_z \cos\theta_z \\ \bar{F}_2 &= \bar{i}F'_x \sin\theta_x + \bar{j}F'_y \sin\theta_y + \bar{k}F'_z \sin\theta_z\end{aligned} \tag{6.5-15}$$

It follows from (6.5-12) and (6.5-14) that

$$\bar{F} = \bar{F}_1 \cos\omega t - \bar{F}_2 \sin\omega t \tag{6.5-16}$$

so that, in general, the end point of $\bar{F}$ traces out an elliptical path with time. $\bar{F}$ is then said to be elliptically polarised in the plane of $\bar{F}_1$ and $\bar{F}_2$.

---

4. There should be no confusion between the unit vector $\bar{j}$ and $j = (-1)^{\frac{1}{2}}$.

When $\overline{F}_1$ and $\overline{F}_2$ are of equal magnitude and mutually perpendicular, the end point of $\overline{F}$ traces out a circle with a constant angular velocity ω, and $\overline{F}$ is said to be circularly polarised[5]. It follows from (6.5-15) that this condition obtains when the following simultaneous equations are satisfied:

$$\begin{aligned} F'^2_x \sin 2\theta_x + F'^2_y \sin 2\theta_y + F'^2_z \sin 2\theta_z &= 0 \\ F'^2_x \cos 2\theta_x + F'^2_y \cos 2\theta_y + F'^2_z \cos 2\theta_z &= 0 \end{aligned} \qquad (6.5\text{-}17)$$

When $\overline{F}_1$ and $\overline{F}_2$ are everywhere collinear, $\overline{F}$ maintains a fixed, but not necessarily equal, orientation at each point and is said to be linearly polarised. From (6.5-15) the required condition is

$$\tan\theta_x = \tan\theta_y = \tan\theta_z$$

in which case

$$\overline{F} = \hat{\overline{F}}_1\,(F_1^2+F_2^2)^{\frac{1}{2}}\cos(\omega t+\theta) = \hat{\overline{F}}_1\,(F'^2_x+F'^2_y+F'^2_z)^{\frac{1}{2}}\cos(\omega t+\theta)$$

or

$$\overline{F} = \overline{F}'\cos(\omega t+\theta)$$

where

$$\tan\theta = \tan\theta_x = \tan\theta_y = \tan\theta_z \qquad \left(-\frac{\pi}{2} \le \theta \le +\frac{\pi}{2}\right)$$

We see also that

$$\tilde{\overline{F}} = \overline{F}'\,e^{j\theta}$$

This simple relationship obtains only in the case of linear polarisation.

Since

$$\text{Re}\,\{\tilde{g}\,\tilde{\overline{F}}\,e^{j\omega t}\} = \text{Re}\,\{g\,\tilde{\overline{F}}\,e^{j(\omega t+\alpha)}\}$$

---

5. Polarisation in this sense has, of course, no connection with polarisation as treated previously.

where

$$\tilde{g} = g\, e^{j\alpha}$$

it is evident that multiplication of the vector space phasor $\tilde{\overline{F}}$ by $\tilde{g}$ increases the magnitude of $\overline{F}$ by the factor g and advances it in time by α/ω, but leaves the shape of the polarisation ellipse unaltered. Similarly, differentiation with respect to time has no effect on the shape of the ellipse but increases its size ω times and produces a time advance of $\pi/2\omega$. On the other hand space differentiation, in general, alters both the shape and size of the ellipse.

As in the scalar case, each real field relationship involving space or time differentiation is paralleled by a vector space phasor relationship.

We have also

$$\mathrm{div}\,\overline{F} = \mathrm{Re}\,\{(\mathrm{div}\,\tilde{\overline{F}})\, e^{j\omega t}\}$$

$$[\mathrm{div}\,\overline{F}] = \mathrm{Re}\,\{(\mathrm{div}\,\tilde{\overline{F}})\, e^{-j\omega r/c}\, e^{j\omega t}\}$$

$$\mathrm{div}\,[\overline{F}] = \mathrm{Re}\,\{(\mathrm{div}\,\tilde{\overline{F}}e^{-j\omega r/c})\, e^{j\omega t}\}$$

with similar expressions for the curl, and

$$\frac{\partial\overline{F}}{\partial t} = \mathrm{Re}\,\{j\omega\tilde{\overline{F}}e^{j\omega t}\}$$

$$\left[\frac{\partial\overline{F}}{\partial t}\right] = \frac{\partial[\overline{F}]}{\partial t} = \mathrm{Re}\,\{j\omega\tilde{\overline{F}}e^{-j\omega r/c}\, e^{j\omega t}\}$$

6.6 Time-Averaged Products of Time-Harmonic Quantities

Let

$$V = \mathrm{Re}\,\{\tilde{V}e^{j\omega t}\} \quad \text{where} \quad \tilde{V} = V_1 + jV_2 \tag{6.6-1}$$

Then it is easily seen that

$$V = \mathrm{Re}\,\{\tilde{V}^*e^{-j\omega t}\} \quad \text{where} \quad \tilde{V}^* = V_1 - jV_2 \tag{6.6-2}$$

It follows that

$$V = \tfrac{1}{2}\,\mathrm{Re}\,\{\tilde{V}e^{j\omega t}+\tilde{V}^*e^{-j\omega t}\} = \tfrac{1}{2}\,(\tilde{V}e^{j\omega t}+\tilde{V}^*e^{-j\omega t}) \tag{6.6-3}$$

Similarly, if

$$U = \text{Re}\ \{\tilde{U}e^{j\omega t}\}$$

then

$$U = \tfrac{1}{2}\ (\tilde{U}e^{j\omega t} + \tilde{U}^*e^{-j\omega t})$$

and

$$VU = \tfrac{1}{4}\ (\tilde{V}\tilde{U}e^{2j\omega t} + \tilde{V}\tilde{U}^* + \tilde{V}^*\tilde{U} + \tilde{V}^*\tilde{U}^*e^{-2j\omega t})$$

The time average of the first and last terms within brackets is zero so that the time average of VU is given by

$$\overline{VU} = \tfrac{1}{4}\ (\tilde{V}\tilde{U}^* + \tilde{V}^*\tilde{U}) = \tfrac{1}{2}\ \text{Re}\ \{\tilde{V}\tilde{U}^*\} = \tfrac{1}{2}\ \text{Re}\ \{\tilde{V}^*\tilde{U}\} \tag{6.6-4}$$

Hence

$$\overline{V^2} = \tfrac{1}{2}\ (\tilde{V}\tilde{V}^*) = \tfrac{1}{2}\ \text{Re}\ \{\tilde{V}\tilde{V}^*\} = \tfrac{1}{2}\ (V_1^2+V_2^2) \tag{6.6-5}$$

We may proceed in a similar manner with vector quantities. Thus, if

$$\overline{F} = \text{Re}\ \{\tilde{\overline{F}}e^{j\omega t}\} \quad \text{where} \quad \tilde{\overline{F}} = \overline{F}_1 + j\overline{F}_2 \tag{6.6-6}$$

then

$$\overline{F} = \text{Re}\ \{\tilde{\overline{F}}^*e^{-j\omega t}\} \tag{6.6-7}$$

and

$$\overline{F} = \tfrac{1}{2}\ (\tilde{\overline{F}}e^{j\omega t} + \tilde{\overline{F}}^*e^{-j\omega t}) \tag{6.6-8}$$

Similarly, if

$$\overline{G} = \text{Re}\ \{\tilde{\overline{G}}e^{j\omega t}\} \quad \text{where} \quad \tilde{\overline{G}} = \overline{G}_1 + j\overline{G}_2$$

then

$$\overline{G} = \tfrac{1}{2}\ (\tilde{\overline{G}}e^{j\omega t} + \tilde{\overline{G}}^*e^{-j\omega t})$$

and

$$\overline{\overline{F}.\overline{G}} = \tfrac{1}{2}\,(\tilde{\overline{F}}.\tilde{\overline{G}}* + \tilde{\overline{F}}*.\tilde{\overline{G}}) = \tfrac{1}{2}\,\mathrm{Re}\,\{\tilde{\overline{F}}.\tilde{\overline{G}}*\} = \tfrac{1}{2}\,\mathrm{Re}\,\{\tilde{\overline{F}}*.\tilde{\overline{G}}\}$$
$$= \tfrac{1}{2}\,(\overline{F}_1.\overline{G}_1 + \overline{F}_2.\overline{G}_2) \qquad (6.6\text{-}9)$$

Hence

$$\overline{\overline{F}.\overline{F}} = \tfrac{1}{2}\,(\tilde{\overline{F}}.\tilde{\overline{F}}*) = \tfrac{1}{2}\,(F_1^2+F_2^2) \qquad (6.6\text{-}10)$$

Further,

$$\overline{\overline{F}\times\overline{G}} = \tfrac{1}{2}\,(\tilde{\overline{F}}\times\tilde{\overline{G}}* + \tilde{\overline{F}}*\times\tilde{\overline{G}}) = \tfrac{1}{2}\,\mathrm{Re}\,\{\tilde{\overline{F}}\times\tilde{\overline{G}}*\} = \tfrac{1}{2}\,\mathrm{Re}\,\{\tilde{\overline{F}}*\times\tilde{\overline{G}}\}$$
$$= \tfrac{1}{2}\,(\overline{F}_1\times\overline{G}_1 + \overline{F}_2\times\overline{G}_2) \qquad (6.6\text{-}11)$$

6.7 Uniqueness Criteria for Time-Harmonic Fields

6.7a The time-harmonic scalar field

Uniqueness criteria for scalar fields of unspecified time dependence have already been discussed in Sec. 5.4a. When consideration is restricted to time-harmonic fields a rather different set of permissible conditions emerges.

Suppose that a scalar point function V' has continuous second space derivatives within the region $\underline{R}$ bounded by the surfaces $S_{1..n}\Sigma$, and that its associated space phasor is represented by $\tilde{V}'$. Then from (6.1-1) with $\tilde{V}$ primed and $\tilde{U} = \tilde{V}'*$ we see that

$$\oint_{S_{1..n}\Sigma} \tilde{V}'*\,\frac{\partial\tilde{V}'}{\partial n}\,dS = \int_\tau \tilde{V}'*\,\nabla^2\tilde{V}'\,d\tau + \int_\tau \mathrm{grad}\,\tilde{V}'*.\mathrm{grad}\,\tilde{V}'\,d\tau$$
$$= \int_\tau \tilde{V}'*\,(\nabla^2\tilde{V}'-p\tilde{V}'-j\omega q\tilde{V}'+r\omega^2\tilde{V}')\,d\tau \qquad (6.7\text{-}1)$$
$$+ \int_\tau \mathrm{grad}\,\tilde{V}'*.\mathrm{grad}\,\tilde{V}'\,d\tau + \int_\tau (p+j\omega q-r\omega^2)\,\tilde{V}'\tilde{V}'*\,d\tau$$

where p, q, r are real functions of position, or constants.

Let V_1 be such that[6]

$$\left.\begin{aligned}&(1)\quad \nabla^2 V_1 - pV_1 - q\frac{\partial V_1}{\partial t} - r\frac{\partial^2 V_1}{\partial t^2} = f\,(x,y,z,t) \text{ in } \underline{R}\\ &(2)\quad V_1 \text{ or } \frac{\partial V_1}{\partial n} = g(x,y,z,t) \text{ on } S_{1..n}\Sigma\end{aligned}\right\} \qquad (6.7\text{-}2)$$

and let V_2 satisfy the same conditions.

On writing $V_1 - V_2 = V'$ we obtain the homogeneous equations

$$\nabla^2 V' - pV' - q\frac{\partial V'}{\partial t} - r\frac{\partial^2 V'}{\partial t^2} = 0 \text{ in } \underline{R} \qquad (6.7\text{-}3)$$

and

$$V' \text{ or } \frac{\partial V'}{\partial n} = 0 \text{ on } S_{1..n}\Sigma$$

so that

$$\nabla^2\widetilde{V'} - p\widetilde{V'} - j\omega q\widetilde{V'} + r\omega^2\widetilde{V'} = 0 \text{ in } \underline{R}$$

and (6.7-4)

$$\widetilde{V'} \text{ or } \frac{\partial\widetilde{V'}}{\partial n} = 0 \text{ on } S_{1..n}\Sigma$$

Substitution in equation (6.7-1) yields

$$\int_\tau \text{grad}\,\widetilde{V'}^*.\text{grad}\,\widetilde{V'}\,d\tau + \int_\tau (p+j\omega q-r\omega^2)\,\widetilde{V'}\widetilde{V'}^*\,d\tau = 0$$

Since the first volume integral of the above equation and the factor $\widetilde{V'}\widetilde{V'}^*$ are real and positive, or zero, it follows that if q is everywhere positive or everywhere negative in $\underline{R}$, $\widetilde{V}$ must be everywhere zero, in which case $V_1 = V_2$ throughout $\underline{R}$. If q does not satisfy this condition but $p-r\omega^2$ is everywhere positive in $\underline{R}$, the same result obtains.

When p and q are zero and r is a positive constant $\left(\text{say } \frac{1}{c^2}\right)$ the proof of uniqueness is seen to fail. This is the case, discussed in Sec. 6.1a, where the Helmholtz equation reduces to $\left(\nabla^2 + \frac{\omega^2}{c^2}\right)\widetilde{V} = 0$ and where $\widetilde{k}\left(= \pm\frac{\omega}{c}\right)$ may be an eigenvalue. The corresponding real equation is

6. In physical applications the coefficients will, in general, be constant, and one or more will be zero.

$$\nabla^2 V - \frac{1}{c^2}\frac{\partial^2 V}{\partial t^2} = \text{dal } V = f(x,y,z,t)$$

and is known as the undamped wave equation.

Nevertheless, the considerations of Sec. 6.1c demonstrate that $\tilde{V}'$ vanishes in the exterior case if $R\tilde{V}'$ is bounded and $R\left(\frac{\partial \tilde{V}'}{\partial R} - j\left(\pm \frac{\omega}{c}\right)\tilde{V}'\right) \to 0$ uniformly in all directions as $R\to\infty$. Hence V will be uniquely determined in the exterior case provided that

(1) $\nabla^2 V - \frac{1}{c^2}\frac{\partial^2 V}{\partial t^2}$ is a specified function of position and time in $\underline{R}$

(2) V or $\frac{\partial V}{\partial n}$ is a specified function of position and time upon $S_{1..n}$

(3) RV is bounded, and either $R\left(\frac{\partial V}{\partial R} - \frac{1}{c}\frac{\partial V}{\partial t}\right)$ or $R\left(\frac{\partial V}{\partial R} + \frac{1}{c}\frac{\partial V}{\partial t}\right)$

approaches zero uniformly in all directions as $R\to\infty$.

6.7b The time-harmonic vector field

The results for the scalar field are applicable to each of the rectangular components of the vector field, whether linearly polarised or not, so that the specification of $\nabla^2\overline{F} - p\overline{F} - q\frac{\partial \overline{F}}{\partial t} - r\frac{\partial^2\overline{F}}{\partial t^2}$ throughout $\underline{R}$ together with the specification of $\overline{F}$ or $\frac{\partial \overline{F}}{\partial n}$ upon $S_{1..n}\Sigma$ will render $\overline{F}$ unique within $\underline{R}$ provided that p, q, r satisfy the conditions laid down in the previous sub-section.

For the exterior case with $p = q = 0$ and $r = \frac{1}{c^2}$ the Sommerfeld conditions become

$$R\overline{F} \text{ bounded;} \quad R\left(\frac{\partial \overline{F}}{\partial R} \mp \frac{1}{c}\frac{\partial \overline{F}}{\partial t}\right) \to 0$$

uniformly in all directions as $R\to\infty$.

Alternatively, by proceeding along the lines of Sec. 6.1b, we find that $\overline{F}$ will be unique within $\underline{R}$ provided that during one time-harmonic period

(1) div $\overline{F}$ is specified throughout $\underline{R}$

(2) $\nabla^2\overline{F} - p\overline{F} - q\frac{\partial \overline{F}}{\partial t} - r\frac{\partial^2\overline{F}}{\partial t^2}$ is specified throughout $\underline{R}$

(3) $\hat{n}\times\overline{F}$ or $\hat{n}\times \text{curl }\overline{F}$ is specified upon $S_{1..n}\Sigma$

and provided that $p - r\omega^2$ is everywhere positive, or q is everywhere positive or everywhere negative in $\underline{R}$.

CHAPTER 7

EXPONENTIAL POTENTIAL THEORY

7.1 Introduction

The exponential potentials of point, line, surface and volume sources are derived from the unretarded potentials of Ch. 4 by the substitution of $\frac{1}{r}e^{j\tilde{k}r}$ for $\frac{1}{r}$, where $\tilde{k}$ is a real or complex constant. The corresponding scalar and vector source densities are treated as complex and time-invariant.

In these circumstances it is possible to proceed as in Ch. 4 to develop expressions for the gradient, divergence, curl and higher-order derivatives of the appropriate potential functions. The procedure is straightforward and has already been required of the reader, in some measure, in the solution of certain exercises appearing in earlier sections. We may take the subject further by postulating a nexus between the scalar and vector source densities along the lines of equations (5.17-15), (5.17-18) and (5.17-19), viz

$$\text{div}\,\tilde{\underline{J}} = j\tilde{k}c\tilde{\rho}$$

$$\text{divs}\,\tilde{\underline{K}} + \Delta\tilde{\underline{J}}.\hat{\underline{n}} = j\tilde{k}c\tilde{\sigma}$$

$$\frac{d\tilde{I}}{ds} + \Delta\tilde{\underline{K}}.\hat{\underline{n}}' = j\tilde{k}c\tilde{\lambda}$$

It is then possible to show that

$$\text{div}\,\tilde{\underline{A}} = j\tilde{k}\,\tilde{\phi}$$

where $\tilde{\phi}$ is the scalar potential and $\tilde{\underline{A}}$ the vector potential of a mixed source system involving $\tilde{\rho}$, $\tilde{\sigma}$, $\tilde{\lambda}$ and $\tilde{\underline{J}}$, $\tilde{\underline{K}}$, $\tilde{\underline{I}}$, $\tilde{\underline{P}}$, $\tilde{\underline{M}}$. Further, if we define

$$\tilde{\underline{E}} = -\,\text{grad}\,\tilde{\phi} + j\tilde{k}\,\tilde{\underline{A}}$$

$$\tilde{\underline{B}} = \text{curl}\,\tilde{\underline{A}}$$

we can develop sets of equations which are closely allied to Maxwell's equations and the associated boundary relationships.

While this type of self-contained treatment of the exponential potentials is possible, their primary value lies in the considerable simplification which they can introduce into the treatment of the retarded potentials of time-harmonic sources. For this purpose the complex densities are identified with the space phasor equivalents of the time-harmonic densities, as discussed in Sec. 6.5, and $\tilde{k}$ is made equal to $-\frac{\omega}{c}$. Thus

$$\int_\tau \frac{[\rho]}{r}\,d\tau = \mathrm{Re}\left\{\int_\tau \frac{\tilde{\rho}}{r} e^{j\omega(t-r/c)}\,d\tau\right\} = \mathrm{Re}\left\{\int_\tau \frac{\tilde{\rho}}{r} e^{j\tilde{k}r}\,d\tau\; e^{j\omega t}\right\}$$

where

$$\tilde{k} = -\frac{\omega}{c}$$

and

$$\mathrm{grad}\int_\tau \frac{[\rho]}{r}\,d\tau = \mathrm{grad}\,\mathrm{Re}\left\{\int_\tau \frac{\tilde{\rho}}{r} e^{j\tilde{k}r}\,d\tau\; e^{j\omega t}\right\} = \mathrm{Re}\left\{\mathrm{grad}\int_\tau \frac{\tilde{\rho}}{r} e^{j\tilde{k}r}\,d\tau\; e^{j\omega t}\right\}$$

Similarly

$$\mathrm{curl}\int_\tau \frac{[\bar{J}]}{r}\,d\tau = \mathrm{Re}\left\{\mathrm{curl}\int_\tau \frac{\tilde{\bar{J}}}{r} e^{j\tilde{k}r}\,d\tau\; e^{j\omega t}\right\}$$

Also

$$\frac{\partial}{\partial t}\int_\tau \frac{[\rho]}{r}\,d\tau = \frac{\partial}{\partial t}\mathrm{Re}\left\{\int_\tau \frac{\tilde{\rho}}{r} e^{j\tilde{k}r}\,d\tau\; e^{j\omega t}\right\} = \mathrm{Re}\left\{j\omega\int_\tau \frac{\tilde{\rho}}{r} e^{j\tilde{k}r}\,d\tau\; e^{j\omega t}\right\}$$

It is seen that in carrying out space operations upon the retarded potentials it is necessary only to perform these operations on the equivalent exponential potentials and to extract the real part of the final expression subsequent to the restoration of the factor $e^{j\omega t}$.

Time differentiation and integration involve multiplication and division of the exponential potentials by $j\omega$ followed by the same procedure.

With this interpretation the nexus equations set out above appear as logical consequences of equations (5.17-15), (5.17-18) and (5.17-19), rather than as independent postulates.

While this particular approach to retarded potential theory involves considerably less effort than that required in Ch. 5 because of the more mechanical nature of the accompanying manipulations, it should be borne in mind that its application is restricted to single-frequency systems. Whether or not a Fourier extension to non-repetitive time functions is permissible will be dependent upon the possibility of accommodating an infinite set of values of $\tilde{k}$ in the case under consideration. Indeed, the single-frequency treatment raises some philosophical difficulty when a bounding surface lies at infinite distance because of the infinite time required for the establishment of the time-harmonic variation there.

We will proceed to treat the exponential potentials independently of the retarded time interpretation - at least in the first instance - because the restriction of $\tilde{k}$ to a real and negative value is unnecessary and because, in any case, complex values of $\tilde{k}$ are required in the exponential solutions of the general form of the Helmholtz equation. Since the procedure largely follows that presented in Ch. 4 it will be unnecessary to dwell overmuch upon detail.

7.2 The Scalar Exponential Potential and its Derivatives

7.2a The point singlet source

For a source of strength $\tilde{a}$ located at P we have

$$\tilde{\phi}_O = \tilde{a}\left(\frac{1}{R}\, e^{j\tilde{k}R}\right)_O \equiv \tilde{a}\,\tilde{\gamma}'_O$$

or (7.2-1)

$$\tilde{\phi}_O = \tilde{a}\left(\frac{1}{r}\, e^{j\tilde{k}r}\right)_P \equiv \tilde{a}\,\tilde{\gamma}_P$$

where R is distance measured from P and r is distance measured from O.

Then

$$(\nabla\tilde{\phi})_O = \left(\hat{R}\,\frac{d\tilde{\gamma}'}{dR}\right)_O = \left\{\hat{R}\,\tilde{a}\left(-\frac{1}{R^2} + \frac{j\tilde{k}}{R}\right) e^{j\tilde{k}R}\right\}_O = \left\{\hat{r}\,\tilde{a}\left(\frac{1}{r^2} - \frac{j\tilde{k}}{r}\right) e^{j\tilde{k}r}\right\}_P \tag{7.2-2}$$

and

$$(\nabla^2\tilde{\phi})_O = \left\{\frac{1}{R^2}\,\frac{d}{dR}\left(R^2\,\frac{d\tilde{\phi}}{dR}\right)\right\}_O = \left\{\frac{\tilde{a}}{R^2}\,\frac{d}{dR}\left((-1 + j\tilde{k}R)\, e^{j\tilde{k}R}\right)\right\}_O = -\tilde{k}^2\tilde{\phi}_O$$

so that

$$(\nabla^2+\tilde{k}^2)\tilde{\phi} = 0 \quad \text{outside the source.} \tag{7.2-3}$$

7.2b The point doublet source

Suppose that the doublet is located at the origin of spherical coordinates and aligned with the positive sense of the z axis (θ=0). Then it follows from Sec. 4.1 that

$$\tilde{\phi}_O = -\left\{\tilde{\underline{p}}.\nabla\,\frac{1}{R}\, e^{j\tilde{k}R}\right\}_O = \tilde{p}\left\{\cos\theta\left(\frac{1}{R^2} - \frac{j\tilde{k}}{R}\right) e^{j\tilde{k}R}\right\}_O \tag{7.2-4}$$

Then

$$(\nabla\phi)_o = \left\{\overline{R}\,\frac{\partial\phi}{\partial R} + \overline{\theta}\,\frac{1}{R}\,\frac{\partial\phi}{\partial\theta}\right\}_o$$

$$= \left\{\overline{R}\ p\cos\theta\left(-\frac{2}{R^3} + \frac{2jk}{R^2} + \frac{k^2}{R}\right)e^{jkR} - \overline{\theta}\ p\sin\theta\left(\frac{1}{R^3} - \frac{jk}{R^2}\right)e^{jkR}\right\}_o$$

On bearing in mind that $\overline{p} = (\overline{R}\ p\cos\theta - \overline{\theta}\ p\sin\theta)$ we find by collection of terms that

$$(\nabla\phi)_o = \left\{\left(\frac{\overline{p}}{R^3} - \frac{3\ p\cos\theta\ \overline{R}}{R^3} - \frac{jk\ \overline{p}}{R^2} + \frac{3\ jk\ p\cos\theta\ \overline{R}}{R^2} + \frac{k^2 p\cos\theta\ \overline{R}}{R}\right)e^{jkR}\right\}_o \tag{7.2-5}$$

and this may be written in the general form

$$\nabla\phi = \left(\frac{\overline{p}}{r^3} - \frac{3\overline{p}.\overline{r}\ \overline{r}}{r^5} - \frac{jk\ \overline{p}}{r^2} + \frac{3jk\ \overline{p}.\overline{r}\ \overline{r}}{r^4} + \frac{k^2\overline{p}.\overline{r}\ \overline{r}}{r^3}\right)e^{jkr} \tag{7.2-6}$$

Whereas only inverse-cube terms appear in the expression for the non-exponential gradient in Sec. 4.6, the present expression includes, in addition, inverse-square and inverse-first-power terms. This behaviour carries over into higher-order derivatives.

As would be expected from (7.2-3)

$$(\nabla^2+k^2)\phi = 0 \quad \text{outside the doublet} \tag{7.2-7}$$

This may be demonstrated by applying (2.6-8) (with R replacing r) to (7.2-4), or by working in Cartesian coordinates with transposition of partial derivatives along the lines of the derivation of $\nabla^2 \int_\Gamma \overline{L}.\nabla\left(\frac{1}{r}\right) ds$ on p. 263.

The interpretation of the exponential potential of singlets and doublets in terms of their retarded counterparts calls for some comment.

Expressions (7.2-1) and (7.2-4) imply the presence of one or two fixed point sources of time-dependent magnitude. Since discrete sources of this type have not been admitted to the model on which the present work is based, it is evident that the singlet source must be identified with that point of a line source where the current phasor I is discontinuous. The doublet potential, in turn, may be taken to correspond with the retarded scalar potential of the end points of a Hertzian dipole (Ex.5-103., p. 572). However, reference to (5.13-9) reveals that it may also be interpreted as the complex form of the retarded potential of a point source pair having a time-dependent vector moment in virtue of oscillatory motion.

Correspondingly, the vector exponential potential of a doublet is taken to be the complex form of the retarded potential of a Hertzian dipole or of an oscillatory pair of sources as given by equation (5.13-10). Similar considerations apply to the exponential potential of a whirl.

7.2c Line singlets and doublets

For a simple line source of density $\tilde{\lambda}$

$$\tilde{\phi} = \int_{\Gamma} \tilde{\gamma}\,\tilde{\lambda}\,ds \tag{7.2-8}$$

As in the non-exponential case, the potential becomes logarithmically infinite as the source is approached.

At exterior points

$$\text{grad}\,\tilde{\phi} = -\int_{\Gamma} \tilde{\lambda}\,\nabla\tilde{\gamma}\,ds \tag{7.2-9}$$

$$\text{div grad}\,\tilde{\phi} = \int_{\Gamma} \tilde{\lambda}\,\nabla^2\tilde{\gamma}\,ds = -\tilde{k}^2 \int_{\Gamma} \tilde{\gamma}\,\tilde{\lambda}\,ds$$

or

$$(\nabla^2+\tilde{k}^2) \int_{\Gamma} \tilde{\gamma}\,\tilde{\lambda}\,ds = 0 \tag{7.2-10}$$

For a line doublet of scalar density $\tilde{L}$, in the notation of Sec. 4.2,

$$\tilde{\phi} = \int_{\Gamma} \tilde{L}\,(\hat{n}'.\nabla\tilde{\gamma})\,ds \tag{7.2-11}$$

$$\text{grad}\,\tilde{\phi} = -\int_{\Gamma} \tilde{L}\,(\hat{n}'.\nabla)\nabla\tilde{\gamma}\,ds \tag{7.2-12}$$

$$\text{div grad}\,\tilde{\phi} = \int_{\Gamma} \tilde{L}\,(\hat{n}'.\nabla)\nabla^2\tilde{\gamma}\,ds = -\tilde{k}^2 \int_{\Gamma} \tilde{L}\,(\hat{n}'.\nabla\tilde{\gamma})\,ds$$

or

$$(\nabla^2+\tilde{k}^2) \int_{\Gamma} \tilde{L}\,(\hat{n}'.\nabla\tilde{\gamma})\,ds = 0 \tag{7.2-13}$$

7.2d Singlet and doublet surface sources

The exponential potential of a simple surface source of density $\tilde{\sigma}$ is given by

$$\tilde{\phi} = \int_S \tilde{\gamma}\tilde{\sigma}\, dS \qquad (7.2\text{-}14)$$

The potential is everywhere convergent and continuous since the density of the equivalent non-exponential system (viz $\tilde{\sigma}e^{j\tilde{k}r}$) can be split into real and imaginary parts, each of which may be treated by the arguments of Sec. 4.3.

$$\text{grad}\,\tilde{\phi} = -\int_S \tilde{\sigma}\,\nabla\tilde{\gamma}\, dS \quad \text{at points outside the surface} \qquad (7.2\text{-}15)$$

Because $\tilde{\sigma}e^{j\tilde{k}r}$ is continuous wherever $\tilde{\sigma}$ is continuous and approaches $\tilde{\sigma}$ as $r \to 0$, the increment in the normal derivative of the potential on passing through the surface at an interior point where $\tilde{\sigma}$ is continuous is related to the local surface density (for a common arbitrary sense of the normal) by

$$\Delta \frac{\partial\tilde{\phi}}{\partial n} = -4\pi\tilde{\sigma} \qquad (7.2\text{-}16)$$

We have also

$$\text{div grad}\,\tilde{\phi} = \int_S \tilde{\sigma}\,\nabla^2\tilde{\gamma}\, dS = -\tilde{k}^2\int_S \tilde{\gamma}\tilde{\sigma}\, dS$$

or

$$(\nabla^2+\tilde{k}^2)\int_S \tilde{\gamma}\tilde{\sigma}\, dS = 0 \quad \text{at exterior points} \qquad (7.2\text{-}17)$$

For a normally-orientated surface doublet of scalar density $\tilde{\mu}$

$$\tilde{\phi} = \int_S \tilde{\mu}(d\bar{S}.\nabla\tilde{\gamma}) \qquad (7.2\text{-}18)$$

where the positive sense of $\bar{S}$ is taken to correspond with the positive sense of the doublet moment.

As in the non-exponential case there is a discontinuity of potential of $4\pi\tilde{\mu}$ on passing positively through the surface at an interior point where $\tilde{\mu}$ is continuous.

$$\text{grad}\,\tilde{\phi} = -\int_S \tilde{\mu}(d\bar{S}.\nabla)\nabla\tilde{\gamma} \qquad (7.2\text{-}19)$$

The normal derivative of the potential approaches the same value from either side (for a common sense of the normal) but, like the potential itself, is undefined on the surface.

$$\text{div grad}\ \tilde{\phi} = \int_S \tilde{\mu}(d\bar{S}.\nabla)\nabla^2\tilde{\gamma} = -\tilde{k}^2 \int_S \tilde{\mu}(d\bar{S}.\nabla\tilde{\gamma}) \tag{7.2-20}$$

or

$$(\nabla^2+\tilde{k}^2) \int_S \tilde{\mu}(d\bar{S}.\nabla\tilde{\gamma}) = 0 \quad \text{at exterior points.}$$

7.2e Volume sources

For reasons discussed in Sec. 4.4 the exponential potential of a volume source of piecewise-continuous density is everywhere convergent and continuous. The gradient of the potential is likewise convergent and continuous (see p. 278).

The field-slipping technique developed in Sec. 4.8 continues to apply, so that in numerous instances the expressions for the derivatives of the exponential potential differ from their non-exponential counterparts only in the substitution of $\tilde{\gamma}$ for $\frac{1}{r}$.

However, since $\nabla^2\left(\frac{1}{r}\right) = 0$ and $\nabla^2\tilde{\gamma} = -\tilde{k}^2\tilde{\gamma}\ (r\neq 0)$, terms involving $\tilde{k}^2$ may appear as additional components in the exponential expressions. The various formulae have been brought together in Table 9, p. 614. The derivatives of the cavity potential, which have not been included, are identical with the corresponding derivatives of the complete potential at points outside the source, with τ and $S_{1..n}\Sigma$ replaced by $\tau-\tau_\delta$ and $S_{1..n}\Sigma$, S_δ as required. (This also holds for Tables 10 to 12.)

The potential function $\int \tilde{\tilde{P}}.\nabla\tilde{\gamma}\,d\tau$ and its derivatives are set out in Table 10, p. 615. In all cases the volume integrals are convergent unless expressed as limits, in which case τ' replaces τ_δ and S' replaces S_δ.

The derivations of the various expressions in these and subsequent tables are left as exercises for the reader, who should endeavour, in particular, to develop the multiple derivatives of the partial potential by alternative methods.

TABLE 9

The Scalar Exponential Potential Function $\int \tilde{\gamma}\,\tilde{\rho}\,d\tau$ and its Derivatives

(1)

$$\text{pot}\,\tilde{\rho} = \int_\tau \tilde{\gamma}\,\tilde{\rho}\,d\tau \qquad \text{(interior and exterior points of } \tau)$$

(2)

$$\text{partial pot}\,\tilde{\rho} = \int_{\tau-\tau_\delta} \tilde{\gamma}\,\tilde{\rho}\,d\tau \quad \text{(evaluated at centre of moving } \delta \text{ sphere within } \tau)$$

(3)

$$\text{cavity pot}\,\tilde{\rho} = \int_{\tau-\tau_\delta} \tilde{\gamma}\,\tilde{\rho}\,d\tau \quad \text{(defined throughout fixed } \delta \text{ sphere within } \tau)$$

(4)

$$\left.\begin{aligned} \text{grad pot}\,\tilde{\rho} &= \int_\tau \tilde{\gamma}\,\text{grad}\,\tilde{\rho}\,d\tau - \oint_{S_{1..n}^{\Sigma}} \tilde{\gamma}\,\tilde{\rho}\,d\bar{S} \\ &= -\int_\tau \tilde{\rho}\,\nabla\tilde{\gamma}\,d\tau \end{aligned}\right\} \text{(interior and exterior points of } \tau)$$

(5)

$$\begin{aligned} \text{grad partial pot}\,\tilde{\rho} &= \int_{\tau-\tau_\delta} \tilde{\gamma}\,\text{grad}\,\tilde{\rho}\,d\tau - \oint_{S_{1..n}^{\Sigma}} \tilde{\gamma}\,\tilde{\rho}\,d\bar{S} \\ &= -\int_{\tau-\tau_\delta} \tilde{\rho}\,\nabla\tilde{\gamma}\,d\tau + \oint_{S_\delta} \tilde{\gamma}\,\tilde{\rho}\,d\bar{S} \end{aligned}$$

(6)

$$\nabla^2\,\text{pot}\,\tilde{\rho} = \oint_{S_{1..n}^{\Sigma}} \left(\tilde{\rho}\,\frac{\partial\tilde{\gamma}}{\partial n} - \tilde{\gamma}\,\frac{\partial\tilde{\rho}}{\partial n}\right) dS + \text{pot}\,\nabla^2\tilde{\rho} \quad \text{(interior and exterior points of } \tau)$$

(7)

$$(\nabla^2+\tilde{k}^2)\,\text{pot}\,\tilde{\rho} = 0 \qquad \text{at exterior points of } \tau$$

(8)

$$(\nabla^2+\tilde{k}^2)\,\text{pot}\,\tilde{\rho} = -4\pi\tilde{\rho} \quad \text{at interior points of } \tau$$

(9)

$$\nabla^2\,\text{partial pot}\,\tilde{\rho} = \oint_{S_{1..n}^{\Sigma}} \left(\tilde{\rho}\,\frac{\partial\tilde{\gamma}}{\partial n} - \tilde{\gamma}\,\frac{\partial\tilde{\rho}}{\partial n}\right) dS + \text{partial pot}\,\nabla^2\tilde{\rho}$$

TABLE 9(CONTD.).

(10)

$$(\nabla^2+\tilde{k}^2)\ \text{partial pot}\ \tilde{\rho} = \oint_{S_\delta} \left(\tilde{\gamma}\frac{\partial\tilde{\rho}}{\partial n} - \tilde{\rho}\frac{\partial\tilde{\gamma}}{\partial n}\right) dS$$

TABLE 10

The Scalar Exponential Potential Function $\int \tilde{\bar{P}}.\nabla\tilde{\gamma}\, d\tau$ and its Derivatives

(1)

$$\int_\tau \tilde{\bar{P}}.\nabla\tilde{\gamma}\, d\tau = \int_\tau \tilde{\gamma}\,(-\operatorname{div}\tilde{\bar{P}})\, d\tau + \oint_{S_{1..n}\Sigma} \tilde{\gamma}\,\tilde{\bar{P}}.d\bar{S}$$ (interior and exterior points of τ)

(2)

$$\text{partial}\int_{\tau-\tau_\delta} \tilde{\bar{P}}.\nabla\tilde{\gamma}\, d\tau = \int_{\tau-\tau_\delta} \tilde{\gamma}\,(-\operatorname{div}\tilde{\bar{P}})\, d\tau + \oint_{S_{1..n}\Sigma,\ S_\delta} \tilde{\gamma}\,\tilde{\bar{P}}.d\bar{S}$$ (evaluated at centre of moving δ sphere within τ)

(3)

$$\text{cavity}\int_{\tau-\tau_\delta} \tilde{\bar{P}}.\nabla\tilde{\gamma}\, d\tau = \int_{\tau-\tau_\delta} \tilde{\gamma}\,(-\operatorname{div}\tilde{\bar{P}})\, d\tau + \oint_{S_{1..n}\Sigma,\ S_\delta} \tilde{\gamma}\,\tilde{\bar{P}}.d\bar{S}$$ (defined throughout fixed δ sphere within τ)

(4)

$$\operatorname{grad}\int_\tau \tilde{\bar{P}}.\nabla\tilde{\gamma}\, d\tau$$

$$= \int_\tau \nabla\tilde{\gamma}\,\operatorname{div}\tilde{\bar{P}}\, d\tau - \oint_{S_{1..n}\Sigma} \nabla\tilde{\gamma}\,\tilde{\bar{P}}.d\bar{S}$$ (interior and exterior points of τ)

$$= \int_\tau (\nabla\tilde{\gamma}\times\operatorname{curl}\tilde{\bar{P}})\, d\tau - \oint_{S_{1..n}\Sigma} \nabla\tilde{\gamma}\times(d\bar{S}\times\tilde{\bar{P}}) + \tilde{k}^2\int_\tau \tilde{\gamma}\,\tilde{\bar{P}}\, d\tau$$

$$= -\operatorname{curl}\int_\tau (\tilde{\bar{P}}\times\nabla\tilde{\gamma})\, d\tau + \tilde{k}^2\int_\tau \tilde{\gamma}\,\tilde{\bar{P}}\, d\tau$$

$$= -\int_\tau (\tilde{\bar{P}}.\nabla)\nabla\tilde{\gamma}\, d\tau$$

(exterior points of τ)

TABLE 10(CONTD.).

$$= \int_{\tau} (\nabla\tilde{\gamma} \times \text{curl}\, \tilde{\bar{P}})\, d\tau - \oint_{S_{1..n}^{\Sigma}} \nabla\tilde{\gamma} \times (d\bar{S}\times\tilde{\bar{P}}) + \tilde{k}^2 \int_{\tau} \tilde{\gamma}\, \tilde{\bar{P}}\, d\tau + 4\pi\tilde{\bar{P}}$$

$$= -\,\text{curl} \int_{\tau} (\tilde{\bar{P}}\times\nabla\tilde{\gamma})\, d\tau + \tilde{k}^2 \int_{\tau} \tilde{\gamma}\, \tilde{\bar{P}}\, d\tau + 4\pi\tilde{\bar{P}}$$

$$= \lim_{\tau'\to 0} \int_{\tau-\tau'} -(\tilde{\bar{P}}.\nabla)\nabla\tilde{\gamma}\, d\tau + \lim_{S'\to 0} \oint_{S'} d\bar{S}\, \tilde{\bar{P}}.\nabla\tilde{\gamma}$$

(interior points of τ)

(5)

$$\text{grad (partial)} \int_{\tau-\tau_\delta} \tilde{\bar{P}}.\nabla\tilde{\gamma}\, d\tau$$

$$= \int_{\tau-\tau_\delta} \nabla\tilde{\gamma}\, \text{div}\, \tilde{\bar{P}}\, d\tau - \oint_{S_{1..n}^{\Sigma}} \nabla\tilde{\gamma}\, \tilde{\bar{P}}.d\bar{S} + \oint_{S_\delta} \tilde{\bar{P}} \times (d\bar{S}\times\nabla\tilde{\gamma})$$

$$= \int_{\tau-\tau_\delta} (\nabla\tilde{\gamma} \times \text{curl}\, \tilde{\bar{P}})\, d\tau + \tilde{k}^2 \int_{\tau-\tau_\delta} \tilde{\gamma}\, \tilde{\bar{P}}\, d\tau - \oint_{S_{1..n}^{\Sigma}} \nabla\tilde{\gamma} \times (d\bar{S}\times\tilde{\bar{P}}) + \oint_{S_\delta} \tilde{\bar{P}}\, \nabla\tilde{\gamma}.d\bar{S}$$

$$= -\,\text{curl (partial)} \int_{\tau-\tau_\delta} (\tilde{\bar{P}}\times\nabla\tilde{\gamma})\, d\tau + \tilde{k}^2 \int_{\tau-\tau_\delta} \tilde{\gamma}\, \tilde{\bar{P}}\, d\tau + \oint_{S_\delta} \{\tilde{\bar{P}}\, d\bar{S}.\nabla\tilde{\gamma} + \tilde{\bar{P}} \times (d\bar{S}\times\nabla\tilde{\gamma})\}$$

$$= -\int_{\tau-\tau_\delta} (\tilde{\bar{P}}.\nabla)\nabla\tilde{\gamma}\, d\tau + \oint_{S_\delta} d\bar{S}\, \tilde{\bar{P}}.\nabla\tilde{\gamma}$$

(6)

$$(\nabla^2+\tilde{k}^2) \int_{\tau} \tilde{\bar{P}}.\nabla\tilde{\gamma}\, d\tau = 0 \quad \text{at exterior points of } \tau$$

(7)

$$(\nabla^2+\tilde{k}^2) \int_{\tau} \tilde{\bar{P}}.\nabla\tilde{\gamma}\, d\tau = 4\pi\, \text{div}\, \tilde{\bar{P}} \quad \text{at interior points of } \tau$$

(8)

$$(\nabla^2+\tilde{k}^2)\ \text{(partial)} \int_{\tau-\tau_\delta} \tilde{\bar{P}}.\nabla\tilde{\gamma}\, d\tau = \oint_{S_\delta} \{\text{div}\, \tilde{\bar{P}}\, d\bar{S}.\nabla\tilde{\gamma} + (\text{curl}\, \tilde{\bar{P}}).(d\bar{S}\times\nabla\tilde{\gamma}) + \tilde{k}^2\tilde{\gamma}\tilde{\bar{P}}.d\bar{S}\}$$

7.3 The Vector Exponential Potential and its Derivatives

7.3a Line singlets and doublets

Whirls

The exponential potential of a line singlet of density $\tilde{\bar{I}}$ (not necessarily tangential to the contour) is given by

$$\tilde{\bar{A}} = \int_{\Gamma} \tilde{\gamma}\,\tilde{\bar{I}}\,ds \tag{7.3-1}$$

At points exterior to the contour

$$\operatorname{div}\tilde{\bar{A}} = -\int_{\Gamma} \tilde{\bar{I}}.\nabla\tilde{\gamma}\,ds \tag{7.3-2}$$

$$\operatorname{grad}\operatorname{div}\tilde{\bar{A}} = \int_{\Gamma} (\tilde{\bar{I}}.\nabla)\nabla\tilde{\gamma}\,ds \tag{7.3-3}$$

$$\operatorname{curl}\tilde{\bar{A}} = \int_{\Gamma} \tilde{\bar{I}} \times \nabla\tilde{\gamma}\,ds \tag{7.3-4}$$

$$\operatorname{curl}\operatorname{curl}\tilde{\bar{A}} = \tilde{k}^2 \int_{\Gamma} \tilde{\gamma}\,\tilde{\bar{I}}\,ds + \int_{\Gamma} (\tilde{\bar{I}}.\nabla)\nabla\tilde{\gamma}\,ds \tag{7.3-5}$$

$$(\nabla^2+\tilde{k}^2)\int_{\Gamma} \tilde{\gamma}\,\tilde{\bar{I}}\,ds = \bar{0} \tag{7.3-6}$$

The corresponding expressions for a vector line doublet, in the notation of Sec. 4.12a, are

$$\tilde{\bar{A}} = \int_{\Gamma} \tilde{\bar{L}}_1\,(\hat{\bar{n}}'.\nabla\tilde{\gamma})\,ds \tag{7.3-7}$$

$$\operatorname{div}\tilde{\bar{A}} = -\int_{\Gamma} (\tilde{\bar{L}}_1.\nabla)(\hat{\bar{n}}'.\nabla\tilde{\gamma})\,ds \tag{7.3-8}$$

$$\operatorname{grad}\operatorname{div}\tilde{\bar{A}} = \int_{\Gamma} (\tilde{\bar{L}}_1.\nabla)\nabla(\hat{\bar{n}}'.\nabla\tilde{\gamma})\,ds \tag{7.3-9}$$

$$\operatorname{curl} \tilde{\bar{A}} = \int_{\Gamma} \tilde{\bar{L}}_1 \times \nabla(\hat{\bar{n}}'.\nabla\tilde{\gamma})\, ds \qquad (7.3\text{-}10)$$

$$\operatorname{curl}\operatorname{curl} \tilde{\bar{A}} = \tilde{k}^2 \int_{\Gamma} \tilde{\bar{L}}_1 (\hat{\bar{n}}'.\nabla\tilde{\gamma})\, ds + \int_{\Gamma} (\tilde{\bar{L}}_1.\nabla)\nabla(\hat{\bar{n}}'.\nabla\tilde{\gamma})\, ds \qquad (7.3\text{-}11)$$

$$(\nabla^2+\tilde{k}^2) \int_{\Gamma} \tilde{\bar{L}}_1 (\hat{\bar{n}}'.\nabla\tilde{\gamma})\, ds = \bar{0} \qquad (7.3\text{-}12)$$

A closed, uniform, tangential line source gives rise to the vector potential

$$\tilde{\bar{A}} = \oint_{\Gamma} \tilde{\bar{I}}\, \tilde{\gamma}\, ds = \tilde{I} \oint_{\Gamma} \tilde{\gamma}\, d\bar{r} = \tilde{I} \int_{S} d\bar{S} \times \nabla\tilde{\gamma}$$

so that the vector potential of a whirl is given at exterior points by

$$\tilde{\bar{A}} = \tilde{\bar{m}} \times \nabla\tilde{\gamma} \qquad (7.3\text{-}13)$$

where

$$\tilde{\bar{m}} = \lim_{\substack{\tilde{I}\to\infty \\ \bar{S}\to\bar{0}}} \tilde{I}\bar{S}$$

Then

$$\operatorname{div} \tilde{\bar{A}} = 0 \qquad (7.3\text{-}14)$$

$$\operatorname{grad}\operatorname{div} \tilde{\bar{A}} = \bar{0} \qquad (7.3\text{-}15)$$

$$\operatorname{curl} \tilde{\bar{A}} = (\tilde{\bar{m}}.\nabla)\nabla\tilde{\gamma} + \tilde{k}^2\, \tilde{\gamma}\, \tilde{\bar{m}} \qquad (7.3\text{-}16)$$

$$\operatorname{curl}\operatorname{curl} \tilde{\bar{A}} = \tilde{k}^2(\tilde{\bar{m}} \times \nabla\tilde{\gamma}) \qquad (7.3\text{-}17)$$

$$(\nabla^2+\tilde{k}^2)\, \tilde{\bar{A}} = \bar{0} \qquad (7.3\text{-}18)$$

7.3b Singlet and doublet surface sources

The exponential potential of a simple surface source of density $\tilde{\bar{K}}$ (not necessarily tangential to the surface) is given by

$$\tilde{\bar{A}} = \int_S \tilde{\gamma}\,\tilde{\bar{K}}\,dS \qquad (7.3\text{-}19)$$

$\tilde{\bar{A}}$ is convergent and continuous everywhere.

The increment in the normal derivative of $\tilde{\bar{A}}$, on passing through the surface at an interior point where $\tilde{\bar{K}}$ is continuous, is related to the local source density (for a common arbitrary sense of the normal) by

$$\Delta\,\frac{\partial\tilde{\bar{A}}}{\partial n} = -4\pi\tilde{\bar{K}} \qquad (7.3\text{-}20)$$

We have also, at points outside the surface,

$$\operatorname{div}\tilde{\bar{A}} = -\int_S \tilde{\bar{K}}.\nabla\tilde{\gamma}\,dS \qquad (7.3\text{-}21)$$

$$\Delta\,\operatorname{div}\tilde{\bar{A}} = -4\pi\,\hat{\bar{n}}.\tilde{\bar{K}} \qquad (7.3\text{-}22)$$

$$\operatorname{curl}\tilde{\bar{A}} = \int_S \tilde{\bar{K}}\times\nabla\tilde{\gamma}\,dS \qquad (7.3\text{-}23)$$

$$\Delta\,\operatorname{curl}\tilde{\bar{A}} = -4\pi\,(\hat{\bar{n}}\times\tilde{\bar{K}}) \qquad (7.3\text{-}24)$$

$$\operatorname{grad}\operatorname{div}\tilde{\bar{A}} = \int_S (\tilde{\bar{K}}.\nabla)\nabla\tilde{\gamma}\,dS \qquad (7.3\text{-}25)$$

$$\operatorname{curl}\operatorname{curl}\tilde{\bar{A}} = \tilde{k}^2\int_S \tilde{\gamma}\,\tilde{\bar{K}}\,dS + \int_S (\tilde{\bar{K}}.\nabla)\nabla\tilde{\gamma}\,dS \qquad (7.3\text{-}26)$$

$$(\nabla^2+\tilde{k}^2)\int_S \tilde{\gamma}\,\tilde{\bar{K}}\,dS = \bar{0} \qquad (7.3\text{-}27)$$

When $\tilde{\tilde{K}}$ is tangential to the surface it follows from (2.12-27a) that

$$\int_S \mathrm{divs}\,\tilde{\gamma}\,\tilde{\tilde{K}}\,dS = \oint_\Gamma \tilde{\gamma}\,\tilde{\tilde{K}}.\hat{n}'\,ds$$

But from equation (2.12-7)

$$\mathrm{divs}\,\tilde{\gamma}\,\tilde{\tilde{K}} = \tilde{\gamma}\,\mathrm{divs}\,\tilde{\tilde{K}} + \tilde{\tilde{K}}.\mathrm{grads}\,\tilde{\gamma}$$

and from equation (2.12-12)

$$\mathrm{grad}\,\tilde{\gamma} = \mathrm{grads}\,\tilde{\gamma} + \hat{n}\,\frac{\partial\tilde{\gamma}}{\partial n}$$

hence

$$\begin{aligned}\mathrm{div}\int_S \tilde{\gamma}\,\tilde{\tilde{K}}\,dS &= -\int_S \tilde{\tilde{K}}.\nabla\tilde{\gamma}\,dS = -\int_S \tilde{\tilde{K}}.\mathrm{grads}\,\tilde{\gamma}\,dS \\ &= \int_S \tilde{\gamma}\,\mathrm{divs}\,\tilde{\tilde{K}}\,dS - \oint_\Gamma \tilde{\gamma}\,\tilde{\tilde{K}}.\hat{n}'\,ds\end{aligned} \quad (7.3\text{-}28)$$

In the notation of Sec. 4.12b the exponential potential of a vector source doublet of density $\tilde{\tilde{\mu}}_1$ is given, at points outside the surface, by

$$\tilde{\tilde{A}} = \int_S \tilde{\tilde{\mu}}_1\,(d\bar{S}.\nabla\tilde{\gamma}) \quad (7.3\text{-}29)$$

whence

$$\mathrm{div}\,\tilde{\tilde{A}} = -\int_S (\tilde{\tilde{\mu}}_1.\nabla)d\bar{S}.\nabla\tilde{\gamma} \quad (7.3\text{-}30)$$

$$\mathrm{grad}\,\mathrm{div}\,\tilde{\tilde{A}} = \int_S (\tilde{\tilde{\mu}}_1.\nabla)\nabla(d\bar{S}.\nabla\tilde{\gamma}) \quad (7.3\text{-}31)$$

$$\mathrm{curl}\,\tilde{\tilde{A}} = \int_S \tilde{\tilde{\mu}}_1 \times \nabla(d\bar{S}.\nabla\tilde{\gamma}) \quad (7.3\text{-}32)$$

$$\text{curl curl } \tilde{\tilde{\bar{A}}} = \tilde{k}^2 \int_S \tilde{\tilde{\bar{\mu}}}_1 \, (d\bar{S}.\nabla\tilde{\gamma}) + \int_S (\tilde{\tilde{\bar{\mu}}}_1.\nabla)\nabla(d\bar{S}.\nabla\tilde{\gamma}) \qquad (7.3\text{-}33)$$

$$(\nabla^2+\tilde{k}^2) \int_S \tilde{\tilde{\bar{\mu}}}_1 \, (d\bar{S}.\nabla\tilde{\gamma}) = \bar{0} \qquad (7.3\text{-}34)$$

7.3c Volume sources

The comments made in Sec. 7.2e in connection with the scalar potentials $\int \tilde{\gamma}\,\tilde{\rho}\,d\tau$ and $\int \tilde{\tilde{\bar{P}}}.\nabla\tilde{\gamma}\,d\tau$ apply equally to the treatment of the corresponding vector potentials, and need not be repeated. Formulae relating to $\int \tilde{\gamma}\,\tilde{\tilde{\bar{J}}}\,d\tau$ will be found in Table 11 below, while the function $\int \tilde{\tilde{\bar{M}}} \times \nabla\tilde{\gamma}\,d\tau$ and its derivatives are set out in Table 12, p. 625.

TABLE 11

The Vector Exponential Potential Function $\int \tilde{\gamma}\,\tilde{\tilde{\bar{J}}}\,d\tau$ and its Derivatives

(1)

$$\text{pot } \tilde{\tilde{\bar{J}}} = \int_\tau \tilde{\gamma}\,\tilde{\tilde{\bar{J}}}\,d\tau \quad \text{(interior and exterior points of } \tau)$$

(2)

$$\text{partial pot } \tilde{\tilde{\bar{J}}} = \int_{\tau-\tau_\delta} \tilde{\gamma}\,\tilde{\tilde{\bar{J}}}\,d\tau \quad \text{(evaluated at centre of moving } \delta \text{ sphere within } \tau)$$

(3)

$$\text{cavity pot } \tilde{\tilde{\bar{J}}} = \int_{\tau-\tau_\delta} \tilde{\gamma}\,\tilde{\tilde{\bar{J}}}\,d\tau \quad \text{(defined throughout fixed } \delta \text{ sphere within } \tau)$$

(4)

$$\left.\begin{aligned} \text{div pot } \tilde{\tilde{\bar{J}}} &= \int_\tau \tilde{\gamma}\,\text{div}\,\tilde{\tilde{\bar{J}}}\,d\tau - \oint_{S_{1..n}\Sigma} \tilde{\gamma}\,\tilde{\tilde{\bar{J}}}.d\bar{S} \\ &= -\int_\tau \tilde{\tilde{\bar{J}}}.\nabla\tilde{\gamma}\,d\tau \end{aligned}\right\} \text{(interior and exterior points of } \tau)$$

TABLE 11(CONTD.).

(5)

$$\text{div partial pot } \tilde{\bar{J}} = \int_{\tau-\tau_\delta} \tilde{\gamma}\,\text{div}\,\tilde{\bar{J}}\,d\tau - \oint_{S_{1..n}^{\Sigma}} \tilde{\gamma}\,\tilde{\bar{J}}.d\bar{S}$$

$$= -\int_{\tau-\tau_\delta} \tilde{\bar{J}}.\nabla\tilde{\gamma}\,d\tau + \oint_{S_\delta} \tilde{\gamma}\,\tilde{\bar{J}}.d\bar{S}$$

(6)

$$\left.\begin{aligned} \text{curl pot } \tilde{\bar{J}} &= \int_{\tau} \tilde{\gamma}\,\text{curl}\,\tilde{\bar{J}}\,d\tau - \oint_{S_{1..n}^{\Sigma}} \tilde{\gamma}(d\bar{S}\times\tilde{\bar{J}}) \\ &= \int_{\tau} \tilde{\bar{J}} \times \nabla\tilde{\gamma}\,d\tau \end{aligned}\right\} \text{(interior and exterior points of } \tau\text{)}$$

(7)

$$\text{curl partial pot } \tilde{\bar{J}} = \int_{\tau-\tau_\delta} \tilde{\gamma}\,\text{curl}\,\tilde{\bar{J}}\,d\tau - \oint_{S_{1..n}^{\Sigma}} \tilde{\gamma}(d\bar{S}\times\tilde{\bar{J}})$$

$$= \int_{\tau-\tau_\delta} \tilde{\bar{J}} \times \nabla\tilde{\gamma}\,d\tau + \oint_{S_\delta} \tilde{\gamma}(d\bar{S}\times\tilde{\bar{J}})$$

(8)

$(\nabla^2+\tilde{k}^2)$ pot $\tilde{\bar{J}} = \bar{0}$ at exterior points of τ

(9)

$(\nabla^2+\tilde{k}^2)$ pot $\tilde{\bar{J}} = -4\pi\tilde{\bar{J}}$ at interior points of τ

(10)

$$\nabla^2 \text{ pot } \tilde{\bar{J}} = \oint_{S_{1..n}^{\Sigma}} \left(\tilde{\bar{J}}\,\frac{\partial\tilde{\gamma}}{\partial n} - \tilde{\gamma}\,\frac{\partial\tilde{\bar{J}}}{\partial n}\right) dS + \text{pot } \nabla^2\tilde{\bar{J}} \text{ (interior and exterior points of } \tau\text{)}$$

(11)

$$\nabla^2 \text{ partial pot } \tilde{\bar{J}} = \oint_{S_{1..n}^{\Sigma}} \left(\tilde{\bar{J}}\,\frac{\partial\tilde{\gamma}}{\partial n} - \tilde{\gamma}\,\frac{\partial\tilde{\bar{J}}}{\partial n}\right) dS + \text{partial pot } \nabla^2\tilde{\bar{J}}$$

TABLE 11(CONTD.).

(12)

$$(\nabla^2+\tilde{k}^2)\ \text{partial pot}\ \tilde{\bar{J}} = \oint_{S_\delta} \left(\tilde{\gamma}\,\frac{\partial\tilde{\bar{J}}}{\partial n} - \tilde{\bar{J}}\,\frac{\partial\tilde{\gamma}}{\partial n}\right) dS$$

(13)

grad div pot $\tilde{\bar{J}}$

$$= \int_\tau (-\,\text{div}\ \tilde{\bar{J}})\ \nabla\tilde{\gamma}\ d\tau + \oint_{S_{1..n}\Sigma} \nabla\tilde{\gamma}\ \tilde{\bar{J}}.d\bar{S} \quad \text{(interior and exterior points of } \tau\text{)}$$

interior points of τ:

$$= \int_\tau (\text{curl}\ \tilde{\bar{J}}) \times \nabla\tilde{\gamma}\ d\tau + \oint_{S_{1..n}\Sigma} \nabla\tilde{\gamma} \times (d\bar{S}\times\tilde{\bar{J}}) - \tilde{k}^2 \int_\tau \tilde{\gamma}\ \tilde{\bar{J}}\ d\tau - 4\pi\tilde{\bar{J}}$$

$$= \lim_{\tau'\to 0} \int_{\tau-\tau'} (\tilde{\bar{J}}.\nabla)\nabla\tilde{\gamma}\ d\tau - \lim_{S'\to 0} \oint_{S'} \nabla\tilde{\gamma}\ \tilde{\bar{J}}.d\bar{S}$$

exterior points of τ:

$$= \int_\tau (\text{curl}\ \tilde{\bar{J}}) \times \nabla\tilde{\gamma}\ d\tau + \oint_{S_{1..n}\Sigma} \nabla\tilde{\gamma} \times (d\bar{S}\times\tilde{\bar{J}}) - \tilde{k}^2 \int_\tau \tilde{\gamma}\ \tilde{\bar{J}}\ d\tau$$

$$= \int_\tau (\tilde{\bar{J}}.\nabla)\nabla\tilde{\gamma}\ d\tau$$

TABLE 11(CONTD.).

(14)

grad div partial pot $\tilde{\bar{J}}$

$$= \int_{\tau-\tau_\delta} (-\operatorname{div} \tilde{\bar{J}})\, \nabla\tilde{\gamma}\, d\tau + \oint_{S_{1..n}\Sigma} \nabla\tilde{\gamma}\, \tilde{\bar{J}}.d\bar{S} + \oint_{S_\delta} \tilde{\gamma} \operatorname{div} \tilde{\bar{J}}\, d\bar{S}$$

$$= \int_{\tau-\tau_\delta} (\operatorname{curl} \tilde{\bar{J}}) \times \nabla\tilde{\gamma}\, d\tau + \oint_{S_{1..n}\Sigma} \nabla\tilde{\gamma} \times (d\bar{S}\times\tilde{\bar{J}}) - \tilde{k}^2 \int_{\tau-\tau_\delta} \tilde{\gamma}\, \tilde{\bar{J}}\, d\tau$$

$$+ \oint_{S_\delta} (\tilde{\gamma} \operatorname{div} \tilde{\bar{J}}\, d\bar{S} + \tilde{\bar{J}}\times(d\bar{S}\times\nabla\tilde{\gamma}) - \tilde{\bar{J}}\, \nabla\tilde{\gamma}.d\bar{S})$$

$$= \int_{\tau-\tau_\delta} (\tilde{\bar{J}}.\nabla)\nabla\tilde{\gamma}\, d\tau + \oint_{S_\delta} (\tilde{\gamma} \operatorname{div} \tilde{\bar{J}}\, d\bar{S} - \nabla\tilde{\gamma}\, \tilde{\bar{J}}.d\bar{S})$$

(15)

curl curl pot $\tilde{\bar{J}}$

$$= \int_{\tau} (\operatorname{curl} \tilde{\bar{J}}) \times \nabla\tilde{\gamma}\, d\tau + \oint_{S_{1..n}\Sigma} \nabla\tilde{\gamma} \times (d\bar{S}\times\tilde{\bar{J}}) \quad \text{(interior and exterior points of } \tau)$$

$$\left.\begin{aligned} &= \int_{\tau} (-\operatorname{div} \tilde{\bar{J}})\, \nabla\tilde{\gamma}\, d\tau + \oint_{S_{1..n}\Sigma} \nabla\tilde{\gamma}\, \tilde{\bar{J}}.d\bar{S} + \tilde{k}^2 \int_{\tau} \tilde{\gamma}\, \tilde{\bar{J}}\, d\tau + 4\pi\tilde{\bar{J}} \\ &= \lim_{\tau'\to 0} \int_{\tau-\tau'} (\tilde{\bar{J}}.\nabla)\nabla\tilde{\gamma}\, d\tau + \tilde{k}^2 \int_{\tau} \tilde{\gamma}\, \tilde{\bar{J}}\, d\tau + \lim_{S'\to 0} \oint_{S'} \nabla\tilde{\gamma} \times (\tilde{\bar{J}}\times d\bar{S}) \end{aligned}\right\} \text{interior points of } \tau$$

$$\left.\begin{aligned} &= \int_{\tau} (-\operatorname{div} \tilde{\bar{J}})\, \nabla\tilde{\gamma}\, d\tau + \oint_{S_{1..n}\Sigma} \nabla\tilde{\gamma}\, \tilde{\bar{J}}.d\bar{S} + \tilde{k}^2 \int_{\tau} \tilde{\gamma}\, \tilde{\bar{J}}\, d\tau \\ &= \int_{\tau} (\tilde{\bar{J}}.\nabla)\nabla\tilde{\gamma}\, d\tau + \tilde{k}^2 \int_{\tau} \tilde{\gamma}\, \tilde{\bar{J}}\, d\tau \end{aligned}\right\} \text{exterior points of } \tau$$

TABLE 11(CONTD.).

(16)

curl curl partial pot $\tilde{\tilde{J}}$

$$= \int_{\tau-\tau_\delta} (\text{curl}\ \tilde{\tilde{J}}) \times \nabla\tilde{\gamma}\, d\tau + \oint_{S_{1..n}\Sigma} \nabla\tilde{\gamma} \times (d\bar{S}\times\tilde{\tilde{J}}) + \oint_{S_\delta} \tilde{\gamma}(d\bar{S} \times \text{curl}\ \tilde{\tilde{J}})$$

$$= \int_{\tau-\tau_\delta} (-\text{div}\ \tilde{\tilde{J}})\nabla\tilde{\gamma}\, d\tau + \oint_{S_{1..n}\Sigma} \nabla\tilde{\gamma}\, \tilde{\tilde{J}}.d\bar{S} + \tilde{k}^2 \int_{\tau-\tau_\delta} \tilde{\gamma}\, \tilde{\tilde{J}}\, d\tau$$

$$+ \oint_{S_\delta} (\tilde{\tilde{J}}\ \nabla\tilde{\gamma}.d\bar{S} - \tilde{\tilde{J}}\times(d\bar{S}\times\nabla\tilde{\gamma}) + \tilde{\gamma}(d\bar{S}\times\text{curl}\ \tilde{\tilde{J}})$$

$$= \int_{\tau-\tau_\delta} (\tilde{\tilde{J}}.\nabla)\nabla\tilde{\gamma}\, d\tau + \tilde{k}^2 \int_{\tau-\tau_\delta} \tilde{\gamma}\, \tilde{\tilde{J}}\, d\tau + \oint_{S_\delta} \{\tilde{\gamma}(d\bar{S} \times \text{curl}\ \tilde{\tilde{J}}) + \nabla\tilde{\gamma}\times(\tilde{\tilde{J}}\times d\bar{S})\}$$

(17)

$$(\nabla^2+\tilde{k}^2)\ \text{partial pot}\ \tilde{\tilde{J}} = \oint_{S_\delta} \{\tilde{\gamma}(\text{div}\ \tilde{\tilde{J}}\ d\bar{S} - d\bar{S} \times \text{curl}\ \tilde{\tilde{J}}) - \tilde{\tilde{J}}\ d\bar{S}.\nabla\tilde{\gamma} + \tilde{\tilde{J}}\times(d\bar{S}\times\nabla\tilde{\gamma})\}$$

TABLE 12

The Vector Exponential Potential Function $\int (\tilde{\tilde{M}}\times\nabla\tilde{\gamma})\, d\tau$ and its Derivatives

(1)

$$\int_\tau (\tilde{\tilde{M}}\times\nabla\tilde{\gamma})\, d\tau = \int_\tau \tilde{\gamma}\ \text{curl}\ \tilde{\tilde{M}}\, d\tau - \oint_{S_{1..n}\Sigma} \tilde{\gamma}\ (d\bar{S}\times\tilde{\tilde{M}})$$ (interior and exterior points of τ)

(2)

$$\text{partial} \int_{\tau-\tau_\delta} (\tilde{\tilde{M}}\times\nabla\tilde{\gamma})\, d\tau = \int_{\tau-\tau_\delta} \tilde{\gamma}\ \text{curl}\ \tilde{\tilde{M}}\, d\tau - \oint_{S_{1..n}\Sigma, S_\delta} \tilde{\gamma}\ (d\bar{S}\times\tilde{\tilde{M}})$$ (evaluated at centre of moving δ sphere within τ)

(3)

$$\text{cavity} \int_{\tau-\tau_\delta} (\tilde{\tilde{M}}\times\nabla\tilde{\gamma})\, d\tau = \int_{\tau-\tau_\delta} \tilde{\gamma}\ \text{curl}\ \tilde{\tilde{M}}\, d\tau - \oint_{S_{1..n}\Sigma, S_\delta} \tilde{\gamma}\ (d\bar{S}\times\tilde{\tilde{M}})$$ (defined throughout fixed δ sphere within τ)

TABLE 12(CONTD.).

(4)

$$\operatorname{div} \int_\tau (\tilde{\bar{M}} \times \nabla\tilde{\gamma})\, d\tau = 0 \quad \text{(interior and exterior points of } \tau)$$

(5)

$$\operatorname{div}\ (\text{partial}) \int_{\tau-\tau_\delta} (\tilde{\bar{M}} \times \nabla\tilde{\gamma})\, d\tau = \oint_{S_\delta} \tilde{\bar{M}} . (\nabla\tilde{\gamma} \times d\bar{S})$$

(6)

$$\operatorname{curl} \int_\tau (\tilde{\bar{M}} \times \nabla\tilde{\gamma})\, d\tau$$

$$= \int_\tau (\operatorname{curl} \tilde{\bar{M}}) \times \nabla\tilde{\gamma}\, d\tau + \oint_{S_{1..n}\Sigma} \nabla\tilde{\gamma} \times (d\bar{S} \times \tilde{\bar{M}}) \quad \text{(interior and exterior points of } \tau)$$

$$\left.\begin{aligned}
&= \int_\tau \nabla\tilde{\gamma}\, (-\operatorname{div} \tilde{\bar{M}})\, d\tau + \tilde{k}^2 \int_\tau \tilde{\gamma}\, \tilde{\bar{M}}\, d\tau + \oint_{S_{1..n}\Sigma} \nabla\tilde{\gamma}\, \tilde{\bar{M}} . d\bar{S} \\
&= -\operatorname{grad} \int_\tau \tilde{\bar{M}} . \nabla\tilde{\gamma}\, d\tau + \tilde{k}^2 \int_\tau \tilde{\gamma}\, \tilde{\bar{M}}\, d\tau \\
&= \int_\tau (\tilde{\bar{M}} . \nabla)\nabla\tilde{\gamma}\, d\tau + \tilde{k}^2 \int_\tau \tilde{\gamma}\, \tilde{\bar{M}}\, d\tau
\end{aligned}\right\} \text{exterior points of } \tau$$

$$\left.\begin{aligned}
&= \int_\tau \nabla\tilde{\gamma}\, (-\operatorname{div} \tilde{\bar{M}})\, d\tau + \tilde{k}^2 \int_\tau \tilde{\gamma}\, \tilde{\bar{M}}\, d\tau + \oint_{S_{1..n}\Sigma} \nabla\tilde{\gamma}\, \tilde{\bar{M}} . d\bar{S} + 4\pi\tilde{\bar{M}} \\
&= -\operatorname{grad} \int_\tau \tilde{\bar{M}} . \nabla\tilde{\gamma}\, d\tau + \tilde{k}^2 \int_\tau \tilde{\gamma}\, \tilde{\bar{M}}\, d\tau + 4\pi\tilde{\bar{M}} \\
&= \lim_{\tau'\to 0} \int_{\tau-\tau'} (\tilde{\bar{M}} . \nabla)\nabla\tilde{\gamma}\, d\tau + \tilde{k}^2 \int_\tau \tilde{\gamma}\, \tilde{\bar{M}}\, d\tau + \lim_{S'\to 0} \oint_{S'} d\bar{S} \times (\tilde{\bar{M}} \times \nabla\tilde{\gamma})
\end{aligned}\right\} \text{interior points of } \tau$$

TABLE 12(CONTD.).

(7)

$$\text{curl (partial)} \int_{\tau-\tau_\delta} (\tilde{\tilde{M}}\times\nabla\tilde{\gamma})\, d\tau$$

$$= \int_{\tau-\tau_\delta} (\text{curl } \tilde{\tilde{M}}) \times \nabla\tilde{\gamma}\, d\tau + \oint_{S_{1..n}\Sigma} \nabla\tilde{\gamma} \times (d\bar{S}\times\tilde{\tilde{M}}) + \oint_{S_\delta} \tilde{\tilde{M}} \times (d\bar{S}\times\nabla\tilde{\gamma})$$

$$= \int_{\tau-\tau_\delta} \nabla\tilde{\gamma}\, (-\, \text{div } \tilde{\tilde{M}})\, d\tau + \tilde{k}^2 \int_{\tau-\tau_\delta} \tilde{\gamma}\, \tilde{\tilde{M}}\, d\tau + \oint_{S_{1..n}\Sigma} \nabla\tilde{\gamma}\, \tilde{\tilde{M}}.d\bar{S} + \oint_{S_\delta} \tilde{\tilde{M}}\, d\bar{S}.\nabla\tilde{\gamma}$$

$$= -\, \text{grad (partial)} \int_{\tau-\tau_\delta} \tilde{\tilde{M}}.\nabla\tilde{\gamma}\, d\tau + \tilde{k}^2 \int_{\tau-\tau_\delta} \tilde{\gamma}\, \tilde{\tilde{M}}\, d\tau + \oint_{S_\delta} \{\tilde{\tilde{M}}\, d\bar{S}.\nabla\tilde{\gamma} + \tilde{\tilde{M}}\times(d\bar{S}\times\nabla\tilde{\gamma})\}$$

$$= \int_{\tau-\tau_\delta} (\tilde{\tilde{M}}.\nabla)\nabla\tilde{\gamma}\, d\tau + \tilde{k}^2 \int_{\tau-\tau_\delta} \tilde{\gamma}\, \tilde{\tilde{M}}\, d\tau + \oint_{S_\delta} d\bar{S} \times (\tilde{\tilde{M}}\times\nabla\tilde{\gamma})$$

(8)

$$\text{grad div} \int_\tau (\tilde{\tilde{M}}\times\nabla\tilde{\gamma})\, d\tau = \bar{0} \quad \text{(interior and exterior points of } \tau)$$

(9)

$$\text{grad div (partial)} \int_{\tau-\tau_\delta} (\tilde{\tilde{M}}\times\nabla\tilde{\gamma})\, d\tau = \oint_{S_\delta} (\text{curl } \tilde{\tilde{M}}) \times (d\bar{S}\times\nabla\tilde{\gamma})$$

(10)

$$\text{curl curl} \int_\tau (\tilde{\tilde{M}}\times\nabla\tilde{\gamma})\, d\tau = \tilde{k}^2 \int_\tau (\tilde{\tilde{M}}\times\nabla\tilde{\gamma})\, d\tau \quad \text{at exterior points of } \tau$$

(11)

$$\text{curl curl} \int_\tau (\tilde{\tilde{M}}\times\nabla\tilde{\gamma})\, d\tau = \tilde{k}^2 \int_\tau (\tilde{\tilde{M}}\times\nabla\tilde{\gamma})\, d\tau + 4\pi \text{ curl } \tilde{\tilde{M}} \quad \text{at interior points of } \tau$$

TABLE 12(CONTD.).

(12)

$$\text{curl curl (partial)} \int_{\tau-\tau_\delta} (\tilde{\overline{M}}\times\nabla\tilde{\gamma})\, d\tau$$

$$= \tilde{k}^2 \int_{\tau-\tau_\delta} (\tilde{\overline{M}}\times\nabla\tilde{\gamma})\, d\tau + \oint_{S_\delta} \{\text{curl } \tilde{\overline{M}}\, (d\overline{S}.\nabla\tilde{\gamma}) - \text{div } \tilde{\overline{M}}\, (d\overline{S}\times\nabla\tilde{\gamma}) + \tilde{k}^2\tilde{\gamma}(d\overline{S}\times\tilde{\overline{M}})\}$$

(13)

$$(\nabla^2+\tilde{k}^2) \int_\tau (\tilde{\overline{M}}\times\nabla\tilde{\gamma})\, d\tau = \overline{0} \quad \text{at exterior points of } \tau$$

(14)

$$(\nabla^2+\tilde{k}^2) \int_\tau (\tilde{\overline{M}}\times\nabla\tilde{\gamma})\, d\tau = -\, 4\pi \text{ curl } \tilde{\overline{M}} \quad \text{at interior points of } \tau$$

(15)

$$(\nabla^2+\tilde{k}^2) \text{ (partial)} \int_{\tau-\tau_\delta} (\tilde{\overline{M}}\times\nabla\tilde{\gamma})\, d\tau$$

$$= \oint_{S_\delta} \{\text{div } \tilde{\overline{M}}\, (d\overline{S}\times\nabla\tilde{\gamma}) + (\text{curl } \tilde{\overline{M}})\times(d\overline{S}\times\nabla\tilde{\gamma}) - \text{curl } \tilde{\overline{M}}\, (d\overline{S}.\nabla\tilde{\gamma}) - \tilde{k}^2\tilde{\gamma}(d\overline{S}\times\tilde{\overline{M}})\}$$

7.4 The Representation of a Complex Field as the Exponential Potential of Surface and Volume Sources

At interior points of τ, (6.1-7) yields

$$4\pi\tilde{V}_o = \oint_{S_{1..n}\Sigma} \left\{\frac{1}{r}\, e^{j\tilde{k}r}\, \frac{\partial\tilde{V}}{\partial n} - \tilde{V}\frac{\partial}{\partial n}\left(\frac{1}{r}\, e^{j\tilde{k}r}\right)\right\} dS - \int_\tau \frac{1}{r}\, e^{j\tilde{k}r}\, (\nabla^2+\tilde{k}^2)\tilde{V}\, d\tau \qquad (7.4\text{-}1)$$

It follows that $\tilde{V}$ may be considered to be the exponential potential of a simple surface source of density $\frac{1}{4\pi}\frac{\partial\tilde{V}}{\partial n}$ and a surface doublet of density $-\frac{\tilde{V}}{4\pi}$ upon $S_{1..n}\Sigma$, together with a volume source of density $-\frac{1}{4\pi}(\nabla^2+\tilde{k}^2)\tilde{V}$ throughout τ. The value of $\tilde{k}$ is arbitrary.

Since $\hat{\overline{n}}$ is directed out of τ, the negative side of the double layer faces τ when the doublet density is positive.

More generally, if $\tilde{V}$ is defined throughout all space and Sommerfeld conditions hold at infinity (the real part of $j\tilde{k}$ being negative or zero) we find from the arguments of Sec. 4.5 that at all points removed from the surfaces

$$4\pi\tilde{V}_o = \oint_{S_{1..n}\Sigma} \left\{\frac{1}{r} e^{j\tilde{k}r}\left(-\Delta\frac{\partial\tilde{V}}{\partial n}\right) + \Delta\tilde{V}\frac{\partial}{\partial n}\left(\frac{1}{r}e^{j\tilde{k}r}\right)\right\} dS - \int_{\infty} \frac{1}{r} e^{j\tilde{k}r} (\nabla^2+\tilde{k}^2)\tilde{V}\, d\tau \tag{7.4-2}$$

There are, of course, an infinite number of density functions giving rise to a prescribed field within a finite region. Thus, if $\tilde{U}_1..\tilde{U}_n$ are well-behaved point functions defined throughout the regions $\tau_1..\tau_n$ bounded by $S_1..S_n$, then within τ

$$\begin{aligned}4\pi\tilde{V}_o = &\oint_{S_{1..n}} \left\{\frac{1}{r} e^{j\tilde{k}r}\frac{\partial}{\partial n}(\tilde{V}-\tilde{U}) - (\tilde{V}-\tilde{U})\frac{\partial}{\partial n}\left(\frac{1}{r}e^{j\tilde{k}r}\right)\right\} dS - \int_{\tau} \frac{1}{r} e^{j\tilde{k}r} (\nabla^2+\tilde{k}^2)\tilde{V}\, d\tau \\ &+ \oint_{\Sigma} \left\{\frac{1}{r} e^{j\tilde{k}r}\frac{\partial\tilde{V}}{\partial n} - \tilde{V}\frac{\partial}{\partial n}\left(\frac{1}{r}e^{j\tilde{k}r}\right)\right\} dS - \int_{\tau_{1..n}} \frac{1}{r} e^{j\tilde{k}r} (\nabla^2+\tilde{k}^2)\tilde{U}\, d\tau\end{aligned} \tag{7.4-3}$$

The complex vector field may be treated in exactly the same way via the vectorial form of (6.1-7).

7.5 Equivalent Layers in Scalar Exponential Potential Theory

Let simple and double layer surface sources of density $\tilde{\sigma}$ and $\tilde{\mu}$ be spread over open or closed non-intersecting surfaces S which lie within the region τ' bounded by the closed surface Σ, and let a volume source of density $\tilde{\rho}$ occupy τ'. Then at any interior point 0 of τ' not coincident with S, the exponential potential of the system is given by

$$\tilde{\phi}_o = \int_S \tilde{\sigma}\frac{1}{r} e^{j\tilde{k}r}\, dS + \int_S \tilde{\mu}\, d\bar{S}.\nabla\left(\frac{1}{r}e^{j\tilde{k}r}\right) + \int_{\tau'} \tilde{\rho}\frac{1}{r}e^{j\tilde{k}r}\, d\tau \tag{7.5-1}$$

where the positive sense of doublet orientation corresponds with the positive sense of $d\bar{S}$

It follows from previous considerations that we may also express $\tilde{\phi}_o$ as

$$\begin{aligned}\tilde{\phi}_o = &\frac{1}{4\pi}\int_S \left\{\frac{1}{r} e^{j\tilde{k}r}\left(-\Delta\frac{\partial\tilde{\phi}}{\partial n}\right) + \Delta\tilde{\phi}\frac{\partial}{\partial n}\left(\frac{1}{r}e^{j\tilde{k}r}\right)\right\} dS - \frac{1}{4\pi}\int_{\tau'} \frac{1}{r} e^{j\tilde{k}r} (\nabla^2+\tilde{k}^2)\tilde{\phi}\, d\tau \\ &+ \frac{1}{4\pi}\oint_{\Sigma} \left\{\frac{1}{r} e^{j\tilde{k}r}\frac{\partial\tilde{\phi}}{\partial n} - \tilde{\phi}\frac{\partial}{\partial n}\left(\frac{1}{r}e^{j\tilde{k}r}\right)\right\} dS\end{aligned} \tag{7.5-2}$$

where Δ represents the increment associated with movement through S in an arbitrarily-defined positive sense common to both sides of the surface.

But it has been shown that in these circumstances

$$\Delta\frac{\partial\tilde{\phi}}{\partial n} = -4\pi\tilde{\sigma} \quad ; \quad \Delta\tilde{\phi} = 4\pi\tilde{\mu} \quad ; \quad (\nabla^2+\tilde{k}^2)\tilde{\phi} = -4\pi\tilde{\rho} \tag{7.5-3}$$

hence equation (7.5-2) reduces to

$$\tilde{\phi}_o = \int_S \frac{\tilde{\sigma}}{r} e^{j\tilde{k}r}\, dS + \int_S \tilde{\mu}\, d\bar{S}.\nabla\left(\frac{1}{r} e^{j\tilde{k}r}\right) + \int_{\tau'} \frac{\tilde{\rho}}{r} e^{j\tilde{k}r}\, d\tau$$
$$+ \frac{1}{4\pi}\oint_\Sigma \left\{\frac{1}{r} e^{j\tilde{k}r} \frac{\partial\tilde{\phi}}{\partial n} - \tilde{\phi}\frac{\partial}{\partial n}\left(\frac{1}{r} e^{j\tilde{k}r}\right)\right\} dS$$

whence

$$\oint_\Sigma \left\{\frac{1}{r} e^{j\tilde{k}r} \frac{\partial\tilde{\phi}}{\partial n} - \tilde{\phi}\frac{\partial}{\partial n}\left(\frac{1}{r} e^{j\tilde{k}r}\right)\right\} dS = 0 \tag{7.5-4}$$

The argument is readily extended to cover the case in which volume polarisation is present at interior points of τ; $\tilde{\rho}$ is replaced by $\tilde{\rho}$ - div $\tilde{\bar{P}}$, $\tilde{P}_n$ appears upon surfaces of discontinuity of $\tilde{\bar{P}}$, and the relationships embodied in (7.5-3) are modified accordingly.

Equation (7.5-4) holds for all configurations of Σ which enclose the source system. As a consequence, we see that the surface integral

$$\oint_\infty \left\{\frac{1}{r} e^{j\tilde{k}r} \frac{\partial\tilde{V}}{\partial n} - \tilde{V}\frac{\partial}{\partial n}\left(\frac{1}{r} e^{j\tilde{k}r}\right)\right\} dS$$

may be equated to zero whenever $\tilde{V}$ is the exponential potential of a finite source system, without reference to Sommerfeld conditions or the nature of the potential exponent, provided only that $\tilde{k}$ is made equal to it.

Equivalent layer expressions may now be developed in the manner described in Sec. 4.10 for the non-exponential case. It is easily shown that if $\tilde{\phi}_i$ and $\tilde{\phi}_e$ are respectively the exponential potential of sources within and without the region $\underline{R}$ bounded by the closed surfaces $S_{1..n}\Sigma$, and if $\tilde{\phi} = \tilde{\phi}_i + \tilde{\phi}_e$, then

(1) at an interior point 0 of $\underline{R}$

$$4\pi\tilde{\phi}_{e_o} = \oint_{S_{1..n}\Sigma} \left\{\frac{1}{r} e^{j\tilde{k}r} \frac{\partial\tilde{\phi}_e}{\partial n} - \tilde{\phi}_e\frac{\partial}{\partial n}\left(\frac{1}{r} e^{j\tilde{k}r}\right)\right\} dS \tag{7.5-5}$$

or

$$4\pi\tilde{\phi}_{e_o} = \oint\limits_{S_{1..n}\Sigma} \left\{\frac{1}{r}\, e^{j\tilde{k}r}\, \frac{\partial\tilde{\phi}}{\partial n} - \tilde{\phi}\,\frac{\partial}{\partial n}\left(\frac{1}{r}\, e^{j\tilde{k}r}\right)\right\} dS \qquad (7.5\text{-}6)$$

where $\hat{n}$ is directed out of $\underline{R}$ and $\tilde{k}$ is the constant in the exponent of the potential function.

Simple and double equivalent layer densities are, correspondingly, $\frac{1}{4\pi}\frac{\partial\tilde{\phi}_e}{\partial n}$ and $\frac{-1}{4\pi}\tilde{\phi}_e$. The doublet density has a positive value when doublet alignment corresponds with $\hat{n}$, ie when the negative side side of the double layer faces $\underline{R}$.

It is seen from equation (7.5-6) that the above densities may be replaced by $\frac{1}{4\pi}\frac{\partial\tilde{\phi}}{\partial n}$ and $-\frac{1}{4\pi}\tilde{\phi}$.

(2) at an exterior point 0 of $\underline{R}$ (ie within $\tau_1..\tau_n$ or outside Σ)

$$4\pi\tilde{\phi}_{i_o} = \oint\limits_{S_{1..n}\Sigma} \left\{\frac{1}{r}\, e^{j\tilde{k}r}\, \frac{\partial\tilde{\phi}_i}{\partial n} - \tilde{\phi}_i\,\frac{\partial}{\partial n}\left(\frac{1}{r}\, e^{j\tilde{k}r}\right)\right\} dS \qquad (7.5\text{-}7)$$

or

$$4\pi\tilde{\phi}_{i_o} = \oint\limits_{S_{1..n}\Sigma} \left\{\frac{1}{r}\, e^{j\tilde{k}r}\, \frac{\partial\tilde{\phi}}{\partial n} - \tilde{\phi}\,\frac{\partial}{\partial n}\left(\frac{1}{r}\, e^{j\tilde{k}r}\right)\right\} dS \qquad (7.5\text{-}8)$$

where $\hat{n}$ is directed into $\underline{R}$.

Simple and double equivalent layer densities are now $\frac{1}{4\pi}\frac{\partial\tilde{\phi}_i}{\partial n}$ and $-\frac{1}{4\pi}\tilde{\phi}_i$, the latter having a positive value when the positive side of the double layer faces $\underline{R}$.

These densities may be replaced by $\frac{1}{4\pi}\frac{\partial\tilde{\phi}}{\partial n}$ and $-\frac{1}{4\pi}\tilde{\phi}$.

As in the non-exponential case, the equivalent layer densities (defined by (7.5-6) and (7.5-8)) reverse sign when the exterior problem replaces the interior.

The equivalence of sources (both scalar and vector) are commonly considered from the point of view of the $\tilde{\underline{E}}$ and $\tilde{\underline{B}}$ fields which derive from them rather than of the potentials themselves. This subject is treated of in Sec. 7.8.

EXERCISES

7-1. It has been tacitly assumed that if $\bar{J}$, $\bar{K}$, $\bar{I}$ are time-harmonic then ρ, σ, λ are also time-harmonic (supposing that their mean values are zero). Similarly, it has been assumed that if the above density functions, together with $\bar{P}$ and $\bar{M}$, are time-harmonic, then the potentials are time-harmonic. Prove these assumptions.

7-2. Confirm the relationship

$$\operatorname{div}\,[\bar{F}] = [\operatorname{div}\,\bar{F}] - \frac{\bar{r}}{cr}\cdot\left[\frac{\partial\bar{F}}{\partial t}\right]$$

by transforming from real to complex notation and back again.

7-3. By expanding $(\tilde{\bar{J}}.\nabla)$ in rectangular coordinates show that

$$(\tilde{\bar{J}}.\nabla)\nabla\tilde{\gamma} = \left\{-\frac{\tilde{\bar{J}}}{r^3} + \frac{3\tilde{\bar{J}}.\bar{r}\,\bar{r}}{r^5} + \frac{j\tilde{k}\,\tilde{\bar{J}}}{r^2} - \frac{3j\tilde{k}\,\tilde{\bar{J}}.\bar{r}\,\bar{r}}{r^4} - \frac{\tilde{k}^2\tilde{\bar{J}}.\bar{r}\,\bar{r}}{r^3}\right\} e^{j\tilde{k}r}$$

Use this result to show that if $\tilde{k} = -\frac{\omega}{c}$ and $\bar{J} = \operatorname{Re}\,\{\tilde{\bar{J}}\,e^{j\omega t}\}$, the real part of $(\tilde{\bar{J}}.\nabla)\nabla\tilde{\gamma}\,e^{j\omega t}$ is given by

$$-\frac{[\bar{J}]}{r^3} + \frac{3[\bar{J}].\bar{r}\,\bar{r}}{r^5} - \frac{1}{cr^2}\left[\frac{\partial\bar{J}}{\partial t}\right] + \frac{3\bar{r}}{cr^4}\left[\frac{\partial\bar{J}}{\partial t}\right].\bar{r} + \frac{\bar{r}}{c^2r^3}\left[\frac{\partial^2\bar{J}}{\partial t^2}\right].\bar{r}$$

$$= ([\bar{J}].\nabla)\operatorname{grad}\frac{1}{r} - \left(\left[\frac{\partial\bar{J}}{\partial t}\right].\nabla\right)\frac{\bar{r}}{cr^2} + \frac{\bar{r}}{cr^4}\left[\frac{\partial\bar{J}}{\partial t}\right].\bar{r} + \frac{\bar{r}}{c^2r^3}\left[\frac{\partial^2\bar{J}}{\partial t^2}\right].\bar{r}$$

Hence show that the following expressions for curl curl partial pot $\tilde{\bar{J}}$ and curl curl partial pot $\bar{J}$ which appear in Table 11 and in Ex.5-21., p. 435 are in agreement:

$$\int_{\tau-\tau_\delta}(\tilde{\bar{J}}.\nabla)\nabla\tilde{\gamma}\,d\tau + \tilde{k}^2\int_{\tau-\tau_\delta}\tilde{\gamma}\,\tilde{\bar{J}}\,d\tau + \oint_{S_\delta}\{\tilde{\gamma}(d\bar{S}\times\operatorname{curl}\tilde{\bar{J}}) + \nabla\tilde{\gamma}\times(\tilde{\bar{J}}\times d\bar{S})\}$$

and

$$\int_{\tau-\tau_\delta}\left\{([\bar{J}].\nabla)\operatorname{grad}\frac{1}{r} - \left(\left[\frac{\partial\bar{J}}{\partial t}\right].\nabla\right)\frac{\bar{r}}{cr^2} + \frac{\bar{r}}{cr^4}\left[\frac{\partial\bar{J}}{\partial t}\right].\bar{r} + \frac{\bar{r}}{c^2r^3}\left[\frac{\partial^2\bar{J}}{\partial t^2}\right].\bar{r}\right\}d\tau$$

$$-\frac{1}{c^2}\frac{\partial^2}{\partial t^2}\int_{\tau-\tau_\delta}\frac{[\bar{J}]}{r}\,d\tau + \oint_{S_\delta}\left\{d\bar{S}\times\frac{1}{r}\,[\operatorname{curl}\bar{J}] + \operatorname{grad}\frac{1}{r}\times([\bar{J}]\times d\bar{S}) - \frac{\bar{r}}{cr^2}\times\left(\left[\frac{\partial\bar{J}}{\partial t}\right]\times d\bar{S}\right)\right\}$$

Note the considerable simplification introduced into the expressions (and the working) by the complex notation. (Dyadic notation effects a further simplification since $\int_{\tau-\tau_\delta} ((\tilde{\vec{J}}.\nabla)\nabla\tilde{\gamma} + \tilde{k}^2\,\tilde{\gamma}\,\tilde{\vec{J}})\,d\tau$ is replaced by $\tilde{k}^2 \int_{\tau-\tau_\delta} \bar{\bar{\Gamma}}.\tilde{\vec{J}}\,d\tau$).

7-4. It was pointed out in Ex.5-13., p. 424, that the computed value of $\Delta \frac{\partial}{\partial n} \operatorname{grad} \oint_S \frac{\sigma}{r}\,dS$, which accompanies movement through the surface, will be in error if S is spherical and it is assumed, for the purpose of calculation, that a sufficiently small portion of the surface may be treated as plane.

Show that this assumption does not lead to error in the case of $\Delta \operatorname{grad} \tilde{\phi}$, $\Delta \operatorname{div} \tilde{\vec{A}}$ and $\Delta \operatorname{curl} \tilde{\vec{A}}$ where $\tilde{\phi}$ and $\tilde{\vec{A}}$ are exponential potentials (and consequently include retarded and non-retarded potentials) by deriving values for the general case and demonstrating their agreement with previously-determined values based upon a planar analysis.

[Hint: Let S be a regular closed surface having a continuous simple source density $\tilde{\sigma}$. Show that

$$4\pi\tilde{\phi}_o = \oint_S \left\{\frac{1}{r}\,e^{j\tilde{k}r}\left(-\Delta\frac{\partial\tilde{\phi}}{\partial n}\right) + \Delta\tilde{\phi}\,\frac{\partial}{\partial n}\left(\frac{1}{r}\,e^{j\tilde{k}r}\right)\right\} dS$$

where O lies within or without the enclosure.

Hence show that $\Delta \frac{\partial\tilde{\phi}}{\partial n} = -4\pi\tilde{\sigma}$ at all points of the surface. Similarly, demonstrate that $\Delta \frac{\partial\tilde{\vec{A}}}{\partial n} = -4\pi\tilde{\vec{K}}$.

Then substitute these values in equations (2.12-12/14)].

7-5. Confirm that $\Delta \frac{\partial\tilde{\phi}}{\partial n} = -4\pi\tilde{\sigma}$ when the local surface comprises a spherical cap (ie the portion of a spherical surface cut off by a plane) by calculating the limiting values of $\frac{\partial\tilde{\phi}}{\partial n}$ when the centre of the surface is approached normally from each side. Show that the result is consistent with equation (4.7-8).

Show further that the normal derivative of grad $\tilde{\phi}$ is discontinuous by $8\pi\tilde{\sigma}/R$, where R is the radius of curvature of the surface.

Ans:

$$\frac{\partial\tilde{\phi}}{\partial n} = 2\pi\tilde{\sigma}\left\{\pm 1 + \frac{R}{d}(1-\cos\psi)e^{j\tilde{k}d} - \frac{1}{j\tilde{k}R}(e^{j\tilde{k}d}-1)\right\}$$

where d is the length of the chord joining the centre of the cap to its periphery and ψ is the angle subtended by the chord at the centre of curvature.

7-6. Show that

$$\left.\begin{matrix}-4\pi \\ 0\end{matrix}\right\} = \oint_S \frac{\partial}{\partial n}\left(\frac{1}{r}\, e^{j\tilde{k}r}\right) dS$$

according as the origin of r lies within or without the region enclosed by S.

Hence deduce that there will be an increase of $4\pi\tilde{\mu}$ in the exponential potential on passing positively through an interior point of an open surface doublet where the local density is continuous and equal to $\tilde{\mu}$.

Then employ the procedure of Ex.7-4. to show that the associated value of $\Delta \frac{\partial\tilde{\phi}}{\partial n}$ is zero.

7-7. If the spherical cap of Ex.7-5. now comprises a doublet source of constant density $\tilde{\mu}$, show that as the surface is approached centrally along the normal from each side

$$\tilde{\phi} \to 2\pi\tilde{\mu}\left\{\pm 1 + \frac{1}{j\tilde{k}R} + e^{j\tilde{k}d}\left(\frac{d}{2R} - \frac{1}{j\tilde{k}R}\right)\right\}$$

$$\frac{\partial\tilde{\phi}}{\partial n} \to \frac{-2\pi\tilde{\mu}}{R}\left\{\frac{1}{j\tilde{k}R} - j\tilde{k}R + e^{j\tilde{k}d}\left(\frac{d}{2R} - \frac{1}{j\tilde{k}R} + \frac{R}{d} + \frac{1}{2}\,(1-j\tilde{k}d)\,\frac{\partial d}{\partial n}\right)\right\}$$

for a common positive sense of the normal.

Show that for $\tilde{k} = 0$ and $R\to\infty$ these results reduce to equations (4.7-11/12).

7-8. Consider a uniform, spherical volume source of radius a and density $\tilde{\rho}$. Show by integration that the exponential potential at exterior points is identical with that deriving from a point source at the centre, of strength

$$\frac{4\pi\tilde{\rho}}{\tilde{k}^3}\,(\sin\tilde{k}a - \tilde{k}a\cos\tilde{k}a)$$

Show further that at interior points, at a distance R from the centre,

$$\tilde{\phi} = \frac{4\pi\tilde{\rho}}{\tilde{k}^2}\left\{e^{j\tilde{k}a}\,(1-j\tilde{k}a)\,\frac{\sin\tilde{k}R}{\tilde{k}R} - 1\right\}$$

Hence demonstrate that both $\tilde{\phi}$ and grad $\tilde{\phi}$ are continuous through the surface.

Finally, prove that

$$\nabla^2\tilde{\phi} + \tilde{k}^2\tilde{\phi} = -4\pi\tilde{\rho}$$

at interior points of the sphere (including the centre).

7-9. A spherical surface S of radius a is centred upon the point P. A point function $\tilde{V}$ is defined by

$$\tilde{V} = \frac{\sin \alpha R}{R} \qquad (R \le a)$$

and

$$\tilde{V} = e^{j\alpha a} (\sin \alpha a) \frac{e^{-j\alpha R}}{R} \qquad (R \ge a)$$

where R is distance measured from P and α is a real constant.

Show that $\tilde{V}$ may be represented within and without the sphere by

$$\tilde{V} = \frac{\alpha e^{j\alpha a}}{4\pi a} \oint_S \frac{1}{r} e^{-j\alpha r} \, dS$$

where r is distance measured from the point of evaluation.

Now suppose that

$$\tilde{V} = \frac{e^{-j\alpha a}}{\sin \alpha a} \frac{\sin \alpha R}{R} \qquad (R \le a; \quad \sin \alpha a \ne 0)$$

and

$$\tilde{V} = \frac{1}{R} e^{-j\alpha R} \qquad (R \ge a)$$

Show that $\tilde{V}$ may be represented within and without the sphere by

$$\tilde{V} = \frac{\alpha}{4\pi a \sin \alpha a} \oint_S \frac{1}{r} e^{-j\alpha r} \, dS$$

Deduce that $\oint_S \frac{1}{r} e^{-j\alpha r} \, dS = 0$ when the origin of r lies outside the sphere and α is such as to make $\sin \alpha a$ zero, and confirm this by direct integration.

7-10. A rectilinear source of constant axial density $\tilde{I}$ subtends an angle 2θ at a point 0 on the perpendicular bisector at a distance d. Show by expansion in a power series that the curl of the exponential vector potential at 0 is given by

$$(\text{curl } \tilde{\bar{A}})_o = -\tilde{I}\int_\Gamma d\bar{r} \times \left(\frac{\bar{r}}{r^3} + \frac{\tilde{k}^2\bar{r}}{2r} + \frac{j\tilde{k}^3\bar{r}}{3} ----\right)$$

where the positive sense of $d\bar{r}$ corresponds with the direction of $\tilde{\bar{I}}$.

Hence show that

$$(\text{curl } \tilde{\bar{A}})_o = (\hat{\bar{r}}\times d\hat{\bar{r}})\left\{\frac{2\tilde{I}}{d}\sin\theta + \frac{\tilde{k}^2\tilde{I}}{2} d\ln\left(\frac{1+\sin\theta}{1-\sin\theta}\right) + \frac{2j\tilde{k}^3\tilde{I}}{3} d^2\tan\theta ---\right\}$$

Now evaluate $\oint_{\Gamma'} \text{curl } \tilde{\bar{A}}.d\bar{r}'$ around a circle of radius d in the mid plane (designated Γ') and show that for right-handed movement relative to $\tilde{\bar{I}}$

$$\oint_{\Gamma'} \text{curl } \tilde{\bar{A}}.d\bar{r}' \to 4\pi\tilde{I} \quad \text{as} \quad d\to 0$$

Show further that for any closed contour Γ'' which does not embrace the source (although possibly coaxial with it)

$$\oint_{\Gamma''} \text{curl } \tilde{\bar{A}}.d\bar{r}''\to 0$$

as the dimensions of the contour shrink to zero.

Hence deduce that the limiting value of the line integral of curl $\tilde{\bar{A}}$ around a closed contour which shrinks about an interior point of a line source with axially directed density $\tilde{\bar{I}}$ is dependent only upon the local value of $\tilde{I}$ provided that $\tilde{\bar{I}}$ is continuous in a neighbourhood of the point.

7-11. Let $\tilde{V}$ be a point function having continuous second derivatives throughout the infinite region τ bounded internally by the closed surfaces $S_{1..n}$, and let $(\nabla^2+\tilde{k}^2)\tilde{V}$ be zero outside a sphere of finite radius. If $\tilde{V}$ satisfies the Sommerfeld conditions when the real part of $j\tilde{k}$ is negative or zero, show that $\tilde{V}$ is given everywhere within τ by

$$4\pi\tilde{V}_o = \oint_{S_{1..n}}\left\{\frac{1}{r}e^{j\tilde{k}r}\frac{\partial\tilde{V}}{\partial n} - \tilde{V}\frac{\partial}{\partial n}\left(\frac{1}{r}e^{j\tilde{k}r}\right)\right\}dS - \int_\tau \frac{1}{r}e^{j\tilde{k}r}(\nabla^2+\tilde{k}^2)\tilde{V}\,d\tau$$

[Hint: For the case $j\tilde{k} = 0$ make use of the result of Ex.6-2. to show that $\tilde{V} = \tilde{\phi}$ throughout τ, where $\tilde{\phi}$ is the exponential potential of singlet and doublet sources upon $S_{1..n}$ of densities $\frac{1}{4\pi}\frac{\partial\tilde{V}}{\partial n}$ and $-\frac{\tilde{V}}{4\pi}$, and of volume sources in τ of density $-\frac{1}{4\pi}(\nabla^2+\tilde{k}^2)\tilde{V}$.]

7-12. O and P are interior points of a region τ bounded externally by the closed surface S. A point doublet of moment $\tilde{\bar{p}}$ is located at P and gives rise to the exponential potential $\tilde{\phi} = -\tilde{\bar{p}}.\nabla \frac{e^{j\tilde{k}r'}}{r'}$ where r' is distance from P. Show by direct integration over a δ sphere centred on P that for all finite values of $\tilde{k}$

$$\oint_S \left\{ \frac{1}{r} e^{j\tilde{k}r} \frac{\partial\tilde{\phi}}{\partial n} - \tilde{\phi}\frac{\partial}{\partial n}\left(\frac{1}{r} e^{j\tilde{k}r}\right)\right\} dS = 0$$

where r is distance measured from O.

7-13. If, in the previous exercise , a point whirl is substituted for the doublet, and if $\tilde{\bar{A}}$ is the vector potential of the whirl, show that

$$\oint_S \left\{ \frac{1}{r} e^{j\tilde{k}r} \frac{\partial\tilde{\bar{A}}}{\partial n} - \tilde{\bar{A}}\frac{\partial}{\partial n}\left(\frac{1}{r} e^{j\tilde{k}r}\right)\right\} dS = 0$$

Show further that this relationship continues to hold when $\tilde{\bar{A}}$ is the vector potential of piecewise continuous sources of densities $\tilde{\bar{I}}$, $\tilde{\bar{K}}$ and $\tilde{\bar{J}}$, provided that all sources lie within τ.

7-14. Derive the following forms of the generalised grad-curl theorem by expanding

$$\text{curl curl} \int_\tau \tilde{\gamma}\,\tilde{\bar{F}}\,d\tau = \text{grad div} \int_\tau \tilde{\gamma}\,\tilde{\bar{F}}\,d\tau - \nabla^2 \int_\tau \tilde{\gamma}\,\tilde{\bar{F}}\,d\tau$$

where $\tilde{\bar{F}}$ is any well-behaved vector field.

$$\left.\begin{matrix} 4\pi\tilde{\bar{F}}_O \\ \bar{0} \end{matrix}\right\} = -\,\text{grad} \int_\tau \tilde{\gamma}\,\text{div}\,\tilde{\bar{F}}\,d\tau + \text{grad} \oint_{S_{1..n}\Sigma} \tilde{\gamma}\,\tilde{\bar{F}}.d\bar{S}$$

$$+\,\text{curl} \int_\tau \tilde{\gamma}\,\text{curl}\,\tilde{\bar{F}}\,d\tau - \text{curl} \oint_{S_{1..n}\Sigma} \tilde{\gamma}\,(d\bar{S}\times\tilde{\bar{F}}) - \tilde{k}^2 \int_\tau \tilde{\gamma}\,\tilde{\bar{F}}\,d\tau$$

$$= \int_\tau \nabla\tilde{\gamma}\,\text{div}\,\tilde{\bar{F}}\,d\tau - \oint_{S_{1..n}\Sigma} \nabla\tilde{\gamma}\,\tilde{\bar{F}}.d\bar{S}$$

$$-\int_\tau \nabla\tilde{\gamma}\times\text{curl}\,\tilde{\bar{F}}\,d\tau + \oint_{S_{1..n}\Sigma} \nabla\tilde{\gamma}\times(d\bar{S}\times\tilde{\bar{F}}) - \tilde{k}^2 \int_\tau \tilde{\gamma}\,\tilde{\bar{F}}\,d\tau$$

7-15. Let $\bar{F}$ be a time-harmonic vector field which is well-behaved within the region τ bounded by the surfaces $S_{1..n}\Sigma$. Since we may write $\bar{F} = \mathrm{Re}\{\tilde{\bar{F}}\,e^{j\omega t}\}$ it follows from the previous exercise that

$$\left.\begin{matrix}4\pi\bar{F}_o\\ \bar{0}\end{matrix}\right\} = \mathrm{Re}\left\{\left(\int_\tau \nabla\tilde{\gamma}\,\mathrm{div}\,\tilde{\bar{F}}\,d\tau - \oint_{S_{1..n}\Sigma} \nabla\tilde{\gamma}\,\tilde{\bar{F}}.d\bar{S} - \int_\tau \nabla\tilde{\gamma}\times \mathrm{curl}\,\tilde{\bar{F}}\,d\tau \right.\right.$$

$$\left.\left. + \oint_{S_{1..n}\Sigma}\nabla\tilde{\gamma}\times(d\bar{S}\times\tilde{\bar{F}}) - \frac{\omega^2}{c^2}\int_\tau \tilde{\gamma}\,\tilde{\bar{F}}\,d\tau\right)e^{j\omega t}\right\}$$

where $\tilde{\gamma} = \frac{1}{r}\,e^{-j\omega r/c}$

Transform this equation into the retarded form of the grad-curl theorem as expressed in Ex.5-1., p. 403.

7.6 The Complex Form of Maxwell's Equations

The exponential equivalents of the general expressions for the retarded scalar and vector potentials deriving from the mixed source configuration discussed in Sec. 5.19a are given by

$$\tilde{\phi} = \int_\infty \tilde{\rho}\,\tilde{\gamma}\,d\tau + \int_S \tilde{\sigma}\,\tilde{\gamma}\,dS + \int_\Gamma \tilde{\lambda}\,\tilde{\gamma}\,ds + \int_\infty \tilde{\bar{P}}.\nabla\tilde{\gamma}\,d\tau + \int_S \tilde{\bar{P}}'.\nabla\tilde{\gamma}\,dS \tag{7.6-1}$$

and

$$\tilde{\bar{A}} = \frac{1}{c}\int_\infty \tilde{\bar{J}}\,\tilde{\gamma}\,d\tau + \frac{1}{c}\int_S \tilde{\bar{K}}\,\tilde{\gamma}\,dS + \frac{1}{c}\int_\Gamma \tilde{\bar{I}}\,\tilde{\gamma}\,ds + \frac{1}{c}\int_\infty j\omega\tilde{\bar{P}}\,\tilde{\gamma}\,d\tau + \frac{1}{c}\int_S j\omega\tilde{\bar{P}}'\tilde{\gamma}\,dS$$

$$+ \int_\infty \tilde{\bar{M}}\times\nabla\tilde{\gamma}\,d\tau + \int_S \tilde{\bar{M}}'\times\nabla\tilde{\gamma}\,dS \tag{7.6-2}$$

Here, the surface sources, which need not be coincident, are designated S. $\tilde{\bar{K}}$ and $\tilde{\bar{I}}$ are tangential and $\tilde{k}$ is restricted in value to $-\frac{\omega}{c}$, so that $\tilde{\gamma} \equiv \frac{1}{r}\,e^{-j\omega r/c}$.

The corresponding Hertzian vectors are

$$\tilde{\bar{\Pi}}_e = \int_\infty \tilde{\bar{P}}\,\tilde{\gamma}\,d\tau + \int_S \tilde{\bar{P}}'\tilde{\gamma}\,dS \tag{7.6-3}$$

$$\tilde{\overline{\Pi}}_m = \int_\infty \tilde{\overline{M}}\,\tilde{\gamma}\,d\tau + \int_S \tilde{\overline{M}}'\tilde{\gamma}\,dS \tag{7.6-4}$$

whence the associated values of $\tilde{\phi}$ and $\tilde{\overline{A}}$ are given by

$$\tilde{\phi} = -\operatorname{div} \tilde{\overline{\Pi}}_e \tag{7.6-5}$$

$$\tilde{\overline{A}} = \operatorname{curl} \tilde{\overline{\Pi}}_m + \frac{j\omega}{c}\tilde{\overline{\Pi}}_e \tag{7.6-6}$$

It is seen from (5.17-15), (5.17-18) and (5.17-19) that the complex equations of continuity take the form

$$\operatorname{div} \tilde{\overline{J}} = -j\omega\,\tilde{\rho} \tag{7.6-7}$$

$$\operatorname{divs} \tilde{\overline{K}} + \Delta\,\tilde{\overline{J}}.\hat{n} = -j\omega\,\tilde{\sigma} \tag{7.6-8}$$

$$\frac{d\tilde{I}}{ds} + \Delta\,\tilde{\overline{K}}.\hat{n}' = -j\omega\,\tilde{\lambda} \tag{7.6-9}$$

It then follows that for any combination of complete sources

$$\operatorname{div} \tilde{\overline{A}} = -\frac{j\omega}{c}\tilde{\phi} \quad \text{at exterior points of } S \tag{7.6-10}$$

This is most easily demonstrated by considering each type of source separately and making use, if required, of equation (7.3-28) and its line integral counterpart, viz

$$\int_\Gamma \frac{d}{ds}(\tilde{\overline{I}}\,\tilde{\gamma})\,ds = \int_\Gamma \tilde{\gamma}\frac{d\tilde{\overline{I}}}{ds}\,ds + \int_\Gamma \tilde{\overline{I}}.\nabla\tilde{\gamma}\,ds$$

in closed or open form.

From equations (5.11-19) and (5.11-20)

$$\tilde{\overline{E}} = -\operatorname{grad}\tilde{\phi} - \frac{j\omega}{c}\tilde{\overline{A}} \tag{7.6-11}$$

$$\tilde{\overline{B}} = \operatorname{curl} \tilde{\overline{A}} \tag{7.6-12}$$

Then

$$\operatorname{div} \tilde{\overline{E}} = -\nabla^2\tilde{\phi} - \frac{j\omega}{c}\operatorname{div} \tilde{\overline{A}} = -\left(\nabla^2 + \frac{\omega^2}{c^2}\right)\tilde{\phi}$$

whence, from Sec. 7.2 and Tables 9 and 10,

$$\operatorname{div} \tilde{\overline{E}} = 4\pi (\tilde{\rho} - \operatorname{div} \tilde{\overline{P}}) \tag{7.6-13}$$

Also

$$\operatorname{curl} \tilde{\overline{B}} = \operatorname{grad} \operatorname{div} \tilde{\overline{A}} - \nabla^2 \tilde{\overline{A}} = \frac{j\omega}{c} (-\operatorname{grad} \tilde{\phi}) - \nabla^2 \tilde{\overline{A}}$$

$$= \frac{j\omega}{c} \tilde{\overline{E}} - \left(\nabla^2 + \frac{\omega^2}{c^2}\right) \tilde{\overline{A}}$$

whence, from Sec. 7.3 and Tables 11 and 12,

$$\operatorname{curl} \tilde{\overline{B}} = \frac{4\pi}{c} \tilde{\overline{J}} + \frac{4\pi}{c} j\omega \tilde{\overline{P}} + 4\pi \operatorname{curl} \tilde{\overline{M}} + \frac{j\omega}{c} \tilde{\overline{E}} \tag{7.6-14}$$

On putting

$$\tilde{\overline{D}} = \tilde{\overline{E}} + 4\pi \tilde{\overline{P}} \tag{7.6-15}$$

$$\tilde{\overline{H}} = \tilde{\overline{B}} - 4\pi \tilde{\overline{M}} \tag{7.6-16}$$

and taking the curl of (7.6-11) and divergence of (7.6-12) we arrive at the complex form of Maxwell's equations, viz

$$\operatorname{div} \tilde{\overline{D}} = 4\pi \tilde{\rho} \tag{7.6-17}$$

$$\operatorname{curl} \tilde{\overline{E}} = \frac{-j\omega}{c} \tilde{\overline{B}} \tag{7.6-18}$$

$$\operatorname{div} \tilde{\overline{B}} = 0 \tag{7.6-19}$$

$$\operatorname{curl} \tilde{\overline{H}} = \frac{4\pi}{c} \tilde{\overline{J}} + \frac{j\omega}{c} \tilde{\overline{D}} \tag{7.6-20}$$

In the notation of Sec. 5.21 the corresponding boundary conditions are found to be

$$\hat{\overline{n}}_1 . \tilde{\overline{D}}_1 + \hat{\overline{n}}_2 . \tilde{\overline{D}}_2 = -4\pi\tilde{\sigma} \tag{7.6-21}$$

$$\hat{\overline{n}}_1 \times \tilde{\overline{E}}_1 + \hat{\overline{n}}_2 \times \tilde{\overline{E}}_2 = \overline{0} \tag{7.6-22}$$

$$\hat{n}_1 . \tilde{\bar{B}}_1 + \hat{n}_2 . \tilde{\bar{B}}_2 = 0 \tag{7.6-23}$$

$$\hat{n}_1 \times \tilde{\bar{H}}_1 + \hat{n}_2 \times \tilde{\bar{H}}_2 = -\frac{4\pi}{c} \tilde{\bar{K}} \tag{7.6-24}$$

7.7 The Macroscopic Vector Fields $\tilde{\bar{E}}, \tilde{\bar{D}}, \tilde{\bar{B}}, \tilde{\bar{H}}$

By combining (7.6-10) with (7.6-11) we may express $\tilde{\bar{E}}$ in terms of $\tilde{\bar{A}}$ alone. Thus

$$\tilde{\bar{E}} = \frac{c}{j\omega}\left(\text{grad div } \tilde{\bar{A}} + \frac{\omega^2}{c^2}\tilde{\bar{A}}\right) = \frac{c}{j\omega}\left\{\text{curl curl } \tilde{\bar{A}} + \left(\nabla^2 + \frac{\omega^2}{c^2}\right)\tilde{\bar{A}}\right\} \tag{7.7-1}$$

Since (7.6-10) holds only for a complete source system, this limitation is imposed upon (7.7-1) also.

Substitution in (7.7-1) of the various expressions derived in Sec. 7.3 for curl curl $\tilde{\bar{A}}$ and $\left(\nabla^2 + \frac{\omega^2}{c^2}\right)\tilde{\bar{A}}$, and listed in Tables 11 and 12, lead to the following formulae for $\tilde{\bar{E}}, \tilde{\bar{D}}, \tilde{\bar{B}}$ and $\tilde{\bar{H}}$ at points exterior to surfaces and lines of discontinuity. To simplify the presentation, the different parent current systems are treated individually.

7.7a Line current (and associated values of $\tilde{\lambda}$ and $\tilde{a}$)

$$\tilde{\bar{E}} = \tilde{\bar{D}} = \frac{1}{j\omega}\int_\Gamma \left\{(\tilde{\bar{I}}.\nabla)\nabla\tilde{\gamma} + \frac{\omega^2}{c^2}\tilde{\gamma}\tilde{\bar{I}}\right\} ds = -\frac{j\omega}{c^2}\int_\Gamma \bar{\bar{\Gamma}}.\tilde{\bar{I}}\, ds \tag{7.7-2}$$

$$\tilde{\bar{B}} = \tilde{\bar{H}} = \frac{1}{c}\int_\Gamma \tilde{\bar{I}} \times \nabla\tilde{\gamma}\, ds \tag{7.7-3}$$

7.7b Surface current (and associated values of $\tilde{\sigma}$ and $\tilde{\lambda}$)

$$\tilde{\bar{E}} = \tilde{\bar{D}} = \frac{1}{j\omega}\int_S \left\{(\tilde{\bar{K}}.\nabla)\nabla\tilde{\gamma} + \frac{\omega^2}{c^2}\tilde{\gamma}\tilde{\bar{K}}\right\} dS = -\frac{j\omega}{c^2}\int_S \bar{\bar{\Gamma}}.\tilde{\bar{K}}\, dS \tag{7.7-4}$$

$$\tilde{\bar{B}} = \tilde{\bar{H}} = \frac{1}{c}\int_S \tilde{\bar{K}} \times \nabla\tilde{\gamma}\, dS \tag{7.7-5}$$

7.7c <u>Volume current</u> (and associated values of $\tilde{\rho}$ and $\tilde{\sigma}$)

(1) at interior points of τ.

$$\tilde{\bar{E}} = \tilde{\bar{D}} = \frac{1}{j\omega}\left\{\int_{\tau} (-\operatorname{div}\tilde{\bar{J}})\nabla\tilde{\gamma}\,d\tau + \oint_{S_{1..n}^{\Sigma}} \nabla\tilde{\gamma}\,\tilde{\bar{J}}.d\bar{S} + \frac{\omega^2}{c^2}\int_{\tau} \tilde{\gamma}\,\tilde{\bar{J}}\,d\tau\right\} \quad (7.7\text{-}6)$$

$$= \frac{1}{j\omega}\left\{\int_{\tau} (\operatorname{curl}\tilde{\bar{J}}) \times \nabla\tilde{\gamma}\,d\tau + \oint_{S_{1..n}^{\Sigma}} \nabla\tilde{\gamma} \times (d\bar{S}\times\tilde{\bar{J}}) - 4\pi\tilde{\bar{J}}\right\} \quad (7.7\text{-}7)$$

$$= \frac{1}{j\omega}\left\{\lim_{\tau'\to 0}\int_{\tau-\tau'} (\tilde{\bar{J}}.\nabla)\nabla\tilde{\gamma}\,d\tau - \lim_{S'\to 0}\oint_{S'} \nabla\tilde{\gamma}\,\tilde{\bar{J}}.d\bar{S} + \frac{\omega^2}{c^2}\int_{\tau} \tilde{\gamma}\,\tilde{\bar{J}}\,d\tau\right\} \quad (7.7\text{-}8)$$

$$= -\frac{j\omega}{c^2}\lim_{\tau'\to 0}\int_{\tau-\tau'} \bar{\bar{\Gamma}}.\tilde{\bar{J}}\,d\tau - \frac{1}{j\omega}\lim_{S'\to 0}\oint_{S'} \nabla\tilde{\gamma}\,\tilde{\bar{J}}.d\bar{S} \quad (7.7\text{-}9)$$

(2) at exterior points of τ.

$$\tilde{\bar{E}} = \tilde{\bar{D}} = \frac{1}{j\omega}\left\{\int_{\tau} (-\operatorname{div}\tilde{\bar{J}})\nabla\tilde{\gamma}\,d\tau + \oint_{S_{1..n}^{\Sigma}} \nabla\tilde{\gamma}\,\tilde{\bar{J}}.d\bar{S} + \frac{\omega^2}{c^2}\int_{\tau} \tilde{\gamma}\,\tilde{\bar{J}}\,d\tau\right\} \quad (7.7\text{-}10)$$

$$= \frac{1}{j\omega}\left\{\int_{\tau} (\operatorname{curl}\tilde{\bar{J}}) \times \nabla\tilde{\gamma}\,d\tau + \oint_{S_{1..n}^{\Sigma}} \nabla\tilde{\gamma} \times (d\bar{S}\times\tilde{\bar{J}})\right\} \quad (7.7\text{-}11)$$

$$= \frac{1}{j\omega}\left\{\int_{\tau} (\tilde{\bar{J}}.\nabla)\nabla\tilde{\gamma}\,d\tau + \frac{\omega^2}{c^2}\int_{\tau} \tilde{\gamma}\,\tilde{\bar{J}}\,d\tau\right\} \quad (7.7\text{-}12)$$

$$= \frac{-j\omega}{c^2}\int_{\tau} \bar{\bar{\Gamma}}.\tilde{\bar{J}}\,d\tau \quad (7.7\text{-}13)$$

(3) at interior and exterior points of τ.

$$\tilde{\bar{B}} = \tilde{\bar{H}} = \frac{1}{c}\int_{\tau} \tilde{\gamma}\operatorname{curl}\tilde{\bar{J}}\,d\tau - \frac{1}{c}\oint_{S_{1..n}^{\Sigma}} \tilde{\gamma}\,(d\bar{S}\times\tilde{\bar{J}}) \quad (7.7\text{-}14)$$

$$= \frac{1}{c}\int_\tau \tilde{\bar{J}} \times \nabla\tilde{\gamma}\, d\tau \qquad (7.7\text{-}15)$$

7.7d <u>Surface doublet</u> (not necessarily normal to surface)

$$\tilde{\bar{E}} = \tilde{\bar{D}} = \int_S \left\{(\tilde{\bar{P}}'.\nabla)\nabla\tilde{\gamma} + \frac{\omega^2}{c^2}\tilde{\gamma}\,\tilde{\bar{P}}'\right\} dS = \frac{\omega^2}{c^2}\int_S \bar{\bar{\Gamma}}.\tilde{\bar{P}}'\, dS \qquad (7.7\text{-}16)$$

$$\tilde{\bar{B}} = \tilde{\bar{H}} = \frac{j\omega}{c}\int_S \tilde{\bar{P}}' \times \nabla\tilde{\gamma}\, dS \qquad (7.7\text{-}17)$$

7.7e <u>Volume doublet</u>

(1) at interior points of τ.

$$\tilde{\bar{E}} = \int_\tau (-\operatorname{div}\tilde{\bar{P}})\nabla\tilde{\gamma}\, d\tau + \oint_{S_{1..n}\Sigma} \nabla\tilde{\gamma}\,\tilde{\bar{P}}.d\bar{S} + \frac{\omega^2}{c^2}\int_\tau \tilde{\gamma}\,\tilde{\bar{P}}\, d\tau \qquad (7.7\text{-}18)$$

$$= \int_\tau (\operatorname{curl}\tilde{\bar{P}}) \times \nabla\tilde{\gamma}\, d\tau + \oint_{S_{1..n}\Sigma} \nabla\tilde{\gamma} \times (d\bar{S}\times\tilde{\bar{P}}) - 4\pi\tilde{\bar{P}} \qquad (7.7\text{-}19)$$

$$= \lim_{\tau'\to 0}\int_{\tau-\tau'} (\tilde{\bar{P}}.\nabla)\nabla\tilde{\gamma}\, d\tau + \lim_{S'\to 0}\oint_{S'} \nabla\tilde{\gamma}\,\tilde{\bar{P}}.d\bar{S} + \frac{\omega^2}{c^2}\int_\tau \tilde{\gamma}\,\tilde{\bar{P}}\, d\tau \qquad (7.7\text{-}20)$$

$$= \frac{\omega^2}{c^2}\lim_{\tau'\to 0}\int_{\tau-\tau'} \bar{\bar{\Gamma}}.\tilde{\bar{P}}\, d\tau - \lim_{S'\to 0}\oint_{S'} \nabla\tilde{\gamma}\,\tilde{\bar{P}}.d\bar{S} \qquad (7.7\text{-}21)$$

$$\tilde{\bar{D}} = \tilde{\bar{E}} + 4\pi\tilde{\bar{P}}$$

(2) at exterior points of τ.

$$\tilde{\bar{E}} = \tilde{\bar{D}} = \int_\tau (-\operatorname{div}\tilde{\bar{P}})\nabla\tilde{\gamma}\, d\tau + \oint_{S_{1..n}\Sigma} \nabla\tilde{\gamma}\,\tilde{\bar{P}}.d\bar{S} + \frac{\omega^2}{c^2}\int_\tau \tilde{\gamma}\,\tilde{\bar{P}}\, d\tau \qquad (7.7\text{-}22)$$

$$= \int_\tau (\operatorname{curl}\tilde{\bar{P}}) \times \nabla\tilde{\gamma}\, d\tau + \oint_{S_{1..n}\Sigma} \nabla\tilde{\gamma} \times (d\bar{S}\times\tilde{\bar{P}}) \qquad (7.7\text{-}23)$$

$$= \int_\tau (\tilde{\overline{P}}.\nabla)\nabla\tilde{\gamma}\, d\tau + \frac{\omega^2}{c^2}\int_\tau \tilde{\gamma}\,\tilde{\overline{P}}\, d\tau \qquad (7.7\text{-}24)$$

$$= \frac{\omega^2}{c^2}\int_\tau \overline{\overline{\Gamma}}.\tilde{\overline{P}}\, d\tau \qquad (7.7\text{-}25)$$

(3) at interior and exterior points of τ.

$$\tilde{\overline{B}} = \tilde{\overline{H}} = \frac{j\omega}{c}\int_\tau \tilde{\gamma}\,\text{curl}\,\tilde{\overline{P}}\, d\tau - \frac{j\omega}{c}\oint_{S_{1..n}\Sigma} \tilde{\gamma}\,(d\overline{S}\times\tilde{\overline{P}}) \qquad (7.7\text{-}26)$$

$$= \frac{j\omega}{c}\int_\tau \tilde{\overline{P}}\times\nabla\tilde{\gamma}\, d\tau \qquad (7.7\text{-}27)$$

7.7f <u>Surface whirl (not necessarily normal to surface)</u>

$$\tilde{\overline{E}} = \tilde{\overline{D}} = -\frac{j\omega}{c}\int_S \tilde{\overline{M}}'\times\nabla\tilde{\gamma}\, dS \qquad (7.7\text{-}28)$$

$$\tilde{\overline{B}} = \tilde{\overline{H}} = \int_S \left\{(\tilde{\overline{M}}'.\nabla)\nabla\tilde{\gamma} + \frac{\omega^2}{c^2}\tilde{\gamma}\,\tilde{\overline{M}}'\right\} dS = \frac{\omega^2}{c^2}\int_S \overline{\overline{\Gamma}}.\tilde{\overline{M}}'\, dS \qquad (7.7\text{-}29)$$

7.7g <u>Volume whirl</u>

(1) at interior and exterior points of τ.

$$\tilde{\overline{E}} = \tilde{\overline{D}} = -\frac{j\omega}{c}\int_\tau \tilde{\overline{M}}\times\nabla\tilde{\gamma}\, d\tau \qquad (7.7\text{-}30)$$

(2) at interior points of τ.

$$\tilde{\overline{B}} = \int_\tau (\text{curl}\,\tilde{\overline{M}})\times\nabla\tilde{\gamma}\, d\tau + \oint_{S_{1..n}\Sigma} \nabla\tilde{\gamma}\times(d\overline{S}\times\tilde{\overline{M}}) \qquad (7.7\text{-}31)$$

$$= \int_\tau (-\,\text{div}\,\tilde{\overline{M}})\nabla\tilde{\gamma}\, d\tau + \oint_{S_{1..n}\Sigma} \nabla\tilde{\gamma}\,\tilde{\overline{M}}.d\overline{S} + \frac{\omega^2}{c^2}\int_\tau \tilde{\gamma}\,\tilde{\overline{M}}\, d\tau + 4\pi\tilde{\overline{M}} \qquad (7.7\text{-}32)$$

$$= -\operatorname{grad}\int_{\tau} \widetilde{\overline{\overline{M}}}.\nabla\tilde{\Upsilon}\, d\tau + \frac{\omega^2}{c^2}\int_{\tau} \tilde{\Upsilon}\,\widetilde{\overline{\overline{M}}}\, d\tau + 4\pi\widetilde{\overline{\overline{M}}} \qquad (7.7\text{-}33)$$

$$= \lim_{\tau'\to 0}\int_{\tau-\tau'} (\widetilde{\overline{\overline{M}}}.\nabla)\nabla\tilde{\Upsilon}\, d\tau + \lim_{S'\to 0}\oint_{S'} d\bar{S} \times (\widetilde{\overline{\overline{M}}}\times\nabla\tilde{\Upsilon}) + \frac{\omega^2}{c^2}\int_{\tau} \tilde{\Upsilon}\,\widetilde{\overline{\overline{M}}}\, d\tau \qquad (7.7\text{-}34)$$

$$= \frac{\omega^2}{c^2}\lim_{\tau'\to 0}\int_{\tau-\tau'} \overline{\overline{\Gamma}}.\widetilde{\overline{\overline{M}}}\, d\tau + \lim_{S'\to 0}\oint_{S'} d\bar{S} \times (\widetilde{\overline{\overline{M}}}\times\nabla\tilde{\Upsilon}) \qquad (7.7\text{-}35)$$

$$\widetilde{\overline{\overline{H}}} = \widetilde{\overline{\overline{B}}} - 4\pi\widetilde{\overline{\overline{M}}}$$

(3) at exterior points of τ.

$$\widetilde{\overline{\overline{B}}} = \widetilde{\overline{\overline{H}}} = \int_{\tau} (\operatorname{curl}\widetilde{\overline{\overline{M}}}) \times \nabla\tilde{\Upsilon}\, d\tau + \oint_{S_{1..n}^{\Sigma}} \nabla\tilde{\Upsilon} \times (d\bar{S}\times\widetilde{\overline{\overline{M}}}) \qquad (7.7\text{-}36)$$

$$= \int_{\tau} (-\operatorname{div}\widetilde{\overline{\overline{M}}})\nabla\tilde{\Upsilon}\, d\tau + \oint_{S_{1..n}^{\Sigma}} \nabla\tilde{\Upsilon}\,\widetilde{\overline{\overline{M}}}.d\bar{S} + \frac{\omega^2}{c^2}\int_{\tau} \tilde{\Upsilon}\,\widetilde{\overline{\overline{M}}}\, d\tau \qquad (7.7\text{-}37)$$

$$= -\operatorname{grad}\int_{\tau} \widetilde{\overline{\overline{M}}}.\nabla\tilde{\Upsilon}\, d\tau + \frac{\omega^2}{c^2}\int_{\tau} \tilde{\Upsilon}\,\widetilde{\overline{\overline{M}}}\, d\tau \qquad (7.7\text{-}38)$$

$$= \int_{\tau} (\widetilde{\overline{\overline{M}}}.\nabla)\nabla\tilde{\Upsilon}\, d\tau + \frac{\omega^2}{c^2}\int_{\tau} \tilde{\Upsilon}\,\widetilde{\overline{\overline{M}}}\, d\tau \qquad (7.7\text{-}39)$$

$$= \frac{\omega^2}{c^2}\int_{\tau} \overline{\overline{\Gamma}}.\widetilde{\overline{\overline{M}}}\, d\tau \qquad (7.7\text{-}40)$$

It should be borne in mind that when the formulae of Sec. 7.7a to 7.7c are applied to incomplete current sources, the $\widetilde{\overline{\overline{E}}}$ fields, as evaluated, necessarily include contributions from such point, line or surface singlets as would develop were there no transgression of the boundaries by the associated currents.

EXERCISES

7-16. Prove the relationship (7.6-10) for the most general system of sources.

7-17. Confirm equations (7.6-21) to (7.6-24).

7-18. Show that
$$\operatorname{div} \tilde{\bar{D}}^* = 4\pi\tilde{\rho}^*$$
$$\operatorname{curl} \tilde{\bar{E}}^* = \frac{j\omega}{c} \tilde{\bar{B}}^*$$
$$\operatorname{div} \tilde{\bar{B}}^* = 0$$
$$\operatorname{curl} \tilde{\bar{H}}^* = \frac{4\pi}{c} \tilde{\bar{J}}^* - \frac{j\omega}{c} \tilde{\bar{D}}^*$$

7-19. Show that the flux of $\tilde{\bar{B}}$ through a closed contour Γ_1 due to a current of constant magnitude $\tilde{I}$ in a closed contour Γ_2 is equal to the flux of $\tilde{\bar{B}}$ through Γ_2 due to $\tilde{I}$ in Γ_1. [For this purpose, an arbitrary positive sense of current in Γ_1 is defined to produce a positive flux through Γ_2.]

7-20. Make use of the expansion of $(\tilde{\bar{J}}.\nabla)\nabla\tilde{\gamma}$ in Ex.7-3., p. 632 to show that at exterior points of a volume distribution of density $\tilde{\bar{J}}$

$$\tilde{\bar{E}} = \frac{1}{j\omega} \int_\tau \left\{ -\frac{\tilde{\bar{J}}_t}{r^3} + 2\frac{\tilde{\bar{J}}_r}{r^3} - \frac{j\omega}{cr^2}\tilde{\bar{J}}_t + \frac{2j\omega}{cr^2}\tilde{\bar{J}}_r + \frac{\omega^2}{c^2 r}\tilde{\bar{J}}_t \right\} e^{-j\omega r/c}\, d\tau$$

where $\tilde{\bar{J}}_r = \hat{r}\,\tilde{\bar{J}}.\hat{r}$ and $\tilde{\bar{J}}_t = \tilde{\bar{J}} - \tilde{\bar{J}}_r$

Show that the same value of $\tilde{\bar{E}}$ obtains at both exterior and interior points (for the same limiting process) when the singlet distribution is replaced by a doublet distribution of density $\tilde{\bar{P}} = \frac{1}{j\omega}\tilde{\bar{J}}$.

7-21. Show that the exponential potential of a tangentially-orientated, non-uniform, open surface distribution of doublets of density $\tilde{\bar{P}}'$ is equal at exterior points to the potential of a peripheral singlet source of density $\tilde{\bar{P}}'.\hat{n}'$ together with a singlet surface source of density $-\operatorname{divs}\tilde{\bar{P}}'$. $[\hat{n}' = \hat{s} \times \hat{n}]$

7-22. Show that an open non-uniform surface distribution of normally-orientated whirls of density $\tilde{M}'$ gives rise at exterior points to an exponential vector potential which is equal to that deriving from a peripheral line source of density $\tilde{M}'\,\hat{s}$ together with a singlet surface source of density $(\operatorname{grads}\tilde{M}') \times \hat{n}$.

7-23. Show that the exponential vector potential of a tangentially-orientated, non-uniform, plane open surface distribution of whirls of density $\tilde{\bar{M}}'$ is duplicated at exterior points by that of a peripheral line source of density $\tilde{\bar{M}}' \times \hat{n}'$ together with a surface singlet of density curl$_s$ $\tilde{\bar{M}}'$ and a double layer surface source of density $\tilde{\bar{M}}' \times \hat{n}$.

7-24. A time-harmonic point doublet is located at the origin of spherical coordinates and aligned with the z axis ($\theta = 0$). Show that the $\tilde{\bar{E}}$ field at the point R, θ, Φ is elliptically polarised and that

$$E_{R_{max}} \,/\, E_{\theta_{max}} = \frac{2c}{R\omega}\cot\theta\left(1 + \frac{c^2}{R^2\omega^2}\right)^{\frac{1}{2}} \bigg/ \left(1 - \frac{c^2}{R^2\omega^2} + \frac{c^4}{R^4\omega^4}\right)^{\frac{1}{2}}$$

7-25. Suppose that $\tilde{\bar{J}}$ is well-behaved within the region τ bounded by $S_{1..n}\Sigma$ and is zero outside it. Let the value of $\tilde{\sigma}$ upon $S_{1..n}\Sigma$ be defined by $\tilde{\bar{J}}.\hat{n} = j\omega\tilde{\sigma}$. Then the source system is complete and $\tilde{\bar{E}}$ is given at interior points of τ by (7.7-9), viz

$$\tilde{\bar{E}}_o = -\frac{j\omega}{c^2}\lim_{\tau'\to 0}\int_{\tau-\tau'} \bar{\bar{\Gamma}}.\tilde{\bar{J}}\,d\tau - \frac{1}{j\omega}\lim_{S'\to 0}\oint_{S'} \nabla\tilde{\gamma}\,\tilde{\bar{J}}.d\bar{S}$$

By identifying $\tilde{\bar{E}}$ with $\tilde{\bar{F}}$ in equation (6.4-12), arrive at the following alternative expression for $\tilde{\bar{E}}_o$.

$$\tilde{\bar{E}}_o = \frac{1}{4\pi}\oint_{S_{1..n}\Sigma} (\nabla\times\bar{\bar{\Gamma}}).(d\bar{S}\times\tilde{\bar{E}}) - \bar{\bar{\Gamma}}.(d\bar{S}\times\text{curl}\,\tilde{\bar{E}})$$

$$-\frac{j\omega}{c^2}\lim_{\tau'\to 0}\int_{\tau-\tau'} \bar{\bar{\Gamma}}.\tilde{\bar{J}}\,d\tau - \frac{1}{j\omega}\lim_{S'\to 0}\oint_{S'} \nabla\tilde{\gamma}\,\tilde{\bar{J}}.d\bar{S}$$

It is evident that in the present case the surface integral over $S_{1..n}\Sigma$ is zero. Provide an independent proof of this.

[Hint: Show that when volume integration is carried out over the regions bounded by $S_{1..n}$ and by Σ and a surface at infinity, the previous equation is replaced by

$$0 = \oint_{S_{1..n}\Sigma,\,\infty} (\nabla\times\bar{\bar{\Gamma}}).(d\bar{S}\times\tilde{\bar{E}}) - \bar{\bar{\Gamma}}.(d\bar{S}\times\text{curl}\,\tilde{\bar{E}})$$

where the positive normals over $S_{1..n}\Sigma$ are directed into τ and that over the surface at infinity is directed outwards.

By combining the two equations and evaluating $\Delta(\hat{\bar{n}}\times\tilde{\bar{E}})$ and $\Delta(\hat{\bar{n}} \times \text{curl } \tilde{\bar{E}})$, eliminate the local surface integrals. Then show that the surface integral at infinity is zero by making use, inter alia, of the expansion in Ex.7-3.]

7-26. Extend the analysis of the previous exercise to the case in which surface currents are permitted upon $S_{1..n}\Sigma$, provided that $\tilde{\bar{J}}.\hat{\bar{n}} - \text{divs } \tilde{\bar{K}} = j\omega\tilde{\sigma}$. (The surface integrals should be evaluated just inside τ.)

7-27. By working along the lines of Ex.7-25/6 show that $\tilde{\bar{B}}$ may be expressed at interior points of τ as

$$\tilde{\bar{B}}_o = \frac{1}{4\pi}\oint_{S_{1..n}\Sigma} (\nabla\times\bar{\bar{\Gamma}}).(d\bar{S}\times\tilde{\bar{B}}) - \bar{\bar{\Gamma}}.(d\bar{S} \times \text{curl } \tilde{\bar{B}})$$

$$+ \frac{1}{c}\lim_{\tau'\to 0}\int_{\tau-\tau'} \bar{\bar{\Gamma}}.\text{curl } \tilde{\bar{J}}\, d\tau - \frac{c}{\omega^2}\lim_{S'\to 0}\oint_{S'} \nabla\tilde{\gamma}\, \text{curl } \tilde{\bar{J}}.d\bar{S}$$

Show further that this expression may be transformed into (7.7-14) when $\tilde{\bar{J}}.\hat{\bar{n}} = j\omega\tilde{\sigma}$ and into (7.7-14) plus (7.7-5) when $\tilde{\bar{J}}.\hat{\bar{n}} - \text{divs } \tilde{\bar{K}} = j\omega\tilde{\sigma}$.

7-28. If $\tilde{\bar{J}} = \bar{0}$ and $\tilde{\bar{M}}$ is well-behaved throughout τ and zero outside it, show that $\tilde{\bar{B}}$ may be expressed at interior points of τ as

$$\tilde{\bar{B}}_o = \frac{1}{4\pi}\oint_{S_{1..n}\Sigma} (\nabla\times\bar{\bar{\Gamma}}).(d\bar{S}\times\tilde{\bar{B}}) - \bar{\bar{\Gamma}}.(d\bar{S} \times \text{curl } \tilde{\bar{B}})$$

$$+ \lim_{\tau'\to 0}\int_{\tau-\tau'} \bar{\bar{\Gamma}}.\text{curl curl } \tilde{\bar{M}}\, d\tau - \frac{c^2}{\omega^2}\lim_{S'\to 0}\oint_{S'} \nabla\tilde{\gamma}\, \text{curl curl } \tilde{\bar{M}}.d\bar{S}$$

Transform this expression into (7.7-31) by a procedure similar to that adopted in the previous exercises.

7.8 The Diffraction Integrals

We now show that in a finite, source-free region of space $\tilde{\bar{E}}$ and $\tilde{\bar{B}}$ may be expressed as the sum of integrals taken over the bounding surfaces. Expressions of this type are known as diffraction integrals.

It follows from equations (7.6-17), (7.6-18) and (7.6-20) that in a source-free region ($\tilde{\bar{D}} = \tilde{\bar{E}}$, $\tilde{\bar{H}} = \tilde{\bar{B}}$)

$$\text{div } \tilde{\bar{E}} = 0; \quad \text{curl curl } \tilde{\bar{E}} - \frac{\omega^2}{c^2}\tilde{\bar{E}} = \bar{0}; \quad \left(\nabla^2 + \frac{\omega^2}{c^2}\right)\tilde{\bar{E}} = \bar{0} \qquad (7.8\text{-}1)$$

Then on substituting $\tilde{\bar{E}}$ for the general vector point function $\tilde{\bar{F}}$ in (6.4-2) or (6.4-3) with $\tilde{k}^2 = \frac{\omega^2}{c^2}$ we obtain

$$4\pi\tilde{\bar{E}}_o = \oint_{S_{1..n}\Sigma} \left\{\nabla\tilde{\gamma}\times(d\bar{S}\times\tilde{\bar{E}}) + \frac{j\omega}{c}\tilde{\gamma}(d\bar{S}\times\tilde{\bar{B}}) - \nabla\tilde{\gamma}\, d\bar{S}.\tilde{\bar{E}}\right\} \tag{7.8-2}$$

Similarly

$$\text{div}\,\tilde{\bar{B}} = 0; \quad \text{curl curl}\,\tilde{\bar{B}} - \frac{\omega^2}{c^2}\tilde{\bar{B}} = \bar{0}; \quad \left(\nabla^2 + \frac{\omega^2}{c^2}\right)\tilde{\bar{B}} = \bar{0} \tag{7.8-3}$$

and

$$4\pi\tilde{\bar{B}}_o = \oint_{S_{1..n}\Sigma} \left\{\nabla\tilde{\gamma}\times(d\bar{S}\times\tilde{\bar{B}}) - \frac{j\omega}{c}\tilde{\gamma}(d\bar{S}\times\tilde{\bar{E}}) - \nabla\tilde{\gamma}\, d\bar{S}.\tilde{\bar{B}}\right\} \tag{7.8-4}$$

It then follows from equations (7.2-15) and (7.3-23) that

$$4\pi\tilde{\bar{E}}_o = -\,\text{curl}\oint_{S_{1..n}\Sigma}\tilde{\gamma}\,(d\bar{S}\times\tilde{\bar{E}}) + \frac{j\omega}{c}\oint_{S_{1..n}\Sigma}\tilde{\gamma}\,(d\bar{S}\times\tilde{\bar{B}}) + \text{grad}\oint_{S_{1..n}\Sigma}\tilde{\gamma}\,(d\bar{S}.\tilde{\bar{E}}) \tag{7.8-5}$$

and

$$4\pi\tilde{\bar{B}}_o = -\,\text{curl}\oint_{S_{1..n}\Sigma}\tilde{\gamma}\,(d\bar{S}\times\tilde{\bar{B}}) - \frac{j\omega}{c}\oint_{S_{1..n}\Sigma}\tilde{\gamma}\,(d\bar{S}\times\tilde{\bar{E}}) + \text{grad}\oint_{S_{1..n}\Sigma}\tilde{\gamma}\,(d\bar{S}.\tilde{\bar{B}}) \tag{7.8-6}$$

In retarded potential form these equations become

$$4\pi\bar{E}_o = -\,\text{curl}\oint_{S_{1..n}\Sigma}\frac{1}{r}\,[\hat{\bar{n}}\times\bar{E}]\,dS - \frac{1}{c}\frac{\partial}{\partial t}\oint_{S_{1..n}\Sigma}\frac{1}{r}\,[-\hat{\bar{n}}\times\bar{B}]\,dS - \text{grad}\oint_{S_{1..n}\Sigma}\frac{1}{r}\,[-\hat{\bar{n}}.\bar{E}]\,dS \tag{7.8-7}$$

and

$$4\pi\bar{B}_o = \text{curl}\oint_{S_{1..n}\Sigma}\frac{1}{r}\,[-\hat{\bar{n}}\times\bar{B}]\,dS - \frac{1}{c}\frac{\partial}{\partial t}\oint_{S_{1..n}\Sigma}\frac{1}{r}\,[\hat{\bar{n}}\times\bar{E}]\,dS - \text{grad}\oint_{S_{1..n}\Sigma}\frac{1}{r}\,[-\hat{\bar{n}}.\bar{B}]\,dS \tag{7.8-8}$$

Equations (7.8-7) and (7.8-8) are known as the Larmor-Tedone formulae.

We may obtain the following expressions for $\tilde{\bar{E}}_o$ and $\tilde{\bar{B}}_o$ solely in terms of the tangential components of $\tilde{\bar{E}}$ and $\tilde{\bar{B}}$ upon $S_{1..n}\Sigma$, either by substitution of $\tilde{\bar{B}}$ and $\tilde{\bar{E}}$ for $\tilde{\bar{F}}$ in equations (6.4-7/8), or by transformation of the final terms in equations (7.8-5/6).

$$4\pi\tilde{\bar{E}}_o = -\operatorname{curl}\oint_{S_{1..n}\Sigma}\tilde{\gamma}\,(d\bar{S}\times\tilde{\bar{E}}) + \frac{j\omega}{c}\oint_{S_{1..n}\Sigma}\tilde{\gamma}\,(d\bar{S}\times\tilde{\bar{B}}) - \frac{c}{j\omega}\oint_{S_{1..n}\Sigma}((d\bar{S}\times\tilde{\bar{B}}).\nabla)\,\nabla\tilde{\gamma} \tag{7.8-9}$$

$$4\pi\tilde{\bar{B}}_o = -\operatorname{curl}\oint_{S_{1..n}\Sigma}\tilde{\gamma}\,(d\bar{S}\times\tilde{\bar{B}}) - \frac{j\omega}{c}\oint_{S_{1..n}\Sigma}\tilde{\gamma}\,(d\bar{S}\times\tilde{\bar{E}}) + \frac{c}{j\omega}\oint_{S_{1..n}\Sigma}((d\bar{S}\times\tilde{\bar{E}}).\nabla)\,\nabla\tilde{\gamma} \tag{7.8-10}$$

whence, from equation (6.4-10),

$$4\pi\tilde{\bar{E}}_o = \oint_{S_{1..n}\Sigma}(\nabla\times\bar{\bar{\Gamma}}).(d\bar{S}\times\tilde{\bar{E}}) + \frac{j\omega}{c}\oint_{S_{1..n}\Sigma}\bar{\bar{\Gamma}}.(d\bar{S}\times\tilde{\bar{B}}) \tag{7.8-11}$$

$$4\pi\tilde{\bar{B}}_o = \oint_{S_{1..n}\Sigma}(\nabla\times\bar{\bar{\Gamma}}).(d\bar{S}\times\tilde{\bar{B}}) - \frac{j\omega}{c}\oint_{S_{1..n}\Sigma}\bar{\bar{\Gamma}}.(d\bar{S}\times\tilde{\bar{E}}) \tag{7.8-12}$$

From equation (7.3-25) we see that (7.8-9) and (7.8-10) may be replaced by

$$4\pi\tilde{\bar{E}}_o = -\operatorname{curl}\oint_{S_{1..n}\Sigma}\tilde{\gamma}\,(d\bar{S}\times\tilde{\bar{E}}) + \frac{j\omega}{c}\oint_{S_{1..n}\Sigma}\tilde{\gamma}\,(d\bar{S}\times\tilde{\bar{B}}) - \frac{c}{j\omega}\operatorname{grad}\operatorname{div}\oint_{S_{1..n}\Sigma}\tilde{\gamma}\,(d\bar{S}\times\tilde{\bar{B}}) \tag{7.8-13}$$

$$4\pi\tilde{\bar{B}}_o = -\operatorname{curl}\oint_{S_{1..n}\Sigma}\tilde{\gamma}\,(d\bar{S}\times\tilde{\bar{B}}) - \frac{j\omega}{c}\oint_{S_{1..n}\Sigma}\tilde{\gamma}\,(d\bar{S}\times\tilde{\bar{E}}) + \frac{c}{j\omega}\operatorname{grad}\operatorname{div}\oint_{S_{1..n}\Sigma}\tilde{\gamma}\,(d\bar{S}\times\tilde{\bar{E}}) \tag{7.8-14}$$

These expressions, in turn, may be easily transformed into

$$4\pi\tilde{\bar{E}}_o = -\operatorname{curl}\oint_{S_{1..n}\Sigma}\tilde{\gamma}\,(d\bar{S}\times\tilde{\bar{E}}) - \frac{c}{j\omega}\operatorname{curl}\operatorname{curl}\oint_{S_{1..n}\Sigma}\tilde{\gamma}\,(d\bar{S}\times\tilde{\bar{B}}) \tag{7.8-15}$$

$$4\pi\tilde{\bar{B}}_o = -\operatorname{curl}\oint_{S_{1..n}\Sigma}\tilde{\gamma}\,(d\bar{S}\times\tilde{\bar{B}}) + \frac{c}{j\omega}\operatorname{curl}\operatorname{curl}\oint_{S_{1..n}\Sigma}\tilde{\gamma}\,(d\bar{S}\times\tilde{\bar{E}}) \tag{7.8-16}$$

We will have occasion subsequently to make use of variants of equations (7.8-5/6) which substitute surface divergences for the normal components of $\tilde{\bar{E}}$ and $\tilde{\bar{B}}$ in the final terms.

Let $\tilde{\bar{E}} = \hat{\xi}\,\tilde{E}_\xi + \hat{\zeta}\,\tilde{E}_\zeta + \hat{n}\,E_n$, where ξ and ζ are surface curvilinear coordinates and $\hat{\xi}\,\hat{\zeta}\,\hat{n}$ form a right-handed set.

Then

$$\hat{n} \times \tilde{\bar{E}} = \hat{\zeta}\,\tilde{E}_\xi - \hat{\xi}\,\tilde{E}_\zeta$$

whence

$$\text{divs}\,(\hat{n}\times\tilde{\bar{E}}) = \frac{1}{h_\xi h_\zeta}\left\{-\frac{\partial}{\partial\xi}(h_\zeta\tilde{E}_\zeta) + \frac{\partial}{\partial\zeta}(h_\xi\tilde{E}_\xi)\right\} \qquad \text{(from (2.12-25))}$$

$$= -\hat{n}.\text{curls}\,\tilde{\bar{E}} \qquad \text{(from (2.12-23))}$$

$$= -\hat{n}.\left(\text{curl}\,\tilde{\bar{E}} - \hat{n}\times\frac{\partial\tilde{\bar{E}}}{\partial n}\right)$$

$$= -\hat{n}.\text{curl}\,\tilde{\bar{E}}$$

$$\text{or}\quad \text{divs}\,(\hat{n}\times\tilde{\bar{E}}) = \frac{j\omega}{c}\,\hat{n}.\tilde{\bar{B}} \qquad (7.8\text{-}17)$$

$$\text{Similarly,}\quad \text{divs}\,(\hat{n}\times\tilde{\bar{B}}) = -\frac{j\omega}{c}\,\hat{n}.\tilde{\bar{E}} \qquad (7.8\text{-}18)$$

Then from equations (7.8-5) and (7.8-6)

$$4\pi\tilde{\bar{E}}_o = -\,\text{curl}\oint_{S_{1..n}\Sigma}\tilde{\gamma}\,(d\bar{S}\times\tilde{\bar{E}}) + \frac{j\omega}{c}\oint_{S_{1..n}\Sigma}\tilde{\gamma}\,(d\bar{S}\times\tilde{\bar{B}}) - \frac{c}{j\omega}\,\text{grad}\oint_{S_{1..n}\Sigma}\tilde{\gamma}\,\text{divs}\,(\hat{n}\times\tilde{\bar{B}})\,dS \qquad (7.8\text{-}19)$$

$$4\pi\tilde{\bar{B}}_o = -\,\text{curl}\oint_{S_{1..n}\Sigma}\tilde{\gamma}\,(d\bar{S}\times\tilde{\bar{B}}) - \frac{j\omega}{c}\oint_{S_{1..n}\Sigma}\tilde{\gamma}\,(d\bar{S}\times\tilde{\bar{E}}) + \frac{c}{j\omega}\,\text{grad}\oint_{S_{1..n}\Sigma}\tilde{\gamma}\,\text{divs}\,(\hat{n}\times\tilde{\bar{E}})\,dS \qquad (7.8\text{-}20)$$

When all sources lie within the closed surfaces $S_{1..n}$ the surface integrals over Σ may be deleted in each of the above expressions. This is a consequence of the fact that the surface integrals over Σ vanish at infinity (See Ex.7-29/30., p. 654). Hence if O be exterior to the regions containing sources, then $\tilde{\bar{E}}_o$ and $\tilde{\bar{B}}_o$ may be expressed in terms of integrals over the surfaces bounding these sources. A particular case is that in which O lies within a source-free region bounded by a single surface.

We now enquire whether the exterior sources which give rise to $\tilde{\bar{E}}$ and $\tilde{\bar{B}}$ at interior points of τ can be replaced by surface sources upon $S_{1..n}^{\Sigma}$.

Suppose that we postulate a macroscopic current source of density $\tilde{\bar{K}} = -\frac{c}{4\pi}(\hat{n}\times\tilde{\bar{B}})$ upon the surfaces. From equation (7.6-8) the associated scalar surface density is $-\frac{1}{j\omega}\operatorname{divs}\tilde{\bar{K}}$ so that the value of $\tilde{\bar{E}}_o$ deriving from the combination is given by

$$\tilde{\bar{E}}_o = -\operatorname{grad}\oint_{S_{1..n}^{\Sigma}} \tilde{\gamma}\left(-\frac{1}{j\omega}\operatorname{divs}\tilde{\bar{K}}\right) dS - \frac{j\omega}{c}\frac{1}{c}\oint_{S_{1..n}^{\Sigma}} \tilde{\gamma}\,\tilde{\bar{K}}\, dS$$

whence

$$4\pi\tilde{\bar{E}}_o = -\frac{c}{j\omega}\operatorname{grad}\oint_{S_{1..n}^{\Sigma}} \tilde{\gamma}\operatorname{divs}(\hat{n}\times\tilde{\bar{B}})\, dS + \frac{j\omega}{c}\oint_{S_{1..n}^{\Sigma}} \tilde{\gamma}\,(d\bar{S}\times\tilde{\bar{B}})$$

Thus we have accounted for two of the three terms in (7.8-19).

Similarly, the associated value of $\tilde{\bar{B}}$ is given by

$$\tilde{\bar{B}}_o = \operatorname{curl}\frac{1}{c}\oint_{S_{1..n}^{\Sigma}} \tilde{\gamma}\left\{-\frac{c}{4\pi}(\hat{n}\times\tilde{\bar{B}})\right\} dS$$

whence

$$4\pi\tilde{\bar{B}}_o = -\operatorname{curl}\oint_{S_{1..n}^{\Sigma}} \tilde{\gamma}\,(d\bar{S}\times\tilde{\bar{B}})$$

This accounts for the first term of (7.8-20).

It is not possible, with the surface sources at our disposal, to account for the remaining terms, nor indeed is there any reason to suppose that it should be possible. Nevertheless, this can be accomplished if we are prepared to accept the more complex model considered in Ex.5-93., p. 559 (See Ex.7-33., p. 656). However, this is an ad hoc procedure which, arguably, is of doubtful value in any other context.

In the practical application of the diffraction formulae we are usually concerned with evaluation of the integrals over a single surface upon which, in part at least, the values of $\tilde{\bar{\bar{E}}}$ and $\tilde{\bar{\bar{B}}}$ or their components are not accurately known. In attempting to arrive at reasonable values it must be borne in mind that the field vectors are coupled via the relationships (7.8-17/18) and cannot be assigned independent values. Again, it should not be forgotten that the diffraction integrals were developed on the assumption of continuity of $\tilde{\bar{\bar{E}}}$ and $\tilde{\bar{\bar{B}}}$ and their derivatives. The error which may result from a neglect of this fact becomes evident if the surface S is split into two parts with a common boundary Γ, and it is supposed that $\tilde{\bar{\bar{E}}}$ and $\tilde{\bar{\bar{B}}}$ are zero over one part and non-zero over the other. On applying equations (7.8-5/6) to this case we find that the computed values of $\tilde{\bar{\bar{E}}}$ and $\tilde{\bar{\bar{B}}}$ fail to satisfy Maxwell's equations whenever the components of $\tilde{\bar{\bar{E}}}$ and $\tilde{\bar{\bar{B}}}$ parallel to Γ are discontinuous through Γ. (Ex.7-31., p. 655). It is easily seen that this is the condition for an unbounded surface divergence upon Γ, and consequently unbounded normal values of $\tilde{\bar{\bar{E}}}$ and $\tilde{\bar{\bar{B}}}$. On the other hand, an application of (7.8-9/10) or (7.8-11/12) to the case under consideration yields values of $\tilde{\bar{\bar{E}}}$ and $\tilde{\bar{\bar{B}}}$ which satisfy Maxwell's equations (See Ex.7-32., p. 655). These expressions do not involve the normal components of $\tilde{\bar{\bar{E}}}$ and $\tilde{\bar{\bar{B}}}$. It would appear that if a case should arise in which the tangential values of $\tilde{\bar{\bar{E}}}$ and $\tilde{\bar{\bar{B}}}$ are bounded while the components parallel to Γ are discontinuous through Γ, the expression for $\tilde{\bar{\bar{E}}}$ and $\tilde{\bar{\bar{B}}}$ involving only $\underline{\hat{n}} \times \tilde{\bar{\bar{E}}}$ and $\underline{\hat{n}} \times \tilde{\bar{\bar{E}}}$ continue to be valid.

Problems associated with discontinuities are common in the applications of diffraction theory because the surface of integration is often constrained to coincide in part with discontinuities of the source system.

The subject is a difficult one and further considerations are beyond the scope of this work. The interested reader is referred to the extensive literature on the subject[1].

1. See, for example, the review by C.J. Bouwkamp, "Diffraction Theory" in "Reports on Progress in Physics", 17, pp. 35-100, The Physical Society, London (1954).

EXERCISES

7-29. Deduce the value of the surface integral over Σ in equation (7.8-2), as Σ recedes to infinity, in the following way.

Let Σ be a spherical surface of radius R centred upon a point P located at finite distance from a system of sources and from a point of evaluation O. If $\Delta\tilde{\tilde{E}}$ and $\Delta\tilde{\tilde{B}}$ are the incremental fields deriving from a source element at Q (supposed complete), in accordance with expressions developed in Sec. 5.16 or 5.22, show that

$$\oint_\Sigma \{\Delta\tilde{\gamma}\times(d\bar{S}\times\Delta\tilde{\tilde{E}})-\nabla\tilde{\gamma}\; d\bar{S}.\Delta\tilde{\tilde{E}}\} = \frac{j\omega}{c}\oint_\Sigma \frac{R}{r^2}\, e^{-j\omega r/c}\, \Delta\tilde{\tilde{E}}\, dS + ---$$

and

$$\oint_\Sigma \frac{j\omega}{c}\tilde{\gamma}\,(d\bar{S}\times\Delta\tilde{\tilde{B}}) = -\frac{j\omega}{c}\oint_\Sigma \frac{r'}{rR}\, e^{-j\omega r/c}\, \Delta\tilde{\tilde{E}}\, dS + ---$$

where r' is distance measured from Q.

Hence show that the total surface integral for a complete source vanishes at infinity as $\frac{1}{R}$.

Extend the argument to the surface integral over Σ in equation (7.8-4).

7-30. Let a system of sources be contained within a region τ bounded by the closed surface Σ such that the macroscopic densities $\tilde{\rho}$, $\tilde{\tilde{J}}$, $\tilde{\tilde{P}}$ and $\tilde{\tilde{M}}$ are everywhere continuous and fall smoothly to zero before reaching the bounding surface; and let O be an interior point of τ. By substitution of $\tilde{\tilde{E}}$ for $\tilde{\tilde{F}}$ in equation (6.4-2) show that

$$\tilde{\tilde{E}}_o = \frac{1}{4\pi}\oint_\Sigma \left\{\nabla\tilde{\gamma}\times(d\bar{S}\times\tilde{\tilde{E}}) + \frac{j\omega}{c}\tilde{\gamma}(d\bar{S}\times\tilde{\tilde{B}})-\nabla\tilde{\gamma}\; d\bar{S}.\tilde{\tilde{E}}\right\}$$

$$- \operatorname{grad}\int_\tau \tilde{\gamma}\,(\tilde{\rho} - \operatorname{div}\tilde{\tilde{P}})\, d\tau - \frac{j\omega}{c^2}\int_\tau \tilde{\gamma}\,(\tilde{\tilde{J}} + j\omega\tilde{\tilde{P}} + c\operatorname{curl}\tilde{\tilde{M}})\, d\tau$$

Hence deduce that the surface integral is zero for all configurations of the surface provided that it encloses the sources and the point O, and observe that O is not constrained, as in the previous exercise, to remain at finite distance from the source system as Σ recedes to infinity.

Now develop a similar treatment for $\tilde{\tilde{B}}$, and devise a plausible means of including discontinuous sources of all types in the argument. In this way show that if all sources are contained within a region bounded by the closed surface S, the $\tilde{\tilde{E}}$ and $\tilde{\tilde{B}}$ fields at exterior points may be expressed in terms of $\hat{n}\times\tilde{\tilde{E}}$ and $\hat{n}\times\tilde{\tilde{B}}$ upon S.

7-31. Let equations (7.8-5/6) be applied to a single surface S which encloses a source-free region τ, and let $\tilde{\bar{E}}$ and $\tilde{\bar{B}}$ be well-behaved upon S. Confirm that at interior points of τ

$$\operatorname{div} \tilde{\bar{E}} = \operatorname{div} \tilde{\bar{B}} = 0$$

$$\operatorname{curl} \tilde{\bar{E}} = -\frac{j\omega}{c} \tilde{\bar{B}} \quad ; \quad \operatorname{curl} \tilde{\bar{B}} = \frac{j\omega}{c} \tilde{\bar{E}}$$

Now suppose that $\tilde{\bar{E}}'$ and $\tilde{\bar{B}}'$ are defined by (7.8-5/6) when the integration is carried out over the portion S_A of S having the boundary Γ. Show that in these circumstances

$$\operatorname{div} \tilde{\bar{E}}' = \frac{j\omega}{4\pi c} \oint_{\Gamma} \tilde{\gamma}\, \tilde{\bar{B}}.d\bar{r} \quad ; \quad \operatorname{div} \tilde{\bar{B}}' = -\frac{j\omega}{4\pi c} \oint_{\Gamma} \tilde{\gamma}\, \tilde{\bar{E}}.d\bar{r}$$

$$\operatorname{curl} \tilde{\bar{E}}' = -\frac{j\omega}{c} \tilde{\bar{B}}' - \frac{1}{4\pi} \operatorname{grad} \oint_{\Gamma} \tilde{\gamma}\, \tilde{\bar{E}}.d\bar{r}$$

$$\operatorname{curl} \tilde{\bar{B}}' = \frac{j\omega}{c} \tilde{\bar{E}}' - \frac{1}{4\pi} \operatorname{grad} \oint_{\Gamma} \tilde{\gamma}\, \tilde{\bar{B}}.d\bar{r}$$

7-32. Let $\tilde{\bar{E}}_o''$ and $\tilde{\bar{B}}_o''$ be defined by

$$4\pi \tilde{\bar{E}}_o'' = -\operatorname{curl} \int_{S_A} \tilde{\gamma}\, (d\bar{S} \times \tilde{\bar{E}}) + \frac{j\omega}{c} \int_{S_A} \tilde{\gamma}\, (d\bar{S} \times \tilde{\bar{B}}) + \operatorname{grad} \int_{S_A} \tilde{\gamma}\, d\bar{S}.\tilde{\bar{E}} - \frac{c}{j\omega} \operatorname{grad} \oint_{\Gamma} \tilde{\gamma}\, \tilde{\bar{B}}.d\bar{r}$$

$$4\pi \tilde{\bar{B}}_o'' = -\operatorname{curl} \int_{S_A} \tilde{\gamma}\, (d\bar{S} \times \tilde{\bar{B}}) - \frac{j\omega}{c} \int_{S_A} \tilde{\gamma}\, (d\bar{S} \times \tilde{\bar{E}}) + \operatorname{grad} \int_{S_A} \tilde{\gamma}\, d\bar{S}.\tilde{\bar{B}} + \frac{c}{j\omega} \operatorname{grad} \oint_{\Gamma} \tilde{\gamma}\, \tilde{\bar{E}}.d\bar{r}$$

Make use of the results of the previous exercise to show that $\tilde{\bar{E}}''$ and $\tilde{\bar{B}}''$ satisfy Maxwell's equations.

These expressions are known as Kottler's formulae.

Show that they are identical with the expressions developed previously in terms of Green's dyadic (7.8-11/12) when applied to an open surface.

7-33. Show that the diffraction integrals for $\tilde{\bar{E}}$ and $\tilde{\bar{B}}$ preserve the same form when the source system of Ex.5-93., p. 559 is employed. Show further that the importation of an additional tangential surface source of density $\tilde{\bar{K}}_m = \frac{c}{4\pi}(\hat{n}\times\tilde{\bar{E}})$, as permitted by the model, allows $\tilde{\bar{E}}$ and $\tilde{\bar{B}}$ to be expressed , at interior points, wholly in terms of surface sources. Extend considerations to the time-invariant case, noting that divs $\bar{K}$ and divs $\bar{K}_m$ are zero, and that $\bar{E}$ and $\bar{B}$, ϕ and $\bar{A}$, and ϕ_m and $\bar{A}_m$ are uncoupled.

7.9 Introduction to the Auxiliary Potentials[2]

Let sources be distributed throughout a finite region of space in such a way that Maxwell's equations hold at all points of a subregion τ bounded by the closed surfaces $S_{1..n}\Sigma$.

Since $\tilde{\bar{B}}$ is solenoidal in τ it is possible, in accordance with the results of Sec. 4.17, to determine a point function $\tilde{\bar{A}}_1$ in τ such that

$$\tilde{\bar{B}} = \text{curl}\,\tilde{\bar{A}}_1 \qquad (7.9\text{-}1)$$

$\tilde{\bar{A}}_1$ is one of an infinite set of possible functions which differ from each other by a gradient function.

Further, curl $\tilde{\bar{E}} = \frac{-j\omega}{c}\tilde{\bar{B}}$ so that $\tilde{\bar{E}} + \frac{j\omega}{c}\tilde{\bar{A}}_1$ is irrotational in τ, hence

$$\tilde{\bar{E}} = -\,\text{grad}\,\tilde{\phi}_1 - \frac{j\omega}{c}\tilde{\bar{A}}_1 \qquad (7.9\text{-}2)$$

where $\tilde{\phi}_1$ is one of an infinite number of scalar point functions which differ from each other by a constant.

Equations (7.9-1) and (7.9-2) may be replaced by

$$\tilde{\bar{B}} = \text{curl}\,\tilde{\bar{A}}' \qquad (7.9\text{-}3)$$

$$\tilde{\bar{E}} = -\,\text{grad}\,\tilde{\phi}' - \frac{j\omega}{c}\tilde{\bar{A}}' \qquad (7.9\text{-}4)$$

where

$$\tilde{\bar{A}}' = \tilde{\bar{A}}_1 - \text{grad}\,\tilde{\psi} \qquad (7.9\text{-}5)$$

---

2. While the analysis of the present section is carried out in complex form, an explicitly retarded treatment may be adopted for the non-time-harmonic case.

$$\tilde{\phi}' = \tilde{\phi}_1 + \frac{j\omega}{c}\tilde{\psi} \qquad (7.9\text{-}6)$$

$\tilde{\psi}$ being any well-behaved scalar function.

We will refer to $\tilde{\bar{A}}'$ and $\tilde{\phi}'$ as auxiliary potentials.

The replacement of $\tilde{\bar{A}}_1$ and $\tilde{\phi}_1$ by $\tilde{\bar{A}}'$ and $\tilde{\phi}'$ is said to be a gauge transformation; correspondingly, $\tilde{\bar{B}}$ and $\tilde{\bar{E}}$ are said to be gauge-invariant.

From equations (7.9-5) and (7.9-6) we see that

$$\operatorname{div}\tilde{\bar{A}}' + \frac{j\omega}{c}\tilde{\phi}' = \operatorname{div}\tilde{\bar{A}}_1 + \frac{j\omega}{c}\tilde{\phi}_1 - \left(\nabla^2\tilde{\psi} + \frac{\omega^2}{c^2}\tilde{\psi}\right) \qquad (7.9\text{-}7)$$

Hence if $\tilde{\psi}$ is so chosen as to make

$$\nabla^2\tilde{\psi} + \frac{\omega^2}{c^2}\tilde{\psi} = \operatorname{div}\tilde{\bar{A}}_1 + \frac{j\omega}{c}\tilde{\phi}_1$$

ie if, for example, we put

$$\tilde{\psi} = -\frac{1}{4\pi}\int_\tau \tilde{\Upsilon}\left(\operatorname{div}\tilde{\bar{A}}_1 + \frac{j\omega}{c}\tilde{\phi}_1\right)d\tau$$

then

$$\operatorname{div}\tilde{\bar{A}}' = \frac{-j\omega}{c}\tilde{\phi}'$$

In this case it follows from (9.7-5), (9.7-6) and Maxwell's equations that

$$\nabla^2\tilde{\phi}' + \frac{\omega^2}{c^2}\tilde{\phi}' = -4\pi\,(\tilde{\rho} - \operatorname{div}\tilde{\bar{P}}) \qquad (7.9\text{-}8)$$

$$\nabla^2\tilde{\bar{A}}' + \frac{\omega^2}{c^2}\tilde{\bar{A}}' = -\frac{4\pi}{c}\,(\tilde{\bar{J}} + j\omega\tilde{\bar{P}} + c\operatorname{curl}\tilde{\bar{M}}) \qquad (7.9\text{-}9)$$

The relationship $\operatorname{div}\tilde{\bar{A}}' = \frac{-j\omega}{c}\tilde{\phi}'$, as mentioned in Sec. 5.19c in real form, is known as the Lorentz gauge. Whereas in the present case it is the result of choice, it was then a logical deduction based upon the definitions of retarded scalar and vector potentials.

We may choose a different relationship, say

$$\operatorname{div}\tilde{\bar{A}}' = 0$$

This is known as the Coulomb gauge and requires that $\nabla^2\tilde{\psi} = \operatorname{div}\tilde{\bar{A}}_1$ or, say,

$$\tilde{\psi} = -\frac{1}{4\pi}\int_\tau \frac{\operatorname{div}\tilde{\bar{A}}_1}{r}\,d\tau$$

We easily find that in this case

$$\nabla^2\tilde{\phi}' = -4\pi\,(\tilde{\rho} - \operatorname{div}\tilde{\bar{P}}) \qquad (7.9\text{-}10)$$

$$\nabla^2\tilde{\bar{A}}' = -\frac{4\pi}{c}\,(\tilde{\bar{J}} + j\omega\tilde{\bar{P}} + c\operatorname{curl}\tilde{\bar{M}}) - \frac{j\omega}{c}\tilde{\bar{E}} \qquad (7.9\text{-}11)$$

On the other hand, if we put $\tilde{\psi} = -\frac{c}{j\omega}\tilde{\phi}_1$ then

$$\tilde{\phi}' = 0 \quad ; \quad \tilde{\bar{E}} = -\frac{j\omega}{c}\tilde{\bar{A}}' \qquad (7.9\text{-}12)$$

and

$$\operatorname{div}\tilde{\bar{A}}' = \operatorname{div}\tilde{\bar{A}}_1 + \frac{c}{j\omega}\nabla^2\tilde{\phi}_1$$

Further,

$$\operatorname{div}\tilde{\bar{E}} = -\frac{j\omega}{c}\operatorname{div}\tilde{\bar{A}}' = 4\pi\,(\tilde{\rho} - \operatorname{div}\tilde{\bar{P}})$$

and

$$\operatorname{curl}\tilde{\bar{B}} = \operatorname{grad}\operatorname{div}\tilde{\bar{A}}' - \nabla^2\tilde{\bar{A}}'$$

whence

$$\nabla^2\tilde{\bar{A}}' + \frac{\omega^2}{c^2}\tilde{\bar{A}}' = -\frac{4\pi}{c}\,(\tilde{\bar{J}} + j\omega\tilde{\bar{P}} + c\operatorname{curl}\tilde{\bar{M}}) + \operatorname{grad}\operatorname{div}\tilde{\bar{A}}' \qquad (7.9\text{-}13)$$

For the particular case in which $\tilde{\rho}$ and $\tilde{\bar{P}}$ are zero in τ, equation (7.9-13) reduces to

$$\nabla^2\tilde{\bar{A}}' + \frac{\omega^2}{c^2}\tilde{\bar{A}}' = -\frac{4\pi}{c}\,(\tilde{\bar{J}} + c\operatorname{curl}\tilde{\bar{M}}) \qquad (7.9\text{-}14)$$

Corresponding to the differential equation (7.9-8), the general form of $\tilde{\phi}'$ at an interior point of τ is given by (6.1-7) as

$$\tilde{\phi}_o' = \frac{1}{4\pi}\oint_{S_{1..n}\Sigma}\left\{\frac{1}{r}e^{j\tilde{k}r}\frac{\partial\tilde{\phi}'}{\partial n} - \tilde{\phi}'\frac{\partial}{\partial n}\left(\frac{1}{r}e^{j\tilde{k}r}\right)\right\}dS + \int_{\tau}\frac{1}{r}e^{j\tilde{k}r}(\tilde{\rho} - \text{div}\,\tilde{\bar{P}})\,d\tau \tag{7.9-15}$$

where $\tilde{k} = \pm\frac{\omega}{c}$

Thus $\tilde{\phi}_o'$ may be expressed in terms of advanced or retarded integrals, or combinations of the two. Boundary conditions, however, remain unknown.

Suppose that the complete source system comprises a set of subregions within which the density functions are well-behaved. Then $\tilde{\phi}'$ may be expressed at all points removed from the surfaces by

$$\tilde{\phi}_o' = \frac{1}{4\pi}\oint_{S}\left\{\frac{1}{r}e^{\pm j\omega r/c}\left(-\Delta\frac{\partial\tilde{\phi}'}{\partial n}\right)+\Delta\tilde{\phi}'\frac{\partial}{\partial n}\left(\frac{1}{r}e^{\pm j\omega r/c}\right)\right\}dS + \int_{\infty}\frac{1}{r}e^{\pm j\omega r/c}(\tilde{\rho} - \text{div}\,\tilde{\bar{P}})d\tau$$

$$+\frac{1}{4\pi}\oint_{\infty}\left\{\frac{1}{r}e^{\pm j\omega r/c}\frac{\partial\tilde{\phi}'}{\partial n} - \tilde{\phi}'\frac{\partial}{\partial n}\left(\frac{1}{r}e^{\pm j\omega r/c}\right)\right\}dS \tag{7.9-16}$$

where the Δ notation is that employed previously, and S represents all paired surfaces of integration defined by the interfaces of the juxtaposed regions.

The corresponding solution of equation (7.9-9) is[3]

$$\tilde{\bar{A}}_o' = \frac{1}{4\pi}\oint_{S}\left\{\frac{1}{r}e^{\pm j\omega r/c}\left(-\Delta\frac{\partial\tilde{\bar{A}}'}{\partial n}\right)+\Delta\tilde{\bar{A}}'\frac{\partial}{\partial n}\left(\frac{1}{r}e^{\pm j\omega r/c}\right)\right\}dS$$

$$+\frac{1}{c}\int_{\infty}\frac{1}{r}e^{\pm j\omega r/c}(\tilde{\bar{J}} + j\omega\tilde{\bar{P}} + c\,\text{curl}\,\tilde{\bar{M}})\,d\tau \tag{7.9-17}$$

$$+\frac{1}{4\pi}\oint_{\infty}\left\{\frac{1}{r}e^{\pm j\omega r/c}\frac{\partial\tilde{\bar{A}}'}{\partial n} - \tilde{\bar{A}}'\frac{\partial}{\partial n}\left(\frac{1}{r}e^{\pm j\omega r/c}\right)\right\}dS$$

---

3. Although the subscript ∞ is employed in conjunction with the volume integrals to indicate that all subregions are included, it is supposed that $\tilde{\sigma}$, $\tilde{\rho}$, $\tilde{\bar{J}}$, $\tilde{\bar{P}}$ and $\tilde{\bar{M}}$ are zero beyond a finite region of space.

In virtue of the boundary conditions expressed by equations (5.21-12) to (5.21-15), the discontinuities of $\tilde{\phi}'$ and $\tilde{\bar{A}}'$ and their derivatives through bounding surfaces are not entirely arbitrary. Thus if $\tilde{\phi}'$ and $\tilde{\bar{A}}'$ are postulated to be continuous between subregions, the following relationships must hold (see Ex.7-35., p. 663):

$$\Delta \frac{\partial \tilde{\phi}'}{\partial n} = -4\pi \left(\tilde{\sigma} + \hat{n}_1.\tilde{\bar{P}}_1 + \hat{n}_2.\tilde{\bar{P}}_2\right) \qquad (7.9\text{-}18)$$

$$\Delta \frac{\partial \tilde{\bar{A}}'}{\partial n} = \frac{-4\pi}{c} \left(\tilde{\bar{K}} + c(\tilde{\bar{M}}_1 \times \hat{n}_1) + c(\tilde{\bar{M}}_2 \times \hat{n}_2)\right) \qquad (7.9\text{-}19)$$

Substitution of (7.9-18) and (7.9-19) in (7.9-16) and (7.9-17), and transformation in accordance with equation (1) of Tables 10 and 12 then lead to

$$\tilde{\phi}' = \int_\infty \frac{\tilde{\rho}}{r} e^{\pm j\omega r/c} d\tau + \int_S \frac{\tilde{\sigma}}{r} e^{\pm j\omega r/c} dS + \int_\infty \tilde{\bar{P}}.\nabla \left(\frac{1}{r} e^{\pm j\omega r/c}\right) d\tau$$
$$+ \frac{1}{4\pi} \oint_\infty \left\{\frac{1}{r} e^{\pm j\omega r/c} \frac{\partial \tilde{\phi}'}{\partial n} - \tilde{\phi}' \frac{\partial}{\partial n}\left(\frac{1}{r} e^{\pm j\omega r/c}\right)\right\} dS \qquad (7.9\text{-}20)$$

and

$$\tilde{\bar{A}}' = \frac{1}{c}\int_\infty \frac{\tilde{\bar{J}}}{r} e^{\pm j\omega r/c} d\tau + \frac{j\omega}{c}\int_\infty \frac{\tilde{\bar{P}}}{r} e^{\pm j\omega r/c} d\tau + \frac{1}{c}\int_S \frac{\tilde{\bar{K}}}{r} e^{\pm j\omega r/c} dS$$
$$+ \int_\infty \tilde{\bar{M}} \times \nabla \left(\frac{1}{r} e^{\pm j\omega r/c}\right) d\tau + \frac{1}{4\pi} \oint_\infty \left\{\frac{1}{r} e^{\pm j\omega r/c} \frac{\partial \tilde{\bar{A}}'}{\partial n} - \tilde{\bar{A}}' \frac{\partial}{\partial n}\left(\frac{1}{r} e^{\pm j\omega r/c}\right)\right\} dS \qquad (7.9\text{-}21)$$

More generally,

$$\tilde{\phi}' = a\tilde{\phi}_1' + (1-a)\,\tilde{\phi}_2'$$

where $\tilde{\phi}_1'$ is the advanced (positive exponent) component of (7.9-20), $\tilde{\phi}_2'$ is the retarded component, and a is an arbitrary constant[4].

A corresponding expression holds for $\tilde{\bar{A}}'$.

---

4. Note that these two components have identical values so long as the surface integrals at infinity are retained.

For a possibly independent solution see A.J. Carr, Phil. Mag. 6, p. 241 (1928).

If, now, a is equated to zero and Sommerfeld conditions stated in the form

$$\begin{aligned} &R\tilde{\phi}' \text{ bounded} \quad ; \quad R\left(\frac{\partial\tilde{\phi}'}{\partial R} + \frac{j\omega}{c}\tilde{\phi}'\right) \to 0 \quad \text{as} \quad R\to\infty \\ &R\tilde{\bar{A}}' \text{ bounded} \quad ; \quad R\left(\frac{\partial\tilde{\bar{A}}'}{\partial R} + \frac{j\omega}{c}\tilde{\bar{A}}'\right) \to 0 \quad \text{as} \quad R\to\infty \end{aligned} \tag{7.9-22}$$

the advanced components of the potentials are eliminated together with the retarded surface integrals at infinity, and $\tilde{\phi}'$ and $\tilde{\bar{A}}'$ reduce to those forms of the retarded potentials employed elsewhere in this work (with contributions from surface doublets and whirls omitted)[5].

The foregoing treatment $\left(\text{for the case div } \tilde{\bar{A}}' = \frac{-j\omega}{c}\tilde{\phi}'\right)$, by means of which the retarded potentials in externally unbounded regions are developed from Maxwell's equations, is representative of the approach invariably adopted (if only implicitly) in conventional works on electromagnetic theory. Its unsatisfactory nature is at once evident. Not only must Maxwell's equations be treated as postulates (without prior definition of $\bar{E}$ and $\bar{B}$), but the associated boundary conditions likewise[6]. A further postulate must then be adopted concerning the continuity of the potential functions through surfaces of discontinuity. Finally, the retarded surface integrals at infinity are arbitrarily equated to zero[7] while the advanced solutions are conceded to be incompatible with experience and consequently rejected.

---

5. The complementary functions, ie the solutions of the homogeneous Helmholtz equations, should be added to the particular integrals. However, Sommerfeld conditions, as stated, eliminate the retarded components. For the non-time-harmonic case the analysis of Sec. 5.4a reveals that the complementary functions will be zero everywhere and at all times if they and their time derivatives are everywhere zero at any one instant.

6. Boundary conditions cannot be derived legitimately from Maxwell's equations.

7. Some writers endeavour to justify the procedure on the ground that, when the volume integral is taken to be the complete solution, the computed value of the corresponding surface integral at infinity is zero. This argument is invalid, for once the solution is preempted in this way the surface integral necessarily vanishes.

The reverse path, in which the retarded potentials are treated as primary concepts, is logical, unequivocal[8] and flexible[9].

This criticism is not, of course, intended to decry the use of auxiliary potentials when developed for finite closed regions, within which the constitutive relationships for linear material media may lead to surprising simplifications; nevertheless, the secondary nature and limitations of these potentials, and of the correspondingly modified Maxwellian equations, should not be forgotten. Such applications, however, properly belong to the applied mathematics of electromagnetic theory and, as such, are beyond the scope of this work.

EXERCISES

7-34. Supposing that the solutions of (7.9-8), (7.9-9) and of (7.9-10), (7.9-11) may be expressed entirely as volume integrals and that the potentials are designated $\tilde{\phi}_1'$, $\tilde{\tilde{A}}_1'$ and $\tilde{\phi}_2'$, $\tilde{\tilde{A}}_2'$ respectively, show that

$$\tilde{\phi}_1' = \int \frac{1}{r} e^{-j\omega r/c} (\tilde{\rho} - \text{div } \tilde{\tilde{P}})\, d\tau$$

$$\tilde{\tilde{A}}_1' = \frac{1}{c} \int \frac{1}{r} e^{-j\omega r/c} (\tilde{\tilde{J}} + j\omega\tilde{\tilde{P}} + c \text{ curl } \tilde{\tilde{M}})\, d\tau$$

or

$$\tilde{\tilde{A}}_1' = \frac{1}{c} \int \frac{1}{r} \left(\tilde{\tilde{J}} + \frac{1}{4\pi} j\omega\tilde{\tilde{D}} + c \text{ curl } \tilde{\tilde{M}}\right) d\tau + \frac{j\omega}{4\pi c} \int \frac{1}{r} \text{ grad } \tilde{\phi}_1'\, d\tau$$

and

$$\tilde{\phi}_2' = \int \frac{1}{r} (\tilde{\rho} - \text{div } \tilde{\tilde{P}})\, d\tau$$

$$\tilde{\tilde{A}}_2' = \frac{1}{c} \int \frac{1}{r} \left(\tilde{\tilde{J}} + \frac{1}{4\pi} j\omega\tilde{\tilde{D}} + c \text{ curl } \tilde{\tilde{M}}\right) d\tau$$

---

8. In particular, attention is necessarily focussed from the outset upon the kinematics of retardation. (Needless to say, the conventional view has been adopted in the present work.)

9. Thus, if doublet surface sources were admitted to the system, the modification of the boundary conditions for $\tilde{\phi}$ and $\tilde{\tilde{A}}$, as derived in an earlier chapter, could be written down at once. But in the alternative approach a further postulate would be required.

or

$$\tilde{\tilde{A}}_2' = \frac{1}{c}\int \frac{1}{r} e^{-j\omega r/c} (\tilde{\tilde{J}} + j\omega\tilde{\tilde{P}} + c \text{ curl } \tilde{\tilde{M}}) \, d\tau - \frac{j\omega}{4\pi c}\int \frac{1}{r} e^{-j\omega r/c} \text{ grad } \tilde{\phi}_2' \, d\tau$$

Note that both $\tilde{\tilde{A}}_1'$ and $\tilde{\tilde{A}}_2'$ may be expressed in unretarded or retarded (exponential) form, but that only in the retarded form of $\tilde{\tilde{A}}_1'$ is the integrand expressed directly in terms of density functions alone.

7-35. Let a surface of discontinuity be characterised by the curvilinear coordinates ξ and ζ and let $\hat{\xi}$, $\hat{\zeta}$, $\hat{n}$ form a right-handed set. By substitution of (7.9-4) in (7.6-21) and (7.6-22) show that

$$\frac{\partial\tilde{\phi}_1}{\partial n_1} + \frac{\partial\tilde{\phi}_2}{\partial n_2} + \frac{j\omega}{c} \hat{n}_1.(\tilde{\tilde{A}}_1 - \tilde{\tilde{A}}_2) = 4\pi \, (\tilde{\sigma} + \hat{n}_1.\tilde{\tilde{P}}_1 + \hat{n}_2.\tilde{\tilde{P}}_2)$$

$$\frac{1}{h_\xi}\left(\frac{\partial\tilde{\phi}_1}{\partial\xi} - \frac{\partial\tilde{\phi}_2}{\partial\xi}\right) + \frac{j\omega}{c} \hat{\xi}.(\tilde{\tilde{A}}_1 - \tilde{\tilde{A}}_2) = 0$$

$$\frac{1}{h_\zeta}\left(\frac{\partial\tilde{\phi}_1}{\partial\zeta} - \frac{\partial\tilde{\phi}_2}{\partial\zeta}\right) + \frac{j\omega}{c} \hat{\zeta}.(\tilde{\tilde{A}}_1 - \tilde{\tilde{A}}_2) = 0$$

Hence conclude that (7.6-21) and (7.6-22) are satisfied by the boundary conditions

$$\tilde{\phi}_1 = \tilde{\phi}_2 \quad ; \quad \tilde{\tilde{A}}_1 = \tilde{\tilde{A}}_2$$

$$\frac{\partial\tilde{\phi}_1}{\partial n_1} + \frac{\partial\tilde{\phi}_2}{\partial n_2} = 4\pi \, (\tilde{\sigma} + \hat{n}_1.\tilde{\tilde{P}}_1 + \hat{n}_2.\tilde{\tilde{P}}_2)$$

Show further that the above conditions also satisfy (7.6-23), and that (7.6-24) requires that

$$\hat{n}_1\left(\hat{n}_1.\frac{\partial\tilde{\tilde{A}}_1}{\partial n_1} - \hat{n}_2.\frac{\partial\tilde{\tilde{A}}_2}{\partial n_2}\right) - \left(\frac{\partial\tilde{\tilde{A}}_1}{\partial n_1} + \frac{\partial\tilde{\tilde{A}}_2}{\partial n_2}\right) = -\frac{4\pi}{c} \, (\tilde{\tilde{K}} + c(\tilde{\tilde{M}}_1\times\hat{n}_1) + c(\tilde{\tilde{M}}_2\times\hat{n}_2))$$

Finally, show that $\hat{n}_1.\frac{\partial\tilde{\tilde{A}}_1}{\partial n_1} - \hat{n}_2.\frac{\partial\tilde{\tilde{A}}_2}{\partial n_2} = 0$ because $\text{div } \tilde{\tilde{A}} = -\frac{j\omega}{c}\tilde{\phi}$,

whence (7.6-24) is satisfied if

$$\frac{\partial\tilde{\tilde{A}}_1}{\partial n_1} + \frac{\partial\tilde{\tilde{A}}_2}{\partial n_2} = \frac{4\pi}{c} \, (\tilde{\tilde{K}} + c(\tilde{\tilde{M}}_1\times\hat{n}_1) + c(\tilde{\tilde{M}}_2\times\hat{n}_2))$$

7-36. Suppose that $(\tilde{\rho} - \text{div}\,\tilde{\bar{P}})$ is well-behaved everywhere and is zero outside the externally-bounded region τ'. Let the closed surface Σ, which bounds the region τ, lie everywhere beyond τ'. Then if, at any interior point O of τ, we take the complete solution of (7.9-8) to be

$$\tilde{\phi}_o' = \int_{\tau'} \frac{1}{r} e^{-j\omega r/c} (\tilde{\rho} - \text{div}\,\tilde{\bar{P}})\, d\tau$$

it follows from (7.9-15) that

$$\oint_{\Sigma} \left\{\frac{1}{r} e^{-j\omega r/c} \frac{\partial \tilde{\phi}'}{\partial n} - \tilde{\phi}' \frac{\partial}{\partial n}\left(\frac{1}{r} e^{-j\omega r/c}\right)\right\} dS = 0$$

Confirm this, when Σ recedes to infinity, by direct evaluation of that component of the integral deriving from an arbitrary source element.

7-37. Since the retarded and advanced forms of equation (7.9-15) represent a single solution of (7.9-8), the retarded volume integral, as set down in the previous exercise, must satisfy the advanced form of (7.9-15), ie with $\tilde{k} = +\frac{\omega}{c}$. Demonstrate this for the case in which Σ recedes to infinity.

Show also that the Sommerfeld conditions would have to be modified if a solution involving a combination of retarded and advanced volume integrals alone were to be admitted.

APPENDICES

As mentioned in a footnote to Appendix 1 below, the Lorentz force expression is employed in the following pages. In conformity with our expressed intention to prescind from the applied mathematics of electromagnetic theory, we do not seek to stress the physical interpretation of the various analyses; this will readily be supplied by the reader with a physical background.

The analyses have been included here because of their intrinsic interest and because - as far as the author is aware - they have not appeared elsewhere.

A.1 The Activity Equation for Point Sources

Consider a set of point sources of strengths a_1---a_p which move in any manner (with velocity < c) within the region bounded by a closed surface S. The sources are surrounded by regular closed surfaces S_1---S_p which share their velocities and which, in conjunction with S, bound the region T. Sources of strengths a_q---a_n lie outside S. At some particular instant the interior sources occupy the positions P_1---P_p and move with velocities $\bar{v}_1$---$\bar{v}_p$.

We now proceed to show that if $\bar{E}_i$ and $\bar{B}_i$ are the $\bar{E}$ and $\bar{B}$ fields deriving from the ith source and $\bar{E}_i'$ and $\bar{B}_i'$ are the combined fields of all other sources, and if

$$\bar{F}_i = a_i \left\{ \bar{E}_i' + \frac{1}{c} (\bar{v}_i \times \bar{B}_i') \right\}_{P_i} \qquad \text{(A.1-1)}$$

then[1]

$$\bar{F}_1.\bar{v}_1 \text{ --- } + \bar{F}_p.\bar{v}_p + \oint_S \frac{c}{4\pi}(\bar{E}_1\times\bar{B}_1' \text{ --- } + \bar{E}_n\times\bar{B}_n').d\bar{S}$$

$$= -\frac{d}{dt}\int_\tau \frac{1}{8\pi}(\bar{E}_1.\bar{E}_1' \text{ --- } + \bar{E}_n.\bar{E}_n' + \bar{B}_1.\bar{B}_1' \text{ --- } + \bar{B}_n.\bar{B}_n')\,d\tau \qquad \text{(A.1-2)}$$

where τ is the limiting configuration of T as the surfaces S_1----S_p shrink uniformly about their respective sources.

The surface integrand of (A.1-2), which may be expressed as

$$\sum_{i=1}^{n}\frac{c}{4\pi}(\bar{E}_i\times\bar{B}_i') \quad \text{or} \quad \sum_{i=1}^{n}\frac{c}{4\pi}(\bar{E}_i'\times\bar{B}_i) \quad \text{or} \quad \sum_{i=1}^{n}\frac{c}{8\pi}(\bar{E}_i\times\bar{B}_i' + \bar{E}_i'\times\bar{B}_i),$$

will be referred to as 'the Poynting vector for point sources'.

On substituting $\bar{E}_1$ and $\bar{B}_1'$ and then $\bar{E}_1'$ and $\bar{B}_1$ for $\bar{F}$ and $\bar{G}$ in (1.16-7), adding the resulting equations, and applying (5.11-24) and (5.11-27), we obtain

$$\text{div}\,(\bar{E}_1\times\bar{B}_1' + \bar{E}_1'\times\bar{B}_1) = \bar{B}_1'.\text{curl}\,\bar{E}_1 - \bar{E}_1.\text{curl}\,\bar{B}_1' + \bar{B}_1.\text{curl}\,\bar{E}_1' - \bar{E}_1'.\text{curl}\,\bar{B}_1$$

$$= -\frac{1}{c}\left\{\bar{B}_1'.\frac{\partial\bar{B}_1}{\partial t} + \bar{E}_1.\frac{\partial\bar{E}_1'}{\partial t} + \bar{B}_1.\frac{\partial\bar{B}_1'}{\partial t} + \bar{E}_1'.\frac{\partial\bar{E}_1}{\partial t}\right\}$$

$$= -\frac{1}{c}\frac{\partial}{\partial t}(\bar{E}_1.\bar{E}_1' + \bar{B}_1.\bar{B}_1')$$

---

1. Students of electromagnetics will recognise equation (A.1-1) as the Lorentz expression for the force exerted on a point charge of strength a_i at the point P_i by sources giving rise to $\bar{E}_i'$ and $\bar{B}_i'$, so that $\sum_{i=1}^{p}\bar{F}_i.\bar{v}_i$ may be interpreted as the activity of the discrete electrical forces of interaction which appear in the electrical model within the region bounded by S.

Then

$$\frac{\partial}{\partial t}\int_T (\bar{E}_1.\bar{E}_1'+\bar{B}_1.\bar{B}_1')\,d\tau = -c\oint_{S_{1..p}S} (\bar{E}_1\times\bar{B}_1'+\bar{E}_1'\times\bar{B}_1).d\bar{S}$$

when the surfaces $S_{1..p}$ are fixed in position, and

$$\begin{aligned}\frac{d}{dt}\int_T (\bar{E}_1.\bar{E}_1'+\bar{B}_1.\bar{B}_1')\,d\tau = & -c\oint_{S_{1..p}S} (\bar{E}_1\times\bar{B}_1'+\bar{E}_1'\times\bar{B}_1).d\bar{S} \\ & + \oint_{S_1} (\bar{E}_1.\bar{E}_1'+\bar{B}_1.\bar{B}_1')\,\bar{v}_1.d\bar{S} \\ & + \oint_{S_2} (\bar{E}_1.\bar{E}_1'+\bar{B}_1.\bar{B}_1')\,\bar{v}_2.d\bar{S} \\ & + \text{----}\end{aligned} \qquad \text{(A.1-3)}$$

when the surfaces $S_{1..p}$ move with their associated sources.

In the limit[2] as S_1 shrinks about P_1 the surface integrals over S_1 reduce to

$$c(\bar{B}_1')_{P_1}.\oint_{S_1} (\bar{E}_1\times d\bar{S}) - c(\bar{E}_1')_{P_1}.\oint_{S_1} (\bar{B}_1\times d\bar{S})$$

$$+ (\bar{E}_1')_{P_1}.\oint_{S_1} \bar{E}_1\ \bar{v}_1.d\bar{S} + (\bar{B}_1')_{P_1}.\oint_{S_1} \bar{B}_1\ \bar{v}_1.d\bar{S}$$

and this may be replaced by

$$c(\bar{B}_1')_{P_1}.\oint_{S_1}\left(\bar{E}_1\times d\bar{S} + \frac{1}{c}\,\bar{B}_1\ \bar{v}_1.d\bar{S}\right) + c(\bar{E}_1')_{P_1}.\oint_{S_1}\left(d\bar{S}\times\bar{B}_1 + \frac{1}{c}\,\bar{E}_1\ \bar{v}_1.d\bar{S}\right) \qquad \text{(A.1-4)}$$

2. 'Lim' signs have been omitted in the following pages to simplify the notation.

By making use of the identity

$$d\bar{S} \times (\bar{v}_1 \times \bar{B}_1) = \bar{v}_1\, \bar{B}_1 . d\bar{S} - \bar{B}_1\, \bar{v}_1 . d\bar{S}$$

the first integral of (A.1-4) may be expressed as

$$\oint_{S_1} \left\{ \bar{E}_1 + \frac{1}{c} (\bar{v}_1 \times \bar{B}_1) \right\} \times d\bar{S} + \frac{1}{c} \bar{v}_1 \oint_{S_1} \bar{B}_1 . d\bar{S}$$

$$= \oint_{S_1} \left\{ \bar{E}_1 + \frac{1}{c} (\bar{v}_1 \times \bar{B}_1) \right\} \times d\bar{S} \qquad \text{(from Ex.5-44., p. 485)}$$

Now from equation (1.16-6)

$$\bar{v}_1 \times \bar{B}_1 = \bar{v}_1 \times \text{curl}\, \bar{A}_1 = \text{grad}(\bar{v}_1 . \bar{A}_1) - (\bar{v}_1 . \nabla)\bar{A}_1 - (\bar{A}_1 . \nabla)\bar{v}_1 - \bar{A}_1 \times \text{curl}\, \bar{v}_1$$

$$= \text{grad}(\bar{v}_1 . \bar{A}_1) - (\bar{v}_1 . \nabla)\bar{A}_1 \qquad \text{since } \bar{v}_1 \text{ is constant in this context}$$

hence

$$\bar{E}_1 + \frac{1}{c} (\bar{v}_1 \times \bar{B}_1) = - \text{grad}\left(\phi_1 - \frac{1}{c} \bar{v}_1 . \bar{A}_1\right) - \frac{1}{c}\left(\frac{\partial \bar{A}_1}{\partial t} + (\bar{v}_1 . \nabla)\bar{A}_1\right)$$

and from equation (1.17-2)

$$\oint_{S_1} \left\{ \bar{E}_1 + \frac{1}{c} (\bar{v}_1 \times \bar{B}_1) \right\} \times d\bar{S} = \frac{1}{c} \oint_{S_1} d\bar{S} \times \left(\frac{\partial \bar{A}_1}{\partial t} + (\bar{v}_1 . \nabla)\bar{A}_1\right)$$

$$= \frac{1}{c} \oint_{S_1} d\bar{S} \times \frac{d\bar{A}_1}{dt}$$

where $\frac{d\bar{A}_1}{dt}$ is the rate of change of $\bar{A}_1$ upon the moving surface S_1.

But since the retarded positions of the source appropriate to evaluation of the field quantities upon S_1 converge upon P_1 as S_1 shrinks about P_1, we may take the retarded and instantaneous velocities to be identical in the limit, in which case the retarded distance associated with an element of S_1 does not change with time and $\frac{d\bar{A}_1}{dt}$ involves inverse-distance terms at most. (In the unaccelerated case $\frac{d\bar{A}_1}{dt}$ will be zero.) It then follows on dimensional grounds that the above surface integral, and therefore the first term of (A.1-4), vanishes in the limit.

The integrand in the second term of (A.1-4) may be replaced by

$$d\bar{S} \times \left\{\bar{B}_1 + \frac{1}{c}\,(\bar{E}_1 \times \bar{v}_1)\right\} + \frac{1}{c}\,\bar{v}_1\,\bar{E}_1.d\bar{S}$$

$$= d\bar{S} \times \left\{\left(\frac{\bar{R}}{R} - \frac{\bar{v}_1}{c}\right) \times \bar{E}_1\right\} + \frac{1}{c}\,\bar{v}_1\,\bar{E}_1.d\bar{S} \qquad \text{from (5.11-23)}$$

Bearing in mind that inverse-distance terms in $\bar{E}_1$ do not contribute to the surface integral in the limit and that $\bar{v}_1$ may be identified with the retarded velocity, we find from equation (5.11-21) that

$$\left(\frac{\bar{R}}{R} - \frac{\bar{v}_1}{c}\right) \times \bar{E}_1 \text{ may be replaced by} \left(\frac{\bar{R}}{R} - \frac{\bar{v}_1}{c}\right) \times \left(\frac{\bar{R}}{R} - \frac{\bar{v}_1}{c}\right)\left(1 - \frac{v_1^2}{c^2}\right)\frac{a_1}{\alpha^3 R^2};$$

and this is zero.

Hence

$$\oint\limits_{S_1} \left(d\bar{S}\times\bar{B}_1 + \frac{1}{c}\,\bar{E}_1\,\bar{v}_1.d\bar{S}\right) = \frac{1}{c}\,\bar{v}_1 \oint\limits_{S_1} \bar{E}_1.d\bar{S}$$

$$= -\frac{1}{c}\,4\pi a_1\bar{v}_1 \qquad \text{from Ex.5-42., p. 485}$$

and the sum of the surface integrals over S_1 in equation (A.1-3) is equal to $-4\pi a_1\bar{v}_1.(\bar{E}_1{}')_{P_1}$.

The surface integrals over S_2 in equation (A.1-3) have the limiting value

$$- c\ (\bar{E}_1)_{P_2}.\oint_{S_2} \{\bar{B}_2\times d\bar{S} + (\bar{B}_3+\bar{B}_4---)\times d\bar{S}\}$$

$$+ c\ (\bar{B}_1)_{P_2}.\oint_{S_2} \{\bar{E}_2\times d\bar{S} + (\bar{E}_3+\bar{E}_4---)\times d\bar{S}\}$$

$$+ (\bar{E}_1)_{P_2}.\oint_{S_2} \{\bar{E}_2\ \bar{v}_2.d\bar{S} + (\bar{E}_3+\bar{E}_4---)\bar{v}_2.d\bar{S}\}$$

$$+ (\bar{B}_1)_{P_2}.\oint_{S_2} \{\bar{B}_2\ \bar{v}_2.d\bar{S} + (\bar{B}_3+\bar{B}_4---)\bar{v}_2.d\bar{S}\}$$

$$= c\ (\bar{E}_1)_{P_2}.\oint_{S_2} \left(d\bar{S}\times\bar{B}_2 + \frac{1}{c}\ \bar{E}_2\ \bar{v}_2.d\bar{S}\right) + c\ (\bar{B}_1)_{P_2}.\oint_{S_2} \left(\bar{E}_2\times d\bar{S} + \frac{1}{c}\ \bar{B}_2\ \bar{v}_2.d\bar{S}\right)$$

$$= - 4\pi a_2\bar{v}_2.(\bar{E}_1)_{P_2} \qquad \text{from previous considerations.}$$

It then follows from equation (A.1-3) that

$$\frac{d}{dt}\int_\tau (\bar{E}_1.\bar{E}_1{}'+\bar{B}_1.\bar{B}_1{}')\ d\tau = - c \oint_S (\bar{E}_1\times\bar{B}_1{}'+\bar{E}_1{}'\times\bar{B}_1).d\bar{S} - 4\pi a_1\bar{v}_1.(\bar{E}_1{}')_{P_1}$$

$$- 4\pi a_2\bar{v}_2.(\bar{E}_1)_{P_2} ---- - 4\pi a_p\bar{v}_p.(\bar{E}_1)_{P_p}$$

If we proceed as above with $\bar{E}_2$, $\bar{E}_2{}'$, $\bar{B}_2$, $\bar{B}_2{}'$ replacing $\bar{E}_1$, $\bar{E}_1{}'$, $\bar{B}_1$, $\bar{B}_1{}'$ we obtain

$$\frac{d}{dt}\int_\tau (\bar{E}_2.\bar{E}_2{}'+\bar{B}_2.\bar{B}_2{}')\ d\tau = - c \oint_S (\bar{E}_2\times\bar{B}_2{}'+\bar{E}_2{}'\times\bar{B}_2).d\bar{S} - 4\pi a_1\bar{v}_1.(\bar{E}_2)_{P_1}$$

$$- 4\pi a_2\bar{v}_2.(\bar{E}_2{}')_{P_2} ----- - 4\pi a_p\bar{v}_p.(\bar{E}_2)_{P_p}$$

The sum of n equations of this type reduces to

$$a_1\bar{v}_1.(\bar{E}_1')_{P_1} \text{ ---- } + a_p\bar{v}_p.(\bar{E}_p')_{P_p} + \oint_S \frac{c}{4\pi}(\bar{E}_1\times\bar{B}_1' \text{----} + \bar{E}_n\times\bar{B}_n').d\bar{S}$$

$$= -\frac{d}{dt}\int_\tau \frac{1}{8\pi}(\bar{E}_1.\bar{E}_1' \text{----} + \bar{E}_n.\bar{E}_n' + \bar{B}_1.\bar{B}_1' \text{---} + \bar{B}_n.\bar{B}_n')\, d\tau$$

whence equation (A.1-2) follows.

It should be noted that

(a) no restriction (other than $v < c$) has been imposed upon source motion in the above analysis;

(b) the Poynting vector for point sources vanishes when only a single source is present;

(c) for n equal clustered sources sharing a common motion the Poynting vector, at sufficient distance from sources, is given by $\left(1 - \frac{1}{n}\right)\frac{c}{4\pi}(\bar{E}\times\bar{B})$ where $\bar{E}$ and $\bar{B}$ are total fields. Thus for large values of n the vector approaches $\frac{c}{4\pi}(\bar{E}\times\bar{B})$, while for $n = 2$ it becomes $\frac{c}{8\pi}(\bar{E}\times\bar{B})$.

A.2 The Linear Momentum Equation for Point Sources[3]

We now derive a further relationship for the system of point sources considered in Sec. A.1, viz

$$\bar{F}_1 \text{ ---- } + \bar{F}_p + \frac{d}{dt}\int_\tau \frac{1}{4\pi c}(\bar{E}_1\times\bar{B}_1' \text{----} + \bar{E}_n\times\bar{B}_n')\, d\tau$$

$$= \oint_S \frac{1}{4\pi}\{(\bar{E}_1\ \bar{E}_1'.d\bar{S} \text{----} + \bar{E}_n\ \bar{E}_n'.d\bar{S}) - \tfrac{1}{2}\, d\bar{S}\,(\bar{E}_1.\bar{E}_1' \text{----} + \bar{E}_n.\bar{E}_n') \qquad \text{(A.2-1)}$$

$$+ (\bar{B}_1\ \bar{B}_1'.d\bar{S} \text{----} + \bar{B}_n\ \bar{B}_n'.d\bar{S}) - \tfrac{1}{2}\, d\bar{S}\,(\bar{B}_1.\bar{B}_1' \text{----} + \bar{B}_n.\bar{B}_n')\}$$

A combination of equations (1.17-15) and (1.17-17) yields

$$\int(\bar{F}\ \text{div}\ \bar{G} + \bar{G}\ \text{div}\ \bar{F} - \bar{F}\times\text{curl}\ \bar{G} - \bar{G}\times\text{curl}\ \bar{F})\, d\tau = \oint\{\bar{G}\ \bar{F}.d\bar{S} - \bar{G}\times(d\bar{S}\times\bar{F})\} \qquad \text{(A.2-2)}$$

---

3. The name derives from the interpretation of the volume integrand of (A.2-1) as the 'density of linear momentum of the field' in electromagnetic applications. This interpretation does not concern us here.

where $\bar{F}$ and $\bar{G}$ are any point functions well-behaved within the region of integration[4]

On identifying $\bar{F}$ with $\bar{E}_1$ and $\bar{G}$ with $\bar{E}_1'$, and integrating over the region T as defined in Sec. A.1, we obtain

$$\int_T (\bar{E}_1 \operatorname{div} \bar{E}_1' + \bar{E}_1' \operatorname{div} \bar{E}_1 - \bar{E}_1 \times \operatorname{curl} \bar{E}_1' - \bar{E}_1' \times \operatorname{curl} \bar{E}_1)\, d\tau$$

$$= \oint_{S_{1..p}S} \{\bar{E}_1'\bar{E}_1.d\bar{S} - \bar{E}_1'\times(d\bar{S}\times\bar{E}_1)\}$$

whence from equations (5.11-26) and (5.11-27)

$$\frac{1}{c}\int_T \left\{\bar{E}_1 \times \frac{\partial\bar{B}_1'}{\partial t} + \bar{E}_1' \times \frac{\partial\bar{B}_1}{\partial t}\right\} d\tau = \oint_{S_{1..p}S} \{\bar{E}_1'\bar{E}_1.d\bar{S} - \bar{E}_1'\times(d\bar{S}\times\bar{E}_1)\}$$

Similarly, substitution of $\bar{B}_1$ for $\bar{F}$ and $\bar{B}_1'$ for $\bar{G}$ in (A.2-2) yields

$$-\frac{1}{c}\int_T \left\{\bar{B}_1 \times \frac{\partial\bar{E}_1'}{\partial t} + \bar{B}_1' \times \frac{\partial\bar{E}_1}{\partial t}\right\} d\tau = \oint_{S_{1..p}S} \{(\bar{B}_1'\bar{B}_1.d\bar{S} - \bar{B}_1'\times(d\bar{S}\times\bar{B}_1)\}$$

so that, in the notation of Sec. A.1,

$$\frac{1}{c}\frac{\partial}{\partial t}\int_T (\bar{E}_1\times\bar{B}_1' + \bar{E}_1'\times\bar{B}_1)\, d\tau$$

$$= \oint_{S_{1..p}S} \{\bar{E}_1'\bar{E}_1.d\bar{S} - \bar{E}_1'\times(d\bar{S}\times\bar{E}_1) + \bar{B}_1'\bar{B}_1.d\bar{S} - \bar{B}_1'\times(d\bar{S}\times\bar{B}_1)\}$$

4. $\bar{F}$ in (A.2-2) has, of course, no connection with $\bar{F}_1$ --- + $\bar{F}_p$ in (A.2-1).

and

$$\frac{1}{c}\frac{d}{dt}\int_T (\bar{E}_1\times\bar{B}_1'+\bar{E}_1'\times\bar{B}_1)\,d\tau$$

$$= \oint_{S_1..{}_pS} \{\bar{E}_1'\bar{E}_1.d\bar{S} - \bar{E}_1'\times(d\bar{S}\times\bar{E}_1) + \bar{B}_1'\bar{B}_1.d\bar{S} - \bar{B}_1'\times(d\bar{S}\times\bar{B}_1)\} \tag{A.2-3}$$

$$+ \oint_{S_1} (\bar{E}_1\times\bar{B}_1'+\bar{E}_1'\times\bar{B}_1)\,\frac{1}{c}\,\bar{v}_1.d\bar{S} + \oint_{S_2} (\bar{E}_1\times\bar{B}_1'+\bar{E}_1'\times\bar{B}_1)\,\frac{1}{c}\,\bar{v}_2.d\bar{S} \text{ ----}$$

In the limit as S_1 shrinks about P_1 the surface integrals over S_1 reduce to

$$(\bar{E}_1')_{P_1}\oint_{S_1}\bar{E}_1.d\bar{S} + (\bar{E}_1')_{P_1}\times\oint_{S_1}\left(\bar{E}_1\times d\bar{S} + \frac{1}{c}\,\bar{B}_1\;\bar{v}_1.d\bar{S}\right) + (\bar{B}_1')_{P_1}\oint_{S_1}\bar{B}_1.d\bar{S}$$

$$- (\bar{B}_1')_{P_1}\times\oint_{S_1}\left(d\bar{S}\times\bar{B}_1 + \frac{1}{c}\,\bar{E}_1\;\bar{v}_1.d\bar{S}\right)$$

which, from the considerations of Sec. A.1, is equal to

$$-\,4\pi a_1\left\{\bar{E}_1' + \frac{1}{c}\,(\bar{v}_1\times\bar{B}_1')\right\}_{P_1}$$

The surface integral over S_2 may be re-arranged as

$$\oint_{S_2}\left\{\bar{E}_1\times(\bar{E}_1'\times d\bar{S}) + \bar{E}_1\;\bar{E}_1'.d\bar{S} + \bar{B}_1\times(\bar{B}_1'\times d\bar{S}) + \bar{B}_1\;\bar{B}_1'.d\bar{S} + \frac{1}{c}\,(\bar{E}_1\times\bar{B}_1'+\bar{E}_1'\times\bar{B}_1)\bar{v}_2.d\bar{S}\right\}$$

In the limit this becomes

$$(\bar{E}_1)_{P_2} \times \oint_{S_2} \left(\bar{E}_2 \times d\bar{S} + \frac{1}{c}\,\bar{B}_2\,\bar{v}_2.d\bar{S}\right) + (\bar{E}_1)_{P_2} \oint_{S_2} \bar{E}_2.d\bar{S} + (\bar{B}_1)_{P_2} \oint_{S_2} \bar{B}_2.d\bar{S}$$

$$- (\bar{B}_1)_{P_2} \times \oint_{S_2} \left(d\bar{S} \times \bar{B}_2 + \frac{1}{c}\,\bar{E}_2\,\bar{v}_2.d\bar{S}\right)$$

$$= -4\pi a_2 \left\{\bar{E}_1 + \frac{1}{c}\,(\bar{v}_2 \times \bar{B}_1)\right\}_{P_2}$$

Then from equation (A.2-3)

$$\frac{1}{c}\frac{d}{dt}\int_\tau (\bar{E}_1 \times \bar{B}_1' + \bar{E}_1' \times \bar{B}_1)\,d\tau$$

$$= \oint_S \{\bar{E}_1'\bar{E}_1.d\bar{S} - \bar{E}_1' \times (d\bar{S} \times \bar{E}_1) + \bar{B}_1'\bar{B}_1.d\bar{S} - \bar{B}_1' \times (d\bar{S} \times \bar{B}_1)\}$$

$$- 4\pi a_1 \left\{\bar{E}_1' + \frac{1}{c}\,(\bar{v}_1 \times \bar{B}_1')\right\}_{P_1} - 4\pi a_2 \left\{\bar{E}_1 + \frac{1}{c}\,(\bar{v}_2 \times \bar{B}_1)\right\}_{P_2} \quad ---$$

On carrying through the above analysis with $\bar{E}_2$, $\bar{E}_2'$, $\bar{B}_2$, $\bar{B}_2'$ replacing $\bar{E}_1$, $\bar{E}_1'$, $\bar{B}_1$, $\bar{B}_1'$ we find that

$$\frac{1}{c}\frac{d}{dt}\int_\tau (\bar{E}_2 \times \bar{B}_2' + \bar{E}_2' \times \bar{B}_2)\,d\tau$$

$$= \oint_S \{\bar{E}_2'\bar{E}_2.d\bar{S} - \bar{E}_2' \times (d\bar{S} \times \bar{E}_2) + \bar{B}_2'\bar{B}_2.d\bar{S} - \bar{B}_2' \times (d\bar{S} \times \bar{B}_2)\}$$

$$- 4\pi a_1 \left\{\bar{E}_2 + \frac{1}{c}\,(\bar{v}_1 \times \bar{B}_2)\right\}_{P_1} - 4\pi a_2 \left\{\bar{E}_2' + \frac{1}{c}\,(\bar{v}_2 \times \bar{B}_2')\right\}_{P_2} \quad ----$$

Summation of n equations of the above type yields

$$\frac{2}{c}\frac{d}{dt}\int_\tau (\bar{E}_1\times\bar{B}_1{}'\text{----}+\bar{E}_n\times\bar{B}_n{}')\,d\tau$$

$$=\oint_S \{(\bar{E}_1{}'\bar{E}_1.d\bar{S}\text{---}+\bar{E}_n{}'\bar{E}_n.d\bar{S}+\bar{E}_1\,\bar{E}_1{}'.d\bar{S}\text{---}+\bar{E}_n\bar{E}_n{}'.d\bar{S})-d\bar{S}(\bar{E}_1.\bar{E}_1{}'\text{---}+\bar{E}_n.\bar{E}_n{}')\}$$

$$+\oint_S \{(\bar{B}_1{}'\bar{B}_1.d\bar{S}\text{---}+\bar{B}_n{}'\bar{B}_n.d\bar{S}+\bar{B}_1\bar{B}_1{}'.d\bar{S}\text{---}+\bar{B}_n\bar{B}_n{}'.d\bar{S})-d\bar{S}(\bar{B}_1.\bar{B}_1{}'\text{---}+\bar{B}_n\bar{B}_n{}')\}$$

$$-8\pi\left\{a_1\left(\bar{E}_1{}'+\frac{1}{c}(\bar{v}_1\times\bar{B}_1{}')\right)_{P_1}\text{----}+a_p\left(\bar{E}_p{}'+\frac{1}{c}(\bar{v}_p\times\bar{B}_p{}')\right)_{P_p}\right\}$$

whence (A.2-1) follows.

A.3 The Angular Momentum Equation for Point Sources

Corresponding to the linear momentum equation of Sec. A.2 we have the following angular momentum equation:

$$\bar{r}_1\times\bar{F}_1\text{----}+\bar{r}_p\times\bar{F}_p+\frac{d}{dt}\int_\tau\left\{\bar{r}\times\frac{1}{4\pi c}(\bar{E}_1\times\bar{B}_1{}'\text{----}+\bar{E}_n\times\bar{B}_n{}')\right\}d\tau$$

$$=\oint_S\left\{\bar{r}\times\frac{1}{4\pi}\left((\bar{E}_1\bar{E}_1{}'.d\bar{S}\text{----}+\bar{E}_n\bar{E}_n{}'.d\bar{S})-\tfrac{1}{2}d\bar{S}(\bar{E}_1.\bar{E}_1{}'\text{----}+\bar{E}_n.\bar{E}_n{}')\right.\right. \qquad (A.3\text{-}1)$$

$$\left.\left.+(\bar{B}_1\bar{B}_1{}'.d\bar{S}\text{----}+\bar{B}_n\bar{B}_n{}'.d\bar{S})-\tfrac{1}{2}d\bar{S}(\bar{B}_1.\bar{B}_1{}'\text{----}+\bar{B}_n.\bar{B}_n{}')\right)\right\}$$

where $\bar{r}$ is the position vector drawn from an interior or exterior point Q of τ (not coincident with any source) and $\bar{r}_i = \vec{QP}_i$.

A combination of equations (1.17-16) and (1.17-18) yields

$$\int\{\bar{r}\times\bar{F}\text{ div }\bar{G}+\bar{r}\times\bar{G}\text{ div }\bar{F}-\bar{r}\times(\bar{F}\times\text{curl }\bar{G})-\bar{r}\times(\bar{G}\times\text{curl }\bar{F})\}\,d\tau$$

$$=\oint\{(\bar{r}\times\bar{G})\bar{F}.d\bar{S}-\bar{r}\times(\bar{G}\times(d\bar{S}\times\bar{F}))\} \qquad (A.3\text{-}2)$$

where $\overline{F}$ and $\overline{G}$ are any point functions well-behaved within the region of integration.

It will be observed that the individual terms of the integrands are the vector products of the corresponding terms in (A.2-2) and the position vector $\overline{r}$. This, and the fact that the value of $\overline{r}$ associated with any volume element is independent of time, permits of the development of a proof of (A.3-1) which is entirely analogous to that of (A.2-1) and which, in consequence, need not be presented here.

USEFUL TRANSFORMATIONS

$$\int \operatorname{div} \overline{F}\, d\tau = \oint \overline{F}.d\overline{S}$$

$$\int \operatorname{curl} \overline{F}\, d\tau = \oint d\overline{S} \times \overline{F}$$

$$\int (\operatorname{curl} \overline{F}).d\overline{S} = \oint \overline{F}.d\overline{r}$$

$$\oint (\operatorname{curl} \overline{F}).d\overline{S} \equiv 0$$

$$\int d\overline{S} \times \operatorname{grad} V = \oint V\, d\overline{r}$$

$$\oint d\overline{S} \times \operatorname{grad} V \equiv \overline{0}$$

$$\int \operatorname{grad} V\, d\tau = \oint V\, d\overline{S}$$

$$\int \overline{F} \times \operatorname{grad} V\, d\tau = \int V \operatorname{curl} \overline{F}\, d\tau + \oint V\, \overline{F} \times d\overline{S}$$

$$\int (\overline{F} \times \operatorname{grad} V).d\overline{S} = \int V\, (\operatorname{curl} \overline{F}).d\overline{S} - \oint V\, \overline{F}.d\overline{r}$$

$$\oint (\overline{F} \times \operatorname{grad} V).d\overline{S} = \oint V\, (\operatorname{curl} \overline{F}).d\overline{S}$$

$$\int \overline{F}.\operatorname{grad} V\, d\tau = \int V\, (-\operatorname{div} \overline{F})\, d\tau + \oint V\, \overline{F}.d\overline{S}$$

$$\oint V\, (\operatorname{grad} U).d\overline{S} = \int V\, \nabla^2 U\, d\tau + \int \operatorname{grad} V.\operatorname{grad} U\, d\tau$$

$$\oint (V \operatorname{grad} U - U \operatorname{grad} V).d\overline{S} = \int (V\, \nabla^2 U - U\, \nabla^2 V)\, d\tau$$

$$\oint VW \operatorname{grad} U.d\overline{S} = \int V \operatorname{div}(W \operatorname{grad} U)\, d\tau + \int W \operatorname{grad} V.\operatorname{grad} U\, d\tau$$

$$\oint (VW \text{ grad } U - UW \text{ grad } V).d\overline{S} = \int (V \text{ div}(W \text{ grad } U) - U \text{ div}(W \text{ grad } V))\, d\tau$$

$$\oint d\overline{r} \times \overline{F} = \int (d\overline{S}.\nabla)\overline{F} - \int \text{div } \overline{F}\, d\overline{S} + \int d\overline{S} \times \text{curl } \overline{F}$$

$$\overline{0} \equiv \oint (d\overline{S}.\nabla)\overline{F} - \oint \text{div } \overline{F}\, d\overline{S} + \oint d\overline{S} \times \text{curl } \overline{F}$$

$$\oint \overline{r} \times (d\overline{r}\times\overline{F}) = \int \overline{r} \times (d\overline{S}.\nabla)\overline{F} - \int \overline{r} \times \text{div } \overline{F}\, d\overline{S} + \int \overline{r} \times (d\overline{S} \times \text{curl } \overline{F}) + \int d\overline{S} \times \overline{F}$$

$$\overline{0} \equiv \oint \overline{r} \times (d\overline{S}.\nabla)\overline{F} - \oint \overline{r} \times \text{div } \overline{F}\, d\overline{S} + \oint \overline{r} \times (d\overline{S} \times \text{curl } \overline{F}) + \oint d\overline{S} \times \overline{F}$$

$$\int (\overline{F}.\nabla)\overline{G}\, d\tau = \int (- \text{div } \overline{F})\, \overline{G}\, d\tau + \oint \overline{G}\, \overline{F}.d\overline{S}$$

$$\int \overline{r} \times (\overline{F}.\nabla)\overline{G}\, d\tau = \int \overline{r} \times (- \text{div } \overline{F})\, \overline{G}\, d\tau + \oint (\overline{r}\times\overline{G})\, \overline{F}.d\overline{S} - \int \overline{F} \times \overline{G}\, d\tau$$

$$\int (\overline{F}.\nabla)\overline{G}\, d\tau = \int \overline{F} \text{ div } \overline{G}\, d\tau - \int \overline{F} \times \text{curl } \overline{G}\, d\tau - \int \overline{G} \times \text{curl } \overline{F}\, d\tau + \oint \overline{G} \times (d\overline{S}\times\overline{F})$$

$$\int \overline{r} \times (\overline{F}.\nabla)\overline{G}\, d\tau = \int \overline{r}\times\overline{F} \text{ div } \overline{G}\, d\tau - \int \overline{r} \times (\overline{F} \times \text{curl } \overline{G})d\tau - \int \overline{r} \times (\overline{G} \times \text{curl } \overline{F})d\tau$$
$$+ \oint \overline{r} \times (\overline{G}\times(d\overline{S}\times\overline{F})) - \int \overline{F} \times \overline{G}\, d\tau$$

$$\int \text{divs } \overline{F}\, dS = \oint \overline{F}.\hat{\overline{n}}'\, ds + \int (\text{divs } \hat{\overline{n}})\, \overline{F}.d\overline{S}$$

$$\int \text{curls } \overline{F}\, dS = \oint (\hat{\overline{n}}'\times\overline{F})\, ds + \int \text{divs } \hat{\overline{n}}\, (d\overline{S}\times\overline{F})$$

$$\int d\overline{S} \times \text{grads } V = \oint V\, d\overline{r}$$

$$\oint d\overline{S} \times \text{grads } V \equiv \overline{0}$$

$$\int \text{grads } V\, dS = \oint V\, \hat{\overline{n}}'\, ds + \int (\text{divs } \hat{\overline{n}})\, V\, d\overline{S}$$

ADDENDA TO TABLES

Table 2

grad div partial pot $\bar{J}$

$$= \int_{\tau-\tau_\delta} (\text{curl } \bar{J}) \times \text{grad } \frac{1}{r}\, d\tau + \oint_{S_{1..n}\Sigma} \text{grad } \frac{1}{r} \times (d\bar{S}\times\bar{J})$$

$$+ \oint_{S_\delta} \left\{ \frac{1}{r} (d\bar{S} \times \text{curl } \bar{J}) + \frac{1}{r} \frac{\partial \bar{J}}{\partial n}\, dS - \bar{J} \text{ grad } \frac{1}{r}.d\bar{S} \right\}$$

grad div pot $\bar{J}$

$$= \int_{\tau} (\text{curl } \bar{J}) \times \text{grad } \frac{1}{r}\, d\tau + \oint_{S_{1..n}\Sigma} \text{grad } \frac{1}{r} \times (d\bar{S}\times\bar{J}) - 4\pi\bar{J} \quad \text{at interior points}$$

curl curl partial pot $\bar{J}$

$$= \int_{\tau-\tau_\delta} (-\text{ div } \bar{J}) \text{ grad } \frac{1}{r}\, d\tau + \oint_{S_{1..n}\Sigma} \text{grad } \frac{1}{r}\, \bar{J}.d\bar{S} + \oint_{S_\delta} \left(\frac{1}{r} \text{ div } \bar{J}\, d\bar{S} - \frac{1}{r} \frac{\partial \bar{J}}{\partial n} dS + \bar{J} \text{ grad } \frac{1}{r}.d\bar{S} \right)$$

curl curl pot $\bar{J}$

$$= \int_{\tau} (-\text{ div } \bar{J}) \text{ grad } \frac{1}{r}\, d\tau + \oint_{S_{1..n}\Sigma} \text{grad } \frac{1}{r}\, \bar{J}.d\bar{S} + 4\pi\bar{J} \quad \text{at interior points}$$

Table 3

$$\text{grad (partial)} \int_{\tau-\tau_\delta} \bar{P}.\text{grad}\,\frac{1}{r}\,d\tau = -\int_{\tau-\tau_\delta} (\bar{P}.\nabla)\text{grad}\,\frac{1}{r}\,d\tau + \oint_{S_\delta} \bar{P}.\text{grad}\,\frac{1}{r}\,d\bar{S}$$

$$\text{grad} \int_{\tau} \bar{P}.\text{grad}\,\frac{1}{r}\,d\tau = -\lim_{\tau'\to 0}\int_{\tau-\tau'} (\bar{P}.\nabla)\text{grad}\,\frac{1}{r}\,d\tau + \lim_{S'\to 0}\oint_{S'} \bar{P}.\text{grad}\,\frac{1}{r}\,d\bar{S}$$

Table 4

$$\text{curl (partial)} \int_{\tau-\tau_\delta} \bar{M}\times\text{grad}\,\frac{1}{r}\,d\tau = \int_{\tau-\tau_\delta} (\bar{M}.\nabla)\text{grad}\,\frac{1}{r}\,d\tau + \oint_{S_\delta} d\bar{S}\times\left(\bar{M}\times\text{grad}\,\frac{1}{r}\right)$$

$$\text{curl} \int_{\tau} \bar{M}\times\text{grad}\,\frac{1}{r}\,d\tau = \lim_{\tau'\to 0}\int_{\tau-\tau'} (\bar{M}.\nabla)\text{grad}\,\frac{1}{r}\,d\tau + \lim_{S'\to 0}\oint_{S'} d\bar{S}\times\left(\bar{M}\times\text{grad}\,\frac{1}{r}\right)$$

Table 6

curl curl pot $[\bar{J}]$

$$= \int_{\tau} \left\{\left(-\left[\frac{\partial\bar{J}}{\partial t}\right].\nabla\right)\frac{\bar{r}}{cr^2} + \frac{\bar{r}}{cr^4}\,\bar{r}.\left[\frac{\partial\bar{J}}{\partial t}\right] + \frac{\bar{r}}{c^2r^3}\,\bar{r}.\left[\frac{\partial^2\bar{J}}{\partial t^2}\right]\right\} d\tau - \frac{1}{c^2}\frac{\partial^2}{\partial t^2}\int_{\tau} \frac{[\bar{J}]}{r}\,d\tau$$

$$+ \lim_{\tau'\to 0}\int_{\tau-\tau'} ([\bar{J}].\nabla)\text{grad}\,\frac{1}{r}\,d\tau + \lim_{S'\to 0}\oint_{S'} ([\bar{J}]\times d\bar{S})\times\frac{\bar{r}}{r^3}$$

INDEX

Lecture Notes in Engineering

Edited by C.A. Brebbia and S.A. Orszag

Vol. 21: B. F. Spencer, Jr.
Reliability of Randomly
Excited Hysteretic Structures
XIII, 138 pages. 1986.

Vol. 22: A. Gupta, R. P. Singh
Fatigue Behaviour
of Offshore Structures
XXI, 299 pages. 1986.

Vol. 23: P. Hagedorn, K. Kelkel,
J. Wallaschek
Vibrations and Impedances
of Rectangular Plates
with Free Boundaries
V, 152 pages. 1986.

Vol. 24: Supercomputers
and Fluid Dynamics
Proceedings of the First
Nobeyama Workshop
September 3-6, 1985
VIII, 200 pages. 1986.

Vol. 25: B. Hederson-Sellers
Modeling of Plume Rise
and Dispersion –
The University of Salford
Model: U. S. P. R.
VIII, 113 pages. 1987.

Vol. 26: Shell and Spatial Structures:
Computational Aspects
Proceeding of the International Symposium
July 1986, Leuven, Belgium
Edited by G. De Roeck, A. Samartin Quiroga,
M. Van Laethem and E. Backx
VII, 486 pages. 1987.

Vol. 27: Th. V. Hromadka, Ch.-Ch. Yen
G. F. Pinder
The Best Approximation Method
An Introduction
XIII, 168 pages. 1987.

Vol. 28: Refined Dynamical Theories
of Beams, Plates and Shells and
Their Applications
Proceedings of the Euromech-Colloquim 219
Edited by I. Elishakoff and H. Irretier
IX, 436 pages. 1987.

Vol. 29: G. Menges, N. Hövelmanns,
E. Baur (Eds.)
Expert Systems in Production Engineering
Proceedings of the International Workshop
Spa, Belgium, August 18-22, 1986
IV, 245 pages. 1987.

Vol. 30: R. Doležal
Simulation of Large State Variations
in Steam Power Plants
Dynamics of Large Scale Systems
X, 110 pages. 1987.

Vol. 31: Y. K. Lin, G. I. Schueller (Eds.)
Stochastic Structural Mechanics
U.S.-Austria Joint Seminar, May 4-5, 1987
Boca Raton, Florida, USA
XI, 507 pages. 1987.

Vol. 32: Y. K. Lin, R. Minai (Eds.)
Stochastic Approaches
in Earthquake Engineering
U.S.-Japan Joint Seminar, May 6-7, 1987
Boca Raton, Florida, USA
XI, 457 pages. 1987.

Vol. 33: P. Thoft-Christensen (Editor)
Reliability and Optimization
of Structural Systems
Proceedings of the First IFIP WG 7.5
Working Conference
Aalborg, Denmark, May 6-8, 1987
VIII, 458 pages. 1987.

Vol. 34: M. B. Allen III, G. A. Behie,
J. A. Trangenstein
Multiphase Flow in Porous Media
Mechanics, Mathematics, and Numerics
IV, 312 pages. 1988.

Vol. 35: W. Tang
A New Transformation
Approach in BEM
A Generalized Approach for
Transforming Domain Integrals
VI, 210 pages. 1988.

Vol. 36: R. H. Mendez, S. A. Orszag
Japanese Supercomputing
Architecture, Algorithms, and Applications
IV, 160 pages. 1988.

Vol. 37: J. N. Reddy, C. S. Krishnamoorthy,
K. N. Seetharamu (Eds.)
Finite Element Analysis
for Engineering Design
XIV, 869 pages. 1988.

Vol. 38: S. J. Dunnett, D. B. Ingham
The Mathematics of Blunt Body Sampling
VIII, 213 pages. 1988

Vol. 39: S.L. Koh, C.G. Speziale (Eds.)
Recent Advances in Engineering Science
A Symposium dedicated to A. Cemal Eringen
June 20–22, 1988, Berkeley, California
XVIII, 268 pages. 1989

Made in the USA
Las Vegas, NV
10 November 2024

11356587R00393